International Association of Geodesy Symposia

Fernando Sansò, Series Editor

Springer-Verlag Berlin Heidelberg GmbH

International Association of Geodesy Symposia

Fernando Sansò, Series Editor

Symposium 101: Global and Regional Geodynamics
Symposium 102: Global Positioning System: An Overview
Symposium 103: Gravity, Gradiometry, and Gravimetry
Symposium 104: Sea SurfaceTopography and the Geoid
Symposium 105: Earth Rotation and Coordinate Reference Frames
Symposium 106: Determination of the Geoid: Present and Future
Symposium 107: Kinematic Systems in Geodesy, Surveying, and Remote Sensing
Symposium 108: Application of Geodesy to Engineering
Symposium 109: Permanent Satellite Tracking Networks for Geodesy and Geodynamics
Symposium 110: From Mars to Greenland: Charting Gravity with Space and Airborne Instruments
Symposium 111: Recent Geodetic and Gravimetric Research in Latin America
Symposium 112: Geodesy and Physics of the Earth: Geodetic Contributions to Geodynamics
Symposium 113: Gravity and Geoid
Symposium 114: Geodetic Theory Today
Symposium 115: GPS Trends in Precise Terrestrial, Airborne, and Spaceborne Applications
Symposium 116: Global Gravity Field and Ist Temporal Variations
Symposium 117: Gravity, Geoid and Marine Geodesy
Symposium 118: Advances in Positioning and Reference Frames
Symposium 119: Geodesy on the Move
Symposium 120: Towards an Integrated Global Geodetic Observation System (IGGOS)
Symposium 121: Geodesy Beyond 2000: The Challenges of the First Decade
Symposium 122: IV Hotine-Marussi Symposium on Mathematical Geodesy
Symposium 123: Gravity, Geoid and Geodynamics 2000
Symposium 124: Vertical Reference Systems
Symposium 125: Vistas for Geodesy in the New Millennium
Symposium 126: Satellite Altimetry for Geodesy, Geophysics and Oceanography
Symposium 127: V Hotine Marussi Symposium on Mathematical Geodesy

V Hotine-Marussi Symposium on Mathematical Geodesy

Matera, Italy
June 17-21, 2003

Edited by
Fernando Sansò

Springer

Volume Editor

Prof. Dr. Fernando Sansò
Polytechnic of Milan
D.I.I.A.R. - Surveying Section
Piazza Leonardo da Vinci, 32
20133 Milan
Italy

Series Editor

Prof. Dr. Fernando Sansò
Polytechnic of Milan
D.I.I.A.R. - Surveying Section
Piazza Leonardo da Vinci, 32
20133 Milan
Italy

ISSN 0939-9585
ISBN 978-3-642-06028-1 ISBN 978-3-662-10735-5 (eBook)
DOI 10.1007/978-3-662-10735-5

Cataloging-in-Publication Data

Library of Congress Control Number: 2004104537

springeronline.com

Originally published by Springer-Verlag Berlin Heidelberg New York in 2004
Softcover reprint of the hardcover 1st edition 2004

Camera ready by Editors
Cover design: design & production GmbH, Heidelberg

Printed on acid-free paper 32/3141/as 5 4 3 2 1 0

Contents

Foreword

The Hotine-Marussi symposia are, in the life of the International Association of Geodesy, a moment of pause to stop and think how the geodetic theory, the arsenal of mathematical, physical and geodetic tools, is finding an assessment, a systematization which all of us could agree upon and on the same time the craddle of new ideas and new methods boosted by the fantastic achievements of modern technology, from space geodesy to electronics and informatics, and brought into the arena of theoretic debates.

Traditional items in the Hotine-Marussi's are the geometric description of the gravity field, its mathematical analysis and its approximation methods, the dynamic theory of satellite motion, the inverse relation between gravity field and mass distributions, the positioning problems, nowadays inextricably joined to satellite geodesy, the general concept supporting positioning, namely the definition of reference systems and their practical implementation in frames; across all these items we find a specific geodetic research on numerical methods for the solution of very large systems and on statistical methods in handling data.

All these items were touched during the symposium although some of them have got more relevance in the presentations.
In particular the figure and theoretical achievements of Marussi (the initiatior of the series of these symposia that now bear his name too) have been discussed in a paper by J. Zund, which is worth mentioning because it was not presented orally but has been hosted in the proceedings because of its argument.
As it has been stated, also in a discussion with the reviewers, the paper does not attempt to give a global judgement on the role of geometric thinking in geodesy, but certainly provides an interesting view-angle by one of the authors that has traditionally accompanied our reflections in this area.

Another aspect of this V Hotine-Marussi that I like to underline, is the good interaction that has been established with scientists from geophysical disciplines naturally interfaced with geodesy, like geodynamics of the solid earth, related to the geodetic monitoring of deformation, and to the inverse gravimetric problem, ice and ocean dynamics, so much related to geoid studies, angular momentum of the earth and global rotational dynamics, so strictly linked to the definition and maintenance of a global reference system.

During the symposium we also had time to visit the Bepi Colombo Observatory of the Italian Space Agency (ASI) in Matera, where our classroom was placed, and we would like very much to thank ASI for the hospitality and the visit.

Also some representatives of firms of geodetic instrumentation, in particular Codevintec, Geotop, Leica, have shown up in Matera and discussed some technical items with us;

beyond that they have contributed to our budget with funds that have been used for social activities, during the Symposium period, and we would like to thank them heartly for their generosity. It has to be mentioned too that, as customary with sponsored Symposia, IAG has also provided financial support, which has been used to fund the participation of young scientists.
A special thank is due to Mrs. Elena Raguzzoni, that took care of the organization of the whole event, which ran perfectly with no drawbacks.

To conclude, if I look back to the V Hotine-Marussi and to the results presented, I really think it was worthwhile to have it and hope that such series will last long, to the benefit of the International Association of Geodesy.

Fernando Sansò

Report on the Symposium

The V Hotine-Marussi Symposium on Mathematical Geodesy has been held in Matera from June 17th to June 21st, 2002.

Geodesists interested in the theoretical and methodological aspects of their discipline discuss these issues in a symposium that usually takes place every four years and that is hosted in Italy. Martin Hotine and Antonio Marussi have initiated this tradition in 1959 under the title of "Symposium on Three Dimensional Geodesy".

More than forty papers have been presented in five sessions dedicated to "New missions", "The analysis of geodetic data: methods and numerics", "The theory of geodetic reference systems", "Deterministic and stochastic field theory in geodesy and sister sciences" and "Inverse problems in geodesy and geodynamics".

Some relevant topics re-appeared in several sessions; this shows not only their special importance but also the unitary character of the geodetic problems. For example the topic of boundary value problems has been treated in several sessions under different viewpoints and with applications to different but related fields. The topic of stochastic models for data analysis appeared in all the different sessions.

Stochastic methods have been applied to several situations including potential theory, integer estimation, testing theory for control applications and inverse problems.

Some papers have been dedicated to numerical methods, specially to the method for least squares estimation applied to the huge amount of data coming from space gravimetric missions, spherical harmonic analysis and synthesis, multiresolution analysis and integer estimation.

It must be stressed that the problems treated in the symposium have been mainly analyzed from the theoretical point of view and with a strong attention to methodology, that is the very purpose of the Hotine-Marussi symposia: nevertheless practical problems of real-life geodesy have been quite present.
Most of the presented papers are related to problems quite relevant in the applications: in this respect I want to specially mention the reduction of space-born gravimetry (in a general sense), the positioning with GNSSs, and the intricate difficulties of establishing a reference system.

Some animated discussions completed the topics presented by the speakers, as it is usual in these symposia. Some papers and most of the discussions have highlighted not only the new results obtained but also some important open problems that are relevant both for the applications and interesting from a general theoretical point of view.
The theory of integer estimation, for example, is now well established, but a satisfactory testing theory for the estimates still need to be developed. The problem of a "smooth transition" between real and integer estimations (or float and fixed solution, to stay with the customary expressions) has been also discussed. The optimal weighting of regularization conditions in

inverse problem, and the characterization of the noise in the measuring data are just two other examples of the relevant open problems raised in the discussions.

About seventy persons have attended the symposium from twelve countries. Their active participation is a good invitation to look forward to the next symposium on mathematical geodesy.

The symposium has been held at "Centro di Geodesia Spaziale Giuseppe Colombo". This is the largest Italian facility for space geodetic observations; it is managed by ASI (Agenzia Spaziale Italiana) with some activities operated by the private firm Telespazio. The participants have had the opportunity to visit the instrumentations of the center including the quite impressive device for satellite and lunar laser ranging and the giant VLBI antenna.
Some free time has been dedicated to the visit of the archeological site in Metaponto. Participants also enjoyed the unique and peculiar environment of the town of Matera, characterized by the famous houses and churches caved in the underground of the mountain and named "Sassi".

Battista Benciolini

Previous symposia on Mathematical Geodesy

1. 1st Symposium on Three Dimensional Geodesy, Venezia 1959
2. 2nd Symposium on Three Dimensional Geodesy, Cortina d'Ampezzo 1962
3. 3rd Symposium on Mathematical Geodesy, Torino 1965
4. 1st Hotine Symposium on Mathematical Geodesy, Trieste 1969
5. 2nd Hotine Symposium on Mathematical Geodesy, Firenze 1972
6. 3rd Hotine Symposium on Mathematical Geodesy, Siena 1976
7. 4th Hotine Symposium on Mathematical Geodesy, Assisi 1978
8. 5th Hotine Symposium on Mathematical Geodesy, Como 1981
9. 1st Hotine-Marussi Symposium on Mathematical Geodesy, Roma 1985
10. 2nd Hotine-Marussi Symposium on Mathematical Geodesy, Pisa 1990
11. 3rd Hotine-Marussi Symposium on Mathematical Geodesy, L'Aquila 1994
12. 4th Hotine-Marussi Symposium on Mathematical Geodesy, Trento 1998

Gaussian Differential Geometry and Differential Geodesy

Joseph D. Zund
New Mexico State University, Las Cruces, NM 88003

Abstract. This paper presents an appreciation of the work of Marussi and Hotine, and gives a survey of my investigations of Gaussian differential geometry which are required in formulating the generalized Marussi-Hotine approach to differential geodesy. It is not intended to be either a comprehensive survey, or a status report, on the beautiful contributions of other authors in different approaches to differential geodesy.

1 Introduction

At the General Assembly of the International Association of Geodesy (Oslo, 1948), Antonio Marussi gave a presentation of his seminal ideas on how differential geometry could be applied to the gravity field of the Earth. At the time, he was forty years of age and had published no more than eight papers on classical geodesy and cartography. In the audience was a fifty-year-old retired British brigadier, Martin Hotine, who had a distinguished military career in the service of his country, in the triangulation of East Africa and Great Britain, and had been a pioneer in aerial photography. Hotine's previous experience in theoretical geodesy and cartography had been classical, and although he had considerable mathematical skills, these had yet to be tested in the mathematical arena of theoretical geodesy. Indeed, nothing in his background – apart from his audacity – would seem to suggest that he would be prepared to embark on a new career of theoretical research.

By Marussi's own recollections, nobody understood his presentation, (see [5] for an expanded version). But it immediately struck a chord with the youthful-minded Hotine, who said he understood only very little about it, but that it broke with crystallized tradition and that it must therefore be important, (see page 14 of [13] and the references cited there).

Thus, began a collaboration and friendship, which would last until Hotine's death twenty years later. However, Hotine's entrance into the field was delayed until 1963 since he held a full time position as Director of the Directorate of Overseas Surveys, (formerly the Directorate of Colonial Surveys), in the British government's Ministry of Overseas Development. In the meantime, he mastered the tensor calculus on his own, and his first papers on differential geodesy began to appear in 1956 – 57, (see items 2, 3, and 4 in [4]). Ironically, Hotine's debut occurred about the same time Marussi was becoming a Professor of Geodesy and Geophysics at the University of Trieste. This new position offered him the opportunity to pursue his diverse interests, and in a sense – Marussi's primary activity in differential geodesy essentially ended when Hotine's began. However, together they organized a series of International Symposia on Mathematical Geodesy (Venice, 1959; Cortina d'Ampezzo, 1962; and Turin, 1965), which were the ancestors of the current Hotine-Marussi Symposium. In 1965 Hotine joined the U.S. Coast and Geodetic Survey as a research scientist, and his treatise, [3], was a result of this activity.

It should be emphasized that while they shared the common goal of creating a geometric picture of theoretical geodesy, their approaches and methodology were distinctly different. Marussi was guided by general principles and employed the homographic calculus (an abstract version of tensors due to C. Burali-Forti and R. Marcolongo, which is lucidly discussed by W. I. Reilly in an Appendix in [6], pages 190 – 195). In contrast, Hotine was more computationally minded and primarily employed the traditional tensor calculus (e.g., as presented in the text of J. A. McConnell, [8]).

As far as is known, they never worked together on a particular problem, but there is no doubt that they stimulated each other via their independent – and highly individual – contributions. In effect, they left two distinct views of essentially the same theory, which we call the Marussi-Hotine theory of differential geodesy. Roughly speaking, (see page

152 of [12]), Marussi sketched the conceptual structure of the theory and indicated the possibility of a particular coordinate system, while Hotine attempted to construct a hierarchy $\mathfrak{H}_i$, (i = 0, 1, 2, 3, 4), of suitable coordinate systems, (where $\mathfrak{H}_0$ was Marussi's coordinate system – i.e. the (ω, Φ, N) system in the terminology of [3]). All their work began from a particular coordinate system and then attempted to 'fit' geodesy into this picture.

I first became involved in differential geodesy in 1986, and as a geometer / mathematical physicists, I quickly fell under the magical spell of their writings. My immediate goal was to understand what they had done, had wanted to do, and to attempt to merge their approaches into a coherent single theory. I subsequently edited a set of Marussi's 1952 lectures in America, [7]; a volume of Hotine's papers on differential geodesy, [4], (which was intended to be a companion to Marussi's volume, [6]), and finally I wrote my own monograph on differential geodesy, [14]. While their magic is partially dispelled, the wonderment remains.

My principle conclusion was that Marussi-Hotine theory was quite restrictive and that the hierarchy proposed by Hotine failed to include any admissible coordinate systems other than $\mathfrak{H}_0$, (which was Marussi's original choice). Indeed, it seems likely that $\mathfrak{H}_0$, (or more precisely the equivalence class defined by $\mathfrak{H}_0$), is the *only* possible coordinate system allowed in the Marussi-Hotine theory. Consequently, I proposed that the Marussi-Hotine process be reversed, i.e. given a particular geodetic scenario *construct* a coordinate system which 'describes' this physical situation. In chapter X of [14], I called such an approach the generalized Marussi-Hotine theory, since it faithfully incorporated their ideas, but is independent of any particular coordinate system.

As a byproduct of these investigations, in 1996 I began writing a series of essays on the foundations of Gaussian differential geometry for the *Bolletino di Geodesia e Scienze Affini*, [15] – [22]. My purpose today is to give an overview and status report on these essays, and their implications for differential geodesy.

One might think that after almost two centuries Gaussian differential geometry is a closed and finished chapter in mathematics. However, I believe that this is a hasty judgement, and that in reality only the 'simplest' things have been worked out, and well understood. In particular, the generalized Marussi-Hotine theory offers numerous mathematical puzzles.

2 Gaussian Differential Geometry

The origin of Gaussian differential geometry may be traced back to work of Gauss in the geodetic survey of Kingdom of Hannover during 1821 – 1825. His first paper, [1], was a solution to a prize problem posed by the Danish Academy in Copenhagen. It contained his solution of the conformal representation of one surface onto another, and has immediate applications to cartography. This lead to his great paper, [2], *Disquisitiones generales circa superficies curves*, which essentially marks the birth of Gaussian differential geometry *per se*. The title invites a comparison with his masterwork on number theory, *Disquisitiones arithmeticae*, of 1801, however it is decidedly less polished and more tentative. The viewpoint was primarily methodological and marked the first time that the parametrization of a surface was advocated as a fundamental new point of departure. Gauss introduced the first and second basic forms, the Gaussian curvature K, proved his *theorema egregium*, derived the *curvature integra*, and investigated geodesics, and geodesic triangles. However, his notation was cumbersome and frequently his reasoning was less than transparent. Moreover, he missed the notion of geodesic curvature, and failed to obtain the integrability conditions of a system of partial differential equations which led to the Codazzi equations, (Delfino Codazzi, 1867 / 68; Gaspare Mainardi, 1856).

Although Gauss lived another seventeen years, he published nothing more on differential geometry, and indeed he taught the material only once in his lectures at Gottingen! The task of completing the theory and devising the appropriate analytic formalism was left to others; and clarified by the magnificent treatises of Luigi Bianchi (1886), *Lezioni di geometria differenziale*, and of Gaston Darboux (1887), *Leçons sur la théorie générale des surfaces et les applications géométriques du calcul infinitésimal* respectively.

3 Preliminary Assumptions

My eight essays are based on the following general assumptions which we briefly review. First, almost all of the discussion is local, i.e. although we will frequently refer to a surface S, in reality – as in the usual classical literature – we mean a 'piece' of the surface. Likewise, except in the second essay, the smoothness requirements of S and its imbedding in

Euclidean 3-space $\mathbf{E}_3$ are unspecified. The local coordinates x^r, (r = 1, 2, 3), are generally taken to be the ordinary rectangular Cartesian coordinates, except in general tensor equations, or cases where the context is evident, or explicitly indicated. Apart from the first essay, the order of the subsequent essays is simply that in which they were done. In general, unless specified otherwise, the notation follows that of [14], which in turn is largely that employed in [3] and [8].

4 First Essay: Completeness Questions

The basic motivation for my essays was two-fold. My first goal was to examine the structure of the Gaussian theory with the aim of determining whether it is a complete theory. Today, one commonly regards classical differential geometry as being synonymous with Gaussian differential geometry. But, is this really true, or is it illusory?

A perusal of the standard monographs shows that the usual standard 'textbook' examples normally deal with quadric surfaces. The notable exception is the 2-torus $\mathbf{T}_2$ which is a quartic surface, but higher order surfaces are seldom considered. For such reasonably 'simple' surfaces, the choice of a parametrization is either obvious, or a result of how the surface was geometrically generated, (as in the case of a $\mathbf{T}_2$). However, in geodesy the choice of a parametrization for the geopotential **N** surfaces is far from obvious, and without a parametrization Gaussian differential geometry is inapplicable. Moreover, in practice one wants not merely a parametrization, but one which is susceptible of a geodetic / geometric interpretation!

This suggests that 'new' mathematical tools are required in order to make the Gaussian theory applicable to the physical demands of differential geodesy. Typically, these include the third and fourth basic tensors, and the Darboux tensor. The third basic tensor is useful in the spherical representation of a surface, while the fourth tensor occurs in the bending of surfaces. The third order Darboux tensor vanishes identically for quadric surfaces, but is non-zero – and presumably non-trivial for higher order surfaces. Each of these tensors is defined in terms of the first and second basic tensors, and perfectly consistent with the structural features of Gaussian differential geometry.

My second goal was to recast Gaussian differential geometry into a form which was suitable for the generalized Marussi-Hotine theory. This raises the issue of what is the appropriate mathematical formalism for the Gaussian theory. Gauss himself introduced – without explanation – a multitude of new symbols, and apparently got 'lost' in the resulting melee. However, there is a mathematical formalism which remedies the confusion, namely the tensor calculus of Gregorio Ricci-Curbastro (1853 – 1924). Indeed, in his lectures, [9], Ricci made the first attempt to derive Gaussian differential geometry using the invariant-theoretic processes of the tensor calculus, rather than merely 'translating' the theory into tensors. Today, with the modern notion of tensors, one can do considerably more by employing the leg–calculus, (see Chapters I through IV of [14]).

The first essay contains a 'fundamental lemma' which states that first and second basic tensors are proportional, with a constant proportionality factor, if and only if the surface **S** is a 2-sphere $\mathbf{S}_2$. This may be restated by saying that while the first basic tensor is adequate to describe the geometry on $\mathbf{S}_2$, the second basic tensor is needed to handle more general surfaces.

5 Second Essay: Imbedding Questions

As noted before, the basic problem confronting differential geodesy is the "Parametrization Problem," namely the problem of obtaining a parameter net $\mathfrak{R}$ for a surface S. Gauss gave an ingenious solution: *ab initio* he always assumed that such a parametrization was *a priori* given!

Hence, in Gaussian differential geometry, the question of whether a prescribed surface can always be 'placed' in a 3-dimensional Euclidean space $\mathbf{E}_3$ is trivialized, and rarely considered in the textbooks. On the other hand, in geodetic situations one might construct a 'surface' via measurements and observations. The resulting operationally determined 'surface,' need not be realizable in $\mathbf{E}_3$! Without going into the specifics, in 1900 Hilbert proved that if **S** had constant negative Gaussian curvature K, then it could not be isometrically imbedded in $\mathbf{E}_3$. Geometrically speaking, there is an obvious explanation: the geometry of **S** is non-Euclidean, so why should one expect it to be a subspace of $\mathbf{E}_3$? While the details remain somewhat murky, it seems that this difference in geometries requires an $\mathbf{E}_5$!

Thus the equation

$$a_{\alpha\beta} = g_{rs} x_\alpha^r x_\beta^s,$$

has a two-fold meaning. Assuming the metric tensor g_{rs} of $\mathbf{E}_3$ is given, if a Gaussian parametrization $\mathfrak{G}$ of S is known, then this equation defines the first basic tensor. Conversely, if $a_{\alpha\beta}$ is known, (in terms of some parametrization of S), then this equation is a system of non-linear partial differential equations for the x^r which yields an isometric imbedding of S in E_3.

I then discuss the smoothness requirements for the imbedding problem to be solvable, viz. Hölder continuity, and the basic Cartan – Janet (1926 / 1927) imbedding theorem. The latter has been proven only for the rather restrictive case when the metric tensor is real-analytic, viz. of class C^ω, and this seems to be unavoidable since all known proofs involve the Cauchy-Kowaleskaya theorem (1825, 1875) which is false when the functions are not of class C^ω.

Then I present a set of operational rules for surface covariant differentiation which leads to the Gauss formulas, the Weingarten equations, the Gauss equations and finally the Codazzi-Mainardi equations which yield the Fundamental theorem of surface theory of Ossian Bonnet (1867). An argument is given which clarifies the role of the Weingarten and Gauss equations in the imbedding process.

The essay ends with a discussion of the corresponding global imbedding theorems of Nash, Greene, Gromov and Rokhlin.

6 Third Essay: Sphere Geometry

This essay returns to the 'fundamental lemma' of the first essay, and is devoted to considering the various specializations which occur in Gaussian differential geometry when S is required to be a S_2. At first glance this may seem to be a step backwards since one certainly wants to study surfaces which are more general than 2-spheres. On the other hand, if a result holds only on a sphere, then it will not hold on a more general curved surface S!

The principal results are five corollaries which contrast the situations on S_2, and S. On S_2 one has

1) $K = H^2$,
2) dim $\mathfrak{S} = 1$,
3) $d_{\alpha\beta} = f_{\alpha\beta} = 0$,
4) principal directions are non-trivial,
5) the Weingarten formula and the Codazzi equations are identically satisfied; while for S one has

1) $K = H^2 + K_E$,
2) dim $\mathfrak{S} = 3$,
3) $d_{\alpha\beta}$ & $f_{\alpha\beta}$ are non-zero,
4) any direction is a principal direction,
5) both the Weingarten formula and the Codazzi equations are non-trivial.

In these expressions, $K_E = (\kappa_1 - \kappa_2)^2 / 4$ is the Euler curvature of S, $d_{\alpha\beta}$ and $f_{\alpha\beta}$ are respectively the fourth and fifth basic tensors, and $\mathfrak{S}$ is the linear space of symmetric second order surface tensors, (see [10] and [11]).

In effect for differential geodesy this discussion exhibits the complexities which must be faced when we attempt to discard our *idée fixe* of attempting to model the equipotential field of the Earth on a sphere.

7 Fourth Essay: The Parametrization Problem

In this essay I made my first attempt to understand what was really involved in the parametrization problem for Gaussian differential geometry. Our principle results include a reformulation of the rank conditions for a 3 × 2 matrix of the components of the Gauss operator x^r_α.

Denoting the three 2 × 2 determinants of $\left\|x^r_\alpha\right\|$ by $\mathfrak{D}_i$, (i = 1, 2, 3), one has

1) the *weak rank condition* whenever at least *one* of the $\mathfrak{D}_i$ is non-zero; and
2) the *strong rank condition* when all *three* of the $\mathfrak{D}_i$ are non-zero.

In the literature, the former condition is commonly adopted, however, in practice the latter condition is the appropriate condition. Indeed, except for isolated points, the strong rank condition is satisfied for all of the 'simple' surfaces commonly considered in Gaussian differential geometry, and this essentially my Theorem A.

Theorem B, is concerned with the question of whether there exists a curvilinear analogue $\mathfrak{M}$* of the Monge net $\mathfrak{M}$:

$$x = x,\ y = y,\ z = f(x, y).$$

The answer is in the affirmative, however $\mathfrak{M}$* is useful only in rather restrictive situations, e.g. in the case of a surface of revolution.

The essay concludes with some comments on the Clairaut net $\mathfrak{C}$, and the Tschebysheff net $\mathfrak{T}$ which are useful in special circumstances.

8 Fifth Essay: Operator Methods

This essay begins with a derivation of the two Cesàro equations which yield new tensorial equations involving the curvatures K and H. The original expanded non-tensorial versions of these equations were discovered by Ernesto Cesàro (1859 – 1906) *ante* 1904, and do not appear to be well-known.

The main goal of the essay is to introduce a pair of tensorial operators which can be employed to map vectors, or more generally tensors, onto the tangent plane and unit normal to **S**. For brevity, I respectively call them the Π and Σ projectors. The general theory of these projectors is developed, and then applied to derive the resolution of spatial vectors, and second order tensors, into their tangential and normal parts relative to a given surface **S**. The essay concludes with a new derivation of the local imbedding equation, (which involves the tangential part of g_{rs}), and the McConnell equation,

$$g^{rs} = a^{\alpha\beta} x_\alpha^r x_\beta^s - \nu^r \nu^s .$$

McConnell's derivation of this equation is mysterious, (see Example 5, page 197 of [8]), and depends on a particular choice of coordinate system. Hotine employed it in [3], and gave a proof which it is very neat, but made extensive use of his leg calculus.

9 Sixth Essay: The Reciprocal Gauss Operator

My point of departure was the question as to whether analogous to

$$a_{\alpha\beta} = g_{rs} x_\alpha^r x_\beta^s ,$$

there existed an analogous space tensor ψ_{rs} that

$$b_{\alpha\beta} = \psi_{rs} x_\alpha^r x_\beta^s \, ?$$

The answer would be readily obtained *if* the matrix $\left\| x_\alpha^r \right\|$ was invertible, however this is not possible. On the other hand this matrix admits a generalized inverse in the sense of Moore and Smith. I call the tensorial version of this generalized inverse the reciprocal Gauss operator. By definition,

$$\breve{x}_r^\alpha = g_{rs} a^{\alpha\beta} x_\beta^s ,$$

and it is then easy to obtain the spatial and surface 'orthogonality conditions.' The final answer turns out to be

$$\psi_{rs} = b_{\alpha\beta} \breve{x}_r^\alpha \breve{x}_s^\beta$$

where in terms of the orthonormal Hotine vectorial 3-leg has the representation

$$\psi_{rs} = k_1 \lambda_\alpha \lambda_\beta + k_2 \mu_\alpha \mu_\beta + t_1 \left(\lambda_\alpha \mu_\beta + \mu_\alpha \lambda_\beta \right)$$

where k_1, k_2 are the normal curvatures and t_1 is the geodesic torsion when this expression is restricted to **S**. I call ψ_{rs} the Meusnier tensor in honor of Jean Baptiste Meusnier, (1754 – 1793), who in 1776 almost discovered the second basic tensor $b_{\alpha\beta}$.

If one considers the general differentiation problem of relating the spatial derivatives of a smooth function to its surface derivatives then the Meusnier tensor naturally occurs. More specifically, if the surface **S** is an equipotential **N** surface, then the Meusnier tensor appears as *part* of the Marussi tensor. Indeed, this essay concludes by giving a new derivation of the Marussi tensor by using the reciprocal Gauss operator!

10 Seventh Essay: The Parametrization Problem Revisited

This essay begins with a historical discussion of how surfaces were originally expressed analytically by Antoine Parent (1713). Then we show that the Monge net $\mathfrak{M}$ is applicable to a line element of diagonal form if only when the surface is a generalized cylinder. We then indicate to what extent Alexis Clairaut and Leonhard Euler partially anticipated Gauss differential geometry, and give a self-contained proof of Euler's 'No-Go Theorem' that a 2-sphere $\mathbf{S}_2$ cannot be locally isometrically mapped on a Euclidean plane $\mathbf{E}_2$. Euler's argument is interesting in that he showed that the integrability conditions of such a mapping are not satisfied, however he failed to notice that these furnish an expression for the – then unkown – Gaussian curvature K which is non-zero for $\mathbf{S}_2$!

11 Eighth Essay: The Leg Integrability Conditions.

The leg calculus was present in some detail in [14], (for a brief introduction see the Appendix of [20]), and the present essay is devoted to its use in the fundamental theorem of Gaussian differential geometry. Recall that by definition, a *leg* is a linearly independent system of vectors, or differential forms. It is quite remarkable that although implicitly Hotine used the *leg calculus* in his treatise [3], he did not bother to connect it with Ricci's theory of rotation coefficients, or relate it to Cartan's theory of differential forms (see Chapters II and III of [4]). In this eighth essay my primary concern was to investigate the leg-theoretic version of the Codazzi equations (**C**), and their integrability conditions. This involves developing certain mathematical preliminaries on the leg calculus which occupy the first half of our essay. Then, as Hotine showed on page 44 or [3], (**C**) involves a pair of leg derivatives $\kappa_{1/2}$, $\kappa_{2/1}$ of the principal curvatures κ_1, κ_2 where the slash denotes a directional derivative in the indicated leg directionl It is noteworthy that although the other two leg derivatives, $\kappa_{1/1}$, $\kappa_{2/2}$, are not specified by C, they are restricted by the leg integrability conditions of the Codazzi equations. The essay concludes an interpretation of these integrability conditions and Bonnet's fundamental theorem.

12 Conclusion

I conclude with some impressions on the contributions of Marussi and Hotine to differential geodesy. However, I must emphasize that these are merely 'impressions,' and that the evidence for them is largely circumstantial, and indirect.

Marussi, following his artistic inclinations, was mainly interested in adumbrating the broad structural features of differential geodesy. His passion was for ideas, not lengthy calculations, and he did not fully elaborate the $\mathfrak{H}_0$. Perhaps this was due to the pressure of other activities, or the promise of pursuing other challenges in geophysics. However, I personally suspect that he did not view his contributions to differential geodesy as more than a set of intriguing and very promising ideas. Certainly, in re-reading his presentation to the Venice Symposium in 1959, (the first essay in [6], pages 3 – 13), it is difficult to conclude that he regarded any of it to be in final, or definitive, form. Indeed, one explanation for his neglect of his beloved *geodesia intrinseca* after 1957, may have stemmed from a recognition of the restricted nature of $\mathfrak{H}_0$, and the need for a more general mathematical framework which he hoped Hotine would supply. This is supported by his enthusiastic praise for the contributions of his younger colleagues, (see page xi of [6],) however, at least in his publications, his interests seem to lie elsewhere.

Hotine, during the last five years of his life extensively investigated the mathematical details of *his version* of Marussi's ideas, and in the months before his death was enthusiastically learning Cartan's exterior differential calculus. I suspect that this augurs a possible enlargement of the underlying structure of the Marussi-Hotine theory. In fact, I believe that – apart from $\mathfrak{H}_0$ – he considered his other coordinate systems, $\mathfrak{H}_i$ (i = 1, 2, 3, 4), as merely *possibilities* worthy of careful consideration. Note that in his treatise, [3], his discussion of $\mathfrak{H}_0$ was *twice* as long as that of each of his new systems. However, in reality – had they been admissible – their intricacy and subtlety would certainly have required a much more lengthy elaboration. Hence, there also seems to be a tentativeness to Hotine's work.

Thus, while Marussi and Hotine saw their theory as through a glass darkly, they nevertheless glimpsed wonderful things. Things which they left us as a challenge to complete, and perfect. I think neither of them would be really surprised at what I have called the generalized Marussi-Hotine theory.

If anything might surprise them, I think it would be that, as commonly conceived, traditional Gaussian differential geometry is only partially adequate as a framework for their theory.

It has been my privilege in these essays to take the first steps toward remedying these gaps in the Gaussian picture, and much still remains to be done. However, I remain optimistic and inspired by the elegance and magnificence of Marussi and Hotine's vision of differential geodesy.

The words of the great French physiologist, Claude Bernard (1813 – 1878) well describes them:

"Great men may be compared to torches shining at long intervals, to guide the advance of science. They light up their time, either by discovering unexpected and fertile phenomena which open up new paths and reveal unknown horizons, or by generalizing acquired scientific facts and disclosing truths which their predecessors had not perceived."

Truly, Antonio Marussi and Martin Hotine were such men!

In conclusion, I would like to express my profound gratitude to Professor Fernando Sansò, President of the International Association of Geodesy, for graciously permitting me to make this contribution *in absentia*.

References

1. Gauss, C. F. (1825). Allgemeine Auflösung der Aufgabe: Die Theile einer gegebnen Fläche auf einer andern

gegebnen Fläche so abzubilden, dass die Abbildung der Abgebildeteen in den kleinsten Theilen ähnlich wird, *Astronomischen Nachrichten*. 3: 1 – 30 = Werke, Band 4: 189 – 216.
2. Gauss, C. F. (1828). Disquisitiones generales circa superficies curves. *Commentationes societatis regiae scientiarum Göttingensis, Commentationes classis mathematicae* 6: 99 – 146 = Werke, Band 4: 217 – 258.
3. Hotine, M. (1970). *Mathematical Geodesy*, U. S. Department of Commerce, Washington, D. C.
4. Hotine, M. (1991). *Differential Geodesy*, Zund J. D., Nolton J., Chovitz B. H., Whitten C. A. (eds) Springer-Verlag, Berlin.
5. Marussi, A. (1949). Fondements de géométrie différentielle absolue du champ potentiel terrestre. *Bulletin Géodésique*, 14, pp. 411 – 439.
6. Marussi, A (1985). *Intrinsic Geodesy*. (translated by W. I. Reilly), Springer-Verlag, Berlin.
7. Marussi, A. (1988). Intrinsic Geodesy, (a revised and edited version of his 1952 lectures prepared by J. D. Zund), *Report Number 390*, Department of Geodetic Science and Surveying, The Ohio State University, Columbus.
8. McConnell, J. A. (1931). *Applications of the absolute differential calculus*. Blackie & Son, Limited, London (reprinted as *Applications of tensor analysis*, by Dover Publications, Inc., New York, (1957)).
9. Ricci, G. (1898). *Lezioni sulla teoria della superficie*. Fratelli Drucker – Editore, Verona – Padova.
10. Zund, J. D. (1988). Tensorial methods in classical differential geometry I: basic principles. *Tensor N. S.* 47, pp. 73 – 82.
11. Zund, J. D. (1988). Tensorial methods in classical differential geometry II: basic surface tensors. *Tensor N. S.* 47, pp. 83 – 92.
12. Zund, J. D. (1990). An essay on the mathematical foundations of the Marussi-Hotine approach to geodesy. *Bolletino di Geodesia e Scienze Affini* XLIX, pp. 133 – 179.
13. Zund, J. D. (1991). The work of Antonio Marussi in theoretical geodesy. *Proceedings of the Geodetic Day in honor of Antonio Marussi*. Accademia Nazionale dei Lincei, Rome.
14. Zund, J. D. (1994). *Foundations of differential geodesy*. Springer-Verlag, Berlin.
15. Zund, J. D. (1996). An essay on the foundations of Gaussian differential geometry I: completeness questions. *Bolletino di Geodesia e Scienze Affini* LV, pp. 377 – 384.
16. Zund, J. D. (1997). An essay on the foundations of Gaussian differential geometry II: imbedding questions. *Bolletino di Geodesia e Scienze Affini* LVI, pp. 407 – 424.
17. Zund, J. D. (1998). An essay on the foundations of Gaussian differential geometry III: sphere geometry. *Bolletino di Geodesia e Scienze Affini* LVII, pp. 309 – 321.
18. Zund, J. D. (199). An essay on the foundations of Gaussian differential geometry IV: the parametrization problem. *Bolletino di Geodesia e Scienze Affini* LVIII, pp. 325 – 352.
19. Zund, J. D. (2000). An essay on the foundations of Gaussian differential geometry V: operator methods. *Bolletino di Geodesia e Scienze Affini* LIX, pp. 351 – 368.
20. Zund, J. D. (2001). An essay on the foundations of Gaussian differential geometry VI: the reciprocal Gauss operator. *Bolletino di Geodesia e Scienze Affini* LX, pp. 191 – 212.
21. Zund, J. D. (2003). An essay on the foundations of Gaussian differential geometry VII: the parametrization problem revisited, *Bolletino di Geodesia e Scienze Affini*, LXII, 63 – 76.
22. Zund, J. D. (2003). An essay on the foundations of Gaussian differential geometry VIII: the leg integrability conditions, *Bolletino di Geodesia e Scienze Affini*, LXII, in press.

Numerical techniques for large least-squares problems with applications to GOCE

R. Klees, P. Ditmar and J. Kusche
Physical, Geometric and Space Geodesy (FMR)
Delft University of Technology, 2629 JA Delft, Thijsseweg 11, The Netherlands

Abstract. The development of space-borne measurement sensors and powerful computer hardware, in combination with improvements in mathematical and physical modelling, allow us to determine the Earth's gravity field with increasing resolution and accuracy. The huge amount of data and unknowns and the stringent accuracy requirements pose the limits for simplifications of the functional and stochastic model and for the numerical and approximation errors introduced in the course of the least-squares data processing. This requires the design of sophisticated numerical algorithms despite of the increasing computer power. This paper addresses a number of computational problems that one frequently encounters in least-squares gravity field determination: i) how to set-up and solve the normal equations efficiently, ii) how to deal with instabilities, iii) how to model the observation noise correctly, iv) how to find optimal weights for different types or groups of observations? Possible solutions to these problems are presented and supported by simulations. The gravity field determination from satellite gravity gradients serves as an example, but the presented solutions may be used for other types of satellite data, as well. Moreover, some developments may also be relevant for other data acquisition techniques such as airborne gravimetry.

Keywords. Numerical methods, large least-squares problems, GOCE, satellite gravity gradiometry

1 Introduction

The new gravity field missions CHAMP (Challenging Mini-Satellite Payload for Geophysical Research and Application), GRACE (Gravity Recovery and Climate Experiment) and GOCE (Gravity Field and Steady-State Ocean Circulation Explorer) are triggering a lot of scientific research, among them activities devoted to the least-squares estimation of the gravity field from the huge number of observations collected in the course of these missions. In particular in preparation of the European Space Agency's GOCE mission, to be launched in 2006, a tremendous progress has been achieved in the development of new methods and algorithms, which lead to significant performance improvements in terms of accuracy and CPU-time. Recent numerical simulations indicate that the least-squares estimation of the Earth's gravity field from GOCE SGG data may be done in only a couple of days on a Pentium IV PC (cf. Ditmar and Klees 2002).

One of the main characteristics of space-borne sensors is that they collect a huge amount of data over short periods of time. Sometimes, the noise characteristics are not exactly known and one has to deal with inaccurate stochastic models. Often, the data noise is highly correlated along the orbit. The amount of data yields large design matrices and the high spatial resolution of today's gravity field models yields large normal equations. Then, the proper handling of colored observation noise becomes non-trivial, and may require non-traditional approaches to achieve the highest accuracy. Data gaps such as spikes may further complicate the situation. Finally, the normal equations are often ill-conditioned due to the attenuation effect of the satellite altitude, the high truncation degree of the spherical harmonic expansion (e.g. 300), the orbit geometry, and the noise characteristics of the measurement sensors. Similar problems may also arise in other acquisition techniques such as airborne gravimetry.

The aim of this paper is to present some recent developments in numerical methods that may be of interest for least-squares gravity field modelling. The material is not completely new; parts have been published recently or will be published soon. Some specific items to be addressed are also the subject of contributions to these proceedings. The reader is referred to the contributions of e.g. Klees, Ditmar, Kusche, Schuh, Cai et al. and others.

Section 2 is devoted to the solution of the normal equations and emphasizes the benefit of iterative solvers in least-squares gravity field modelling. The major argument is that in gravity field modelling iterative methods open the door for a number of additional approaches, which may reduce the numerical complexity significantly or improve the accuracy of the least-squares solution. In section 3 it is shown that the (implicit) assembly of the normal equations, which is even more demanding than the solution, can benefit from sophisticated algorithms exploiting the efficiency of FFT methods and 3D interpolation. Section 4 addresses the problem of pre-conditioning, which mainly determines the convergence rate of iterative methods. When satellite data are used we are often in a favorable position, because the physics of the problem allows to design suitable pre-conditioners. In section 5 the issue of colored noise in large least-squares problems is discussed. It is shown that ARMA filters allow to model colored noise very efficiently, provided that the assumption of stationarity is justified. Then, the numerical costs of colored noise modelling are almost negligible compared with the overall costs of the least-squares processing. Section 6 is devoted to the choice of the regularization matrix. An alternative to the classical Kaula regularization is proposed, which is as simple as Kaula, but offers some advantages for high-resolution gravity field modelling. In section 7 an approach to automatically select the regularization parameter is discussed. It exploits a fast Monte-carlo algorithm, which allows us to compute efficiently the trace of large matrices using stochastic trace estimators. Another interesting application of this algorithm are variance-component estimation techniques, which are useful in order to find the optimal weighting for different groups or types of observations.

In the following sections, results of simulations are presented to demonstrate the performance of the various algorithms. The simulations are related to the determination of the Earth's gravity field from satellite gravity gradients, which are the primary observations of the GOCE mission. The data are generated along a realistic 6-month non-repeat GOCE orbit with a mean inclination of 96.6° and an average elevation above the equator of about 262 km. The three diagonal elements of the SGG tensor are used, the sampling rate is 1-s. The disturbing potential to be recovered is the difference between the OSU91a model truncated at degree 300 and the GRS80 normal field. Digital versions of the noise PSD functions published in ESA (1999) are used. The frequency-depending weighting procedure to be discussed in section 5 is applied.

2 Direct vs. iterative solution of normal equations

The traditional approach to solve normal equations in gravity field modelling is the Cholesky method. This requires the explicit computation of the (positive definite) normal matrix and may become very expensive in terms of computation time and memory space if the number of unknowns is large. The computational complexity is often reduced by making some approximations in the course of the assembly and/or the solution of the normal equations. For instance, one may neglect noise correlations or use a simplified noise model; the regularization parameter may be fixed a priori instead of using methods to determine it explicitly from the data; last but not least, simple data weighting schemes are used, since optimal data weighting techniques are forbidden for numerical reasons. Sub-optimal solutions may be a consequence of these approximations.

Iterative methods are often more economic in terms of computation time. For instance, when the normal equations are computed explicitly, the amount of work per iteration is of $O(r^2)$, where r denotes the number of unknowns. If the condition number is of $O(r^\tau)$, the number of iteration steps is of $O(r^{\tau/2})$. Then, the total work estimate is roughly of $O(r^{2+\tau/2})$. When $\tau < 2$, this is less than the amount of work of Cholesky's method, which requires $O(r^3)$ operations. The most expensive part when using Cholesky is the assembly of the normal matrix, which takes $O(nr^2)$ operations. This is not needed when iterative methods are used; it is sufficient to compute the application of the normal matrix to a vector. This requires only $O(2nr)$ operations per iteration, where n is the number of observations.

In situations when both classes of methods are more or less comparable from a computational complexity point of view, the argument in favor of using iterative methods is that they typically require orders of magnitude less memory space, provided that the normal matrix is not computed explicitly. Of course, there are many more aspects that have to be taken into account when comparing direct and iterative methods. For instance, the question whether iterative methods can take advantage of parallel architectures. Another argument often used in favor of Cholesky's method is the need to compute the inverse of the normal matrix to assess the accuracy of the estimated parameters. We believe, however, that this cannot justify the computation of a suboptimal solution, but should be considered as an independent task. The benefit of using iterative methods may be so stringent that in future they may become the primary tool for computing the least-squares solution, whereas the application of direct methods will be limited to the computation of the inverse of the normal matrix.

The iterative method of choice to solve the (symmetric) normal equations, is the preconditioned conjugate gradient method (PCCG) (Hestenes and Stiefel 1952). We refer also to Schuh (1996), who used PCCG for the gravity field determination from satellite gravity gradients. It has some nice features, which makes it particularly suitable for least-squares gravity field modelling: i) the convergence is guaranteed for positive semi-definite normal matrices; ii) it converges to the pseudo-inverse solution in the indefinite case; iii) the speed of convergence is primary a matter of proper pre-conditioning. Maybe the most convincing advantage is that it may be combined with very fast algorithms for i) the application of the design matrix and the transposed design matrix to a vector, and ii) the accurate handling of colored noise. This, in turn, is a pre-requisite for the application of i) advanced regularization parameter choice rules and ii) variance-component estimation techniques for optimal data weighting.

To explain the basic idea, it is assumed that the Tikhonov-regularized normal equations are given by

$$(A^T C^{-1} A + \alpha R)\, x_\alpha = A^T C^{-1} y, \qquad (1)$$

where $y \in \mathcal{R}^n$ denotes the stochastic vector of observations, $A \in \mathcal{R}^{n \times r}$ is the design matrix, $C \in \mathcal{R}^{n \times n}$ is the noise covariance matrix, x_α is the vector of estimated potential coefficients with length r, α a regularization parameter, and $R \in \mathcal{R}^{r \times r}$ denotes the regularization matrix. When PCCG is used to solve Eq. (1), the basic operations to be performed at each iteration is i) the application of $N_\alpha := A^T C^{-1} A + \alpha R$ to a vector, i.e.

$$p = N_\alpha\, q, \qquad (2)$$

and ii) the pre-conditioning

$$s = M^{-1} w. \qquad (3)$$

The matrix M is called the pre-conditioner, which is an approximation to N_α. The better this approximation the better the eigenvalues of the preconditioned matrix $M^{-1/2} N_\alpha M^{-1/2}$ are clustered and the faster the convergence of the iterations is.

In large least-squares problems and in the presence of colored observation noise, it may be forbidden to compute the normal matrix N_α explicitly. This is not needed if Eq. (2) is computed as a sequence of matrix-vector multiplications:

1. $u = A q$.
2. $v = C^{-1} u$.
3. $\tilde{p} = A^T v$.
4. $p = \tilde{p} + \alpha R\, q$.

Often, R is diagonal, and step 4 does not need any special pre-caution. The first three steps may be computationally very demanding and need special attention.

3 Fast spherical harmonic synthesis and co-synthesis

The computation of $u = A q$ is a spherical harmonic synthesis (SHS) at each observation point. Efficient SHS algorithms based on FFT have been developed for observation points located on a regular spherical grid (e.g. Colombo 1981). In real applications, the observation points are distributed along a non-circular orbit, and the sampling along the orbit may be irregular. However, even then it is possible to design very fast algorithms for SHS. An idea is to perform the SHS as a 3D-interpolation at the observation points using the nodes of a number of spherical grids as data (Figure 1). The 3D interpolation is performed in spherical coordinates using Oberhauser tensor product splines. The data

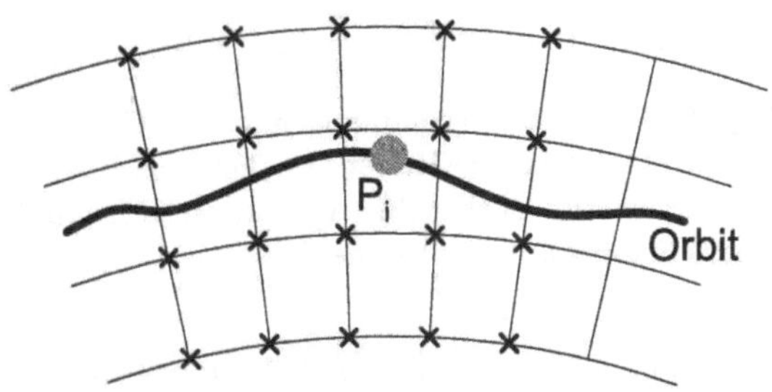

Fig. 1: SHS along a satellite orbit by 3D interpolation.

points on the spherical grids are generated using FFT. The major advantage of this approach is a reduced numerical complexity. SHS at irregular distributed points requires $O(n\,N_{max}^2)$ operations, where N_{max} is the maximum spherical harmonic degree. The fast SHS reduces this to $O(n) + O(N_{max}^2)$, since now the loops over spherical harmonic degrees and orders are de-coupled from the loops over the data. Interpolation errors are controlled by the grid size, i.e. the size of the equiangular cells on each sphere, and the number of spheres. The minimum number of spheres is four for Overhauser splines. This is sufficient for almost circular orbits. Ditmar and Klees (2002) have implemented this technique in the software package GOCESOFT, which is being designed to process SST and SGG data of the GOCE satellite. They showed that a 3D grid with cells of size 0.18° × 0.18° × 24 km is sufficient to make the interpolation error a factor 10 smaller than the expected measurement accuracies of the GOCE gradiometer over the whole measurement bandwidth.

If gravity gradients are to be interpolated the SHS has to be completed by a rotation of the gravity gradient tensor to the adopted frame, and a selection of the components that are included in the SHS. A rotation is also necessary if the gravity vector has been observed.

From a linear algebra point of view, the SHS on a 3D grid, the 3D spline interpolation, the tensor rotation and the component selection are equivalent with a decomposition of the design matrix A into the product of four matrices. This implies that also the transposed operation, i.e. the so-called spherical harmonic co-synthesis (SHCS), can be expressed as the product of four matrices. Each matrix is the transposed of the corresponding matrix of the SHS. Correspondingly, the SHCS can be performed along the same lines as the SHS with the same numerical complexity. For more details see (Ditmar and Klees 2002 and Ditmar et al. 2003). To demonstrate the performance, the fast SHS and SHCS have been performed for three gravity gradient tensor components at about 15.6 million data points distributed along a non-circular GOCE orbit. The maximum degree of the spherical harmonic expansion is $N_{max} = 300$. The fast SHS takes about 41 seconds, the fast SHCS about 45 seconds. This is a factor 1800 faster than the time needed for the classical SHS and SHCS at the irregularly-distributed data points (SGI Origin 3800, 32 processor elements). We want to emphasize that the computational complexity of the fast SHS and SHCS scales linearly with the number of observations n provided that $n \gg N_{max}^2$, which is fulfilled in gravity field modelling from satellite data. Therefore, the higher n the better the fast SHS and SHCS perform.

4 Preconditioning

The number of iterations of the PCCG method is governed by the clustering of the eigenvalues of the normal matrix. This clustering can be improved by pre-conditioning. This means that the matrix N_α, Eq. (1), can be thought of to be scaled using the inverse of some suitable approximation M. The pre-conditioner M is constructed such that the pre-conditioned normal matrix $M^{-1/2} N_\alpha M^{-1/2}$ has a clustered spectrum. A suitable pre-conditioner should also keep the operation count per iteration low. That is, the system $M\,x = z$, with some $z \in \mathcal{R}^r$, must be solvable much easier than $N_\alpha\,x = z$, and the construction of M should not require too many operations.

What makes pre-conditioning difficult is that there is no black-box method that works fine in all applications, i.e. the construction of an efficient pre-conditioner is very much problem-dependent (e.g. Schuh 1996). However, in gravity field determination from satellite data, we are sometimes in a much better position. For instance, Colombo (1986) has shown that for a class of geodetic observables, the normal matrix would become perfectly block-diagonal if the orbit is circular repeat and slowly precessing due to the Earth's flattening, and if the sampling rate is constant. This block-diagonal matrix may be used as a suitable pre-conditioner. Sometimes, appropriate refinements can improve the performance of the pre-conditioner. For instance, Ditmar et al. (2001) relaxed the condition of a circular orbit. They designed a pre-conditioner for the inversion of satellite gravity gradients, which accelerates the convergence 2-3 times compared with the conventional block-diagonal approximation of the normal matrix.

5 Colored noise

The noise of sensors on moving platforms is often colored, i.e. non-uniformly distributed over the frequency. Figure 2 shows an example of the noise in satellite gravity gradients as expected for the GOCE gradiometer. In particular there is a

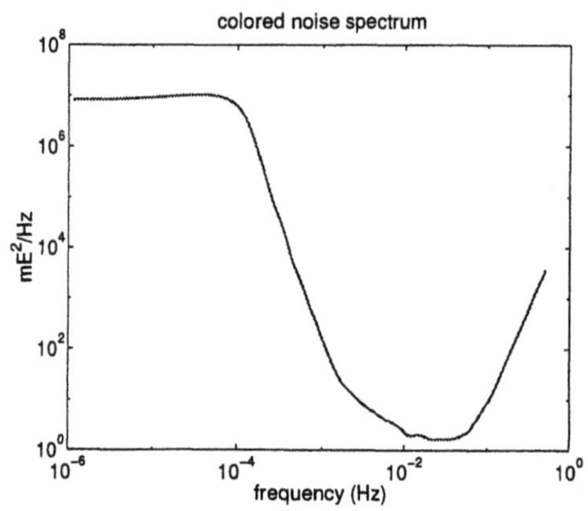

Fig. 2: Typical noise PSD function for one component of the tensor of gravity gradients as expected from the gradiometer on board GOCE

high noise power at low frequencies, which is responsible for strong correlations over longer periods of time. The dependence of the noise on frequency implies a non-diagonal noise covariance matrix. Then, the matrix C^{-1}, which appears in the normal equations, Eq. (1), cannot be easily computed if one deals with millions of observations. When PCCG is used, this inverse is not needed and only the product $C^{-1}u$ has to be computed (see section 2). So far, the problem of colored noise in large least-squares problems has not been solved entirely. However, if the assumption of stationarity is justified, several algorithms have been proposed, which exploit the Toeplitz structure of C in this case. These algorithms have a numerical complexity ranging from $O(n^2)$ to $O(n)$. One example is the $O(n)$-algorithm proposed by Klees et al. (2003). It uses an ARMA model of the colored noise to approximate the solution of the system of linear equations $Cv = u$

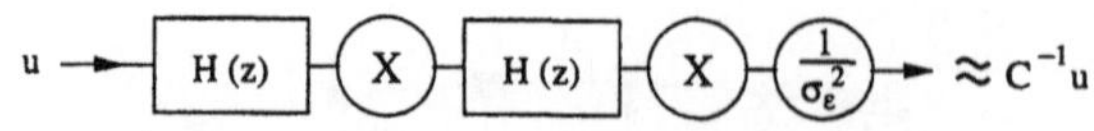

Fig. 3: Approximate computation of $v = C^{-1}u$ using ARMA filtering (Klees et al. 2003). $H(z)$ is the transfer function of the ARMA filter, Eq. (5), and X means flipping a vector upside-down.

(step 2 in section 2). This ARMA model is

$$\xi_n = \varepsilon_n - \sum_{k=1}^{p} a_{p,k}\,\xi_{n-k} + \sum_{i=1}^{q} b_{q,i}\,\varepsilon_{n-i}, \quad n \in \mathcal{Z}, \tag{4}$$

where $\{\xi_n\}$ is the colored noise process, $\{\varepsilon_n\}$ is a white noise process with zero mean and variance σ_ε^2, and $\mathcal{Z}$ denotes the set of integer values (see e.g. Brockwell and Davis 1991). The coefficients $\{a_{p,k} : k = 1,\dots,p\}$ and $\{b_{q,i} : i = 1,\dots,q\}$ are the model parameters, and the pair (p,q) describes the order of the ARMA model. Klees and Broersen (2002) developed methods to compute a best-fitting ARMA model from a given PSD or noise realization. The approximate solution of $Cv = u$ can be split in five steps (cf. figure 3): 1. filter u, 2. flip the result of 1., 3. filter the result of 2., 4. flip the result of 3., and 5. scale the result of step 4 with $1/\sigma_\varepsilon^2$. The filter applied at step 1 and step 3 has the transfer function

$$H(z) = \frac{1 + \sum_{k=1}^{p} a_{p,k}\,z^k}{1 + \sum_{i=1}^{q} b_{q,i}\,z^i}, \quad |z| \le 1 \tag{5}$$

(ibid.). The computation requires approximately $2(p+q)n$ operations per iteration of the PCCG method instead of the classical $O(n^2)$ algorithm of Levinson (1946). Since the order (p,q) of the ARMA process is independent of n, the numerical complexity of the algorithm is essentially of $O(n)$. A different algorithm, which also exploits an ARMA model of colored noise, has been proposed by Schuh (1996).

6 Regularization

Usually, the matrix $A^T C^{-1} A$ is ill-conditioned. This fact may be due to various reasons:

- the downward continuation from satellite altitude to the Earth's surface, which is implicitly included in the estimation procedure;
- colored observation noise, which may not allow to determine certain frequencies of the Earth's gravity field very well;

- data gaps, e.g. due to the orbit geometry.

The ill-posedness can further increase due to the choice of the maximum degree of the harmonic expansion of the gravitational field beyond the resolution limit. This is often done to avoid the loss of information contained in the data.

All this implies that the spherical harmonic coefficients cannot be accurately determined from the data solely. Then, the application of a suitable regularization becomes important. Improper regularization may not only lead to an inaccurate model of the Earth's gravity field, but can also cause divergence of the PCCG method. Usually, a Tikhonov regularization is applied to regularize the normal equations, i.e. a positive definite regularization matrix R times a regularization parameter α is added to the matrix $A^T C^{-1} A$, cf. Eq. (1).

Regularization involves two aspects: 1. the choice of the regularization matrix R, and 2. the choice of the regularization parameter α. A frequently used method in gravity field determination from satellite data is the Kaula regularization. The matrix $R = R^K$ is diagonal with the signal variances given by Kaula's rule of thumb, i.e. $R^K_{ij} = \delta_{ij}\, l_i^4$, where l_i is the degree of the i-th unknown potential coefficient, and the regularization parameter is fixed a priori equal to $\alpha = 10^{10}$. The condition number of R^K is equal to $l_{max}^4/16$ if potential coefficients from degree 2 to l_{max} are estimated. Thus, R^K becomes ill-conditioned for large l_{max}, which may cause divergence of the PCCG method.

An alternative is the so-called first-order Tikhonov regularization (e.g. Zhdanov 2002). This can be interpreted physically as the minimization of the horizontal gradient of the gravitational potential over the sphere σ_R of radius R, i.e. the functional to be minimized is

$$\Phi^{FOT} = \int_{\sigma_R} (\nabla_H V)^2 \, d\sigma_R, \qquad (6)$$

where ∇_H is the surface-gradient operator. Using Green's second formula and taking the orthogonality property of surface spherical harmonics into account, one obtains

$$\Phi^{FOT} = 4\pi \left(\frac{GM}{R^2} \right)^2 x^T R^{FOT} x,$$

where the matrix R^{FOT} is defined as

$$R^{FOT}_{ij} = \delta_{ij}\, l_i\, (l_i + 1).$$

The condition number of R^{FOT} is approximately equal to $l_{max}^2/6$ if coefficients of degrees $l = 2, \ldots, l_{max}$ are solved for. Thus, R^{FOT} may also become ill-conditioned for large l_{max}, but the conditioning is a factor $\frac{3}{8} l_{max}^2$ better than the conditioning of R^K. Ditmar and Klees (2002) obtained very good results in simulations of the GOCE mission even up to degree $l_{max} = 300$.

Sometimes additional precautions have to be taken in order to obtain reasonable results when applying regularization. One example is the inversion of satellite gravity gradients to be measured by the gradiometer on board GOCE. The expected strong low-frequency noise in the measured gravity gradients implies that the low-frequency information in the columns of the design matrix is suppressed, when frequency-dependent weighting (see section 5) is applied. This makes the elements of $A^T C^{-1} A$ that are related to low degrees very small, i.e. these coefficients become biased towards the corresponding elements of αR. This low-frequency bias may significantly corrupt the solution. Therefore, coefficients related to low frequencies should not be regularized. This is achieved by the following modification of the FOT regularization matrix:

$$R^{FOT}_{ij} = \begin{cases} 0 & \text{if } l \le l_{thr}, \\ \delta_{ij}\, l_i (l_i + 1) & \text{otherwise.} \end{cases}$$

The choice of the threshold l_{thr} depends on the noise PSD function. For GOCE it is known that the noise PSD function starts increasing proportional to $1/f^2$ (f is the frequency) below 0.005 Hz. Thus a choice of $l_{thr} \approx 30$ may be appropriate in this case (cf. Ditmar and Klees 2002).

Figure 4 shows a map of geoid height errors obtained from a full inversion of simulated gravity gradients, when FOT regularization is applied. The regularization parameter was chosen to minimize the global RMS geoid height error over the area between ±80° latitude.

Figure 5 shows the RMS geoid height error as a function of the regularization parameter for Kaula and FOT regularization. Both Kaula and FOT regularization yield the same minimum RMS geoid height error of about 0.15 m. The choice $\alpha = 10^{10}$ in combination with Kaula regularization, which is often made in gravity field solutions from satellite data, yields a slightly higher RMS geoid height error. Therefore, when the Kaula matrix is used it makes sense to use a parameter choice rule (e.g. generalized cross valida-

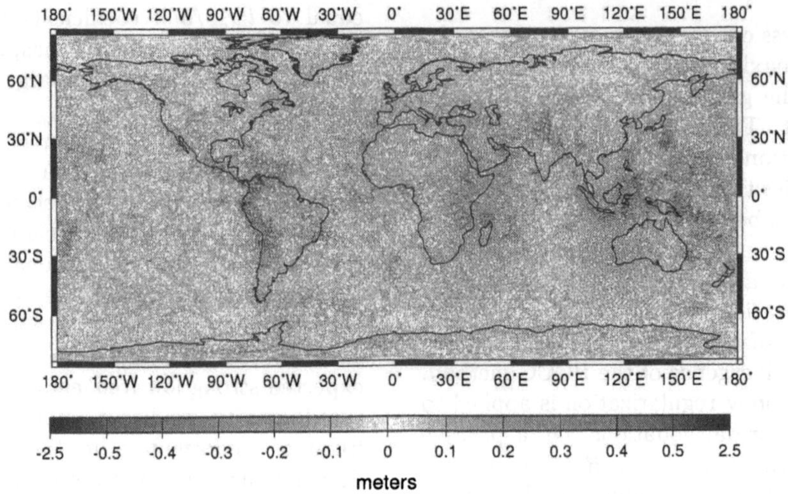

Fig. 4: Map of geoid height errors after inversion of SGG observations using FOT regularization. The regularization parameters is equal to 10^{15}. Only potential coefficients with degrees above 30 are subject to regularization. The RMS geoid height error is 0.15 m.

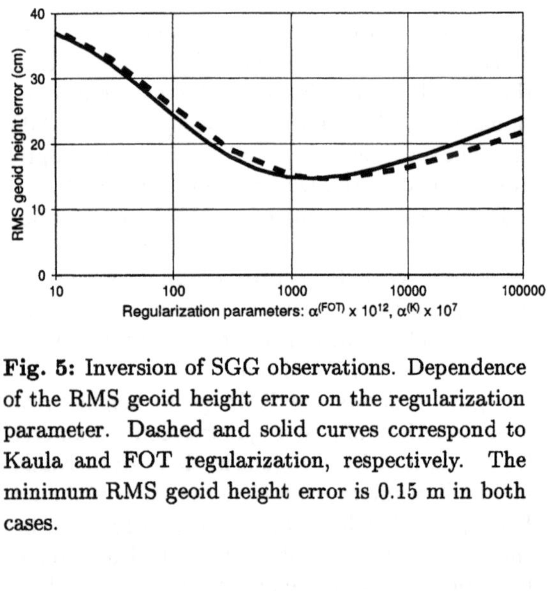

Fig. 5: Inversion of SGG observations. Dependence of the RMS geoid height error on the regularization parameter. Dashed and solid curves correspond to Kaula and FOT regularization, respectively. The minimum RMS geoid height error is 0.15 m in both cases.

tion, cf. Kusche and Klees 2002) instead of fixing the regularization parameter a priori.

7 Choice of the regularization parameter and variance component estimation

In the previous section the regularization parameter was chosen such as to minimize the global RMS geoid height error. The latter has been computed by comparing the estimated gravity field model with the true one. When processing real data, the true model is not available, i.e. the selection of the regularization parameter must be based on some other criterion (see Hanke and Hansen (1993) for an overview of regularization parameter choice rules). One of the problems one frequently encounters when dealing with large least-squares problems, is the numerical complexity of the parameter choice rule. That is the reason why sometimes methods, although very promising from a conceptual point of view, cannot be applied to large least-squares problems. Examples are the methods of Schwintzer (1990), Cai et al. (this volume), variance component estimation (e.g. Koch 1990), and generalized cross-validation (Golub et al. 1979). These parameter choice rules often involve the computation of the trace of one or several matrices of dimension $r \times r$ or even $n \times n$, where r and n is the number of unknowns and observations, respectively.

For instance, generalized cross validation (GCV) selects the optimal α according to

$$\alpha_{gcv} = \arg\min_{\alpha} \frac{n\|Ax_\alpha - y\|^2_{C^{-1}}}{(n - \text{trace } Q_\alpha)^2}, \quad (7)$$

$$Q_\alpha = AN_\alpha^{-1}A^TC^{-1}. \quad (8)$$

This involves the computation of the trace of the $n \times n$ matrix Q_α for each α, which is prohibitively expensive in large least-squares problems.

One faces a similar problem when (possibly multiple) parameters are to be estimated or validated from the data, which govern the stochastic models of the observation, like variance components or correlation coefficients. This is especially important, when, in a least-squares estimation, weights have to be assigned to the different (heterogeneous) data sets and to the prior constraints. Formal errors, based on prior assessment of instrument noise, may be over-optimistic by orders of magnitude. In reality, these covariance parameters also account for inaccuracies of the functional models. Examples are the methods developed by Lerch (1991) for down-weighting data sets, by Schwintzer (1990) for weighting prior information, and the variance component estimation technique recently proposed by Koch and Kusche (2002).

Lerch's method, essentially, requires repeatedly the traces of (two different) variance covariance matrices to be compared, whereas variance component estimation techniques (Koch 1990) usually require trace terms of the data influence matrix or even products of this matrix to be evaluated within each iteration step. In Maximum-Likelihood estimation, for example, the group redundancy number for the i-th data set,

$$\rho_i = n_i - 1/\sigma_i^2 \text{ trace } (A_iN^{-1}A_i^TC_i^{-1}),$$

has to be computed (see Kusche, this volume).

A way out is the use of a fast 'Monte-Carlo' algorithm for the computation of reliable estimates of the trace of the involved matrix. This general algorithm has been proposed by Girard (1989) and goes as follows. Assume that the $s \times s$ matrix B is symmetric and that trace(B^{-1}) has to be computed: i) generate a vector w consisting of s pseudo-random values $w_1, \ldots, w_s$ obeying the standard normal distribution, ii) solve the system of linear equations $Bw_\alpha = w$, iii) take w^Tw_α as an approximation to trace(B^{-1}). This estimate is unbiased (Girard 1989). Thus,

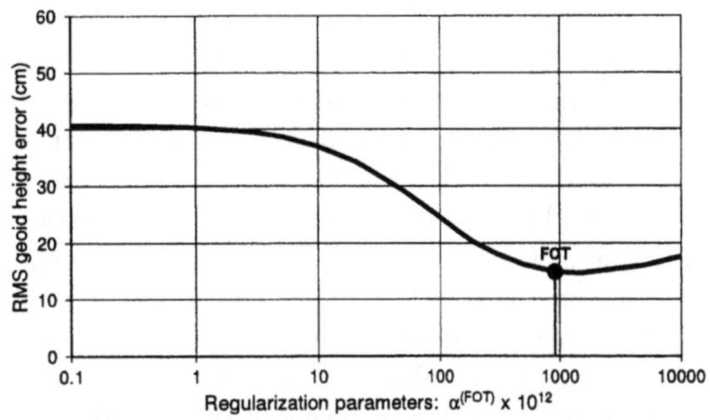

Fig. 6: Gravity field determination from simulated satellite gravity gradients. RMS geoid height error as function of the regularization parameter for FOT. The vertical line indicates the regularization parameter found by GCV.

the arithmetic mean of m of such estimates, produced by running this procedure m times, converges with probability one to trace(B^{-1}), as $m \to \infty$. Girard (ibid.) has also proved that for large s only one run, i.e. $m = 1$, is needed to get a reliable estimate of the trace. Kusche and Klees (2002) performed simulations concerning the gravity field determination from satellite gravity gradiometry and come to the same conclusion.

The algorithm proposed by Girard (ibid.) may be used to compute trace Q_α, Eq. (7), as follows:

1. compute $w_1 = A^TL^{-1}w$ (whitening filter and SHCS, where $C = L^TL$)
2. for each α
   (a) solve $N_\alpha w_\alpha = w_1$,
   (b) compute trace $Q_\alpha \approx w_1^Tw_\alpha$.

Step 1 is independent of α, i.e. it has to be performed only once. The most time consuming step here is 2a, which is nothing but solving the system of normal equations with the right-hand side vector w_1. Finally, the computation of the nominator of Eq. (7) requires the solution of the normal equations, Eq. (1), with $\alpha = \alpha_{gcv}$, one SHS and one filter operation. In the same way we can compute the group redundancy numbers ρ_i when variance components are to be estimated. However, for the evaluation of Eq. (7) with diagonal regularization matrix $R = S^TS$, it is even possible to simplify step 1 by applying Girard's algorithm to trace $(SN_\alpha^{-1}S^T)$ instead, and finally derive trace $Q_\alpha \approx r - \alpha\, w_1^Tw_\alpha$.

Figure 6 shows the global RMS geoid height error as function of the regularization parameter for FOT, and the regularization parameter α_{gcv}, which minimizes the GCV functional, Eq. (7). The selected regularization parameter is $\alpha_{gcv} = 0.92 \cdot 10^{15}$, the associated global RMS geoid height error is 0.149 m. The regularization parameter that minimizes the global RMS geoid height error is $\alpha = 1.50 \cdot 10^{15}$, the corresponding global RMS geoid height error is 0.147 m. Thus, GCV provides a solution, which is very close to the optimal solution in terms of the global RMS geoid height error. This has also been found in many other simulations related to the least-squares gravity field determination from satellite gravity gradients. The GCV function, Eq. (7), was computed at 8 points before the minimum was found. Thus, in total 16 different 'systems of normal equations' were solved in order to find the optimal regularization parameter. The 17th system was solved to find the optimal solution $x_{\alpha_{gcv}}$ itself. The total wall-clock time was 6508 seconds (SGI 3800 Origin, 32 processing elements).

8 Summary

We proposed efficient numerical algorithms to solve a couple of problems one frequently encounters in least-squares gravity field determination from space-borne measurements. The preconditioned conjugate gradient method may be a real alternative to the Cholesky method. The major advantage is that it does not require to compute the normal matrix explicitly, and offers fast algorithms for the spherical harmonic synthesis, co-synthesis, and proper handling of colored observation noise. This accelerates the computation of the least-squares solution dramatically without introducing significant distortions into the solution. Consequently, it may allow to solve large least-squares problems even on PC's. Efficient pre-conditioners can often be found for satellite data. Frequency-depending weighting in combination with ARMA models allow the numerically efficient and accurate treatment of colored noise. Moreover, this approach can easily be integrated into the preconditioned conjugate gradient method. Regularization is routinely applied to get physically meaningful gravity field models. First-order Tikhonov regularization proofed to work well in combination with generalized cross validation, although the improvement relative to Kaula is minor. Randomized trace estimators provide reliable estimates of the trace of large matrices, which is a basic operation to be performed when regularization parameter choice rules are applied such as generalized cross validation, and in optimal weighting of different data sources such as variance component estimation techniques. A pre-requisite to apply advanced parameter choice rules and data weighting techniques, is the fast assembly and solution of the normal equations. The proposed approaches for spherical harmonic synthesis, co-synthesis, and frequency-dependent weighting, provide the necessary tools.

Acknowledgment

Computing resources were provided by Stichting Nationale Computerfaciliteiten (NCF), grant SG-027.

References

Brockwell PJ, Davis RA (1991) Time series: theory and methods. Second Edition, Springer Series in Statistics, New York.

Cai J, Grafarend EW, Schaffrin B (2002) The A-optimal regularization parameter in uniform Tykhonov-Phillips regularization - α-weighted BLE. These proceedings.

Colombo O (1981) Numerical methods for harmonic analysis on the sphere. Reports of the Department of Geodetic Science, Report No. 310, The Ohio State University, Columbus, Ohio, USA.

Colombo O (1986) Notes on the mapping of the gravity field using satellite data. In Sünkel H, editor, *Mathematical and Numerical Techniques in Physical Geodesy*, Lecture Notes in Earth Sciences, volume 7, Springer, Berlin, Germany.

P. Ditmar, E. Schrama, R. Klees (2001). Improved block diagonal approximation of the normal matrix in the inversion of satellite gravity gradiometry data. Proc. IAG 2001 Scientific Assembly, Budapest, Hungary, 2001, CD-ROM, 6 pages.

Ditmar P, Klees R (2002) A method to compute the earth's gravity field from SGG/SST data to be acquired by the GOCE satellite. Delft University Press (DUP) Science, Delft, The Netherlands, 64 pages.

Ditmar P, Klees R, Kostenko F (2003) Fast and accurate computation of spherical harmonic coefficients from satellite gravity gradiometry data. *Journal of Geodesy* 76: 690-705.

ESA (1999) Gravity field and steady-state ocean circulation mission. Reports for Mission Selection - The Four Candidate Earth Explorer Core Missions, ESA SP-1233 (1), Noordwijk, The Nether-

lands.
Girard DA (1989) A fast 'Monte-Carlo cross-validation' procedure for large least squares problems with noisy data. *Numer Math* 56: 1-23.
Golub GH, Heath M, Wahba G (1979) Generalized cross-validation as a method for choosing a good ridge parameter. *Technometrics* 21: 215-223.
Hanke M, Hansen, PC (1993) Regularization methods for large-scale problems. *Surv Math Ind* 3: 253-315.
Hestenes MR, Stiefel E (1952) Methods of conjugate gradients for solving linear systems. *Journal of Research of the National Bureau of Standards* 49: 409-436.
Klees R, Ditmar P, Broersen P (2003) How to handle colored noise in large least-squares problems. *Journal of Geodesy* 76: 629-640.
Klees R, Broersen P (2002) How to handle colored observation noise in large least-squares problems? - Building the optimal filter. Delft University Press, Science series, Delft, The Netherlands, 29 pages.
Koch R (1990) Bayesian inference with geodetic applications. Springer Berlin, Heidelberg, New York.
Kusche J, Klees R (2002) Regularization of gravity field estimation from satellite gravity gradients. *Journal of Geodesy* 76: 359-368.
Koch KR, Kusche J (2002) Regularization of geopotential determination from satellite data by variance components. *Journal of Geodesy* 76: 259-268.
Lerch FL (1991) Optimum data weighting and error calibration for estimation of gravitational parameters. *Bull Géod* 65: 44-52.
Levinson N (1946) The Wiener RMS (root mean square) error criterion in filter design and prediction. *Journal on Mathematical Physics* 25: 261-278.
Schuh WD (1996) Tailored numerical solution strategies for the global determination of the earth's gravity field. Mitteilungen der geodätischen Institute der Technischen Universität Graz, Folge 81, Graz, Austria.
Schwintzer P (1990) Sensitivity analysis in least squares gravity field modelling, redundancy decomposition of stochastic a priori information. Internal Report, Deutsches Geodätisches Forschungsinstitut, München, Germany, 14 pages.
Zhdanov MS (2002) Geophysical inverse theory and regularization problems. Methods in Geochemistry and Geophysics, volume 36, Elsevier, Amsterdam, The Netherlands.

The numerical treatment of the downward continuation problem for the gravity potential

W.-D. Schuh, B. Kargoll
Institute of Theoretical Geodesy (ITGB),
University of Bonn, Nussallee 17, D-53115 Bonn, Germany

Abstract. This paper discusses numerical and statistical techniques used to recover the gravity potential from GOCE mission data. In particular, in a closed loop simulation, it is shown that two completely different and independent solution strategies, i.e. the direct method and the semi-analytic approach, lead to essentially identical results. Both methods can give only a finite representation of the gravity potential. The truncation of the infinite series leads to a special type of regularization, denoted as spectral leakage. The size of this effect is estimated in a closed loop simulation. To verify the quality of the gravity field recovery a detailed analysis of the whole adjustment procedure is necessary. Therefore, a step-by-step statistical test strategy is introduced to validate the deterministic model as well as the stochastic assumptions by analyzing the residuals of the decorrelated adjustment problem. First of all a test of randomness is applied to check the assumption of uncorrelated residuals. The assumption of stationarity is checked by variance analysis of different observation groups. Finally, the autocorrelation function and the periodogram of the transformed residuals are tested for significant correlations. To eliminate the remaining correlations the filter model and thus the stochastic model is improved.

Keywords. Spherical harmonic analysis, spectral leakage, gradiometry mission GOCE, diagnostic checking, model validation

1 Introduction

Present and planned satellite missions are dedicated to recover the detailed structures of the Earth's gravity field, which are represented by a set of the order of fifty to one hundred thousand parameters. This large number of parameters must be derived from an even larger number of correlated observations. But only the fusion of different sensors and information guarantees an accurate solution over the whole spectral range.

Therefore, the gravity gradiometry mission GOCE will deliver essentially two types of measurements: GPS satellite-to-satellite tracking (SST) data reflecting orbit perturbations and representing first order derivatives of the gravitational potential, as well as gravity gradients (SGG) measured by the 3-axes gravity gradiometer representing second order derivatives. Both data types have pros and cons. While the SST measurements determine the low frequencies accurately, the smoothing effect with respect to the satellite altitude filters out the high-frequency contents. In contrast, the SGG observations provide detailed information up to a spatial resolution of 80 km, because the additional differentiation reinforces the high-frequency information. However, the measurement bandwidth of the gradiometer decreases the accuracy of the low-frequency information and yields to a correlated measurement series. Specific strategies must be developed to handle this huge, ill-posed inverse problem and to model the correlations which link millions of observations.

In the last decade, several approaches have been developed to solve this huge system of equations. While the *direct* method directly processes the gravity field observations in the space domain, the *time-wise approach* considers the observations along a satellite track as a time series (Rummel et al., 1993; Sneeuw, 2000). Applying Fourier techniques to the time series opens a very efficient way to handle the correlations in the frequency domain. On the other hand signal processing tools allows the handling of the measurement series also in the time domain. The causality of the discrete filters used clears a path for the application of a least squares adjustment by direct or iterative techniques with sequential access to the observations (Schuh, 1996). Although the observation equations are not identical for the *direct* and the *time-wise* approach, and the correlations are treated in a totally different manner, both processing strategies lead to nearly identical results in terms of the gravity field recovery (cf. Section 3).

Section 2 of this paper introduces the deterministic and stochastic model. This will be followed by a discussion of a closed-loop-simulation in section 3, which shows the performance of the different strategies to reproduce a priori introduced gravity field. In addition to consistent models, the results with truncated models are also summarized to demonstrate the truncation error (spectral leakage effects) on the results. Yet the main part of this paper deals with the interpretation of the residuals of the adjustment procedure. Statistical hypotheses are introduced to find objective criteria for the validation of the model. The (internal) validation of the deterministic and the stochastic model is the heart of the adjustment procedure, but there is no absolute algorithm for approaching it. It is wise to be equipped with a variety of different tools with which we are able to evaluate model qualities. Section 4 describes some techniques and gives examples with respect to the modeling of SGG data.

2 Deterministic and stochastic model

The Earth's gravity field is parameterized in terms of spherical harmonics as base functions,

$$f(\phi,\lambda,r) = \frac{GM}{r}\left\{1+\sum_{\ell=2}^{\infty}\sum_{m=0}^{\ell}\left(\frac{R}{r}\right)^{\ell}\cdot\right.$$

$$\left.\bar{P}_{\ell m}(\sin\phi)\left(\bar{C}_{\ell m}\cos m\lambda + \bar{S}_{\ell m}\sin m\lambda\right)\right\}. \quad (1)$$

G, M, and R, respectively, are the gravitational constant, the Earth's mass and the Earth's reference radius, respectively, while $\bar{P}_{lm}$ denotes the fully normalized Legendre polynomials depending on the harmonic degree ℓ and order m.

The goal is now to determine the corresponding harmonic coefficients $\bar{C}_{lm}$ and $\bar{S}_{lm}$ as accurately as possible.

The dedicated gravity gradiometry mission GOCE will essentially deliver two types of measurements: GPS satellite-to-satellite tracking (SST) data reflecting orbit perturbations and representing first order derivatives of the gravitational potential, as well as gravity gradients (SGG) measured by the 3-axes gravity gradiometer representing second order derivatives, denoted as SGG tensor. In the present paper we will concentrate on the latter measurement type.

Four components (the along track component ΔV_{xx}, the across track component ΔV_{yy}, the zenith component ΔV_{zz}, and the component ΔV_{xz}) will be recorded simultaneously as time series with a time step of $1\,s$ along the satellite orbit. Over the course of the mission (two six-month periods) about 12 million observations will be collected to determine the unknown spherical harmonic coefficients $C_{\ell m}$ and $S_{\ell m}$. This inverse problem is denoted as spherical harmonic analysis.

Gathering all measurements in an observation vector $\boldsymbol{\ell}$ and the corresponding error covariances in the matrix $\boldsymbol{\Sigma}$, and using a standard Gauss-Markov model applying the best linear uniformly unbiased estimation with respect to the $\boldsymbol{\Sigma}^{-1}$-norm leads to the normal equation system

$$(\boldsymbol{A}^T\boldsymbol{\Sigma}^{-1}\boldsymbol{A})\,\tilde{\boldsymbol{x}} = \boldsymbol{A}^T\boldsymbol{\Sigma}^{-1}\boldsymbol{\ell} \quad (2)$$

for the parameter vector $\tilde{\boldsymbol{x}} = \{\bar{C}_{lm};\bar{S}_{lm}\}$, where the design matrix $\boldsymbol{A}$ represents the linear relation between the observations $\boldsymbol{\ell}$ and the parameters $\boldsymbol{x}$, and thus contains the base functions of the series expansion (1). This approach is in principle applied by all *direct methods* to solve for the coefficients.

Due to the fact that measurements taken along a satellite's track are not homogeneously distributed with respect to the spatial domain of the sphere, the orthogonality properties of the base functions are lost, and thus the normal equation matrix $\boldsymbol{A}^T\boldsymbol{\Sigma}^{-1}\boldsymbol{A}$ is a dense matrix. However, if the gravity gradients are measured without interruption and a constant step size along a circular repeat orbit is used, it can be shown that all coefficients of different orders m as well as *cos*- and *sin*-related coefficients become uncorrelated. This leads to a block-diagonal structure of the equation system (Sneeuw, 2000).

In reality, these simplifying conditions cannot be strictly fulfilled; consequently the normal equation system will deviate from the block-diagonal structure. However, due to the large number of unknown parameters and thus the huge dimension of the normal equation system $\boldsymbol{A}^T\boldsymbol{\Sigma}^{-1}\boldsymbol{A}$ (about 90000 unknowns for a maximum resolution of $l_{max} = 300$, corresponding to 70 km half wavelength), the system can not be inverted directly.

Therefore, for the processing of real mission data the *direct method* was implemented in terms of the `pcgma` (**P**reconditioned **C**onjugate **G**radient **M**ultiple **A**djustment) algorithm (Schuh, 1996), using a block-diagonal preconditioner and iteratively applying a conjugate gradient method to successively improve the parameter estimates.

While in the *direct method* the observations are regarded as functions of the geographical location (r,ϕ,λ), they can also be considered as a periodic time series for one repeat period (Rummel et al., 1993; Sneeuw, 2000).

Assuming a circular orbit, (1) can be rewritten as

$$f = \frac{GM}{R} \sum_l \left(\frac{R}{r}\right)^{l+1} \sum_m \sum_k \bar{F}_{lm}^k(i) \cdot$$

$$(\alpha_{lm} \cos \psi_{km}(t) + \beta_{lm} \sin \psi_{km}(t)) \tag{3}$$

with $\bar{F}_{lm}^k(i)$ denoting the inclination function which depends on the orbit inclination i, and the coefficients α_{lm}, β_{lm} related to the harmonic coefficients $\bar{C}_{lm}$, $\bar{S}_{lm}$:

$$\left\{ \begin{array}{c} A_{km} \\ B_{km} \end{array} \right\} = \frac{GM}{R} \sum_l \left(\frac{R}{r}\right)^{l+1} \bar{F}_{lm}^k(i) \left\{ \begin{array}{c} \alpha_{lm} \\ \beta_{lm} \end{array} \right\}. \tag{4}$$

(3) is a Fourier series

$$f^{s_t} = \sum_m \sum_k [A_{km} \cos \psi_{km}(t) + B_{km} \sin \psi_{km}(t)] \tag{5}$$

where $\psi_{km}(t)$ is related to two fundamental frequencies ω_o (orbit revolution) and ω_e (Earth's rotation).

In this *time-wise approach*, also frequently denoted as *semi-analytic approach*, the Fourier coefficients ('lumped coefficients') A_{km}, B_{km} are computed in the first step using FFT methods. In the second step the harmonic coefficients $\bar{C}_{lm}$, $\bar{S}_{lm}$ are estimated order by order, assuming strict block-diagonal structure of the system. The deviations from this property are incorporated by means of an iterative procedure.

As mentioned, the measurements are correlated within each of the four data streams. As *a priori* information, the error budget of the gradiometer output (ESA, 1999) is given in terms of the power spectral density curves (Figure 1).

The semi-analytic approach can directly take advantage of this frequency-domain information, since the frequency-domain coefficients A_{km} and B_{km} can be multiplied frequency-by-frequency. The direct approach works with the covariance matrix $\mathbf{\Sigma}$. However, it is still impossible to handle this huge matrix, and therefore a step-by-step decorrelation process is introduced to circumvent this problem. The main idea is to split up the covariance matrix into two triangular matrices. These triangular Toeplitz systems can be interpreted as linear discrete filters, which allow us to decorrelate the measurement series step-by-step. According to this sequential approach, the spherical harmonic analysis of huge data sets can be handled by current technology, and furthermore the computations can be parallelized easily.

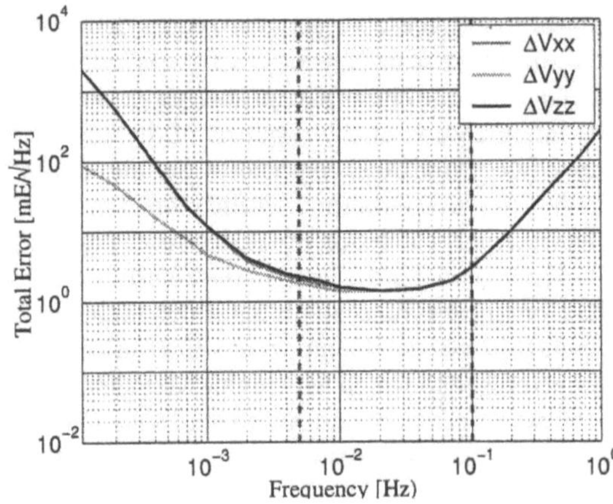

Fig. 1 Square root of the spectra of the gravity-gradient measurement error budget for the components ΔV_{xx}, ΔV_{yy}, and ΔV_{zz} (*Granada*-SGG error budget, cf. ESA (1999), pp. 177, digitized fig. 8.2). The error level in the measurement bandwidth between $5\times10^{-3}\,Hz$ and $0.1\,Hz$ is below $\pm3\,mE/\sqrt{Hz}$.

3 Closed-loop simulations

The current closed loop simulation (Figure 2) starts from an *a priori* known gravity field parameterized by spherical harmonics (OSU91a) up to degree 180. Two data sets were deduced. The geoid heights on a regular grid ($0.25° \times 0.25°$) on the Earth's surface (altitude zero) were computed as a reference field. The observation functionals (three main diagonal elements of the gradient tensor ΔV_{xx}, ΔV_{yy}, ΔV_{zz}) were computed in a sun-synchronous repeat orbit of 29 days with an initial altitude of about $250\,km$ and an inclination of $96.6°$. The simulated measurements, denoted as *Delft-data-set* (cf. ESA (2000), Section 3.6.2), are regularly distributed along the orbit with a sampling interval of $5\,s$. These measurements were superimposed by colored noise generated by an ARMA process[1], which reflects the proposed spectral behavior (cf. Figure 1). This results in more than 1.5 million observations containing information about the Earth's gravity field up to degree and order 180.

[1]The same simulations were also performed with colored noise generated by a harmonic process. The results of these simulations are summarized in Schuh *et al.* (2001) and show a similar performance to the simulations with ARMA noise. Therefore, we leave this aspect beyond the consideration.

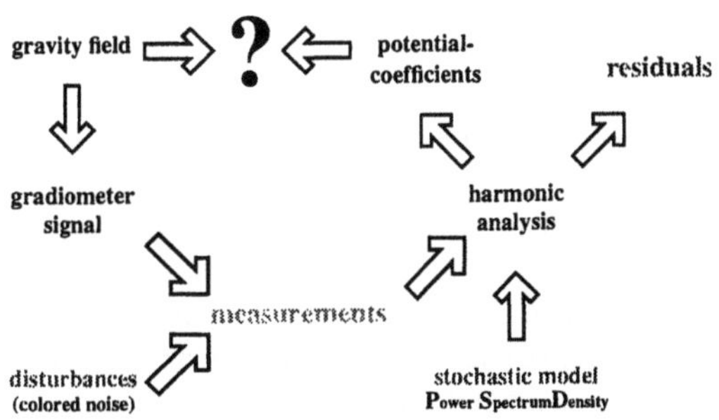

Fig. 2 Closed-loop simulation

The *direct approach* (DA) as well as the *semi-analytic approach* (SA) were applied to estimate the set of 32757 coefficients describing the Earth's gravity field up to a spatial resolution of about 110 km half wavelength. Due to the fact that the input data set is based on a realistic, perturbed orbit, for both methods an iterative solution strategy was applied; at each case 15 iterations were computed.

The numerical results of the closed-loop simulations are summarized in Table 1. Aside from the global behavior (excluding polar regions) the local evaluation shows the capability of reconstructing the high-frequency behavior. For the local representation the data were reduced by a moving average using a Gaussian type window function with a variance of 5°. The first two lines of Table 1 give a comparison between the *semi-analytic approach* and the *direct approach*. Our results show clearly that both techniques lead to essentially identical results.

Further investigations are also summarized in Table 1. As a first aspect, the mission duration of the simulated data set was extended to 58 days. The external validation reflects exactly the same behavior that is expected from the variance propagation law. If we double the number of (independent) observations, the variances decrease by a factor of $\sqrt{2}$ (cf. Table 1, third row).

In contrast to this consistent simulation, in which the maximal degree of the simulated input data corresponds exactly with the maximal degree of the resolved data, we performed further simulations with truncated series, to outline the spectral leakage effect. To get a feeling of the size of this effect, we first computed only the section between 181 to 360 of the measurement signal, which varies within the range of $\pm 6\,mE$. This systematic effect was twice as large as the noise in the measurement bandwidth ($\pm 3\,mE$). But, in contrast to the random characteristic of the noise, the truncation effect has a deterministic behavior. Because of the non-orthogonality of the base functions caused by the distribution of the discrete measurement points, we can expect this high-frequency information also to exert some influence on the low-degree spherical harmonic coefficients. A detailed theoretical foundation of these effects can be found in Wasle (2001).

To quantify this effect, we used the signal between the degrees 181 and 360 as the input signal and performed a spherical harmonic analysis, resolving only the degrees 2 to 180. The computed geoid heights, which should have been zero, varied between $\pm 0.396m$ globally, and $\pm 0.060m$ locally. The variances with respect to the local area averaged to $\pm 0.012m$ (cf. Table 1, second group). This effect is approximately two times smaller than the reconstruction error of the consistent simulation. This field was used to reconstruct the measurement signal at satellite altitude, and it can be seen that 10% of the input signal is absorbed by the adjusted coefficients. A simulation series with realistic mission data showed that for the GOCE mission the coefficients of the last two degrees were largely influenced by this truncation effect. One strategy to minimize the truncation effect is to use the coefficients of the last two degrees as a 'buffer', but in addition, low pass filters can also be used to extenuate the high-frequency part of the signal. A detailed description of these simulations can be found in ESA (2002), Section 2.3.3.

For further investigations the combination of random noise and the spectral leakage effect were of interest. Therefore, a measurement series including both effects was generated. The signal based on the

Table 1. Reconstruction of second level information on a 0.25° × 0.25° grid; geoid heights [m]: Numerical comparison between different approaches (*semi-analytic* (SA) versus *direct*(DA)), different mission durations (29 days and 58 days), and different input signals (consistent model plus noise 2-180+n, truncated signal 181-360, and an inconsistent model 2-360+n). The residuals between the true input model (OSU) and the computed models are summarized by the maximal deviations, the mean value and the variance σ. The factor +2 in the third column points out that this model was solved up to a higher degree, but the last coefficients were used as a buffer to absorb model inconsistencies.

	duration	input signal	solved	global max	local max	local mean	local σ
SA	29	2-180+n	180	0.284	0.082	-0.001	0.020
DA	29	2-180+n	180	0.272	0.082	-0.001	0.019
DA	58	2-180+n	180	0.217	0.053	-0.001	0.013
DA	29	181-360	180	0.396	0.060	0.000	0.012
DA	29	181-360	178+2	0.396	0.042	0.000	0.009
DA	29	2-360+n	180+2	0.510	0.099	-0.001	0.021
DA	58	2-360+n	180+2	0.568	0.060	-0.001	0.016

OSU reference field up to degree 360 was computed, and an ARMA noise was superimposed to this signal. The spherical harmonic analysis was resolved up to degree 182, and the last two degrees were seen as a buffer. The last two lines in Table 1 summarize the results. It can be seen that for the 29-day model, the systematic truncation effects had no large influence on the result. The variance of the disturbances rose minimally, from $\pm 0.020m$ to $\pm 0.021m$. This situation changed, however, when we extended the duration of the mission in our simulation. The influence of the random noise decreased (variance propagation law), whereas the contribution of the deterministic truncation effect remained equal. Therefore, in the case of the 58 day model, our calculations showed an increase in variance of the inconsistent model with respect to the consistent model (from $\pm 0.013m$ to $\pm 0.016m$). It can be stated that the truncation effect worsens the result significantly. With respect to the real measurement duration of the GOCE mission of two six-month periods, an additional reduction of the random noise by a factor of approximately $\sqrt{12} \approx 3.5$ is expected, but also the truncation effects are much smaller if higher degree/order global models are estimated. As a rule of thumb the size of the truncation error is halved if the model is extended by 20 degrees. Therefore, if the resolved degrees are extended by a factor of 40 (maximal degree/order 220), the truncation signal is reduced by a factor of four. Thus, it can be expected that similar random (error : deterministic) error ratios appear for a one-month model developed up to degree 180 and a 12-month model resolved up to degree 220.

These closed-loop simulations are very useful to study individual effects and to compare different solution strategies. Unfortunately, reality does not permit us to fall back to an errorless reference field. Therefore, it is necessary to develop new strategies to validate the results of a spherical harmonic analysis, and to find better ways to evaluate the deterministic model as well as the stochastic model.

4 Model validation

The observations made by the gradiometer will be correlated, because the instrument will work optimally only within a certain measurement bandwidth, while the spectral density will decrease considerably outside this bandwidth (Figure 1). Thus the residuals of the Gauss-Markov model will be correlated *a priori*. To obtain optimal estimations from this least-squares adjustment the *a priori* variance/covariance matrix of the model residuals is required. This approach is equivalent to decorrelating the residuals/observations and then estimating the parameters from an unweighted least squares adjustment. This transformation can be done efficiently by filtering the observations in the time domain by means of an ARMA model. In many cases a parsimonious ARMA model can be used to transform an (uncorrelated) series of a white noise process into a (correlated) series of a colored noise process and vice versa. The coefficients of this parametric ARMA model were estimated from the inverse *Granada-SGG* error budget represented by a spectral density function (cf. Figure 1). Since no real data is available yet, a simulated data set, namely the above mentioned *Delft-data-set* was used instead.

If the given stochastic model is inaccurate the model residuals will not be perfectly uncorrelated, and the assumptions regarding the optimality properties will not be strictly fulfilled. In order to check this assumption, one has to analyze the estimated residuals after the adjustment of the transformed, i.e. filtered observations, denoted as *filtered residuals*. Important goals are to find out whether the filtered residuals are still significantly correlated, indicating model inaccuracies, what kind of pattern this dependence possibly shows, and how one can make use of this information to improve the stochastic model. At this point it must be mentioned, though, that autocorrelations may also be caused by non-vanishing expectations of the residuals, i.e. by a trend function possibly varying in time and space which was not expressed by the deterministic model within the adjustment procedure. However, within this closed-loop simulation study we may assume to have parameterized the observations sufficiently well.

In order to detect significant systematic effects still inherent in the filtered residuals we suggest a strategy consisting of different statistical tests as well as graphical examination of the data beforehand. We will take a look at the residuals both in the time and the frequency domain allowing the detection of certain characteristics from different perspectives, thus producing more reliability.

A quick examination of the filtered residuals (Figure 3) reveals some kind of randomness regarding the distribution around its expectation. To put this into a rather strict mathematical context, Moore and Wallis (1943) proposed several hypothesis tests based on the signs of the first differences in order to test the alteration rate of the series. We preferred the turning point test, since its distribution is known to follow, for large time series, the Normal distribution. This test is made by calculating the differences between adjacent elements, taking the signs of these differences, and extracting the turning points, i.e. points that mark a reversal of the sign. Thus, at a turning point, a series stops rising and starts declining, or vice versa.

For a random series, three consecutive values may equally likely occur in six different orders with four cases containing a turning point (Figure 4). Keeping in mind that the very first and last point of a series may not be a turning point, we can expect $\frac{2}{3}(N-2)$ turning points (N: number of values within the time series); the variance can be shown to be $\frac{16N-29}{90}$. Setting the confidence level to 95% yielded, for the filtered residuals, a test value which is outside the region of acceptance of the white noise hypothesis. In fact, there were significantly more turning points in the data than were to be expected for 'real' white noise.

```
>> turning_point_test(signal,0.95)
+---------------------------------+
|        turning-point test       |
+---------------------------------+
|  testvalue     =          3.643 |
|  fractile      =          1.960 |
|  p-value       =          0.000 |
+---------------------------------+
ans =    false
```

Since the hypothesis had to be rejected, we still have to assume dependence of the data. In order to get a better understanding of these correlations, the filtered residuals series was thinned out by computing the test value again considering only every second, every third, etc., value. As a result the approximate correlation length as the sample interval, after which no further correlations could be verified (Figure 5) was obtained. The p-value, which gives the two-tailed area (standardized between 0 and 1) beneath the probability density function outside the calculated test value, was less than the chosen α-value. In other words, no dependence for sample intervals higher than 25 could be detected. Thus, independence of the data was assumed and tests for inhomogeneity were applied. In doing so, the thinned out filtered residuals were divided into M segments of equal length N_i, in our case one revolution within a repeat orbit.

Then it was tested, whether every group with its individual mean and variance belongs to the same basic population. Rejection of this test would provide information about remaining non-stationarities in the data. As before, a graphical examination should precede the strict hypothesis test in order to detect any obvious phenomena in the data.

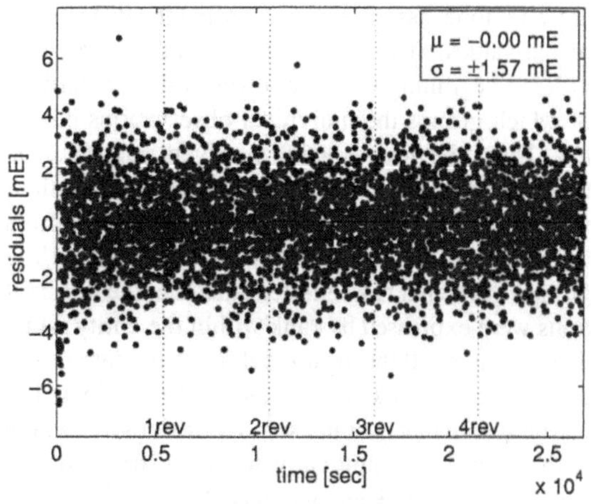

Fig. 3 Filtered residuals of the ΔV_{zz} gradiometer component with expectation μ and standard deviation σ.

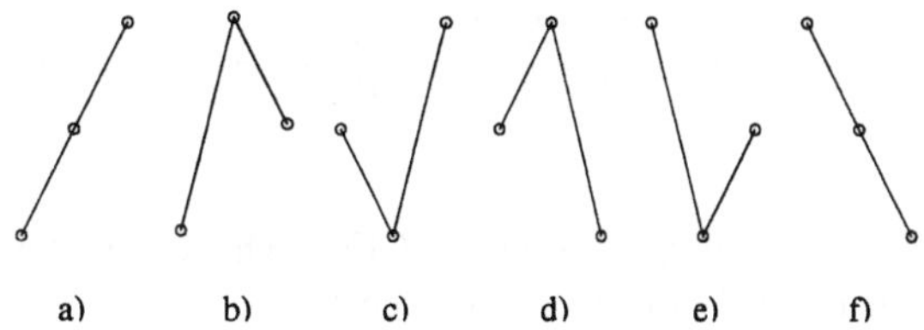

Fig. 4 Equally likely sequences of three consecutive points within a random series with b), c), d) and e) being turning points.

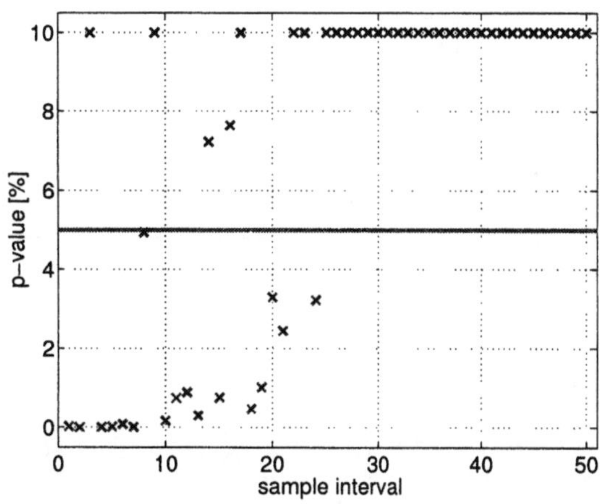

Fig. 5 p-values for different sampling intervals for the turning point test.

Figure 6 does not reveal any extreme outliers and deviations of the median or quartile values. In order to investigate any significant differences between at least two groups, first the Kruskal-Wallis test (Taubenheim, 1969) was applied, which is free of any additional assumptions regarding the distribution of the values. The observations of all groups were ordered beginning with the lowest value, regardless of which groups the individual observations belong to. Thus, the lowest value is given rank# 1, and the highest value rank# N (N: total observations within all groups). In order to be independent from the distribution, only the ranks were considered. The groups were then reassembled such that the observations were expressed by rank within the entity rather than by values. If the groups belong to the same basic population, the ranks will be randomly distributed. It can be shown that the rank sum R_i of elements within each group is asymptotically normal distributed with expectation $N_i \frac{N+1}{2}$ and variance $N_i \frac{N(N+1)}{12}$ if N_i is large. Computing the squared sum of the normalized rank sums yields the test value

$$\frac{1}{M-1} \sum_{i=1}^{M} \frac{(R_i - N_i(M+1)/2)^2}{N_i N(N+1)/12} \sim \chi^2_{M-1} \quad (6)$$

With a confidence level of 95% the hypothesis for the Kruskal-Wallis test was accepted:

```
>> kruskal_wallis_test(signal,0.95)
+-------------------------------------+
|        Kruskal \& Wallis test       |
+-------------------------------------+
|  testvalue      =         497.734   |
|  fractile       =         516.272   |
|  p-value        =           0.142   |
+-------------------------------------+
ans =    true
```

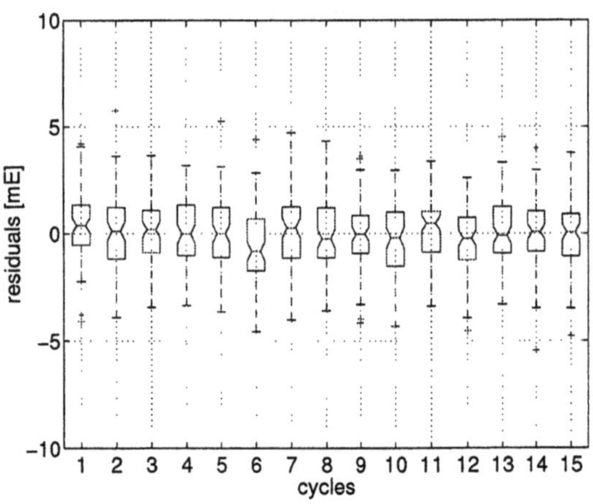

Fig. 6 Grouped data with lines at the lower/upper quartile and the median value. The vertical lines comprise the rest of the data; values outside these lines are outliers.

Thus it could not be proven that two (or more) groups of thinned out filtered residuals show significant differences with respect to the mean value. An alternative, parametric test is provided by the well known variance analysis (cf., for instance, Kreyszig (1998)) based on a decomposition of the squared sum of the deviations of all sample values from the mean value of the series. For this test one must assume that all M groups emanate from M normally distributed populations, all of which have the same variance (which may be unknown). It was tested whether the mean values of all groups are equal as well. In order to derive the test value one must compute the mean value within every group ($\bar{x}_i$) and of all observations ($\bar{x}$), followed by the above mentioned decomposition into the squared sum *within the groups* and the squared sum *between different groups*

$$\sum_{i=1}^{M} \sum_{k=1}^{N_i} (x_{ik} - \bar{x})^2 = q_1 + q_2 = \sum_{i=1}^{M} \sum_{k=1}^{n_i} (x_{ik} - \bar{x}_i)^2 + \sum_{i=1}^{M} N_i (\bar{x}_i - \bar{x})^2. \quad (7)$$

The ratio $\frac{q_1/(M-1)}{q_2/(N-M)}$ defines a Fisher distributed quantity with $f_1 = M - 1$ and $f_2 = N - M$ degrees of freedom. The confidence level was once again set to 95% and the test value was computed for the filtered residuals as described. The hypothesis that the mean values do not differ significantly between the different groups could not be rejected again.

```
>> variance_analysis_test(signal,0.95)
+-------------------------------------+
| Variance-Analysis: vtv    f   sigma |
+-------------------------------------+
| total:            84006.3 33551 1.582 |
| between groups:    1235.5   465 1.630 |
| within group:     82770.8 33086 1.582 |
+-------------------------------------+
|  testvalue     =      1.062          |
|  fractile      =      1.111          |
|  p-value       =      0.173          |
+-------------------------------------+
ans =    true
```

As it is possible now to verify inhomogeneities in the long term behavior of the filtered residuals, the estimated autocorrelation function can be introduced into the test procedure. For a white noise series, the autocorrelations should vanish for all lags greater than zero. Therefore, the autocorrelations of white noise are distributed normally with zero mean and variance $1/N$. This variance can only be used as an approximate upper limit for the confidence interval of the estimated autocorrelations. Considering that 5% of the autocorrelations may be allowed to be outside the region of confidence, far too many outliers for the first one to two thousand lags were found in the graphical representation (Figure 7).

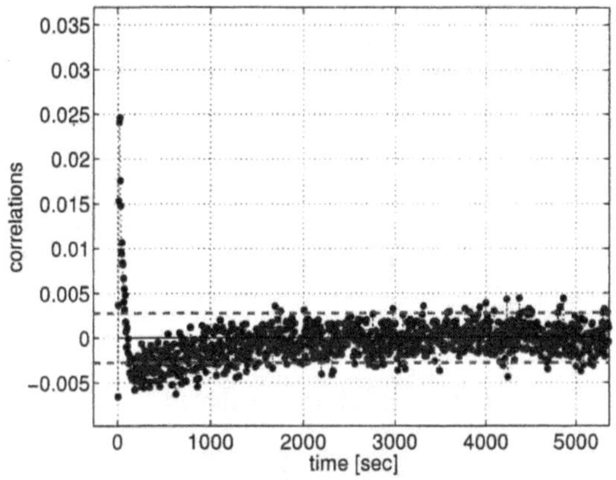

Fig. 7 Autocorrelations of the filtered residuals compared to the variance of a white noise series (95%).

At this point, another hypothesis test was introduced to validate this visual impression, since the usage of $1/N$ as in the graphic above is only accurate for high lags (Box and Jenkins, 1976). In order to avoid the problems arising with lower lags the modified Box-Pierce test was applied using the statistic

$$Q^*(\tilde{r}) = N(N+2)\sum_{k=1}^{M}(N-k)^{-1}\tilde{r}_k^2 \qquad (8)$$

by conjointly testing the first M estimated autocorrelations $\tilde{r}_k^2$ for significance. The quantity Q^* is approximately χ^2 distributed with $M-p-q$ degrees of freedom; p and q denote the orders of the ARMA model. M was chosen according to the length of the impulse response function of the ARMA model. The hypothesis of white noise due to insignificant autocorrelations thus had to be rejected. This result had been previously suspected on the basis of the graphical examination of the lower lags.

```
>> box_pierce_test(signal,0.95)
+-----------------------------------+
|          Box-Pierce test          |
+-----------------------------------+
| testvalue      =        1838.722  |
| fractile       =         137.701  |
| p-value        =           0.000  |
+-----------------------------------+
ans =    false
```

As before with the turning point test, this significance test of the autocorrelations allowed us to obtain some information on the correlation length by thinning out the autocorrelation function. This is equivalent to resampling the filtered residuals. Figure 8 shows that the autocorrelations are certainly significant for lags up to 20.

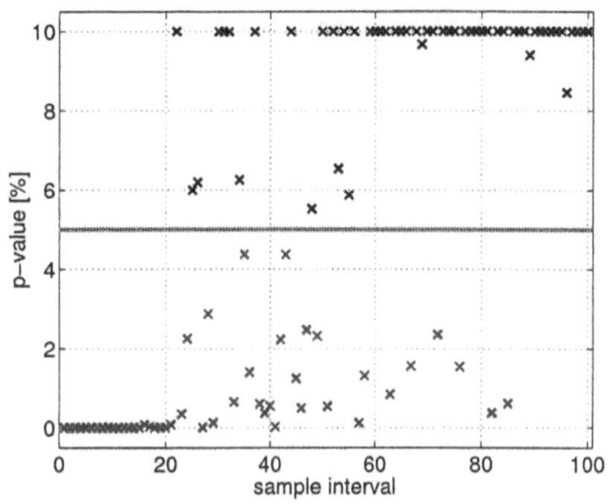

Fig. 8 p-values for different sampling intervals for the significance test of the estimated autocorrelations.

Finally, remaining correlations within the filtered residuals were analyzed in the frequency domain, where periodic patterns of the series were revealed better than they were by the autocorrelation function. In order to detect such periodic non-randomness, the periodogram $I(f_k)$ of the time series, with f_k denoting the Fourier frequency, was calculated. The quantities

$$CP_r = \frac{\sum_{k=1}^{r} I(f_k)}{\sum_{k=1}^{n} I(f_k)} \qquad (9)$$

with $r = 1, 2, \ldots, n$ and n as the number of unique Fourier frequencies define the normalized cumulative periodogram of the series (Box and Jenkins, 1976). A correct ARMA model would produce a white noise series with equal powers for all frequencies, shown as a line with constant rise in Figure 9. An imperfect model would result in unequal powers in the spectrum, so that its integration over the frequencies would cause deviations from this straight line. The significance of cumulative periodogram was checked by a comparison with the $(1-\alpha)$-

quantile

$$c = \frac{\sqrt{-\frac{1}{2}\ln\frac{\alpha}{2}}}{\sqrt{n-1}+0.2+\frac{0.68}{\sqrt{n-1}}} - \frac{0.4}{n-1} \qquad (10)$$

(Schlittgen and Streitberg, 1991) shown as lines parallel to the cumulative white noise periodogram. The bump as the integrated excess of power (for the sake of a clearer illustration plotted inflatedly in comparison to the confidence interval) indicates periodicities in the residuals that were distributed over a few frequencies. Since the estimated curve for our data exceeded the 95% confidence interval, the white noise hypothesis was rejected.

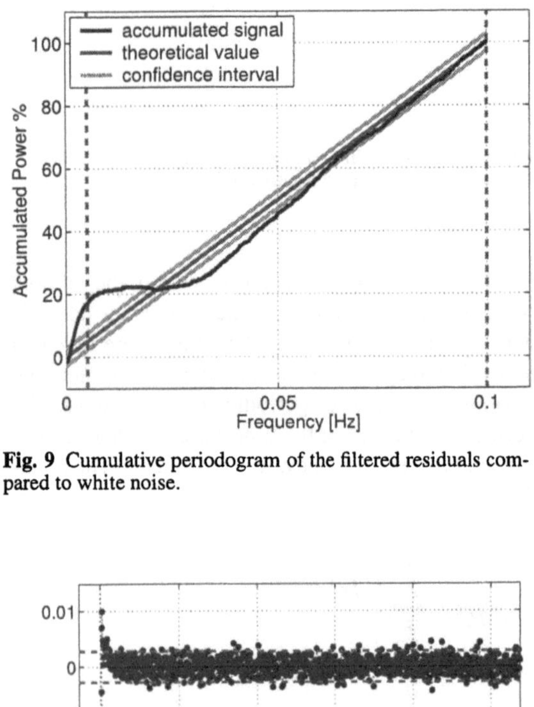

Fig. 9 Cumulative periodogram of the filtered residuals compared to white noise.

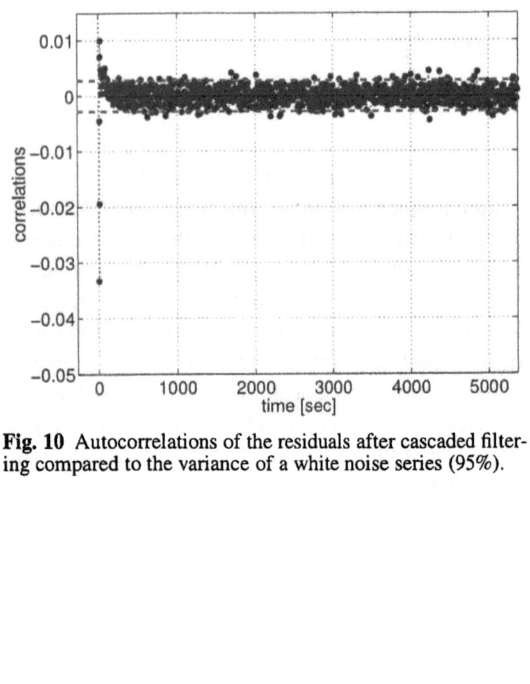

Fig. 10 Autocorrelations of the residuals after cascaded filtering compared to the variance of a white noise series (95%).

Having gained insight into the characteristics of the filtered residuals, it was possible to improve the model by adjusting another model to the residuals that remained after the first filter step. Thus, a cascade of filters applied sequentially to the residuals from the adjustment process was determined taking into account especially the model inadequacies for low frequencies, which were obvious in the representation of the cumulative periodogram. As an example, the autocorrelations of the residuals produced by an improved, cascaded filter are illustrated in Figure 10. It is clearly shown that the range of significant autocorrelations has decreased. However, there remained some systematic effects in the data resulting in significant correlations within the first 100 lags.

5 Summary and Conclusions

This study focusses on the modeling of a realistic scenario to recover the Earth's gravitational potential from satellite gravity gradiometry data. Independent solution strategies were applied and a closed loop simulation showed that a local geoid reconstruction with an accuracy on cm-level is possible. Truncation effects were discussed, the size of the effects was estimated, and special attention was paid to validate the model assumptions. Statistical test strategies were combined to define a systematic way to analyze the residuals of time series. First, non-randomness was detected with the turning point test. Determining the correlation length by thinning out the data was useful to find that the dependence has a long-periodic characteristic. Variance analysis of groups of resampled data allowed to prove stationarity which was mandatory for the application of both the autocorrelation function and the periodogram. Both functions expressed a long-periodic correlation structure, too. Further investigations are necessary to improve these hypothesis tests and to find tailor-made strategies to validate real mission scenarios. We believe that these strategies have the potential to fine-tune the deterministic and stochastic models, as well as to impact the calibration process of the gradiometer.

Acknowledgments

Parts of the work described in this report were done under the ESA/ESTEC contract No. 14287/00/NL/DC (*'From Eötvös to Milligal+'*). Additional funding is provided by the GOCE-GRAND project within the German 'Geotechnologien Projekt'. The coefficients of the semi-analytic model were kindly provided by R. Pail, TU Graz. We would like to thank Jeramy Flora for proof-reading and suggesting ways to improve the presentation of the material.

References

Box, G.E.P. and G.M. Jenkins (1976). *Time Series Analysis*. Holden Day, Oakland.

ESA (1999). *The four candidate earth explorer core missions - gravity field and steady-state ocean circulation mission*. ESA Report SP-1233(1), Granada.

ESA (2000). *From Eötvös to mGal*. Technical report, ESA-Project, ESA/ESTEC Contract No. 13392/98/NL/GD.

ESA (2002). *From Eötvös to mGal+*. Technical report, ESA-Project, ESA/ESTEC Contract No. 14287/00/NL/DC.

Kreyszig, E. (1998). *Statistische Methoden und ihre Anwendungen*. Vandenhoeck & Ruprecht, Göttingen.

Moore, G. and W. Wallis (1943). *Time series significance tests based on signs of differences*. Journal of the American Statistical Association, 38.

Rummel, R., F. Sansò, M. vanGelderen, M. Brovelli, R. Koop, F. Migliaccio, E. Schrama, and F. Sacerdote (1993). *Spherical harmonic analysis of satellite gradiometry*. Netherlands Geodetic Commission, New Series, 39.

Schlittgen, R. and B. Streitberg (1991). *Zeitreihenanalyse*. Oldenbourg Verlag, München - Wien.

Schuh, W.-D. (1996). *Tailored numerical solution strategies for the global determination of the earth's gravity field*. Mitteilungen der Geodätischen Institute der TU, Graz. Folge 81.

Schuh, W.-D., R. Pail, and G. Plank (2001). *Assessment of different numerical solution strategies for gravity field recovery*. In: Proceedings of the "International GOCE User Workshop", pp. 87–95. ESTEC, Noordwijk, The Netherlands (April 23-24, 2001).

Sneeuw, N. (2000). *A semi-analytical approach to gravity field analysis from satellite observations*. Reihe C, 527. Deutsche Geodätische Kommission, München.

Taubenheim, J. (1969). *Statistische Auswertung geophysikalischer und meteorologischer Daten*. Akademische Verlagsgesellschaft, Leipzig.

Wasle, E. (2001). *The significance of orthogonality in the framework of digital signal analysis*. Master's thesis, TU Graz.

Wiener filters and collocation in satellite gradiometry

A. Albertella, F. Migliaccio, M. Reguzzoni, F. Sansò
DIIAR – Sezione Rilevamento
Politecnico di Milano – Piazza Leonardo da Vinci, 32 – 20133 MILANO – Italy

Abstract. Many years ago the problem was addressed in satellite geodesy whether the collocation scheme based on covariances estimated on the rotation group, under the hypothesis of stochastic invariance of the underlying random field, could be directly applied to the time-wise data analysis, which should be typically based on a prior hypothesis of time stationarity.

The answer was indeed in the negative, for the very simple reason that couples of points separated by constant time differences do not have equal spherical distances, not even in the case of a perfectly circular motion, due to earth rotation.

Nevertheless a direct computation based on simple examples shows that the variations of such distances are at most of a few percent up to half a cycle and that the covariances based on spherical models do agree quite well with those computed along the orbit if the estimation time is long enough, e.g. 1 month.

This has been verified in particular for functionals of the anomalous potential that are of direct interest for a gradiometric mission like GOCE; for instance this is true for T itself, for T_r and T_{rr}.

Keywords. Collocation, covariance function, Wiener filter.

1 The problem of time-wise (Wiener filter) versus space-wise collocation

When using the Wiener filter in a time-wise approach for the treatment of a time-series of data, covariances are estimated under the hypothesis of time stationarity. On the other hand, when working in the frame of a space-wise approach (collocation), covariances are estimated under the hypothesis of rotation invariance of the underlying random field.

The question arises whether we can compare these covariance functions, and how the two schemes can be combined in satellite gradiometry.

Let us briefly describe the two concepts.

1.1 The time-wise approach: modelling and filtering the observables

The problem is to model correctly the observational functionals, including the correlation along the orbit. This can be done exploiting the Wiener filter theory.

We model the observables $Q_0(t)$ as a signal plus a noise $\nu(t)$

$$\begin{cases} Q_0(t) = Q(t) + \nu(t) \\ Q(t) = L(t,T) = L_0(T)\big|_{\underline{x}(t)} \end{cases} \tag{1}$$

where $L_0(T)|_{\underline{x}(t)}$ is a local functional of the gravitational potential T and $\underline{x}(t)$ is the orbit point at time t.

Note that under the hypothesis of L_0 being a local functional, we can take $Q(t)$ as stationary and, assuming ergodicity, we can calculate the covariance function from one realization only (Papoulis, 1984)

$$C(\tau) = \lim_{A\to\infty} \frac{1}{A} \int_0^A Q(t+\tau)Q(t)dt\,. \tag{2}$$

Although this is very much a matter of "belief", let us assume that the noise too is stationary and ergodic, so that by the same formula (2) one can compute the noise covariance $C_\nu(\tau)$. We call $S(p)$ and $S_\nu(p)$ the Fourier transforms of $C(\tau)$ and $C_\nu(\tau)$, i.e. the power spectra of $Q(t)$ and $\nu(t)$.

When the above statistics are known, it is possible to smooth the observations Q_0. This can be done in principle by filtering Q_0 in the time domain

$$\hat{Q}(t) = \sum_{i,k} C(t-t_i)\{C(t_i-t_k) + C_\nu(t_i-t_k)\}^{-1} Q_0(t_k) \tag{3}$$

and in the frequency domain

$$\hat{Q}(t) = F^{-1}\left\{\frac{S(p)}{S(p)+S_\nu(p)}\,\tilde{Q}_0(p)\right\} \tag{4}$$

where

$$\begin{cases} \tilde{Q}_0(p) = F\{Q_0(t)\} \\ S(p) = F\{C(\tau)\} \\ S_\nu(p) = F\{C_\nu(\tau)\} \end{cases}$$

and $\dfrac{S(p)}{S(p)+S_\nu(p)}$ represents the Wiener filter (Papoulis, 1984).

Moreover, according to the well known Wiener theory, we can estimate any time-wise functional of $Q(t)$ (e.g. $\frac{d}{dt}Q(t)$, etc.), but not different space-wise functionals, unless we also measure them in such a way that cross-covariances become available through realization-wise averages as in (2).

As an example, let us suppose we have measured the following quantity

$$Q(t) = \frac{\partial^2 T}{\partial r^2}\Big|_{\underline{x}(t)}$$

which is therefore "known". It is however not possible to derive an estimate of another space-wise functional of T, e.g. it is not possible to apply the Wiener filter to obtain

$$Y(t) = \frac{\partial T}{\partial r}\Big|_{\underline{x}(t)}$$

unless this quantity is observed, too, so that the cross-covariance between T_{rr} and T_r can be derived from empirical data. When this is not the case, one is led to introduce the space-wise stochastic structure of the function (or random field) $T(\underline{x})$.

1.2 The space wise approach: modelling and filtering/predicting the observables

In this case $T(\underline{x})$ is to be considered a "stationary" (rotation invariant) harmonic random field. In formulas, this is translated into

$$\begin{aligned} & T(\underline{x}) \leftrightarrow \{T_{\ell m}\} \\ & E[T_{\ell m}] = 0 \\ & E[T_{\ell m} T_{jk}] = \sigma_\ell^2\, \delta_{\ell j}\, \delta_{mk} \end{aligned} \qquad (5)$$

where $T_{\ell m}$ are the harmonic coefficients, and σ_ℓ^2 are the degree variances.

When using this model, we are able to compute all the covariances (and cross-covariances) of bounded functionals; in fact, from one "space-wise covariance" only (Ψ being the spherical angle between $\underline{x}$ and $\underline{y}$)

$$\begin{aligned} C(\underline{x},\underline{y}) &\equiv C(\Psi_{\underline{x},\underline{y}}) = \sum_{\ell,m} \sigma_\ell^2\, Y_{\ell m}(\underline{x}) Y_{\ell m}(\underline{y}) = \\ &= \sum_\ell \sigma_\ell^2 (2\ell+1) P_\ell(\cos\Psi_{\underline{x},\underline{y}}) \end{aligned} \qquad (6)$$

we can compute at arbitrary points $\underline{x}$, $\underline{y}$ in space for any couple of functionals $F_{\underline{x}}$, $G_{\underline{y}}$

$$\begin{aligned} & C\left[F_{\underline{x}}(T), G_{\underline{y}}(T)\right] = \\ & = \sum_{\ell,m} \sigma_\ell^2\, F_{\underline{x}}\left[Y_{\ell m}(\underline{x})\left(\frac{R}{r_{\underline{x}}}\right)^{\ell+1}\right] G_{\underline{y}}\left[Y_{\ell m}(\underline{y})\left(\frac{R}{r_{\underline{y}}}\right)^{\ell+1}\right]. \end{aligned} \qquad (7)$$

As an example, we can compute $C_{Trr\,Trr}(\underline{x},\underline{y})$, $C_{Tr\,Trr}(\underline{x},\underline{y})$ and then obtain by collocation $Y(t) = \frac{\partial T}{\partial r}\Big|_{\underline{x}(t)}$ from $Q(t) = \frac{\partial^2 T}{\partial r^2}\Big|_{\underline{x}(t)}$ (Moritz, 1989).

2 The almost equivalence of time-wise and space-wise covariances

The two schemes described in Section 1 are obviously not equivalent, and it must be remarked that it is so even when it is possible to apply both.

This happens because one must take into account the fact that the earth is rotating in space while measurements are taken in time, so, even for a circular orbit lying naturally on a sphere, it is (Betti and Sansò, 1989)

$$\begin{aligned} & C_{SW}(\underline{x}(t), \underline{x}(t')) \equiv C_{SW}(\Psi_{\underline{x}(t),\underline{x}(t')}) \neq \\ & \neq C_{TW}(t-t') \equiv C_{TW}(\tau)\,. \end{aligned} \qquad (8)$$

In other words, couples of points separated by constant time differences along the orbit do not have equal spherical distances, not even in the case of a perfectly circular motion, due to earth rotation.

On the other hand, two remarks are due here:

- the collocation solution robustly depends on the shape of the covariance function (Sansò et al., 1999);
- the shape of time-wise covariance functions follow very strictly that of space-wise functions on circular orbits (which is a useful consideration in the case of gravity-geodetic applications).

As a matter of fact, it happens that $\Psi_{\underline{x}(t),\underline{x}(t')}$ is almost invariant with $\tau = t - t'$, at least up to half a cycle. Direct computations based on simple examples show that the variations of $\Psi_{\underline{x}(t),\underline{x}(t')}$ are at most of a few percent up to half a cycle and that the covariances based on spherical models do agree quite well with those computed along the orbit, if the time span for which the covariance function is estimated is long enough, e.g. 1 month.

A simulation has been performed, with data which are typical of a gradiometric mission like GOCE, with a circular orbit, altitude $h = 250\ km$ and inclination $i = 97°$.

The variation of Ψ along the orbit has been computed for different values of τ and the results in terms of extremely relative differences are reported in Table 1.

Table 1. Results of the simulation of values of Ψ: differences are expressed in relative (percentage) values

τ (sec)	$\frac{(\Psi_{max} - \Psi_{min})}{\overline{\Psi}}$
10	0.19 %
50	0.19 %
100	0.19 %
500	0.21 %
1000	0.25 %

Afterwards, covariances have been computed according to the formulas

$$C_{TW}(\tau) = \frac{1}{N-\tau}\sum_{i=1}^{N-\tau} Q(t_i)\,Q(t_i+\tau) \tag{9}$$

$$C_{SW}(\overline{\Psi}) = \sum_{\ell=0}^{L} a_\ell\,\sigma_\ell^2\,(2\ell+1)\,P_\ell(\cos\overline{\Psi}) \tag{10}$$

where TW and SW denote the time-wise and space-wise computations respectively and a_ℓ represents the action of the different functionals on the basis of spherical harmonics.

Covariance and cross-covariance functions have been computed in particular for functionals of the anomalous potential that are of direct interest for a mission like GOCE, for instance T itself, T_r and T_{rr}, giving rise to the use of the a_ℓ coefficients reported in Table 2 into Equation (10).

Table 2. Coefficients for the computation of the covariances of different functional of T

Functional	a_ℓ
T	$\left(\frac{R}{R_{sat}}\right)^{2\ell+2}$
T_r	$\frac{(\ell+1)^2}{R^2}\left(\frac{R}{R_{sat}}\right)^{2\ell+4}$
T_{rr}	$\frac{(\ell+1)^2(\ell+2)^2}{R^4}\left(\frac{R}{R_{sat}}\right)^{2\ell+6}$

The EGM96 gravity model has been used, with degrees from 25 to 300, both for simulating "data" along the orbit and then computing C_{TW}, and for computing $\sigma_\ell^2 = \frac{1}{2\ell+1}\sum_m T_{\ell m}^2$ and then C_{SW} according to (10).

The results are shown in Fig. 1 to Fig. 3 for covariances and in Fig. 4 to Fig. 6 for cross-covariances.

One comment is that when dealing with cross-covariances we find a marked difference between space-wise and time-wise formulas. The former, as far as the functionals are represented by simple multiplying factors in the spectral domain of spherical harmonics, are functions of the spherical distance $\Psi_{\underline{x}\underline{y}}$ which is symmetric with respect to the exchange $\underline{x} \leftrightarrow \underline{y}$.

On the contrary, the latter is not symmetric due to the well known antisymmetric relation holding for any two stationary random functions $H(t)$, $K(t)$

$$C_{HK}(\tau) = C_{KH}(-\tau)\ . \tag{11}$$

It is for this reason that, in order to verify the closeness of the space-wise stochastic structure to the corresponding time-wise counterpart, we have compared C_{SW} with the symmetric part of C_{TW} and verified that the antisymmetric part of C_{TW} was small.

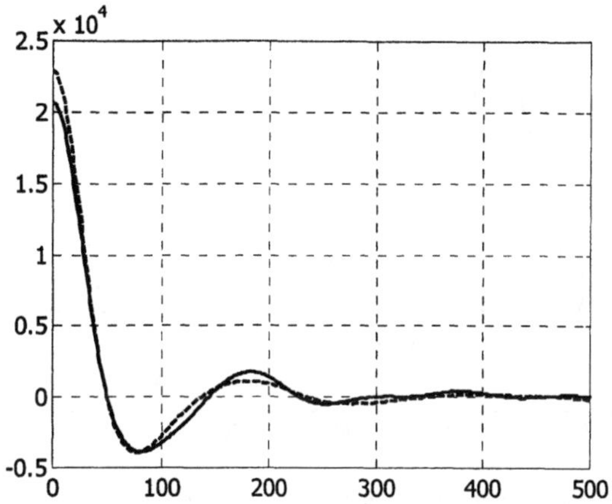

Fig. 1 $C_{T_{rr}T_{rr}}$ [mE^2] covariance as a function of time [sec]. Time-wise in solid line, space-wise in dashed line

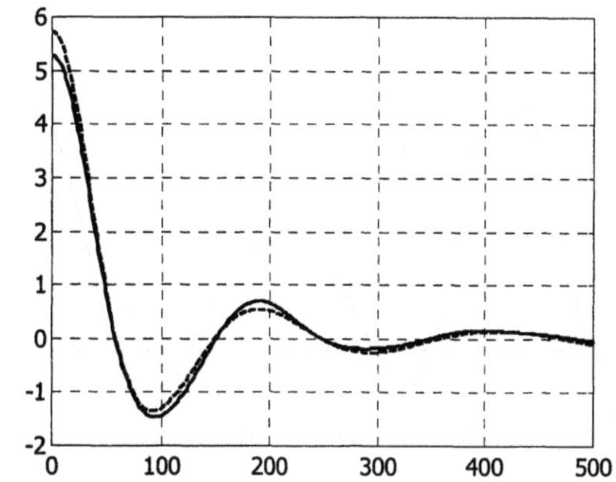

Fig. 2 $C_{T_rT_r}$ [mGal^2] covariance as a function of time [sec]. Time-wise in solid line, space-wise in dashed line

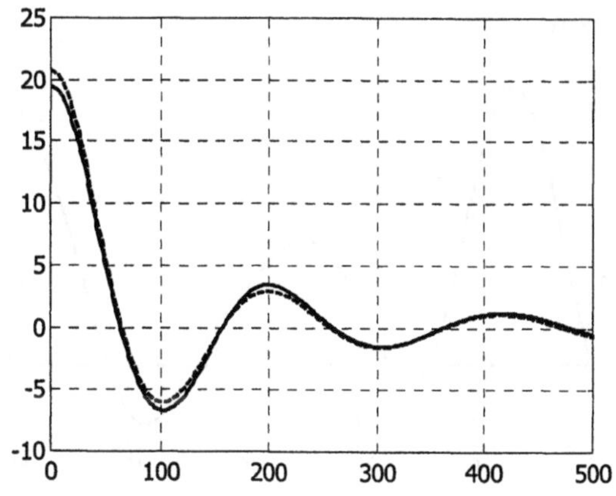

Fig. 3 C_{TT} [m^4/s^4] covariance as a function of time [sec]. Time-wise in solid line, space-wise in dashed line

3 Time-wise and space-wise filtering and prediction schemes

As a first and general conclusion of the numerical experiments which have been carried out, we can state that a scheme can be built for the point reduction of several functionals of the field in a Wiener filter mode, where not only covariances are made consistent with one another but also cross-covariances, usually more difficult to estimate, are rigorously derived from a unique model.

As a consequence, a prediction scheme can be set up, which is consistent between the time-wise and space-wise approach, though the two schemes are not equivalent. In this way we can exploit the advantages of both the time-wise approach (Wiener filter), which exploits the correlations along the orbit, and the space-wise approach, which exploits the correlations between points close in space but far in time (e.g. cross-over points).

In other words, the time-wise approach can only exploit the correlation along the orbit up to say half a cycle, but when the orbit winds up the space-wise approach is able to describe the correlation between points close in space though far in time.

According to this scheme, the time-wise approach can be used to reduce the estimation error along the orbit and in particular we can use a multiple-input single-output approach to process several types of data (e.g. T_{zz}, T_{yy}, T_{xx}) and estimate the "best" time-wise value T_{zz}. Afterwards we have to apply some algorithm that exploits the spatial correlation, if we want to extract all the information from the data. The basic scheme for this filtering procedure is shown in Fig. 7.

Here one has to recall that the instrument triad $(\underline{e}_x,\underline{e}_y,\underline{e}_z)$ along which second derivatives are measured, is not aligned to the global geographic triad $(\underline{e}_r,\underline{e}_\varphi,\underline{e}_\lambda)$ so that a small rotation has to be applied even to go from T_{zz} to T_{rr}.

Furthermore, using the space-wise cross-covariances in the Wiener filter we can directly estimate $\hat{T}_{rr}$ from time-wise data such as T_{zz}, T_{yy}, T_{xx}, without introducing rotations, according to the prediction scheme shown in Fig. 8. This is the best observation we can exploit for further data analysis in the space-wise approach.

Naturally the same idea can be further applied to provide the best estimate of another second derivative along the orbit, in order to catch all the information contained in the data.

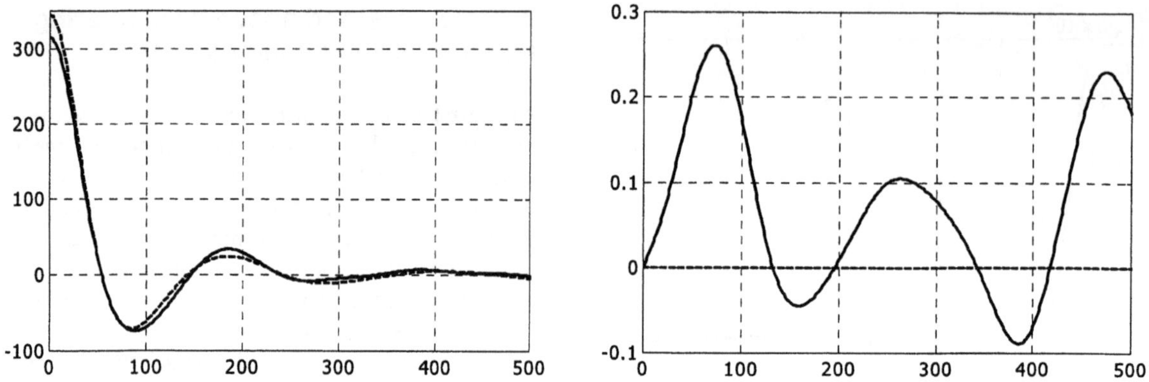

Fig. 4 $C_{Tr\,Trr}$ [mE mGal] cross-covariance as a function of time [sec]. Time-wise in solid line, space-wise in dashed line.

On the left: $\dfrac{C_{TrTrr}+C_{TrrTr}}{2}$. On the right: $\dfrac{C_{TrTrr}-C_{TrrTr}}{2}$

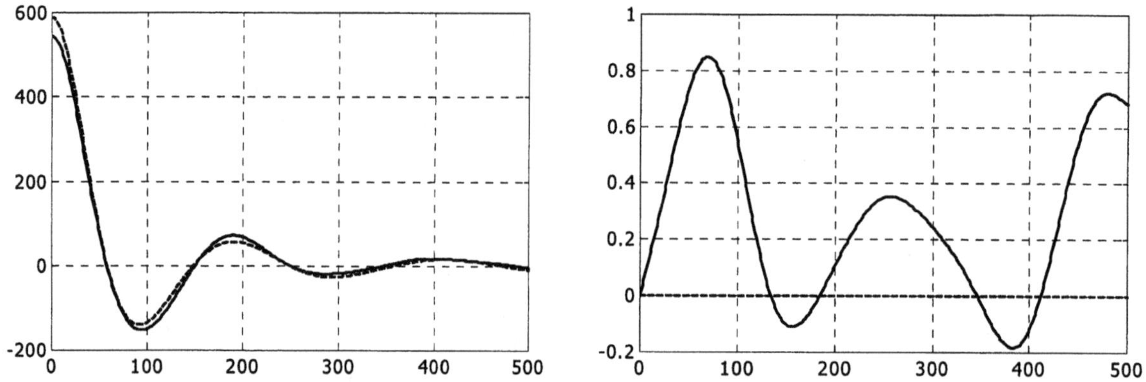

Fig. 5 $C_{T\,Trr}$ [mE m^2/s^2] cross-covariance as a function of time [sec]. Time-wise in solid line, space-wise in dashed line.

On the left: $\dfrac{C_{T\,Trr}+C_{TrrT}}{2}$. On the right: $\dfrac{C_{T\,Trr}-C_{TrrT}}{2}$

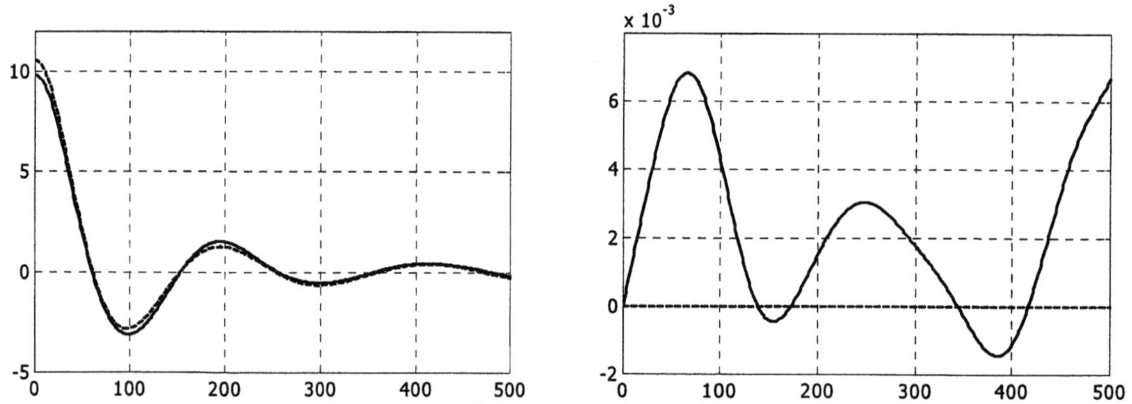

Fig. 6 $C_{T\,Tr}$ [mGal m^2/s^2] cross-covariance as a function of time [sec]. Time-wise in solid line, space-wise in dashed line.

On the left: $\dfrac{C_{T\,Tr}+C_{Tr\,T}}{2}$. On the right: $\dfrac{C_{T\,Tr}-C_{Tr\,T}}{2}$

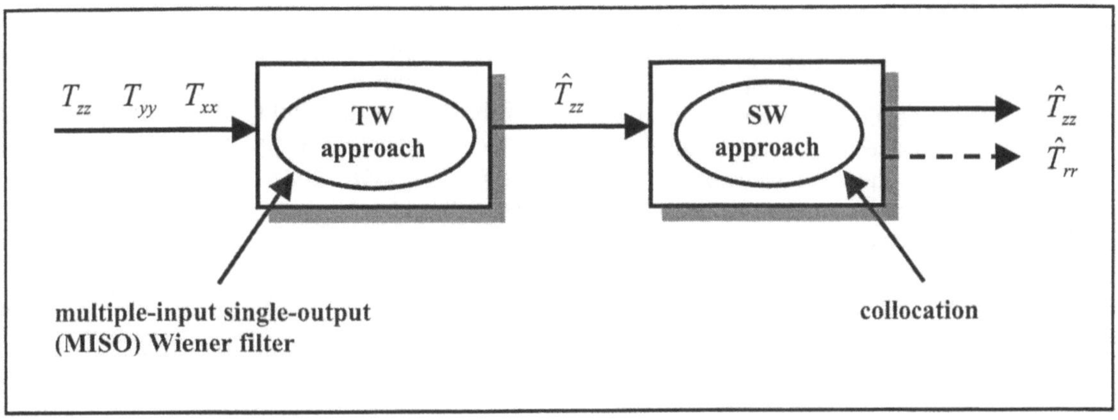

Fig. 7 The time-wise and space-wise filtering scheme

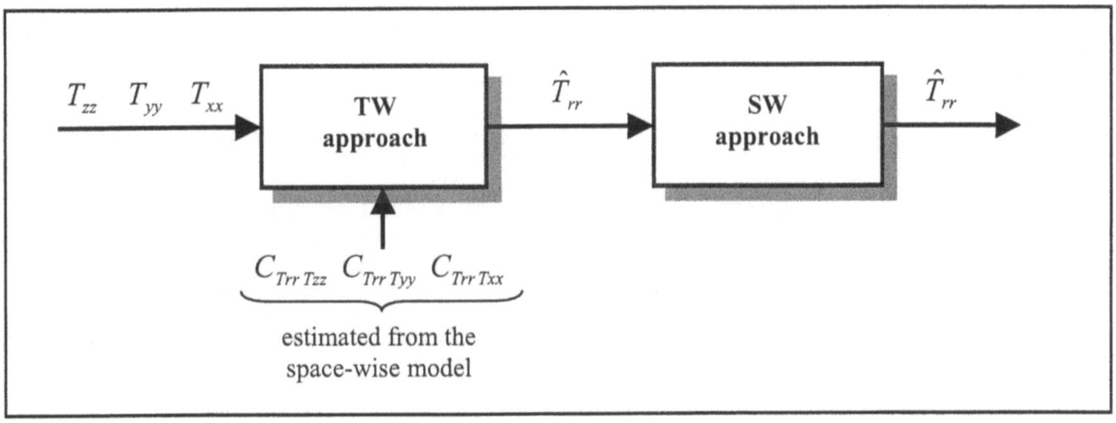

Fig. 8 The time-wise and space-wise prediction scheme

Whichever approach we use, an important point has to be underlined, that according to the above scheme we do not need to introduce the other noisy components of the gravity tensor to compute for instance T_{rr}, since in this case the rotation is applied only to the covariance structure and not to the tensor itself.

4 The error spectrum calibration

There is another possible interesting application of the above ideas.

The empirical (time-wise) covariance function provides us information only when we assume that the measurement error spectrum is perfectly known.

But since the full covariance model of the relevant signals can be safely computed from an a priori model, without going through the empirical covariances, this opens the possibility of calibrating the error spectrum from true data.

The covariance function of the noise can be simply estimated by

$$\begin{aligned} C_\nu &\cong C_{Q_0,TW} - C_{Q,SW} \\ S_\nu &= F\{C_\nu\} \end{aligned} \tag{12}$$

since Q, ν are assumed to be independent.

Indeed this reasoning can be applied only on condition that the shapes of space-wise and time-wise covariances do not depend too strongly on the coefficients, so that the use of some model can already provide useful covariance information. As a check of this hypothesis we have simulated, starting from T computed with EGM96, another fictitious model $\widetilde{T}$ by modifying the coefficients $T_{\ell m}$ with a random noise of the order of *20%* of σ_ℓ.

Recomputing the space-wise and time-wise covariances of $\tilde{T}$ we find quite a good agreement (see Fig. 9).

Another interesting point is that this can be applied as well to the flow of data, which are a by-product of the precise orbit determination, like T_r or T: in fact though these estimates have indeed an error spectrum which cannot be completely controlled a priori, this can be estimated a posteriori so that they could be used too in a multiple-input single-output mode.

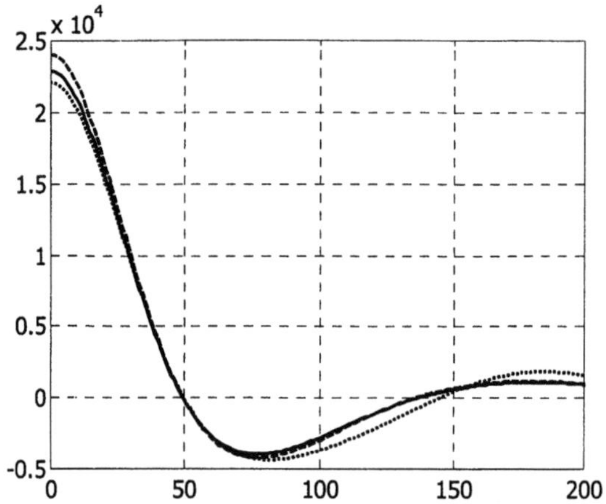

Fig. 9 $C_{T_{rr}T_{rr}}$ [mE^2] covariance as a function of time [sec]. Space-wise T in solid line; space-wise $\tilde{T}$ in dashed line; time-wise $\tilde{T}$ in dotted line.

5 Conclusions

The work presented in this paper was actually aiming at verifying whether the space-wise covariance structure of a random field T can be compared with the corresponding time-wise covariance, computed along a circular orbital motion. The question has been positively answered on the basis of simulations.

As a consequence, the way has been opened to perform direct estimates of different functionals of T from orbital data, for instance going from the observations T_{xx}, T_{yy}, T_{zz} to the prediction of T_{rr} or to any other directional second derivative along the orbit. In fact what is needed to perform this estimation is the time-wise cross-covariance between predictions and observations, which in turn can be approximately estimated from a space-wise computation applied to an a priori model.

The same reasoning can also be applied to calculate the noise covariance (or spectrum) by using the covariance of the observations and a model to compute the signal covariance.

This opens the possibility of exploiting data with a fairly uncertain error structure, like data derived during the orbit processing, which are indeed functionals of T.

References

Betti B. and Sansò F. (1989). "The integrated approach to satellite geodesy", in: *Lecture notes in earth sciences*, vol. 25 "Theory of satellite geodesy and gravity field determination", F. Sansò and R. Rummel eds., Springer-Verlag, Berlin.

ESA (1999). "Gravity Field and Steady-State Ocean Circulation Mission", ESA SP-1233 (1).

Moritz H. (1989). "Advanced physical geodesy", Wichmann, Karlsruhe.

Papoulis A. (1984). "Signal analysis", McGraw-Hill, New York.

Reguzzoni M. (2002). "From the time-wise to space-wise GOCE observables", *EGS General Assembly*, Nice, 21-26 April 2002.

Sansò F., Venuti G. and Tscherning C.C. (1999). "A theorem of insensitivity of the collocation solution to variations of the metric of the interpolation space", in: *International Association of Geodesy Symposia*, vol. 121 "Geodesy beyond 2000. The challenges of the first decade", K.P. Schwarz ed., Springer-Verlag, Berlin, pag. 233-240.

How to handle colored noise in large least-squares problems in the presence of data gaps?

R. Klees and P. Ditmar
Department of Geodesy, Physical, Geometric and Space Geodesy (FMR)
Delft University of Technology, 2629 JA Delft, Thijsseweg 11, The Netherlands

Abstract. An approach to handle stationary colored noise in least-squares problems in the presence of data gaps is presented. The presence of colored noise implies that a non-diagonal covariance matrix has to be inverted. The problem is reduced to the solution of systems of linear equations with the covariance matrix as equation matrix. Each system is solved using a pre-conditioned conjugate gradient method. An ARMA representation of the colored noise is used both to design an efficient pre-conditioner and to compute the product of the covariance matrix with a vector. This results in an algorithm that has a numerical complexity of $O(N)$ operations, where N is the number of observations. This makes the algorithm particularly suited for large least-squares problems. It is shown that the approach works perfectly in the sense that no edge effects show up and no observations have to be thrown away.

Keywords. Colored noise, ARMA filters, data gaps, satellite gravity gradiometry (SGG)

1 Introduction

Satellite sensors often provide measurements that are contaminated with colored noise. Prominent examples are the accelerometers on board CHAMP and GRACE and the gradiometer on board GOCE. This implies that the covariance matrix describing the measurement noise is non-diagonal. If the number of observations is moderate this would not cause any problem from a computational point of view: the covariance matrix could be set up and inverted explicitly. When, however, millions of observations have to be processed this is not possible anymore for computational reasons.

In many situations it is justified to consider the colored noise as one realization of a stationary random process. Stationarity implies that the covariance matrix is Toeplitz. A number of approaches have been proposed in geodetic literature that exploit the Toeplitz structure of the covariance matrix, see e.g. Colombo (1979), Eren (1980, 1982), Bottoni & Barzaghi (1993), Schuh (1996), and Klees et al. (2003). These approaches, however, cannot properly handle colored noise in the presence of data gaps as the covariance matrix loses the Toeplitz structure. Data gaps (particularly caused by spikes) are not exceptional in measurement time-series. A well-known example is the accelerometer on board CHAMP, which produces a significant amount of spikes.

The objective of this paper is to present an approach to handle colored noise, which can be applied to data with gaps. In section 2 it is shown that the question of how to handle colored noise reduces to the efficient solution of a system of linear equations with the covariance matrix as equation matrix. The approach of Klees et al. (2003) is taken as the starting point, because it is essentially of the order of $O(N)$. It is shown in section 3 that this approach corrupts a few minutes of data before and after each data gap, no matter whether the gaps are ignored or filled with zeroes, interpolated observations or even exact observations. The same holds if the data segments separated by data gaps are treated as uncorrelated. In fact, simply removing these corrupted observations from the further data processing may solve the problem. Unfortunately, this simple approach has two drawbacks: first, it may lead to a significant reduction of redundancy if many gaps are present in the data. Secondly, it violates the symmetry of the covariance matrix and, therefore, that of the normal matrix, which makes the least-squares inversion more difficult. Therefore, in section 4 a new approach is presented, which handles data gaps properly and has a numerical complexity of $O(N)$. In section 5 the results of some numerical experiments are presented. They demonstrate the accuracy and efficiency of the pursued approach.

2 Colored noise, Toeplitz systems, and ARMA filters

It is assumed that the normal equations are given by

$$A^T C^{-1} A x = A^T C^{-1} y, \quad (1)$$

where $y \in \mathcal{R}^N$ denotes the stochastic vector of observations, $A \in \mathcal{R}^{N\times r}$ is the design matrix, $C \in \mathcal{R}^{N\times N}$ is the noise covariance matrix, and x is the least-squares estimator of length r. For simplicity, regularization is not considered though it is always applied in practice. If the covariance matrix is non-diagonal, the computation of the normal matrix $A^T C^{-1} A$ may be extremely time-consuming. When we consider the product $C^{-1} A$ as the application of C^{-1} to the columns of A, the problem requires the solution of r systems of linear equations with equation matrix C.

It is much more efficient to solve the normal equations iteratively, e.g. using the method of preconditioned conjugate gradients (PCCG) (Hestenes and Stiefel 1952). Per iteration of the PCCG method the application of the normal matrix $A^T C^{-1} A$ to a vector d has to be computed. This can be done in three steps: (1) $u = A d$, (2) $v = C^{-1} u$, and (3) $A^T C^{-1} A d = A^T v$. In order to take colored noise properly into account one has to solve the linear system of equations $C v = u$ once per iteration. Provided that a good pre-conditioner is found, the number of iterations will be much less than the number of unknown parameters r. Therefore, the problem to be addressed now is how the system $C v = u$ can be solved efficiently.

If the colored noise can be seen as a realization of a stationary random process, and if there are no data gaps in the time series of observations, the covariance matrix C is Toeplitz. A very efficient method to solve Toeplitz systems of equations has been proposed by Klees et al. (2003). It is essentially of the order of $O(N)$ and uses an ARMA (autoregressive moving average) representation of the colored noise:

$$\xi_n = \varepsilon_n - \sum_{k=1}^{p} a_{p,k}\, \xi_{n-k} + \sum_{i=1}^{q} b_{q,i}\, \varepsilon_{n-i}, \quad n \in \mathcal{Z}, \quad (2)$$

where $\{\xi_n\}$ is the colored noise process, $\{\varepsilon_n\}$ is a white noise process with zero mean and variance σ_ε^2, and $\mathcal{Z}$ denotes the set of integer values (see e.g. Brockwell and Davis 1991). The coefficients $\{a_{p,k} : k = 1, \ldots, p\}$ and $\{b_{q,i} : i = 1, \ldots, q\}$ are the model parameters, and the pair (p, q) describes the order of the ARMA process. Equation (2) allows one to build a 'whitening' filter, i.e. a filter that transforms a colored noise time series into a white noise time series:

$$\varepsilon_n = \xi_n + \sum_{k=1}^{p} a_{p,k}\, \xi_{n-k} - \sum_{i=1}^{q} b_{q,i}\, \varepsilon_{n-i}, \quad n \in \mathcal{Z}. \quad (3)$$

The solution of the system $C v = u$ can be obtained *approximately* by two whitening operations, provided that the second filtering is performed backwards in time followed by a scaling with the factor $1/\sigma_\varepsilon^2$ (Klees et al. 2003). In terms of linear algebra this may be written as

$$\tilde{v} = \frac{1}{\sigma_\varepsilon^2} X F X F u, \quad (4)$$

where X is the exchange matrix (i.e. the matrix that is equal to the unit matrix but with the columns in reverse order), and F is a lower triangular Toeplitz matrix, which is a function of the ARMA coefficients. The matrix F relates the output $s = (s_1, \ldots, s_N)^T$ of the filter with transfer function

$$H(z) = \frac{1 + \sum_{k=1}^{p} a_{p,k}\, z^k}{1 + \sum_{i=1}^{q} b_{q,i}\, z^i}, \quad |z| \le 1, \quad (5)$$

to the input $t = (t_1, \ldots, t_N)^T$ according to $s = F t$. Thus, the *approximate* solution of $C v = u$ can be split in five steps (cf. Figure 1):

Step 1 Filter u (filter with transfer function $H(z)$).
Step 2 Flip result of step 1 (apply exchange matrix).
Step 3 Filter result of step 2 (filter with transfer function $H(z)$).
Step 4 Flip result of step 3 (apply exchange matrix).
Step 5 Scale result of step 4 with $1/\sigma_\varepsilon^2$.

This requires approximately $2(p+q)N$ floating point operations (flops). If n_{it} iterations of the PCCG algorithm have to be performed, the overall costs are $2n_{it}(p+q)N$ flops. Since n_{it} and the order of the ARMA process (p, q) are independent of N, and usually much lower than N, the

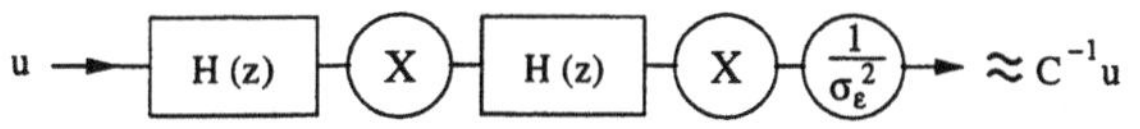

Fig. 1: Approximate computation of $C^{-1}u$ using ARMA filtering according to Klees et al. (2003). $H(z)$ is the transfer function of the ARMA filter, Eq. (5), and X is the exchange matrix.

numerical complexity of the algorithm is essentially of $O(N)$. In the following we refer to this approach as the 'ARMA filtering approach'. -1

3 Performance of the ARMA filtering approach in the presence of data gaps

When the data contain gaps, ARMA filtering can be applied in different ways:

Procedure 1 The available observations form one data vector, which is then filtered. This implies that correlations between data segments separated by a data gap are not correctly taken into account.

Procedure 2 Each data segment between two data gaps is filtered separately. This implies that correlations between data segments separated by a data gap are neglected.

Procedure 3 Data gaps are filled with zeroes and the data vector (including the zero-filled data gaps) is filtered. After filtering the data vector is reduced to the original length by removing the corresponding elements. This implies that the correlation between data segments separated by a gap are properly taken into account.

Procedure 4 Initially, procedure 3 is followed. After the least-squares solution has been obtained, adjusted gravity gradients are computed and used to fill the data gaps. Then, a new least-squares solution is computed. This approach is iterated until convergence is achieved. This means that the data vector is extended by a number of 'adjusted' observations, which are handled as 'observations'. Thus, the stochastic properties of the 'adjusted' observations are assumed to be identical with the stochastic properties of the missing data, which is of course not correct.

The first three procedures are investigated further. The following experiment is designed: a 74-hour time series of gravity gradients (zz-component) is computed along a realistic GOCE orbit from the difference between the OSU91a model (Rapp et al. 1991) complete up to degree and order 300 and the GRS80 Somigliana-Pizzetti model (Moritz 1980). The sampling rate is one second. A realistic noise PSD function is taken (cf. Figure 2, solid line) and approximated by an ARMA model according to the procedure proposed in (Klees et al. 2003). The selected order of the model is (41,40). A 74-hour colored noise time series is generated using this ARMA(41,40) model and added to the exact gravity gradients. Then, two data gaps of length 1 hour are simulated: one after 24 hours, the other after 49 hours. The corresponding observations were removed from the data vector. The resulting data vector contains 259200 observations. Then, the product of C^{-1} with the data vector is computed according to the procedures 1, 2 and 3 using the ARMA filtering approach. Next, $C^{-1}y$ is exactly computed by solving the system of linear equations $Cz = y$. Finally, the differences between the solutions obtained with the procedures 1-3 and the exact solution z are computed. They are shown in Figure 3. The differences are almost everywhere negligible but not before and after a data gaps occurs and not at the beginning of the time series. A closer look at the filter output after the second data gap (bottom row of Figure 3) shows that about 2-3 minutes of data are corrupted. The errors go gradually down to zero, and are completely negligible after about 2-3 minutes. That is what we call 'edge effects'. Obviously ARMA filtering cannot handle data gaps properly, independently of which

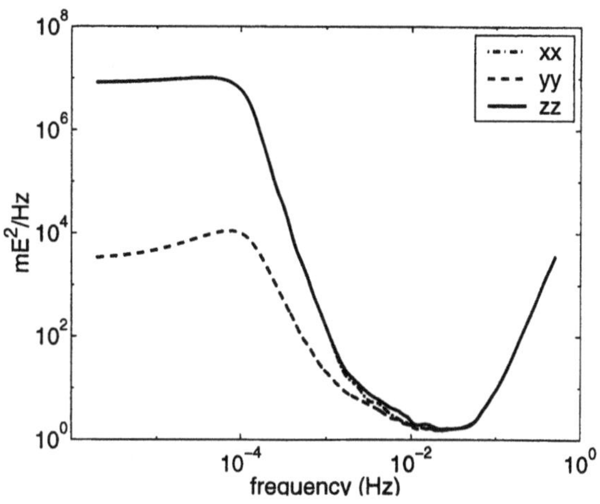

Fig. 2: Noise PSD functions of the three diagonal components of the tensor of gravity gradients. The total noise variance is $\sigma^2_{xx} = 3029$ mE^2, $\sigma^2_{yy} = 760$ mE^2, and $\sigma^2_{zz} = 3029$ mE^2, respectively (1 E = $10^{-9}\ s^{-2}$).

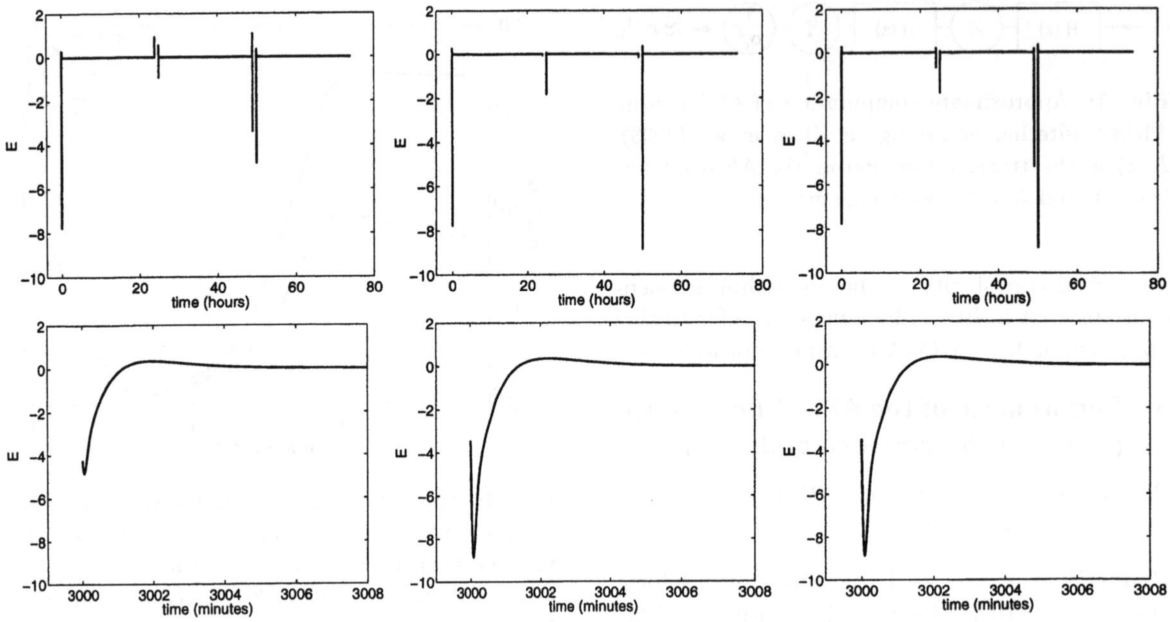

Fig. 3: Errors of the ARMA filtering approach in the presence of data gaps. From left to right: procedure 1, procedure 2, and procedure 3. The filter output is scaled by a factor of $1.5 \cdot 10^{-4}$ E^2. Note that the errors are of the order of the signal. Top: the whole time series. Bottom: the first eight minutes after the end of the second data gap.

procedure is followed. The edge effects may exceed the signal. Procedure 1 seems to perform better than the other two procedures, because the amplitudes of the edge effects are smaller. It is conspicuous that procedure 2 causes edge effects only at the end of a data gap, whereas the procedures 1 and 3 produce edge effects also at the beginning of a data gap. In all cases ARMA filtering generates edge effects at the beginning of the time series but not at the end.

A natural solution to this problem is to remove the erroneous data in the course of the data processing. The major disadvantage of this procedure is that good data would be eliminated from the data processing. If many data gaps occur this could lead to a significant reduction of the data set, which may reduce the quality of the least-squares solution. Besides, the symmetry of the normal matrix is violated when this approach is followed.

4 Proper handling of data gaps

The simulations done in the previous section have shown that ARMA filtering produces edge effects at the beginning of the time series and in the vicinity of data gaps. If one does not want to throw away good data an alternative approach to handle colored noise in the presence of data gaps is needed. Such an approach should have three properties: (1) the solution of $C v = u$ should have about the same numerical complexity as the ARMA filtering approach, i.e. it should be of the order of $O(N)$; (2) the approach should not produce any edge effects; (3) the approach should not rely on excluding 'good' data in the course of the data processing.

The pursued approach fulfils these requirements. It combines the advantages of an 'exact' solution of $C v = u$ with the numerical efficiency of the ARMA filtering approach. The basic idea is to solve the system $C v = u$ iteratively, using the PCCG method. The two most time-consuming operations per iteration of the PCCG method are the pre-conditioning and the application of the matrix C to a vector, say d. Both have to be designed to need no more than $O(N)$ operations.

The only task of the pre-conditioner is to reduce the number of iterations of the PCCG method such that convergence is achieved after $n_{it} \ll N$ iterations. As pre-conditioner, the ARMA filtering approach of Section 2, i.e. the inverse of $1/\sigma_\varepsilon^2 XFXF$, is proposed. As $\tilde{v}$, Eq. (4), is a good approximation to $C^{-1}u$, the product $1/\sigma_\varepsilon^2 XFXF\,C$ is about equal to the identity. Therefore, the number of iterations of the PCCG method is expected to be small. The numerical complexity of this type of pre-conditioning is of the order of $O(N)$, i.e. it meets the requirements. Note that the selection of a pre-conditioner influences only the convergence rate, not the solution itself.

It remains to perform the computation of $C\,d$ in $O(N)$ operations. We want to emphasize that this matrix-vector product has to be computed exactly, otherwise the PCCG method does not converge to the solution of $C\,v = u$. We propose to perform this operation also by exploiting the ARMA model of the colored noise. The basic idea can be explained easily when assuming that the time series does not contain any data gap. Thereafter, it is shown what modifications are needed when there are gaps in the data.

Let us assume that an infinite-length realization of a white noise random process $\{\varepsilon_n\}$ with variance σ_ε^2 is filtered by a filter with transfer function $H^{-1}(z)$, Eq. 5. Defining

$$\xi = (\ldots, \xi_0, \xi_1, \ldots, \xi_N, \xi_{N+1}, \ldots)^T, \qquad (6)$$

$$\varepsilon = (\ldots, \varepsilon_0, \varepsilon_1, \ldots, \varepsilon_N, \varepsilon_{N+1}, \ldots)^T \qquad (7)$$

we may write $\xi = G\,\varepsilon$, where G is an infinite-dimensional lower-triangular Toeplitz matrix. Then, the covariance matrix of the random vector ξ is

$$C_\xi = \sigma_\varepsilon^2 GG^T = \sigma_\varepsilon^2 GXGX, \qquad (8)$$

since G is per-symmetric, i.e. symmetric with respect to the North-East South-West diagonal. Thus, the computation of the product $C_\xi\,d_e$, where the vector d_e contains an infinitely long sequence of real numbers, can be split in five steps (cf. Figure 4):

Step 1 Flip the vector d_e (apply the exchange matrix).
Step 2 Filter the result of step 1 (filter with transfer function $H^{-1}(z)$, cf. Eq. (5)).
Step 3 Flip the result of step 2 (apply the exchange matrix).
Step 4 Filter the result of step 3 (filter with transfer function $H^{-1}(z)$).
Step 5 Scale the result of step 4 with σ_ε^2.

Consider now the finite-dimensional case. Let C be the covariance matrix of the sub-vector $(\xi_1, \ldots, \xi_N)^T$. Thus, C is a $N \times N$ sub-matrix of

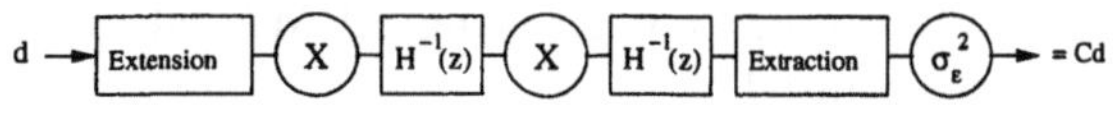

Fig. 4: Exact computation of $C\,d$ using ARMA filtering according to Klees et al. (2003).

C_ξ. Moreover, let d be the $N \times 1$ sub-vector of d_e. Then, we may write

$$C_\xi\,d_e = \begin{pmatrix} \cdots & \cdots & \cdots & \cdots & \cdots \\ \cdots & \cdots & \cdots & \cdots & \cdots \\ \cdots & \cdots & C & \cdots & \cdots \\ \cdots & \cdots & \cdots & \cdots & \cdots \\ \cdots & \cdots & \cdots & \cdots & \cdots \end{pmatrix} \begin{pmatrix} \vdots \\ d \\ \vdots \end{pmatrix}, \qquad (9)$$

and, therefore,

$$\begin{pmatrix} \vdots \\ \vdots \\ C\,d \\ \vdots \\ \vdots \end{pmatrix} = \begin{pmatrix} \cdots & \cdots & \cdots & \cdots & \cdots \\ \cdots & \cdots & \cdots & \cdots & \cdots \\ \cdots & \cdots & C & \cdots & \cdots \\ \cdots & \cdots & \cdots & \cdots & \cdots \\ \cdots & \cdots & \cdots & \cdots & \cdots \end{pmatrix} \begin{pmatrix} 0 \\ \vdots \\ 0 \\ d \\ 0 \\ \vdots \\ 0 \end{pmatrix}. \qquad (10)$$

Thus, when we want to compute $C\,d$ using ARMA filtering, we simply extend d on both sides by a sufficiently large number of zeros and filter the extended vector according to the procedure outlined above. Of course, the exact computation of $C\,d$ would require to pad the vector d with infinitely many zeros. However, when the ARMA model of the colored noise is causal and invertible (cf. Brockwell and Davis 1991), the filter coefficients decay quickly. In practice, the number of zeros needed depends on the desirable accuracy of the computation.

In the presence of data gaps, this procedure has to be modified by (1) adding zeros for the missing elements in the vector d prior to filtering, and (2) setting equal to zero the corresponding elements in the vector $C\,d$ after filtering. This yields the following conjugate gradient algorithm to compute the solution of $C\,v = u$:

1. $v_0 = 0$, $r_0 = u$, compute $p_0 = C^{-1}v_0$ by ARMA filtering (cf. Figure 1), $\tilde{p}_0 = p_0$, $k = 0$.

2. $b_k = \tilde{p}_k^T r_k$

3. Extend p_k by adding zeros for the missing data: new vector p_k

4. Compute $a_k = C p_k$ using ARMA filtering (cf. Figure 4)
5. Replace elements of a_k at the locations were data were missing by zeros: new vector a_k
6. $\gamma_k = \frac{r_k^T p_k}{a_k^T p_k}$
7. $v_{k+1} = v_k + \gamma_k p_k$
8. $r_{k+1} = r_k - \gamma_k a_k$
9. If $\|r_{k+1}\| < \epsilon_1$ and difference $(v_k, v_{k+1}) < \epsilon_2$ stop
10. Compute $\tilde{p}_{k+1} = C^{-1} r_{k+1}$ by ARMA filtering (cf. Figure 1)
11. $\beta_{k+1} = \frac{r_{k+1}^T \tilde{p}_{k+1}}{r_k^T \tilde{p}_k}$
12. $p_{k+1} = \tilde{p}_{k+1} + \beta_{k+1} p_k$
13. $k = k + 1$, go to item (2).

If n_0 denotes the number of zeros, the overall costs of performing the operation $C\,d$ is $2(p+q)(N+M+n_0)$, where N is the number of observations, and M is the number of missing observations due to data gaps. Provided that $M \ll N$ and $n_0 \ll N$ the algorithm is essentially of the order of $O(N)$.

5 Simulations

The following set-up has been chosen: along a non-repeat 60-day GOCE orbit with a mean altitude of 262 km, and a mean inclination of 96.6°, the three diagonal components of the tensor of gravity gradients were generated with a sampling rate of 1 second. This yields $N \approx 1.5 \cdot 10^7$ observations. The difference between the OSU91a gravity field model (Rapp et al. 1991) complete up to degree and order 300 and the GRS80 Somigliana-Pizetti model (Moritz 1980) defines the disturbing potential that has to be estimated from the observations by least-squares. The simulated gravity gradients were deliberately corrupted by colored noise according to the noise PSD functions for the diagonal components published in (ESA 1999). Below 10^{-4} Hz, which is the minimum frequency considered in (ESA 1999), the PSD was extrapolated by a natural cubic spline. The corresponding PSD functions for the three diagonal components are shown in Figure 2. Next, appropriate ARMA models were estimated from the given PSD functions according to the procedure in (Klees et al. 2003). Then, one-hour data gaps with frequency once per 25 hours have been generated in the 60-day time series of noisy gravity gradients for each diagonal component of the gravity gradient tensor. Consequently, a least-squares inversion was performed with the GOCE data processor GOCE-SOFT (Ditmar and Klees 2002) using the different strategies to cope with the data gaps. From the estimated potential coefficients global maps of geoid height errors were computed in a latitudinal band of ±80°.

In the first example, procedure 3 (see section 3) was followed, i.e. data gaps were filled with zeros. Moreover, a second run was performed with a different data-gap frequency: 1-hour data gaps with frequency once per 23 hours instead of frequency once per 25 hours. Both solutions exhibit noticeable distortions, especially at lower spatial frequencies (Figure 5). However, the magnitudes of the distortions differ significantly. The maximum geoid height error is 5.7 m and 7.1 m, the RMS geoid height error is 0.33 m and 1.08 m, respectively. Moreover, the geographical pattern of the geoid error map is completely different. This indicates that the propagation of edge effects into the solution strongly depends on the location of the data gaps. Moreover, the amplitude of each particular edge effect seems to be proportional to the strength of the noisy signal at the corresponding location. As soon as the geographical distribution of data gaps is very inhomogeneous, the error amplitude and geographical pattern may vary very significantly from case to case. If most of the data gaps begin and end in areas where the signal is close to zero, the errors in the geoid heights will be modest and only due to the measurement noise. Vice versa, if data gaps are located in areas with large signal, the edge effects generate large geoid-height errors.

In the second example, the computations are carried out according to procedure 1. It turns out that this procedure performs even worse than procedure 3: the the maximum error reaches 7.4 m, the RMS geoid height error is 0.68 m. When the data-gap frequency is changed, similar observations are made as in case of procedure 3.

In the third example, the gaps are filled with the exact (i.e. noise-free) gravity gradients (top panel in Figure 6). In this way, the best possible result obtainable according to procedure 4 is simulated. Unfortunately, the geoid height errors are only slightly better compared with procedure 3. A computation with the data set exhibiting data gaps with frequency ones per 23 hours

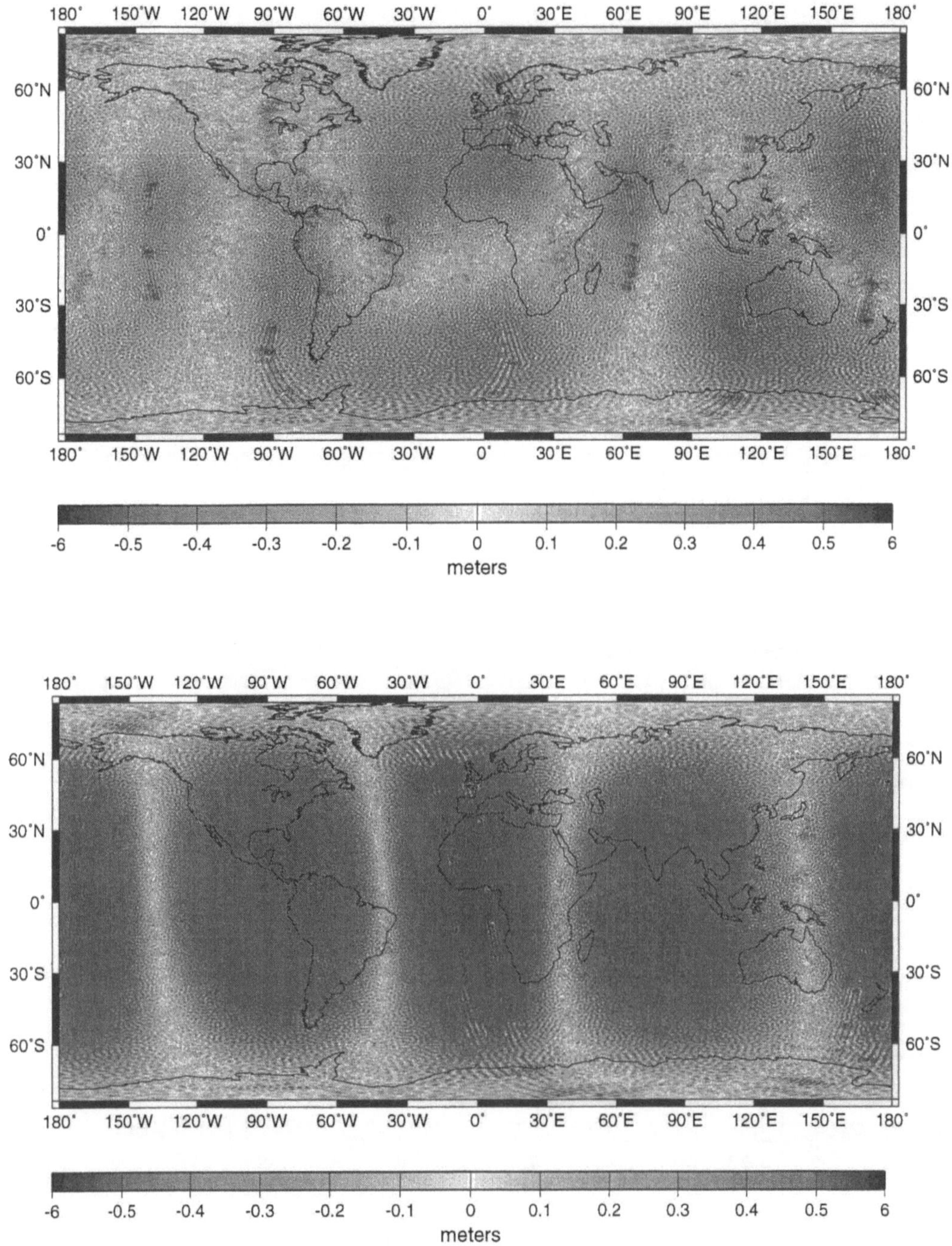

Fig. 5: Map of geoid height errors below ±80° latitude in the presence of 'data gaps' when procedure 3 of the ARMA filtering approach is used (cf. section 3). Top panel: 1-hour data gaps evenly distributed with frequency once per 25 hours. The maximum geoid height error is 5.7 m, the RMS geoid height error is 0.33 m. Bottom panel: 1-hour data gaps evenly distributed with frequency once per 23 hours. The maximum geoid height error is 7.1 m, the RMS geoid height error is 1.08 m.

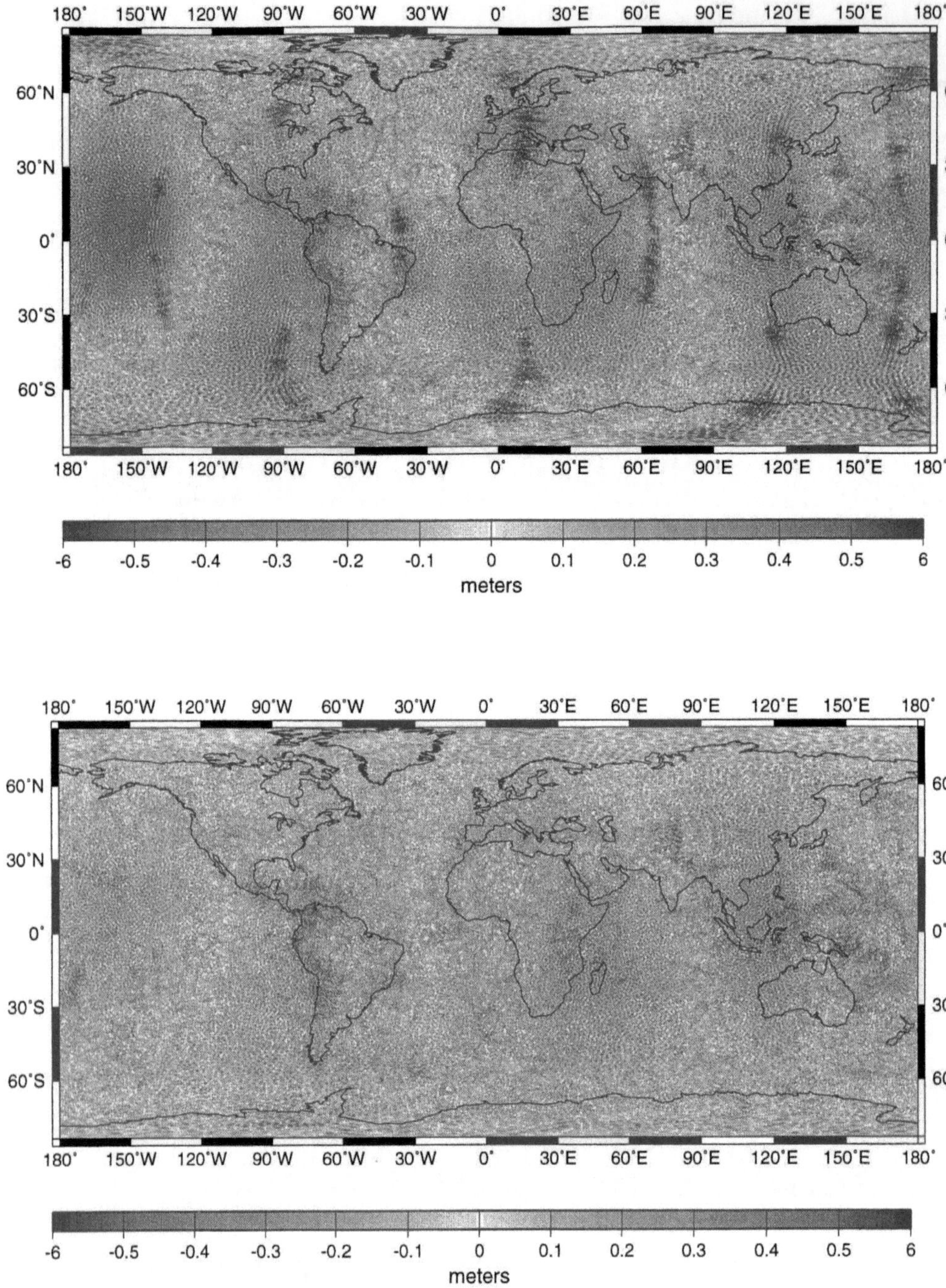

Fig. 6: Map of geoid height errors below ±80° latitude in the presence of 'data gaps'. Top panel: The ARMA filtering approach is followed and the data gaps have been filled with the exact gravity gradients (limit of procedure 4). The maximum geoid height error is 5.2 m, the RMS geoid height error is 0.19 m. Bottom panel: The PCCG approach is followed. The maximum geoid height error is 1.5 m, the RMS geoid height error is 0.20 m.

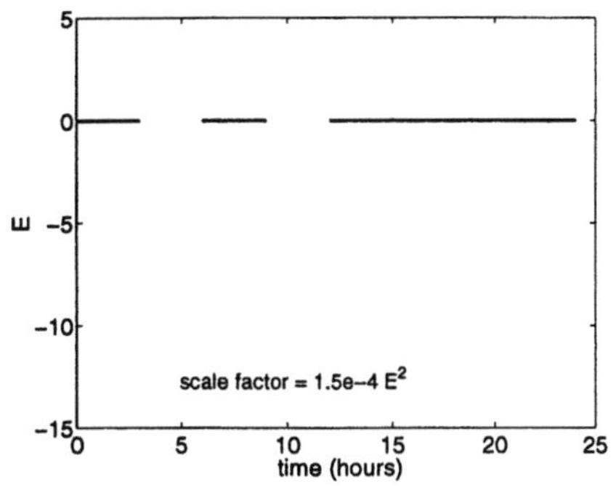

Fig. 7: Exact computation of $C^{-1}y$ versus PCCG approach in the presence of data gaps. Only the first 24 hours of the 74-hour time series are shown.

(not shown) gives very similar error statistics, although the geographical distribution of the errors is different. This is in agreement with the interpretation given before.

In the last example the computations are repeated using the proposed PCCG method to handle colored noise in the presence of data gaps. Figure 5 shows that the filter output does not suffer from any edge effects unlike the ARMA filtering approach (cf. Figure 3). This is confirmed by the solution in terms of geoid height errors (bottom panel in Figure 6). The quality of this solution is noticeably better than the one obtainable with the approximate filtering approaches considered before. In fact, this solution is almost identical to that obtained in the absence of data gaps. This is not surprising because differences to the solution in the absence of data gaps are only due to the loss of data caused by the gaps, which is less than 5%.

The price that has to be payed for an accurate treatment of data gaps is a higher computation time. The approximate procedures 1-3 require not more than half an hour to process the data set (SGI Origin 3800, 8 processing elements). The proposed PCCG-based procedure takes about 20 hours. This is a degradation of the performance in terms of the wall-clock time by a factor 40! The reason is that we did not succeed yet to parallelize the operation $a_k = Cp_k$, which has to be performed in each PCCG iteration and for which ARMA filtering according to Figure 4 is used.

6 Conclusions

The ARMA filtering approach, which was developed by Klees et al. (2003) in order to efficiently handle stationary colored noise in large least-squares problems, may cause significant distortions in the estimated gravity field model if the time series of observations exhibits gaps. A PCCG-based approach overcomes these problems. ARMA filters should be used to improve the performance of the PCCG-based approach in terms of wall-clock time. Nevertheless, the computation time is much higher than the computation time for the ARMA filtering approach, since an essential part of the PCCG iteration could not be parallelized yet. Whether this will be possible without significant modification of the overall data processing strategy has still to be investigated. Anyway, the PCCG-based approach can handle any 'type' of data gap whether short, long, frequent, or incidental etc. Unlike the PCCG-based approach, the magnitude of the distortions introduced by the ARMA filtering approach may depend on the 'type' of the data gap. So far, relatively long ('long' compared with the noise correlation length) data gaps have been considered and it was concluded that the distortions are significant. Many more scenarios, among them very short gaps of a few seconds, have still to be investigated. It is expected that the ARMA filtering approach performs better for some 'types' of data gaps than for others. Then, it could be used as an alternative to the more time-consuming PCCG-based approach on a case-by-case basis.

Acknowledgment

Computing resources were provided by Stichting Nationale Computerfaciliteiten (NCF), grant SG-027.

References

Bottoni GP, Barzaghi R (1993) Fast collocation. *Bulletin Géodesique* 67: 119-126.

Brockwell PJ, Davis RA (1991) Time series: theory and methods. Second Edition, Springer Series in Statistics, New York.

Colombo OL (1979) Optimal estimation from data regularly sampled on a sphere with applications in geodesy. Reports of the Department of Geodetic Science, Report No. 291, The Ohio State University, Columbus, Ohio, USA.

Eren K (1980) Spectral analysis of GEOS-3 altimeter

data and frequency domain collocation. Reports of the Department of Geodetic Science, Report No. 297, The Ohio State University, Columbus, Ohio, USA.

Eren K (1982) Toeplitz matrices and frequency domain collocation. *Manuscripta Geodaetica* 7: 85-118.

ESA (1999) Gravity field and steady-state ocean circulation mission. Reports for Mission Selection - The Four Candidate Earth Explorer Core Missions, ESA SP-1233 (1), Noordwijk, The Netherlands.

Ditmar P, Klees R (2002) A method to compute the earth's gravity field from SGG/SST data to be acquired by the GOCE satellite. Delft University Press (DUP) Science, Delft, The Netherlands, 64 pages.

Hestenes MR, Stiefel E (1952) Methods of conjugate gradients for solving linear systems. *Journal of Research of the National Bureau of Standards* 49: 409-436.

Klees R, Ditmar P, Broersen P (2003) How to handle colored observations noise in large least-squares problems? *Journal of Geodesy* 76: 629-640.

Moritz H (1980) *Geodetic reference system 1980. Bulletin Géodesique* 54: 395-405.

Rapp R, Wang Y, Pavlis N (1991) The Ohio State 1991 geopotential and sea surface topography harmonic coefficient models. Rep 410, Department of Geodetic Science and Surveying, The Ohio State University, Columbus, USA.

Schuh WD (1996) Tailored numerical solution strategies for the global determination of the earth's gravity field. Mitteilungen der geodätischen Institute der Technischen Universität Graz, Folge 81, Graz, Austria.

Statistical hypothesis tests in case of imprecise data

Hansjörg Kutterer
Deutsches Geodätisches Forschungsinstitut, Marstallplatz 8, D-80539 München, Germany

Abstract. Various types of uncertainty may occur in geodetic observation and modelling such as stochasticity (random variability) of the data or imprecision. Stochasticity is a consequence of uncontrollable effects during the observation procedure. Imprecision is caused by remaining systematic deviations between data and model due to imperfect knowledge or simplifications in order to make a model practicable. In the applications either stochasticity or imprecision can dominate the uncertainty budget. In the applications, imprecision is usually modelled and treated by means of fuzzy-theory.

Statistical hypothesis tests are designed for stochastic quantities only. In case of test statistics which are both stochastic and imprecise, the traditional test strategy needs to be extended. In the paper a solution of this problem is presented for one-dimensional Normal test statistics. The procedure is based on a precise decision criterion. Both precise and imprecise regions of acceptance and rejection can be handled. The respective probabilities for type I and type II errors are given. The case of L-fuzzy test statistics and precise regions of acceptance and rejection, respectively, is given as an example.

Keywords. Hypothesis tests, imprecision, fuzzy numbers, fuzzy data analysis

1 Introduction

Geodetic data analysis has to cope with uncertainty caused by various sources. Stochasticity and imprecision which are the most prominent types of data uncertainty are subject of this paper. They are explained below. The complete error budget may contain additional uncertainties such as the imprecision of the geodetic model or deficiencies due to fragmentary or contradictory data which are not considered in the following. For a comprehensive view see Kutterer (2001a, 2002).

Stochasticity or random variability describes deviations of the results of repeated observations due to the laws of probability. It corresponds with the random errors which are assumed to be Normal with zero expectation. *Imprecision* as it is considered throughout the paper is a non-stochastic type of uncertainty. It is due to remaining systematic deviations in the data which could not be eliminated.

Depending on the particular scientific interest and the corresponding model-data relation either stochasticity or imprecision can dominate the uncertainty budget. Real-time differential GPS may serve as an example: if two GPS antennas are very close to each other, usually all systematics are cancelled out by observation double-differencing leaving pure stochasticity. However, this strategy does not work if two GPS antennas are far away from each other. In this case imprecision due to imperfect knowledge of, e.g., atmospheric effects is the prevailing type of uncertainty.

Fuzzy-theory has proven to be an adequate mathematical tool for the handling and assessment of imprecise data (Bandemer and Näther, 1992; Viertl, 1996). It is used here to jointly treat stochasticity and imprecision of the data.

The purpose of the paper is to introduce a general concept for statistical hypothesis tests for such cases when both stochasticity and imprecision have to be taken into account. The geodetic model is considered as exact. The concept is worked out in detail for a one-dimensional Normal test statistics with known variance.

In the following, some fuzzy-theoretical basics are given. Then a strategy for the extension of statistical hypothesis tests is outlined which allows to take imprecision into account. The main focus lies on the development of an adequate test criterion. The probabilities of type I and type II errors are derived to bridge the gap between traditional hypothesis tests and the new concept. Finally, an example is given in order to illustrate the practicability of the theoretical derivations.

2 Basics of Fuzzy Data Analysis

Fuzzy-theory was initiated by Zadeh (1965). It extends the classical set theory by describing gradually the degree of membership that a certain element belongs to a set. In classical set theory the membership degrees are either 1 (is element) or 0 (is not element). In fuzzy set theory the range of membership degrees is [0,1]. Thus, a *fuzzy set* is defined as

$$\tilde{A} := \{(x, m_{\tilde{A}}(x)) \mid x \in X\}, \; m_{\tilde{A}}: X \to [0,1] \quad (1)$$

with $m_{\tilde{A}}(x)$ denoting the *membership function* and X a classical set like, e.g., the real numbers.

The *support* of a fuzzy set $\tilde{A}$ is defined as the classical set of elements of $\tilde{A}$ with positive degrees of membership, the *height* of a fuzzy set as the maximum membership degree value, the *core* of a fuzzy set as the classical set of elements of $\tilde{A}$ with membership degree equal to 1, and the *α-cut* of a fuzzy set as the classical set of elements of $\tilde{A}$ with membership degree greater equal $\alpha \in [0,1]$.

The basic operation for the analysis of fuzzy data is the *intersection* of fuzzy sets. It is defined through the resulting membership function

$$m_{\tilde{A}\cap\tilde{B}} = \min(m_{\tilde{A}}, m_{\tilde{B}}) \quad (2)$$

This definition is the one which is mostly used but there are others which are also consistent extensions with respect to classical set theory; see, e.g., Dubois and Prade (1980). The *complement* $\tilde{A}^c$ of a fuzzy set $\tilde{A}$ is defined by its membership function as

$$m_{\tilde{A}^c} = 1 - m_{\tilde{A}} \quad (3)$$

Fuzzy numbers are the basic quantities of fuzzy data analysis. They can be defined as a special case of a fuzzy set: a *fuzzy number* is a fuzzy set over the set of real numbers R with a single element core and compact α-cuts. The mostly used fuzzy numbers are the *LR-fuzzy numbers* defined by Dubois and Prade (1980). Their membership functions are given by monotonously decreasing (left and right) *reference functions* L and R, respectively. The range of values is [0,1]. Due to the single element core, L(0) = R(0) = 1. For a graphical sketch see Figure 1.

LR-fuzzy numbers can be represented by $\tilde{X} = (x_m, x_l, x_r)_{LR}$. The *mean point* is denoted by x_m. The (left and right) *spreads* x_l and x_r serve as measures of fuzziness. Totally symmetric LR-fuzzy numbers (identical reference functions and spreads) are called *L-fuzzy numbers*. They are represented by $\tilde{X} = (x_m, x_s)_L$.

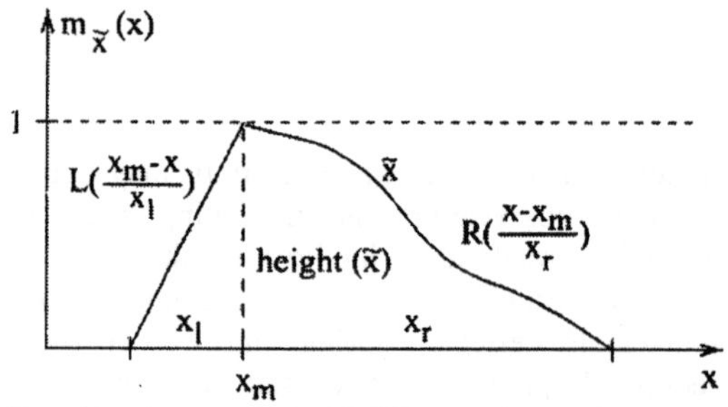

Fig. 1: LR-fuzzy number with different reference functions

Arithmetic operations for fuzzy numbers can be defined based on *Zadeh's extension principle* which allows the generalization of functions with real arguments to functions with fuzzy arguments:

$$\tilde{B} = \tilde{g}(\tilde{A}_1, \ldots, \tilde{A}_n) :\Leftrightarrow$$

$$m_{\tilde{B}}(y) = \sup_{\substack{(x_1,\ldots,x_n)\in X_1\times\ldots\times X_n \\ g(x_1,\ldots,x_n)=y}} \ldots \quad (4)$$

$$\ldots \min\left((m_{\tilde{A}_1}(x_1), \ldots, m_{\tilde{A}_n}(x_n))\right) \; \forall \; y \in Y$$

For L-fuzzy numbers, the resulting arithmetic rules are, e.g.,

$$\tilde{X} + \tilde{Y} = (x_m + y_m, x_s + y_s)_L \quad \text{Addition}$$

$$a\tilde{X} = (a\,x_m, |a|\,x_s)_L \quad \text{Multiplication by a real number}$$

(5a, b)

Obviously, the type of the reference function is preserved. And: the arithmetic operations can be based simply on the mean points and the spreads. Eqs. (5a, b) indicate the linear propagation of imprecision. This is a direct consequence of Eqs. (2) and (4), respectively. Generalized formulations of the extension principle based on a t-norm yield different types of propagation of imprecision.

For a more detailed introduction see standard references like, e.g., Dubois and Prade (1980), Kaufmann and Gupta (1991), Bandemer and Näther (1992), and Kruse et al. (1994). Studies of fuzzy data analysis in the geodetic context are presented by Kutterer (2001a, b, c; 2002).

3 Test scenario and strategy in case of precise but stochastic data

In case of precise but stochastic data the procedure of statistical hypothesis testing is well known; see, e.g., Koch (1999). It is briefly compiled here to illustrate its extension to the imprecise case which is presented in the following section. Please note that "precise" is used here as the opposite of "imprecise".

A precise one-dimensional and Normal *test statistics* is given as

$$\underline{T} \sim N(\mu, 1) \tag{6}$$

with the expectation value μ and (without loss of generality) the variance 1. The underline "_" denotes a random variable. The null hypothesis H_0 and alternative hypothesis H_a, respectively, are formulated with respect to μ and read as

$$\begin{aligned} H_0 &: \; E(\underline{T}) = \mu = \mu_0 = 0 \\ H_a &: \; E(\underline{T}) = \mu = \mu_0 + \delta \neq 0 \end{aligned} \tag{7a, b}$$

with the noncentrality parameter δ. For the sake of the simplicity $\mu_0 = 0$ is assumed furtheron. Based on the probability α of a type I error, a critical value

$$k := z_{1-\frac{\alpha}{2}} \tag{8}$$

can be defined with $z_{1-\frac{\alpha}{2}}$ the $\left(1-\frac{\alpha}{2}\right)$-fractile of the standardized normal distribution. Hence, the *region of acceptance*

$$A = [-k, k] \tag{9}$$

and the *region of rejection*

$$R = A^c = (-\infty, -k) \cup (k, \infty) \tag{10}$$

are obtained. The superscript "c" denotes the set-theoretical complement. The two sets A and R are called *regions of decision.*

The test decision is now straightforward: if the actual value T of the test statistics $\underline{T}$ is element of R, then H_0 is rejected, otherwise it is accepted.

Fig. 2 shows a typical situation in the precise case. The value of the test statistics is either element of A or element of R. For this reason a clear and unique decision is achievable in any case.

The probability of a type I error (rejection of H_0 despite it is valid) is defined as

$$\alpha = P(T \in R \mid H_0) \tag{11}$$

and the probability of a type II error (acceptance of H_0 in case it is not valid) as

$$1-\beta = P(T \in A \mid H_a). \tag{12}$$

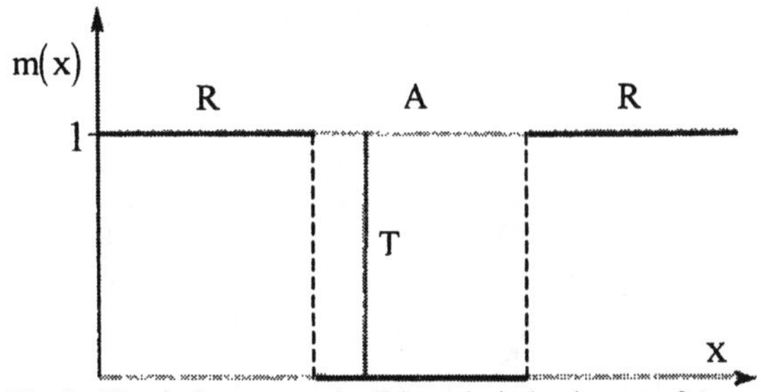

Fig. 2: Classical test scenario· T is exclusively element of A or R; a clear decision is always possible

4 Test scenario in case of imprecise and stochastic data

An imprecise one-dimensional test statistics with Normal mean point is given as

$$\tilde{\underline{T}} = (\underline{T}_m, T_l, T_r)_{LR} \tag{13a}$$

$$\text{with} \quad \underline{T}_m \sim N(\mu, 1) \tag{13b}$$

$$\text{and} \quad \begin{cases} \mu = 0 & H_0 \\ \mu = \delta, \delta \neq 0 & H_a \end{cases} \tag{13c}$$

This represents a direct and consistent extension of the case described in Section 3. Please note that the variance of $\underline{T}_m$ is assumed to be known precisely. The concept which is presented in the following principally permits to handle imprecise variances. However, this is not discussed here for the sake of a compact and clear presentation.

It is possible to consider additional imprecision due to the fuzziness of linguistic expressions which originates from the formulation of hypotheses such as "observation i contains an *outlier*". Hence, the general definition of a region of acceptance can based on a LR-fuzzy interval

$$\tilde{A} = (A_m, A_n, A_l, A_r)_{LR} \tag{14}$$

with the core $[A_m, A_n]$, the left spread A_l and the right spread A_r. The corresponding region of rejection (fuzzy-theoretical complement) is given as

$$\tilde{R} = \tilde{A}^c \quad \Rightarrow \quad m_{\tilde{R}} = 1 - m_{\tilde{A}} \tag{15}$$

Fig. 3 illustrates exemplarily the problems which have to be solved in case of imprecise test statistics and imprecision regions of decision. A unique test decision is only possible if the value $\tilde{T}$ of the imprecise test statistics $\tilde{\underline{T}}$ is completely contained in $\tilde{A}$ or in $\tilde{R}$, respectively. A more general decision criterion is needed which is precise and unique if the test statistics overlaps the regions of decision. A dedicated strategy which uses an adequate criterion is developed in the next section.

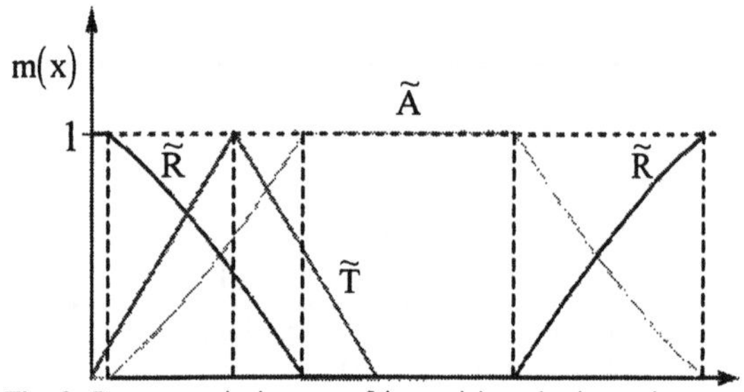

Fig. 3: Test scenario in case of imprecision: the imprecise test statistics is in general element of both the region of acceptance and the region of rejection

5 Test and decision strategy

From a pragmatic point of view the basic conditions for a general test strategy in case of imprecise data are: the imprecise test statistics and the regions of decision have to be compared quantitatively. The final decision criterion for acceptance or rejection has to be precise. A probabilistic evaluation and interpretation (derivation of the probability for a type I and a type II error) should be possible. The steps of the proposed test are:

1. Define the regions of acceptance and rejection: in case of precise regions of decision they are given by classical intervals A and R. In case of imprecise regions fuzzy intervals $\tilde{A}$ and $\tilde{R}$ have to be used. In the context of fuzzy sets the membership function of a classical interval Y is given by
$$m_Y(y) = \begin{cases} 1, & y \in Y \\ 0 & \text{else} \end{cases} \tag{16}$$
This holds since classical intervals are special cases of fuzzy sets. Thus, the discussion can be given completely in terms of fuzzy-theory.
2. Calculate the value $\tilde{T}$ of the imprecise test statistics through fuzzy extension of the classical expression according to Eqs. (5a,b).
3. Evaluate the decision criterion
$$\rho_{\tilde{R}}(\tilde{T}) \begin{Bmatrix} \leq \\ > \end{Bmatrix} \rho_{crit} \in [0,1] \Rightarrow \begin{cases} \text{Do not reject } H_0 \\ \text{Reject } H_0 \end{cases} \tag{17}$$
which will be defined below in this section.
4. Calculate the probability of a type I error
$$\alpha = P\left(\rho_{\tilde{R}}(\tilde{T}) > \rho_{crit} \mid H_0\right) \tag{18}$$
and the probability of a type II error
$$1-\beta = P\left(\rho_{\tilde{R}}(\tilde{T}) \leq \rho_{crit} \mid H_a\right) \tag{19}$$

The decision criterion used in Step 3 is based on a quantitive comparison of $\tilde{T}$ with $\tilde{A}$ and with $\tilde{R}$, respectively. The idea is to assess the rejectability of $\tilde{T}$ by its agreement with $\tilde{R}$ and its disagreement with $\tilde{A}$. In this section a particular choice is presented to illustrate the procedure. A more rigorous treatment is given in Section 6.

The *degree of agreement* of $\tilde{T}$ and $\tilde{R}$ can be defined as
$$\gamma_{\tilde{R}}(\tilde{T}) := \gamma(\tilde{T}, \tilde{R}) = \text{height}(\tilde{T} \cap \tilde{R}) \tag{20}$$
with the intersection of the two fuzzy sets according to Eq. (2) and the height as the maximum membership degree. If $\tilde{T}$ lies completely inside $\tilde{R}$, we obtain $\gamma_{\tilde{R}}(\tilde{T}) = 1$. The *degree of disagreement* of $\tilde{T}$ and $\tilde{A}$ can be defined as
$$\delta_{\tilde{A}}(\tilde{T}) := 1 - \gamma_{\tilde{A}}(\tilde{T}) \tag{21}$$
which reflects the complement of the degree of agreement. Finally, the *degree of rejectability* of $\tilde{T}$ with respect to $\tilde{R}$ (precisely: the rejectability of the null hypothesis H_0 under the condition of $\tilde{T}$) is defined as
$$\rho_{\tilde{R}}(\tilde{T}) := \min\left(\gamma_{\tilde{R}}(\tilde{T}), \delta_{\tilde{R}^c}(\tilde{T})\right). \tag{22}$$

Hence, $\tilde{T}$ is only rejected if it agrees with $\tilde{R}$ *and* it does not agree with $\tilde{A}$. The test decision is now based on the comparison of $\rho_{\tilde{R}}(\tilde{T})$ with a (new) critical value $\rho_{crit} \in [0,1]$ (Step 3).

The presented strategy is based on Römer and Kandel (1995), Viertl (1996), Steinmetz (1998) and Kutterer (2002). However, it extends the given references in significant parts. Its key issue is the unique and easy-to-handle test decision criterion which enables the probabilistic assessment of the strategy according to Step 4.

6 A generalized decision criterion

Actually, the test decision criterion presented in the previous section can be embedded in a more general context. The necessary definitions are now given.

The *degree of agreement* $\gamma : \mathcal{F}(\mathbb{R} \times \mathbb{R}) \rightarrow [0,1]$ of a non-empty fuzzy set $\tilde{M} \in \mathcal{F}(\mathbb{R})$ with a fuzzy set $\tilde{N} \in \mathcal{F}(\mathbb{R})$ is defined as

$$\gamma_{\tilde{N}}(\tilde{M}) := \gamma(\tilde{M}, \tilde{N}) = \frac{h(\tilde{M} \cap \tilde{N})}{h(\tilde{M})} \tag{23}$$

with $\mathcal{F}(\mathbb{R})$ denoting the space of fuzzy sets over the real numbers $\mathbb{R}$ and $\mathcal{F}(\mathbb{R} \times \mathbb{R})$ the space of fuzzy sets over $\mathbb{R} \times \mathbb{R}$. The degree of agreement is a relative measure which relates $\tilde{M}$ to $\tilde{N}$. This is also indicated by the notation $\gamma_{\tilde{N}}(\tilde{M})$. The class of functions

$h: \quad \mathcal{F}(\mathbb{R}) \to [0, \infty)$

is defined by the conditions

(i) $\tilde{U} = \emptyset \Leftrightarrow h(\tilde{U}) = 0,$

(ii) $\tilde{U} \subseteq \tilde{V} \Rightarrow h(\tilde{U}) \le h(\tilde{V})$

for the fuzzy sets $\tilde{U}, \tilde{V} \in \mathcal{F}(\mathbb{R})$. Condition (i) reflects minimum agreement which is equivalent to an empty set as argument, condition (ii) guarantees monotonicity for fuzzy-theoretical set inclusion. The two conditions lead to some obvious properties of the degree of agreement γ

(a) $\tilde{M} \cap \tilde{N} = \emptyset \Leftrightarrow \gamma_{\tilde{N}}(\tilde{M} \cap \tilde{N}) = 0$

(b) $\tilde{N} \neq \emptyset \Rightarrow \gamma_{\tilde{N}}(\tilde{N}) = 1$

(c) $\tilde{M} \subseteq \tilde{P} \wedge \tilde{N} \subseteq \tilde{Q} \Rightarrow \gamma_{\tilde{N}}(\tilde{M}) \le \gamma_{\tilde{Q}}(\tilde{P})$

Hence, the minimum agreement of the two argument sets is equivalent to empty intersection. Maximum agreement is obtained for $\tilde{M} \cap \tilde{N} = \tilde{N}$.

One example for h is the height of a fuzzy set

$$h(\tilde{U}) = \text{height}(\tilde{U}). \tag{24}$$

As the height of a fuzzy number equals 1, we find

$$\gamma_{\tilde{N}}(\tilde{M}) = \text{height}(\tilde{M} \cap \tilde{N}) \tag{25}$$

which is symmetric with respect to $\tilde{M}$ and $\tilde{N}$. A second example for h is the cardinality

$$h(\tilde{U}) = \text{card}(\tilde{U}) := \int_{\mathbb{R}} m_{\tilde{U}}(x)\,dx. \tag{26}$$

The definition of a *degree of disagreement* $\delta: \mathcal{F}(\mathbb{R} \times \mathbb{R}) \to [0, 1]$ of a non-empty fuzzy set $\tilde{M} \in \mathcal{F}(\mathbb{R})$ with a fuzzy set $\tilde{N} \in \mathcal{F}(\mathbb{R})$ as

$$\delta_{\tilde{N}}(\tilde{M}) := 1 - \gamma_{\tilde{N}}(\tilde{M}) \tag{27}$$

is straightforward from the definition of γ. It leads to the corresponding properties

(a) $\tilde{M} \cap \tilde{N} = \emptyset \Leftrightarrow \delta_{\tilde{N}}(\tilde{M} \cap \tilde{N}) = 1$

(b) $\tilde{N} \neq \emptyset \Rightarrow \delta_{\tilde{N}}(\tilde{N}) = 0$

(c) $\tilde{M} \subseteq \tilde{P} \wedge \tilde{N} \subseteq \tilde{Q} \Rightarrow \delta_{\tilde{N}}(\tilde{M}) \ge \delta_{\tilde{Q}}(\tilde{P})$

The remarks in the context of $\gamma_{\tilde{N}}(\tilde{M})$ can be transferred analogously to $\delta_{\tilde{N}}(\tilde{M})$.

Finally, the *degree of approval* $\rho: \mathcal{F}(\mathbb{R}) \to [0, 1]$ of a fuzzy set $\tilde{N} \in \mathcal{F}(\mathbb{R})$ by a non-empty fuzzy set $\tilde{M} \in \mathcal{F}(\mathbb{R})$ is defined as

$$\rho_{\tilde{N}}(\tilde{M}) := \bar{\rho}\left(\gamma_{\tilde{N}}(\tilde{M}), \delta_{\tilde{N}^c}(\tilde{M})\right). \tag{28}$$

The function $\bar{\rho}: [0, 1] \times [0, 1] \to [0, 1]$ can be defined by the following properties

(a) $\bar{\rho}(x, y) = \bar{\rho}(y, x)$

(b) $\bar{\rho}(x, \bar{\rho}(y, z)) = \bar{\rho}(\bar{\rho}(x, y), z)$

(c) $x \le u \wedge y \le v \Rightarrow \bar{\rho}(x, y) \le \bar{\rho}(u, v)$

(d) $\bar{\rho}(x, 1) = x, \ \bar{\rho}(x, 0) = 0$

These are the defining properties of a t-norm and correspond with the fuzzy extension of the intersection of classical sets. See, e.g., Bandemer and Näther (1992) for t-norms and the s-norms which are considered below. Hence, a suitable choice for the t-norm is the minimum:

$$\bar{\rho}(x, y) = \min(x, y). \tag{29}$$

Alternatively, a s-norm could be chosen for the definition of $\bar{\rho}$ instead of a t-norm. This corresponds with the fuzzy extension of the union of classical sets. In this case the criterion for rejection would be more sensitive as then just one of the arguments "$\tilde{M}$ agrees with $\tilde{N}$" and "$\tilde{M}$ disagrees with $\tilde{N}^c$" has to be fulfilled.

Please note that in case of

$$\rho_{\tilde{R}}(\tilde{T}) = \bar{\rho}\left(\gamma_{\tilde{R}}(\tilde{T}), \delta_{\tilde{A}}(\tilde{T})\right) \tag{30}$$

a degree of rejectability of $\tilde{T}$ with respect to $\tilde{R}$ is defined in analogy to the presentation in Section 5.

7 Example: L-fuzzy test statistics and precise regions of decision

In order to illustrate the theoretical concept, an exemplary application is presented now. In case of precise regions of decision A and R, we have

$$m_A(x) = \begin{cases} 1, & -k \le x < k \\ 0 & \text{else} \end{cases}$$
$$m_R(x) = \begin{cases} 0, & -k \le x < k \\ 1 & \text{else} \end{cases} \qquad (31a, b)$$

for the regions of acceptance and rejection, respectively (see Section 5, Step 1). The imprecise value of a L-fuzzy test statistics is given by

$$m_{\tilde{T}}(x) = \begin{cases} L\left(\dfrac{T_m - x}{T_s}\right), & x < T_m \\ L\left(\dfrac{x - T_m}{T_s}\right), & T_m \le x \end{cases} \qquad (32)$$

with the mean point T_m as in the classical case and the reference function L according to Section 2 (see Section 5, Step 2).

In order to prepare the evaluation of the test decision criterion (Section 5, Step 3), the degree of rejectability $\rho_{\tilde{R}}(\tilde{T})$ has to be computed based on the degree of agreement $\gamma_{\tilde{R}}(\tilde{T})$ and the degree of disagreement $\delta_{\tilde{A}}(\tilde{T})$. This yields

$$\left.\begin{aligned} \gamma_R(\tilde{T}) &= \text{height}(\tilde{T} \cap R) \\ \delta_A(\tilde{T}) &= 1 - \text{height}(\tilde{T} \cap A) \end{aligned}\right\} \qquad (33a, b)$$

$$\Rightarrow \quad \rho_R(\tilde{T}) = \min\left(\gamma_R(\tilde{T}), \delta_A(\tilde{T})\right) \qquad (34)$$

and leads to

$$\rho_R(\tilde{T}) = \begin{cases} 0, & T_m \in A \\ 1 - L\left(\dfrac{|T_m| - k}{T_s}\right), & T_m \in R \end{cases} \qquad (35)$$

Figures 4 and 5 illustrate the derivation of the results for the two relevant cases $T_m \in A$ and $T_m \in R$. For the sake of simplicity a linear membership function has been chosen for the graphical representations. It is obvious from both figures that the imprecise value of the test statistics is in general element of both the region of acceptance A and the region of rejection R. The cases which have to be distinguished are controlled by the relation of the mean point T_m and the regions of decision.

If T_m is an element of A, then the degree of agreement between $\tilde{T}$ and A equals one and, consistently, their degree of disagreement equals zero. As the degree of rejectability is defined according to Eq. (34), it is obtained in this case as the minimum value zero; see also the first line of Eq. (35).

If T_m is an element of R, then the degree of agreement between $\tilde{T}$ and R equals one. Therefore it is maximum. Correspondingly, the degree of disagreement between $\tilde{T}$ and A equals the degree of rejectability. It is given according to the second line of Eq. (35).

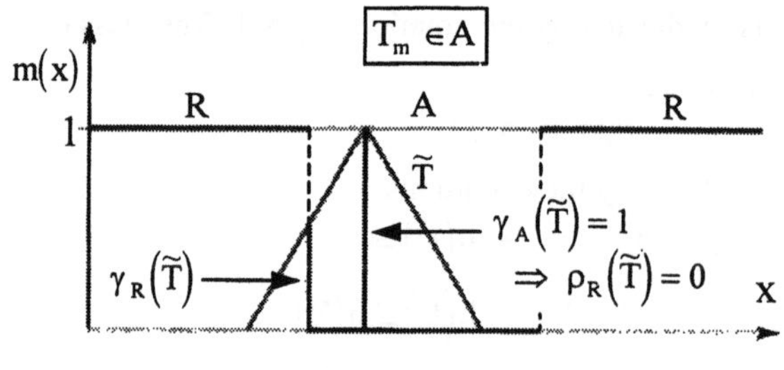

Fig. 4: T_m is element of A: the degree of agreement of $\tilde{T}$ and A is one, hence their degree of disagreement is zero .

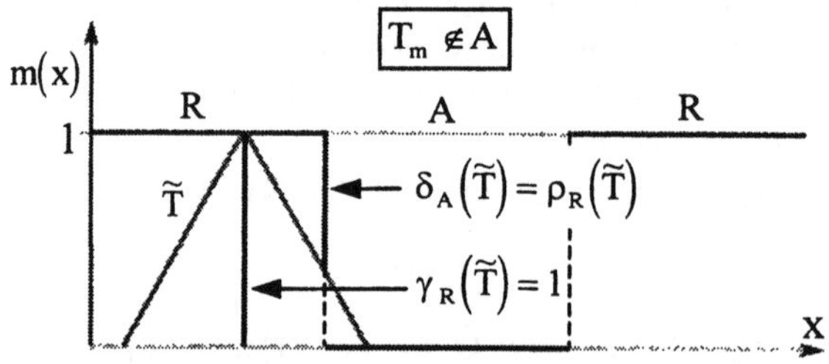

Fig. 5: T_m is element of R: the degree of disagreement of $\tilde{T}$ and A equals the degree of rejectability.

So the criterion for the rejection of the null hypothesis H_0 reads as

$$\rho_R(\tilde{T}) = 1 - L\left(\frac{|T_m| - k}{T_s}\right) > \rho_{crit} \qquad (36)$$
$$\Rightarrow \quad \text{reject } H_0$$

with a suitable $\rho_{crit} \in [0, 1]$. Obviously, $T_m \in R$ is necessary but not sufficient for rejection. If $T_m \in A$, then H_0 cannot be rejected.

As Eq. (36) represents a precise and unique rule, the probabilistic evaluation in terms of the probabilities for type I and type II errors is easy to handle. In case of the type I error its probability is defined as

$$\alpha_{impr} = P\left(\rho_{\tilde{R}}(\tilde{T}) > \rho_{crit} \mid H_0\right), \qquad (37)$$

see Eq. (18). The index "impr" denotes the case of imprecision. Now Eq. (36) has to be introduced. This yields after some calculation steps the

$\left(1-\frac{\alpha_{impr}}{2}\right)$ – fractile value of the standard normal distribution

$$z_{1-\frac{\alpha_{impr}}{2}} = k + T_s \cdot L^{-1}(1-\rho_{crit}), \qquad (38)$$

with L^{-1} denoting the inverse function of L. The fractile value depends on the size of A (controlled by k), on the spread T_s of the imprecise test statistics, on the particular reference function L, and on the chosen value of ρ_{crit}. An equivalent representation can be found for the value ρ_{crit} which reads as

$$\rho_{crit} = 1 - L\left(\frac{z_{1-\frac{\alpha_{impr}}{2}} - k}{T_s}\right) \qquad (39)$$

Two properties are mentioned for the case that A represents the classical region of acceptance ($k = z_{1-\frac{\alpha_{prec}}{2}}$), with the index "prec" denoting the precise case. The first one is related with the spread of the imprecise test statistics: the test situation of the classical case ($\alpha_{impr} = \alpha_{prec}$) is obtained, if imprecision is absent ($T_s = 0$) or $\rho_{crit} = 0$. The second one is related to the value of ρ_{crit}: if $\rho_{crit} \to 1$, then the last term of Eq. (38) yields

$$L^{-1}(1-\rho_{crit}) \to L^{-1}(0) = \begin{cases} v < \infty, & \text{Case I: L has finite support} \\ \infty, & \text{Case II: L has infinite support} \end{cases}$$

(borders of the support) and the fractile value

$$z_{1-\frac{\alpha_{impr}}{2}} \to \begin{cases} z_{1-\frac{\alpha_{prec}}{2}} + v \cdot T_s, & \text{Case I} \\ \infty, & \text{Case II} \end{cases}$$

Obviously in this case the imprecise test statistics needs a finite support. Otherwise the hypothesis test is insensitive. In any case it is obvious from Eq. (38) that $\alpha_{impr} \le \alpha_{prec}$ because

$$T_s \cdot L^{-1}(1-\rho_{crit}) \ge 0.$$

Hence, for the discussed test scenario a type I error is more likely in the precise case than in the imprecise case.

The probability of a type II error can be derived analogously. The starting formula is

$$1-\beta_{impr} = P\left(\rho_{\tilde{R}}(\tilde{T}) \le \rho_{crit} \mid H_a\right) \qquad (40)$$

according to Eq. (19). This is equivalent with

$$1-\beta_{impr} = \Phi\left(z_{1-\frac{\alpha_{impr}}{2}} - \delta_{impr}\right) - \Phi\left(-z_{1-\frac{\alpha_{impr}}{2}} - \delta_{impr}\right) \qquad (41)$$

with Φ denoting the standard normal distribution function and δ_{impr} the non-centrality parameter in the case of imprecision. A direct evaluation of Eq. (41) is possible in two cases. First, an equivalence relation holds which is known from the precise case:

$$\delta_{impr} = 0 \Leftrightarrow \alpha_{impr} = \beta_{impr}$$

Second, if $|\delta_{impr}| >> 0$, then

$$z_{1-\beta_{impr}} \approx k + T_s \cdot L^{-1}(1-\rho_{crit}) - |\delta_{impr}|$$

and

$$|\delta_{impr}| \approx k - z_{1-\beta_{impr}} + T_s \cdot L^{-1}(1-\rho_{crit}),$$

respectively. Obviously, the non-centrality parameter in the imprecise case contains a positive additive term which vanishes in case of $T_s = 0$ or $\rho_{crit} = 0$. If additionally

$$\beta_{prec} = \beta_{impr} \wedge k = z_{1-\frac{\alpha_{prec}}{2}},$$

then

$$|\delta_{impr}| = |\delta_{prec}| + T_s \cdot L^{-1}(1-\rho_{crit}).$$

Hence the non-centrality parameter in the imprecise case is greater than the one in the precise case which indicates reduced sensitivity regarding the rejection of the null hypothesis.

8 Conclusions

Stochasticity and imprecision of the data are two typical kinds of uncertainty which have to be taken into account in geodetic data analysis. The focus of this paper is on test decisions in such cases when both stochasticity and imprecision are relevant. It was shown that their joint treatment is possible. The construction principle is as follows: classical real-valued formulas which are common use in statistics are superposed with imprecision which is modelled in terms of fuzzy-theory.

The presented test strategy is based on the comparison of an imprecise test value and the regions of acceptance and rejection, respectively, which are considered as fuzzy sets. Additionally, this approach permits imprecise regions of decision; the extension of the example given in Section 7 to this case is natural and straightforward.

Two further generalizations have to be mentioned. On the one hand the test strategy allows to handle multi-dimensional test statistics which play a key role in Geodesy when the positions of points in a two-dimensional or three-dimensional Euclidean space have to be tested. On the other hand tests statistics based on estimated variances could be formulated. However, this extension is rather complicated because the extension principle has to be applied to a quadratic form.

It can be stated regarding Sections 5, 6 and 7 (in particular the given example) that the consideration of imprecision tends to reduce the sensitivity of the classical test results. This is plausible and expected as the uncertainty is increased when imprecision is additionally taken into account. However, reality is then met better in the interpretation step of data analysis since the results do not only depend on the assessment of random variations (which may be of minor relevance in some cases) but also on the quality of the knowledge about the observation process.

References

Bandemer H.; Näther W. (1992): Fuzzy Data Analysis. Kluwer Academic Publishers, Dordrecht.

Dubois D.; Prade H. (1980): Fuzzy Sets and Systems. Academic Press, New York.

Kaufmann A.; Gupta M. M. (1991): Introduction to Fuzzy Arithmetic – Theory and Applications. Van Nostrand Reinhold, New York.

Koch K.-R. (1999): Parameter Estimation and Hypothesis Testing in Linear Models (2nd Ed.). Springer, Berlin.

Kruse R.; Gebhardt J.; Klawonn F. (1994): Foundations of Fuzzy Systems. Wiley, Chichester.

Kutterer, H. (2001a): Uncertainty assessment in geodetic data analysis. In: Carosio, A. und H. Kutterer (Eds.): Proceedings of the First International Symposium on Robust Statistics and Fuzzy Techniques in Geodesy and GIS. Swiss Federal Institute of Technology Zurich, Institute of Geodesy and Photogrammetry - Report No. 295, S. 7-12.

Kutterer, H. (2001b): An imprecise search space for GPS phase ambiguity parameters. In: Carosio, A. und H. Kutterer (Ed.): Proceedings of the First International Symposium on Robust Statistics and Fuzzy Techniques in Geodesy and GIS. Swiss Federal Institute of Technology Zurich, Institute of Geodesy and Photogrammetry - Report No. 295, S. 215-220.

Kutterer, H. (2001c): Some considerations on Fuzzy Least-Squares. In: Sideris, M. (Ed.): Gravity, Geoid and Geodynamics 2000. Springer, Berlin Heidelberg, S. 73-78.

Kutterer, H. (2002): Zum Umgang mit Ungewissheit in der Geodäsie - Bausteine für eine neue Fehlertheorie. Deutsche Geodätische Kommission, Reihe C, Nr. 553, München.

Römer Chr.; Kandel A. (1995): Statistical tests for fuzzy data. Fuzzy Sets and Systems 72 (1995) 1-26.

Steinmetz V. (1998): Schließende Statistik mit unscharfen Stichproben. Latex-Version der Nr. A 9802, Diskussionsbeiträge Fachbereich Wirtschaftswissenschaft, Universität des Saarlandes, Saarbrücken, http://www.uni-saarland.de/fak1/fr12/steinmetz/fustat.ps

Viertl R. (1996): Statistical Methods for Non-Precise Data. CRC Press, Boca Raton New York London Tokyo.

Zadeh L. A. (1965): Fuzzy sets. Information Control 8 (1965): 338 - 353.

Multiresolution Data Analysis – Numerical Realization by Use of Domain Decomposition Methods and Fast Multipole Techniques

Willi Freeden, Carsten Mayer
Geomathematics Group, Department of Mathematics
University of Kaiserslautern, P.O. Box 3049, 67653 Kaiserslautern, Germany

Abstract. This survey paper deals with multiresolution analysis of geodetically relevant data and its numerical realization for functions harmonic outside a (Bjerhammar) sphere inside the Earth. Harmonic wavelets are introduced within a suitable framework of a Sobolev-like Hilbert space. Scaling functions and wavelets are defined by means of convolutions. A pyramid scheme provides efficient implementation and economical computation. Essential tools are the multiplicative Schwarz alternating algorithm (providing domain decomposition procedures) and fast multipole techniques (accelerating iterative solvers of linear systems).

Keywords. Multiresolution analysis, harmonic wavelets, reconstruction formula, pyramid schemes, domain decomposition methods, fast multipole techniques.

1 Introduction

Gravity field modelling, although it is always governed by the same classical Newton's law, changes its nature when it is seen from different spatial scales. To be more specific, if one looks at gravity field determination on the basis of an increasing spatial magnification and accuracy, we have to go from something that is suitably characterized by a simple mass point, on astronomical scale, to what is described by a global truncated multipole (i.e., outer harmonic) model, at scales corresponding to satellite altimetry, down to wavelengths of about 100 km. By further *zooming in* we can reach a spatial resolution of about 1 km showing a very complicated pattern, strongly related to the shape of the Earth and to irregular masses inside the Earth's crust. Simultaneously the error in the knowledge of the gravity field model goes from 5 Gal, the flattening effect, down to 10 mGal in a today's global model, down to about 10^{-1} mGal at the future 1 km spatial resolution. There is also a change of the gravity field in the time scale depending on the time interval under consideration, but this aspect will not be discussed here. Thus, gravity field modelling as scientific object to be investigated in this approach is by definition the stationary gravity field with a spatial resolution from a worldwide to about 1 km scale and from about 1000 Gal of the full field down to, at least 10^{-1} mGal in future.

What we would like to present in this note are multiscale structures by which the gravitational part of the gravity field can be approximated progressively better and better, reflecting an increasing flow of observations of terrestrial, airborne and/or satellite type (e.g., terrestrial gravimetry, airborne gravimetry, satellite-to-satellite tracking (SST), satellite gravity gradiometry (SGG), etc). More precisely, we shall try to outline the canonical bridge of gravitational field determination from the well-established global outer harmonic approximation corresponding to a spherical Earth to modern wavelet methods based on the actual geometry of the Earth's surface and geodetically relevant observables (thereby always neglecting the small effect of the atmosphere in the outer space).

At this stage some remarks should be made: First, every geodetic measurement is a functional which may be assumed to be suitably linearizable in case of nonlinearity when the disturbing potential is considered. In other words, the relation between the object function, i.e., the disturbing potential and the data, may be supposed to be linear (for more details see, for example, W.A. Heiskanen, H. Moritz (1967)). Second, more and more measurements refer to satellites and cannot be modelled as functionals of the gravitational potential on the boundary, i.e. the Earth's surface. The satellite observations, which certainly increase in importance in the future, are much more difficult to handle, since they are exponentially smoothed while moving to the outer space. As a consequence, essential knowledge of the gravitational potential should be based on the combination of ground observations with satellite information.

observables	location	measurement method
terrestrial observables		
$\vert\nabla V(x)\vert$	$x \in$ continent	gravimetry
$\frac{\nabla V(x)}{\vert\nabla V(x)\vert}$	$x \in$ continent	geometric–astronomic levelling
$V(x)$	$x \in$ ocean	satellite altimetry
x, y	$x, y \in$ surface of the Earth	Doppler, GPS, LASER, RADAR, VLBI techniques
spaceborne techniques		
$\nabla V(x)$	$x \in$ outer space of the Earth	satellite–to–satellite tracking
$(\nabla \otimes \nabla)V(x)$	$x \in$ outer space of the Earth	satellite gravity gradiometry
seismic information		
$\rho(x)$	$x \in$ Earth's crust	seismic tomography, etc

Table 1 Gravitational field observables and their location

A very rough overview of what we consider nowadays as relevant geodetic observables can be taken from a classification due to H. Nutz (2002) (see Table 1).

The main interest in modelling data associated to the disturbing potential is a mathematical model that is characterized by two aspects, on the one hand side by realistic geodetic assumptions and on the other hand side by an acceptable numerical complexity. It is very helpful from the point of approximation theory to require the class $\mathcal{H}$ of object functions to constitute a Hilbert space of potentials F being harmonic down to an internal sphere A (with radius α around the origin). In doing so, we are led to a sphere-oriented Runge-Walsh approach (see W. Freeden, F. Schneider (1998)) which tells us that mean squares approximation on the internal ("Bjerhammar") sphere implies uniform approximation on and outside the Earth's surface Σ, a feature that is of exceptional significance. The price that must be paid when approximately replacing the actual gravitational potential V of the Earth by an object function $F \in \mathcal{H}$ is the smoothness of the gravitational potential V on the Earth's surface which is clearly acceptable seen from the viewpoint of numerical analysis. In addition, we are confronted with the fact that the Hilbert space $\mathcal{H}$ of potentials in $\overline{A_{\text{ext}}}$ is infinite-dimensional, briefly formulated (for more details see W. Freeden (1999)) $\mathcal{H}$ is the space of potentials in $\overline{A_{\text{ext}}}$ of the form $\sum_{n=0}^{\infty}\sum_{k=1}^{2n+1}\frac{1}{A_n}F^\wedge(n,k)H_{-n-1,k}(\alpha;\cdot)$ satisfying $\sum_{n=0}^{\infty}\sum_{k=1}^{2n+1}(F^\wedge(n,k))^2 < \infty$, where $\{A_n\}_{n\in\mathbb{N}_0}$ is a sequence of non-vanishing real numbers and $\{F^\wedge(n,k)\}_{\substack{n=0,1,\ldots\\k=1,\ldots,2n+1}}$ given by

$$F^\wedge(n,k) = \int\limits_A F(x)H_{-n-1,k}(\alpha;x)\,d\omega(x)$$

($d\omega$: the surface element on A) is the table of Fourier coefficients $F^\wedge(n,k)$ of F on A with respect to the outer harmonics $H_{-n-1,k}(\alpha;\cdot)$ of degree n and order k. (Note that A_{ext} is the outer space of the sphere A with radius α around the origin 0; $\overline{A_{\text{ext}}} = A_{\text{ext}} \cup A$). In other words, the information about the object function $V \in \mathcal{H}$ is incomplete, since only a finite number of data is available. Moreover, the data are of heterogeneous type relating to different positions in $\overline{A_{\text{ext}}}$.

2 Outer Harmonic Multiresolution

It is fortunate that the Hilbert space $\mathcal{H}$ contains as subspaces all $(2j+1)$-dimensional spaces $Harm_j$ of outer harmonics $H_{-j-1,l}(\alpha;\cdot)$ of degree j and order l. More explicitly, we are able to make profit of the following (in physical geodesy well-known) *multiresolution analysis in terms of outer harmonics:*

- $\mathcal{V}_0 \subset \mathcal{V}_1 \subset \cdots \subset \mathcal{H}$,
- $\overline{\bigcup_{j=0}^{\infty} \mathcal{V}_j}^{\|\cdot\|_{\mathcal{H}}} = \mathcal{H}$,
- $\mathcal{V}_{j+1} = \mathcal{V}_0 \oplus \bigoplus_{l=0}^{j} \mathcal{W}_l = \mathcal{V}_{J_0} \oplus \bigoplus_{l=J_0}^{j} \mathcal{W}_l$, $0 \le J_0 \le j$,

where the *scale spaces* $\mathcal{V}_j$ and the *detail spaces* $\mathcal{W}_j$ are given by

$$\mathcal{V}_j = \bigoplus_{l=0}^{j} Harm_l = Harm_{0,\ldots,j}$$

and

$$\mathcal{W}_j = \mathcal{V}_{j+1} \ominus \mathcal{V}_j = Harm_{j+1},$$

respectively. Advantages may be characterized as follows: Any member F of $\mathcal{V}_j$ (that means any band-limited element of $\mathcal{H}$) is representable as a finite Fourier (orthogonal) *expansion*

$$F \in \mathcal{V}_j \Longleftrightarrow \sum_{n=0}^{j} \sum_{l=1}^{2n+1} F^{\wedge}(n,l) H_{-n-1,l}(\alpha;\cdot)$$

in terms of the $\mathcal{L}^2(A)$-orthonormal outer harmonics of order $\leq j$. It is well–known that each outer harmonic localizes ideally in the frequency space; the "polynomial structure" of the outer harmonics produces a good global "low frequency" approximation; each Fourier coefficient (i.e., potential coefficient) of the orthogonal expansion may be expressed exactly as finite linear combination of data functionals applied to outer harmonics. In conclusion, multiresolution analysis in terms of outer harmonics has proven to be very useful in physical geodesy for the purpose of global modelling. Nowadays, all updated global models include orthogonal coefficients up to the degree 360 (for example, EGM96). The Fourier coefficients are derivable in different ways on certain spectral domains (by global combination of terrestrial and spaceborne data).

3 Wavelet Multiresolution Analysis

The fundamental difficulty of multiresolution in terms of outer harmonics, however, is that we have to deal with trial potentials which do not show any phenomenon of space localization. Consequently, any local change of the potential affects the whole table of Fourier coefficients. In consequence, local modelling is hardly treatable. The essential reason is that the consecutive scale spaces $\mathcal{V}_j$ and $\mathcal{V}_{j+1}$ in outer harmonic multiresolution are not related by a dilation and shifting (translation) procedure. In other words, the spaces $\mathcal{V}_j$ do not consist of dilated and shifted copies of one fixed function, i.e., there is *no "mother outer harmonic"* from which all other outer harmonics may be generated by dilation and translation. For these reasons we are convinced that local modelling cannot be performed adequately within the framework of outer harmonics. A new component of approximation has to come into play, viz. harmonic wavelets (see W. Freeden (1999)).

The power of wavelets is based on a multiresolution analysis which enables both frequency as well as space localization. In fact, wavelets may be used as mathematical means for breaking up the complicated structure of the Earth's disturbing field into many simple pieces at different scales and positions. Basically this is done by filtering, i.e. forming convolutions $\Psi_j * F$, $j \in \mathbb{N}_0$, of a harmonic function F such as the disturbing potential against "dilated" and "shifted" versions Ψ_j of one fixed function, viz. the *mother wavelet* Ψ. In consequence, the disturbing potential is represented by a two-parameter family reflecting the different localization and different levels of resolution. The definition of wavelets has to be given in close connection with a so-called scaling function $\{\Phi_j\}_{j\in\mathbb{N}_0}$, approaching formally in the limit case the "Dirac function(al)" Φ_∞ given by

$$\Phi_\infty(x,y) = \sum_{n=0}^{\infty} \frac{1}{A_n^2} \sum_{k=1}^{2n+1} H_{-n-1,k}(\alpha;x) H_{-n-1,k}(\alpha;y),$$

$x, y \in \overline{A_{\text{ext}}}$, of the Sobolev space $\mathcal{H}$ under consideration. Observing the addition theorem of spherical harmonics (see, for example, W. Freeden et al. (1998)) we get

$$\Phi_\infty(x,y) = \sum_{n=0}^{\infty} \frac{2n+1}{4\pi\alpha^2} \frac{1}{A_n^2} \left(\frac{\alpha^2}{|x||y|}\right)^{n+1} P_n\left(\frac{x}{|x|}\cdot\frac{y}{|y|}\right),$$

where P_n is the Legendre polynomial of degree n.

The scaling (kernel) function is given as series expansion in terms of outer harmonics as follows:

$$\Phi_j(x,y) = \sum_{n=0}^{\infty} \frac{\varphi_j(n)}{A_n^2} \sum_{k=1}^{2n+1} H_{-n-1,k}(\alpha;x) H_{-n-1,k}(\alpha,y)$$

$x, y \in \overline{A_{\text{ext}}}$, where the family $\{\{\varphi_j(n)\}_{n\in\mathbb{N}_0}\}_{j\in\mathbb{N}_0}$ is (usually) assumed to satisfy the following properties:

(i) for all $j \in \mathbb{N}_0$,

$$\varphi_j(0) = 1$$

(ii) for all $j, j' \in \mathbb{N}_0$ with $j \leq j'$ and all $n \in \mathbb{N}$

$$\varphi_j(n) \leq \varphi_{j'}(n)$$

(iii) for all $n \in \mathbb{N}$

$$\lim_{j\to\infty} \varphi_j(n) = 1$$

(iv) for all $j \in \mathbb{N}_0$,

$$\sum_{n=1}^{\infty} \frac{\varphi_j(n)}{A_n^2} \frac{2n+1}{4\pi\alpha^2} < \infty.$$

The aforementioned convolutions in $\mathcal{H}$ are understood in the following way:

$$F * G = \sum_{n=0}^{\infty} \sum_{k=1}^{2n+1} A_n^2 F^\wedge(n,k) G^\wedge(n,k)$$

for $F, G \in \mathcal{H}$. In particular, for all $x \in \overline{A_{\text{ext}}}$ and all $F \in \mathcal{H}$,

$$(\Phi_j * F)(x) = \sum_{n=0}^{\infty} \sum_{k=1}^{2n+1} A_n^2 \varphi_j(n) F^\wedge(n,k) H_{-n-1,k}(\alpha; x).$$

Wavelet function $\{\Psi_j\}_{j\in\mathbb{N}_0}$ and scaling function $\{\Phi_j\}_{j\in\mathbb{N}_0}$ are related via their symbols $\{\varphi_j(n)\}_{n\in\mathbb{N}_0}$, $\{\psi_j(n)\}_{n\in\mathbb{N}_0}$ via the *scaling equation*

$$\psi_j(n) = \varphi_{j+1}(n) - \varphi_j(n), \quad n \in \mathbb{N}_0 .$$

Starting from the mother kernels Φ_0 and Ψ_0, respectively, the j-th level kernels Φ_j and Ψ_j are obtained by dilation. The scale index j serves as a "measure for decreasing frequency localization" or equivalently as a "measure for increasing space localization".

To be more specific, observing the addition theorem of spherical harmonics the kernels Φ_j and Ψ_j, respectively, can be written as follows:

$$\Phi_j(x,y) = \sum_{n=0}^{\infty} \frac{\varphi_j(n)}{A_n^2} \frac{2n+1}{4\pi\alpha^2} \left(\frac{\alpha^2}{|x||y|}\right)^{n+1} P_n\left(\frac{x}{|x|} \cdot \frac{y}{|y|}\right),$$

$$\Psi_j(x,y) = \Phi_{j+1}(x,y) - \Phi_j(x,y) = \sum_{n=0}^{\infty} \frac{\psi_j(n)}{A_n^2} \frac{2n+1}{4\pi\alpha^2} \left(\frac{\alpha^2}{|x||y|}\right)^{n+1} P_n\left(\frac{x}{|x|} \cdot \frac{y}{|y|}\right),$$

$(x,y) \in \overline{A_{\text{ext}}} \times \overline{A_{\text{ext}}}$.

The so–called *dilation operator* D_j is defined in the following way. $D_j : \Phi_0(\cdot,\cdot) \mapsto D_j\Phi_0(\cdot,\cdot) = \Phi_j(\cdot,\cdot)$, $j \in \mathbb{N}_0$. If y is a point of $\overline{A_{\text{ext}}}$, then the *shift* operator S_y is defined by $S_y\Phi_0(\cdot,\cdot) = \Phi_0(\cdot,y)$, $y \in \overline{A_{\text{ext}}}$. Combining both operators we find $\Phi_j(\cdot,y) = S_y\Phi_j(\cdot,\cdot) = S_yD_j\Phi_0(\cdot,\cdot)$ and $\Psi_j(\cdot,y) = S_y\Psi_j(\cdot,\cdot) = S_yD_j\Psi_0(\cdot,\cdot)$.

The *multiresolution analysis in terms of harmonic wavelets* for a potential $F \in \mathcal{H}$ (such as the disturbing potential) is illustrated by the scheme below:

- $\Phi_j * F \in \mathcal{V}_j$, $j = 0, 1, \ldots$,
- $\Phi_j * F \underset{j\to\infty}{\longrightarrow} F$,
- $\mathcal{V}_0 \subset \cdots \subset \mathcal{V}_j \subset \mathcal{V}_{j+1} \subset \mathcal{V}_{j+2} \subset \cdots \subset \mathcal{H}$,
- $\mathcal{V}_{j+1} = \mathcal{V}_j + \mathcal{W}_j$,

  $$\mathcal{V}_{j+1} = \mathcal{V}_{J_0} + \sum_{l=J_0}^{j} \mathcal{W}_l, \quad J_0 \geq 0,$$
- $\overline{\bigcup_{j=0}^{\infty} \mathcal{V}_j}^{\|\cdot\|_{\mathcal{H}}} = \mathcal{H}$,
- $\Psi_j * F \in \mathcal{W}_j$,
- $\Phi_{J_0} * F + \sum_{l=J_0}^{J} \Psi_l * F \underset{J\to\infty}{\longrightarrow} F$.

The last result is known in the wavelet terminology as *reconstruction formula*. Explicitly written out it reads:

$$\lim_{J\to\infty} \left\| \Phi_{J_0} * F + \sum_{l=J_0}^{J} \Psi_l * F - F \right\|_{\mathcal{H}} = 0.$$

In terms of filtering, the sequences $\{\Phi_j\}_{j\in\mathbb{N}_0}$ and $\{\Psi_j\}_{j\in\mathbb{N}_0}$, respectively, may be interpreted as *low-pass filter* and *band-pass filter*. The convolutions $\Psi_j * F$ may be understood as a version of F blurred to the scale j. It describes the "detail structure" of F at scale j. The *detail space* $\mathcal{W}_j$ contains the detail information needed to go from an approximation at resolution j to an approximation at resolution $j+1$. For more theoretical details confer W. Freeden (1999); for graphical illustrations see e.g. Figure 2 of this paper.

4 Some Wavelet Examples

Next some examples of scaling functions and associated wavelets will be presented.

Example 1. Let $\{\gamma_j\}_{j\in\mathbb{N}_0}$ be a strict monotonically decreasing sequence of positive real numbers with $\lim_{j\to\infty} \gamma_j = 0$. The generating symbol $\{\{\varphi_j(n)\}_{n\in\mathbb{N}_0}\}_{j\in\mathbb{N}_0}$ of the *Shannon scaling function* $\{\varphi_j\}_{j\in\mathbb{N}_0}$ is given by

$$\varphi_j(n) = \begin{cases} 1 & , \quad n \in [0, \gamma_j^{-1}) \\ 0 & , \quad n \in [\gamma_j^{-1}, \infty). \end{cases}$$

The Shannon scaling functions constitutes a bandlimited wavelet function.

The following example is of non-bandlimited nature.

Example 2. Let $\{\gamma_j\}_{j\in\mathbb{N}_0}$ be a strict monotonically decreasing sequence of positive real numbers with $\lim_{j\to\infty} \gamma_j = 0$. The generating symbol

$\{\{\varphi_j(n)\}_{n\in\mathbb{N}_0}\}_{j\in\mathbb{N}_0}$ of the *Tikhonov scaling function* $\{\Phi_j\}_{j\in\mathbb{N}_0}$ is given by

$$\varphi_j(n) = \frac{\sigma_n^2}{\sigma_n^2 + \gamma_j},$$

where the sequence $\{\sigma_n\}_{n\in\mathbb{N}_0}$ has to satisfy $\sigma_n \neq 0$, $n \in \mathbb{N}_0$, and $\sum_{n=0}^{\infty}(2n+1)\sigma_n^2 < \infty$.

In particular, the sequence $\{\gamma_j\}_{j\in\mathbb{N}_0}$ may be chosen in dyadic way: $\gamma_j = 2^{-j}\alpha$, $j \in \mathbb{N}_0$, for some constant $\alpha > 0$.

Finally, another type of a non–bandlimited scaling function should be listed that turns out to be of particular importance in numerical computations.

Example 3. Let $\{\gamma_j\}_{j\in\mathbb{N}_0}$ be a strict monotonically decreasing sequence of positive real numbers with $\lim_{j\to\infty}\gamma_j = 0$. Let $Q : [0,\infty) \to [0,\infty)$, $t \mapsto Q(t)$, be a function with the following properties:

(α) $Q \in C^{(\infty)}[0,\infty)$,

(β) $Q(0) = 0$,

(γ) $Q(t) > 0, t > 0$

(δ) $Q(t) < Q(t'), 0 < t < t'$.

Then the generating symbol $\{\{\varphi_j(n)\}_{n\in\mathbb{N}_0}\}_{j\in\mathbb{N}_0}$ of the *exponential scaling function* $\{\Phi_j\}_{j\in\mathbb{N}_0}$ is given by

$$\varphi_j(n) = h^{n/2}e^{-\gamma_j Q(n)}, \quad h \in (0,1].$$

Of particular significance are kernels Φ_j that are available as elementary functions.

Abel–Poisson kernel: $A_n = 1, Q(t) = Rt, R > 0$.

$$\Phi_j(x,y) = \frac{1}{4\pi}\frac{|x|^2|y|^2 - h^2\alpha^4e^{-2\gamma_j R}}{(|x|^2|y|^2 + h^2\alpha^4e^{-2\gamma_j R} - 2h\alpha^2e^{-\gamma_j R}(x,y))^{3/2}}.$$

Singularity kernel: $A_n = (n+1/2)^{1/2}$, $Q(t) = Rt, R > 0$.

$$\Phi_j(x,y) = \frac{1}{2\pi}\frac{1}{(|x|^2|y|^2 + h^2\alpha^4e^{-2\gamma_j R} - 2h\alpha^2e^{-\gamma_j R}(x\cdot y))^{1/2}}.$$

A fundamental aspect that should be mentioned is that a combined concept of outer harmonic and wavelet expansion may be formulated, the outer harmonic expansion being responsible for the global modelling and the wavelet expansion being appropriate for local modelling involving a zooming-in procedure. (For more details the reader is referred to W. Freeden (1999)).

5 A Tree Algorithm

For computational purposes it is useful to discretize the convolutions by approximate formulas based on bounded linear functionals $L_i^{N_j}$, $i = 1,\dots,N_j$, on $\mathcal{H}$ characterizing the observables of the disturbing potential. What we are going to realize is a tree algorithm (*pyramid scheme*) with the following ingredients: Starting from a sufficiently large J such that

$$F(x) \simeq (\Phi_J * F)(x) \simeq \sum_{i=1}^{N_J} L_i^{N_J}\Phi_J(x,\cdot)a_i^{N_J},$$

$x \in \overline{A_{\text{ext}}}$, we are interested in showing that the coefficient vectors $a^{N_j} = (a_1^{N_j},\dots,a_{N_j}^{N_j})^T \in \mathbb{R}^{N_j}$, $j = J_0,\dots,J-1$ (being, of course, dependent on the potential $F \in \mathcal{H}$ under consideration) can be calculated such that the following statements hold true:

(i) The vectors a^{N_j}, $j = J_0,\dots,J-1$, are obtainable by recursion from the values a^{N_J}.

(ii) For $j = J_0,\dots,J$,

$$(\Phi_j * F)(x) \simeq \sum_{i=1}^{N_j} L_i^{N_j}\Phi_j(x,\cdot)a_i^{N_j},$$

$x \in \overline{A_{\text{ext}}}$. Analogously, we find for $j = J_0,\dots,J-1$,

$$(\Psi_j * F)(x) \simeq \sum_{i=1}^{N_j} L_i^{N_j}\Psi_j(x,\cdot)a_i^{N_j},$$

$x \in \overline{A_{\text{ext}}}$.

Our considerations leading to a tree algorithm are divided into two parts, viz. the initial step concerning the scale level J and the pyramid step establishing the recursion relation for $J-1,\dots,J_0$:

The Initial Step: For a suitably large integer J, $\Phi_J * F$ is sufficiently close to F on the whole space $\overline{A_{\text{ext}}}$. Formally understood, the kernel Φ_J replaces the "Dirac kernel" Φ_∞. In other words,

$$L_i^{N_J}(\Phi_J * F) \simeq L_i^{N_J}F, \quad i = 1,\dots,N_J.$$

It remains to determine the coefficients $a_i^{N_J}$, $i = 1,\dots,N_J$. Different strategies can be applied. Dependent on the topology of the Hilbert space $\mathcal{H}$ and the type of the linear functionals the coefficients $a_i^{N_J}$ can be calculated by numerical integration (as, for example, in the case of a spherical Earth) or by solving certain linear equations (for example, relating to points on the actual surface of the Earth or a satellite orbit).

Applying a general *interpolation procedure* we may base our initial step on the ansatz

$$a_i^{N_J} = \sum_{k=1}^{N_J} w_{i,k}^{N_J} L_k^{N_J} F, \quad i = 1, \ldots, N_J,$$

where the coefficients $w_{i,k}^{N_J}$ satisfy the linear equations

$$\sum_{l=1}^{N_J} w_{l,k}^{N_J} L_i^{N_J} L_l^{N_J} \Phi_J(\cdot,\cdot) = \delta_{i,k}; i,k = 1, \ldots, N_J. \tag{1}$$

(It is worth mentioning that using numerical integration formulas, for example, in case of evaluation functionals $L_i^{N_J} : F \to F(x_i^{N_J})$, $x_i^{N_J} \in A$, we are canonically led to coefficients of the form

$$a_i^{N_J} = w_i^{N_J} F(x_i^{N_J}), \quad i = 1, \ldots, N_J,$$

where $w_i^{N_J}$ are the weights and $x_i^{N_J}$ the knots of the approximate integration formula. This approach also leads to a tree algorithm; however, it will not be discussed here (see W. Freeden (1999)).

The Pyramid Step. The essential idea for the development of a pyramid scheme is the existence of a reproducing kernel function $K_{\mathcal{V}_j}$, $j = J_0, \ldots, J$, of the detail spaces $\mathcal{V}_j$ such that

$$\Phi_j \simeq K_{\mathcal{V}_j} * \Phi_j$$

and

$$K_{\mathcal{V}_j} \simeq K_{\mathcal{V}_{j+1}} * K_{\mathcal{V}_j}$$

for $j = J_0, \ldots, J$. Observing our concept of discretizing convolutions by approximate formula based on interpolation we are led to

$$\begin{aligned}(\Phi_j * F)(x) &= (\Phi_j * K_{\mathcal{V}_j} * F)(x) \qquad (2)\\ &\simeq \sum_{i=1}^{N_j} L_i^{N_j} \Phi_j(x,\cdot) a_i^{N_j}, \quad x \in \overline{A_{\text{ext}}},\end{aligned}$$

where

$$a_i^{N_j} = \sum_{k=1}^{N_j} w_{i,k}^{N_j} L_k^{N_j} (K_{\mathcal{V}_j} * F), \quad i = 1, \ldots, N_j,$$

and the coefficients $w_{i,k}^{N_j}$ satisfy the linear equations

$$\sum_{l=1}^{N_j} w_{l,k}^{N_j} L_i^{N_j} L_l^{N_j} \Phi_j(\cdot,\cdot) = \delta_{ik}; i,k = 1, \ldots, N_j. \tag{3}$$

Now it follows by use of our approximate (interpolation based) formulae that

$$\begin{aligned}a_i^{N_j} &= \sum_{k=1}^{N_j} w_{i,k}^{N_j} L_k^{N_j} (K_{\mathcal{V}_j} * F)\\ &\simeq \sum_{k=1}^{N_j} w_{i,k}^{N_j} L_k^{N_j} (K_{\mathcal{V}_j} * K_{\mathcal{V}_{j+1}} * F)\\ &\simeq \sum_{k=1}^{N_j} w_{i,k}^{N_j} L_k^{N_j} \sum_{l=1}^{N_{j+1}} L_l^{N_{j+1}} K_{\mathcal{V}_j}(\cdot,\cdot) a_l^{N_{j+1}}\\ &= \sum_{k=1}^{N_j} \sum_{l=1}^{N_{j+1}} w_{i,k}^{N_j} a_l^{N_{j+1}} L_k^{N_j} L_l^{N_{j+1}} K_{\mathcal{V}_j}(\cdot,\cdot)\\ &= a_i^{N_{j+1}}\\ &\quad + \sum_{k=1}^{N_j} \sum_{l=N_j+1}^{N_{j+1}} w_{i,k}^{N_j} a_k^{N_{j+1}} L_k^{N_j} L_l^{N_{j+1}} K_{\mathcal{V}_j}(\cdot,\cdot).\end{aligned}$$

In other words, the coefficients $a_i^{N_{J-1}}$ can be calculated recursively starting from the values $a_i^{N_J}$ for the initial level J, $a_i^{N_{J-2}}$ can be deduced recursively from $a_i^{N_{J-1}}$, etc. Moreover, it is easy to see that the coefficients are independent of the chosen kernel. Consequently we have

$$(\Psi_j * F)(x) \simeq \sum_{i=1}^{N_j} L_i^{N_j} \Psi_j(x,\cdot) a_i^{N_j}, \quad x \in \overline{A_{\text{ext}}},$$

with the coefficients $a_i^{N_j}$, $i = 1, \ldots, N_j$, given by

$$\begin{aligned}a_i^{N_j} &\simeq a_i^{N_{j+1}}\\ &\quad + \sum_{k=1}^{N_j} \sum_{l=N_j+1}^{N_{j+1}} w_{i,k}^{N_j} a_k^{N_{j+1}} L_k^{N_j} L_l^{N_{j+1}} K_{\mathcal{V}_j}(\cdot,\cdot).\end{aligned}$$

F	$\to$	a^{N_J}	$\to$	$a^{N_{J-1}}$	$\to$	$\ldots$	$\to$	$a^{N_{J_0}}$
		$\downarrow$		$\downarrow$				$\downarrow$
		$\Phi_J * F$		$\Phi_{J-1} * F$				$\Phi_{J_0} * F$
		$\Psi_J * F$		$\Psi_{J-1} * F$				$\Psi_{J_0} * F$

Table 2 Decomposition Scheme

$a^{N_{J_0}}$				$a^{N_{J_0+1}}$				$a^{N_{J_0+2}}$		
$\downarrow$										
$\Psi_{J_0} * F$				$\Psi_{J_0+1} * F$				$\Psi_{J_0+2} * F$		
	$\searrow$				$\searrow$				$\searrow$	
$\Phi_{J_0} * F$	$\rightarrow$	$+$	$\rightarrow$	$\Phi_{J_0+1} * F$	$\rightarrow$	$+$	$\rightarrow$	$\Phi_{J_0+2} * F$	$\rightarrow$ $+$ $\rightarrow$	$\ldots$

Table 3 Decomposition Scheme

From numerical point of view we have to deal with *two essential problems*: (i) Large linear systems have to be solved to determine the coefficients $w_{i,k}^{N_j}$, $j = J_0, \ldots, J$. (ii) Large summations $\sum_{i=1}^{N_j} \ldots$ must be performed during the solution process within the system of linear equations.

6 The Schwarz Alternating Algorithm: A Domain Decomposition Method

The pyramid scheme leads to a system of linear equations of the form (1) with a positive definite symmetric matrix $\mathbf{A} = A_{i,l} = (L_i^{N_j} L_l^{N_j} \Phi_j(\cdot,\cdot))_{i,l=1,\ldots,N_j}$, as far as the linear functionals under consideration are linearly independent. In other words, we have to solve a linear system

$$\mathbf{A}x = b \tag{4}$$

where $\mathbf{A} \in \mathbb{R}^{N \times N}$, $\mathbf{A}^T = \mathbf{A}$, $(\mathbf{A}y, y) > 0$ for all $y \in \mathbb{R}^N \backslash \{0\}$, and $x, b \in \mathbb{R}^N$. Such a linear equation system can be solved with iterative solvers or direct solvers, but if N is large (for example, $N \geq 10000$), the runtime of an iterative solver without a suitable pre-conditioner or of a direct solver increases tremendously. Therefore, we need a more sophisticated method to solve (4) for problems with a large number of given data. One such method is a multiplicative variant of the Schwarz alternating algorithm, a domain decomposition method, which allows to split the matrix $\mathbf{A}$ in (4) into several smaller submatrices relating the linear functionals (see Table 1) to subdomains of the entire domain and thus providing a division in certain domains, which may (and will in numerical implementations) in general overlap. This multiplicative variant of the Schwarz alternating algorithm is an iterative method which solves in each iteration step successively a linear system with the matrices obtained from the splitting. This fact reduces both runtime and memory requirement drastically. (A further speed-up can be achieved if an additive variant of the Schwarz alternating algorithm is used which runs on parallel computers (see, for example, M. Griebel, P. Oswald (1995) and the references therein)).
The Schwarz alternating algorithm dates back to H.A. Schwarz' work, published in 1890, and has been investigated by many authors since then. A revived interest in variants of the Schwarz alternating method arose since 1985 due to the availability of fast modern and parallel computers. Roughly formulated, there are mainly two types of the Schwarz alternating algorithm: multiplicative variants (like the one used in this paper) and additive variants, which can be implemented on parallel computers and which are usually faster. For more information about the Schwarz alternating algorithm, the reader is referred to W. Frommer, H. Schwandt (1997), M. Griebel, P. Oswald (1995), and W. Freeden, K. Hesse (2002) and the references therein. In the last few years, a great interest has also been taken in the relation between the Schwarz alternating algorithm, multisplittings, multigrid methods, preconditioned conjugate gradient methods, as well as other iterative schemes.

The solution of (4) with a multiplicative variant of the Schwarz alternating algorithm, which will from now on be called briefly the multiplicative Schwarz alternating algorithm, is based on two facts: (i) every positive definite symmetric matrix is a Gram matrix, (ii) the convergence proof of the multiplicative Schwarz alternating algorithm is based on its formulation in terms of orthogonal projectors.
The matrix $\mathbf{A} = (A_{i,j})_{i,j=1,\ldots N}$ in (4) is positive definite and symmetric. Within the context of Cholesky factorization, there exists a uniquely determined invertible lower triangular matrix $\mathbf{L}$ with positive diagonal entries, such that

$$\mathbf{A} = \mathbf{L}\,\mathbf{L}^T\,. \tag{5}$$

Denote the row vectors of $\mathbf{L}$ by $v_1, \ldots, v_N$. Then (5) implies that

$$A_{ij} = v_i \cdot v_j = (v_i, v_j), \quad i, j = 1, \ldots, N\,.$$

Thus, $\mathbf{A}$ is the Gram matrix of the basis $\{v_1, \ldots, v_N\}$ of $\mathbb{R}^N$, and the solution $x = (x_1, \ldots, x_N)^T$ of the linear equation system (4) is the solution of the following orthogonal projection problem: Find $x = (x_1, \ldots, x_N)^T \in \mathbb{R}^N$ such that $f \in \mathbb{R}^N$ with $(f, v_i) = b_i$, $i = 1, \ldots, N$, has the representation

$$f = \sum_{i=1}^{N} x_i\, v_i\,. \tag{6}$$

Indeed, the solution of this problem demands the solution of the linear system

$$\sum_{i=1}^{N} x_i\,(v_i, v_j) = (f, v_j) = b_j\,, \qquad j = 1, \dots, N\,, \tag{7}$$

which is just the linear system (4). The orthogonal projection operator corresponding to (6) is, of course, the identity operator: We seek a representation of $f = \mathrm{Id}_{\mathbb{R}^N} f$ with respect to the basis $\{v_1, \dots, v_N\}$. Now we split the basis $\{v_1, \dots, v_N\}$ into several smaller possibly overlapping subsets $\Xi_r^{N_r} = \{v_1^r, \dots, v_{N_r}^r\} \subset \{v_1, \dots, v_N\}, r = 1, \dots, M$, such that

$$\bigcup_{r=1}^{M} \Xi_r^{N_r} = \{v_1, \dots, v_N\}\,.$$

This union will, in general, not be a disjoint, and we speak of overlapping subsets, if there are at least two subsets $\Xi_r^{N_r}$, $\Xi_k^{N_k}$ with $\Xi_r^{N_r} \cap \Xi_k^{N_k} \neq \emptyset$ and $k \neq r$. Denote the orthogonal projector from $\mathbb{R}^N$ onto $\mathrm{span}(\Xi_r^{N_r})$ by

$$P_r : \mathbb{R}^N \to \mathrm{span}\{v_1^r, \dots, v_{N_r}^r\}, g \mapsto P_r g, \tag{8}$$

i.e., $P_r = P_r \circ P_r$ and $(P_r v, w) = (v, P_r w)$ for all $v, w \in \mathbb{R}^N$. In order to compute $P_r g$, we assume again that (g, v_i), $i = 1, \dots, N$, is known. We want to calculate the coefficient vector $y = (y_1, \dots, y_{N_r})$ of the representation

$$P_r g = \sum_{i=1}^{N_r} y_i\, v_i^r\ .$$

Taking the inner product with $v_1^r, \dots, v_{N_r}^r$ successively leads to the linear equation system

$$\sum_{i=1}^{N_r} y_i\,(v_i^r, v_j^r) \stackrel{!}{=} (P_r g, v_j^r) = (g, P_r v_j^r) = (g, v_j^r), \tag{9}$$

$j = 1, \dots, N_r$. Clearly the matrix $\mathbf{A}_r = ((v_i^r, v_j^r))_{i,j=1,\dots,N_r}$ is a submatrix of the matrix $\mathbf{A}$ of the linear equation system (4).

Algorithm 6.1 *(Multiplicative Schwarz Alternating Algorithm)*
set $f_0 = f \in \mathbb{R}^N$ *and* $s_0^f = 0$
for $n = 0, 1, 2, \dots$ *do*
  for $r = 1, \dots, M$ *do*
    calculate $s_{nM+r}^f = s_{nM+(r-1)}^f + P_r(f_{nM+(r-1)})$
    update $f_{nM+r} = f_{nM+(r-1)} - P_r(f_{nM+(r-1)})$
until

$$\frac{|((f_{(n+1)M}, v_1), \dots, (f_{(n+1)M}, v_N))^T|}{|((f, v_1), \dots, (f, v_N))^T|} \leq \varepsilon$$

Recently, W. Freeden, K. Hesse (2002) have shown that the sequence of iterates $\{s_{nM}^f\}_{n \in \mathbb{N}_0}$ converges to f as $n \to \infty$.

Algorithm 6.1 will now be transformed into a matrix formulation via (7) and (9). For this purpose, we need the following restriction operators $R_r : \mathbb{R}^N \to \mathbb{R}^{N_r}, w \mapsto R_r(w) = ((R_r(w))_1, \dots, (R_r(w))_{N_r})^T$, and embedding operators $I_r : \mathbb{R}^{N_r} \to \mathbb{R}^N$, $z \mapsto I_r(z) = ((I_r(z))_1, \dots, (I_r(z))_N)^T$, corresponding to the subspaces $\mathbb{R}^{N_r}$ of the subproblems (9). They are defined by

$$\begin{aligned} &(R_r(w))_i \\ &= w_j \text{ for the index } i \in \{1, \dots, N_r\} \text{ with } v_i^r = v_j, \\ &(I_r(z))_i \\ &= \begin{cases} z_j & \text{if there is } j \in \{1, \dots, N_r\} \text{ with } v_j^r = v_i \\ 0 & \text{else}\ . \end{cases} \end{aligned}$$

Algorithm 6.2
(matrix formulation of Algorithm 6.1)
Define the matrices $\mathbf{A}_r = ((v_i^r, v_j^r))_{i,j=1,\dots,N_r}$, $r = 1, \dots M$
set $\tilde{f}_0 = ((f, v_1), \dots, (f, v_N))^T$, $a_0 = (0, \dots, 0)^T \in \mathbb{R}^N$, *where* $f \in \mathbb{R}^N$
for $n = 0, 1, 2 \dots$ *do*
  for $r = 1, \dots, M$ *do*
    solve $\mathbf{A}_r d = R_r(\tilde{f}_{nM+(r-1)})$, $d = (d_1, \dots, d_{N_r})^T \in \mathbb{R}^{N_r}$
    update $a_{nM+r} = a_{nM+(r-1)} + I_r(d)$
    update $\tilde{f}_{nM+r} = \tilde{f}_{nM+(r-1)} - \left(\left(\sum_{i=1}^{N_r} d_i\,(v_i^r, v_k) \right)_{k=1,\dots,N} \right)^T$
until

$$\frac{|\tilde{f}_{(n+1)M}|}{|\tilde{f}_0|} \leq \varepsilon$$

compute

$$s_{(n+1)M}^f = \sum_{i=1}^{N} (a_{(n+1)M})_i\, v_i.$$

We stress that all the computations (except the computation of $s_{(n+1)M}^f$) in Algorithm 6.2 can be performed without actually computing $v_1, \dots, v_N \in \mathbb{R}^N$, i.e., we do not need the Cholesky factorization of $\mathbf{A}$: The matrices $\mathbf{A}_r$ are available as submatrices of $\mathbf{A}$, and the update involves a matrix vector multiplication with the matrix $((v_k, v_i^r))_{\substack{k=1,\dots,N;\\ i=1,\dots,N_r}}$, which is also a submatrix of $\mathbf{A}$.

Summarizing our result we obtain the following corollary.

COROLLARY 6.3 *Let the notation and the assumptions be the same as in Algorithm 6.2. Then the sequence $\{a_{nM}\}_{n\in\mathbb{N}_0} \subset \mathbb{R}^N$ converges to the solution $x \in \mathbb{R}^N$ of the linear equation system $\mathbf{A}x = b$, where $\mathbf{A} = ((v_i, v_j))_{i,j=1,\ldots,N}$ and $b = ((f, v_1), \ldots, (f, v_N))^T$.*

Finally, we want to give some comments concerning the implementation of Algorithm 6.2 for the solution of the aforementioned linear equation systems. In case of Example 2 or 3 in section 4 the matrix entries are available as elementary function. This enables us to evaluate a matrix entry with small computational effort (in comparison to the generation of kernels by aid of series expansions in terms of outer harmonics). In an implementation of Algorithm 6.2 we will generate only the small matrices A_r in advance, compute, for example, their Cholesky factorization in a preprocessing step, and keep the matrices of the Cholesky factorizations of the A_r in the memory. The other matrix entries of A, which will be needed for the update (computation of the new residual), are generated while the update is performed. The update is the time-consuming task, whereas the smaller equation systems can now be solved extremely fast.

If fast multipole methods (fast summation techniques) are available for the type of kernel, which determines the matrix entries, the update can even be accelerated. In what follows this numerical method will be explained in more detail.

7 A Fast Multipole Technique

Although the fast multipole technique may also be applied to other types of kernels we restrict our considerations here to the singularity kernel (see Example 3) which is the reproducing kernel of the Sobolev space $\mathcal{H}$ associated to the sequence $\{A_n\}_{n\in\mathbb{N}_0}$ given by

$$A_n = (n + \frac{1}{2})^{1/2} h^{-n/2}; n = 0, 1, \ldots; 0 < h < 1.$$

This particular choice yields the following kernel representation for the singularity kernel

$$K_{\mathcal{H}}(x,y) = \frac{1}{2\pi\alpha^2} \sum_{n=0}^{\infty} h^n \left(\frac{\alpha^2}{|x||y|}\right)^{n+1} P_n\left(\frac{x}{|x|} \cdot \frac{y}{|y|}\right),$$

$x, y \in \overline{A_{\text{ext}}}$. Applying the Kelvin transform of potential theory (see, for example, O.D. Kellogg (1929)) with respect to the variable y and the sphere A we obtain by setting $y^* = h\frac{\alpha^2}{|y|^2}y$

$$\begin{aligned} K_{\mathcal{H}}(x,y) &= \frac{1}{2\pi\alpha^2}\frac{|y*|}{h}\sum_{n=0}^{\infty}\frac{|y^*|^n}{|x|^{n+1}}P_n\left(\frac{x}{|x|}\cdot\frac{y}{|y|}\right) \\ &= \frac{1}{2\pi\alpha^2}\frac{|y^*|}{h}\frac{1}{|x-y^*|}. \end{aligned}$$

This shows us, for example, in case of *evaluation functionals* $L_iF = F(x_i)$, $x_i \in \overline{A_{\text{ext}}}$,

$$L_iL_jK_{\mathcal{H}}(\cdot,\cdot) = \frac{1}{2\pi\alpha^2}\frac{|x_j^*|}{h}\frac{1}{|x_i - x_j^*|}, \tag{10}$$

and for functionals of *oblique derivatives* given by $L_iF = ((x \cdot \nabla F)(x))\big|_{x=x_i}$

$$\begin{aligned} &L_iL_jK_{\mathcal{H}}(\cdot,\cdot) \qquad (11) \\ &= \frac{1}{2\pi\alpha^2}\frac{|y^*|}{h}\left[(y^*\cdot\nabla_{y*})^2\left(\frac{1}{|x-y^*|}\right)\right. \\ &\left.+ 2(y^*\cdot\nabla_{y^*})\left(\frac{1}{|x-y^*|}\right) + \frac{1}{|x-y^*|}\right]\Big|_{(x_i,x_j^*)}. \end{aligned}$$

Since the entries of the matrix consist of the fundamental solution of the Laplace equation and directional derivatives of it, the solution of the linear system by a fast multipole accelerated solver becomes possible.

In order to reduce the computational effort and the storage requirements for solving the linear system (4) the Geomathematics Group Kaiserslautern adapted the fast multipole technique to accelerate the matrix multiplication in an iterative (GMRES) solver. Theory and algorithmic aspects of the FMM are very sophisticated and their description would go beyond the scope of this article. In this approach we merely outline the ideas and the basic aspects of the fast multipole method (FMM).

The subject of interest is the fast evaluation of a matrix multiplication of the form $\sum_{i=1}^N a_iL_iL_jK_{\mathcal{H}}(\cdot,\cdot)$, $j = 1, \ldots, N$, N large. The idea of the FMM is based on the localization property of the kernel defining the entries $L_iL_jK_{\mathcal{H}}(\cdot,\cdot)$, i.e. on the fact that most of the energy is contained within a small vicinity of x_j^*. The sum is split up into a near-field part (of course, defined with respect to the linear functional associated to x_j^*), which is evaluated explicitly, and a far-field part, for which we seek a "fast" approximation with "sufficient accuracy".
The algorithm is started by embedding the computational domain into an (initial) cube, followed by an adaptive and hierarchical subdivision of each cube into eight child cubes. Adaptivity means that a cube is only subdivided if it contains a minimum number of nodal points and is essential for the efficiency.

The approximation of the far-field is performed on the basis of the well-known multipole expansion of $1/|x - y|$ in terms of inner and outer harmonics. The kernel is expanded for both variables with respect to the centers x_0 and y_0 of the cubes (instead of the origin) and the expansion is truncated at a certain (low) degree p. Conditions for the convergence of the expansion can easily be derived. For example, an expansion for $(y \cdot \nabla_y)\, K_{\mathcal{H}}(x, y)$ can be found indirectly by applying $(y \cdot \nabla_y)$ to the expansion of $K_{\mathcal{H}}(x, y)$. Since the error of the expansion is controllable only in a small vicinity of (x_0, y_0), we have to perform several expansions with respect to different centers (x_0, y_0) to obtain a global approximation. A lot of work can be done in an a-priori step independent of the number and the location of the target points, resulting in a small set of so-called far-field coefficients containing the information from distant measurements. The hierarchical subdivision in connection with the translation and conversion theorems of spherical harmonics allow to calculate the far-field coefficients in level l recursively from those of the parent cube in level $l + 1$ so that an explicit calculation is only needed for the finest level. Altogether we obtain an algorithm which performs a matrix multiplication in $\mathcal{O}(N)$ operations and with order N storage requirements. Details and extensive numerical experiments concerning the efficiency of the method for satellite problems in physical geodesy can be found in the Diploma thesis D. Michel (2001) and the PhD thesis O. Glockner (2002).

8 Some Numerical Results

In the foregoing we have seen that harmonic wavelets are numerically efficient "building blocks" that enable fast modelling of geopotential data such as oblique derivative data of the disturbing potential. Now we discuss in more detail the concept of multiresolution data analysis. This method "looks at" the Earth's disturbing potential through a microscope, whose resolution gets finer and finer. Thus it associates to the disturbing potential a sequence of smoothed versions, labelled by the scale parameter j. This aspect is illustrated by the figures below (see Figure 2, (a)-(f)) from oblique derivative data on the actual Earth derived from the EGM96-model.
More explicitly, in our approach, we use the so-called TerrainBase data model of the National Geodetic Data Center in Boulder, Colorado, to have a representation of the actual surface of the Earth.

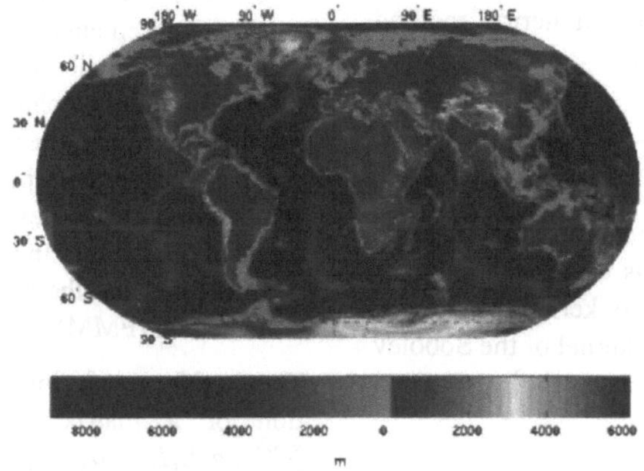

Figure 1 The topography of the Earth as given by TerrainBase data.

The EGM96 model (F.G. Lemoine et al. (1996)) led us to create gradients of the disturbing potential on the actual Earth's surface, i.e. the TerrainBase (data) model. Hence, in an *a priori step*, oblique derivatives became available on a grid for the TerrainBase model of the actual Earth. Based on this material the interpolation oriented multiscale procedure described above could be used to model the disturbing potential on the Earth's surface and in the outer space of the Earth from prescribed oblique derivatives on the Earth's surface.
Actually, the interpolation procedure was executed on a (regular) grid of about 160 000 points, where the oblique derivatives were derived from the EGM96–model. It turned out that the multiplicative Schwarz alternating algorithm had to perform 45 iterations to get the results of the multiresolution data analysis (illustrated in Figure 4). For more computational details including an error analysis the reader is referred to the Diploma thesis due to M. Gutting (2002).

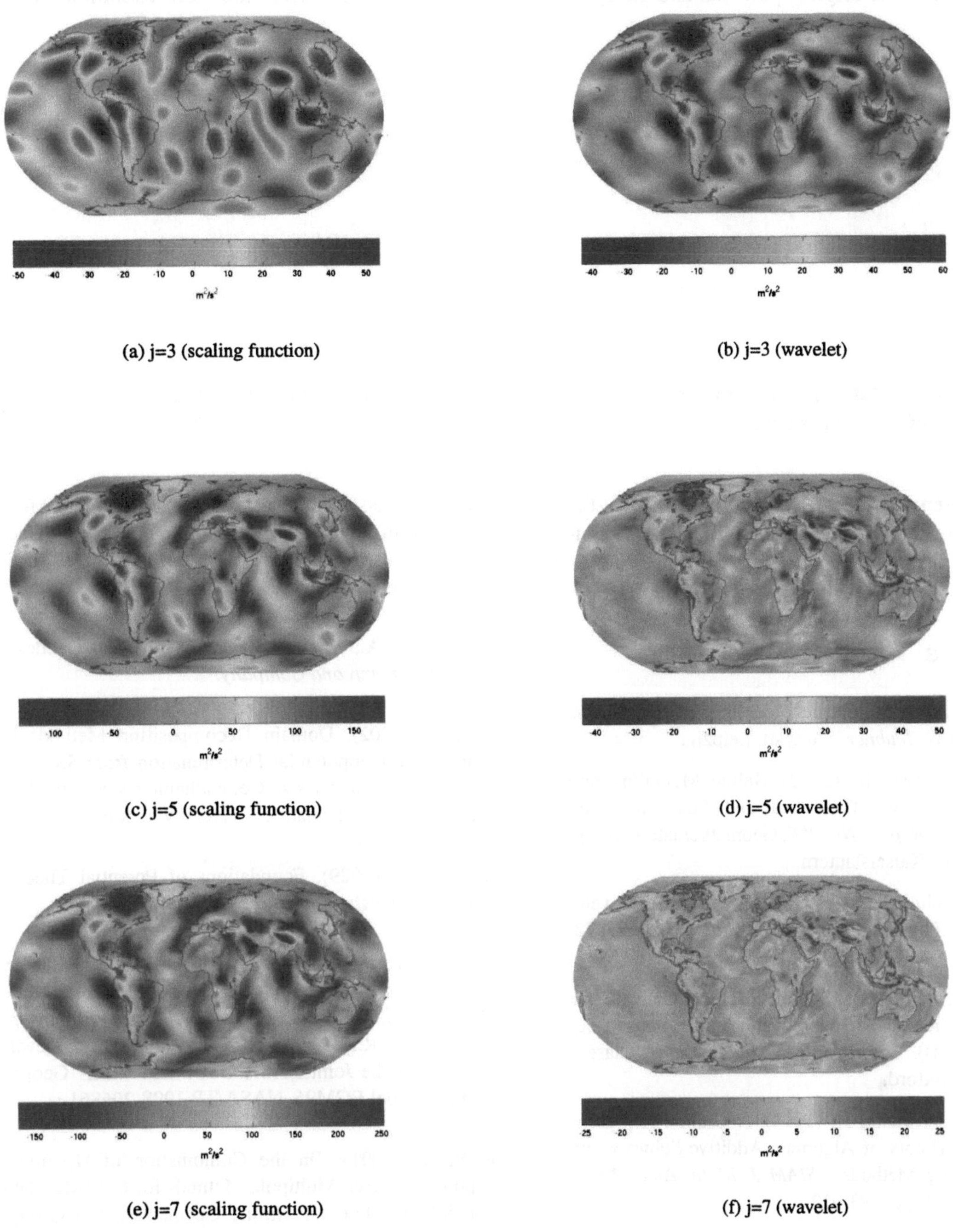

(a) j=3 (scaling function)

(b) j=3 (wavelet)

(c) j=5 (scaling function)

(d) j=5 (wavelet)

(e) j=7 (scaling function)

(f) j=7 (wavelet)

Figure 2 Dyadic multiresolution data analysis on the actual Earth's surface for the disturbing potential corresponding to oblique derivatives (derived from the EGM96–model). On the left hand side you can see the scaling function reconstructions while on the right hand the corresponding wavelet reconstructions are shown. The scaling function reconstruction at scale $j + 1$ is calculated by adding the wavelet reconstruction at scale j to the scaling function reconstruction at scale j.

Finally, we show the original potential and the approximate potential resulting from the multiresolution procedure.

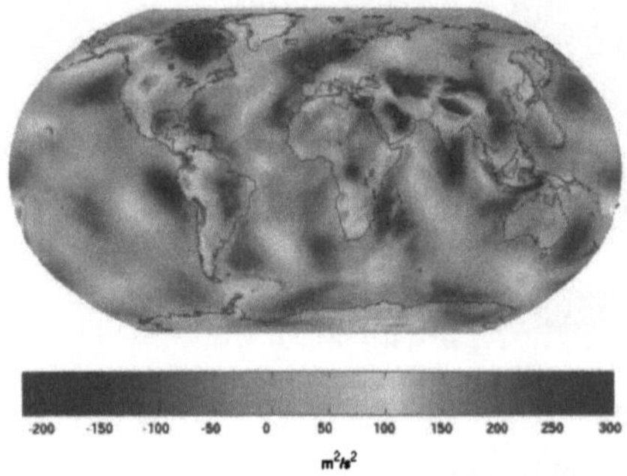

Figure 3 Original EGM96–potential on the actual topography of the Earth as given by TerrainBase

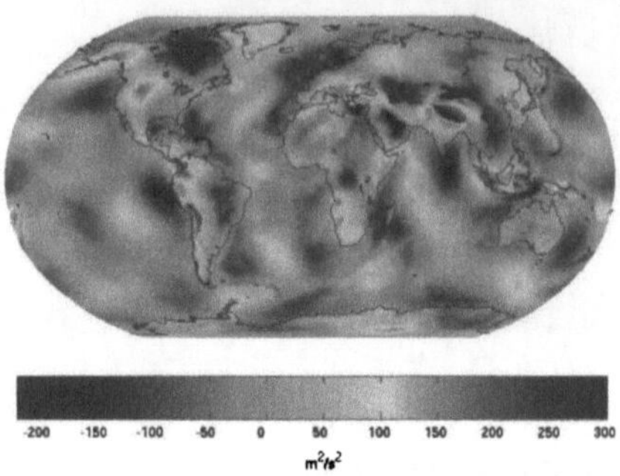

Figure 4 Approximate EGM96–potential from oblique derivatives on the actual Earth

Acknowledgement. The authors thank the Graduiertenkolleg "Mathematik und Praxis", University of Kaiserslautern and the "Stiftung Innovation für Rheinland-Pfalz" for financial support.

References

W. Freeden. (1999) Multiscale Modelling of Spaceborne Geodata. *B.G. Teubner*, Stuttgart, Leipzig.

Freeden, W., Hesse, K. (2002). Spline Modelling and Geostrophic Flow: Theoretical and Algorithmic Aspects. *AGTM-Report, No. 250*, Geomathematics Group, University of Kaiserslautern.

Freeden, W., Schneider, F. (1998). An Integrated Wavelet Concept of Physical Geodesy. *Journal of Geodesy*, 72, 259-281.

Freeden W., T. Gervens, M. Schreiner (1998). Constructive Approximation on the Sphere (With Applications to Geomathematics). *Oxford Science Publications*, Clarendon Press, Oxford.

Frommer,W., Schwandt, H. (1997). A Unified Representation and Theory of Algebraic Additive Schwarz and Multisplitting Methods. *SIAM J. Math. Anal. Appl.*, Vol. 18, No. 4, pp. 893-912.

Glockner O. (2002). On Numerical Aspects of Gravitational Field Modeling from SST and SGG by Harmonic Splines and Wavelets (With Application to CHAMP Data). Doctoral Thesis, Department of Mathematics, Geomathematics Group, University of Kaiserslautern.

Griebel, M., Oswald, P. (1995). On the abstract theory of additive and multiplicative Schwarz algorithms. *Numer. Math.*, 70, pp. 163-180.

Gutting M. (2002). Multiscale Gravitational Field Modelling From Oblique Derivatives. Diploma Thesis, Geomathematics Group, Department of Mathematics, University of Kaiserslautern.

Heiskanen, W.A., Moritz, H. (1967). Physical Geodesy. *W.H. Freemann and Company.*

Hesse, K. (2002). Domain Decomposition Methods in Multiscale Geopotential Determination from SST and SGG. Doctoral Thesis, Geomathematics Group, Department of Mathematics, University of Kaiserslautern.

Kellogg O.D. (1929). Foundations of Potential Theory. *Frederick Ungar Publishing Company.*

Lemoine F.G., S.C. Kenyon, J.K. Factor, R.G. Trimmer, N.K. Pavlis, D.S. Chinn, C.M. Cox, S.M. Klosko, S.B. Luthcke, M.H. Torrence, Y.M. Wang, R.G. Williamson, E.C. Pavlis, R.H. Rapp, T.R. Olson (1998). The Development of the Joint NASA GSFC and NIMA Geopotential Model EGM96, NASA/TP-1998-206861.

Michel, D. (2001). On the Combination of Harmonic Splines and Fast Multipole Methods for CHAMP Data Modelling. Diploma Thesis, Geomathematics Group, Department of Mathematics, University of Kaiserslautern.

Nutz, H. (2002). A Unified Setup of Gravitational Field Observables. PhD-thesis, Geomathematics Group, Department of Mathematics, University of Kaiserslautern.

Schwarz, H.A. (1890). Gesammelte Mathematische Abhandlungen. Vol. 2, *Springer Verlag*, Berlin.

Integer Least-Squares

P.J.G. Teunissen

Department of Mathematical Geodesy and Positioning
Delft University of Technology, Thijsseweg 11, 2629 JA Delft, The Netherlands
Fax: ++ 31 15 278 3711

Abstract. Carrier phase ambiguity resolution is the key to high precision Global Navigation Satellite System (GNSS) positioning and navigation. It applies to a great variety of current and future models of GPS, modernized GPS and Galileo. A proper handling of carrier phase ambiguity resolution requires a proper understanding of the underlying theory of integer inference. In this contribution a brief review is given of the probabilistic theory of integer ambiguity estimation with special emphasis on the integer least-squares principle. We describe the concept of ambiguity pull-in regions, introduce the class of admissible integer estimators, determine their probability mass functions and show how their variability affect the uncertainty in the so-called 'fixed' baseline solution. The theory is worked out in more detail for integer least-squares and integer bootstrapping. It is shown that the integer least-squares principle maximizes the probability of correct integer estimation. Sharp and easy-to-compute bounds are given for both the ambiguity success rate and the baseline's probability of concentration.

Keywords. GNSS, ambiguity resolution, integer least-squares

1 Introduction

Global Navigation Satellite System (GNSS) ambiguity resolution is the process of resolving the unknown cycle ambiguities of double difference (DD) carrier phase data as integers. It is the key to high precision GNSS positioning and navigation. The availability of a theory of integer inference is therefore a prerequisite for a proper handling and understanding of the various intricate aspects of carrier phase ambiguity resolution. Although the theory of integer inference as a whole is unfortunately still far from mature, significant progress has been made in the last decade in the area of integer ambiguity estimation. This holds true for the computational aspects of integer estimation as well as for the corresponding probabilistic aspects. We now have a theoretical framework available which enables one to define integer estimators unequivocally, to compare their performance, to single-out optimal estimators, to judge the probabilistic consequences for the so-called 'fixed' baseline solution and to design measurement set-ups in accordance to specifications. In addition the framework has also enabled one to identify pitfalls in some of the earlier proposed methods of ambiguity resolution. It is the purpose of the current invited contribution to give a brief review of this probabilistic framework of integer estimation. The focus will be on integer least-squares estimation and some closely related integer estimation principles. The presentation will be non-Bayesian throughout. For a Bayesian approach to ambiguity resolution we refer to e.g. Betti et al. (1993), Gundlich and Koch (2001), Gundlich and Teunissen (2002). The presentation will also not touch upon the theory of integer validation, a theory which unfortunately is still in its infancy. Some first, but modest steps in this direction are reported in Teunissen (2002) and Verhagen and Teunissen (2002).

The practical importance of carrier phase ambiguity resolution becomes clear when one realizes the great variety of current and future GNSS models to which it applies. These models may differ greatly in complexity and diversity. They range from single-baseline models used for kinematic positioning to multi-baseline models used as a tool for studying geodynamic phenomena. The models may or may not have the relative receiver-satellite geometry included. They may also be discriminated as to whether the slave receiver(s) are stationary or in motion. When in motion, one solves for one or more trajectories, since with the receiver-satellite geometry included, one will have new coordinate unknowns for each epoch. One may also discriminate between the models as to whether or not the differential atmospheric delays (ionosphere and troposphere) are included as unknowns. In the case of sufficiently short baselines they are usually excluded.

Apart from the current Global Positioning System (GPS) models, carrier phase ambiguity resolution also applies to the future modernized GPS and the future European Galileo GNSS. An overview of GNSS models, together with their applications in surveying, navigation, geodesy and geophysics, can be found in textbooks such as Hofmann-Wellenhof et al (2001), Leick (1995), Misra and Enge (2001), Parkinson and Spilker (1996), Strang and Borre (1997) and Teunissen and Kleusberg (1998).

In the present contribution we emphasize the probabilistic aspects of integer ambiguity estimation. This contribution is organized as follows. In section 2 we introduce a general class of integer ambiguity estimators, determine their probability mass functions and show how their variability affect the uncertainty in the computed GNSS baselines. This theory is worked out in sections 3 and 4 for two of the most important integer ambiguity estimators. We refrain from giving proofs of the theorems and corollaries. For these proofs we refer to the referenced literature. In section 3 we discuss the properties of integer bootstrapping and in the final section 4 those of integer least-squares. The properties of these two estimators are compared. It is shown that integer least-squares maximizes the probability of correct integer estimation. We also give sharp and easy-to-compute bounds for the ambiguity success rate.

2 Integer Ambiguity Resolution

2.1 The GNSS model

As our point of departure we will take the following system of linear(ized) observation equations

$$y = Aa + Bb + e \tag{1}$$

where y is the given GNSS data vector of order m, a and b are the unknown parameter vectors respectively of order n and p, and where e is the noise vector. In principle all the GNSS models can be cast in this frame of observation equations. The data vector y will usually consist of the 'observed minus computed' single- or dual-frequency double-difference (DD) phase and/or pseudorange (code) observations accumulated over all observation epochs. The entries of vector a are then the DD carrier phase ambiguities, expressed in units of cycles rather than range. They are known to be *integers*, $a \in Z^n$. The entries of the vector b will consist of the remaining unknown parameters, such as for instance baseline components (coordinates) and possibly atmospheric delay parameters (troposphere, ionosphere). They are known to be real-valued, $b \in R^p$.

The procedure which is usually followed for solving the GNSS model (1), can be divided into three steps. In the *first* step one simply disregards the integer constraints $a \in Z^n$ on the ambiguities and performs a standard least-squares adjustment. As a result one obtains the (real-valued) estimates of a and b, together with their variance-covariance (vc-) matrix

$$\begin{bmatrix} \hat{a} \\ \hat{b} \end{bmatrix}, \begin{bmatrix} Q_{\hat{a}} & Q_{\hat{a}\hat{b}} \\ Q_{\hat{b}\hat{a}} & Q_{\hat{b}} \end{bmatrix} \tag{2}$$

This solution is referred to as the 'float' solution. In the *second* step the 'float' ambiguity estimate $\hat{a}$ is used to compute the corresponding integer ambiguity estimate $\check{a}$. This implies that a mapping $S : R^n \mapsto Z^n$, from the n-dimensional space of reals to the n-dimensional space of integers, is introduced such that

$$\check{a} = S(\hat{a}) \tag{3}$$

Once the integer ambiguities are computed, they are used in the *third* step to finally correct the 'float' estimate of b. As a result one obtains the 'fixed' solution

$$\check{b} = \hat{b} - Q_{\hat{b}\hat{a}} Q_{\hat{a}}^{-1}(\hat{a} - \check{a}) \tag{4}$$

In the present review we will primarily focus our attention on the probabilistic properties of (3) and (4).

2.2 Admissible integer estimation

There are many ways of computing an integer ambiguity vector $\check{a}$ from its real-valued counterpart $\hat{a}$. To each such method belongs a mapping $S : R^n \mapsto Z^n$ from the n-dimensional space of real numbers to the n-dimensional space of integers. Due to the discrete nature of Z^n, the map S will not be one-to-one, but instead a many-to-one map. This implies that different real-valued ambiguity vectors will be mapped to the same integer vector. One can therefore assign a subset $S_z \subset R^n$ to each integer vector $z \in Z^n$:

$$S_z = \{x \in R^n \mid z = S(x)\}, \quad z \in Z^n \tag{5}$$

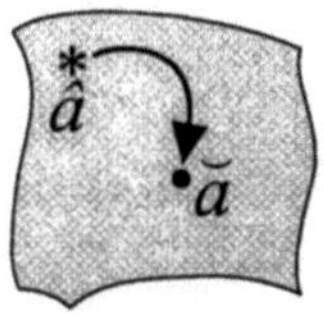

Fig. 1 An ambiguity pull-in region of $z = \check{a}$.

The subset S_z contains all real-valued ambiguity vectors that will be mapped by S to the same integer vector $z \in Z^n$. This subset is referred to as the *pull-in region* of z (see figure 1). It is the region in which all ambiguity 'float' solutions are pulled to the same 'fixed' ambiguity vector z.

Using the pull-in regions, one can give an explicit expression for the corresponding integer ambiguity estimator. It reads

$$\check{a} = \sum_{z \in Z^n} z s_z(\hat{a}) \tag{6}$$

with the indicator function:

$$s_z(\hat{a}) = \begin{cases} 1 & if \;\; \hat{a} \in S_z \\ 0 & otherwise. \end{cases}$$

Since the pull-in regions define the integer estimator completely, one can define classes of integer estimators by imposing various conditions on the pull-in regions. One such class is referred to as the class of admissible integer estimators. This class was introduced in Teunissen (1999a) and it is defined as follows.

Definition 1
The integer estimator $\check{a} = \sum_{z \in Z^n} z s_z(\hat{a})$ is said to be *admissible* if

$$\begin{aligned} &(i) \quad \bigcup_{z \in Z^n} S_z = R^n \\ &(ii) \quad Int(S_{z_1}) \bigcap Int(S_{z_2}) = \emptyset, \quad \forall z_1, z_2 \in Z^n, \; z_1 \neq z_2 \\ &(iii) \quad S_z = z + S_0, \quad \forall z \in Z^n \end{aligned}$$

This definition is motivated as follows (see figure 2). Each one of the above three conditions describes a property of which it seems reasonable that it is possessed by an arbitrary integer ambiguity estimator. The first condition states that the pull-in regions should not leave any gaps and the second that they should not overlap. The absence of gaps is needed in order to be able to map any 'float' solution $\hat{a} \in R^n$ to Z^n, while the absence of overlaps is needed to guarantee that the 'float' solution is mapped to just one integer vector. Note that we allow the pull-in regions to have common boundaries. This is permitted if we assume to have zero probability that $\hat{a}$ lies on one of the boundaries. This will be the case when the probability density function (pdf) of $\hat{a}$ is continuous.

The third and last condition of the definition follows from the requirement that $S(x + z) = S(x) + z, \forall x \in R^n, z \in Z^n$ (see figure 3). Also this condition is a reasonable one to ask for. It states that when the 'float' solution is perturbed by $z \in Z^n$, the corresponding integer solution is perturbed by the same amount. This property allows one to apply the *integer remove-restore* technique: $S(\hat{a} - z) + z = S(\hat{a})$. It therefore allows one to work with the fractional parts of the entries of $\hat{a}$, instead of with its complete entries.

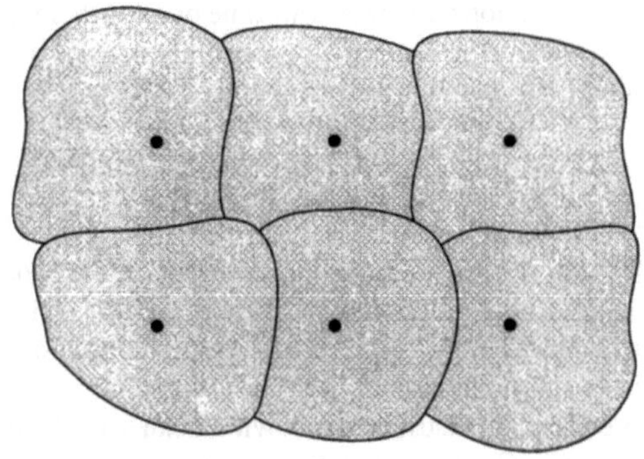

Fig. 2 Pull-in regions that cover R^n without gaps and overlaps.

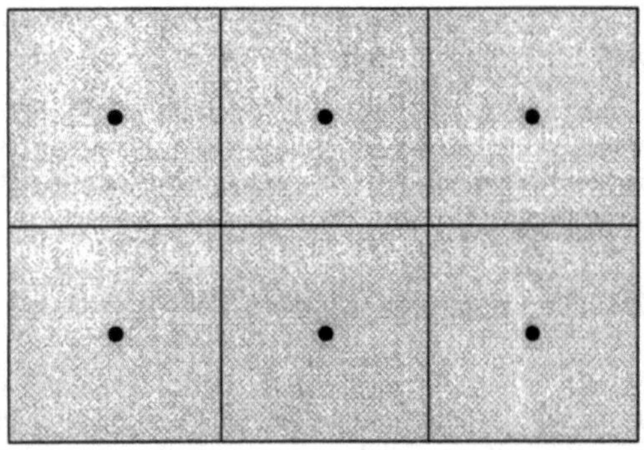

Fig. 3 An example of integer translational invariant pull-in regions that cover R^n without gaps and overlaps.

With the division of R^n into mutually exclusive pull-in regions, we are now in the position to consider the distribution of $\check{a}$. This distribution is of the *discrete* type and it will be denoted as $P(\check{a} = z)$. It is a probability mass function, having zero masses at non-grid points and nonzero masses at some or all grid points. If we denote the *continuous* probability density function of $\hat{a}$ as $p_{\hat{a}}(x)$, the distribution of $\check{a}$ follows as

$$P(\check{a} = z) = \int_{S_z} p_{\hat{a}}(x)dx \ , \ z \in Z^n \tag{7}$$

This expression holds for any distribution the 'float' ambiguities $\hat{a}$ might have. In most GNSS applications however, one assumes the vector of observables y to be normally distributed. The estimator $\hat{a}$ is therefore normally distributed too, with mean $a \in Z^n$ and vc-matrix $Q_{\hat{a}}$. Its probability density function reads

$$p_{\hat{a}}(x) = \frac{1}{\sqrt{\det(Q_{\hat{a}})(2\pi)^{\frac{1}{2}n}}} \exp\{-\frac{1}{2} \parallel x - a \parallel^2_{Q_{\hat{a}}}\} \tag{8}$$

with the squared weighted norm $\parallel \ . \ \parallel^2_{Q_{\hat{a}}} = (.)^T Q_{\hat{a}}^{-1}(.)$. Note that $P(\check{a} = a)$ equals the probability of *correct* integer ambiguity estimation. It describes the expected success rate of GNSS ambiguity resolution.

2.3 The baseline solution

We are now in the position to determine the pdf of the 'fixed' baseline estimator (4). In order to determine this pdf, one needs to propagate the uncertainty of the 'float' solution, $\hat{a}$ and $\hat{b}$, as well as the uncertainty of the integer solution $\check{a}$ through (4). Should one neglect the random character of the integer solution and therefore consider the ambiguity vector $\check{a}$ as deterministic and equal to, say, z, then the pdf of $\check{b}$ would equal the conditional baseline distribution

$$p_{\hat{b}|\hat{a}}(x \mid z) = \frac{1}{\sqrt{\det Q_{\hat{b}|\hat{a}}(2\pi)^{\frac{1}{2}p}}} \exp\{-\frac{1}{2}\|x - b(z)\|^2_{Q_{\hat{b}|\hat{a}}}\} \tag{9}$$

with conditional mean $b(z) = b - Q_{\hat{b}\hat{a}}Q_{\hat{a}}^{-1}(a - z)$, conditional variance matrix $Q_{\hat{b}|\hat{a}} = Q_{\hat{b}} - Q_{\hat{b}\hat{a}}Q_{\hat{a}}^{-1}Q_{\hat{a}\hat{b}}$ and $\|.\|^2_{Q_{\hat{b}|\hat{a}}} = (.)^T Q_{\hat{b}|\hat{a}}^{-1}(.)$. However, since $\check{a}$ is random and not deterministic, the conditional baseline distribution will give a too optimistic description of the quality of the 'fixed' baseline. To get a correct description of the 'fixed' baseline's pdf, the integer ambiguity's pmf needs to be considered. As the following theorem shows this results in a baseline distribution, which generally will be multi-modal.

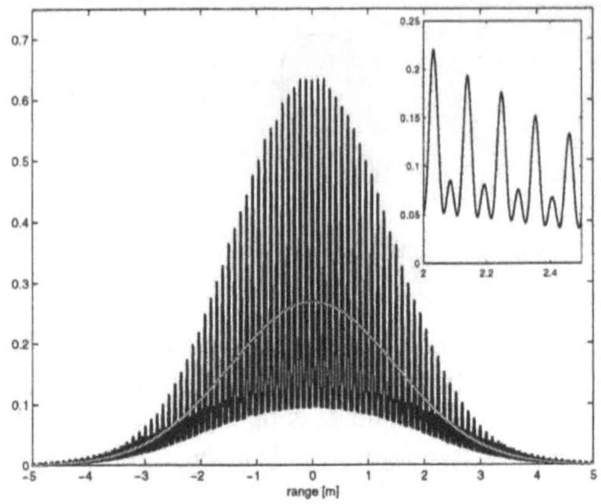

Fig. 4 An example of the multi-modal pdf of $\check{b}$ and corresponding uni-modal pdf of $\hat{b}$.

Theorem 1 (*Pdf of the 'fixed' baseline*)
Let the 'float' solution, $\hat{a}$ and $\hat{b}$, be normally distributed with mean $a \in Z^n$ and mean $b \in R^p$, and vc-matrix (2), let $\check{a}$ be an admissible integer estimator and let the 'fixed' baseline $\check{b}$ be given as in (4). The pdf of $\check{b}$ reads then

$$p_{\check{b}}(x) = \sum_{z \in Z^n} p_{\hat{b}|\hat{a}}(x \mid z)P(\check{a} = z) \tag{10}$$

Note that, although the model (1) is linear and the observables normally distributed, the distribution of the 'fixed' baseline is not normal, but multi-modal (see figure 4). As the theorem shows, the 'fixed' baseline distribution equals an infinite sum of weighted conditional baseline distributions. These conditional baseline distributions $p_{\hat{b}|\hat{a}}(x \mid z)$ are shifted versions of one another. The size and direction of the shift is governed by $Q_{\hat{b}\hat{a}}Q_{\hat{a}}^{-1}z$, $z \in Z^n$. Each of the conditional baseline distributions in the infinite sum is down-weighted. These weights are given by the probability masses of the distribution of the integer ambiguity estimator $\check{a}$. This shows that the dependence of the 'fixed' baseline distribution on the choice of integer estimator is only felt through the weights $P(\check{a} = z)$.

2.4 On the quality of the 'fixed' baseline

In order to describe the quality of the 'fixed' baseline, one would like to know how close one can expect the baseline estimate $\check{b}$ to be to the unknown, but true baseline value b. As a measure of confidence, we take

$$P(\check{b} \in R) = \int_R p_{\check{b}}(x)dx \ \ with \ \ R \subset R^p \tag{11}$$

But in order to evaluate this integral, we first need to make a choice about the shape and location of the subset R. Since it is common practice in GNSS positioning to use the vc-matrix of the conditional baseline estimator as a measure of precision for the 'fixed' baseline, the vc-matrix $Q_{\hat{b}|\hat{a}}$ will be used to define the shape of the confidence region. For its location, we choose the confidence region to be centered at b. After all, we would like to know by how much the baseline estimate $\check{b}$ can be expected to differ from the true, but unknown baseline value b. That is, one would like (11) to be a measure of the baseline's probability of concentration about b.

With these choices on shape and location, the region R takes the form

$$R = \{x \in R^p \mid (x-b)^T Q_{\hat{b}|\hat{a}}^{-1}(x-b) \leq \beta^2\} \quad (12)$$

The size of the region can be varied by varying β. The following theorem shows how the baseline's probability of concentration (11) can be evaluated as a weighted sum of probabilities of noncentral Chi-square distributions.

Theorem 2 *(The 'fixed' baseline's probability of concentration)*
Let $\check{b}$ be the 'fixed' baseline estimator, let R be defined as in (12), and let $\chi^2(p, \lambda_z)$ denote the noncentral Chi-square distribution with p degrees of freedom and noncentrality parameter λ_z. Then

$$P(\check{b} \in R) = \sum_{z \in Z^n} P(\chi^2(p,\lambda_z) \leq \beta^2) P(\check{a} = z) \quad (13)$$

with

$$\lambda_z = \| \nabla \check{b}_z \|^2_{Q_{\hat{b}|\hat{a}}} \quad and \quad \nabla \check{b}_z = Q_{\hat{b}\hat{a}} Q_{\hat{a}}^{-1}(z-a)$$

This result shows that the probability of the 'fixed' baseline lying inside the ellipsoidal region R centered at b equals an infinite sum of probability products. If one considers the two probabilities of these products separately, two effects are observed. First the probabilistic effect of shifting the conditional baseline estimator away from b and secondly the probabilistic effect of the peakedness or nonpeakedness of the ambiguity pmf. The second effect is related to the expected performance of ambiguity resolution, while the first effect has to do with the sensitivity of the baseline for changes in the values of the integer ambiguities. This effect is measured by the noncentrality parameter λ_z. Since the tail of a noncentral Chi-square distribution becomes heavier when the noncentrality parameter increases, while the degrees of freedom remain fixed, $P(\chi^2(p, \lambda_z) \leq \beta^2)$ gets smaller when λ_z gets larger.

The two probabilities in the product reach their maximum values when $z = a$. The following corollary shows how these two maxima can be used to lower bound and to upper bound the probability $P(\check{b} \in R)$. Such bounds are of importance for practical purposes, since it is difficult in general to evaluate (13) exactly.

Corollary 1 *(Lower and upper bounds)*
Let $\check{b}$ be the 'fixed' baseline estimator and let R be defined as in (12). Then

$$P(\hat{b}_{|\hat{a}=a} \in R) P(\check{a} = a) \leq P(\check{b} \in R) \leq P(\hat{b}_{|\hat{a}=a} \in R) \quad (14)$$

with

$$P(\hat{b}_{|\hat{a}=a} \in R) = P(\chi^2(p, 0) \leq \beta^2)$$

Note that the two bounds relate the probability of the 'fixed' baseline estimator to that of the conditional estimator and the ambiguity success rate. The above bounds become tight when the ambiguity success rate approaches one. This shows, although the probability of the conditional estimator always overestimates the probability of the 'fixed' baseline estimator, that the two probabilities are close for large values of the success rate. This implies that in case of GNSS ambiguity resolution, one should first evaluate the success rate $P(\check{a} = a)$ and make sure that its value is close enough to one, before making any inferences on the basis of the distribution of the conditional baseline estimator. In other words, the (unimodal) distribution of the conditional estimator is a good approximation to the (multi-modal) distribution of the bootstrapped baseline estimator, when the success rate is sufficiently close to one.

3 Integer Bootstrapping

3.1 The bootstrapped estimator

The distributional results presented so far hold for any admissible ambiguity estimator. The simplest way to obtain an integer vector from the real-valued 'float' solution is to round each of the entries of $\hat{a}$ to its nearest integer. The corresponding integer estimator reads therefore

$$\check{a}_R = ([\hat{a}_1], \ldots, [\hat{a}_n])^T \quad (15)$$

where '[.]' denotes rounding to the nearest integer. The pull-in region of this integer estimator equals the multivariate version of the unit square (see figure 3).

Another relatively simple integer ambiguity estimator is the bootstrapped estimator. The bootstrapped estimator can be seen as a generalization of the previous estimator, Teunissen (1998b) and Teunissen (2001). It still makes use of integer rounding, but it also takes some of the correlation between the ambiguities into account. The bootstrapped estimator follows from a sequential conditional least-squares adjustment and it is computed as follows. If n ambiguities are available, one starts with the first ambiguity $\hat{a}_1$, and rounds its value to the nearest integer. Having obtained the integer value of this first ambiguity, the real-valued estimates of all remaining ambiguities are then corrected by virtue of their correlation with the first ambiguity. Then the second, but now corrected, real-valued ambiguity estimate is rounded to its nearest integer. Having obtained the integer value of the second ambiguity, the real-valued estimates of all remaining $n-2$ ambiguities are then again corrected, but now by virtue of their correlation with the second ambiguity. This process is continued until all ambiguities are considered. We thus have the following definition.

Definition 2 (*Integer bootstrapping*)
Let $\hat{a} = (\hat{a}_1, \ldots, \hat{a}_n)^T \in R^n$ be the ambiguity 'float' solution and let $\check{a}_B = (\check{a}_{B,1}, \ldots, \check{a}_{B,n})^T \in Z^n$ denote the corresponding integer bootstrapped solution. The entries of the bootstrapped ambiguity estimator are then defined as

$$\begin{aligned} \check{a}_{B,1} &= [\hat{a}_1] \\ \check{a}_{B,2} &= [\hat{a}_{2|1}] = [\hat{a}_2 - \sigma_{21}\sigma_1^{-2}(\hat{a}_1 - \check{a}_{B,1})] \\ &\vdots \\ \check{a}_{B,n} &= [\hat{a}_{n|N}] = [\hat{a}_n - \sum_{j=1}^{n-1} \sigma_{n,j|J}\sigma_{j|J}^{-2}(\hat{a}_{j|J} - \check{a}_{B,j})] \end{aligned} \tag{16}$$

where '[.]' denotes the operation of rounding to the nearest integer, and $\sigma_{i,j|J}$ denotes the covariance between $\hat{a}_i$ and $\hat{a}_{j|J}$, and $\sigma_{j|J}^2$ is the variance of $\hat{a}_{j|J}$. The shorthand notation $\hat{a}_{i|I}$ stands for the ith least-squares ambiguity obtained through a conditioning on the previous $I = \{1, \ldots, (i-1)\}$ sequentially rounded ambiguities.

Note that the bootstrapped estimator is not unique. Changing the order in which the ambiguities appear in vector $\hat{a}$ will already produce a different bootstrapped estimator. Although the principle of bootstrapping remains the same, every choice of ambiguity parameterization has its own bootstrapped estimator.

3.2 The bootstrapped pull-in regions

The pull-in regions for rounding are unit cubes centered at integer grid points. For bootstrapping the shape of the pull-in regions will depend on the vc-matrix of the ambiguities. They will coincide with the unit cubes only in case the vc-matrix is a diagonal matrix. Bootstrapping reduces namely to rounding in the absence of any correlation between the ambiguities. The following theorem gives a description of the bootstrapped pull-in regions in the general case.

Theorem 3 (*Bootstrapped pull-in regions*)
The pull-in regions of the bootstrapped ambiguity estimator $\check{a}_B = (\check{a}_{B,1}, \ldots, \check{a}_{B,n})^T \in Z^n$ are given as

$$S_{B,z} = \{x \in R^n \mid | c_i^T L^{-1}(x-z) | \leq \frac{1}{2}, \; i = 1, \ldots, n\}, \; \forall z \in Z^n \tag{17}$$

where L denotes the unique unit lower triangular matrix of the ambiguity vc-matrix' decomposition $Q_{\hat{a}} = LDL^T$ and c_i denotes the ith canonical unit vector having a 1 as its ith entry and zeros otherwise.

That the bootstrapped estimator is indeed admissible, can now be seen as follows. The first two conditions of Definition 1 are easily verified using the definition of the bootstrapped estimator. Since every real-valued vector $\hat{a}$ will be mapped by the bootstrapped estimator to an integer vector, the pull-in regions $S_{B,z}$ cover R^n without any gaps. There is also no overlap between the pull-in regions, since - apart from boundary ties - any real-valued vector $\hat{a}$ is mapped to not more than one integer vector. To verify the last condition of Definition 1, we make use of (17). From

$$\begin{aligned} S_{B,z} &= \{x \in R^n \mid | c_i^T L^{-1}(x-z) | \leq \frac{1}{2}, \; i = 1, \ldots, n\} \\ &= \{x \in R^n \mid | c_i^T L^{-1} y | \leq \frac{1}{2}, \; x = y + z, \; i = 1, \ldots, n\} \\ &= S_{B,0} + z \end{aligned}$$

it follows that all bootstrapped pull-in regions are translated copies of $S_{B,0}$. All pull-in regions have therefore the same shape and the same volume. Their volumes all equal 1. This can be shown by transforming $S_{B,0}$ to the unit cube centered at the origin.

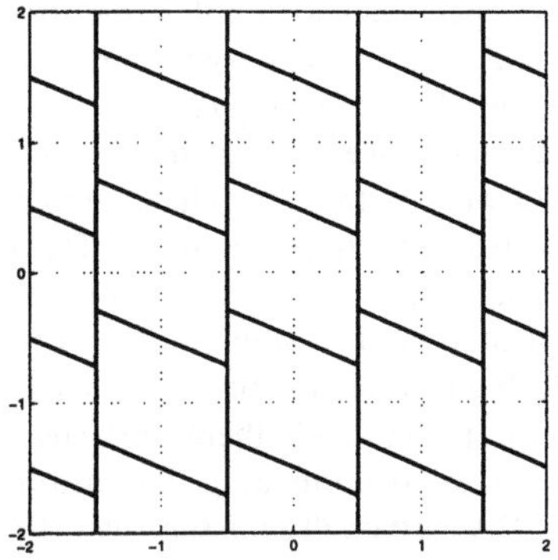

Fig. 5 The two-dimensional pull-in regions of integer bootstrapping.

Consider the linear transformation $y = L^{-1}x$. Then

$$L^{-1}(S_{B,0}) = \{y \in R^n \mid \mid c_i^T y \mid \leq \frac{1}{2}\ , \ i = 1, \ldots, n\}$$

equals the unit cube centered at the origin. Since the determinant of the unit lower triangular matrix L^{-1} equals one and since the volume of the unit cube equals one, it follows that the volume of $S_{B,0}$ must equal one as well. To infer the shape of the bootstrapped pull-in region, we consider the two-dimensional case first. Let the lower triangular matrix L be given as

$$L = \begin{bmatrix} 1 & 0 \\ l & 1 \end{bmatrix}$$

Then

$$\begin{aligned} S_{B,0} &= \{x \in R^2 \mid \mid c_i^T L^{-1} x \mid \leq \frac{1}{2}\ , \ i = 1, 2\} \\ &= \{x \in R^2 \mid \mid x_1 \mid \leq \frac{1}{2}, \mid x_2 - l x_1 \mid \leq \frac{1}{2}\} \end{aligned}$$

which shows that the two-dimensional pull-in region equals a parallelogram (see figure 5). Its region is bounded by the two vertical lines $x_1 = 1/2$ and $x_1 = -1/2$, and the two parallel slopes $x_2 = lx_1 + 1/2$ and $x_2 = lx_1 - 1/2$. The direction of the slope is governed by $l = \sigma_{21}\sigma_1^{-2}$. Hence, in the absence of correlation between the two ambiguities, the parallelogram reduces to the unit square. In higher dimensions the above construction of the pull-in region can be continued. In three dimensions for instance, the intersection of the pull-in region with the x_1x_2-plane remains a parallelogram, while along the third axis the pull-in region becomes bounded by two parallel planes.

3.3 The bootstrapped pmf

Since the integer bootstrapped estimator is defined as $\check{a}_B = z \iff \hat{a} \in S_{B,z}$, it follows that $P(\check{a}_B = z) = P(\hat{a} \in S_{B,z})$. The pmf of $\check{a}_B$ follows therefore as

$$P(\check{a}_B = z) = \int_{S_{B,z}} p_{\hat{a}}(x)dx\ , \ z \in Z^n \tag{18}$$

Hence, the probability that $\check{a}_B$ coincides with z is given by the integral of the pdf $p_{\hat{a}}(x)$ over the bootstrapped pull-in region $S_{B,z} \subset R^n$. The above expression holds for any distribution the 'float' ambiguities $\hat{a}$ might have. In most GNSS applications however, one usually assumes the vector of observables y to be normally distributed. For that case the following theorem gives an exact expression of the bootstrapped pmf (see figure 6).

Theorem 4 (*The integer bootstrapped pmf*)
Let $\hat{a}$ be distributed as $N(a, Q_{\hat{a}})$, $a \in Z^n$, and let $\check{a}_B$ be the corresponding integer bootstrapped estimator. Then

$$\begin{aligned} P(\check{a}_B = z) = \prod_{i=1}^{n} [\Phi(&\frac{1 - 2l_i^T(a - z)}{2\sigma_{\hat{a}_{i|I}}}) \\ &+ \Phi(\frac{1 + 2l_i^T(a - z)}{2\sigma_{\hat{a}_{i|I}}}) - 1]\ , \ z \in Z^n \end{aligned} \tag{19}$$

with

$$\Phi(x) = \int_{-\infty}^{x} \frac{1}{\sqrt{2\pi}} \exp\{-\frac{1}{2}v^2\}dv$$

and with l_i the ith column vector of the unit lower triangular matrix L^{-T} and $\sigma^2_{\hat{a}_{i|I}}$ the variance of the ith least-squares ambiguity obtained through a conditioning on the previous $I = \{1, \ldots, (i-1)\}$ ambiguities.

The bootstrapped pmf equals a product of univariate pmf's and is therefore easy to compute. Note that the bootstrapped pmf is completely governed by the ambiguity vc-matrix $Q_{\hat{a}}$. The pmf follows once the triangular factor L and the diagonal matrix D of the decomposition $Q_{\hat{a}} = LDL^T$ are given. The above result also shows that the bootstrapped pmf is symmetric about the mean of $\hat{a}$. This implies that the bootstrapped estimator $\check{a}_B$ is an unbiased estimator of $a \in Z^n$. Since the 'float' solutions, $\hat{a}$ and $\hat{b}$, are unbiased too, it follows from taking the expectation of (4) that the bootstrapped baseline is also unbiased.

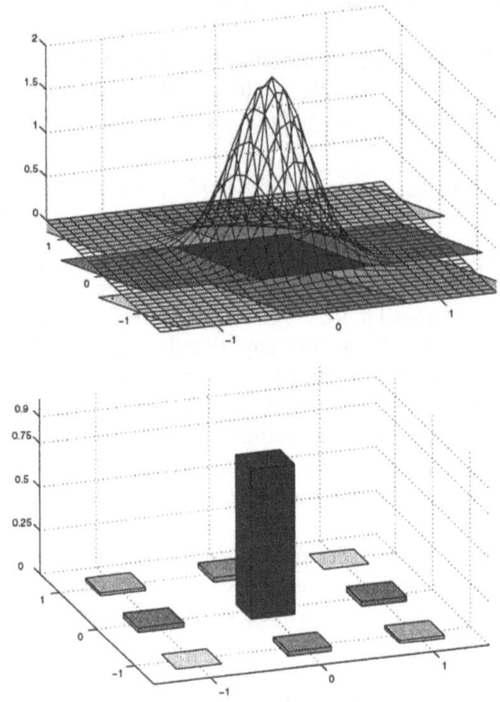

Fig. 6 (top) The two-dimensional pdf of the 'float' solution $\hat{a}$ situated over the bootstrapped pull-in regions; (bottom) the two-dimensional pmf of the integer bootstrapped solution $\check{a}_B$.

For the purpose of predicting the success of ambiguity resolution, the probability of correct integer estimation is of particular interest. For the bootstrapped estimator this success rate is given in the following corollary.

Corollary 2 (*The bootstrapped success rate*)
The bootstrapped probability of correct integer estimation (the success rate) is given as

$$P(\check{a}_B = a) = \prod_{i=1}^{n}[2\Phi(\frac{1}{2\sigma_{\hat{a}_{i|I}}}) - 1] \quad (20)$$

The method of integer bootstrapping is easy to implement and it does not need, as opposed to the method of integer least-squares (see next section), an integer search for computing the sought for integer solution. However, as it was mentioned earlier, the outcome of bootstrapping depends on the chosen ambiguity parameterization. Bootstrapping of DD ambiguities, for instance, will produce an integer solution which generally differs from the integer solution obtained from bootstrapping of reparameterized ambiguities. Since this dependency also holds true for the bootstrapped pmf, one still has some important degrees of freedom left for improving (20).

In order to improve the bootstrapped success rate, one should work with decorrelated ambiguities instead of with the original ambiguities. The method of bootstrapping performs relatively poor, for instance, when applied to the DD ambiguities. This is due to the usually high correlation between the DD ambiguities. Bootstrapping should therefore be used in combination with the decorrelating Z-transformation of the LAMBDA method (Teunissen 1993; 1995). This transformation decorrelates the ambiguities further than the best reordering would achieve and thereby reduces the values of the sequential conditional variances. By reducing the values of the sequential conditional variances, the bootstrapped success rate gets enlarged.

It may however happen that it is simply not possible to resolve the complete vector of ambiguities with sufficient probability. As an alternative of resolving the complete vector of ambiguities, one might then consider resolving only a subset of the ambiguities. The idea of partial ambiguity resolution is based on the fact that the success rate will generally increase when fewer integer constraints are imposed. However, in order to apply partial ambiguity resolution, one first will have to determine which subset of ambiguities to choose. It will be clear that this decision should be based on the precision of the 'float' ambiguities. The more precise the ambiguities, the larger the ambiguity success rate. It is at this point where the decorrelation step of the LAMBDA method and the bootstrapping principle can be applied. Once the transformed and decorrelated ambiguity vc-matrix is obtained, the construction of the subset proceeds in a sequential fashion. One first starts with the most precise ambiguity, say $\hat{z}_1$, and computes its success rate $P(\check{a}_1 = z_1)$. If this success rate is large enough, one continues and determines the most precise pair of ambiguities, say $(\hat{z}_1, \hat{z}_2)$. If their success rate is still large enough, one continues again by trying to extend the set. This procedure continues until one reaches a point where the corresponding success rate becomes unacceptably small. When this point is reached, one can expect that the previously identified ambiguities can be resolved successfully.

Once the subset for partial ambiguity resolution has been identified, one still needs to determine what this will do to improve the baseline estimator. After all, being able to successfully resolve the ambiguities does not necessarily mean that the 'fixed' solution is significantly better than the 'float' solution. The theory presented in the previous sections provides the necessary tools for performing such an evaluation.

4 Integer Least-squares

4.1 The ILS estimator

When using the least-squares principle, the GNSS model can be solved by means of the minimization problem

$$\min_{a,b} \| y - Aa - Bb \|^2_{Q_y} \ , a \in Z^n \ , b \in R^p \quad (21)$$

with Q_y the vc-matrix of the GNSS observables. This type of least-squares problem was first introduced in Teunissen (1993) and has been coined with the term *'integer least-squares'*. It is a nonstandard least-squares problem due to the integer constraints $a \in Z^n$, see Teunissen (1995) and Hassibi and Boyd (1998). The solution of (21) is consistent with the three solution steps of section 1. This can be seen as follows. It follows from the orthogonal decomposition

$$\| y - Aa - Bb \|^2_{Q_y} = \| \hat{e} \|^2_{Q_y} + \| \hat{a} - a \|^2_{Q_{\hat{a}}} + \| \hat{b}(a) - b \|^2_{Q_{\hat{b}|\hat{a}}} \quad (22)$$

with $\hat{e} = y - A\hat{a} - B\hat{b}$ and $\hat{b}(a) = \hat{b} - Q_{\hat{b}\hat{a}} Q_{\hat{a}}^{-1}(\hat{a} - a)$, that the sought for minimum is obtained when the second term on the right-hand side is minimized for $a \in Z^n$ and the last term is set to zero. The integer least-squares (ILS) estimator of the ambiguities is therefore defined as follows.

Definition 3 *(Integer least-squares)*
Let $\hat{a} = (\hat{a}_1, \ldots, \hat{a}_n)^T \in R^n$ be the ambiguity 'float' solution and let $\check{a}_{LS} \in Z^n$ denote the corresponding integer least-squares solution. Then

$$\check{a}_{LS} = \arg \min_{z \in Z^n} \| \hat{a} - z \|^2_{Q_{\hat{a}}} \quad (23)$$

In contrast to integer rounding and integer bootstrapping, an integer search is needed to compute $\check{a}_{LS}$. Although we will refrain from discussing the computational intricacies of ILS estimation, the conceptual steps of the computational procedure will be described briefly. The ILS procedure is mechanized in the GNSS LAMBDA (Least-squares AMBiguity Decorrelation Adjustment) method, which is currently one of the most applied methods for GNSS carrier phase ambiguity resolution. For more information on the LAMBDA method, we refer to e.g. Teunissen (1993), Teunissen (1995) and de Jonge and Tiberius (1996a) or to the textbooks Hofmann-Wellenhof (2001), Strang and Borre (1997), Teunissen and Kleusberg (1998). Practical results obtained with it can be found, for example, in Boon and Ambrosius (1997), Boon et al. (1997), Cox and Brading (1999), de Jonge and Tiberius (1996b), de Jonge et al (1996), Han (1995), Jonkman (1998), Peng et al. (1999), Tiberius and de Jonge (1995), Tiberius et al. (1997).

The main steps as implemented in the LAMBDA method are as follows. One starts by defining the ambiguity search space

$$\Omega_a = \{a \in Z^n \mid (\hat{a}-a)^T Q_{\hat{a}}^{-1} (\hat{a}-a) \leq \chi^2\} \quad (24)$$

with χ^2 a to be chosen positive constant. The boundary of this search space is ellipsoidal. It is centered at $\hat{a}$, its shape is governed by the vc-matrix $Q_{\hat{a}}$ and its size is determined by χ^2. In case of GNSS, the search space is usually extremely elongated, due to the high correlations between the ambiguities. Since this extreme elongation usually hinders the computational efficiency of the search, the search space is first transformed to a more spherical shape,

$$\Omega_z = \{z \in Z^n \mid (\hat{z}-z)^T Q_{\hat{z}}^{-1} (\hat{z}-z) \leq \chi^2\} \quad (25)$$

using the admissible ambiguity transformations $\hat{z} = Z^T \hat{a}$, $Q_{\hat{z}} = Z^T Q_{\hat{a}} Z$. Ambiguity transformations Z are said to be admissible when both Z and its inverse Z^{-1} have integer entries. Such matrices preserve the integer nature of the ambiguities. In order for the transformed search space to become more spherical, the volume-preserving Z-transformation is constructed as a transformation that decorrelates the ambiguities as much as possible. Using the triangular decomposition of $Q_{\hat{z}}$, the left-hand side of the quadratic inequality in (25) is then written as a sum-of-squares:

$$\sum_{i=1}^{n} \frac{(\hat{z}_{i|I} - z_i)^2}{\sigma^2_{i|I}} \leq \chi^2 \quad (26)$$

On the left-hand side one recognizes the conditional least-squares estimator $\hat{z}_{i|I}$, which follows when the conditioning takes place on the integers $z_1, z_2, \ldots, z_{i-1}$. Using the sum-of-squares structure, one can finally set up the n intervals which are used for the search. These sequential intervals are given as

$$\begin{aligned} (\hat{z}_1 - z_1)^2 &\leq \sigma_1^2 \chi^2 \\ (\hat{z}_{2|1} - z_2)^2 &\leq \sigma_{2|1}^2 \left(\chi^2 - \tfrac{(\hat{z}_1 - z_1)^2}{\sigma_1^2}\right) \\ &\vdots \end{aligned} \quad (27)$$

In order for the search to be efficient, one not only would like the vc-matrix $Q_{\hat{z}}$ to be as close as possible to a diagonal matrix, but also that the search space does not contain too many integer grid points. This requires the choice of a small value for χ^2, but one that still guarantees that the search space contains at least one integer grid point. Since the bootstrapped estimator is so easy to compute and at the same time gives a good approximation to the ILS estimator (see section 4.4), the bootstrapped solution is an excellent candidate for setting the size of the ambiguity search space. Following the decorrelation step $\hat{z} = Z^T\hat{a}$, the LAMBDA-method therefore uses, as one of its options, the bootstrapped solution $\check{z}_B$ for setting the size of the ambiguity search space as

$$\chi^2 = (\hat{z} - \check{z}_B)^T Q_{\hat{z}}^{-1} (\hat{z} - \check{z}_B) \tag{28}$$

In this way one can work with a very small search space and still guarantee that the sought for integer least-squares solution is contained in it.

4.2 The ILS pull-in region

The pull-in regions of integer rounding are unit cubes, while those of integer bootstrapping are multivariate versions of parallelograms. To determine the ILS pull-in regions we need to know the set of 'float' solutions $\hat{a} \in R^n$ that are mapped to the same integer vector $z \in Z^n$. This set is described by all $x \in R^n$ that satisfy $z = \arg\min_{u \in Z^n} \| x - u \|^2_{Q_{\hat{a}}}$. The ILS pull-in region that belongs to the integer vector z follows therefore as

$$S_{LS,z} = \{x \in R^n \mid \| x - z \|^2_{Q_{\hat{a}}} \leq \| x - u \|^2_{Q_{\hat{a}}}, \forall u \in Z^n\} \tag{29}$$

It consists of all those points which are closer to z than to any other integer point in R^n. The metric used for measuring these distances is determined by the vc-matrix $Q_{\hat{a}}$. Based on (29), one can give a representation of the ILS pull-in regions that resembles the representation of the bootstrapped pull-in regions. This representation reads as follows.

Theorem 5 (*ILS pull-in regions*)
The pull-in regions of the ILS ambiguity estimator $\check{a}_{LS} \in Z^n$ are given as

$$S_{LS,z} = \bigcap_{c_i \in Z^n} \{x \in R^n \mid \mid c_i^T Q_{\hat{a}}^{-1}(x - z) \mid \leq \frac{1}{2} \| c_i \|^2_{Q_{\hat{a}}}\}, \forall z \in Z^n \tag{30}$$

This shows that the ILS pull-in regions are constructed from intersecting half-spaces. One can also show that at most $2^n - 1$ pairs of such half spaces are needed for constructing the pull-in region. The ILS pull-in regions are convex, symmetric sets of volume 1, which satisfy the conditions of Definition 1. The ILS estimator is therefore admissible. The ILS pull-in regions are hexagons in the two-dimensional case (see figure 7).

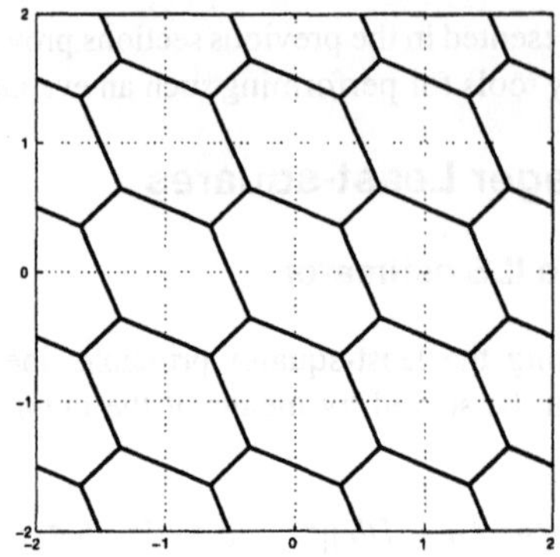

Fig. 7 The two-dimensional ILS pull-in regions.

4.3 Maximizing the success rate

Although various integer estimators exist which are admissible, some may be better than others. Having the problem of GNSS ambiguity resolution in mind, one is particularly interested in the estimator which maximizes the probability of correct integer estimation. This probability equals $P(\check{a} = a)$, but it will differ for different ambiguity estimators. The following theorem, due to Teunissen (1999b), shows that the ILS estimator maximizes the probability of correct integer estimation.

Theorem 6 (*ILS is optimal*)
Let the pdf of the 'float' solution $\hat{a}$ be given as

$$p_{\hat{a}}(x) = \sqrt{\det(Q_{\hat{a}}^{-1})}G(\| x - a \|^2_{Q_{\hat{a}}}) \tag{31}$$

where $G : R \mapsto [0, \infty)$ is decreasing and $Q_{\hat{a}}$ is positive-definite. Then

$$P(\check{a}_{LS} = a) \geq P(\check{a} = a) \tag{32}$$

for any admissible estimator $\check{a}$. This theorem gives a probabilistic justification for using the ILS estimator. For GNSS ambiguity resolution it shows, that one is better off using the ILS estimator than any other admissible integer estimator. The family of distributions defined in (31), is known as the family of elliptically contoured distributions. Several important distributions belong to this family. The multivariate normal distribution can be shown to be a member of this family by choosing $G(x) = (2\pi)^{-\frac{n}{2}} \exp -\frac{1}{2}x, x \in R$. Another member is the multivariate t-distribution.

As a direct consequence of the above theorem we have the following corollary.

Corollary 3 *(The effect of the weight matrix)*
Let Σ be any positive-definite matrix of order n and define

$$\check{a}_{\Sigma} = \arg \min_{z \in Z^n} \parallel \hat{a} - z \parallel^2_{\Sigma} \quad (33)$$

Then $\check{a}_{\Sigma}$ is admissible and

$$P(\check{a}_{LS} = a) \geq P(\check{a}_{\Sigma} = a) \quad (34)$$

In order to prove the corollary, we only need to show that $\check{a}_{\Sigma}$ is admissible. Once this has been established, the stated result (34) follows from theorem 6. The admissibility can be shown as follows. The first two conditions of Definition 1 are satisfied, since the ILS-map produces - apart from boundary ties - a unique integer vector for any 'float' solution $\hat{a} \in R^n$. And since $\check{a}_{\Sigma} = \arg\min_{z \in Z^n} \parallel \hat{a} - u - z \parallel^2_{\Sigma} + u$ holds true for any integer $u \in Z^n$, also the integer remove-restore technique applies.

As the corollary shows, a proper choice of the data weight matrix is also of importance for ambiguity resolution. The choice of weights is optimal when the weight matrix equals the inverse of the ambiguity vc-matrix. A too optimistic precision description or a too pessimistic precision description, will both result in a less than optimal ambiguity success rate. In the case of GNSS, the observation equations (the functional model) are sufficiently known and well documented. However, the same can not yet be said of the vc-matrix of the GNSS data. In the many GNSS textbooks available, we will usually find only a few comments, if any, on this vc-matrix. Examples of studies that have been reported in the literature are: Euler and Goad (1991), Gerdan (1995), Gianniou (1996), and Jin and de Jong (1996), who studied the elevation dependence of the observation variances; Jonkman (1998) and Tiberius (1998), who considered time correlation and cross correlation; and Schaffrin and Bock (1988), Bock (1998) and Teunissen (1998a), who considered the inclusion of stochastic ionospheric constraints.

4.4 Bounding the ILS success rate

A very useful application of theorem 6 is that it shows how one can *lower bound* the ILS probability of correct integer estimation. This is particularly useful since the ILS success rate is usually difficult to compute. This is due to the rather complicated geometry of the ILS pull-in region. The bootstrapped success rate is a good candidate for the ILS success rates' lower bound. The bootstrapped success rate is easy to compute and it becomes a sharp lower bound when applied to the decorrelated ambiguities $\hat{z} = Z^T\hat{a}$. In fact, at present, the bootstrapped success rate is the sharpest available lower bound of the ILS success rate.

Apart from having a lower bound, it is also useful to have an upper bound available. For obtaining an upper bound one can make use of the *geometric mean* of the ambiguity conditional variances. This geometric mean is referred to as the Ambiguity Dilution of Precision (ADOP) and it is given as

$$ADOP = \sqrt{detQ_{\hat{a}}}^{\frac{1}{n}} \ (cycles) \quad (35)$$

Note that this scalar measure of the ambiguity precision is invariant for the admissible volume preserving ambiguity transformations. With the ADOP one can obtain an upper bound by making use of the fact that the probability content of the ILS pull-in region $S_{LS,a}$ would be maximal if its shape would coincide with that of the ambiguity search space, while its volume would still be constrained to 1. We have the following bounds for the ILS success rate.

Theorem 7 *(Bounds on the ILS success rate)*
The ILS success rate $P(\check{a}_{LS} = a)$ is bounded from below and from above as

$$\begin{aligned} P(\check{z}_B = z) &\leq P(\check{a}_{LS} = a) \\ &\leq P\left(\chi^2(n,0) \leq \frac{c_n}{ADOP^2}\right) \quad (36) \\ \text{with:} \quad c_n &= (\frac{n}{2}\Gamma(\frac{n}{2}))^{2/n}/\pi \end{aligned}$$

References

Betti, B., M. Crespi, and F. Sanso (1993): A geometric illustration of ambiguity resolution in GPS theory and a Bayesian approach. *Manuscr Geod*, 18: 317-330.

Bock, Y. (1998): Medium distance GPS measurements. In: Teunissen, P.J.G., and A. Kleusberg (Eds), *GPS for Geodesy*, 2nd edition, Springer Verlag.

Boon, F., B. Ambrosius (1997): Results of real-time applications of the LAMBDA method in GPS based aircraft landings. *Proceedings KIS97*, pp. 339-345.

Boon, F., P.J. de Jonge, C.C.J.M. Tiberius (1997): Precise aircraft positioning by fast ambiguity resolution using improved troposphere modelling. *Proceedings ION GPS-97*, Vol. 2, pp. 1877-1884.

Cox, D.B. and J.D.W. Brading (1999): Integration of LAMBDA ambiguity resolution with Kalman filter for relative navigation of spacecraft. *Proceedings ION NTM-99*, pp. 739-745.

de Jonge P.J., C.C.J.M. Tiberius (1996a): The LAMBDA method for integer ambiguity estimation: implementation aspects. Publications of the Delft Computing Centre, *LGR-Series* No. 12.

de Jonge, P.J., C.C.J.M. Tiberius (1996b): Integer estimation with the LAMBDA method. *Proceedings IAG Symposium No. 115, 'GPS trends in terrestrial, airborne and spaceborne applications'*, G. Beutler et al. (eds), Springer Verlag, pp. 280-284.

de Jonge, P.J., C.C.J.M. Tiberius, P.J.G. Teunissen (1996): Computational aspects of the LAMBDA method for GPS ambiguity resolution. *Proceedings ION GPS-96*, pp. 935-944.

Euler, H.J., C. Goad (1991): On optimal filtering of GPS dual frequency observations without using orbit information. *Bull Geod*, 65: 130- 143.

Frei, E., G. Beutler (1990): Rapid static positioning based on the Fast Ambiguity Resolution Approach FARA: Theory and first results. *Manuscripta Geodaetica*, 15(6): 325-356.

Gerdan, G.P. (1995): A comparison of four methods of weighting double difference pseudo range measurements. *Trans Tasman Surv*, 1(1): 60-66.

Gianniou, M. (1996): Genauigkeitssteigerung bei kurzzeitstatischen und kinematischen Satellitenmessungen bis hin zu Echtzeitanwendung. PhD- thesis, DGK, Reihe C, no. 458, Muenchen.

Gundlich, B., K-R. Koch (2001): Confidence regions for GPS baselines by Bayesian statistics. *J Geod*, 76: 55-62.

Gundlich, B., P.J.G. Teunissen (2002): Multiple Models: Fixed, Switching and Interacting. *Proceedings V Hotine-Marussi Symposium*, Matera, Italy (this volume).

Han, S. (1995): Ambiguity resolution techniques using integer least-squares estimation for rapid static or kinematic positioning. Symposium *Satellite Navigation Technology: 1995 and beyond*, Brisbane, Australia, 10 p.

Hassibi, A., S. Boyd (1998): Integer parameter estimation in linear models with applications to GPS. *IEEE Transactions on Signal Processing*, 46(11): 2938-2952.

Hofmann-Wellenhof, B., H. Lichtenegger, J. Collins (2001): *Global Positioning System: Theory and Practice*. 5th edition. Springer Verlag.

Jin, X.X. and C.D. de Jong (1996): Relationship between satellite elevation and precision of GPS code measurements. *J Navig*, 49: 253-265.

Jonkman, N.F. (1998): Integer GPS ambiguity estimation without the receiver-satellite geometry. Publications of the Delft Geodetic Computing Centre, *LGR-Series*, No. 18.

Leick, A. (1995): *GPS Satellite Surveying*. 2nd edition, John Wiley, New York.

Misra, P., P. Enge (2001): *Global Positioning System: Signals, Measurements, and Performance*. Ganga-Jamuna Press.

Parkinson, B., J.J. Spilker (eds) (1996): *GPS: Theory and Applications*, Vols 1 and 2, AIAA, Washington DC.

Peng, H.M., F.R. Chang, L.S. Wang (1999): Attitude determination using GPS carrier phase and compass data. *Proceedings ION NTM-99*, pp. 727-732.

Schaffrin, B., Y. Bock (1988): A unified scheme for processing GPS dual- band observations. *Bull Geod*, 62: 142-160.

Strang, G., K. Borre (1997): *Linear Algebra, Geodesy, and GPS*, Wellesley-Cambridge Press.

Teunissen, P.J.G., A. Kleusberg (eds) (1998): *GPS for Geodesy*, 2nd enlarged edition, Springer Verlag.

Teunissen, P.J.G. (1993): Least-squares estimation of the integer GPS ambiguities. Invited Lecture, Section IV Theory and Methodology, IAG General Meeting, Beijing, China, August 1993. Also in: *LGR Series*, No. 6, Delft Geodetic Computing Centre.

Teunissen, P.J.G. (1995): The least-squares ambiguity decorrelation adjustment: a method for fast GPS integer ambiguity estimation. *Journal of Geodesy*, 70: 65-82.

Teunissen, P.J.G. (1998a): The ionosphere-weighted GPS baseline precision in canonical form. *Journal of Geodesy*, 72: 107-117.

Teunissen, P.J.G. (1998b): Success probability of integer GPS ambiguity rounding and bootstrapping. *Journal of Geodesy*, 72: 606-612.

Teunissen, P.J.G. (1999a): The probability distribution of the GPS baseline for a class of integer ambiguity estimators. *Journal of Geodesy*, 73: 275-284.

Teunissen, P.J.G. (1999b): An optimality property of the integer least-squares estimator. *Journal of Geodesy*, 73: 587-593.

Teunissen, P.J.G. (2001): The probability distribution of the ambiguity bootstrapped GNSS baseline. *Journal of Geodesy*, 75: 267-275.

Teunissen, P.J.G. (2002): The parameter distributions of the integer GPS model. *Journal of Geodesy*, 2002, 76: 41-48.

Tiberius, C.C.J.M., P.J. de Jonge (1995): Fast positioning using the LAMBDA method. *Proceedings DSNS-95*, paper 30, 8p.

Tiberius, C.C.J.M., P.J.G. Teunissen, P.J. de Jonge (1997): Kinematic GPS: performance and quality control. *Proceedings KIS97*, pp. 289-299.

Tiberius, C.C.J.M. (1998): Recursive data processing for kinematic GPS surveying. Publ Geodesy, no. 45, Netherlands Geodetic Commission, Delft.

Verhagen, S., P.J.G. Teunissen (2002): PDF evaluation of the ambiguity residuals. *Proceedings V Hotine-Marussi Symposium*, Matera, Italy (this volume).

The Use of Wavelets for the Acceleration of Iteration Schemes

W. Keller
Geodetic Institute, The University of Stuttgart
Geschwister-Scholl-Str. 24/D , 70174 Stuttgart, Germany

Abstract. In general, large systems of linear equations cannot be solved directly. Both the storage requirements and the number of $O(n^3)$ necessary operations let direct solvers appear inappropriate.
In many cases iterative solvers can be designed in such a way that only a small fraction of the necessary information is held in the computers memory and the rest can be computed *on-the-fly* during the iteration process.
Unfortunately, iterative solvers as SOR or conjugate gradients have a notoriously slow convergence rate, which in the worst case can prevent convergence at all, due to the unavoidable rounding errors.
The paper aims at a demonstration that wavelet can be used as a systematic tool for the construction of iteration accelerators such as multi-grid solvers or preconditioners for the conjugate gradient iteration scheme.

The paper starts with a theoretical explanation of the links between wavelets multi-grid solvers and preconditioning of conjugate gradient iteration schemes. Two numerical examples will demonstrate the efficiency of wavelet based multi-grid solvers for the planar Stokes problem.

Keywords. multi-grid solver, conjugate gradients, preconditioning,wavelets, Stokes problem, satellite

1 Introduction

Recently launched and upcoming gravity-field satellite missions like CHAMP, GRACE, GOCE will face the geodetic community with the challenge to analyze a huge amount of data. Most of the mathematical models used for data analysis will lead to linear systems of equations with a very high dimension. If the dimension is so high, that the system matrix does not fit into the computers memory, iterative equation solvers are the only option. Unfortunately, these iterative solvers have a very slow convergence rate. Frequently, this slow convergence rate is related to the long-wavelength features in the solution.This relationship between the spectrum and the convergence properties of the iterative solver can be used for an acceleration of the convergence. Two techniques for convergence acceleration will be discussed here:

- Multi-grid solvers and
- preconditioned conjugate gradients iterations.

In many cases it can be observed that an iterative solver smoothes out the short-wavelength errors of the initial guess for the solution but leaves the long-wavelength errors almost untouched.
Therefore, the idea of a multi-grid solver is

- to determine the short-wavelength solution parts by an usual iterative solver on a fine grid and subsequently
- to determine the long-wavelength features by a direct solver on a much coarser grid.

This alternating treatment of the same problem on a fine and on a coarse grid can speed up the convergence considerably.
The treatment of a problem on hierarchical grids of different resolutions has a natural link to the multi-scale analysis of wavelet theory. Therefore, wavelet techniques can be employed for an efficient construction of multi-grid iteration schemes.
For the conjugate gradient iteration the speed of convergence is determined by the ratio of the largest and the smallest eigenvalue. Therefore, the convergence can be accelerated by multiplication of the system by a matrix, which is close to the inverse of the matrix under consideration. Under the assumption that the linear system

stems from a discretization of a continuous problem and that wavelets are used for the discretization, the multi-scale properties of wavelets can be used to construct a good approximation of the inverse. This approximation can be used as preconditioner.
The efficiency of these schemes will be demonstrated by two numerical examples.

2 Linear equations as discretization of PDO equations

We will restrict our considerations to situations where the connection between the given data and the unknown solution can be formulated as a differential or an integral equation. The concept of a pseudo-differential operator (PDO), introduced by L. Svensson into geodesy (c.f. Svensson (1983)), is the generalization both of the concept of a differential and an integral operator. It is defined as an operator of the following structure

$$pu(x) := \int_{-\infty}^{\infty} \hat{u}(\omega) \cdot a(\omega) e^{\imath x \omega} d\omega \tag{1}$$

In the equation (1) the quantity $\hat{u}$ is the Fourier transform of the signal u. This means, the application of a PDO on a signal is a three-step procedure

1. computation of the spectrum $\hat{u}$ of the signal,
2. filtering of the spectrum by the so-called symbol $a(\omega)$ of the PDO and
3. back-transformation of the filtered signal into the space- or time domain.

In general, a PDO equation

$$pu = f \tag{2}$$

cannot be solved analytically. A frequently applied technique for the generation of an approximate solution is Galerkin's method.
The idea of Galerkin's method is, not to look for a solution in the whole definition space of the operator p but in a finite-dimensional subspace. Let $\{\varphi_1, \ldots, \varphi_n\}$ be a base of such an n-dimensional subspace. Then, every candidate for an approximate solution has the following structure

$$u_n(x) = \sum_{i=1}^{n} u_i \cdot \varphi_i(x) \tag{3}$$

No matter how the weights u_i are chosen, one cannot expect that the approximate solution u_n fulfills (1) exactly. There will always remain a residual,

$$r_n := f - pu_n = f - \sum_{i=1}^{n} u_i p\varphi_i(x). \tag{4}$$

Now, the weights u_i have to be chosen so that the residual r_n becomes as small as possible. If $\{\varphi_j\}_{j=1}^{\infty}$ forms a Riesz base of the imapge space of the operator p, one criterion for a small residual is that it is orthogonal to all base functions φ_i. This leads to the so-called Galerkin's equations for the unknown weights u_i.

$$\sum_{j=1}^{n} a_{ij} \cdot u_j = b_i \ , \quad i = 1, \ldots, n \tag{5}$$

with the coefficients a_{ij}, b_i given by

$$a_{ij} := \int_{-\infty}^{\infty} p\varphi_j \cdot \varphi_i dx \ , \quad b_i := \int_{-\infty}^{\infty} f \cdot \varphi_i dx \tag{6}$$

Equation (6) is valid for any kind of an operator p. If the operator p is a PDO the coefficients a_{ij} can also be determined by

$$a_{ij} = \int_{-\infty}^{\infty} a(\omega) \hat{\varphi}_i(\omega) \hat{\varphi}_j(\omega) d\omega \tag{7}$$

This means that for the establishment of the Galerkin's equations it is not necessary to know the PDO exactly, it is already sufficient to know its symbol $a(\omega)$. Additionally, it will be shown later that only a small part of the coefficients a_{ij} has to be stored in the computers memory. The rest can be determined from this core-part during the iteration.

3 Multi-grid solvers

In many cases the base functions φ_i are related to the discretization of the PDO equation on a siutable grid. If the grid is fine enough, the dimension of the Galerkin's equations is so high that it has to be solved iteratively. Generally, it can be observed that an iterative solver reduces the short-wavelength error constituents in the initial guess for the solution but leaves the long-wavelength error constituents almost unchanged. For the elimination of the remaining long-wavelength error constituents a much coarser grid is sufficient. On such a coarse grid the dimension of the remaining problem is so small that the error elimination can be performed by a direct solver.

This alternating treatment of the same problem on a fine grid by an iterative solver and on a coarse grid by a direct solver is the basic principle of a multi-grid method.

Multi-grid solvers discretize a given problem on two hierarchical grids, having the mesh-widths h and H. The discretized problems are

$$A_h u_h = b_h \ , \quad A_H u_H = b_H \tag{8}$$

Let be U_h and U_H the set of all possible solutions of the problem discretized on the fine and on the coarse grid respectively. For the restriction to a coarser and for the uplift to a finer grid two operators are necessary

$$I_h^H : U_h \to U_H \ , \quad I_H^h : U_H \to U_h \tag{9}$$

On the fine grid the problem is solved iteratively.

$$u_h^{(l+1)} = T_h u_h^{(l)} + t_h \ , \quad l = 0, \ldots, n-1 \quad , \tag{10}$$

where T_h denotes the iteration matrix and t_h the constant part of the iterative solver on the fine grid. The iteration will leave a long-wavelength residual

$$d_h = b_h - A_h u_h^{(n)} \tag{11}$$

The cause of the residual can be determined on the coarse grid

$$\nu_H = A_H^{-1}(I_h^H d_h) \tag{12}$$

After uplift of the long-wavelength error ν_H to a finer grid the solution can be corrected for the long-wavelength errors

$$u_h = u_h^{(n)} + I_H^h \nu_H \tag{13}$$

This process has to be repeated several times. Compressed into a single formula a multi-grid solver is of the following structure (cf. Großmann et. al. (1992))

$$u_h^{(k+1)} = T_h^n (I - I_H^h A_H^{-1} I_h^H A_h) T_h^n u_h^{(k)} + s_h \tag{14}$$

For suitable operators p the multi-grid strategy provides a sufficiently high and dimension-independent convergence speed.

4 Multi-grid solvers on consecutive scaling spaces

The solution spaces U_h and U_H on two hierarchical grids strongly remind of the hierarchical scaling spaces of a wavelet multi-scale analysis (MSA). Therefore, it is natural to identify these solution spaces U_h and U_H with two spaces V_0 and V_1 of a MSA. The advantage of this choice is that the restriction operator I_h^H and the uplift operator I_H^h can be identified with Mallat's smoothing operator H and its adjoint H^*. From this setting two important conclusions can be drawn. First, the discretization of the problem on a coarser grid can be efficiently computed from the given discretization on the fine grid by a wavelet transformation

$$A_H = H A_h H^* \tag{15}$$

Instead of a new discretization according to (7) a single step of Mallat's algorithm is sufficient to compute the matrix A_H from the given matrix A_h.

5 Preconditioned Conjugate Gradients

Since for a positive and selfadjoint operator p the Galerkin matrix A is positive definite, the problem (5) is equivalent to the minimization problem

$$\frac{1}{2} u^\top A u - b^\top u + b^\top b \to \min . \tag{16}$$

The minimum of this convex surface is found by moving along directions p^j, which are orthogonal with respect to A

$$(p^l)^\top A p^k = \delta_{lk} . \tag{17}$$

The speed of convergence of the conjugate gradients iteration is governed by the following rule (cf. Großmann et. al. (1992))

$$(u^{k-1} - u)^\top A (u^{k-1} - u) \leq \tag{18}$$
$$2\kappa^k (u^1 - u)^\top A (u^1 - u) \tag{19}$$

with

$$\kappa = \left(\frac{\sqrt{\mu} - \sqrt{\nu}}{\sqrt{\mu} + \sqrt{\nu}} \right) \tag{20}$$

and μ being the largest and ν the smallest eigenvalue.

Obviously, the convergence is the faster the closer μ and ν are to each other. Therefore, if instead of (5) the equivalent system

$$MAu = Mb \tag{21}$$

with $M \approx A^{-1}$ is considered, the eigenvalues of MA will be close to each other. Usually, the matrix M is chosen positive definit

$$M = C^{\top} C \tag{22}$$

and instead of(21) the equation

$$\tilde{A}\tilde{u} = \tilde{b},\ \tilde{A} = CAC^{\top},\ \tilde{u} = (C^{\top})^{-1},\ \tilde{b} = Cb \tag{23}$$

is considered. The matrix $\tilde{A}$ is symmetric, positive definit and has the same spectrum as MA. Therefore, the corresponding conjugate gradients iteration will converge very fast.
If now A stems from a wavelet discretization (7) of a PDO equation, it can be expected that M can be obtained by the same wavelet discretization of the inverse PDO

$$m_{ij} = \int_{-\infty}^{\infty} a^{-1}(\omega)\hat{\varphi}_i(\omega)\hat{\varphi}_j(\omega)d\omega \tag{24}$$

As it is the case for A also for M only a small part has to be stored and the rest can be computed on-the-fly.

6 Planar Stokes problem

The use of wavelet techniques for the construction of a multi-grid solver and preconditioned conjugate gradients iterations will be tested for the Stokes problem in planar approximation The planar Stokes problem looks for a function u to be harmonic in the upper half-space. The normal derivatives of this function have to coincide with given values δg at the $x_3 = 0$ plane.

$$\Delta u(x) = 0\ ,\ x_3 > 0 \tag{25}$$

$$\frac{\partial u}{\partial n}\Big|_{x_3=0} = \delta g \tag{26}$$

Despite the fact that in spherical approximation the Stokes problem is a Robin-type boundary value problem in planar approximation the Stokes problem simplifies to a Neumann problem The PDO formulation of this problem is

$$\delta g = pu = \frac{1}{4\pi}\int\int_{-\infty}^{\infty} a(\omega)\hat{u}(\omega)e^{\imath x\omega}d\omega \tag{27}$$

$$a(\omega) = 4\pi|\omega| \tag{28}$$

For the MSA, generating a multi-grid solver for the planar Stokes problem, the simplest possible choice was made: The two-dimensional Haar tensor wavelet. This wavelet has the following scaling function

$$\varphi(\mathbf{x}) := \varphi(x_1)\cdot\varphi(x_2) \tag{29}$$

$$\varphi(x) := \begin{cases} 1\ , 0 \le x < 1 \\ 0\ , \text{else} \end{cases} \tag{30}$$

$$\hat{\varphi}(\omega) = \frac{1}{\imath\omega}[1 - e^{-\imath\omega}] \tag{31}$$

For this choice Mallat's smoothing operator H becomes (c.f. Lois et. al. (1994))

$$(Hu)_{ij} := \sum_{k=0}^{1}\sum_{l=0}^{1} u_{kl}h_{k-2i}h_{l-2j} \tag{32}$$

$$h_0 = -h_1 = \frac{1}{\sqrt{2}} \tag{33}$$

According to (7) the coefficients of the Galerkin matrix can be computed by

$$\begin{aligned} a^h_{ijkl} &= \int\int_{-\infty}^{\infty} a(\omega)\hat{\varphi}_{ij}\hat{\varphi}_{kl}d\omega \\ &= 4\pi\int\int_{-\infty}^{\infty} |\omega||\hat{\varphi}|^4 e^{-\imath(\omega_1(i-k)+\omega_2(j-l))}d\omega \end{aligned}$$

Obviously, the coefficients a^h_{ijkl} of the Galerkin matrix are the Fourier transform of $|\omega||\hat{\varphi}|^4$ sampled at integer locations. Therefore, these coefficients efficiently can be computed by FFT. For a $N \times N$ grid the Galerkin matrix contains N^4 coefficients. Fortunately, it is sufficient to compute the N^2 coefficients of a core block. Figure 1 shows the core block for the planar Stokes problem

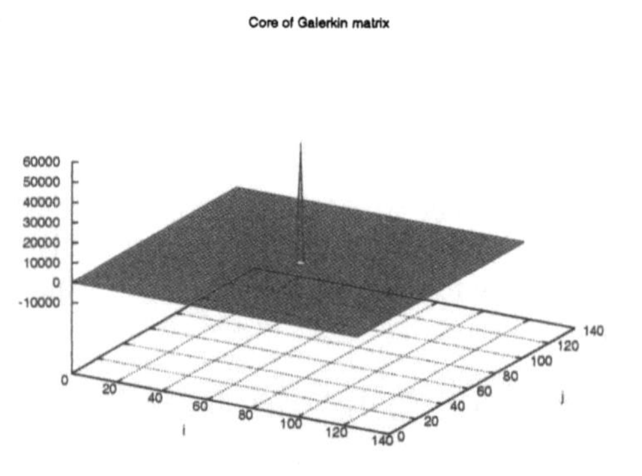

Fig. 1: Core block for the planar Stokes Problem.

Only this core block has to be stored in the computers memory. The remaining coefficients can be determined on-the-fly by scanning this core block row wise. In the examples to be discussed here N always was equal to 128, which means a 16384 × 16384 system had to be solved. In order to test the performance of the resulting multi-grid solver the Galerkin's equations were solved twice: With the usual Gauß-Seidel iteration and by the multi-grid solver. (Of course, also other iterative solvers e.g. CG-solvers could be used. Independently from the iterative solver the multi-grid iteration will have a dimension-independent rate of convergence.) Both iterations started with an identical initial error distribution, which is displayed in figure 2.

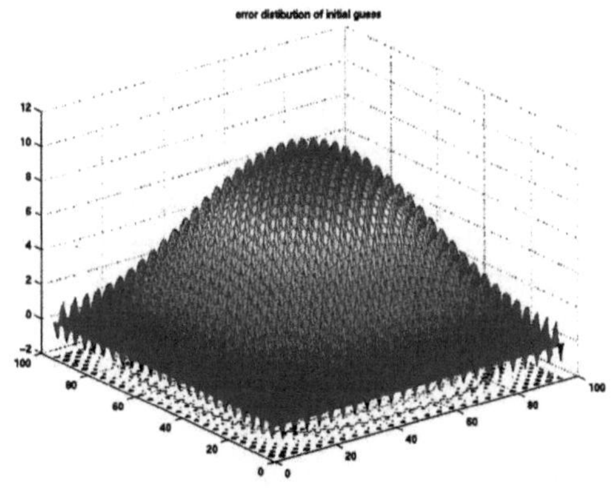

Fig. 2: Errors of the initial guess.

These initial errors consist of a long-wavelength and a short-wavelength part. The remaining errors after 12 steps of Gauß-Seidel iteration are displayed in figure 3.

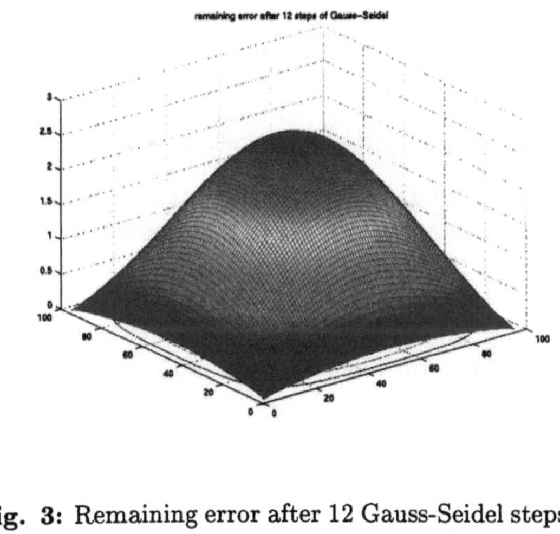

Fig. 3: Remaining error after 12 Gauss-Seidel steps.

The short-wavelength error constituents are completely eliminated but the long-wavelength errors are almost unchanged. Only a slight change in its magnitude can be observed. The situation is quite different for the multi-grid solver. The remaining errors after 12 steps of multi-grid iteration are shown in figure 4.

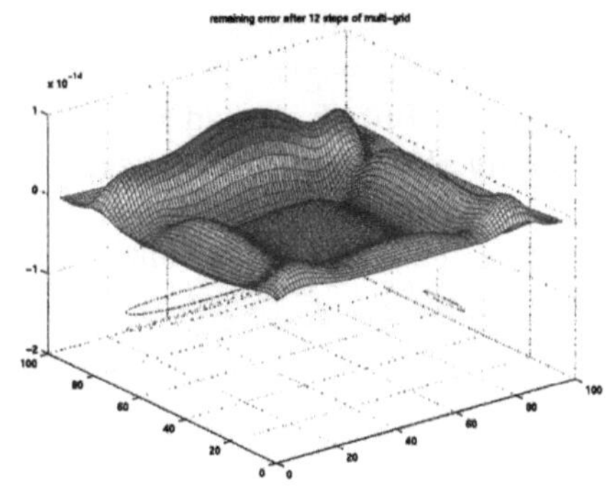

Fig. 4: Remaining errors after 12 multi-grid steps.

Here, the long-wavelength errors are reduced by a factor of about 10^{-14}. The short-wavelength errors are so small that they are not visible in the plot.

7 Geoid in Baden-Württemberg

In order to test the multi-grid solver for a more realistic example, the planar Stokes problem for the region of Baden-Württemberg is solved. As input data a quasi-geoid model for that region is used. It is displayed in figure 5.

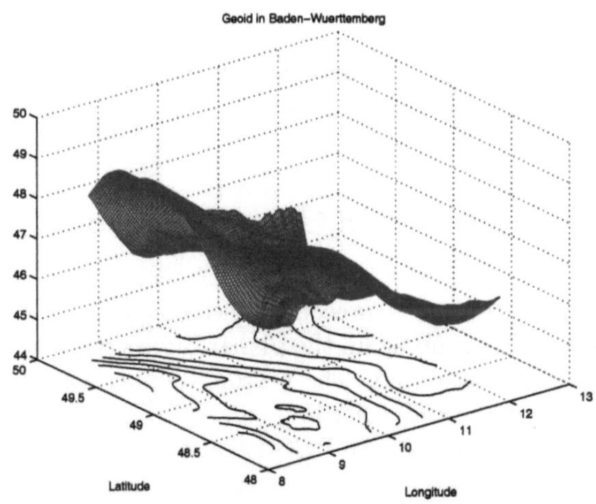

Fig. 5: Quasi-geoid in Baden-Württemberg.

The quasi-geoid heights ζ were tranformed into gravity disturbances δg by standard FFT methods and the so obtained data were used as boundary values in the planar Stokes problem (25),(26).
Due to the unavoidable boundary effects the differences between the original and the the recovered quasi-geoid heights get fairly large at the boundaries. But if instead of the original 128×128 only the inner 96×96 region is considered, the differences in this region are below $1\ mm$. The differences in the inner region are displayed in figure 6.

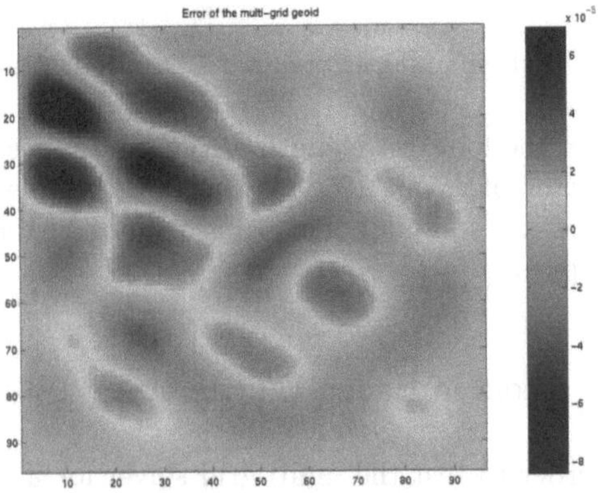

Fig. 6: Differences between original and recovered quasi-geoid in the inner zone.

Hence, besides the boundary region, the quasi-geoid is perfectly recovered.

8 Preconditioned Conjugate Gradients

For the test of the preconditioned conjugate gradients iteration again the planar Stokes problem (25),(26) was used. The preconditioning matrix was constructed according to (24). The convergence behavior of the PCG iteration is displayed in figure 7.

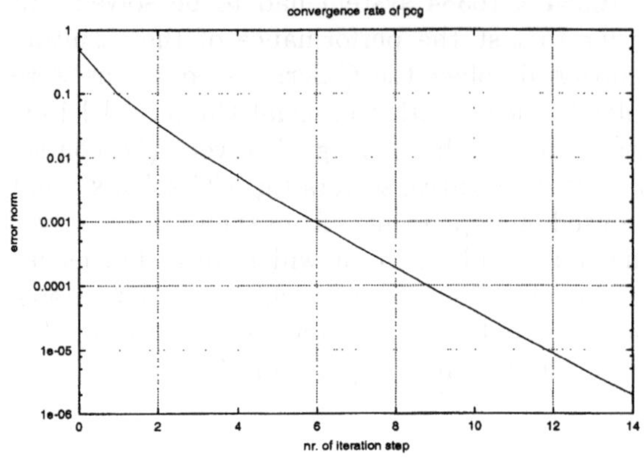

Fig. 7: Convergence behavior of preconditioned conjugate gradients.

Obviously, the PCG iteration has a convergence rate according to (18) with $\kappa \approx 0.215$. This comparatively lage value of κ does not satisfy the expectations. As possible cause for this weak acceleration an improper treatment of the singularity of a^{-1} in (24) can be supposed.

9 Summary

For linear equations stemming from a wavelet discretization of a PDO equation multi-grid solvers and preconditioning of conjugate gradients iteration are efficient tools for the acceleration of convergence.
In the multi-grid context the wavelet discretization enables the use of wavelet techniques for

- the discretization on different grids,
- the restriction from a finer to a coarser grid,
- the uplift from a coarser to a finer grid.

In the conjugate gradients context wavelet discretization allows for an efficient construction of a preconditioner.
All matrices involved can be computed on the fly from much smaller core matrices. In this case really large problems can be tackled even on a PC.

References

Großmann Ch and Roos H-G (1992) *Numerik partieller Differentialgleichungen.* B.G. Teubner, Stuttgart

Hackbusch W (1985) *Multi-Grid Methods and Applications.* Springer Series in Computational Mathematics, Berlin 1985

Kusche J *Implementation of multigrid solvers for satellite gravity anomaly recovery.* Journal of Geodesy 74(2001) pp. 773-782

Kusche J and Rudolph S (2001) *Satellite Gravity Anomaly Recovery Using Multigrid Methods.* In: Sideris M G: (ed.) *Gravity, Geoid and Geodynamics 2000* Springer Verlag Berlin, Heidelberg New York 2001.

Lois A.K. , Maaß P and Rieder A (1994) *Wavelets. Theorie und Anwendungen.* B.G. Teubner, Stuttgart

Svensson L (1983) *Pseudodifferential Operators - A New Approach to the Boundary Problems of Physical Geodesy* mansuscripta geodaetica 8(1983) pp.1-44

Helmert Geometry

T. Krarup
Carolinelundsvej 23, DK-8700 Horsens, Denmark

K. Borre
Aalborg University, Niels Jernes Vej 14, DK-9220 Aalborg Ø, Denmark

Abstract

The Baltic Geodetic Commission was founded in 1924 and remained active till 1940. It was created in order to establish a triangulation network around the Baltic Sea to provide uniform coordinates in neighboring countries. The objectives of the Commission were known to the first author and this makes the starting point for the ideas underlying the present paper.

The simplest version of the problem is described by two r by s matrices X and Y. Their columns contain the r-dimensional coordinates of s points that have been observed in two different coordinate systems with random observational errors. The problem is to estimate in some least-squares sense a transformation matrix H (rotations and scale changes) and an r-dimensional vector $\boldsymbol{t}$ (translation between the two origins); the translational matrix is $\boldsymbol{t}\boldsymbol{e}^{\mathrm{T}}$ where the s-dimensional vector $\boldsymbol{e}$ has mere ones as components:

$$\tilde{Y} = H\tilde{X} - \tilde{\boldsymbol{t}}\,\boldsymbol{e}^{\mathrm{T}}$$

We generalize by introducing weights p, and q. All coordinates in X are observed with weight p and the Y coordinates with weight q.

There is another type of generalization: we want to transform between more than two point sets X and Y. With background in the Baltic Ring Triangulation we may introduce a *total network* that is composed of individual *national networks.* Neighboring national networks have points in common. We focus on these common points which we want to transform into a global frame for the total network by using *similarity transformations,* that is transformations allowing for rotations, translations, and changes of scale.

The common points are the points that really interest us. The rest is named *inner points.* In this way we have defined a natural splitting of points: *common points* (that are common to adjacent national networks) and *inner points* belonging only to one nation.

The national networks also can be perceived as repeatedly measured versions of one and the same network and the problem changes into one of *interpolation* between different coordinate sets for the same point set.

The problem admits a geometrical theory which in the opinion of the authors is interesting and even beautiful, and, what is perhaps more important, it suggests a method for computing the parameters.

As in most geodetic problems we also here use the principle of least squares. In most adjustment problems the unknown quantity is a vector; here it is a vector and two (or more) matrices. The relevant entries of the matrices could be reshaped into vectors; but we want to work with the matrices directly and demonstrate the richer structure of matrices.

The present paper has been several years under its way and has used various clothings. However, we hope this final version is readable, understandable, and useful for a lot of different disciplines like image processing, computer vision, and computer graphics. We include MATLAB routines that directly solve the various formulations of the problem. Our MATLAB algorithms also pay attention to the issue of numerical stability.

The title of the present paper is chosen as to honour the German geodesist Friedrich Robert Helmert (1843-1917). In 1872 he addressed a special case of the present problem.

We use the following *notational convention* throughout the paper: Capital Latin letter for any matrix, small Latin letter for any scalar, capital

Greek letter for any orthogonal matrix, and small Greek letter for any diagonal matrix.

We choose to work in analytic geometry, that is with lines and circles*. Traditionally the derivation of the Helmert Transformation starts from "inside" while we start from "outside".

---

* A specific geodetic question: is the Helmert Transformation a general result of the principle of least squares or do we build on analytical geometry? Do we need further corrections and how is the relationship to Integrated Geodesy?

Multiscale Solution of Oblique Boundary-Value Problems by Layer Potentials

Willi Freeden, Carsten Mayer
Geomathematics Group, Department of Mathematics
University of Kaiserslautern, P.O. Box 3049, 67653 Kaiserslautern, Germany

Abstract. With the aid of classical results of potential theory, the limit- and jump-relations, a multiscale framework on geodetically relevant regular surfaces is established corresponding to oblique derivative data. By the oblique distance to the regular surface a scale factor in the kernel functions of the limit- and jump-operators is introduced, which connects these intergral kernels with the theory of scaling functions and wavelets.
As applications of the wavelet approach some numerical examples are presented on an ellipsoid of revolution. At the end we discuss a fast multiscale representation of the solution of the (exterior) oblique derivative (boundary–value) problem corresponding to geoscientifically relevant surfaces. A local as well as global reconstruction of the gravitational potential model EGM96 on the reference ellipsoid will illustrate the power of this appoach.

Keywords. Scaling functions and wavelets on regular surfaces, potential operators, jump relations, multiscale analysis, (exterior) oblique derivative problem of potential theory

1 Introduction

Wavelets are known as mathematical means for breaking up a complicated function (signal) into many simple pieces at different scales and positions. Thus wavelets have become a powerful and flexible tool for scientific computation and data handling. Basically, wavelet analysis is done by convolving the function under consideration against 'dilated' and 'shifted' versions of one fixed function, viz. the 'mother wavelet'. Traditionally, applications of wavelets have been signal analysis, image processing, noise cancellation, etc, but there is also a growing interest in the numerical treatment of partial differential equations. However, wavelet methods are merely known for unfolding their computational economy and efficiency when applied to problems on Euclidian spaces, the sphere or the torus. The aim of this article is to present a new wavelet approach to geodetically relevant surfaces. Our purpose is to develop a multiscale theory on regular surfaces by using results of classical potential theory.
The outline of this paper is as follows: First we introduce the notations and preliminaries that are needed for our wavelet approach. We specify regular surfaces on which our theory is established. Then we introduce potential operators with respect to an arbitrary non-tangential vector field which are the main ingredients of this work. We develop the oblique limit and jump relations of these potential operators formulated in the framework of the Hilbert space of square–integrable functions. The setup of a multiresolution analysis (i.e. scaling functions, scale spaces, wavelets, detail spaces) is defined by interpreting the kernel functions of the limit and jump integral operators as scaling functions on regular surfaces. The oblique distance to the parallel surfaces of the regular surface under consideration thereby represents the scale level in the scaling function. At the end we deal with the already mentioned discretization of Fredholm integral equations in order to give a multiscale representation of the solution of the (exterior) oblique derivative problem (EODP) in three dimensions corresponding to geoscientifically relevant regular surfaces. Furthermore, we dicuss some numerical examples. In particular we are interested in the zoom–in property and the detection of a high frequency perturbation which are typical features within a wavelet framework.

2 Basic Concept

At first we introduce some settings which are standard in potential theory (see, for example, Kellogg (1929)).

We begin our considerations by introducing the notation of a regular surface:

Definition 2.1 *A surface $\Sigma \subset \mathbb{R}^3$ is called regular, if it satisfies the following properties:*

(i) Σ divides the three-dimensional Euclidean space $\mathbb{R}^3$ into the bounded region Σ_{int} (inner space) and the unbounded region Σ_{ext} (outer

space) defined by $\Sigma_{\text{ext}} = \mathbb{R}^3 \backslash \overline{\Sigma_{\text{int}}}$, $\overline{\Sigma_{\text{int}}} = \Sigma_{\text{int}} \cup \Sigma$,

(ii) Σ_{int} *contains the origin,*

(iii) Σ *is a closed and compact surface free of double points,*

(iv) Σ *is locally of class* $C^{(2,\mu)}$ *(*ν *denotes the unit normal field on* Σ *pointing into the outer space* Σ_{ext}*).*

Geoscientifically relevant regular surfaces are, for example, sphere, ellipsoid, spheroid, geoid, (regular) Earth's surface.
Given a regular surface, then there exist positive constants α, β such that

$$\alpha < \sigma^{\inf} = \inf_{x\in\Sigma} |x| \leq \sup_{x\in\Sigma} |x| = \sigma^{\sup} < \beta. \quad (1)$$

By λ we designate a $c^{(1,\mu)}$–unit vector field on Σ satisfying $\inf_{x\in\Sigma}(\lambda(x) \cdot \nu(x)) > 0$. The set

$$\Sigma^{(\lambda)}(\tau) = \{x \in \mathbb{R}^3 | x = y + \tau\lambda(y), y \in \Sigma\}$$

generates an *oblique (parallel) surface* which is exterior to Σ for $\tau > 0$ and interior for $\tau < 0$. It is well known from differential geometry (see e.g. Freeden and Kersten (1980), Freeden and Kersten (1981)) that if $|\tau|$ is sufficiently small, then the surface $\Sigma^{(\lambda)}(\tau)$ is regular.
As usual, by $\mathrm{C}^{(0,\mu)}(\Sigma)$, $0 < \mu < 1$, we denote the space of all μ-Hölder continuous functions on the regular surface Σ and by $\mathrm{L}^2(\Sigma)$ we denote the space of (Lebesgue) square-integrable functions on the regular surface Σ. $\mathrm{L}^2(\Sigma)$ is a Hilbert space with respect to the inner product $(\cdot,\cdot)_{\mathrm{L}^2(\Sigma)}$ and a Banach space with respect to the norm $\|\cdot\|_{\mathrm{L}^2(\Sigma)}$. $\mathrm{L}^2(\Sigma)$ is the completion of $\mathrm{C}^{(0,\mu)}(\Sigma)$ with respect to the norm $\|\cdot\|_{\mathrm{L}^2(\Sigma)}$:

$$\overline{\mathrm{C}^{(0,\mu)}(\Sigma)}^{\|\cdot\|_{\mathrm{L}^2(\Sigma)}} = \mathrm{L}^2(\Sigma) \ .$$

3 Limit Formulae und Jump Relation

Next we want to formulate the classical limit and jump relations of potential theory in comprehensive manner. They will be the fundamental framework for the development of scaling functions and wavelets on regular surfaces.
For $\tau \neq \sigma$ with $|\tau|, |\sigma|$ sufficiently small, the functions

$$(x,y) \mapsto \frac{1}{|x + \tau\lambda(x) - (y + \sigma\lambda(y))|},$$

for $(x,y) \in \Sigma \times \Sigma$ are continuous. Thus the *potential operators* $P^{(\lambda)}(\tau,\sigma)$ formally defined by

$$P^{(\lambda)}(\tau,\sigma)F(x) = \int_\Sigma F(y) \frac{1}{|x + \tau\lambda(x) - (y + \sigma\lambda(y))|} \, d\omega(y)$$

form mappings from $\mathrm{L}^2(\Sigma)$ into $\mathrm{C}^{(0,\mu)}(\Sigma)$ and are Hölder continuous with respect to $\|\cdot\|_{\mathrm{C}^{(0,\mu)}(\Sigma)}$.
By formal operations we obtain for $F \in \mathrm{C}^{(0,\mu)}(\Sigma)$

$$P^{(\lambda)}(\tau,0)F(x) = \int_\Sigma F(y) \frac{1}{|x + \tau\lambda(x) - y|} \, d\omega(y)$$

($P^{(\lambda)}(\tau,0)$*: operator of the single–layer potential on* Σ *for values on* $\Sigma^{(\lambda)}(\tau)$),

$$P^{(\lambda)}_{|2}(\tau,0)F(x) = \frac{\partial}{\partial\sigma} P^{(\lambda)}(\tau,\sigma)F(x)|_{\sigma=0} = \int_\Sigma F(y) \frac{\lambda(y)\cdot(x + \tau\lambda(x) - y)}{|x + \tau\lambda(x) - y|^3} \, d\omega(y)$$

$\left(P^{(\lambda)}_{|2}(\tau,0)\right.$*: operator of the double–layer potential on* Σ *for values on* $\left.\Sigma^{(\lambda)}(\tau)\right)$.
The notation $P^{(\lambda)}_{|i}$ indicates differentiation with respect to the i-th variable. Analogously we get

$$P^{(\lambda)}_{|1}(\tau,0)F(x) = \frac{\partial}{\partial\tau} P^{\lambda}(\tau,\sigma)F(x)\big|_{\sigma=0}, = -\int_\Sigma F(y) \frac{\lambda(x)\cdot(x + \tau\lambda(x) - y)}{|x + \tau\lambda(x) - y|^3} \, d\omega(y)$$

and

$$P^{(\lambda)}_{|2|1}(\tau,0)F(x) = \frac{\partial^2}{\partial\tau\partial\sigma} P^{(\lambda)}(\tau,\sigma)F(x)\big|_{\sigma=0}$$

for the *operators of the normal derivatives.*
The potential operators now enable us to give concise formulations of the classical *limit formulae* and *jump relations* in potential theory which are given in Table 1 in the appendix. Proofs of these relations are given in Freeden (1980), Kersten (1980).

4 Multiscale Modelling in $\mathrm{L}^2(\Sigma)$

Since we are interested in a reconstruction of a function $F \in \mathrm{L}^2(\Sigma)$ we will only take those cases in Table 1 into account, where the construction of an approximating identity for F is possible. These are the second and the third limit relation and the second and the third jump relation, which were simply deduced from the limit relations. The disadvantage of the two limit relations mentioned above is, that there appear strongly singular integral kernels in the potential operators ($P^{(\lambda)}_{|1}(0,0)$, respectively $P^{(\lambda)}_{|2}(0,0)$ understood in the sense of Cauchy), while the integral kernels of the jump relations are fully regular. This is the reason why we will only take into account the first two jump relations, the jump relation of the normal

derivative of the single layer potential (labeled by 5) and the jump relation of the double layer potential (labeled by 6). It should be noted, that the following theory can be formulated in almost the same manner for the two limit relations.
Writing out the jump relations explicitly we obtain the following theorem.

THEOREM 4.1 *For* $F \in L^2(\Sigma)$ *and* $i = 5, 6$

$$\lim_{\substack{\tau \to 0 \\ \tau > 0}} \int_\Sigma \Phi^i_\tau(\cdot, y)\, F(y)\, d\omega(y) = F$$

holds in the sense of the $\|\cdot\|_{L^2(\Sigma)}$ *–norm, where the kernel functions* Φ^5_τ *and* Φ^6_τ *are known explicitly (see Table 2).*

4.1 Scaling and Wavelet Functions

For $\tau > 0$ the family $\{\Phi^i_\tau\}_{\tau>0}$ of kernels $\Phi^i_\tau : \Sigma \times \Sigma \to \mathbb{R}$ is called an *oblique* Σ*–scaling function of type* i. Moreover, $\Phi^i_1 : \Sigma \times \Sigma \to \mathbb{R}$ (i.e.: $\tau = 1$) is called the *oblique mother kernel of the oblique* Σ*–scaling function of type* i.
Correspondingly, for $\tau > 0$ and $i = 5, 6$, the family $\{\Psi^i_\tau\}_{\tau>0}$ of kernels $\Psi^i_\tau : \Sigma \times \Sigma \to \mathbb{R}$ given by the scaling equation

$$\Psi^i_\tau(x, y) = -\alpha(\tau)^{-1} \frac{d}{d\tau} \Phi^i_\tau(x, y), \qquad (2)$$

for $x, y \in \Sigma$ is called an *oblique* Σ*–wavelet function of type* i.
In the remainder of this paper we particularly choose $\alpha(\tau) = \tau^{-1}$ (of course, other weight functions than $\alpha(\tau) = \tau^{-1}$ can be chosen in (2)). Moreover, $\Psi^i_1 : \Sigma \times \Sigma \to \mathbb{R}$ (i.e.: $\tau = 1$) is called the *oblique mother kernel of the oblique* Σ*–wavelet function of type* i.

DEFINITION 4.2 *Let* $\{\Phi^i_\tau\}_{\tau>0}$ *be a* Σ*–scaling function of type* i. *Then the associated oblique* Σ*–wavelet transform of type* i *is defined by*

$$(WT)^{(i)} : L^2(\Sigma) \to L^2((0,\infty) \times \Sigma)$$
$$(WT)^{(i)}(F)(\tau, x) = \int_\Sigma \Psi^i_\tau(x, y) F(y)\, d\omega(y) \ .$$

Explicit formulations of the oblique Σ–wavelet function of type 5 and type 6 are given in Table 2.

It is not difficult to see that the wavelets Ψ^i_τ, $i = 5, 6$, behave like $O(\tau^{-1})$, hence, the convergence of the integrals in the following *reconstruction theorem* is guaranteed.

THEOREM 4.3 *Let* $\{\Phi^i_\tau\}_{\tau>0}$ *be an oblique* Σ*–scaling function of type* i. *Suppose that* F *is of class* $L^2(\Sigma)$. *Then the reconstruction formula*

$$\int_0^\infty (WT)^i(F)(\tau, \cdot) \frac{d\tau}{\tau} = F$$

holds for $i = 5, 6$ *in the sense of* $\|\cdot\|_{L^2(\Sigma)}$.

Proof. Let $R > 0$ be arbitrary. By observing Fubini's theorem and the identity

$$\Phi^i_R(x, y) = \int_R^\infty \Psi^i_\tau(x, y) \frac{d\tau}{\tau}, \qquad (x, y) \in \Sigma \times \Sigma,$$

we obtain

$$\begin{aligned} &\int_R^\infty (WT)^i(F)(\tau, \cdot) \frac{d\tau}{\tau} \\ &= \int_R^\infty \int_\Sigma \Psi^i_\tau(\cdot, y) F(y)\, d\omega(y) \frac{d\tau}{\tau} \\ &= \int_\Sigma \int_R^\infty \Psi^i_\tau(\cdot, y) F(y)\, \frac{d\tau}{\tau} d\omega(y) \\ &= \int_\Sigma \Phi^i_R(\cdot, y) F(y)\, d\omega(y) \ . \end{aligned}$$

The limit $R \to 0$ in connection with Theorem 4.1 yields the desired result. □
Note that the properties of the oblique Σ–wavelets of type i (analogously to variants of spherical wavelets developed in Freeden et al. (1998)) do not presume the zero–mean property of Ψ^i_τ. The wavelets constructed in this way, therefore, do not satisfy a substantial condition of the Euclidean concept.

4.2 Scale Discretized Reconstruction Formula

In what follows, scale discrete oblique Σ–scaling functions and wavelets of type i will be introduced. We start with the choice of a sequence which divides the continuous scale interval $(0, \infty)$ into discrete subintervals. More explicitly, $(\tau_j)_{j\in\mathbb{Z}}$ denotes a sequence of real numbers satisfying

$$\lim_{j\to\infty} \tau_j = 0 \quad \text{and} \quad \lim_{j\to-\infty} \tau_j = \infty \ . \qquad (3)$$

For example, one may choose $\tau_j = 2^{-j}$, $j \in \mathbb{Z}$ (note that in this case, $2\tau_{j+1} = \tau_j$, $j \in \mathbb{Z}$).
Given an oblique Σ–scaling function $\{\Phi^i_\tau\}_{\tau>0}$ of type i, then we define the (scale) *discretized oblique* Σ*–scaling function* of type i by $\{\Phi^i_{\tau_j}\}_{j\in\mathbb{Z}}$.
In doing so, we immediately get the following result.

THEOREM 4.4 *For* $F \in L^2(\Sigma)$

$$\lim_{j\to\infty} \int_\Sigma \Phi^i_{\tau_j}(\cdot, y) F(y)\, d\omega(y) = F$$

holds for $i = 5, 6$ *in the* $\|\cdot\|_{L^2(\Sigma)}$*–sense.*

Our procedure canonically leads us to the following type of scale discretized wavelets.

DEFINITION 4.5 *Let $\{\Phi^i_{\tau_j}\}_{j\in\mathbb{Z}}$ be a discretized oblique Σ–scaling function of type i. Then the (scale) discretized oblique Σ–wavelet function of type i is defined by*

$$\Psi^i_{\tau_j}(\cdot,\cdot) = \int_{\tau_{j+1}}^{\tau_j} \Psi^i_\tau(\cdot,\cdot)\frac{d\tau}{\tau}, \quad j \in \mathbb{Z}.$$

In connection with (2) it follows that

$$\Psi^i_{\tau_j} = -\int_{\tau_{j+1}}^{\tau_j} \tau\frac{d}{d\tau}\Phi^i_\tau\frac{d\tau}{\tau} = \Phi^i_{\tau_{j+1}} - \Phi^i_{\tau_j}. \quad (4)$$

Formula (4) is called *(scale) discretized Σ–scaling equation of type i.*

Assume now that F is a function of class $L^2(\Sigma)$. Observing the discretized Σ–scaling equation of type i we get for $J \in \mathbb{Z}$ and $N \in \mathbb{N}$

$$\begin{aligned}&\int_\Sigma \Phi^i_{\tau_{J+N}}(\cdot,y)F(y)\,d\omega(y)\\ &= \int_\Sigma \Phi^i_{\tau_J}(\cdot,y)F(y)\,d\omega(y)\\ &+ \sum_{j=J}^{J+N-1}\int_\Sigma \Psi^i_{\tau_j}(\cdot,y)F(y)\,d\omega(y)\ .\end{aligned}$$

The *(scale) discretized oblique Σ–wavelet transform of type i* is defined by

$$(WT)^i : L^2(\Sigma) \mapsto \{H : \mathbb{Z}\times\Sigma \to \mathbb{R}\}$$

$$(WT)^i(F)(\tau_j;x) = \int_\Sigma \Psi^i_{\tau_j}(x,y)F(y)\,d\omega(y)\ .$$

As a final result we are able to formulate the following theorem.

THEOREM 4.6 *Let $\{\Psi^i_{\tau_j}\}_{j\in\mathbb{Z}}$ be a (scale) discretized oblique Σ–wavelet function of type i. Then, for all $F \in L^2(\Sigma)$, the reconstruction formula*

$$\sum_{j=-\infty}^{+\infty} (WT)^i(F)(\tau_j;\cdot) = F$$

holds for $i = 5, 6$ in $\|\cdot\|_{L^2(\Sigma)}$–sense.

4.3 Scale and Detail Spaces

Comparing the above result with the continuous analogue (Theorem 4.3) we notice that the subdivision of the continuous scale interval $(0,\infty)$ into discrete pieces means substitution of the integral over τ by an associated discrete sum.

As in the spherical theory of wavelets (see Freeden et al. (1996a), Freeden, Windheuser (1996b)), the operators $R^i_{\tau_j}$, $P^i_{\tau_j}$ defined by

$$R^i_{\tau_j}(F) = \int_\Sigma \Psi^i_{\tau_j}(\cdot,y)F(y)\,d\omega(y),$$

$$P^i_{\tau_j}(F) = \int_\Sigma \Phi^i_{\tau_j}(\cdot,y)F(y)\,d\omega(y),$$

for $F \in L^2(\Sigma)$ may be understood as band pass and low pass filter, respectively. The *scale spaces* $V^i_{\tau_j}$ and the *detail spaces* $W^i_{\tau_j}$ *of type* i are defined as usual by

$$V^i_{\tau_j} = P^i_{\tau_j}(L^2(\Sigma)) = \left\{P^i_{\tau_j}(F)\middle|F \in L^2(\Sigma)\right\},$$

$$W^i_{\tau_j} = R^i_{\tau_j}(L^2(\Sigma)) = \left\{R^i_{\tau_j}(F)\middle|F \in L^2(\Sigma)\right\},$$

respectively. From the identity

$$\Phi^i_{\tau_{J+1}} = \Phi^i_{\tau_J} + \Psi^i_{\tau_J},$$

i.e.

$$P^i_{\tau_{J+1}}(F) = P^i_{\tau_J}(F) + R^i_{\tau_J}(F)$$

for all $J \in \mathbb{Z}$ it easily follows that

$$V^i_{\tau_{j+1}} = V^i_{\tau_j} + W^i_{\tau_j}\ . \quad (5)$$

However, it should be remarked that the sum (5) generally is neither direct nor orthogonal.

Equation (5) may be interpreted in the following way: The set $V^i_{\tau_j}$ contains a $P^i_{\tau_j}$–filtered version of a function belonging to the class $L^2(\Sigma)$. The lower the scale, the stronger the intensity of filtering. By adding '$R^i_{\tau_j}$–details' contained in the space $W^i_{\tau_j}$ the space $V^i_{\tau_{j+1}}$ is created, which consists of a filtered versions at resolution $j+1$. Obviously, for $i = 5, 6$,

$$\overline{\bigcup_{j=-\infty}^{\infty} V^i_{\tau_j}}^{\|\cdot\|_{L^2(\Sigma)}} = \overline{\bigcup_{j=-\infty}^{\infty} W^i_{\tau_j}}^{\|\cdot\|_{L^2(\Sigma)}} = L^2(\Sigma)\ .$$

4.4 Solving BVPs by Wavelets

In what follows we want to show, how our multiscale approach on geoscientifically relevant regular surfaces can be used to approximate the solution of the exterior oblique derivative problem (EODP). The problem can be formulated briefly as follows:

(EODP) Given a function F of class $C^{(0,\mu)}(\Sigma)$. Find a function $U \in C^{(1,\mu)}(\overline{\Sigma_{ext}})$ with

$$\begin{aligned}\Delta U &= 0, \quad x \in \Sigma_{ext},\\ \frac{\partial U^+}{\partial\lambda}(x) &= \lim_{\substack{\tau\to 0\\ \tau>0}} \lambda(x)\cdot(\nabla U)(x+\tau\lambda(x))\\ &= F(x), \quad x \in \Sigma,.\end{aligned} \quad (6)$$

and U being regular at infinty (that is, $|U(x)| = O(|x|^{-1})$, $|\nabla U(x)| = O(|x|^{-2})$ for $|x| \to \infty$ uniformly with respect to all directions).

If the field λ coincides with the normal field ν on Σ, equation (6) becomes the boundary condition of the

classical exterior Neumann problem.
For given $F \in C^{(0,\mu)}(\Sigma)$, the solution $U \in \mathrm{Pot}^{(1,\mu)}(\overline{\Sigma_{ext}}) = \{U \in C^{(1,\mu)}(\overline{\Sigma_{ext}}) | \Delta U = 0 \text{ in } \Sigma_{ext}, U \text{ regular at infinity}\}$ of the (EODP) can be written as a single layer potential,

$$U(x) = \int_\Sigma Q(y) \frac{1}{|x-y|} \, d\omega(y), \tag{7}$$

where the single layer density $Q \in C^{(0,\mu)}(\Sigma)$ satisfies the Fredholm integral equation of the second kind

$$-2\pi Q(x)\,(\lambda(x)\cdot\nu(x)) + \int_\Sigma Q(y) \frac{\partial}{\partial\lambda(x)} \frac{1}{|x-y|} \, d\omega(y) = F(x) \tag{8}$$

for all $x \in \Sigma$. An approximation of scale J

$$P_{\tau_J}(Q)(x) = \sum_{l=1}^{N_J} a_l^{N_J} \Phi^i_{\tau_J}\left(x, y_l^{N_J}\right), \quad x \in \Sigma$$

(with $i \in \{5,6\}$, $a_l^{N_J} \in \mathbb{R}$, $y_l^{N_J} \in \Sigma$ appropriate point system on Σ, $l = 1, \ldots, N_J$ and $J, N_J \in \mathbb{N}$ sufficiently large) is deducable from (8) by solving a sytem of linear equations obtained by an appropriate approximation method such as collocation, Galerkin procedure, least squares approximation, etc .
For solving the linear systems fast multipole methods (FMM) are applicable (see e.g. Glockner (2001) and the reference therein).

REMARK 4.7 *It should be noted that the singular kernel function* $\frac{\partial}{\partial\lambda(x)} \frac{1}{|x-y|}$ *in the boundary integral equation* (8) *is substituted (regularized) in numerical applications by a regular approximation of the form*

$$\frac{1}{2} \frac{\partial}{\partial\lambda(x)} \left(\frac{1}{|x+\tau_L\lambda(x)-y|} + \frac{1}{|x-\tau_L\lambda(x)-y|} \right)$$

with $L \in \mathbb{N}$ *chosen large enough. For an explicit description of this type of regularisation of the singular integral operator in* (8) *for Neumann boundary conditions the reader is referred to Freeden, Mayer (2001).*

As a final result we can formulate the following

THEOREM 4.8 *For given* $F \in C^{(0,\mu)}(\Sigma)$, *let* U *be the potential of class* $\mathrm{Pot}^{(1,\mu)}(\overline{\Sigma_{ext}})$ *with* $\frac{\partial U^+}{\partial\lambda} = F$ *on* Σ. *Then the function* $U_J \in \mathrm{Pot}^{(0,\mu)}(\overline{\Sigma_{ext}})$ *given by*

$$U_J(x) = \sum_{l=1}^{N_J} a_l^{N_J} \int_\Sigma \Phi^i_{\tau_J}\left(y, y_l^{N_j}\right) \frac{1}{|x-y|} \, d\omega(y) \tag{9}$$

represents a $J-$*scale approximation of* U *in the* $\|\cdot\|_{C^0(\overline{K})}-$*sense for every* $K \subset \Sigma_{ext}$ *with* $dist(\overline{K}, \Sigma) > 0$.

REMARK 4.9 *It should be noted, that in numerical applications a similar regularization as in Remark 4.7 is applied to the kernel* $\frac{1}{|x-y|}$ *in* (9). *Then a fully discrete representation of* (9) *can be achieved by discretizing the integral with a suitable integration rule (for more details see Mayer (2001)).*

Acknowledgement. The authors thank the Graduiertenkolleg "Mathematik und Praxis", University of Kaiserslautern for financial support.

References

Freeden W.(1980). On the Approximation of External Gravitational Potential With Closed Systems of (Trial) Functions, *Bull. Géod.*, 54, 1-20.

Freeden W., H. Kersten (1980). The Geodetic Boundary Value Problem Using the Known Surface of the Earth, Report No. 29, Geod. Inst. RWTH Aachen.

W. Freeden, H. Kersten (1981). A Constructive Approximation Theorem for the Oblique Derivative Problem in Potential Theory, *Math. Meth. Appl. Sci.*, 3, 104-114.

Freeden W., M. Schreiner, R. Franke (1996). A Survey on Spherical Spline Approximation, *Surv. Math.*, 7, 29-85.

Freeden W., U. Windheuser, Spherical Wavelet Transform and Its Discretization, *Adv. Comput. Math.*, 5, 51-94.

Freeden W., T. Gervens, M. Schreiner (1998). Constructive Approximation on the Sphere (With Applications to Geomathematics). Oxford Science Publications, Clarendon.

Freeden W., C. Mayer (2001). Wavelets Generated by Layer Potentials, *Apllied Computational Harmonic Analysis* (accepted for publication).

Glockner O. (2001). On Numerical Aspects of Gravitational Field Modeling from SST and SGG by Harmonic Splines and Wavelets (With Application to CHAMP Data). Doctoral Thesis, Geomathematics Group, Shaker Verlag.

Kellogg O.D. (1929). Foundations of Potential Theory. Frederick Ungar Publishing Company.

Kersten H. (1980). Grenz- und Sprungrelationen für Potentiale mit quadratsummierbarer Flächenbelegung, *Resultate der Mathematik*, 3, 17-24.

Mayer C. (2001). Potential Operators, Jump Relations and Wavelets. Diploma Thesis, Department of Mathematics, Geomathematics Group, University of Kaiserslautern.

$$\lim_{\substack{\tau\to 0\\ \tau>0}} \left\| P^{(\lambda)}(\pm\tau,0)F - P^{(\lambda)}(0,0)F \right\| = 0,$$
$$\lim_{\substack{\tau\to 0\\ \tau>0}} \left\| P^{(\lambda)}_{|1}(\pm\tau,0)F - P^{(\lambda)}_{|1}(0,0)F \pm 2\pi(\lambda(x)\cdot\nu(x))F \right\| = 0,$$
$$\lim_{\substack{\tau\to 0\\ \tau>0}} \left\| P^{(\lambda)}_{|2}(\pm\tau,0)F - P^{(\lambda)}_{|2}(0,0)F \mp 2\pi(\lambda(x)\cdot\nu(x))F \right\| = 0,$$

$$\lim_{\substack{\tau\to 0\\ \tau>0}} \left\| P^{(\lambda)}(\tau,0)F - P^{(\lambda)}(-\tau,0)F \right\| = 0,$$
$$\lim_{\substack{\tau\to 0\\ \tau>0}} \left\| P^{(\lambda)}_{|1}(\tau,0)F - P^{(\lambda)}_{|1}(-\tau,0)F + 4\pi(\lambda(x)\cdot\nu(x))F \right\| = 0,$$
$$\lim_{\substack{\tau\to 0\\ \tau>0}} \left\| P^{(\lambda)}_{|2}(\tau,0)F - P^{(\lambda)}_{|2}(-\tau,0)F - 4\pi(\lambda(x)\cdot\nu(x))F \right\| = 0,$$

Table 1 Classical limit and jump relations of potential theory for $F \in C^{(0,\mu)}(\Sigma)$, respectively $F \in \mathrm{L}^2(\Sigma)$. The norm $\|\cdot\|$ indicates the Hölder-norm, the $\mathrm{C}^{(0)}(\Sigma)-$norm or the $\mathrm{L}^2(\Sigma)-$norm. For proofs of these relations see Freeden and Kersten (1980) and Freeden and Kersten (1981).

$$\begin{aligned}
\Phi^5_\tau(x,y) &= \frac{1}{4\pi(\lambda(x)\cdot\nu(x))}\left(\frac{(x+\tau\lambda(x)-y)\cdot\lambda(x)}{|x+\tau\lambda(x)-y|^3} - \frac{(x-\tau\lambda(x)-y)\cdot\lambda(x)}{|x-\tau\lambda(x)-y|^3}\right),\\
\Phi^6_\tau(x,y) &= \frac{1}{4\pi(\lambda(x)\cdot\nu(x))}\left(\frac{(x+\tau\lambda(x)-y)\cdot\lambda(y)}{|x+\tau\lambda(x)-y|^3} - \frac{(x-\tau\lambda(x)-y)\cdot\lambda(y)}{|x-\tau\lambda(x)-y|^3}\right),\\
\Psi^5_\tau(x,y) &= \frac{-\tau}{4\pi(\lambda(x)\cdot\nu(x))}\left(\frac{1}{|x-\tau\lambda(x)-y|^3} + \frac{1}{|x-\tau\lambda(x)-y|^3}\right)\\
&\quad + \frac{3\tau}{4\pi(\lambda(x)\cdot\nu(x))}\left(\frac{((x+\tau\lambda(x)-y)\cdot\lambda(x))^2}{|x+\tau\lambda(x)-y|^5} + \frac{((x-\tau\lambda(x)-y)\cdot\lambda(x))^2}{|x-\tau\lambda(x)-y|^5}\right),\\
\Psi^6_\tau(x,y) &= \frac{-\tau}{4\pi(\lambda(x)\cdot\nu(x))}\left(\frac{\lambda(x)\cdot\lambda(y)}{|x+\tau\lambda(x)-y|^3} + \frac{\lambda(x)\cdot\lambda(y)}{|x-\tau\lambda(x)-y|^3}\right)\\
&\quad + \frac{3\tau}{4\pi(\lambda(x)\cdot\nu(x))}\left(\frac{((x+\tau\lambda(x)-y)\cdot\lambda(x))((x+\tau\lambda(x)-y)\cdot\lambda(y))}{|s+\tau\lambda(x)-y|^5}\right.\\
&\quad \left. + \frac{((x-\tau\lambda(x)-y)\cdot\lambda(x))((x-\tau\lambda(x)-y)\cdot\lambda(y))}{|x-\tau\lambda(x)-y|^5}\right),
\end{aligned}$$

Table 2 Explicit terms of the oblique $\Sigma-$scaling functions and wavelet functions defined in Theorem 4.1 and equation (2). In the approach presented here if restricted to the sphere, elementary representations in explicit form and spectral representations in terms of spherical harmonics are available for the $\Sigma-$scaling functions and the $\Sigma-$wavelets.

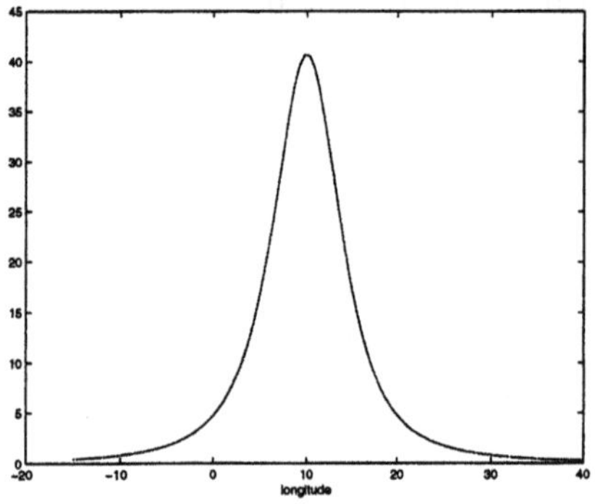

(a) scaling function

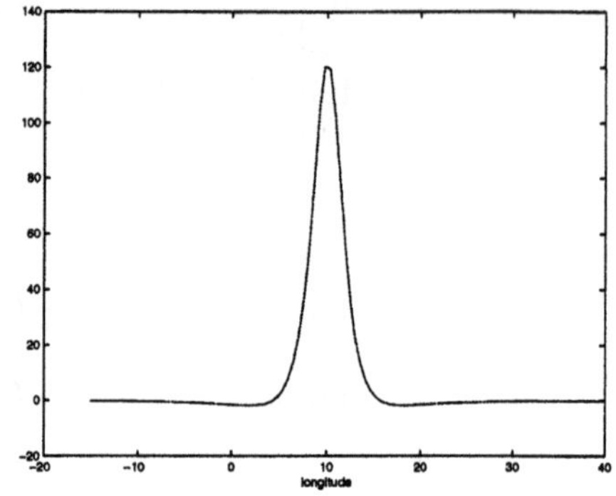

(b) wavelet function

Figure 1 $\Sigma-$scaling function Φ_{τ}^{6} and $\Sigma-$wavelet–function Ψ_{τ}^{6} (sectional illustration) for $\tau = 2^{-4}$.

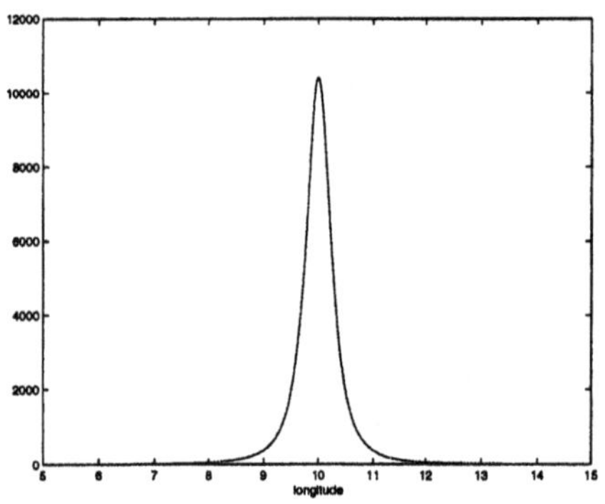

(a) scaling function

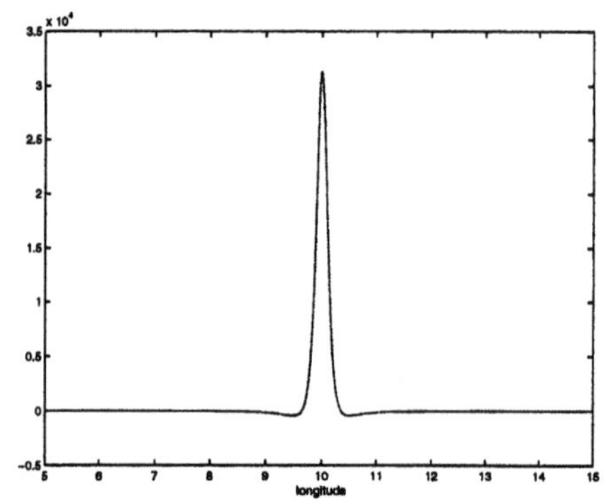

(b) wavelet function

Figure 2 Oblique $\Sigma-$scaling function Φ_{τ}^{6} and $\Sigma-$wavelet–function Ψ_{τ}^{6} (sectional illustration) for $\tau = 2^{-8}$. Comparing to Figure 1 the property of space localization without the appearance of any oszillations which is a main disadvantage of some wavelet functions can clearly be seen.

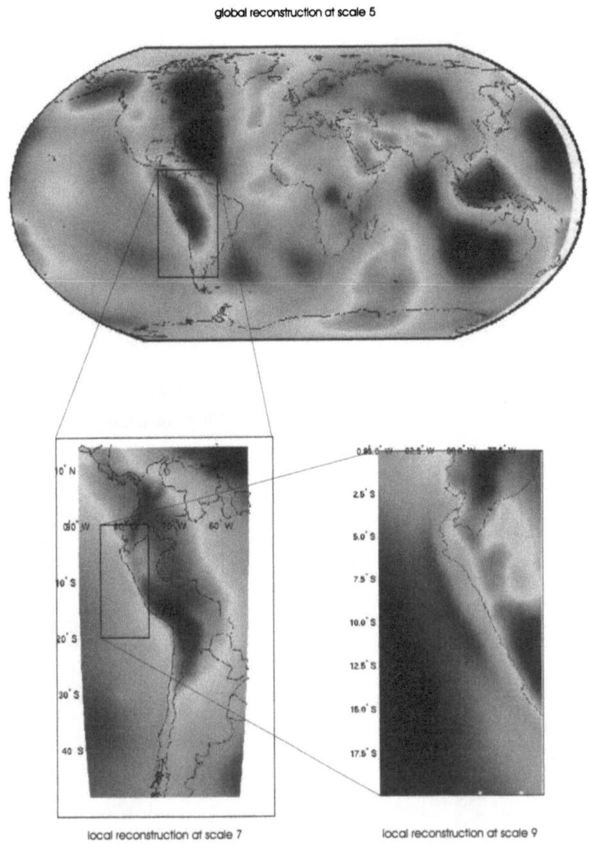

Figure 3 Illustration of the zoom–in property. In order to reconstruct a function on a local area, only data in a certain neighborhood of this area are used. Since global high–scale reconstruction of fine structure is very time–consuming, only the area of interest is reconstructed which can be done with a considerably fewer effort.

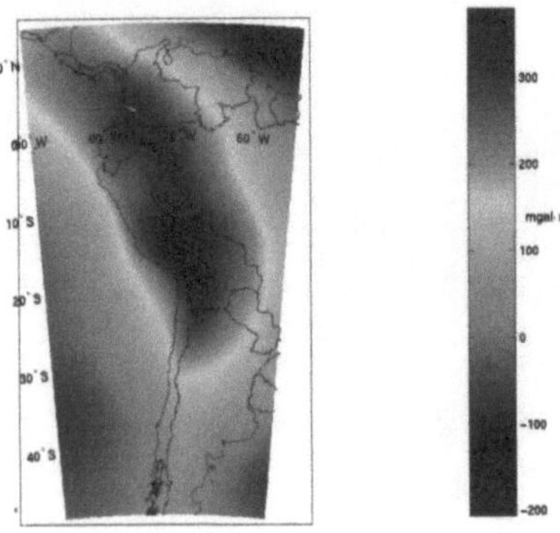

(a) local reconstruction with $\Sigma-$scaling function $\Phi^6_{\tau_j}$ at scale $j = 5$

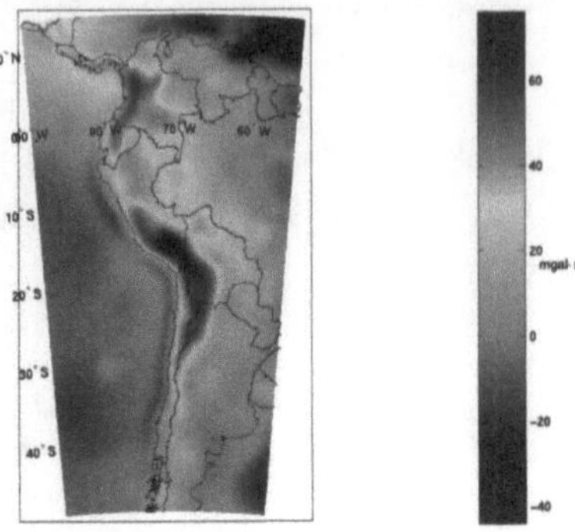

(b) local reconstruction with $\Sigma-$wavelet function $\Psi^6_{\tau_j}$ at scale $j = 5$

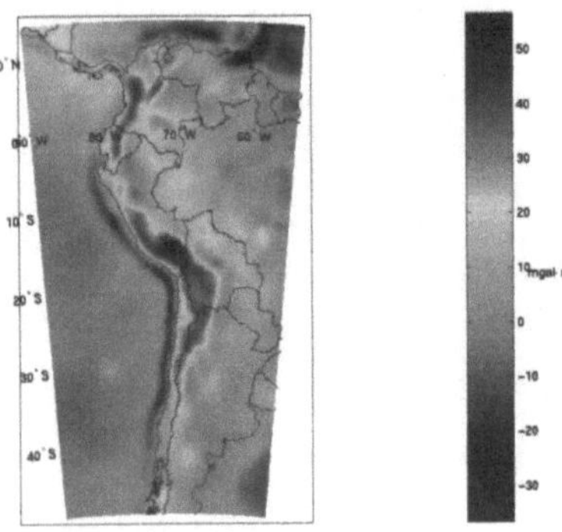

(c) local reconstruction with $\Sigma-$wavelet function $\Psi^6_{\tau_j}$ at scale $j = 6$

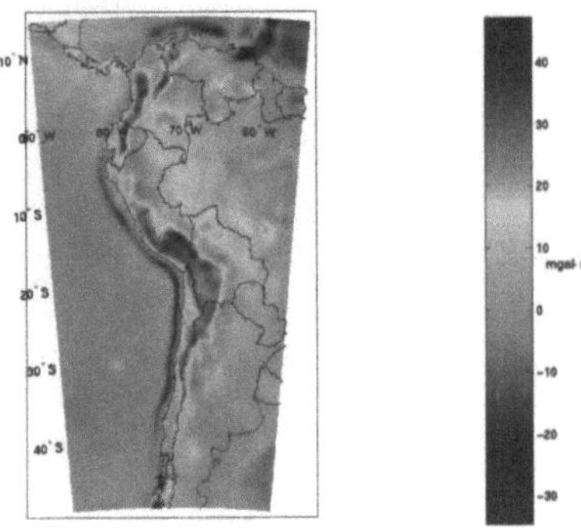

(d) local reconstruction with $\Sigma-$wavelet function $\Psi^6_{\tau_j}$ at scale $j = 7$

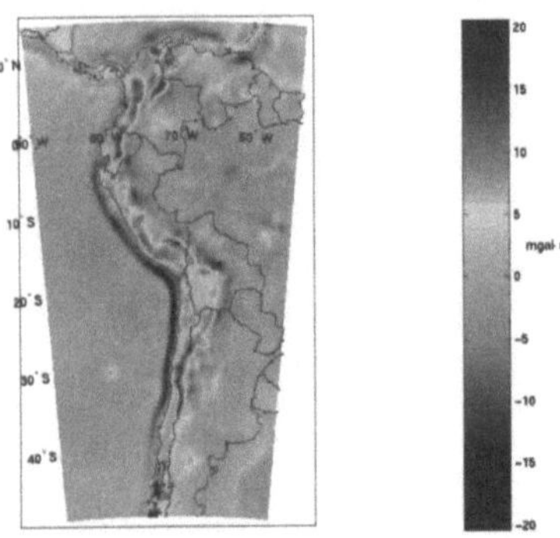

(e) local reconstruction with $\Sigma-$scaling function $\Psi^6_{\tau_j}$ at scale $j = 8$

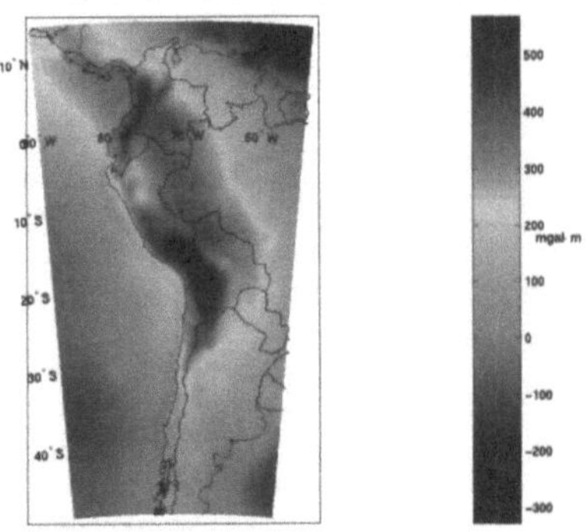

(f) local reconstruction with $\Sigma-$scaling function $\Phi^6_{\tau_j}$ at scale $j = 5$

Figure 4 Detection of high frequency perturbation within a local area of the EGM96–geopotential model. The buried mass point at 80° West, 30° South is clearly detected, especially in the wavelet reconstruction at scale 8.

Fast spherical harmonic synthesis and co-synthesis with applications in gravity field modeling

P. Ditmar, R. Klees, and F. Kostenko
Department of Physical, Geometrical and Space Geodesy (FMR)
Delft University of Technology, 2629 JA Delft, Thijsseweg 11, The Netherlands

Abstract. Fast algorithms for spherical harmonic synthesis and co-synthesis have been developed. Synthesis and co-synthesis are defined as the application to a vector of the design matrix (matrix of partial derivatives, Jacobian matrix) and of the transposed design matrix, respectively. The algorithms are applicable to a variety of functional models including the cases of irregularly distributed data points, arbitrary sensor orientations, and data contaminated by colored noise. The performance of the proposed algorithms is of the order of $O(N)$ for large number N of data points. The fast synthesis and co-synthesis algorithms can be efficiently used for data simulation and, in combination with the pre-conditioned conjugate gradient method (PCCG), for the least-squares inversion of data into model parameters. Importantly, the PCCG method allows one to avoid an assembling of the normal matrix. If, however, the normal matrix has to be computed explicitly, the application of the proposed algorithms can also be beneficial. With the fast synthesis and co-synthesis, the number of operations required for the computation of the normal matrix is of the order of $O\left(NL_{max}^2\right)$ for large N, where L_{max} is the maximum degree of the spherical harmonic expansion. The explicitly computed normal matrix may be used, in particular, for the combination of the estimated model parameters with other data sets (possibly, to be acquired later).

Keywords. Spherical harmonics, gravity field, co-synthesis, normal matrix, data combination.

1 Introduction

Spherical harmonics are widely used base functions to parameterize models of different physical phenomena. In particular, in Earth's sciences they are frequently utilized to represent the global gravity field (Heiskanen and Moritz, 1984) and magnetic field (Sabaka et.al., 2000). Therefore, fast algorithms for carrying out various computations in the context of such models can be broadly applied. Many of such computations include the following two operations: (i) spherical harmonic synthesis, i. e. the multiplication of the design matrix (other names: matrix of partial derivatives; Jacobian matrix) to a vector; (ii) spherical harmonic co-synthesis, i. e. the multiplication of the transposed design matrix to a vector. A fast implementation of these operations (especially, of the fast co-synthesis) in many situations is not obvious (think of an arbitrary distribution of data points, arbitrary sensor orientation at each point, incomplete list of measured components in case of vector/tensor measurements).

Development of fast synthesis and co-synthesis algorithms was originally motivated by inversion of satellite gravity gradiometry (SGG) data to be acquired during the GOCE mission (ESA, 1999). A fast algorithm for the spherical harmonic synthesis in the context of SGG data has been proposed by Koop et al. (2000) and for a fast spherical harmonic co-synthesis by Ditmar et al. (2003). With the present publication, we would like to emphasize that the spectrum of possible applications of the developed algorithms is, in fact, much broader, both in terms of data types and the computation goals. We also consider some practical aspects of the fast spherical harmonic synthesis and co-synthesis such as the numerical complexity, parallelization issues, and the accuracy.

2 Applications of fast spherical harmonic synthesis and co-synthesis

One of the tasks one frequently encounters in the context of spherical harmonics is the data simulation, i. e. the computation of a linear combination of spherical harmonics at a set of points. This task can be directly fulfilled by means of the spherical harmonic synthesis. An example is the computation of the gravitational potential:

$$V(r,\theta,\lambda) = \frac{GM}{R} \sum_{l=L_{min}}^{L_{max}} \left(\frac{R}{r}\right)^{l+1} \times$$

$$\sum_{m=0}^{l} \{C_{lm} \cos m\lambda + S_{lm} \sin m\lambda\} P_{lm}(\cos\theta), \quad (1)$$

where r, θ, λ are the geocentric coordinates (radial distance, co-latitude, and longitude, respectively); G is the universal gravity constant; M is the Earth's mass; R is the semi-major axis of a reference ellipsoid of revolution; $P_{lm}(\cos\theta)$ are the 4π-normalized (or fully normalized) associated Legendre functions; C_{lm}, S_{lm} are 4π-normalized potential coefficients; L_{min} and L_{max} are the minimum and the maximum degree of the spherical harmonic expansion, respectively. Eq. (1) can be re-written in matrix notation as

$$\mathbf{d} = \mathbf{A}\mathbf{x}, \tag{2}$$

where $\mathbf{x}$ is the vector of potential coefficients, $\mathbf{A}$ is the design matrix, and $\mathbf{d}$ is the vector of computed values of the gravitational potential.

Another frequently encountered task is the inversion, i. e. the estimation of the potential coefficients from observations at a set of points in space. For instance, the least-squares solution is obtained as the solution of the system of normal equations:

$$\mathbf{N}\mathbf{x} = \mathbf{y}, \tag{3}$$

where $\mathbf{y}$ is the right-hand side vector:

$$\mathbf{y} = \mathbf{A}^T\mathbf{C_d}^{-1}\mathbf{d}, \tag{4}$$

$\mathbf{N}$ is the normal matrix:

$$\mathbf{N} = \mathbf{A}^T\mathbf{C_d}^{-1}\mathbf{A} + \alpha\mathbf{R}, \tag{5}$$

$\mathbf{C_d}$ is the noise covariance matrix, α is the *regularization parameter*, and $\mathbf{R}$ is the *regularization matrix*. It is not uncommon that noise in the observations is colored (i. e. frequency-dependent), so that the matrix $\mathbf{C_d}$ is non-diagonal.

If the number of data and unknown parameters is large, the explicit computation of the normal matrix may be very time-consuming. An alternative is to solve the system of normal equations iteratively, e. g. by means of the method of conjugate gradients with pre-conditioning (PCCG) (Hestenes & Stiefel, 1952; Bertsekas, 1982). At each iteration of this method, one should apply the normal matrix $\mathbf{N}$ to a certain vector $\mathbf{q}$, which can be done, according to Eq. (5), as a sequence of matrix-to-vector multiplications:

$$\mathbf{N}\mathbf{q} = \mathbf{A}^T(\mathbf{C_d}^{-1}(\mathbf{A}\mathbf{q})) + \alpha\mathbf{R}\mathbf{q}. \tag{6}$$

Thus, an explicit computation of the normal matrix is avoided. The key operations to be performed in this approach are: (1) The application of the design matrix to the vector $\mathbf{q}$, which is nothing but the spherical harmonic synthesis; (2) The application of the matrix $\mathbf{C_d}^{-1}$ to the result of synthesis; this operation can be efficiently implemented as a filtering if the data noise is stationary (Schuh, 1996; Ditmar and Klees, 2002; Klees et al. 2003a; Klees and Ditmar, 2003); (3) The application of the transposed design matrix to the result of filtering, i. e. the *spherical harmonic co-synthesis*.

Though the normal matrix is not needed for the data inversion, there might be some reasons to obtain this matrix explicitly. For instance, one may be interested in the covariance matrix of the estimated model parameters:

$$\mathbf{C_x} = \mathbf{N}^{-1}\left(\mathbf{A}^T\mathbf{C_d}^{-1}\mathbf{A}\right)\mathbf{N}^{-1}. \tag{7}$$

Another likely goal is the joint inversion of the given data set with other sets of data, possibly to be collected in the future. In principle, a joint inversion assumes that the following system of equations has to be solved:

$$\begin{cases} \mathbf{A_1}\mathbf{x} = \mathbf{d_1} \\ \mathbf{A_2}\mathbf{x} = \mathbf{d_2}, \end{cases} \tag{8}$$

where $\mathbf{d_1}$ and $\mathbf{d_2}$ are two data sets to be considered simultaneously; $\mathbf{A_1}$ and $\mathbf{A_2}$ are the corresponding design matrices. In practice, however, usage of the system 8) may be inconvenient. An alternative is to invert the data sequentially: first, one computes the solution using the first data set: $\mathbf{x_1} = \mathbf{N_1}^{-1}\mathbf{A_1}^T\mathbf{C_{d_1}}\mathbf{d_1}$, where $\mathbf{N_1} = \mathbf{A_1}^T\mathbf{C_{d_1}}^{-1}\mathbf{A_1} + \alpha\mathbf{R}$ and $\mathbf{C_{d_1}}$ is the covariance matrix of the first data set. Secondly, the joint solution is obtained from the following system of linear equation:

$$\begin{cases} \mathbf{A_2}\mathbf{x} = \mathbf{d_2} \\ \mathbf{x} = \mathbf{x_1}, \end{cases} \tag{9}$$

provided that the covariance matrix assigned to the "pseudo-observations" $\mathbf{x_1}$ is set equal to $\mathbf{N_1}^{-1}$. Assuming that $\mathbf{C_{d_2}}$ is the covariance matrix of the second data set, we can write the least-square solution of the system (9) as:

$$\begin{aligned} \mathbf{x} = &\left(\mathbf{A_2}^T\mathbf{C_{d_2}}^{-1}\mathbf{A_2} + \mathbf{N_1}\right)^{-1} \times \\ &\left(\mathbf{A_2}^T\mathbf{C_{d_2}}^{-1}\mathbf{d_2} + \mathbf{N_1}\mathbf{x_1}\right) = \\ &\left(\mathbf{A_2}^T\mathbf{C_{d_2}}^{-1}\mathbf{A_2} + \mathbf{N_1}\right)^{-1} \times \\ &\left(\mathbf{A_2}^T\mathbf{C_{d_2}}^{-1}\mathbf{d_2} + \mathbf{A_1}^T\mathbf{C_{d_1}}^{-1}\mathbf{d_1}\right), \end{aligned} \tag{10}$$

which coincides with the (regularized) least-square solution of the original system (8). Naturally, one may repeat such a procedure every time as another new set of data becomes available.

Thus, a fast computation of the normal matrix may be essential. With the fast synthesis and co-synthesis procedures, one can do so column-by-column. In order to compute the m-th column, one has to define a unit vector $\mathbf{e_m}$: all the elements of this vector are equal to 0 except for the element at the position m, this element being equal to 1. A sequential application of the synthesis, filtering, and co-synthesis to this vector results in the m-th column of the (unregularized) normal matrix:

$$\{\mathbf{N}^{(\mathbf{unreg})}\}_m = \mathbf{A}^T(\mathbf{C_d}^{-1}(\mathbf{A}\mathbf{e_m})). \qquad (11)$$

3 Methodology of the fast synthesis and co-synthesis

Gridded data. Fast spherical synthesis and co-synthesis are especially simple if the data points are located at the nodes of a regular spherical grid. Let us return to the computation of the gravitational potential. By interchanging the order of the summation over l and m in Eq. (1), we obtain

$$V(r,\theta,\lambda) = \qquad (12)$$

$$\sum_{m=0}^{L_{\max}} \left(\hat{C}_m(r,\theta)\cos m\lambda + \hat{S}_m(r,\theta)\sin m\lambda\right),$$

with radius- and co-latitude-dependent coefficients $\hat{C}_m(r,\theta)$ and $\hat{S}_m(r,\theta)$ defined as:

$$\left\{ \begin{array}{c} \hat{C}_m(r,\theta) \\ \hat{S}_m(r,\theta) \end{array} \right\} = \frac{GM}{R} \times \qquad (13)$$

$$\sum_{l=\max(m,L_{\min})}^{L_{\max}} \left\{ \begin{array}{c} C_{lm} \\ S_{lm} \end{array} \right\} \left(\frac{R}{r}\right)^{l+1} P_{lm}(\cos\theta).$$

It is now obvious that an efficient way to perform the space-wise synthesis on a spherical grid is to consider circular grid lines situated at different latitudes and elevations one by one. For each circular line one should (1) compute the set of Legendre functions $P_{lm}(\cos\theta)$, for which purpose a recursive approach can be efficiently used (Rummel et.al, 1993); (2) calculate the coefficients $\hat{C}_m(r,\theta)$ and $\hat{S}_m(r,\theta)$; and (3) find the values of the gravitational potential along the circle by means of formula (12), which can be efficiently implemented as an inverse FFT.

Let us consider now the co-synthesis, i. e. the transposed synthesis. The transposition operation changes a summation over a certain index into just a loop, in which this index plays the role of the loop variable. Vice versa, each loop is replaced by the summation. Therefore, the synthesis formula (13) can be transformed into the following expression for the co-synthesis:

$$\left\{ \begin{array}{c} c_{lm}(r,\theta) \\ s_{lm}(r,\theta) \end{array} \right\} = \frac{GM}{R}\left(\frac{R}{r}\right)^{l+1} \times \qquad (14)$$

$$P_{lm}(\cos\theta) \sum_{i=0}^{N_\lambda - 1} \left\{ \begin{array}{c} \cos im\Delta\lambda \\ \sin im\Delta\lambda \end{array} \right\} y(r,\theta,i\,\Delta\lambda),$$

where we take into account that longitudes are considered with a constant step (denoted as $\Delta\lambda$); N_λ is the total number of longitudinal lines in the grid; and $y(r,\theta,i\,\Delta\lambda)$ are the input data for the co-synthesis on the grid. The expression (14) allows for an efficient implementation of the co-synthesis, which is somewhat similar to that of the synthesis. The latitudinal grid lines should be considered one by one at all latitudes and radii r; FFT and computation of Legendre functions should be carried out for each grid line. The final result of the co-synthesis is obtained by the summation of coefficients $(c_{lm}(r,\theta), s_{lm}(r,\theta))$ over all latitudes and radii.

Obviously, the FFT at a certain grid line inside synthesis and co-synthesis takes $O(N_\lambda \log N_\lambda)$ operations, whereas other steps require $O(L_{max})^2$ operations. Assuming that the number of grid nodes in the latitudinal and the longitudinal direction is of the same order, the total number of operations during the synthesis and co-synthesis on the grid can be estimated as $O\left((N_\lambda \log N_\lambda + L_{max}^2)N_\lambda N_r\right)$, where N_r is the number of grid nodes in the radial direction. In particular, $N_r = 1$ if all the observation points lie on a sphere or spheroid (naturally, the grid should be spheroidal rather than spherical in the latter case).

Arbitrarily distributed data. If the data to be inverted are distributed in the 3-D space in an arbitrary way, it is sometimes possible to reduce them to nodes of a regular grid during the pre-processing. In many cases, however, such an approach is not desirable. For example, such a reduction can be difficult due to colored noise in the data or because of a varying sensor orientation (see the next paragraph). Then, the true distribution of observation points should be incorporated into the functional model. In this case, the synthesis on a regular grid should be supplied with an interpolation procedure. It is reasonable to choose the interpolation method such that (1) the result of interpolation is a smooth function, (2) the interpolation is "local", so that only a limited number of nodes in the vicinity of the given point have to be considered, and (3) interpolated values depend on values at grid nodes linearly. A suitable interpolation method, which matches these requirements, is Overhauser spline interpolation (Overhauser, 1968; Klees, 1999). The interpolation with 3-D Overhauser splines is based on $4 \times 4 \times 4$ grid nodes in the vicinity of the current point and results in a function with continuous first derivatives. Such an interpolation can be represented as a matrix-to-vector multiplication, where the interpolation matrix contains $4^3 = 64$ explicitly given non-zero values in each row (Ditmar et al., 2003).

Once the co-synthesis is the transposed synthesis, it can be adapted to an arbitrary distribution of observation points by incorporation of a *transposed* interpolation *before* the co-synthesis on the grid. The matrix describing the transposed interpolation is the transposed copy of the original interpolation matrix. For example, the matrix of the transposed 3-D Overhauser interpolation contains 64 non-zero values in each *column* (Ditmar et al., 2003).

Thus, the numerical complexity of an interpolation and of a transposed interpolation is of a same order. As long as the number of nodes involved in the interpolation at a single data point is very limited (that is always the case when a local interpolation method is used), the total number of operations required by the interpolation and the transposed interpolation can be assessed as $O(N)$, where N is the number of observation points.

Vector and tensor measurements. It is not uncommon that the data to be inverted represent vector or tensor components. Examples are components of the gravity vector and gravity gradients. If the orientation of the sensor varies and the list of measured components is incomplete, rotation of measured data into a uniform frame at the pre-processing stage can be impossible. Then, the functional model should be extended further in order to account for the true sensor orientation at each observation point. In terms of the synthesis, this can be achieved by (1) computing all the vector/tensor components at each observation point in a uniform frame, and (2) rotating the computed vector/tensor according to the true sensor orientation. It is reasonable to define the uniform frame as a local Cartesian frame one, e. g. as the frame with the X-axis pointing to the North, the Y-axis to the West, and the Z-axis upwards. Expressions for the derivatives of the original scalar value (e. g. of the gravitational potential) can be easily obtained analytically in such a frame. Naturally, the interpolation procedure should be applied to each vector/tensor component independently.

As soon as the vector/tensor at a given observation point has been rotated, only the components that have been actually measured there must be selected. Both the rotation and the component selection can be represented as matrix-to-vector multiplications. In particular, the component-selection matrix can be obtained from the unit one by removing the rows corresponding to the not measured components. Thus, the total synthesis procedure in case of vector/tensor measurements performed at arbitrary points with arbitrary sensor orientations can be represented by four matrix-to-vector multiplications:

$$\mathbf{Aq} = \mathbf{A_S}(\mathbf{A_R}(\mathbf{A_I}(\mathbf{A_G q}))), \qquad (15)$$

where $\mathbf{A_G}$ is the matrix describing the synthesis on a grid, $\mathbf{A_I}$ is the interpolation matrix, $\mathbf{A_R}$ is the rotation matrix, and $\mathbf{A_S}$ is the component selection matrix.

Then, the co-synthesis procedure can be represented as a sequence of similar matrix-to-vector multiplication provided that they are understood in the transposed sense and performed in the reverse order:

$$\mathbf{A}^T\mathbf{p} = \mathbf{A_G}^T(\mathbf{A_I}^T(\mathbf{A_R}^T(\mathbf{A_S}^T\mathbf{p}))), \qquad (16)$$

where $\mathbf{p}$ is the input vector for the co-synthesis. In this case, the co-synthesis should start from the *transposed component selection*. The matrix for the transposed component selection at a given point can be obtained from the unit matrix by removing the *columns* corresponding to the not measured components. It is easy to see that the 'transposed component selection' means a restoration of the full vector/tensor in such a way that the measured components are taken from the input vector/tensor and those not measured are set equal to zero (Ditmar et al., 2003). Next, the *transposed rotation* has to be applied, which is nothing other than the rotation in the opposite direction (from the sensor-related frame to the uniform one). In case of vector measurements, this statement is obvious; one can show that the same holds for tensor measurements (Ditmar et al., 2003). Finally, the transposed interpolation and the co-synthesis on the grid have to be applied as outlined above.

As far as the computational complexity is concerned, the rotation and component selection in the synthesis and co-synthesis require not more than $O(N)$ operations.

4 Parallelization aspects

Though the proposed spherical harmonic synthesis and co-synthesis are usually much faster than those based on an explicit computation of the design matrix, one may wish to accelerate the computations even further by making use of a multi-processor computer. If, however, the data to be inverted are collected at arbitrarily distributed points and contaminated by colored noise, the optimal approach to the parallelization is not evident.

As was already mentioned, colored noise requires incorporation of filtering into the inversion procedure. The only practical approach to the parallelization of a filtering is the time-wise one: the data are arranged in the chronological order and then split into fragments, so that each processing element (PE) handles one fragment. The computations on a grid, however, require a space-wise parallelization, so that each PE considers its own interval of latitudes. The interpolation is the procedure where two types of information (the one parallelized in the time-wise fashion and the one parallelized in the space-wise fashion) should interact. At this point, a data transfer between processing elements is needed. There are two ways to organize this data transfer, depending on how the interpolation itself is parallelized: in the time-wise way or in the space-wise way.

Let us consider, for example, the synthesis. The time-wise parallelization of the interpolation requires that each PE is supplied in advance with a whole 3-D set of gridded values (or, at least, with one spherical layer from the 3-D grid). It means that each PE should compute gridded values in the interval of latitudes it is responsible for and then distribute these values over all other PE's. A disadvantage of this approach is a large data traffic between processing elements, which may easily become the dominant contributor to the overall time expenditures. In our practice, for example, we faced situations when the exchange of gridded values took up to 70% of the total time of the synthesis and co-synthesis. Moreover, these time expenditures do not decrease with increasing number of PE's, i. e. the parallelization becomes inefficient. Another drawback of such an approach is a relatively large memory needed to store sets of gridded values.

Another approach is to parallelize the interpolation in the space-wise fashion and then to distribute the results over PE's in accordance with the requirement of the filtering (i. e. in the chronological order). The data traffic in this case is usually much less because each data transfer operation involves only two PEs rather than all of them. Hence the efficiency of parallelization substantially increases. Besides, it is not necessary to store a whole 3-D grid in the computer memory at a time. Instead, one can arrange the observation points in compliance with their geographical locations, so that each time only a narrow interval of latitudes is under consideration. Then, the number of latitudinal grid lines to be stored in the computer memory at a time is determined only by the interpolation method. The Overhauser splines require, for example, only lines situated at four different latitudes.

The only disadvantage of the second approach is a necessity to develop a more complicated programming code in order to establish the necessary "book-keeping", which governs the data transfer.

5 Interpolation errors

Strictly speaking, the proposed algorithms for fast synthesis and co-synthesis are inexact. They substitute the true design matrix $\mathbf{A}$ with its approximation $\tilde{\mathbf{A}}$. The error $(\tilde{\mathbf{A}} - \mathbf{A})$ is caused by the interpolation and depends on the coarseness of the 3-D grid. How to choose the 3-D grid properly? If the computations are aimed at data simulation, so that only the fast synthesis algorithm is used, the answer is simple. One should compare results of the fast synthesis $\tilde{\mathbf{d}} = \tilde{\mathbf{A}}\mathbf{x}$ with those obtained in the traditional way $\mathbf{d} = \mathbf{A}\mathbf{x}$ (or with results of the fast synthesis based on a much finer 3-D grid). The difference $\delta\mathbf{d} = \tilde{\mathbf{d}} - \mathbf{d}$ must not exceed the affordable noise level in the data. As an example, we have investigated the accuracy of the fast synthesis in computations of the disturbing gravitational potential along the equator at zero elevation (R = 6378137 m). The disturbing potential is defined as the difference between the OSU91a model (Rapp et al. 1991) and the GRS80 normal field (Moritz, 1980). The considered grid is 2-D and is composed of cells with size $0.09° \times 0.09°$, the equator crossing the centers of the cells (thus, an interpolation in the vertical direction is excluded). The average error in terms of geoid heights turns out to be 0.2 mm; the maximum error (in the Caribean area) reaches 2.2 mm (Fig. **1**). Such an accuracy is sufficient for most of practical applications.

In case of data inversion, the model error $\delta\mathbf{x}$ caused by the interpolation may be assessed as follows. First of all, the synthesis error vector $\delta\mathbf{d}$ should be computed on the basis of the model obtained by the inversion. Then, the inversion must be repeated with the vector $\delta\mathbf{d}$ substituted for the data. The result of the last computation, up to the terms of the second order of smallness, just describes the propagation of interpolation errors into the model obtained by the inversion (Ditmar et al., 2003):

$$\delta\mathbf{x} = \mathbf{N}^{-1}\mathbf{A}^T\mathbf{C_d}^{-1}\delta\mathbf{d}. \qquad (17)$$

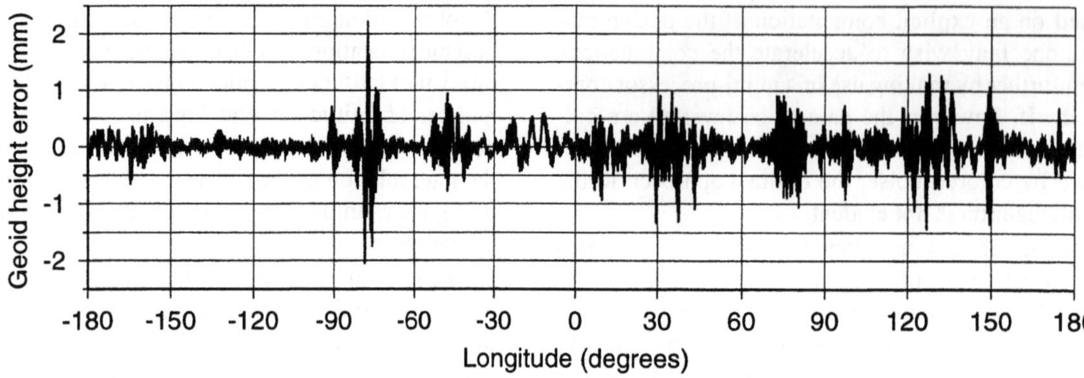

Fig 1. Errors of the fast synthesis in terms of geoid heights. A 2-D grid with cells of size 0.09° × 0.09° is used. Points at zero elevation along the equator are considered. The gravity is defined as difference between the OSU91a model and the GRS80 reference field.

The only necessary condition is that exactly the same 3-D grid is used in all the computations. It is reasonable to relate the selection of the 3-D grid to the data accuracy: the influence of interpolation can be considered as negligible if synthesis errors $\delta\mathbf{d}$ are well below the noise level in the data.

Finally, the fast synthesis and co-synthesis can be used, as was mentioned above, for the computation of the covariance matrix of the estimated parameters. We believe that the 3-D grid in this case can be coarsened, because usually it is sufficient to find the covariance matrix with a relatively low accuracy. A more in-depth consideration of this issue will be presented in a forthcoming paper.

6 Discussion and conclusions

Fast algorithms for spherical harmonic synthesis and co-synthesis have been developed. In general, they can be split into two major parts: (1) computations on a regular spherical grid and (2) other computations. The first part requires $O\left(\left(N_\lambda \log N_\lambda + L_{\max}^2\right) N_\lambda N_r\right)$ operations, where N_λ is the number of grid nodes in the horizontal directions (we do not distinguish between the number of points in the latitudinal and in the longitudinal directions because we assume that they are of the same order), N_r is the number of grid nodes in the vertical direction, and L_{max} is the maximum degree of the spherical harmonic expansion. The second part can be completed in $O\left(N\right)$ operations, where N is the number of observation points. Notice that we prefer to count observation points rather than data because the time of computations at each iteration of the PCCG method is independent of the number of vector/tensor components measured at one point. If the number of observation points is large, computations on a regular grid take relatively small time, so that the asymptotic estimation of total time expenditures becomes just $O\left(N\right)$. On the other hand, the straightforward approach, which is based on the explicit computation of the design matrix, requires $O\left(NL_{max}^2\right)$ operations. Thus, the gain in terms of the required CPU-time may be really dramatic. Furthermore, the proper (space-wise) parallelization of the interpolation procedures makes the fast synthesis and co-synthesis algorithms well suitable for multi-processor computers.

The speed-up offered by the proposed algorithms can be particularly important when the normal matrix is to be computed explicitly. A straightforward assembly of the normal matrix requires $O\left(NL_{max}^4\right)$ operations, so that it may be prohibitively time-consuming in the context of models with large L_{max}. Thanks to the fast synthesis and co-synthesis, the time required for this operation may be asymptotically estimated as only $O\left(NL_{max}^2\right)$. Therefore, it is believed that it will be a common practice in the future to distribute an obtained model of the Earth's gravity field (and of other global phenomena represented in terms of spherical harmonics) and the corresponding normal matrix together, even if L_{max} is of the order of several hundreds. In this way, the successors will be able to combine old models with new data set(s) in the optimal way.

The estimates for the computational performance of the fast synthesis and co-synthesis algorithms can be validated by simulations. As an example, let us consider the inversion of a simulated set of satellite gravity gradients into potential coefficients. The chosen scenario is in line with the future GOCE satellite mission. The set consists of 15 638 401 observation points (1-sec sampling × 6 month); the disturbing potential is defined as the difference between the OSU91a model complete up to degree 300 and the GRS80 normal gravity field. The hardware used is the SGI Origin 3800 super-computer. The size of the 3-D grid is set to $4096 \times 2048 \times 5$ nodes (size of a cell is about $0.09° \times 0.09° \times 12$km) with the middle spherical layer situated at the height of 260 km above the equator. In case of the time-wise parallelization of the interpolation procedure, the wall-clock time for the synthesis and co-synthesis is equal to 112 seconds and 147 seconds, respectively, provided that 8 processing elements are requested. Implementation of the space-wise parallelization of the interpolation procedure allowed us to reduce the time for the synthesis and co-synthesis to 27 seconds and 32 seconds, respectively. Increasing the number of processing elements to 32 reduces the time for synthesis and co-synthesis further, to 9 seconds and 12 seconds, respectively. Thanks to such a high performance, the total time expenditures of the data inversion do not exceed in the latter case 10 minutes (see Klees et al., 2003b). For comparison: the straightforward synthesis and co-synthesis take in the considered example about 20 hours each, provided that 8 processing elements are used.

The presented values can also be used for estimating the time T_N that would be needed for the computation of the full normal matrix. From Eq. (11) it follows that T_N is equal to the time of synthesis and co-synthesis together multiplied by the number of unknowns (the time required by the filtering is assumed to be negligible). In our case: $T_N \approx (9\,\mathrm{sec} + 12\,\mathrm{sec}) \times 300^2 = 1.89 \times 10^6\,\mathrm{sec} \approx 21\,\mathrm{days}$. This value does not seem unrealistic, especially taking into account that the number of processing elements can be increased to 64 or even to 128. Moreover, the computations may be speeded up even further thanks to a rapid progress of processing facilities.

7 Acknowledgments

The research was financially supported by the Netherlands Space Research Organization SRON (grant eo-028). Computing resources were provided by Stichting Nationale Computerfaciliteiten NCF (grant SG-027).

References

Bertsekas D.P. (1982). *Constrained optimization and Lagrange multiplier methods.* Academic Press, New York etc.

Ditmar P., Klees R. (2002). *A Method to Compute the Earth's Gravity Field from SGG / SST data to be Acquired by the GOCE Satellite.* Delft University Press.

Ditmar P., Klees R., Kostenko F. (2003). Fast and accurate computation of spherical harmonic coefficients from satellite gravity gradiometry data. *Journal of Geodesy*, 76, pp. 690-705.

ESA (1999). *Gravity field and steady-state ocean circulation missions. Reports for mission selection. The four candidate Earth explorer core missions, SP-1233(1).* European Space Agency.

Heiskanen W.A. and Moritz H. (1984). *Physical Geodesy.* Institute of Physical Geodesy, Technical University Graz, Austria.

Hestenes M.R. and Stiefel E. (1952). Methods of conjugate gradients for solving linear systems. *Journal of Research of the National Bureau of Standards*, 49, pp. 409-436.

Klees R. (1999). *Numerical analysis.* Lecture Notes, Faculty of Geodetic Engineering, Delft University of Technology.

Klees R. and Ditmar P. (2003). How to handle colored observation noise in large-scale least-squares problems in the presence of gaps? *This volume.*

Klees R., Ditmar P., Broersen P. (2003a). How to handle colored observation noise in large-scale least-squares problems. *Journal of Geodesy*, 76, pp. 629-640.

Klees R., Ditmar P., Kusche J. (2003b). Numerical techniques for large least-square problems with application to GOCE. *This volume.*

Koop R., Visser P., van den IJssel J., Bouman J., van Gelderen M. (2000). Simulation of gravity gradients, in: *GOCE End-to-end Closed Loop Simulation. Final report. NIVR NRT no. 2703 TU.* Instituut voor Aardgericht Ruimteonderzoek Delft, Technische Universiteit Delft. pp. 25-33.

Moritz H. (1980). Geodetic reference system 1980. *Bull. Géod*, 54, pp. 395-405.

Overhauser A. W. (1968). *Analytic definition of curves and surfaces by parabolic blending. Tech. Report No. SL68-40, Scientific research staff publication.* Ford Motor Company, Detroit.

Rapp R., Wang Y., Pavlis N. (1991). *The Ohio State 1991 geopotential and sea surface topography harmonic coefficient model.* Rep 410, Department of Geodetic Science and Surveying, The Ohio State University, Columbus.

Rummel R., van Gelderen M., Koop R., Schrama E., Sansò F., Brovelli M., Miggliaccio F., Sacerdote F. (1993). *Spherical Harmonic Analysis of Satellite Gravity Gradiometry.* Nederlands Geodetic Comission, Publications on Geodesy, New Series, Number 39, Delft, The Netherlands.

Sabaka T.J., Olsen N., Langel R.A. (2000). *A comprehensive model of the near-Earth magnetic field: Phase 3.* NASA, Goddard Space Flight Center.

Schuh W.-D. (1996). *Tailored numerical solution strategies for the global determination of the Earth's gravity field.* Mitteilungen der geodätischen Institute der Technischen Universität Graz. Folge 81, Graz.

Fixed effects, random effects and the mixed model: new results of geodetic estimation and prediction theory

E.W. Grafarend
Department of Geodesy and GeoInformatics, University of Stuttgart
Geschwister-Scholl-Strasse 24D, 70174 Stuttgart, Germany

Abstract

Under the topic "State-of-the Art" we refer to recent topics of geodetic estimation and prediction theory: Integer Least-Squares for GPS Ambiguity Resolution, fuzzy methods, interval analysis, Thykonov-Phillips fixed effect regularization, robust estimations, reweighted Least-Squares, random tensor prediction and hypothesis testing. *Here* we focus on linear and nonlinear models of type

- fixed effects
  (α-weighted BLE)
  Example: (i) generalized Thykonov-Phillips regularization
  (ii) 3d datum transformation and the Procrustes Algorithm
- random effects
  (Bayesian estimation)
  MSPE{ $\tilde{\mathbf{z}}$ }
  Example: bias in nonlinear error propagation
- random coefficient model
  (errors-in-variables)
  Example: straight line fit by *Total Least-Squares*
- mixed model
  estimates and predictors
  Kolmogorov-Wiener prediction with trend, non-parametric prediction
  Example: Collocation

Some Generalized Equivalence Theorems for Least-Squares Adjustment

Burkhard Schaffrin
Dept. of Civil and Environmental Engineering and Geodetic Science
The Ohio State University, Columbus/OH 43210-1275, USA

Abstract. In 1986 an equivalence theorem had been published by B. Schaffrin and E. Grafarend that allows to check whether any form of differenced GPS data still carries the original information insofar as the Least-Squares Solution (LESS) for the parameters of interest remains the same. In its simplest form a (linearized) *Gauss-Markov Model* is compared with a corresponding *Model of Condition Equations*, after eliminating all parameters. The theorem then states that the residual vector from a least-squares adjustment comes out *identical* in both models if and only if two relations hold true, one *orthogonality* and one *rank-additivity* relation. Without the latter, we would have to admit a loss of information that usually results in a biased LESS.
In this contribution we shall generalize the existing equivalence theorem to cases that include (fixed or stochastic) constraints, random effects parameters and, in particular, the *Dynamic Linear Model* where the LESS amounts to "Kalman filtering", with substantial benefits in some GPS ambiguity resolution problems.

---

1 Introduction and Review of the Standard Theorem

It has been a long geodetic tradition to seek the computationally most efficient way to adjust a given data set in the least-squares sense. First comparisons of competing models for this purpose can already be found in the second (and third) edition of the respective monography by *F.R. Helmert (1907)*, based on what is known today as "Helmert's Knack". We shall apply (and explain) this little trick later in the context of the *Random Effects Model* and, in particular, the *Dynamic Linear Model.*
Another long-lasting debate has been on the use of "observation equations" (i.e., the Gauss-Markov Model) versus " condition equations ", or perhaps " condition equations with parameters " (i.e., the Gauss-Helmert Model). Beside the computational burden, it also matters that the adjustment result itself –in form of the residual vector, for instance,– *is not affected* by the chosen model. One of the first systematic investigations into this "equivalence" – after a preliminary study by *J.K.Baksalary and R. Kala (1981)* –was done by *J.K.Baksalary (1984)* from a statistical perspective, and shortly thereafter by *B. Schaffrin and E. Grafarend (1986)* for geodetic application in the context of the Global Positioning System (GPS).
For the *standard result* we assume a Gauss-Markov Model in linear(ized) form, namely

$$y = \underset{n\times m}{A}\,\xi + e\,,\ e \sim \left(0, \sigma_0^2 P^{-1}\right), \tag{1.1a}$$

$$q := rkA \le \min\{m, n\}, \tag{1.1b}$$

where y is the $n\times 1$ vector of observational increments, ξ the $m\times 1$ vector of parameter increments (unknown), e the $n\times 1$ random error vector (unknown), A the $n\times m$ matrix of coefficients, q its rank; e has zero expectation, P^{-1} as $n\times n$ cofactor matrix, P as $n\times n$ weight matrix, and σ_0^2 as variance component (unknown).
After some *linear transformation* with an $r\times n$ matrix B, we obtain the *equivalent* model of "condition equations"

$$w := By = Be\,,\ rkB = r\,,\ e \sim \left(0, \sigma_0^2 P^{-1}\right), \tag{1.2}$$

if and only if the following two relations hold true:

$$BA = 0 \ \text{ and } \ rkA + rkB = n\,; \tag{1.3}$$

$r = n - q$ denotes the "redundancy" (or "degrees of freedom") in the respective models. The proof can be based on the "*decomposition of the identity*" into the sum of two complementary projections

$$I_n = A\left(A^T PA\right)^{-} A^T P + P^{-1}B^T\left(BP^{-1}B^T\right)^{-1} B \tag{1.4}$$

which is valid independently of the g-inverse $(A^T PA)^-$. As a result, we arrive at two representations for both, the *residual vector*

$$\tilde{e} = [I_n - A(A^T PA)^- A^T P]y \quad (1.5a)$$
$$= P^{-1}B^T(BP^{-1}B^T)^{-1}w \quad (1.5b)$$

and – less trivially – the *weighted sum of squared residuals*

$$\Omega := \tilde{e}^T P\tilde{e} = y^T Py - y^T PA(A^T PA)^- A^T Py \quad (1.6a)$$
$$= w^T(BP^{-1}B^T)^{-1}w = (B\tilde{e})^T(BP^{-1}B^T)^{-1}(B\tilde{e}). \quad (1.6b)$$

Particularly (1.6b) permits some reduction of the computational burden whenever $r=rkB$ is comparatively small in view of $q=rkA$. For the derivation of (1.6a), note that $A^TA(A^TPA)^-A^TP=A^TP$ for any g-inverse $(APA)^-$.

Instead of eliminating all parameters at once from the Gauss-Markov Model (1.1) prior to the adjustment let us now consider the case where only the so-called "*nuisance parameters*" are eliminated right away. We thus partition the original model into

$$y = \underset{n\times(m-l)}{A_1}\xi_1 + \underset{n\times l}{A_2}\xi_2 + e,\ e \sim (0, \sigma_0^2 P^{-1}), \quad (1.7a)$$
$$rkA_2 = l < q := rk[A_1, A_2] \le \min\{m, n\}, \quad (1.7b)$$

and transform it into an *equivalent* Gauss-Helmert Model

$$w_2 := \underset{(r+q-l)\times n}{B_2}y = (B_2A_1)\xi_1 + B_2e, , \quad (1.8a)$$
$$rkB_2 = r+q-l,\ e \sim (0, \sigma_0^2 P^{-1}), \quad (1.8b)$$

via a multiplication with a matrix B_2 *if and only if* the two relations

$$B_2A_2 = 0 \text{ and } rkA_2 + rkB_2 = n \quad (1.9)$$

hold true. Obviously, this implies that the rank of the $(n-l)\times(m-l)$ matrix B_2A_1 – but not necessarily that of A_1 itself – will be

$$rk(B_2A_1) = q - l = rk[A_1, A_2] - rkA_2 \le rkA_1 \quad (1.10)$$

with some of its columns potentially also vanishing. This situation is slightly more general than in the case of *B. Schaffrin and E. Grafarend (1986, Corollary 1.2)* where A_1 and A_2 were assumed to be complementary with $rk[A_1, A_2] = rkA_1 + rkA_2$.

The proof is again very similar and will lead to the following representations of the *residual vector*

$$\tilde{e} = [I_n - A(A^T PA)^- A^T P]y \quad (1.11a)$$

$$= P^{-1}B_2^T(B_2P^{-1}B_2^T)^{-1}B_2 \cdot \{I_n - A_1[A_1^TB_2^T(B_2P^{-1}B_2^T)^{-1}B_2A_1]^- \cdot A_1^TB_2^T(B_2P^{-1}B_2^T)^{-1}B_2\}y \quad (1.11b)$$

$$= P^{-1}\{I_n - B_2^T(B_2P^{-1}B_2^T)^{-1}B_2A_1 \cdot [A_1^TB_2^T(B_2P^{-1}B_2^T)^{-1}B_2A_1]^- A_1^T\}B_2^T \cdot (B_2P^{-1}B_2^T)^{-1}w_2 \quad (1.11c)$$

first, and then of the *weighted sum of squared residuals*

$$\Omega := \tilde{e}^T P\tilde{e} = y^T Py - y^T PA(A^T PA)^- A^T Py \quad (1.12a)$$
$$= w_2^T(B_2P^{-1}B_2^T)^{-1}\{I_{r+q-l} - B_2A_1 \cdot [A_1^TB_2^T(B_2P^{-1}B_2^T)^{-1}B_2A_1]^- A_1^TB_2^T(B_2P^{-1}B_2^T)^{-1}\}w_2 \quad (1.12b)$$
$$= (B_2\tilde{e})^T(B_2P^{-1}B_2^T)^{-1}(B_2\tilde{e}). \quad (1.12c)$$

The relations (1.12 a-c) are crucial – and by no means trivial – in reducing the computational costs. We ought to keep in mind, however, that the last identity does not hold for any arbitrary e since, in general, we find

$$e^T Pe \ne e^T B_2^T(B_2P^{-1}B_2^T)^{-1}B_2 e \text{ if } e \notin R(P^{-1}B_2). \quad (1.13)$$

In the following section we shall generalize the above equivalence statements to also cover models with *(stochastic or fixed) constraints*. Having established suitable conditions for this case, we shall apply "Helmert's Knack" to the *Random Effects Model* in order to give the prior information a form similar to stochastic constraints. The ensuing results will then be used to investigate the respective equivalence theorems for the *Dynamic Linear Model* and, consequently, Kalman filtering,

2 The Gauss-Markov Model with Stochastic Constraints

This case can be considered as straightforward extension of the simple Gauss-Markov Model by writing the respective *GM-Model with stochastic constraints* in partitioned form

$$\begin{bmatrix} y \\ z_0 \end{bmatrix} = \begin{bmatrix} \underset{n\times m}{A} \\ \underset{l\times m}{K} \end{bmatrix}\xi + \begin{bmatrix} e \\ e_0 \end{bmatrix}, \quad (2.1a)$$
$$rkK = l \le m = rk[A^T, K^T] < n + l, \quad (2.1b)$$
$$\begin{bmatrix} e \\ e_0 \end{bmatrix} \sim \left(\begin{bmatrix} 0 \\ 0 \end{bmatrix}, \sigma_0^2\begin{bmatrix} \underset{n\times m}{P^{-1}} & 0 \\ 0 & \underset{l\times l}{Q_0} \end{bmatrix}\right), \quad (2.1c)$$

where the $l \times 1$ vector z_0, the $l \times m$ matrix K, and the $l \times 1$ random error vector e_0 (unknown) define the constraints; note that e_0 has zero expectation and Q_0 as –possibly singular– $l \times l$ cofactor matrix.
By partitioning the transformation matrix $B = [B_A, B_K]$ of size $r \times (n+l)$ accordingly, we arrive at the model of "*condition equations*"

$$w_0 := B_A y + B_K z_0 = B_A e + B_K e_0, \qquad (2.2a)$$

$$rk[B_A, B_K] = r, \qquad (2.2b)$$

$$\begin{bmatrix} e \\ e_0 \end{bmatrix} \sim \left(\begin{bmatrix} 0 \\ 0 \end{bmatrix}, \sigma_0^2 \begin{bmatrix} P^{-1} & 0 \\ 0 & Q_0 \end{bmatrix} \right) \qquad (2.2c)$$

which turns out to be *equivalent if and only if* the two relations

$$B_A A + B_K K = 0, \qquad (2.3a)$$

$$rk[B_A, B_K] + rk[A^T, K^T] = n + l, \qquad (2.3b)$$

hold true where $r := n - m + l$ incidentally denotes the *redundancy* of the above model.
For the proof, we introduce the *invertible* matrix $N_K := A^T P A + K^T K$ and establish the matrix identity

$$\sigma_0^2 \begin{bmatrix} A^T P A & K^T \\ K & -Q_0 \end{bmatrix}^{-1}$$

$$= \sigma_0^2 \begin{bmatrix} N_K^{-1} - N_K^{-1} K^T [Q_0 + (I_l - Q_0) K N_K^{-1} K^T]^{-1} (I_e - Q_0) K N_K^{-1} & \vdots \\ [Q_0 + K N_K^{-1} K^T (I_l - Q_0)]^{-1} K N_K^{-1} & \vdots \end{bmatrix}$$

$$\begin{matrix} \vdots & N_K^{-1} K^T [Q_0 + (I_l - Q_0) K N_K^{-1} K^T]^{-1} \\ \vdots & [Q_0 + K N_K^{-1} K^T (I_l - Q_0)]^{-1} [(K N_K^{-1} K^T) - I_l] \end{matrix}$$

$$= \begin{bmatrix} D\{\hat{\xi}\} & \cdots \\ \cdots & -D\{\hat{\lambda}\} \end{bmatrix} \qquad (2.4)$$

where the auxiliary $l \times 1$ vector $\hat{\lambda}$ is given by one of the formulas

$$\hat{\lambda} = -B_K^T (B_A P^{-1} B_A^T + B_K Q_0 B_K^T)^{-1} w_0 \qquad (2.5a)$$

$$= -[Q_0 + K N_K^{-1} K^T (I_l - Q_0)]^{-1} \cdot$$

$$[z_0 - K N_K^{-1} (A^T P y + K^T z_0)], \qquad (2.5b)$$

uncorrelated with the estimated parameter vector $\hat{\xi}$,

$$C\{\hat{\xi}, \hat{\lambda}\} = 0. \qquad (2.6)$$

From $\hat{\lambda}$ we immediately obtain the *residual vector* $\tilde{e}_0$ for the stochastic constraints via

$$\tilde{e}_0 = -Q_0 \hat{\lambda}$$

$$= Q_0 B_K^T (B_A P^{-1} B_A^T + B_K Q_0 B_K^T)^{-1} w_0 \qquad (2.7a)$$

$$= z_0 - K (A^T P A + K^T P_0 K)^{-1} (A^T P y + K^T P_0 z_0)$$

if $P_0 := Q_0^{-1}$ exists, (2.7b)

and, due the *orthogonality relations*

$$0 = A^T P \tilde{e} - K^T \hat{\lambda} \qquad (2.8a)$$

$$= A^T P \tilde{e} + K^T P_0 \tilde{e}_0 \quad \text{(if } P_0 \text{ exist)}, \qquad (2.8b)$$

the further representation

$$\hat{\lambda} = (K N_K^{-1} K^T)^{-1} K N_K^{-1} A^T P \tilde{e} \qquad (2.9)$$

with the (observational) *residual vector*

$$\tilde{e} = P^{-1} B_A^T (B_A P^{-1} B_A^T + B_K Q_0 B_K^T)^{-1} w_0$$

(2.10a) $= y - A (A^T P A + K^T P_0 K)^{-1} \cdot$

$(A^T P y + K^T P_0 z_0)$ if $P_0 := Q_0^{-1}$ exists, (2.10b)

$$= (I_n - A N_K^{-1} A^T P) y$$

$$- A N_K^{-1} K^T [Q_0 + (I_l - Q_0) K N_K^{-1} K^T]^{-1} \cdot$$

$$[z_0 - (I_l - Q_0) K N_K^{-1} A^T P y]. \qquad (2.10c)$$

The *weighted sum of squared residuals* then results in

$$\Omega := \tilde{e}^T P \tilde{e} + \hat{\lambda}^T Q_0 \hat{\lambda}$$

$$= \tilde{e}^T P \tilde{e} + \tilde{e}_0^T P_0 \tilde{e}_0 \quad \text{if } P_0 := Q_0^{-1} \text{ exists}, \qquad (2.11a)$$

$$= w_0^T (B_A P^{-1} B_A^T + B_K Q_0 B_K^T)^{-1} w_0$$

(2.11b)

$$= (B_A \tilde{e} + B_K \tilde{e}_0)^T (B_A P^{-1} B_A + B_K Q_0 B_K^T)^{-1} \cdot$$

$$(B_A \tilde{e} + B_K \tilde{e}_0) \qquad (2.11c)$$

which reflects the main effect in reducing computational costs when using equivalent condition equations instead of the original GM-Model with stochastic constraints.

3 Specialization: The GM-Model with Fixed Constraints

Here we start out with a set-up that resembles the model (2.1a-c), except that the cofactor matrix Q_0 is strictly set to 0; this implies $e_0 = 0$ and consequently

$$y = \underset{n \times m}{A} \xi + e, \; e \sim (0, \sigma_0^2 P^{-1}), \qquad (3.1a)$$

$$\kappa_0 = \underset{l \times m}{K} \xi, \qquad (3.1b)$$

$$rk K = l < rk[A^T, K^T] = m < n + l. \qquad (3.1c)$$

Note that the left hand side vector of the constraints is now *non-random*; hence the symbol z_0 has been replaced by the (greek) κ_0 to indicate this.

By applying our previous results to this special case, we find the model of "*condition equations*"

$$w_0 := \underset{r\times n}{B_A} y + \underset{r\times l}{B_K} \kappa_0 = B_A e, \quad rk[B_A, B_K] = r, \tag{3.2a}$$

$$e \sim (0, \sigma_0^2 P^{-1}), \tag{3.2b}$$

to be *equivalent if and only if* the two relations (2.3a-b) are fulfilled, again with $r := n - m + l$ denoting the *redundancy* of model (3.1a-c).

The key identities to prove this result simplify considerably; with $N_K := A^T PA + K^T K$ as before, we get

$$\sigma_0^2 \begin{bmatrix} A^T PA & K^T \\ K & 0 \end{bmatrix}^{-1} =$$

$$\sigma_0^2 \begin{bmatrix} N_K^{-1} - N_K^{-1} K^T (K N_K^{-1} K^T)^{-1} K N_K^{-1} & N_K^{-1} K^T (K N_K^{-1} K^T)^{-1} \\ (K N_K^{-1} K^T)^{-1} K N_K^{-1} & I_l - (K N_K^{-1} K^T)^{-1} \end{bmatrix} \tag{3.3a}$$

$$= \begin{bmatrix} D\{\hat{\xi}\} & \cdots \\ \cdots & -D\{\hat{\lambda}\} \end{bmatrix} \tag{3.3b}$$

where the auxiliary $l \times 1$ vector $\hat{\lambda}$ is uncorrelated with the estimated parameter vector $\hat{\xi}$ and can be calculated via

$$\hat{\lambda} = \kappa_0 - (K N_K^{-1} K^T)^{-1} (\kappa_0 - K N_K^{-1} A^T P y) \tag{3.4a}$$

$$= -B_K^T (B_A P^{-1} B_A^T)^{-1} w_0 . \tag{3.4b}$$

The *orthogonality relations* reduce to

$$0 = A^T P \tilde{e} - K^T \hat{\lambda}, \tag{3.5a}$$

yielding

$$A^T P \tilde{e} = -K^T B_K^T (B_A P^{-1} B_A^T)^{-1} B_A \tilde{e} \neq 0 \text{ (in general)} \tag{3.5b}$$

and finally

$$\tilde{e} = P^{-1} B_A^T (B_A P^{-1} B_A^T)^{-1} w_0 \tag{3.6a}$$

$$= (I_n - A N_K^{-1} A^T P) y - A N_K^{-1} K^T (K N_K^{-1} K^T)^{-1} (\kappa_0 - K N_K^{-1} A^T P y) \tag{3.6b}$$

$$= (I_n - A N_K^{-1} A^T P) y - A N_K^{-1} K^T (\kappa_0 - \hat{\lambda}) \tag{3.6c}$$

for the *residual vector*. Consequently, the *weighted sum of squared residuals* will be

$$\Omega := \tilde{e}^T P \tilde{e} = w_0^T (B_A P^{-1} B_A^T)^{-1} w_0 \tag{3.7a}$$

$$= (B_A \tilde{e})^T (B_A P^{-1} B_A^T)^{-1} (B_A \tilde{e}) \tag{3.7b}$$

as expected.

By comparing various of the above relations, we may obtain even more "*decompositions of the identity*", for instance

$$I_l = (K N_K^{-1} K^T)^{-1} - B_K^T (B_A P^{-1} B_A^T)^{-1} B_K , \tag{3.8a}$$

$$I_n = A [N_K^{-1} - N_K^{-1} K^T (K N_K^{-1} K^T)^{-1} K N_K^{-1}] A^T P + P^{-1} B_A^T (B_A P^{-1} B_A^T)^{-1} B_A \tag{3.8b}$$

or a "*decomposition of the zero matrix*"

$$0 = (K N_K^{-1} K^T)^{-1} K N_K^{-1} A^T + B_K^T (B_A P^{-1} B_A^T)^{-1} B_A , \tag{3.9}$$

which all prove quite useful to develop formulas for equivalent "condition equations" when "fixed constraint" versions have been provided, or vice versa.

4 Extension: The Random Effects Model

In this section the (incremental) parameter vector contains *random effects*, perhaps as a result of using *stochastic* prior information for the linearization. Hence we have to replace the (greek) ξ symbolically by its counterpart x when writing up the *Random Effects Model*

$$y = \underset{n\times m}{A} x + e, \quad \underset{m\times 1}{\beta_0} = x + e_0, \quad \begin{bmatrix} e \\ e_0 \end{bmatrix} \sim \left(\begin{bmatrix} 0 \\ 0 \end{bmatrix}, \sigma_0^2 \begin{bmatrix} P^{-1} & 0 \\ 0 & Q_0 \end{bmatrix} \right). \tag{4.1}$$

Here β_0 denotes the "bias vector of the prior information" which is assumed to be known (for instance, $\beta_0 := 0$); the corresponding random error vector has zero expectation with the $l \times l$ matrix Q_0 as (possibly singular) cofactor matrix.

In order to use the results from *Section 2*, let us apply "*Helmert's Knack*" which introduces the "random zero vector" $\underset{\sim}{0}$ for the decomposition

$$x = \xi - \underset{\sim}{0}, \text{ or } \xi := x + \underset{\sim}{0}, \tag{4.2a}$$

with

$$E\{\underset{\sim}{0}\} = E\{\xi - x\} = \xi - \beta_0 , \tag{4.2b}$$

$$D\{\underset{\sim}{0}\} = D\{\xi - x\} = D\{x\} = \sigma_0^2 Q_0 . \tag{4.2c}$$

Obviously, ξ is a (totally unknown) non-random vector, in contrast to x (which is random) and to β_0 (which is known). If we now further introduce

$$\underset{\sim}{y} := y + A \cdot \underset{\sim}{0} \text{ and } b_0 := \beta_0 + \underset{\sim}{0}, \quad (4.3)$$

we arrive at the equivalent model

$$\begin{bmatrix} \underset{\sim}{y} \\ b_0 \end{bmatrix} = \begin{bmatrix} A \\ I_m \end{bmatrix} \xi + \begin{bmatrix} e \\ e_0 \end{bmatrix}, \quad (4.4a)$$

$$\begin{bmatrix} e \\ e_0 \end{bmatrix} \sim \left(\begin{bmatrix} 0 \\ 0 \end{bmatrix}, \sigma_0^2 \begin{bmatrix} P^{-1} & 0 \\ 0 & Q_0 \end{bmatrix} \right), \quad (4.4b)$$

which clearly turns out to be of type "*Gauss-Markov with stochastic constraints*", with the special provisions that $K := I_m$ and the redundancy $r = (n+m) - m = n$.
Consequently the first relation in (2.3a) turns into

$B_K := -B_A A$ for *any nonsingular* matrix B_A (4.5a)

so that the second relation of (2.3b) holds true as well:

$$rk(B_A[I_n, -A]) + rk[A^T, K^T] = n + m. \quad (4.5b)$$

We can thus define the whole class of *equivalent* "condition equations" by

$$w_0 := B_A\left(\underset{\sim}{y} - Ab_0 \right) = B_A(y - A\beta_0)$$
$$= B_A(e - Ae_0), \quad \underset{n\times n}{B_A} \text{ nonsingular}, \quad (4.6a)$$

$$\begin{bmatrix} e \\ e_0 \end{bmatrix} \sim \left(\begin{bmatrix} 0 \\ 0 \end{bmatrix}, \sigma_0^2 \begin{bmatrix} P^{-1} & 0 \\ 0 & Q_0 \end{bmatrix} \right). \quad (4.6b)$$

Note that the *Mixed Linear Model* could be treated quite similarly by replacing the second part of (4.1), i.e., the "prior information", by

$$\beta_0 = \underset{l\times m}{K} x + e_0, \quad K := [I_l, 0], \quad l \le m. \quad (4.7)$$

The equivalence between (4.1) and (4.6a-b) could also be directly established by relying on identities (with $N := A^T P A$) such as

$$\sigma_0^2 \begin{bmatrix} N & I_m \\ I_m & -Q_0 \end{bmatrix}^{-1}$$
$$= \sigma_0^2 \begin{bmatrix} Q_0(NQ_0 + I_m)^{-1} & (Q_0 N + I_m)^{-1} \\ (NQ_0 + I_m)^{-1} & -N(Q_0 N + I_m)^{-1} \end{bmatrix} \quad (4.8a)$$

$$= \begin{bmatrix} D\{\tilde{x} - x\} & \cdots \\ \cdots & -D\{\hat{\lambda}\} \end{bmatrix}, \quad C\{\tilde{x} - x, \hat{\lambda}\} = 0. \quad (4.8b)$$

Now the $m \times 1$ vector $\hat{\lambda}$ is uncorrelated with the *difference vector* $\tilde{x} - x$ and may be determined by

$$\hat{\lambda} = A^T B_A^T \left[B_A (P^{-1} + AQ_0 A^T) B_A^T \right]^{-1} w_0 \quad (4.9a)$$

$$= A^T (P^{-1} + AQ_0 A^T)^{-1} (y - A\beta_0) \quad (4.9b)$$

$$= (NQ_0 + I_m)^{-1} A^T P (y - A\beta_c) = A^T P \tilde{e}, \quad (4.9c)$$

so that the *residual vectors* (for $N_I := N + I$) result in

$$\tilde{e}_0 = -Q_0 \hat{\lambda} = -Q_0 A^T P \tilde{e}, \quad (4.10)$$

$$\tilde{e} = P^{-1} B_A^T \left[B_A (P^{-1} + AQ_0 A^T) B_A^T \right]^{-1} w_0 \quad (4.11a)$$
$$= (I_n + AQ_0 A^T P)^{-1} (y - A\beta_0)$$
(4.11b)
$$= (I_n - AN_I^{-1} A^T P) y$$
$$- A(Q_0 N + I_m)^{-1} [\beta_0 - (I_m - Q_0) N_I^{-1} A^T P y], \quad (4.11c)$$

and the *weighted sum of squared residuals* in

$$\Omega := \tilde{e}^T P \tilde{e} + \hat{\lambda}^T Q_0 \hat{\lambda} \quad (4.12a)$$
$$= \tilde{e}^T P \tilde{e} + \tilde{e}_0^{\,T} P_0 \tilde{e}_0 \quad \text{(if } P_0 \text{ exists)} \quad (4.12b)$$
$$= w_0^T \left[B_A (P^{-1} + AQ_0 A^T) B_A^T \right]^{-1} w_0 \quad (4.12c)$$
$$= (y - A\beta_0)^T (P^{-1} + AQ_0 A^T)^{-1} (y - A\beta_0)$$
(4.12d)
$$= (\tilde{e} - A\tilde{e}_0)^T (P^{-1} + AQ_0 A^T)^{-1} (\tilde{e} - A\tilde{e}_0) \quad (4.12e)$$
$$= \tilde{e}^T P (P^{-1} + AQ_0 A^T) P \tilde{e}. \quad (4.12f)$$

Particularly the last sequence of identities for Ω is designed to improve the computational efficiency.

Warning : We have seen that, in order to establish *equivalence* between model (4.1) and (4.6a-b), the matrix B_A ought to be *non-singular*. Thus, the common practice to apply a rectangular $r \times n$ matrix B_1 with $r \le n - rkA$ and $B_1 A = 0$ will automatically lead to *biased results*.
To correct the approach, another *(n-r)×n* matrix B_2 has to be introduced such that $B_A^T = [B_1^T, B_2^T]$ becomes nonsingular with

$$B_k^T := -A^T B_A^T = [0, -A^T B_2^T]. \quad (4.13)$$

5 Application: The Dynamic Linear Model

Let us now apply the results of Section 4 do the case of *Kalman filtering* where we do the least-squares adjustment within a *Dynamic Linear Model,* defined for the time interval from $t=t_0$ to $t=t_1$ by

$$y_1 = \left[\underset{n\times m}{A_1}, \underset{n\times m}{0} \right] \begin{bmatrix} x_1 \\ x_0 \end{bmatrix} + e_1, \tag{5.1a}$$

$$0 = \left[\underset{m\times m}{-I}, \underset{m\times m}{\phi_0} \right] \begin{bmatrix} x_1 \\ x_0 \end{bmatrix} + u_1, \tag{5.1b}$$

$$\tilde{x}_0 = \left[\underset{n\times m}{0}, \underset{m\times m}{I_m} \right] \begin{bmatrix} x_1 \\ x_0 \end{bmatrix} + e_0{}^0, \tag{5.1c}$$

$$\begin{bmatrix} e_1 \\ u_1 \\ e_0^0 \end{bmatrix} \sim \left(\begin{bmatrix} 0 \\ 0 \\ 0 \end{bmatrix}, \begin{bmatrix} \Sigma_1 & 0 & 0 \\ 0 & \Theta_1 & 0 \\ 0 & 0 & \Sigma_0^0 \end{bmatrix} \right). \tag{5.1d}$$

After applying *"Helmert's Knack"* $\tilde{x}_0 = \kappa_0 + \underset{\sim}{0}$ to the *initial condition* (5.1c), we obtain a Random Effects Model for which we can readily state this result, based on the *complete elimination* of the state vector $[x_1^T, x_0^T]^T$:

Theorem: *In the Dynamic Linear Model (5.1a-d), let B_{A_1}, be any $r\times n$ matrix with $rkB_{A_1}=r$ where r denotes the "redundancy" (thus, $r\leq n$). Further, define the r×2m matrix.*

$$B_{K_1} := -B_{A_1}[A_1, 0] \begin{bmatrix} -I_m & \phi_0 \\ 0 & I_m \end{bmatrix} = B_{A_1} A_1 \begin{bmatrix} I_m \\ -\phi_0^T \end{bmatrix}^T \tag{5.2}$$

so that the two required relations

$$rk\left[B_{A_1}, B_{K_1}\right] := rkB_{A_1}\left[I_n, A_1, -A_1\phi_0\right]$$
$$= (n+2m) - 2m, \tag{5.3a}$$

$$B_{A_1}[A_1, 0] + B_{K_1} K_1 = 0, \tag{5.3b}$$

hold true for the 2m×2m matrix

$$K_1 := \begin{bmatrix} -I_m & \phi_0 \\ 0 & I_m \end{bmatrix} = K_1^{-1}. \tag{5.4}$$

Then the model of "condition equations"

$$w_1 := B_{A_1} y_1 + B_{K_1} \begin{bmatrix} 0 \\ \tilde{x}_0 \end{bmatrix} = B_{A_1}(y_1 - A_1\breve{x}_1)$$
$$= B_{A_1} e_1 + B_{K_1} \begin{bmatrix} u_1 \\ e_0^0 \end{bmatrix} = B_{A_1}\left[e_1 + A_1\left(u_1 - \phi_0 e_0^0\right)\right], \tag{5.5a}$$

$$\begin{bmatrix} e_1 \\ u_1 \\ e_0^0 \end{bmatrix} \sim \left(\begin{bmatrix} 0 \\ 0 \\ 0 \end{bmatrix}, \begin{bmatrix} \Sigma_1 & 0 & 0 \\ 0 & \Theta_1 & 0 \\ 0 & 0 & \Sigma_0^0 \end{bmatrix} \right), \; rk \underset{n\times n}{B_{A_1}} = n, \tag{5.5b}$$

is necessarily equivalent.

Obviously, the special choice $B_{A_1} := I_n$ is possible in which case $w_1 := y_1 - A_1\breve{x}_1 = e_1 + A_1\left(u_1 - \phi_0 e_0^0\right)$ with $\breve{x}_1 := \phi_0 \tilde{x}_0$ becomes the vector of *"innovations"*, and the *weighted sum of squared residuals* at time epoch $t=t_1$ can be computed by one of the formulas

$$\Omega_1 := \tilde{e}_1^T \Sigma_1^{-1} \tilde{e}_1 + \tilde{u}_1^T \Theta_1^{-1} \tilde{u}_1 + \left(\tilde{e}_0^0\right)^T \left(\Sigma_0^0\right)^{-1} \left(\tilde{e}_0^0\right) \tag{5.6a}$$

$$= \left[\tilde{e}_1 + A_1\left(\tilde{u}_1 - \phi_0 \tilde{e}_0^0\right)\right]^T \left[\Sigma_1 + A_1\left(\Theta_1 + \phi_0 \Sigma_0^0 \phi_0^T\right) A_1^T\right]^{-1} \cdot \left[\tilde{e}_1 + A_1\left(\tilde{u}_1 - \phi_0 \tilde{e}_0^0\right)\right] \tag{5.6b}$$

$$= \left(y_1 - A_1\breve{x}_1\right)^T \left[\Sigma_1 + A_1\left(\Theta_1 + \phi_0 \Sigma_0^0 \phi_0^T\right) A_1^T\right]^{-1} \left(y_1 - A_1\breve{x}_1\right). \tag{5.6c}$$

6 Outlook

The Theorem in *Section 5* can obviously be further generalized to include (bias) parameters beside the state vector. This is relevant in the case of GPS phase observations where integer ambiguities must be resolved. Computationally efficient formulas may be obtained along the same lines as presented in *Section 1*, formulas (1.7a-b) to (1.12a-c), for the GM-Model with "nuisance parameters". By applying identities such as (1.12a-c) and (5.6a-c), the derivations by *A.A. Povalyaev(1999)* for the likelihood function, after introducing the "float ambiguities", could be drastically simplified. This is, however, left for a separate publication.

References

Baksalary, J.K. and R. Kala (1981). Linear transformations preserving best linear unbiased estimators in a general Gauss-Markov Model. *Ann. Statist.*, 9, pp.913-916

Baksalary, J.K.(1984). A study of the equivalence between a Gauss-Markov Model and its augmentation by nuisance parameters. *Math. Operations-Forschung, Ser. Statistics*, 15, pp.1-35.

Helmert, F.R. (1907). *Adjustment Computations by the Method of Least-Squares (in german)*, 2nd Edition, Teubner: Leipzig, 578 pages.

Povalyaev, A.A. (1999). Filtering problem for ambiguous phase measurements, *J. of Communications Technology and Electronics of the Russian Acad. of Sciences (English edition)*, 44, pp.904-912.

Schaffrin, B. and E. Grafarend (1986). Generating classes of equivalent linear models by nuisance parameter elimination. Applications to GPS observations, *Manus. Geodaet.*,11, pp.262-271.

Procrustes Statistics to Test for Significant OLS Similarity Transformation Parameters

A. Beinat, F. Crosilla
Department of Georesources and Territory,
University of Udine, Via Cotonificio 114, I-33100 Udine UD, Italy

Abstract. Ordinary least squares (OLS) similarity transformations are often used in practice to perform datum conversions in a wide spectrum of geodetic and photogrammetric problems.

Traditionally, the estimation of the unknown parameters is computed by solving a non linear transformation model, build on a set of correspondence point coordinates known in both reference systems.

In the past many efforts have been made to provide a direct OLS solution to the parameter estimation problem. Among the solutions investigated (e.g. Sansò 1973), today very promising results are expected from some algorithms of the Procrustes Analysis (e.g. Crosilla 1999; Grafarend and Awange 2000, 2002).

These methods furnish the mathematical tools to perform a direct similarity transformation parameters estimation between two datums (Ordinary Procrustes Analysis), and among n > 2 different datums by an iterative sequence of simultaneous direct solutions (Generalised Procrustes analysis).

In the paper we present a method, inspired to the perturbation theory of the Procrustes Analysis proposed by Langron and Collins (1985), aimed to evaluate the level of significance of the different parameters (translation, rotation and global dilatation) computed by the solution of a classical OLS similarity transformation problem. The procedure has been developed for both the Ordinary Procrustes problem and for the Generalised Procrustes problem.

Keywords. Procrustes statistics, Procrustes analysis, OLS similarity transformation

1 Introduction

Since the fifties, the development of methods capable to provide in direct way the 7-parameter estimation of a similarity transformation between two different frames, starting from the coordinates defined for the same set of points in both reference systems, has attracted the interest of many authors (e.g. Thompson 1958; Sansò 1973; Arun et al. 1987; Horn 1987; Horn et al. 1988).

Just recently, geodetic and photogrammetric literatures have considered a set of alternative algorithms, developed since the seventies in psychometry and statistics, globally known as Procrustes Analysis (Schoenemann 1966; Schoenemann and Carroll 1970; Gower 1975).

These methods furnish the mathematical tools to directly perform the classical OLS similarity transformation parameters estimation between two datums (Ordinary Procrustes Analysis), and among m > 2 different datums by an iterative sequence of simultaneous direct solutions (Generalised Procrustes Analysis) of an original non linear LS problem. Being based on the Singular Value Decomposition of a product matrix, obtained by the data matrices containing the correspondence point coordinates in the two reference systems, they completely avoid the conventional equation system and other computational problems. For this reason the interest on Procrustes applications in geodesy and related disciplines is still growing, as confirmed by recent contributions. Among these, Beinat and Crosilla developed an innovative approach to the global registration of laserscanning range images (2001) and to the photogrammetric block adjustment solution (Crosilla and Beinat, 2002). Their method seems suitable in the geodetic field for the partial network densification and for the definition of a global and common reference frame. Very recently, Grafarend and Awange (2002) proposed a new formulation of the weighted orthogonal Procrustes problem, in which the stochasticity of the point coordinates of both data matrices is considered.

To evaluate the significance of the parameters involved in an OLS similarity transformation problem we present a method inspired to the perturbation theory of the Procrustes analysis proposed by Langron and Collins (1985).

The test is developed for both the Ordinary Procrustes problem (OPA), in which we match a matrix configuration (origin) over another one (target), and for the Generalised Procrustes problem (GPA) where we simultaneously fit a set of $m > 2$ coordinate matrices one to each other.

Procrustes statistics $G_{t,R,c}$ is defined as the (weighted) residual square distance, i.e. the sum of the (weighted) squared positional differences, between two (OPA) or more (GPA) configuration matrices after an OLS similarity transformation has occurred. Under the hypothesis that the errors affecting the coordinates matrices are normally distributed, it can be proved that the Procrustes statistics is well approximated by a chi-squared distribution with degrees of freedom depending on the space dimension and on the rank of the coordinates matrices.

Following Langron and Collins (1985), the residual square distances before and after a Procrustes analysis can be presented as an analysis of variance (ANOVA) and the mentioned effects (translation, rotation and scale variation) tested for by the classical F–ratio.

For this purpose let us define G as the square distance between the original data matrices, G_t the residual square distance after centring translation, $G_{t,R}$ the residual square distance after translation and rotation and $G_{t,R,c}$ the residual square distance after a complete similarity transformation.

Under the null hypothesis, the ratios:

$$\frac{(G-G_t)}{G_{t,R,c}},\ \frac{(G_t-G_{t,R})}{G_{t,R,c}}\ \text{and}\ \frac{(G_{t,R}-G_{t,R,c})}{G_{t,R,c}}$$

all follow an F distribution with their proper degrees of freedom.

In particular the term $G - G_t$ represents the square distance contribution due to the centring translation, $G_t - G_{t,R}$ is the square distance contribution due to the rotation and $G_{t,R} - G_{t,R,c}$ corresponds to the square distance contribution due to the global dilatation.

According to Goodall (1991) the mentioned ratios follow an F distribution also in the case in which the various points are differently weighted. In the opposite case, that is if the null hypothesis is rejected, at least one of the similarity transformation components is significant.

Geodetic literature reports alternative methods to estimate the parameters of a similarity datum transformation. We refer for instance to the so called Total Least Squares (TLS) solution, studied in detail by Golub and van Loan (1980) and also mentioned by Strang and Borre (1997).

This paper represents a first contribution in testing the significance of the OLS similarity transformation parameters by Procrustes Statistics; future research will try to extend these topics also to the so-called total least squares problem solution.

2 Ordinary Procrustes Analysis (OPA)

Let's we recall the basics of the OLS Procrustes algorithms implemented in the test procedure.

Given two $n \times k$ matrices $\mathbf{X}_1$ and $\mathbf{X}_2$, each one containing by rows the k coordinates of n corresponding points, defined in two different datums, the OPA algorithm directly estimates the parameters of the OLS similarity transformation that fits $\mathbf{X}_1$ (origin) respect to $\mathbf{X}_2$ (target).

The basic OPA model can be immediately extended to consider a given set of weights singularly assigned to every i point ($i = 1 \ldots n$) and stored as corresponding row elements of a $n \times n$ weighting matrix $\mathbf{W}$. If $\mathbf{W} = \mathbf{Q}^T\mathbf{Q}$ (Cholesky decomposition) the solution can be easily achieved.

In most geodetic problems k is usually 3, therefore said $\mathbf{t}$, $\mathbf{R}$ and c be the unknown transformation parameters, *i.e.* the 3×1 translation vector, the 3×3 rotation matrix and the global scale factor respectively, and $\mathbf{1}$ a $n \times 1$ predefined auxiliary unitary vector ($1 = [1_1 \ \ldots \ 1_n]$), under the orthogonality condition $\mathbf{R}^T\mathbf{R} = \mathbf{I}$, by substituting $\mathbf{X}_{1w} = \mathbf{QX}_1$, $\mathbf{X}_{2w} = \mathbf{QX}_2$, $\mathbf{1}_w = \mathbf{Q1}$ one defines:

$$\mathbf{B} = \mathbf{I} - \frac{\mathbf{1}_w\mathbf{1}_w^T}{\mathbf{1}_w^T\mathbf{1}_w} \tag{1}$$

idempotent, such that $\mathbf{BB} = \mathbf{B}$. Computing the singular value decomposition of the following matrix product:

$$\mathbf{X}_{1w}^T\mathbf{B}\mathbf{X}_{2w} = \mathbf{V}\mathbf{D}_s\mathbf{W}^T \tag{2}$$

one directly obtains:

- The orthogonal rotation matrix **R**:

$$\mathbf{R} = \mathbf{V}\mathbf{W}^{\mathrm{T}} \tag{3}$$

- The isotropic scale factor c:

$$c = \frac{\operatorname{tr}\left[\mathbf{R}^{\mathrm{T}}\mathbf{X}_{1\mathrm{W}}^{\mathrm{T}}\mathbf{B}\mathbf{X}_{2\mathrm{W}}\right]}{\operatorname{tr}\left[\mathbf{X}_{1\mathrm{W}}^{\mathrm{T}}\mathbf{B}\mathbf{X}_{1\mathrm{W}}\right]} \tag{4}$$

- The translation vector **t**:

$$\mathbf{t} = \left(\mathbf{X}_{2\mathrm{W}} - c\mathbf{X}_{1\mathrm{W}}\mathbf{R}\right)^{\mathrm{T}} \frac{\mathbf{1}_{\mathrm{W}}}{\mathbf{1}_{\mathrm{W}}^{\mathrm{T}}\mathbf{1}_{\mathrm{W}}} \tag{5}$$

Very recently, Grafarend and Awage (2002) proposed a valuable method to define a diagonal weighting matrix **W**, that incorporates the stochasticity of both $\mathbf{X}_1$ and $\mathbf{X}_2$ pseudo-observation matrices.

On the opposite, the solution of the unit weight case $\mathbf{W}=\mathbf{I}$ can be immediately derived simply considering the original terms in the equations without the W subscript.

Note that, although we referred to the R^3 space, the equation formalism is valid for every R^k $(k>3)$. The OPA model can be further extended to the case of incomplete matching components between **Xi** and **Xj**. The missing elements could be entire points (matrix rows), single coordinates and also whole dimensions (matrix columns).

We forward to the references for more details (e.g. Commandeur 1991, ten Berge et al. 1993).

3 Generalised Procrustes Analysis (GPA)

GPA considers the non linear multiple and simultaneous fit of $m>2$ data matrices $\mathbf{X}_1 \ldots \mathbf{X}_m$ of size $n \times k$, each one containing the coordinates of the same set of n corresponding k–dimensional points, defined in m different R^k datums. For simplicity we can assume $k=3$.

Among the infinite $\{\mathbf{t}, \mathbf{R}, c\}_{ij}$ $(i=1 \ldots m;\ i<j)$ similarity transformations linking every possible pair of *i, j* matrices, GPA defines the specific set for which the least squares objective function satisfies the following minimum condition:

$$S = \operatorname{tr}\sum_{i<j}^{m}\left[\left(c_i\mathbf{X}_i\mathbf{R}_i + \mathbf{1}\mathbf{t}_i\right) - \left(c_j\mathbf{X}_j\mathbf{R}_j + \mathbf{1}\mathbf{t}_j\right)\right]^{\mathrm{T}} \cdot \left[\left(c_i\mathbf{X}_i\mathbf{R}_i + \mathbf{1}\mathbf{t}_i\right) - \left(c_j\mathbf{X}_j\mathbf{R}_j + \mathbf{1}\mathbf{t}_j\right)\right] = \min \tag{6}$$

with the orthogonality restriction $\mathbf{R}_i^{\mathrm{T}}\mathbf{R}_i = \mathbf{I}$ and $c_i \neq 0$ $(i=1 \ldots m)$.

The fulfilment of the above condition represents the solution of the well known Generalised Procrustes problem (GPA), described by Gower (1975), Ten Berge (1977) and Goodall (1991).

The problem can be defined as the search of an unknown "optimal" matrix **K**, also named "consensus", related to every $\mathbf{X}_i$ by a proper unknown similarity transformation, unless a random error matrix $\mathbf{E}_i$:

$$\begin{aligned} \mathbf{K} + \mathbf{E}_i &= \mathbf{X}_i^{\mathrm{P}} \\ \mathbf{X}_i^{\mathrm{P}} &= c_i\mathbf{X}_i\mathbf{R}_i + \mathbf{1}\mathbf{t}_i^{\mathrm{T}} \end{aligned} \tag{7}$$

$\mathbf{E}_i$ has the general definition of:

$$\operatorname{vec}(\mathbf{E}_i) \approx N\left\{0, \boldsymbol{\Sigma} = \sigma^2\left(\mathbf{Q}_P \otimes \mathbf{Q}_K\right)\right\} \tag{8}$$

where $\boldsymbol{\Sigma}$ is the covariance matrix, $\mathbf{Q}_P$ is the cofactor matrix of the n points, $\mathbf{Q}_K$ is the cofactor matrix of the k components, $\otimes$ is the Kronecker product and σ^2 is the variance factor.

Based on equation (7), the GPA problem can be concisely rewritten as:

$$S = \sum_{i<j}^{m}\left\|\mathbf{X}_i^{\mathrm{P}} - \mathbf{X}_j^{\mathrm{P}}\right\|^2 = \min \tag{9}$$

Defining a matrix **C**:

$$\mathbf{C} = \frac{1}{m}\sum_{i=1}^{m}\mathbf{X}_i^{\mathrm{P}} \tag{10}$$

as the *geometrical centroid,* the two functions

$$\sum_{i<j}^{m}\left\|\mathbf{X}_i^{\mathrm{P}} - \mathbf{X}_j^{\mathrm{P}}\right\|^2 = \sum_{i<j}^{m}\operatorname{tr}\left(\mathbf{X}_i^{\mathrm{P}} - \mathbf{X}_j^{\mathrm{P}}\right)^{\mathrm{T}}\left(\mathbf{X}_i^{\mathrm{P}} - \mathbf{X}_j^{\mathrm{P}}\right) \tag{11}$$

$$m\sum_{i=1}^{m}\left\|\mathbf{X}_i^{\mathrm{P}} - \mathbf{C}\right\|^2 = m\sum_{i=1}^{m}\operatorname{tr}\left(\mathbf{X}_i^{\mathrm{P}} - \mathbf{C}\right)^{\mathrm{T}}\left(\mathbf{X}_i^{\mathrm{P}} - \mathbf{C}\right) \tag{12}$$

are equivalent as demonstrated e.g. by Kristof & Wingersky (1971) or by Borg & Groenen (1997). Consequently, the GPA problem can be also solved minimising equation (12) instead of equation (11).

Once the solution is determined, under the condition that $\mathbf{Q}_P$ is diagonal and $\mathbf{Q}_K = \mathbf{I}$, *i.e.* when the measurement precision of a specific point is the same among the m different configurations, it can be demonstrated that the centroid **C** corresponds to the least squares estimate $\hat{\mathbf{K}}$ of the true value **K**. Proof of this can be found in literature (Beinat and Crosilla, 2001).

4 Procrustes statistics with errors in one configuration

Let $\mathbf{X}_i$ be, as before, a $n \times k$ matrix containing the k coordinates of n points defined into an arbitrary datum. Let $\mathbf{X}_j$ be a $n \times k$ matrix containing the coordinates of the same points, obtained summing up to $\mathbf{X}_i$ a matrix of random errors $\mathbf{Z}$, of zero mean and unit variance, and where $\mathbf{Z}$ is furthermore premultiplied by a common factor s, i.e.:

$$\mathbf{X}_j = \mathbf{X}_i + s\mathbf{Z} \tag{13}$$

where $\mathbf{Z} \simeq \mathrm{N}(0,1)$, or:

$$\mathrm{vec}(\mathbf{Z}) \simeq N\{0, \mathbf{\Sigma} = \mathbf{Q}_P \otimes \mathbf{Q}_K\}$$

with $\mathbf{Q}_P = \mathbf{I}$ and $\mathbf{Q}_K = \mathbf{I}$.

The Procrustes distance between the two configurations is given by the sum, extended to all the n points, of the square coordinate differences, that is in matrix form:

$$\begin{aligned} \mathrm{G}(\mathbf{X}_i, \mathbf{X}_j) &= \mathrm{tr}(\mathbf{X}_i - \mathbf{X}_j)^{\mathrm{T}}(\mathbf{X}_i - \mathbf{X}_j) \\ &= s^2\,\mathrm{tr}(\mathbf{Z}^{\mathrm{T}}\mathbf{Z}) \end{aligned} \tag{14}$$

The initial quantity $\mathrm{G}(\mathbf{X}_i,\mathbf{X}_j)$, due to the normal distribution of the elements of $\mathbf{Z}$, follows a χ^2 distribution with $n \times k$ degrees of freedom.

To match one matrix with respect to the other, we proceed now to the sequential application to one of the two matrices of a group of geometrical transformations:

- translation;
- translation and rotation;
- translation, rotation and isotropic dilatation;

in such a way to minimise the OLS Procrustes distance between the two matrices, for each kind of transformation.

This condition can be written for the whole transformations by the satisfaction of the following expressions:

$$\mathrm{G}_{\mathbf{t}}(\mathbf{X}_i, \mathbf{X}_j) := \inf\left(\mathrm{G}\left(\mathbf{X}_i, \mathbf{X}_j + \mathbf{1t}^{\mathrm{T}}\right)\right) \tag{15}$$

$$\mathrm{G}_{\mathbf{t,R}}(\mathbf{X}_i, \mathbf{X}_j) := \inf\left(\mathrm{G}\left(\mathbf{X}_i, \mathbf{X}_j\mathbf{R} + \mathbf{1t}^{\mathrm{T}}\right)\right) \tag{16}$$

$$\mathrm{G}_{\mathbf{t,R},c}(\mathbf{X}_i, \mathbf{X}_j) := \inf\left(\mathrm{G}\left(\mathbf{X}_i, c\mathbf{X}_j\mathbf{R} + \mathbf{1t}^{\mathrm{T}}\right)\right) \tag{17}$$

By a series expansion of the generic OLS Procrustes distance $\mathrm{G}_\phi(\mathbf{X}_i, \mathbf{X}_i + s\mathbf{Z})$ with respect to s, where ϕ stands for translation (**t**), translation and rotation (**t**,**R**), translation, rotation and dilatation (**t**,**R**,c), Sibson (1979, lemma 3.3, p. 222) shows, and experimentally verifies, that the distances $\mathrm{G}_{\mathbf{t}}(\mathbf{X}_i,\mathbf{X}_j)$ and $\mathrm{G}_{\mathbf{t,R}}(\mathbf{X}_i,\mathbf{X}_j)$ follow a χ^2 distribution, with respectively, [nk-k] and [nk-1/2k(k+1)] degrees of freedom, also for great values of s, that is:

$$\mathrm{G}_{\mathbf{t}}(\mathbf{X}_i, \mathbf{X}_j) \simeq s^2 \chi^2_{[\mathrm{nk-k}]}$$

$$\mathrm{G}_{\mathbf{t,R}}(\mathbf{X}_i, \mathbf{X}_j) \simeq s^2 \chi^2_{[\mathrm{nk-1/2k(k+1)}]}$$

while the quantity $\mathrm{G}_{\mathbf{t,R},c}(\mathbf{X}_i,\mathbf{X}_j)$ (Sibson 1979, proposition 3.6, p. 224) follows a χ^2 distribution with [nk-1/2k(k+1)-1] degrees of freedom and remains valid just for small values of s, that is:

$$\mathrm{G}_{\mathbf{t,R},c}(\mathbf{X}_i, \mathbf{X}_j) \simeq s^2 \chi^2_{[\mathrm{nk-1/2k(k+1)-1}]}$$

Goodall (1991, p. 312) reports that the χ^2 distribution of $\mathrm{G}_{\mathbf{t}}$, $\mathrm{G}_{\mathbf{t,R}}$ and $\mathrm{G}_{\mathbf{t,R},c}$ holds also for $\mathbf{Q}_P \neq \mathbf{I}$, that is for a different weight distribution of the various points.

5 Procrustes statistics with errors in two configurations

Let us consider now the case in which two matrices $\mathbf{X}_i$ and $\mathbf{X}_j$ are obtained by the same coordinate matrix $\mathbf{X}$ to which two independent matrices $\mathbf{Z}_i$ and $\mathbf{Z}_j$ are summed up; $\mathbf{Z}_i$ and $\mathbf{Z}_j$ contain random errors with mean zero and unit variance and both matrices are premultiplied by the same factor s (Langron and Collins 1985, p. 279), i.e.:

$$\mathbf{X}_i = \mathbf{X} + s\mathbf{Z}_i \tag{18}$$

where $\mathbf{Z}_i \simeq \mathrm{N}(0,1)$,

$$\mathbf{X}_j = \mathbf{X} + s\mathbf{Z}_j \tag{19}$$

where $\mathbf{Z}_j \simeq \mathrm{N}(0,1)$, or, in both cases:

$$\{\mathrm{vec}(\mathbf{Z}_i), \mathrm{vec}(\mathbf{Z}_j)\} \simeq N\{0, \mathbf{\Sigma} = \mathbf{Q}_P \otimes \mathbf{Q}_K\}$$

with $\mathbf{Q}_P = \mathbf{I}$ and $\mathbf{Q}_K = \mathbf{I}$.

The original Procrustes distance between $\mathbf{X}_i$ and $\mathbf{X}_j$ is given by:

$$\begin{aligned} \mathrm{G}(\mathbf{X}_i, \mathbf{X}_j) &= \mathrm{tr}(\mathbf{X}_i - \mathbf{X}_j)^{\mathrm{T}}(\mathbf{X}_i - \mathbf{X}_j) \\ &= s^2\,\mathrm{tr}(\mathbf{Z}_i - \mathbf{Z}_j)^{\mathrm{T}}(\mathbf{Z}_i - \mathbf{Z}_j) \\ &= s^2\,\mathrm{tr}(\mathbf{W}_{ij}^{\mathrm{T}}\mathbf{W}_{ij}) \end{aligned} \tag{20}$$

where the matrix $\mathbf{W}_{ij} = \mathbf{Z}_i - \mathbf{Z}_j$, seen the independence of $\mathbf{Z}_i$ and $\mathbf{Z}_j$, follows a distribution N(0,2).

Analogously to the previous case where just one matrix configuration is modified by a random error matrix, $G(\mathbf{X}_i,\mathbf{X}_j)$ follows a χ^2 distribution with $n \times k$ degrees of freedom (Langron and Collins 1985, p. 281); that is:

$$G(\mathbf{X}_i,\mathbf{X}_j) \approx 2s^2\chi^2_{[nk]}$$

We proceed again to the OLS matching of the two matrices by applying geometrical transformations of the following kind:

- translation
- translation and rotation
- translation rotation and dilatation

Langron and Collins (1985, pp. 279 and 280, theorems 4.2 and 4.3) proof the type of distribution that characterises the following residual distances:

$$G_{\mathbf{t}}(\mathbf{X}_i,\mathbf{X}_j) \approx 2s^2\chi^2_{[nk-k]}$$

$$G_{\mathbf{t,R}}(\mathbf{X}_i,\mathbf{X}_j) \approx 2s^2\chi^2_{[nk-1/2k(k+1)]}$$

$$G_{\mathbf{t,R,c}}(\mathbf{X}_i,\mathbf{X}_j) \approx 2s^2\chi^2_{[nk-1/2k(k+1)-1]}$$

6 Analysis of variance for the OLS Procrustes statistics

We want to verify now the significance of the parameters of each geometrical transformation, i.e. translation, rotation and dilatation, that minimise, independently, the Procrustes distance between matrices $\mathbf{X}_i$ and $\mathbf{X}_j$.

The proposed test is based on the analysis of variance (ANOVA) of the various contributions and consists in the study of the statistical distribution assumed by the numerical values obtained by the ratio of the Procrustes distance, specific for each transformation of kind translation, rotation and dilatation, and the residual distance that remains when a complete transformation is carried out.

Let's consider at first the result of a translation procedure of $\mathbf{X}_i$ and $\mathbf{X}_j$. The Procrustes distance that remains after centring both matrices is given by (Sibson 1978, p. 235):

$$\begin{aligned} G_{\mathbf{t}}(\mathbf{X}_i,\mathbf{X}_j) &= \mathrm{tr}\left(\mathbf{X}_i - \mathbf{1t}_i - \mathbf{X}_j + \mathbf{1t}_j\right)^{\mathrm{T}} \times \\ &\quad \left(\mathbf{X}_i - \mathbf{1t}_i - \mathbf{X}_j + \mathbf{1t}_j\right) \\ &= \mathrm{tr}\left[\left(\mathbf{X}_i - \mathbf{X}_j\right)^{\mathrm{T}}\left(\mathbf{X}_i - \mathbf{X}_j\right)\right] - \\ &\quad n\left(\mathbf{x}_i - \mathbf{x}_j\right)^{\mathrm{T}}\left(\mathbf{x}_i - \mathbf{x}_j\right) \end{aligned} \qquad (21)$$

where $\mathbf{x}_i$ and $\mathbf{x}_j$ are the positional vectors of the centers of gravity of the matrices $\mathbf{X}_i$ and $\mathbf{X}_j$ expressed in the original reference system.
The difference:

$$T(\mathbf{X}_i,\mathbf{X}_j) = G(\mathbf{X}_i,\mathbf{X}_j) - G_{\mathbf{t}}(\mathbf{X}_i,\mathbf{X}_j) \qquad (22)$$

corresponds to the Procrustes distance due to the translation of the k-components of the two matrices. This value will be submitted afterwards to a statistical test in order to verify its significance.

The residual distance $G_{\mathbf{t}}(\mathbf{X}_i,\mathbf{X}_j)$ following the translation can be further partitioned into a sum of squares due to the rotation $R(\mathbf{X}_i,\mathbf{X}_j)$ and in a residual distance defined by $G_{\mathbf{t,R}}(\mathbf{X}_i,\mathbf{X}_j)$:

$$R(\mathbf{X}_i,\mathbf{X}_j) = G_{\mathbf{t}}(\mathbf{X}_i,\mathbf{X}_j) - G_{\mathbf{t,R}}(\mathbf{X}_i,\mathbf{X}_j) \qquad (23)$$

According to the hypothesis made previously, the Procrustes distance $\mathbf{R}(\mathbf{X}_i,\mathbf{X}_j)$, due to the rotation, follows a χ^2 distribution (Langron and Collins, 1985, p. 281)

$$R(\mathbf{X}_i,\mathbf{X}_j) \approx 2s^2\chi^2_{[1/2k(k-1)]}$$

The residual quantity $G_{\mathbf{t,R}}(\mathbf{X}_i,\mathbf{X}_j)$ can be finally partitioned in a Procrustes distance due to the application of a dilatation factor c and in a residual distance $G_{\mathbf{t,R,c}}(\mathbf{X}_i,\mathbf{X}_j)$ that follows from a complete OLS similarity transformation:

$$S(\mathbf{X}_i,\mathbf{X}_j) = G_{\mathbf{t,R}}(\mathbf{X}_i,\mathbf{X}_j) - G_{\mathbf{t,R,c}}(\mathbf{X}_i,\mathbf{X}_j) \qquad (24)$$

The χ^2 distribution of the various distances just considered remains valid in the case of a null hypothesis, that is when the signals of kind translation, rotation and dilatation are not separable from the remaining noise. If the belonging of the various Procrustes distances to a χ^2 distribution is wanted to be tested, it is necessary to exactly know also the value of s^2.

The difficulty to know with enough precision and reliability this value, shows that it is more reasonable to test the rate between the Procrustes distances due to translation, rotation and dilatation and the residual distance of a complete similarity transformation. The value so obtained does not depend on s^2 and, in the case of a null hypothesis, follows an F central distribution (rate of two χ^2 distributions).

The rejection of the null hypothesis, once a value α for a first kind error has been fixed and for the specific degrees of freedom, shows the presence of significant transformation parameters.

In the case in which the translation component is not significant with respect to the remaining white noise, then the condition given by the following null hypothesis is satisfied:

$$H_0 : \frac{T(\mathbf{X}_i, \mathbf{X}_j)}{G_{\mathbf{t},\mathbf{R},c}(\mathbf{X}_i, \mathbf{X}_j)} \leq F_{k,\, nk-1/2k(k+1)-1}$$

In the opposite case, for significant translations, the following alternative hypothesis is satisfied:

$$H_1 : \frac{T(\mathbf{X}_i, \mathbf{X}_j)}{G_{\mathbf{t},\mathbf{R},c}(\mathbf{X}_i, \mathbf{X}_j)} > F_{k,\, nk-1/2k(k+1)-1}$$

In a similar way, we report the null hypothesis conditions of the test when it is applied for the study of the rotation component:

$$H_0 : \frac{R(\mathbf{X}_i, \mathbf{X}_j)}{G_{\mathbf{t},\mathbf{R},c}(\mathbf{X}_i, \mathbf{X}_j)} \leq F_{1/2k(k-1),\, nk-1/2k(k+1)-1}$$

and of the dilation component:

$$H_0 : \frac{S(\mathbf{X}_i, \mathbf{X}_j)}{G_{\mathbf{t},\mathbf{R},c}(\mathbf{X}_i, \mathbf{X}_j)} \leq F_{1,\, nk-1/2k(k+1)-1}$$

7 Generalised Procrustes statistics and the analysis of variance

The analysis carried out for two matrices of point coordinates can be generalised to m configurations, where $m > 2$. In this case the generic coordinate matrix $\mathbf{X}_i$ is given by:

$$\mathbf{X}_i = \mathbf{X} + s\mathbf{Z}_i \tag{25}$$

where, as usual, $\mathbf{Z}_i$ is an error matrix distributed according to N(0,1) and s is a common multiplicative factor.

In the Generalised Procrustes analysis (Gower 1975; Ten Berge 1977), the simultaneous matching of m matrices $\mathbf{X}_i$ is considered.

To all these matrices the following series of common transformations are applied:

- translation
- translation and rotation
- translation, rotation and dilatation

so to minimise, for each transformation, the Procrustes distance between the various matrices and the so called centroid $\mathbf{K}$ (estimated figure of maximal consensus), that is:

$$G_{\mathbf{t}}(\mathbf{X}_1 \dots \mathbf{X}_m) := \sum_{i=1}^{m} \inf\left(G\left(\mathbf{K}; \mathbf{X}_i + \mathbf{1}\mathbf{t}_i^T\right)\right) \tag{26}$$

$$G_{\mathbf{t},\mathbf{R}}(\mathbf{X}_1 \dots \mathbf{X}_m) := \sum_{i=1}^{m} \inf\left(G\left(\mathbf{K}; \mathbf{X}_i\mathbf{R}_i + \mathbf{1}\mathbf{t}_i^T\right)\right) \tag{27}$$

$$G_{\mathbf{t},\mathbf{R},c}(\mathbf{X}_1 \dots \mathbf{X}_m) := \sum_{i=1}^{m} \inf\left(G\left(\mathbf{K}; c_i\mathbf{X}_i\mathbf{R}_i + \mathbf{1}\mathbf{t}_i^T\right)\right) \tag{28}$$

where:

$$\mathbf{K} = \frac{1}{m} \sum_{i=1}^{m} \left(c_i\mathbf{X}_i\mathbf{R}_i + \mathbf{1}\mathbf{t}_i^T\right) \tag{29}$$

represents the mean transformed configuration.

In the article previously cited, Langron and Collins (1985, p. 282-284) apply the analysis of variance (ANOVA) also to the generalised Procrustes solution and propose the extendibility of the hypothesis tests, previously discussed for two configurations of points, to the generalised case of m configurations.

First, these authors show (p. 282, theorems 6.1 and 6.2) that starting from the hypothesis reported in paragraph 4, the Procrustes distances, for the additive property of the χ^2 distribution and with the respective degrees of freedom multiplied by $m-1$, satisfy:

$$G_{\mathbf{t}}(\mathbf{X}_1 \dots \mathbf{X}_m) \simeq s^2 \chi^2_{(m-1)(nk-k)}$$

$$G_{\mathbf{t},\mathbf{R}}(\mathbf{X}_1 \dots \mathbf{X}_m) \simeq s^2 \chi^2_{(m-1)(nk-1/2k(k+1))}$$

$$G_{\mathbf{t},\mathbf{R},c}(\mathbf{X}_1 \dots \mathbf{X}_m) \simeq s^2 \chi^2_{(m-1)(nk-1/2k(k+1)-1)}$$

being m–1 the independent configurations $\mathbf{X}_i$ with respect to their linear combination $\mathbf{K}$.
Let's define therefore:

$$T(\mathbf{X}_1 \dots \mathbf{X}_m) = G(\mathbf{X}_1 \dots \mathbf{X}_m) - G_{\mathbf{t}}(\mathbf{X}_1 \dots \mathbf{X}_m) \tag{30}$$

$$R(\mathbf{X}_1 \dots \mathbf{X}_m) = G_{\mathbf{t}}(\mathbf{X}_1 \dots \mathbf{X}_m) - G_{\mathbf{t},\mathbf{R}}(\mathbf{X}_1 \dots \mathbf{X}_m) \tag{31}$$

$$S(\mathbf{X}_1 \dots \mathbf{X}_m) = G_{\mathbf{t},\mathbf{R}}(\mathbf{X}_1 \dots \mathbf{X}_m) - G_{\mathbf{t},\mathbf{R},c}(\mathbf{X}_1 \dots \mathbf{X}_m) \tag{32}$$

with the same corresponding meaning given in (22), (23) and (24) respectively.

Passing now to the study of the translation component by the application of a hypothesis test, in the case in which the translation components have the same magnitude as the noise, then, the following condition that corresponds to the null hypothesis is satisfied:

$$H_0 : \frac{T(\mathbf{X}_1 \ldots \mathbf{X}_m)}{G_{\mathbf{t},\mathbf{R},c}(\mathbf{X}_1 \ldots \mathbf{X}_m)} \leq F_{(m-1)k,\ (m-1)(nk-1/2k(k+1)-1)}$$

On the contrary, that is when the translation is significant, the null hypothesis is rejected and the following inequality is satisfied:

$$H_1 : \frac{T(\mathbf{X}_1 \ldots \mathbf{X}_m)}{G_{\mathbf{t},\mathbf{R},c}(\mathbf{X}_1 \ldots \mathbf{X}_m)} > F_{(m-1)k,\ (m-1)(nk-1/2k(k+1)-1)}$$

The same can be done for the rotation and for the dilation components. The value satisfying the null hypothesis of the test for the rotation is:

$$H_0 : \frac{R(\mathbf{X}_1 \ldots \mathbf{X}_m)}{G_{\mathbf{t},\mathbf{R},c}(\mathbf{X}_1 \ldots \mathbf{X}_m)} \leq F_{(m-1)/2k(k-1),\ (m-1)(nk-1/2k(k+1)-1)}$$

while the one relative to the scale factor is:

$$H_0 : \frac{S(\mathbf{X}_1 \ldots \mathbf{X}_m)}{G_{\mathbf{t},\mathbf{R},c}(\mathbf{X}_1 \ldots \mathbf{X}_m)} \leq F_{(m-1),(m-1)(nk-1/2k(k+1)-1)}$$

8 Conclusions

Procrustes algorithms constitute valid tools for the direct and simultaneous estimation of the OLS similarity transformation parameters for two (OPA) and for more (GPA) data matrices.

In this paper we have presented a method, based on the perturbation theory from Langron and Collins (1985) to test for significant OLS similarity transformation parameters.

Future research will be devoted to the extension of Procrustes analysis techniques and statistics to the datum transformation model given by Total Least Squares (Golub and van Loan, 1980).

References

Arun K, Huang T and Blostein S (1987) Least-Squares Fitting of Two 3-D Point Sets. *IEEE Trans. on Pattern Analysis and Machine Intelligence*, 9(5), pp 698-700.

Beinat A and Crosilla F (2001) Generalised Procrustes Analysis for Size and Shape 3-D Object Reconstructions. In: *Optical 3-D Measurement Techniques*. Gruen and Kahmen (Eds), Wien, pp. 345-353.

Borg I and Groenen P (1997) *Modern Multidimensional Scaling. Theory and Applications*. Springer-Verlag, New York.

Commandeur JJF (1991) *Matching configurations*. DSWO Press, Leiden University, III, M&T series; 19.

Crosilla F (1999) Procrustes Analysis and Geodetic Sciences. In: *Quo vadis geodesia ... ?*, Krumm and Schwarze (Eds), Festschrift to EW Grafarend on the occasion of his 60th bithday, pp 69-78.

Crosilla F and Beinat A (2002) Use of Generalised Procrustes analysis for photogrammetric block adjustment by independent models. *ISPRS Journal of Photogrammetry and Remote Sensing*, 56(3), pp. 195-209.

Golub G and van Loan CF (1980) An analysis of the total least squares problem, *SIAM Journal on Numerical Analysis*, 17, pp. 883-893.

Goodall C (1991) Procrustes methods in the statistical analysis of shape. *Journal Royal Stat. Soc.*, Part B 53, 2, pp 285-339.

Gower JC (1975) Generalized Procrustes analysis. *Psychometrika*, 40(1), pp 33-51.

Grafarend EW and Awange JL (2000) Determination of vertical deflections by GPS/LPS measurements. *Zeitschrift für Vermessungswesen*, 125, pp 279-288.

Grafarend EW and Awange JL (2002) Nonlinear analysis of the threedimensional datum transformation (conformal group $\mathbb{C}_7(3)$), *Journal of Geodesy*, in print

Horn BKP (1987) Closed Form Solutions of Absolute Orientation Using Unit Quaternions. *Journal of Optical Society of America*, A-4(4), pp 629-642.

Horn BKP, Hilden HM and Negahdaripour S (1988) Closed Form Solution of Absolute Orientation Using Orthonormal Matrices. *Journal of Optical Society of America*, A-5(7), pp 1127-1135.

Langron SP and Collins AJ (1985) Perturbation theory for Generalized Procrustes Analysis. *Journal Royal Statistical Society*, 47(2), pp 277-284.

Kristof W and Wingersky B (1971) Generalization of the orthogonal Procrustes rotation procedure to more than two matrices. *Proc. of the 79-th Annual Conv. of the American Psychological Ass.*, 6, pp 81-90

Sansò F (1973) An exact solution of the roto-translation problem. *Photogrammetria*, 29, pp 203-216

Schönemann PH (1966) A generalized solution of the orthogonal Procrustes problem. *Psychometrika*, 31(1).

Schönemann PH and Carroll RM (1970) Fitting one matrix to another under choice of a central dilation and a rigid motion. *Psychometrika*, 35(2), pp 245-255.

Sibson R (1978) Studies in the Robustness of Multidimensional Scaling: Procrustes Statistics. J. Royal Statistical Society, 40(2), pp 234-238.

Sibson R (1979) Studies in the Robustness of Multidimensional Scaling: perturbational analysis of classical scaling. *J. Royal Statistical Society*, 41(2), pp 217-229.

Strang G and Borre K (1997) *Linear Algebra, Geodesy, and GPS*, Wellesley-Cambridge Press.

ten Berge JMF (1977) Orthogonal Procrustes rotation for two or more matrices. *Psychometrika*, 42, pp 267-276.

ten Berge JMF, Kiers HAL and Commandeur JJF (1993) Orthogonal Procrustes rotation for matrices with missing values. *British Journal of Mathematical and Statistical Psychology*, 46, pp 119-134.

Thompson EH (1958) On exact linear solution of the problem of absolute orientation. *Photogrammetria*, 13(4), pp 163-178.

PDF Evaluation of the Integer Ambiguity Residuals

S. Verhagen and P.J.G. Teunissen

Department of Mathematical Geodesy and Positioning
Delft University of Technology, Thijsseweg 11, 2629 JA Delft, The Netherlands
E-mail: S.Verhagen@GEO.TUDelft.nl

Abstract. A parameter estimation theory is incomplete if no rigorous measures are available for validating the parameter solution. Since the classical theory of linear estimation does not apply to the integer GPS model, rigorous validation is not possible when use is made of the classical results. As with the classical theory, a first step for being able to validate the integer GPS model is to make use of the residuals and their probabilistic properties. The residuals quantify the inconsistency between data and model, while their probabilistic properties can be used to measure the significance of the inconsistency.

In this contribution we will present and evaluate the joint probability density function (PDF) of the multivariate integer GPS carrier phase ambiguity residuals, which are defined as the difference between the real-valued float ambiguity estimates and the integer-valued fixed ambiguity estimates. Since the residuals and their properties depend on the integer estimation principle used, we will present the PDF of the ambiguity residuals for the whole class of admissible integer estimators. This includes the estimation principles of integer rounding, integer bootstrapping and integer least-squares. In order to get a better understanding of the various features of the joint PDF of the ambiguity residuals we will use a step-by-step construction aided by graphical means. Although the results apply for any dimension, the one-dimensional case and the two-dimensional case are highlighted.

Keywords. GPS, integer ambiguity residuals, parameter distributions

1 Introduction

Any GPS model of observation equations that includes carrier phase data of two or more receivers can be parameterized in non-integers and integers. The non-integers refer to the baseline components and additional unknowns like atmospheric delays. The integer parameters refer to the unknown cycle ambiguities of double-differenced carrier phase data. When the integerness of these parameters is explicitily taken into account in the parameter estimation process, we speak of carrier phase ambiguity resolution. It can be applied to a great variety of GPS models that are used in applications like surveying, navigation, and geophysics. An overview of GPS models can be found in textbooks like Hoffmann-Wellenhof et al. (1997), Leick (1995), Parkinson and Spilker (1996), Strang and Borre (1997), Teunissen and Kleusberg (1998).

Parameter estimation provides the estimates of the unknown parameters, together with the corresponding variance-covariance (vc-) matrices. In the classical theory of linear estimation, the vc-matrices provide sufficient information on the precision of the estimated parameters. The reason is that a linear model applied to normally distributed (Gaussian) data, provides linear estimators that are also normally distributed, and the peakedness of the multivariate normal distribution is completely captured by the vc-matrix.

Unfortunately, this relatively simple approach cannot rigorously be applied in case integer parameters are involved in the estimation process, since the integer estimators do not have a Gaussian distribution, even if the model is linear and the data are normally distributed. Instead of the vc-matrices, we therefore have to use the parameter distribution itself in order to obtain the appropriate measures that can be used to validate the integer parameter solution. For that purpose, the probability density function (PDF) of the difference between the real-valued float ambiguity estimates and the integer-valued fixed ambiguity estimates, referred to as the ambiguity residuals, can be used, since these residuals quantify the inconsistency between data and model, while the PDF describes their probabilistic properties, which are a measure for the significance of the inconsistency.

Our goal is to evaluate the PDF of the ambiguity residuals for the whole class of admissible integer estimators, since the residuals and their probabilistic properties depend on the estimation principle that is used.

We will start with the formulation of the integer GPS model in section 2, where also the class of admissible estimators is defined. In section 3 the PDF of the ambiguity residuals for the whole class of admissible integer estimators is presented. In practice, the PDF that is often used is based on the incorrect assumption that the integer estimator is deterministic. It will be shown how the resulting PDF differs from the correct one as presented here.

The evaluation of the PDFs is the subject of section 4. We will focus on the one-dimensional and two-dimensional case. Thereby, we will also look at a realistic GPS model. In order to get a good understanding of the various features of the joint PDF of the residuals, it is shown how the PDF can be constructed step-by-step, aided by graphical means. Furthermore, the second moments of the ambiguity residuals with relation to the precision of the GPS data are shown numerically as well as graphically. The PDFs and second moments will be compared to the ones that are often used in practice, where the randomness of the integer ambiguity estimates is neglected, so that it is assumed that all estimates are normally distributed. Due to this incorrect assumption the ambiguity validation tests that are used are also incorrect.

2 Integer ambiguity resolution

Any GPS observation model can be parameterized in integers and non-integers. This gives the following system of linear(ized) observation equations:

$$E\{\underset{m\times 1}{y}\} = \underset{m\times n}{A}\,\underset{n\times 1}{a} + \underset{m\times p}{B}\,\underset{p\times 1}{b} \tag{2.1}$$

where $E\{.\}$ is the mathematical expectation operator, y is the GPS observation vector of order m, a and b are the unknown parameter vectors of order n and p respectively. The data vector y usually consists of the observed-minus-computed double-difference (DD) phase and/or code observations on one, two or three frequencies and accumulated over all observation epochs. The entries of the parameter vector a will then consist of the unknown integer carrier phase ambiguities, which are expressed in units of *cycles* rather than in units of range. Since it is known that the entries are integers, $a \in Z^n$. The remaining unknown parameters form the entries of the vector b. These parameters may be the unknown baseline components and for instance atmospheric (ionospheric, tropospheric) delays, which are all real-valued, i.e. $b \in R^p$. In this contribution we will refer to the estimator of these real-valued parameters as the baseline estimator, although the vector b may thus contain other parameters than only the baseline components.

The classical linear estimation theory can be applied to models that contain real-valued parameters. However, if the integerness of the ambiguity parameters is taken into account, we have to follow a different approach which includes a separate step for ambiguity resolution. The complete estimation process will then consist of three steps (Euler and Schaffrin 1991, Teunissen 1993). In the first step, the integerness of the vector a is discarded and a float solution is computed with a standard least-squares adjustment. This results in real-valued estimates for a and b and their variance-covariance (vc-) matrix:

$$\begin{pmatrix} \hat{a} \\ \hat{b} \end{pmatrix}, \begin{pmatrix} Q_{\hat{a}} & Q_{\hat{a}\hat{b}} \\ Q_{\hat{b}\hat{a}} & Q_{\hat{b}} \end{pmatrix} \tag{2.2}$$

In the second step the integer ambiguity estimate is computed from the float ambiguity estimate $\hat{a}$:

$$\breve{a} = S(\hat{a}) \tag{2.3}$$

where S: $R^n \rightarrow Z^n$ the mapping from the n-dimensional space of real numbers to the n-dimensional space of integers. The final step is to use integer ambiguity estimates to correct the float estimate of b with

$$\breve{b} = \hat{b} - Q_{\hat{b}\hat{a}} Q_{\hat{a}}^{-1} (\hat{a} - \breve{a}) \tag{2.4}$$

This solution is referred to as the fixed baseline solution. Both eqs.(2.3) and (2.4) depend on the choice of the integer estimator. Different integer estimators are obtained for different choices of the map S: $R^n \rightarrow Z^n$. This implies that also the probability distribution of the estimators depends on the choice of the map.

In order to arrive at a class of integer estimators from which to choose, we will start with the map S: $R^n \rightarrow Z^n$. The space of integers, Z^n, is of a discrete nature, which implies that the map must be a many-to-one map, and not one-to-one. In other words, different real-valued ambiguity vectors a will be mapped to the same integer vector. Therefore, a subset $S_z \subset R^n$ can be assigned to each integer vector $z \in Z^n$:

$$S_z = \{x \in R^n | z = S(x)\},\ z \in Z^n \qquad (2.5)$$

This subset S_z contains all real-valued float ambiguity vectors that will be mapped to the same integer vector z, and it is called the *pull-in region* of z (Jonkman 1998, Teunissen 1998). This implies that $\breve{a}=z \Leftrightarrow \hat{a}\in S_z$. The integer ambiguity estimator can be expressed as:

$$\breve{a} = \sum_{z\in Z^n} z s_z(\hat{a}) \quad \text{with} \quad s_z(x) = \begin{cases} 1 & \text{if } x \in S_z \\ 0 & \text{otherwise} \end{cases} \qquad (2.6)$$

where we used the indicator function $s_z(x)$.

The integer estimator is completely defined by the pull-in region, so that it is possible to define a class of integer estimators by imposing various conditions on the pull-in regions. The class of *admissible* integer estimators is defined as follows.

Definition 2.1 (*Admissible integer estimators*). An integer estimator, $\breve{a} = \sum_{z\in Z^n} z s_z(\hat{a})$, is said to be *admissible* when its pull-in region, $S_z = \{x \in R^n | z = S(x)\},\ z \in Z^n$, satisfies

(*i*) $\bigcup_{z\in Z^n} S_z = R^n$

(*ii*) $\text{Int}\, S_{z_1} \cap \text{Int}\, S_{z_2} = \emptyset, \forall z_1, z_2 \in Z^n, z_1 \neq z_2$

(*iii*) $S_z = z + S_0, \forall z \in Z^n$

where 'Int' denotes the interior of the subset. In Teunissen (1998) the motivation for this definition is given.

Examples of integer estimators that belong to the class of admissible integer estimators are integer rounding (*R*), integer bootstrapping (*B*), and integer least-squares (*LS*). In Teunissen (1999) the corresponding pull-in regions are shown. For a review, see Teunissen (2002b).

3 The Probability Density Function

In Teunissen (2002a) the joint and marginal PDFs of both the integer and non-integer parameters were determined. In order to do so, it was assumed that the float solutions are normally distributed. This implies that the marginal PDF of the float ambiguities is given as:

$$f_{\hat{a}}(x) = \frac{1}{\sqrt{\det Q_{\hat{a}}\,(2\pi)^{\frac{1}{2}n}}} \exp\left\{-\tfrac{1}{2}\|x-a\|^2_{Q_{\hat{a}}}\right\} \qquad (3.1)$$

It can be proven that the joint distribution of $\hat{a}$ and $\breve{a}$ is given by:

$$f_{\hat{a},\breve{a}}(x,z) = f_{\hat{a}}(x) s_z(x),\ x \in R^n, z \in Z^n \qquad (3.2)$$

The distributions of the float and fixed ambiguities can be recovered from this PDF by summing over z or integrating over x respectively. The distribution of the integer ambiguity estimator is given by a probability mass function (PMF), and not by a PDF. This PMF is equal to the integral of the PDF of the float ambiguity over the pull-in region S_z:

$$\int_{R^n} f_{\hat{a},\breve{a}}(x,z)dx = \int_{S_z} f_{\hat{a}}(x)dx = P[\breve{a} = z] \qquad (3.3)$$

Our goal is to determine the PDF of the ambiguity residual, which is defined as:

$$\breve{e} = \hat{a} - \breve{a} \qquad (3.4)$$

In practice it is often incorrectly assumed that the integer estimates are deterministic, which would imply that the PDF of the residuals coincides with the PDF of the float ambiguities. Instead, the PDF can be constructed once the joint distribution of $\breve{e}$ and $\breve{a}$ is known. This PDF can be obtained using the transformation (see Teunissen, 2002a):

$$\begin{bmatrix} \breve{e} \\ \breve{a} \end{bmatrix} = \begin{bmatrix} I_n & -I_n \\ 0 & I_n \end{bmatrix} \begin{bmatrix} \hat{a} \\ \breve{a} \end{bmatrix} \qquad (3.5)$$

Since $f_{\hat{a},\breve{a}}(x)$ is known (eq.(3.2)), the following can be obtained:

$$f_{\breve{e},\breve{a}}(x,z) = f_{\hat{a},\breve{a}}(x+z,z) = f_{\hat{a}}(x+z)s_0(x) \qquad (3.6)$$

The PDF of the ambiguity residuals then follows from summing over all integers:

$$f_{\breve{e}}(x) = \sum_{z\in Z^n} f_{\hat{a}}(x+z)s_0(x),\ x \in R^n, z \in Z^n \qquad (3.7)$$

With eq.(3.1) this finally gives:

$$f_{\breve{e}}(x) = \frac{1}{\sqrt{\det Q_{\hat{a}}\,(2\pi)^{\frac{1}{2}n}}} \times \sum_{z\in Z^n} \exp\{-\tfrac{1}{2}\|x-a+z\|^2_{Q_{\hat{a}}}\} s_0(x) \qquad (3.8)$$

Note that $f_{\breve{e}}(x)$ is only sensitive to the fractional part of a because the summation is over all integers, so that replacing a by a-[a] does not change the result ([a] means that a is rounded to the nearest integer).

Figure 3.1 shows all steps required for the construction of the PDF of the ambiguity residuals in the one-dimensional (1-D) case. The PDF of $\hat{a}$ (top left) is plotted along the x-axis, the PMF of $\breve{a}$ (top right) along the z-axis, and the joint PDF (top

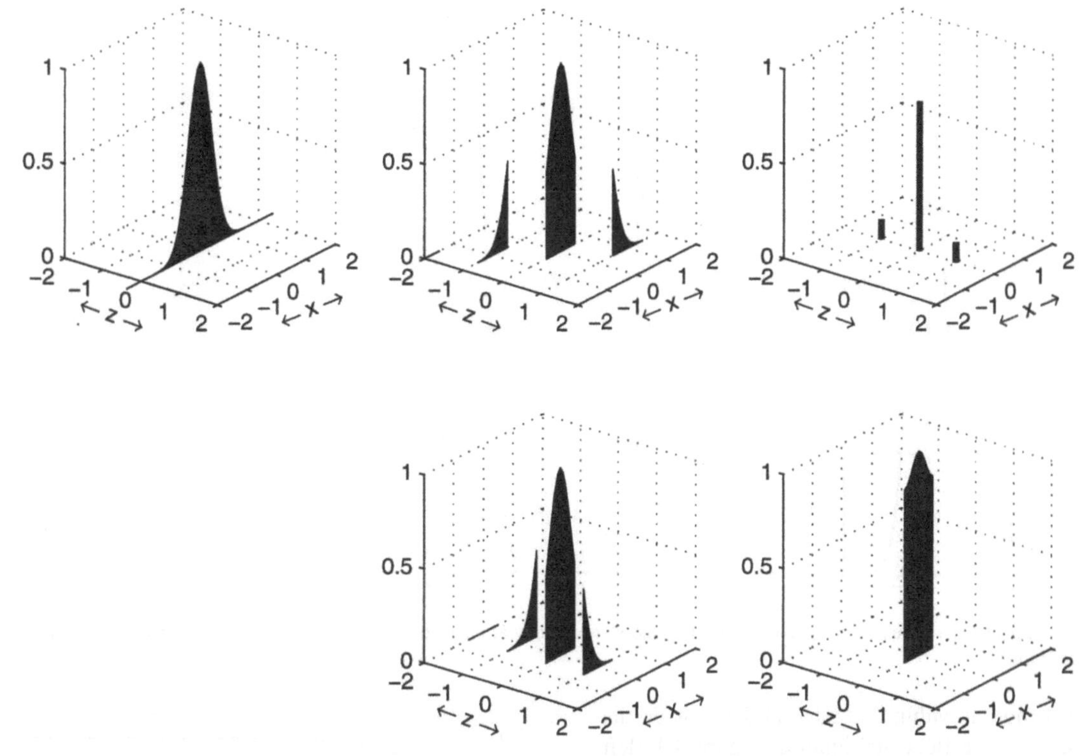

Fig. 3.1 The construction of $f_{\check{\epsilon}}(x)$ from $f_{\hat{a}}(x)$: PDF $f_{\hat{a}}(x)$ (top left); joint PDF $f_{\hat{a},\check{a}}(x,z)$ (top middle); PMF $P[\check{a}=z]$ (top right); joint PDF $f_{\check{\epsilon},\check{a}}(x,z)$ (bottom middle); PDF $f_{\check{\epsilon}}(x)$ (bottom right).

middle) is plotted in the xz-plane. Its construction from the marginal PDF and PMF can be seen as follows. First, the parts of the PDF of $\hat{a}$ are sliced out that correspond to all pull-in regions. For the 1-D case, the pull-in regions simply are intervals with length 1, centered at the integers, $S_z=\{x\in R^n \mid |x-z| \leq \frac{1}{2}\}$. These slices are then translated along the z-axis to the corresponding integers z.

The joint PDF of $\check{\epsilon}$ and $\check{a}$ (bottom left) follows from another translation of the slices, but now along the x-axis, so that they are all centered at the mean value $x = 0$. The PDF of $\check{\epsilon}$ (bottom right) is finally obtained by summing over z, i.e. all slices are again translated along the z-axis to the origin.

The distribution of the ambiguity residuals is clearly non-Gaussian, and it equals zero for all values of x outside the pull-in region. That implies that the norm of the vector of ambiguity residuals is always bounded, irrespective of the value taken by the float solution. A very important implication is that the difference between the float and the fixed baseline solution is then also bounded, as can be seen in eq.(2.4). See also Teunissen (2001).

It can also be seen that the PDF is symmetric around the origin, but the shape depends on the precision of float solution. More examples are given in the next sections. This means that the PDF is independent of the unknown integer ambiguity vector $a\in Z^n$. In other words, the mean of the ambiguity residual equals zero, $E\{\check{\epsilon}\}=0$. Thus, the PDF is completely known once the precision of the float solution is known and the choice of the integer estimator is made.

4 Evaluation of the PDF

The results of the preceding section apply to any dimension. However, for evaluation we focus on the one-dimensional and two-dimensional case.

4.1 One-dimensional case

In the one-dimensional case we do not need to distinguish between the three afore-mentioned admissible estimators, since they are all identical. In figure 3.1 a step-by-step graphical construction of the PDF of the residuals was shown. All probability mass is located within the pull-in interval $(-\frac{1}{2},\frac{1}{2})$.

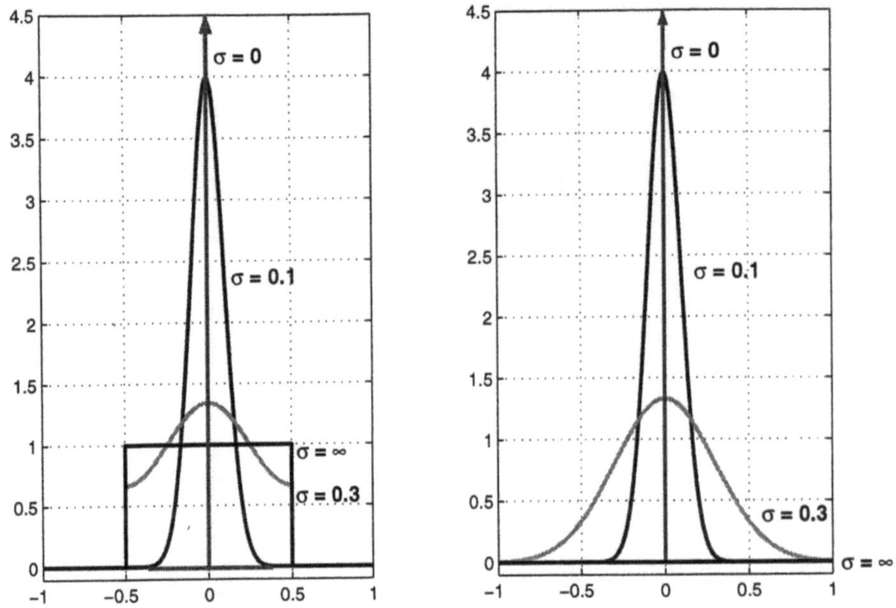

Fig. 4.1 PDF of the residuals for different values of the standard deviation σ: correct PDFs (left); and PDFs as often used in practice (right).

The distribution within this interval depends on the precision of the float ambiguities. Figure 4.1 (left panel) shows the PDF of the residuals for different values of the standard deviation, σ, of the float ambiguities. Also, the extremes $\sigma = 0$ and $\sigma = \infty$ are shown. In the first case, the PDF becomes an impulse function, in the latter case the residuals have a uniform distribution within the pull-in interval. The steepness of the PDF's peak clearly depends on the precision.

The right panel of figure 4.1 shows the PDF of the float ambiguities for the same standard deviations. In practice, often the randomness of the fixed solution is ignored, implying that: $f_{\check{\varepsilon}}(x) := f_{\hat{a}}(x)$. This distribution function has infinite tails. Note that when the precision is high, for example $\sigma = 0.1$ *cycles*, that the PDFs become almost identical. On the other hand, when the precision is low, for example $\sigma = 0.3$ *cycles*, the PDFs are clearly different. If one would use the false PDF by assuming that the fixed solution is deterministic, that may lead to false hypothesis tests on the integerness of the ambiguities.

Knowledge of the PDF also allows us to derive expressions for the second moments, i.e. the variance and covariance, of the ambiguity residuals. The variance can be computed with:

$$\sigma_{\check{\varepsilon}}^2 = \int_{x \in S_0} x^2 f_{\check{\varepsilon}}(x)dx \tag{4.1}$$

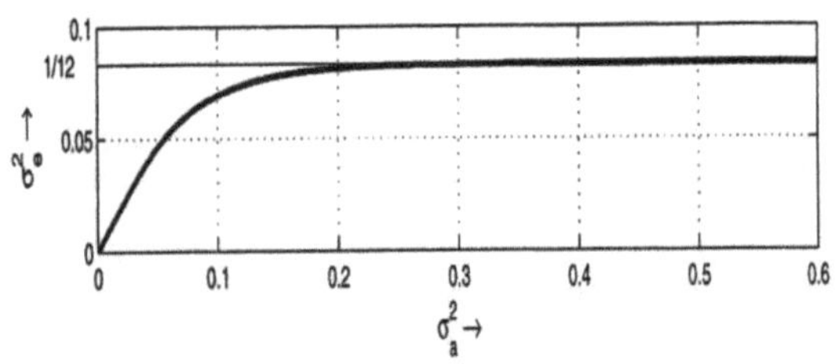

Fig. 4.2 Variance of the residuals, σ_e^2, versus the variance of the float ambiguities, σ_a^2, (1-D).

Figure 4.2 showns the variance of ambiguity residuals, σ_e^2, as function of the variance of the float ambiguities, σ_a^2. As can be seen the variance of the residuals is always smaller than $^1/_{12}$ and it is always lower or equal to the variance of the float ambiguities. Again this shows that it is not correct to assume that the fixed solution is deterministic, so that $\sigma_{\check{\varepsilon}}^2 := \sigma_{\hat{a}}^2$.

4.2 Two-dimensional case

In the two-dimensional case the PDFs of the various admissible estimators may be quite different, especially if there is a high correlation between the float ambiguities. Therefore, we have looked at the resulting PDFs for a vc-matrix that would be obtained by choosing a realistic GPS model. In this example that means that we use the code and phase observations on the L1 and L2 frequency with undifferenced standard deviations of 0.3 *m* and 0.003 *m* for code and phase observations respectively. The resulting vc-matrix of the float ambiguities reads:

$$Q_{\hat{a}} = \begin{bmatrix} 4.9718 & 3.8733 \\ 3.8733 & 3.0188 \end{bmatrix} \tag{4.2}$$

The PDF of the residuals is constructed in a similar way as shown for the 1-D case. This is shown in figure 4.3 in case of rounding: the probability mass of $\hat{a}$ in all pull-in regions, which are squares in this case (top), is added to the mass in the pull-in region centered at the origin (middle).

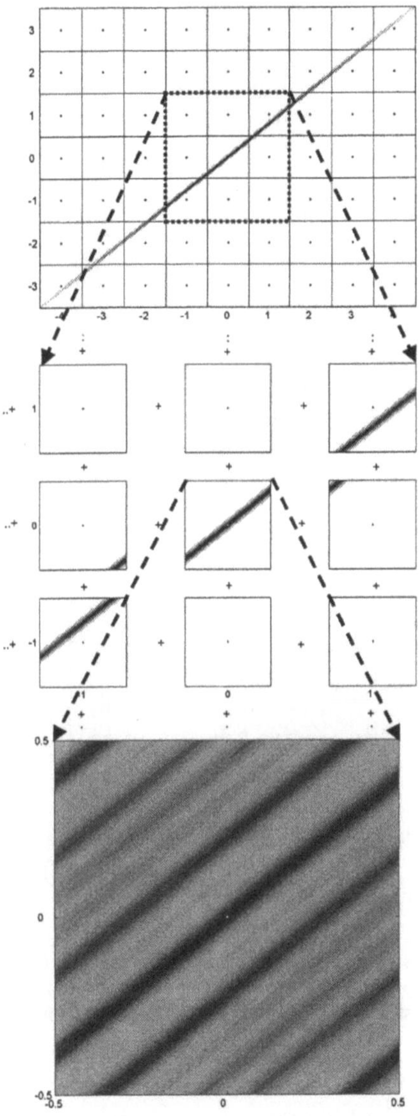

Fig. 4.3 The construction of $f_{\check{\varepsilon}}(x)$ from $f_{\hat{a}}(x)$ in case of rounding (2-D): PDF $f_{\hat{a}}(x)$ (top); summation over all pull-in regions (middle); PDF $f_{\check{\varepsilon}}(x)$ (bottom).

The result is the PDF of the residuals shown in the bottom panel (a dark area corresponds to a large probability mass).

Figure 4.4 shows the PDFs of the float ambiguities (top), and of the ambiguity residuals obtained with the different estimators. In figure 4.5 the PDFs of the corresponding float ambiguities and the residuals after Z-transformation are shown. This Z-transformation is used to decorrelate the components of the float ambiguity vector (see e.g. Teunissen, 1998):

$$\hat{z} = Z^T \hat{a}, \quad Q_{\hat{z}} = Z^T Q_{\hat{a}} Z \tag{4.3}$$

In our example this results in:

$$Q_{\hat{z}} = \begin{pmatrix} 0.0865 & 0.0364 \\ 0.0364 & 0.0847 \end{pmatrix}, \quad Z = \begin{pmatrix} -3 & -4 \\ 4 & 5 \end{pmatrix} \tag{4.4}$$

If we first look at the results obtained for the vc-matrix given in (4.2) we see that the distribution clearly depends on the integer estimator that is used, and that it may be considerably different from the PDF of $\hat{a}$. Firstly, the PDF of $\hat{a}$ is not bounded, whereas the PDFs of the residuals are bounded by the domain of the corresponding pull-in region. Secondly, the PDFs of $\check{\varepsilon}$ in this example are multimodal. This is due to the fact that the PDF of $\hat{a}$ is very elongated and its orientation is not along the a_1-, a_2- or a_1=a_2-axis, where a_1 and a_2 are the values that the two components of the float ambiguity vector may take. This implies that there is also a high probability mass located far away from the origin. So, this confirms that the fixed ambiguities may not always be considered deterministic.

The PDF of the residual for integer LS, $\check{\varepsilon}_{LS}$, is shown in the bottom panel of figure 4.4. The pull-in region of the integer LS estimator follows very much the shape of the PDF of $\hat{a}$ and is thus very elongated. Therefore, it is hard to visually detect any differences between the PDFs of $\hat{a}$ and $\check{\varepsilon}_{LS}$, although there are differences especially near the boundaries of the pull-in region and of course outside the pull-in region. However, this will be easier to see for the residuals of the Z-transformed ambiguities. Note that only for the integer LS estimator the following is true:

$$\check{\varepsilon}_{z,LS} = \hat{z} - \check{z}_{LS} = Z^T(\hat{a} - \check{a}_{LS}) = Z^T \check{\varepsilon}_{LS} \tag{4.5}$$

so that

$$f_{\check{\varepsilon}_{z,LS}}(x) = f_{\check{\varepsilon}_{LS}}(Z^{-T}x) \tag{4.6}$$

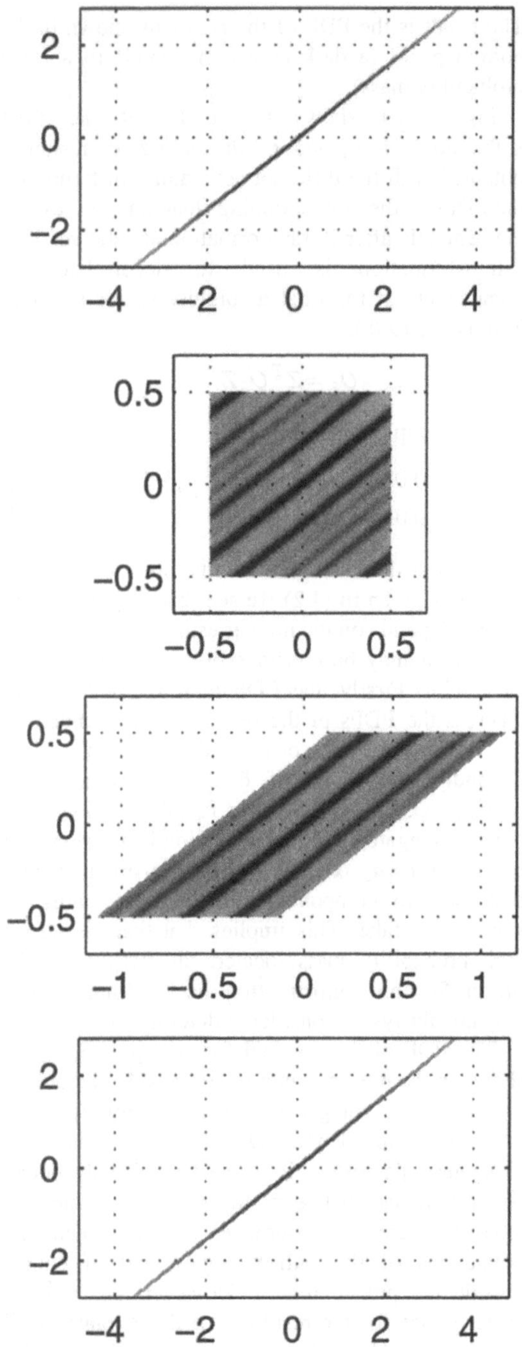

Fig. 4.4 PDFs of the residuals (2-D): PDF $f_{\hat{a}}(x)$ (top); PDF $f_{\check{\epsilon}}(x)$ for rounding (2nd from top); PDF $f_{\check{\epsilon}}(x)$ for bootstrapping (3rd from top); PDF $f_{\check{\epsilon}}(x)$ for integer LS (bottom).

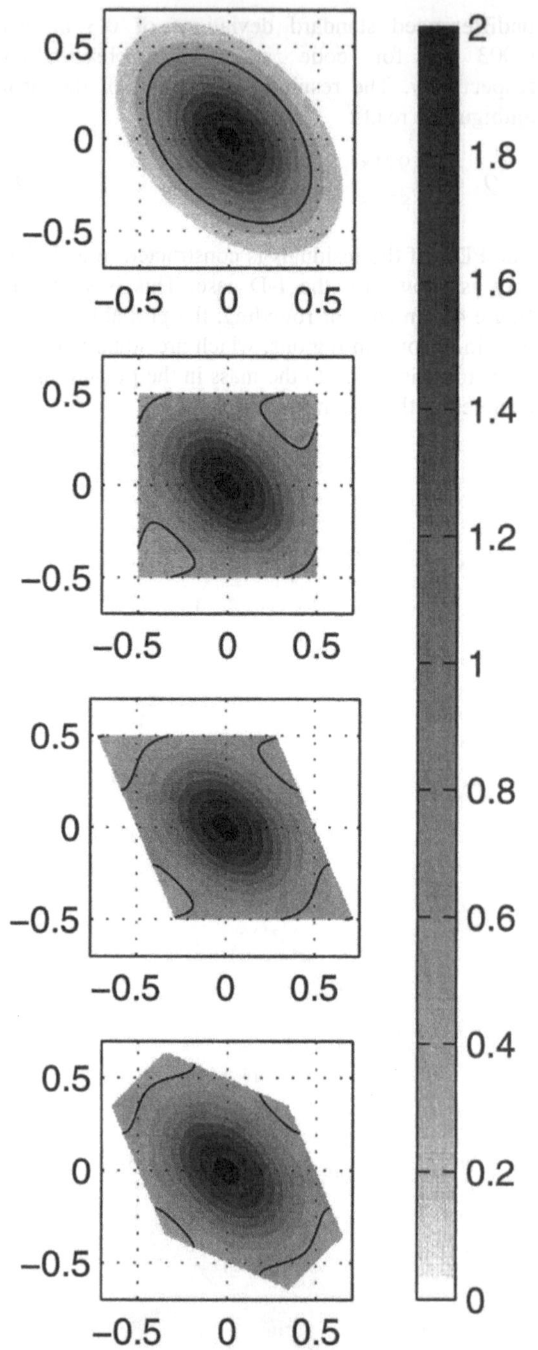

Fig. 4.5 PDFs of the residuals after *Z*-transformation (2-D): PDF $f_{\hat{a}}(x)$ (top); PDF $f_{\check{\epsilon}}(x)$ for rounding (2nd from top); PDF $f_{\check{\epsilon}}(x)$ for bootstrapping (3rd from top); PDF $f_{\check{\epsilon}}(x)$ for integer LS (bottom).

Therefore, the shape of the PDF of the Z-transformed residuals is similar to the PDF of the untransformed residuals, only that it is transformed like the pull-in region.

Figure 4.5 shows that especially for the bootstrapped and integer LS estimator the shape of the PDF of the residuals 'fits' the shape of the pull-in region quite well. Due to the decorrelation, the resulting PDFs are now unimodal. However, the shape near the boundaries of the pull-in regions is clearly different from the shape of the PDF of $\hat{a}$. For clarity this is shown by the black contour lines which are plotted at a value of 0.6. Note that these contour lines are not closed, which means that biases in certain directions will be harder to detect than biases in other directions, depending on the correlation between the float ambiguities.

Although the PDFs of $\hat{a}$ and $\check{e}_{z,LS}$ have the most similar shapes, the figure shows that the distributions are different from each other. Again the conclusion is that ignoring the randomness of the fixed solution leads to false assumptions on the distribution and thus to incorrect tests for integerness of the parameters.

The second moments of the ambiguity residuals in case of rounding, $\check{e}_R$, are computed with:

$$\sigma_{\check{e}_i\check{e}_j} = \iint_{x\in S_0} x_i x_j f_{\check{e}_i\check{e}_j}(x_i, x_j)dx_i dx_j \tag{4.7}$$

The resulting vc-matrices are:

$$Q_{\check{e}_R} = \begin{pmatrix} 0.0833 & -0.0007 \\ -0.0007 & 0.0833 \end{pmatrix}$$
$$Q_{\check{e}_{z,R}} = \begin{pmatrix} 0.0650 & -0.0067 \\ -0.0067 & 0.0643 \end{pmatrix} \tag{4.8}$$

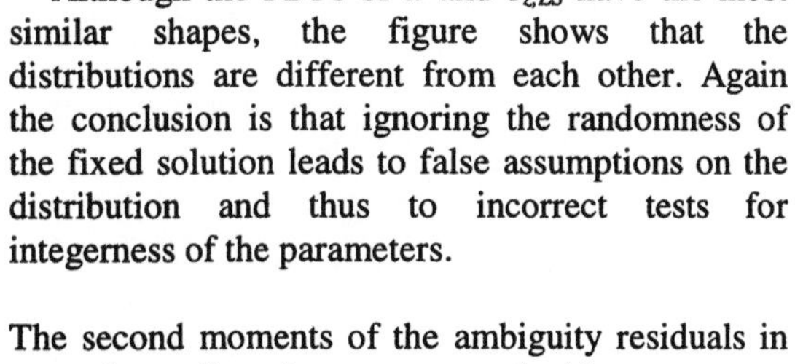

Fig. 4.6 Ellipses corresponding to the vc-matrices of the float ambiguites (light grey) and the residuals (black): for $Q_{\hat{a}}$ (left) and $Q_{\hat{z}}$ (right).

Figure 4.6 shows the ellipses that correspond to these vc-matrices of the residuals without and with Z-transformation. Also, the corresponding vc-matrices of the float ambiguities, $Q_{\hat{a}}$ and $Q_{\hat{z}}$, are shown. These are the second moments of the residuals that are usually used in practice. It can be seen that especially in case the float ambiguities are highly correlated, the deviation is large.

5 Conclusions

In this contribution it is shown how the PDFs of ambiguity residuals can be constructed. This allows for rigorous testing of the integer parameters. The next step will be to formulate hypotheses and determine test statistics sensitive to these hypotheses for the *n*-dimensional case for all admissible integer estimators. Thereby we will have to look at the size and the power of the test and the test should be optimal.

Currently, the randomness of the fixed solution is ignored, which results in wrong PDFs of the residuals, and wrong second moments. This may lead to a false evaluation of the quality of testing the integerness of the ambiguities. The knowledge of the correct PDFs can be used to investigate in which cases the test as used in practice gives a sufficient description, and in which cases it will lead to false conclusions.

References

Euler, H.J. and B. Schaffrin (1990). On a measure of discernibility between different ambiguity solutions in the static-kinematic GPS-mode. In: K.P. Schwarz and G. Lachapelle (eds.), IAG-symp., vol.107, *Kinematic Systems in Geodesy, Surveying, and Remote Sensing*, Springer, New York, pp.285-295.

Hoffmann-Wellenhof, B., H. Lichtenegger, and J. Collins (1997). Global Positioning System: Theory and Practice. 4th edn., Springer, Berlin Heidelberg New York.

Jonkman, N.F. (1998). Integer Ambiguity Estimation without the Receiver-Satellite Geometry. *LGR Series*, no.18, Delft Geodetic Computing Centre, Delft.

Leick, A. (1995). GPS Satellite Surveying. 2nd edn., John Wiley, New York.

Parkinson, B. and J.J. Spilker (eds) (1996). GPS: Theory and Applications. Vols. 1 and 2, AIAA, Washington, DC.

Strang, G. and K. Borre (1997). Linear Algebra, Geodesy, and GPS. Wellesley-Cambridge Press.

Teunissen, P.J.G. (1993). Least-squares Estimation of the Integer GPS Ambiguities. Invited lecture, *Section IV Theory and Methodology, IAG General Meeting*. Beijing, August. Also in: *LGR Series*, No.6, Delft Geodetic Computing Centre, Delft, pp. 59-74.

Teunissen, P.J.G. (1998). On the Integer Normal Distribution of the GPS Ambiguities. *Artific Sat* 33(2): 49-64.

Teunissen, P.J.G. (1999). An Optimality Property of the Integer Least-squares Estimator. *J Geod* 73: 587-593.

Teunissen, P.J.G. (2001). Integer Estimation in the Presence of Biases. *J Geod* 75: 399-407.

Teunissen, P.J.G. (2002a). The Parameter Distributions of the Integer GPS Model. *J Geod* 76: 41-48.

Teunissen, P.J.G. (2002b). Integer Least-Squares. In: *Proc. V Hotine-Marussi Symposium.* Matera, June 17-21 (this volume).

Teunissen, P.J.G. and A. Kleusberg (eds) (1998). GPS for Geodesy. 2nd edn., Springer, Berlin Heidelberg New York.

Multiple Models – Fixed, Switching, Interacting

Brigitte Gundlich, Peter Teunissen
Mathematical Geodesy and Positioning
TU Delft, Thijsseweg 11, 2629 JA Delft, The Netherlands

Abstract. In dynamic models the dynamic and the observation equations are based on a known system model. The multiple model approach introduces uncertainties about the system model by a set of possible system models. In the multiple model approach for fixed models the true system does not change during the whole observation process, wheareas in the approach for switching models a jump from one model to another is allowed. In the later case the state estimation usually has to be approximated, e.g. by so-called interacting multiple models. The multiple model approach for fixed, switching and interacting models are presented and their application for GNSS ambiguity resolution is discussed, but open questions still remain.

Keywords. Bayesian statistics, ambiguity resolution, model uncertainty, recursive estimation, multiple models, interacting multiple models

1 Introduction

The multiple model is a generalization of a dynamic system. In the dynamic system the system model, that defines the dynamic and the observation equations, is considered as known, so that the dynamic system can be interpreted as a single model approach. Compared to it the multiple model approach provides the dynamic system with model uncertainties by offering a whole set of possible system models. In case of fixed models we assume that the true model is one of the offered system models and that it does not change during the measurement process. Further model uncertainty is modeled by the switching model approach that takes dynamic system models into account that can switch from one to another. In figure 1 the possible model sequencies for a single model approach and multiple model approaches with two possible models are shown from time $k-2$ to time k. For the state estimation all possible model sequencies have to be taken into consideration. In case of switching models the number of possibilities rapidly grows with time, so that usually an approximation is necessary, e.g. the interacting multiple model technique. The multiple model approach was originally presented by Magill (1965). The approach for fixed models was extended to switching models in the 70's. The approximation by interacting models was proposed in (Blom 1984) and (Blom and Bar-Shalom 1988). For GNSS ambiguity resolution the integer ambiguities define the set of system models. The technique of fixed multiple models was applied for ambiguity resolution already by Brown and Hwang (1983) as multiple model adaptive estimation or Magill adaptive filter and modified in Henderson (2001). Wolfe et al. (2001) refer to the multiple model approach for fixed models as multiple hypothesis Wald sequential probability ratio test. Estimating the GNSS ambiguities in the multiple model approach for fixed models leads recursively to the same results as the Bayesian approach, see (Betti et al. 1993) and (Gundlich, Koch 2002), that introduces the ambiguities as discrete integer random variables. In the context with ambiguity resolution the multiple model approach for switching models is used to model cycle slips, see Chen and Harigae (2001) and Wolfe et al. (2001). In the following an overview of the multiple model approaches is given in the context of ambiguity resolution.

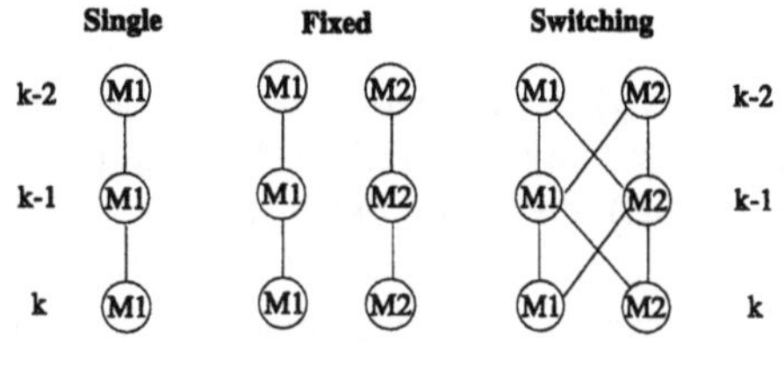

Fig. 1 Single Model, Fixed Models, Switching Models

2 Theory

2.1 Bayesian Statistics

In the Bayesian approach the unknown parameters $\boldsymbol{x}$ are interpreted as random variables. Their probability density function describes the information of the unknown parameters: the prior density $p(\boldsymbol{x})$ expresses the knowledge of the unknown parameters $\boldsymbol{x}$ without considering the observations $\boldsymbol{y}$, the posterior density $p(\boldsymbol{x}|\boldsymbol{y})$ of the parameters $\boldsymbol{x}$ summarizes all information given the observations $\boldsymbol{y}$. The posterior density results from the Bayes' theorem, see for example (Koch 2000, p. 32),

$$p(\boldsymbol{x}|\boldsymbol{y}) = \frac{p(\boldsymbol{x})p(\boldsymbol{y}|\boldsymbol{x})}{p(\boldsymbol{y})} \propto p(\boldsymbol{x})p(\boldsymbol{y}|\boldsymbol{x}) \tag{1}$$

that follows immediately from

$$p(\boldsymbol{x},\boldsymbol{y}) = p(\boldsymbol{y})p(\boldsymbol{x}|\boldsymbol{y}) = p(\boldsymbol{x})p(\boldsymbol{y}|\boldsymbol{x}) .$$

Maximizing the posterior density function leads to the MAP(maximum a posteriori) estimate

$$\hat{\boldsymbol{x}}_{MAP} = \arg \max_{\boldsymbol{x}\in\boldsymbol{X}} p(\boldsymbol{x}|\boldsymbol{y}) . \tag{2}$$

If a quadratic loss function is defined, see e.g. (Koch 2000, p. 65f), the Bayes estimate results in the expectation

$$\hat{\boldsymbol{x}}_B = E\{\boldsymbol{x}|\boldsymbol{y}\} . \tag{3}$$

2.2 Single Model

The linear dynamic system at time k consists of the dynamic equation

$$\boldsymbol{x}(k) = \boldsymbol{\Phi}(k)\boldsymbol{x}(k-1) + \boldsymbol{d}(k-1) \tag{4}$$

and the observation equation

$$\boldsymbol{y}(k) = \boldsymbol{A}(k)\boldsymbol{x}(k) + \boldsymbol{e}(k) \tag{5}$$

with the state $\boldsymbol{x}(.)$, the known transition matrix $\boldsymbol{\Phi}(.)$, the Gaussian system noise $\boldsymbol{d}(.) \sim \mathcal{N}(\boldsymbol{0}, \boldsymbol{Q}_d(.))$, the observations $\boldsymbol{y}(.)$, the known design matrix $\boldsymbol{A}(.)$ and the Gaussian measurement noise $\boldsymbol{e}(.) \sim \mathcal{N}(\boldsymbol{0}, \boldsymbol{Q}(.))$. The Gaussian system and measurement noise are not time correlated and not correlated to each other. State estimation in linear dynamic systems is recursively solved by the Kalman filter, that we shortly derive in a Bayesian approach.

Initialization: Using a non-informative prior, $p(\boldsymbol{x}(1)) \propto const.$, the Bayes' theorem (1) leads to the normal distributed state $\boldsymbol{x}(1)$ given the observations $\boldsymbol{y}(1)$, short $\boldsymbol{x}(1)|\boldsymbol{y}(1)$,

$$\boldsymbol{x}(1)|\boldsymbol{y}(1) \sim \mathcal{N}\Big(\hat{\boldsymbol{x}}(1|1), \boldsymbol{Q}_x(1|1)\Big)$$

with covariance matrix

$$\boldsymbol{Q}_x(1|1) = \Big(\boldsymbol{A}(1)'\boldsymbol{Q}(1)^{-1}\boldsymbol{A}(1)\Big)^{-1}$$

and expectation

$$\hat{\boldsymbol{x}}(1|1) = \boldsymbol{Q}_x(1|1)\boldsymbol{A}(1)'\boldsymbol{Q}(1)^{-1}\boldsymbol{y}(1) .$$

Time update (t-update): For the recursive state estimation the normal distributed state $\boldsymbol{x}(k-1)|\boldsymbol{Y}(k-1)) \sim \mathcal{N}(\hat{\boldsymbol{x}}(k-1|k-1), \boldsymbol{Q}_x(k-1|k-1))$ with $\boldsymbol{Y}(k-1) = \{\boldsymbol{y}(1), \ldots, \boldsymbol{y}(k-1)\}$ is time updated. Because of the linear dynamic equation (4) the prediction leads to the normal density

$$\boldsymbol{x}(k)|\boldsymbol{Y}(k-1) \sim \mathcal{N}\Big(\hat{\boldsymbol{x}}(k|k-1), \boldsymbol{Q}_x(k|k-1)\Big)$$

with expectation and covariance matrix,

$$\hat{\boldsymbol{x}}(k|k-1) = \boldsymbol{\Phi}(k)\hat{\boldsymbol{x}}(k-1|k-1) ,$$

$$\boldsymbol{Q}_x(k|k-1)$$
$$= \boldsymbol{\Phi}(k)\boldsymbol{Q}_x(k-1|k-1)\boldsymbol{\Phi}(k)' + \boldsymbol{Q}_d(k) .$$

Measurement update (m-update): The measurement update is obtained from the recursive Bayes' theorem, see (1),

$$p\Big(\boldsymbol{x}(k)|\boldsymbol{y}(k), \boldsymbol{Y}(k-1)\Big) \propto \tag{6}$$
$$p\Big(\boldsymbol{y}(k)|\boldsymbol{x}(k), \boldsymbol{Y}(k-1)\Big)p\Big(\boldsymbol{x}(k)|\boldsymbol{Y}(k-1)\Big)$$

with the density of the time updated state $p(\boldsymbol{x}(k)|\boldsymbol{Y}(k-1))$ acting as prior and the Gaussian likelihood function $\boldsymbol{y}(k)|\boldsymbol{x}(k) \sim \mathcal{N}(\boldsymbol{A}(k)\boldsymbol{x}(k), \boldsymbol{Q}(k))$, see (5). That leads to the normal distribution

$$\boldsymbol{x}(k)|\boldsymbol{Y}(k) \sim \mathcal{N}\Big(\hat{\boldsymbol{x}}(k|k), \boldsymbol{Q}_x(k|k)\Big) \tag{7}$$

with expectation $\hat{\boldsymbol{x}}(k|k)$ and covariance matrix $\boldsymbol{Q}_x(k|k)$ computed as follows

$$\boldsymbol{v}(k) = \boldsymbol{y}(k) - \boldsymbol{A}(k)\hat{\boldsymbol{x}}(k|k-1) , \tag{8}$$

$$\boldsymbol{Q}_v(k) = \boldsymbol{A}(k)\boldsymbol{Q}_x(k|k-1)\boldsymbol{A}(k)' + \boldsymbol{Q}(k) , \tag{9}$$

$$\boldsymbol{K}(k) = \boldsymbol{Q}_x(k|k-1)\boldsymbol{A}(k)'\boldsymbol{Q}_v(k)^{-1} ,$$

$$\hat{\boldsymbol{x}}(k|k) = \hat{\boldsymbol{x}}(k|k-1) + \boldsymbol{K}(k)\boldsymbol{v}(k) ,$$

$$\boldsymbol{Q}_x(k|k) = \Big(\boldsymbol{I} - \boldsymbol{K}(k)\Big)\boldsymbol{A}(k)\boldsymbol{Q}_x(k|k-1)$$

with the predicted residuals $\boldsymbol{v}(k)$, their covariance matrix $\boldsymbol{Q}_v(k)$ and the filter gain $\boldsymbol{K}(k)$.

Results: Because of the normal distribution (7) the MAP (2) and the Bayes estimate (3) are the same,

$$\hat{\boldsymbol{x}}_{MAP}(k|k) = \hat{\boldsymbol{x}}_B(k|k) = \hat{\boldsymbol{x}}(k|k) \,.$$

2.3 Fixed Models

In the single model case, that is solved via the Kalman filter, the system model defined by the transition and design matrix and noise is known. The multiple model approach deals with a model uncertainty about several system models. In the case of fixed models it is assumed that the true model M is one of the known system models $M_i, i = 1, \ldots, r$,

$$M \in \{M_1, \ldots, M_r\} \,.$$

That means for the linear dynamic model that transition matrix, design matrix and the noise depend on an unknown model M, thus

$$\boldsymbol{x}(k) = \boldsymbol{\Phi}(k, M)\boldsymbol{x}(k-1) + \boldsymbol{u}(k-1, M)$$

and

$$\boldsymbol{y}(k) = \boldsymbol{A}(k, M)\boldsymbol{x}(k) + \boldsymbol{e}(k, M) \,.$$

The common posterior density $p(\boldsymbol{x}(k), M_i|\boldsymbol{Y}(k))$ of state and model has to be derived. It can also be expressed as following

$$p\Big(\boldsymbol{x}(k), M_i|\boldsymbol{Y}(k)\Big)$$

$$= p\Big(\boldsymbol{x}(k)|M_i, \boldsymbol{Y}(k)\Big)P\Big(M_i|\boldsymbol{Y}(k)\Big) \,.$$

Then we get the marginal state density

$$p\Big(\boldsymbol{x}(k)|\boldsymbol{Y}(k)\Big)$$

$$= \sum_{i=1}^{r} p\Big(\boldsymbol{x}(k)|M_i, \boldsymbol{Y}(k)\Big)P\Big(M_i|\boldsymbol{Y}(k)\Big) \qquad (10)$$

as weighted sums of normal densities: Given the model M_i the state $\boldsymbol{x}(k)$ is normal distributed, see the Kalman filter results in section 2.2,

$$\boldsymbol{x}(k)|M_i, \boldsymbol{Y}(k) \sim \mathcal{N}\Big(\hat{\boldsymbol{x}}^i(k|k), \boldsymbol{Q}_x^i(k|k)\Big) \,.$$

The weights are given by the model probabilities, that are estimated recursively.

Initialization: The starting probabilities $P(M_i|\boldsymbol{y}(1)), i = 1, \ldots, r$ have to be chosen. If no other information is given and the state initialization provides no information over the probabilities, usually equal start probabilities are used.

Time update (t-update): No time update of the model probabilities is necessary, because the system models are assumed to be constant.

Measurement update (m-update): Using the Bayes' theorem (6) for the model probabilities we find

$$P\Big(M_i|\boldsymbol{Y}(k)\Big) \propto$$

$$p\Big(\boldsymbol{y}(k)|M_i, \boldsymbol{Y}(k-1)\Big)P\Big(M_i|\boldsymbol{Y}(k-1)\Big) \qquad (11)$$

which is normalized to fulfill the condition

$$\sum_{j=1}^{r} P\Big(M_j|\boldsymbol{Y}(k)\Big) = 1 \,. \qquad (12)$$

The likelihood function $p(\boldsymbol{y}(k)|M_i, \boldsymbol{Y}(k-1))$ is replaced by the corresponding density value of the predicted residuals $\boldsymbol{v}^i(k)$ with, see (8) and (9),

$$\boldsymbol{v}(k)|M_i, \boldsymbol{Y}(k-1) \sim \mathcal{N}\Big(\boldsymbol{0}, \boldsymbol{Q}_v^i(k)\Big) \,.$$

Results: The posterior state density (10) leads to the Bayes estimate (3)

$$\hat{\boldsymbol{x}}_B(k|k) = \sum_{i=1}^{r} \hat{\boldsymbol{x}}^i(k|k)P\Big(M_i|\boldsymbol{Y}(k)\Big)$$

that is a mixture of all single model results weighted by their model probabilities. Maximizing the model probability leads to the MAP estimate (2)

$$\hat{\boldsymbol{x}}_{MAP}(k|k) = \hat{\boldsymbol{x}}^i(k|k),$$

with

$$i = \arg \max_{j=1,\ldots,r} P\Big(M_j|\boldsymbol{Y}(k)\Big) \,.$$

The covariance matrix of the state follows from (10),

$$\boldsymbol{Q}_x(k|k) = \sum_{i=1}^{r} P\Big(M_i|\boldsymbol{Y}(k)\Big)\Big(\boldsymbol{Q}_x^i(k|k) +$$

$$(\hat{\boldsymbol{x}}_B(k|k) - \hat{\boldsymbol{x}}^i(k|k))(\hat{\boldsymbol{x}}_B(k|k) - \hat{\boldsymbol{x}}^i(k|k))'\Big) \,.$$

2.4 Switching Models

The multiple model approach in now extended to switching models. The linear dynamic model is based on a system model that is allowed to change during the observation process. The true unknown system model can vary with time, but still is assumed to be one of a known model set,

$$M(k) \in \{M_1, \ldots, M_r\} \,. \qquad (13)$$

The probability of a jump from one model to another is given by the transition probability of the event $M(k) = M_j$, short $M_j(k)$, given $M(k-1) = M_i$ or $M_i(k-1)$,

$$p_{ij} = P\Big(M_j(k)|M_i(k-1)\Big) \, . \tag{14}$$

It is assumed to be independent of time and state. The linear dynamic model for switching models is

$$\boldsymbol{x}(k) = \boldsymbol{\Phi}(k, M(k))\boldsymbol{x}(k-1) + \boldsymbol{u}(k-1, M(k)),$$

$$\boldsymbol{y}(k) = \boldsymbol{A}(k, M(k))\boldsymbol{x}(k) + \boldsymbol{e}(k, M(k)) \, .$$

In contrast to the multiple model approach for fixed models, we do not only have the uncertainty of one system model but the uncertainty of a whole sequence. At time k r^k different model sequences $S_l(k)$ are possible,

$$S_l(k) = \{M_{l_1}(1), \ldots, M_{l_k}(k)\}$$

with $l_k \in \{1, \ldots, r\}$ and $l = 1, \ldots, r^k$, see figure 2.

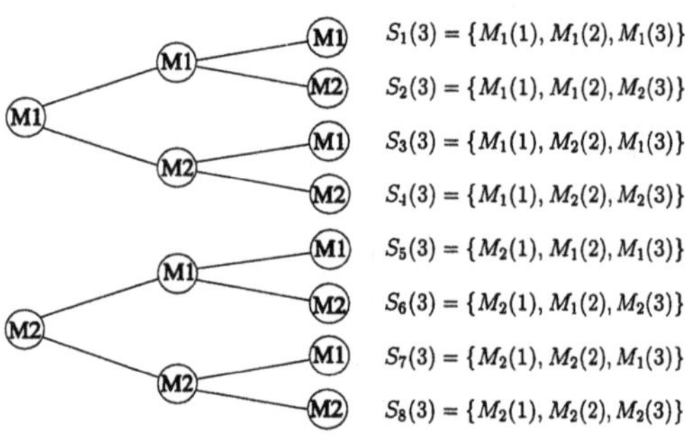

Fig. 2 Possible model sequences for $r = 2$, $k = 1, .., 3$

In analogy to (10) but with respect to all possible model sequences the posterior state density is expressed as marginal density

$$p\Big(\boldsymbol{x}(k)|\boldsymbol{Y}(k)\Big) =$$

$$\sum_{l=1}^{r^k} p\Big(\boldsymbol{x}(k)|S_l(k), \boldsymbol{Y}(k)\Big) P\Big(S_l(k)|\boldsymbol{Y}(k)\Big) \, . \tag{15}$$

The conditional state densities given the model sequence $S_l(k)$ are Gaussian

$$\boldsymbol{x}(k)|S_l(k), \boldsymbol{Y}(k) \sim \mathcal{N}\Big(\hat{\boldsymbol{x}}^l(k|k), \boldsymbol{Q}_x^l(k|k)\Big)$$

whose parameters result from corresponding Kalman filters, see section 2.2. Besides a bank of r^k Kalman filters the update of the probabilities of the model sequences is necessary.

Initialization: In analogy to the fixed model approach the starting probabilities $P(M_i(1)|\boldsymbol{y}(1))$, $i = 1, \ldots, r$ have to be chosen. Further the transition probabilities p_{ij} (14) have to be introduced as priors.

Time update (t-update): The probabilities of sequences $S_l(k-1)$, $l = 1, \ldots, r^{k-1}$ at time $k-1$ are updated with the transition probability, thus

$$P\Big(S_l(k)|\boldsymbol{Y}(k-1)\Big) = p_{ij} P\Big(S_l(k-1)|\boldsymbol{Y}(k-1)\Big)$$

for $S_l(k-1) = \{M_{l_1}(1), \ldots, M_i(k-1)\}$ and $S_l(k) = \{S_l(k-1), M_j(k)\}$.

Measurement update (m-update): In analogy to (11) the Bayes' theorem is applied for the measurement update

$$P\Big(S_l(k)|\boldsymbol{Y}(k)\Big)$$

$$\propto p\Big(\boldsymbol{y}(k)|S_l(k), \boldsymbol{Y}(k-1)\Big) P\Big(S_l(k)|\boldsymbol{Y}(k-1)\Big),$$

which is normalized in analogy to (12). The density $p(\boldsymbol{y}(k)|S_l(k), \boldsymbol{Y}(k-1))$ is obtained from the density of the predicted residuals of the Kalman filter that corresponds to the model sequence $S_l(k)$.

Results: From the posterior state density (15) follows the Bayes estimate (3)

$$\hat{\boldsymbol{x}}_B(k|k) = \sum_{l=1}^{r^k} \hat{\boldsymbol{x}}^l(k|k) P\Big(S_l(k)|\boldsymbol{Y}(k)\Big)$$

and the MAP estimate (2)

$$\hat{\boldsymbol{x}}_{MAP}(k|k) = \hat{\boldsymbol{x}}^l(k|k),$$

with the maximal probability of a sequence

$$l = \arg \max_{i=1,\ldots,r^k} P\Big(S_i(k)|\boldsymbol{Y}(k)\Big) \, .$$

The covariance matrix of the state is also derived from the posterior density (15)

$$\boldsymbol{Q}_x(k|k) = \sum_{l=1}^{r^k} P\Big(S_l(k)|\boldsymbol{Y}(k)\Big) \Big(\boldsymbol{Q}_x^l(k|k) +$$

$$(\hat{\boldsymbol{x}}_B(k|k) - \hat{\boldsymbol{x}}^l(k|k))(\hat{\boldsymbol{x}}_B(k|k) - \hat{\boldsymbol{x}}^l(k|k))'\Big) \, .$$

Because of the fast increasing r^k, see figure 2, approximations of the posterior density and the estimate are necessary. An overview over the update of state and models for a single model, for fixed models and for switching models is shown in figure 3.

	state	model
Single		known
initialization	$p(x(1)\|y(1))$	–
	$p(x(k-1)\|Y(k-1))$	–
t – update	$\downarrow$	
	$p(x(k)\|Y(k-1))$	–
m – update	$\downarrow$	
	$p(x(k)\|Y(k))$	–
	(**KF**)	
Fixed $i=1,\ldots,r$		$M \in \{M_1,\ldots,M_r\}$
initialization	$p(x(1)\|M_i, y(1))$	$P(M_i\|y(1))$
	$p(x(k-1)\|M_i, Y(k-1))$	$P(M_i\|Y(k-1))$
t – update	$\downarrow$	$\|\|$
	$p(x(k)\|M_i, Y(k-1))$	$P(M_i\|Y(k-1))$
m – update	$\downarrow$	$\downarrow$
	$p(x(k)\|M_i, Y(k))$	$P(M_i\|Y(k))$
	(r **KF**)	
Switching $l=1,\ldots r^k$		$M(k) \in \{M_1,\ldots,M_r\}$
prior		p_{ij}
initialization	$p(x(1)\|S_l(1), y(1))$	$P(S_l(1)\|y(1))$
	$p(x(k-1)\|S_l(k-1), Y(k-1))$	$P(S_i(k-1)\|Y(k-1))$
t – update	$\downarrow$	$\downarrow$
	$p(x(k)\|S_l(k), Y(k-1))$	$P(S_l(k)\|Y(k-1))$
m – update	$\downarrow$	$\downarrow$
	$p(x(k)\|S_l(k), Y(k))$	$P(S_l(k)\|Y(k))$
	(l **KF**)	

Fig. 3 Single Model, Fixed Models, Switching Models

2.5 Interacting Models

The multiple model approach for switching models is approximated by interacting models. Instead of equation (15) the posterior state density is derived as marginal density of the common density of state and model $M(k)$,

$$p\Big(x(k)|Y(k)\Big) = \sum_{i=1}^{r} p\Big(x(k)|M_i(k), Y(k)\Big) P\Big(M_i(k)|Y(k)\Big) . \tag{16}$$

In case of switching models, the conditional densities $p(x(k)|M_j(k), Y(k))$, $j = 1, \ldots, r$, are not Gaussian. The idea of interacting multiple models is to approximate the densities by normal distributions. This approximation is also used for the recursive estimation of the model probability. For explanation see the following considerations.

	state	model
IMM $i=1,\ldots r$		$M(k) \in \{M_1,\ldots,M_r\}$
prior		p_{ij}
initialization	$p(x(1)\|M_i(1), y(1))$	$P(M_i(1)\|y(1))$
	$p(x(k-1)\|M_i(k-1), Y(k-1))$	$P(M_i(k-1)\|Y(k-1))$
	interacting	mixing probabilities
	$p(x(k-1)\|M_i(k-1), Y(k-1))$	$P(M_i(k-1)\|Y(k-1))$
t – update	$\downarrow$	$\downarrow$
	$p(x(k)\|M_i(k), Y(k-1))$	$P(M_i(k)\|Y(k-1))$
m – update	$\downarrow$	$\downarrow$
	$p(x(k)\|M_i(k), Y(k))$	$P(M_i(k)\|Y(k))$
	(r interacting **KF**)	

Fig. 4 Interacting Multiple Models (IMM) approximating switching Models

Assume the normal distribution for

$$x(k-1)|M_i(k-1), Y(k-1) \sim \mathcal{N}\Big(\hat{x}^i(k-1|k-1), Q_x^i(k-1|k-1)\Big),$$

which is valid for $k-1=1$.

Time update (t-update): The time update under the condition of $M_j(k)$ is derived with the Kalman filter and leads to the normal distribution

$$x(k)|M_i(k-1), M_j(k), Y(k-1) \sim$$

$$\mathcal{N}\Big(\boldsymbol{x}^{i,j}(k|k-1), \boldsymbol{Q}_x^{i,j}(k|k-1)\Big) . \quad (17)$$

The state density given only $M_j(k)$ is a Gaussian mixture

$$p\Big(\boldsymbol{x}(k)|M_j(k), \boldsymbol{Y}(k-1)\Big) = \quad (18)$$

$$\sum_{i=1}^{r} p\Big(\boldsymbol{x}(k)|M_i(k-1), M_j(k), \boldsymbol{Y}(k-1)\Big)$$

$$P\Big(M_i(k-1)|M_j(k-1), \boldsymbol{Y}(k-1)\Big)$$

with the mixing probabilities $\mu_{i|j}(k-1) = P(M_i(k-1)|M_j(k-1), \boldsymbol{Y}(k-1))$ as weights. In analogy to the Bayes' theorem, see (1), and with $P(M_j(k)|M_i(k-1), \boldsymbol{Y}(k-1)) = p_{ij}$ we find the mixing probabilities

$$P\Big(M_i(k-1)|M_j(k-1), \boldsymbol{Y}(k-1)\Big) \propto$$

$$p_{ij} P\Big(M_i(k-1)|\boldsymbol{Y}(k-1)\Big) .$$

The Gaussian mixture (18) is now approximated by the Gaussian distribution

$$\boldsymbol{x}(k)|M_j(k), \boldsymbol{Y}(k-1)$$

$$\sim \mathcal{N}\Big(\hat{\boldsymbol{x}}^j(k|k-1), \boldsymbol{Q}_x^j(k|k-1)\Big) \quad (19)$$

with parameters

$$\hat{\boldsymbol{x}}^j(k|k-1) = \sum_{i=1}^{r} \hat{\boldsymbol{x}}^{i,j}(k|k-1)\mu_{i|j}(k-1)$$

$$\boldsymbol{Q}_x^j(k|k-1) = \sum_{i=1}^{r} \mu_{i|j}(k-1)\Big(\boldsymbol{Q}_x^{i,j}(k|k-1)+$$

$$(\Delta\boldsymbol{x}^{i,j})(\Delta\boldsymbol{x}^{i,j})'\Big),$$

with $\Delta\boldsymbol{x}^{i,j} = (\hat{\boldsymbol{x}}^{i,j}(k|k-1) - \hat{\boldsymbol{x}}^j(k|k-1))$.

Measurement update (m-update): The measurement update for the state results from the corresponding Kalman filter and leads to the normal distribution

$$\boldsymbol{x}(k)|M_j(k), \boldsymbol{Y}(k) \sim \mathcal{N}\Big(\hat{\boldsymbol{x}}^j(k|k), \boldsymbol{Q}_x^j(k|k)\Big) .$$

Given the model probability $P(M_i(k-1)|\boldsymbol{Y}(k-1))$ the model probability $P(M_j(k)|\boldsymbol{Y}(k))$ is obtained from the time update, see (11),

$$P(M_j(k)|\boldsymbol{Y}(k)) \propto$$

$$p\Big(\boldsymbol{y}(k)|M_j(k), \boldsymbol{Y}(k-1)\Big) P\Big(M_j(k)|\boldsymbol{Y}(k-1)\Big) .$$

As likelihood function $p(\boldsymbol{y}(k)|M_j(k), \boldsymbol{Y}(k-1))$ the normal density of the predicted residuals computed in the Kalman filter is used, given the approximated normal distribution (19).

Results: Finally the posterior density (16) is approximated by a sum of r normal densities that leads to the Bayes estimate

$$\hat{\boldsymbol{x}}_B(k|k) = \sum_{j=1}^{r} \hat{\boldsymbol{x}}^j(k|k) P\Big(M_j(k)|\boldsymbol{Y}(k)\Big)$$

and to the covariance matrix of the state

$$\boldsymbol{Q}_x(k|k) = \sum_{j=1}^{r} P\Big(M_j(k)|\boldsymbol{Y}(k)\Big)\Big(\boldsymbol{Q}_x^j(k|k)+$$

$$(\hat{\boldsymbol{x}}^j(k|k) - \hat{\boldsymbol{x}}_B(k|k))(\hat{\boldsymbol{x}}^j(k|k) - \hat{\boldsymbol{x}}_B(k|k))'\Big) .$$

In the algorithm for interacting models the order of conditional time update (17) for a certain model and the mixing, see (19), is changed. In figure 4 the interacting models technique is summarized.

3 Ambiguity Resolution

We start the ambiguity resolution with the float model. That means, we consider the ambiguities as real valued parameters. The state vector is divided into the real valued ambiguities $\boldsymbol{a}(k)$ and the remaining parameters $\boldsymbol{b}(k)$. If we assume, that during the observation process the same satellites are observed and no cycle slips occur, then the ambiguities are constant. We get the dynamic equations

$$\boldsymbol{a}(k+1) = \boldsymbol{a}(k) = \boldsymbol{a} ,$$

$$\boldsymbol{b}(k+1) = \boldsymbol{\Phi}_b(k) + \boldsymbol{u}_b(k)$$

and the observation equation

$$\boldsymbol{y}(k) = \boldsymbol{A}_a(k)\boldsymbol{a} + \boldsymbol{A}_b(k)\boldsymbol{b}(k) + \boldsymbol{e}(k) .$$

If we treat the ambiguities as real valued parameters, the posterior densities are normal distributed,

$$\boldsymbol{a}(k)|\boldsymbol{Y}(k) \sim \mathcal{N}\Big(\hat{\boldsymbol{a}}(k|k), \boldsymbol{Q}_a(k|k)\Big) , \quad (20)$$

$$\boldsymbol{b}(k)|\boldsymbol{Y}(k) \sim \mathcal{N}\Big(\hat{\boldsymbol{b}}(k|k), \boldsymbol{Q}_b(k|k)\Big) . \quad (21)$$

The state estimate and the covariance matrix are recursively computed with the Kalman filter. In the following single and multiple model approaches integer ambiguities are interpreted as system models.

3.1 Single Model

If the ambiguities are supposed to be the known integers $a \in \mathbb{Z}$, then these integers represent a system model that defines the dynamic and observation equations

$$b(k) = \Phi_b b(k-1) + d_b(k-1) \, ,$$

$$y(k) = A_a(k)a + A_b(k)b(k) + e(k) \, .$$

The remaining parameters $b(k)$ given the ambiguities a are normal distributed

$$b(k)|a, Y(k) \sim \mathcal{N}\Big(\hat{b}_a(k|k), Q_{b|a}(k|k)\Big). \quad (22)$$

If the integer ambiguities are the integer least squares solution,

$$a = \arg\min_{z \in \mathbb{Z}} (\hat{a}(k|k) - z)' Q_a(k|k)^{-1} (\hat{a}(k|k) - z),$$

the posterior distribution leads to the well known fixed solution.

3.2 Fixed Models

For the fixed multiple model approach we introduce an uncertainty for the integer ambiguities. Different integer candidates a_i, $i = 1, \dots r$ define the system models. We assume that the ambiguities a lie in this set,

$$a \in \{a_1, \dots, a_r\} \subset \mathbb{Z} \, .$$

The infinite set of integers $\mathbb{Z}$ has to be restricted to a finite one. This leads to r linear dynamic systems

$$b(k) = \Phi_b b(k-1) + d_b(k-1) \, ,$$

$$y(k) = A_a(k)a_i + A_b(k)b(k) + e(k).$$

with normal distributed state

$$b(k)|a_i, Y(k) \sim \mathcal{N}\Big(\hat{b}^i(k|k), Q_{b|a}(k|k)\Big).$$

The posterior densities under the condition of the different integers are solved by the corresponding Kalman filters or derived as conditional densities from the posterior density of the float solution. They differ only concerning their expectations. In analogy to (10) the posterior density that considers all possible system models is

$$p(b(k)|Y(k)) = \sum_{i=1}^{r} p(b(k)|a_i, Y(k)) P(a_i|Y(k)).$$

The probability of the integers is recursively computed. The initialization of the float solution contains information about the starting probabilities. If the initial real valued ambiguities are normal distributed with

$$a(1)|y(1) \sim \mathcal{N}\Big(\hat{a}(1|1), Q_a(1,1)\Big) \, ,$$

the start probabilities are

$$P(a_i|y(1)) \propto$$

$$\exp\Big(-\frac{1}{2}(\hat{a}(1|1) - a_i)' Q_a(1|1)^{-1} (\hat{a}(1|1) - a_i)\Big) \, ,$$

see (Betti et al. 1993), Gundlich and Koch (2002). The recursive computation of the probabilities according to the fixed model approach leads to identical results as the batch solution

$$P\Big(a_i|Y(k)\Big) \propto$$

$$\exp\Big(-\frac{1}{2}(\hat{a}(k|k) - a_i)' Q_a(k|k)^{-1} (\hat{a}(k|k) - a_i)\Big) \, .$$

Thus the fixed multiple model approach has the same posterior density as the Bayesian approach in (Betti et al. 1993) or (Gundlich, Koch 2002), with the Bayes estimate

$$\hat{b}_B(k|k) = \sum_{i=1}^{r} \hat{b}^i(k|k) P\Big(a_i|Y(k)\Big)$$

and the MAP estimate

$$\hat{b}_{MAP}(k|k) = b^i(k|k)$$

with the maximal probability of the ambiguities

$$i = \arg\max_{j=1,\dots,r} P\Big(a_j|Y(k)\Big) \, .$$

The covariance matrix of $b(k)$ is

$$Q_b(k|k) = \sum_{i=1}^{r} P\Big(a_i|Y(k)\Big)\Big(Q_{b|a}$$

$$+ (\hat{b}^i(k|k) - \hat{b}_B(k|k))(\hat{b}^i(k|k) - \hat{b}_B(k|k))'\Big) \, .$$

The MAP estimate represents the state estimate of the fixed solution (22), whereas the Bayes estimate is a mixture of different solutions that belong to different integer ambiguities. The ambiguity probabilities provide a criterion for the choice between the two estimates. The MAP estimate is used for a high maximal probability. In case of a lower probability, if one does not want to rely on a single integer ambiguity vector, the Bayes estimate as a combination of integer vectors can be chosen. In contrast to the float solution, it still considers the ambiguities as integers.

3.3 Switching and Interacting Models

In analogy to (13) we allow jumps from one integer candidate to another one within the ambiguity set

$$a(k) \in \{a_1, \ldots, a_r\} \in \mathbb{Z} .$$

These jumps are possible integer cycle slips. The probability of a cycle slip has to be defined a priori by the transition probability (14). Using switching models we have not only the problem of approximating an infinite set by a finite set, but the chosen set should also take possible cycle slips into account. We get the linear dynamic system

$$b(k) = \Phi_b b(k-1) + d_b(k-1)$$

$$y(k) = A_a(k)a(k) + A_b(k)b(k) + e(k) .$$

In the posterior density

$$p\Big(b(k)|Y(k)\Big) = \qquad (23)$$

$$\sum_{i=1}^{r} p\Big(b(k)|a_i(k), Y(k)\Big) P\Big(a_i(k)|Y(k)\Big)$$

the probability $P(a_i(k)|Y(k))$ can be updated without an approximation, but the density $p(b(k)|a_i(k), Y(k))$ has to be approximated by interacting models.

4 Conclusions

The multiple model approach enables to take model uncertainties into consideration, that are introduced at the expense of computing time. In case of switching models an approximation is necessary, besides further information in the form of the transition probabilities has to be provided. For ambiguity resolution the fixed model approach is the recursive version of the Bayesian approach that introduces integer ambiguities as random parameters. Cycle slips are taken into account, if the switching model approach is applied. Practical problems arise with the choice of the integer ambiguities as system models. In case of ambiguity resolution with fixed models the integer set of ambiguities can be derived from the initial float solution: All integers within a confidence region of the float solution of the ambiguities form the integer set. An efficient search for those integer ambiguities is possible with the LAMBDA method, see (Teunissen 1994). The definition of the integer sets in a multiple model approach for switching models is more complicated, because the new ambiguities after a cycle slip should be contained in the set. If they belong to the set, then the multiple model algorithm for switching or interacting models is robust to the cycle slips. But we can not expect, that after a cycle slip the new integer ambiguities lie in the once chosen set. Therefore the detection of cycle slips and the use of time dependent integer sets are necessary. For the detection of cycle slips the ambiguity probabilities $P(a_i(k)|Y(k))$ in (23) could be used. This approach however needs further study. Not yet solved is also the question, how the transition probabilities have to be defined.

References

Betti B, Crespi M, Sanso F (1993). A geometric illustration of ambiguity resolution in GPS theory and a Bayesian approach. *Manuscripta Geodaetica* 18: 317–330.

Blom HAP (1984). An efficient filter for abruptly changing systems. In: *Proc. 23rd IEEE Conf. Decision and Control*, Las Vegas, NV, Dec. 1984.

Blom HAP, Bar-Shalom Y (1988). The interacting multiple model algorithm for systems with Markovian switching coefficients. *IEEE Transactions on Automatic Control*, AC-33(8): 780–783, August 1988.

Brown RG, Hwang PYC (1983). A Kalman filter approach to precision GPS geodesy. *Journal of The Institute of Navigation* 30: 338–349.

Chen G, Harigae M (2001). Precise DGPS positioning and carrier phase ambiguity resolution using IMM Kalman filters. In: *Proc. of International Symposium on Kinematic Systems in Geodesy, Geomatics and Navigation (KIS 2001)*, Banff, Canada, June 5–8 2001, 339–346.

Gundlich B, Koch KR (2002). Confidence regions of GPS baselines by Bayesian statistics. *Journal of Geodesy* 76: 55–62.

Henderson P (2001). Flight test of a multiple filter approach for precise DGPS positioning and carrier-phase ambiguity resolution. *ION GPS 2001*, 11-14 September 2001, Salt Lake City UT: 1565–1574.

Koch KR (2000). *Einführung in die Bayes–Statistik.* Springer, Berlin Heidelberg New York.

Magill DT (1965). Optimal adaptive estimation of sampled stochastic processes. *IEEE Trans. Automatic Control*, AC-10: 434–439, 1965.

Teunissen (1994). A new method for fast carrier phase ambiguity estimation. In: *Proceedings IEEE Position Location and Navigation Symposium*, PLANS '94, 562–573, Las Vegas, Nevada.

Wolfe JD, Williamson WR, Speyer JL (2001). Hypothesis testing for resolving integer ambiguity in GPS. *ION GPS 2001*, 11-14 September 2001, Salt Lake City UT: 1522–1531.

A Comparison of Data Weighting Methods for the Combination of Satellite and Local Gravity Data

M. Kern
Department of Geomatics Engineering, The University of Calgary
2500 University Drive N.W., Calgary, T2N 1N4, Canada, Tel: ++ 403 220 8794,
Fax: ++ 403 284 1980, E-mail: mkern@ucalgary.ca

Abstract. The combination of heterogeneous gravity data is among the most involved problems in gravity field modeling. With newly available gravity data from dedicated satellites, such as CHAMP, GRACE and GOCE, additional low-frequency information about the gravity field will be available. Local or regional data collected closer to the Earth's surface allows the resolution of the medium to high frequency components of the gravity field. Although for applications, such as geoid determination, most of the power is contained in the lower frequencies, a geoid at the cm-level, which is required by oceanography and geodesy alike, will require a much wider spectrum than the one that can be resolved by satellite methods. To enhance and widen the gravity field spectrum, a data combination process is essential.

The comparison of data weighting methods for the spectral combination of satellite and local gravity data is the topic of the paper. The quality of each method is evaluated with respect to the requirements, inherent assumptions and errors. Based on the detailed characterization of the methods, the quasi-deterministic method seems to be the most promising for the combination of satellite and local gravity data.

Keywords. Combination methods · Gravity field modeling · Satellite gravity missions · Filters

1 Introduction

The combination of satellite and local gravity data is carried out for calibration and application purposes. A calibration process using terrestrial or airborne gravity data ensures the quality of the satellite solution. Furthermore, the local data are used to regularize the unstable global solutions that primarily occur due to the downward continuation problem, colored data noise and inhomogeneous data coverage over the polar regions. Undoubtedly, many global applications can greatly benefit from the dedicated satellite missions such as CHAMP, GRACE and GOCE. However, regional and local applications often require the combination of satellite and local gravity data, since the global data set may not provide the spectrum required for the specific application. As an example, the demand for a high-resolution geoid at the cm-level in oceanography and geodesy alike is constantly increasing.

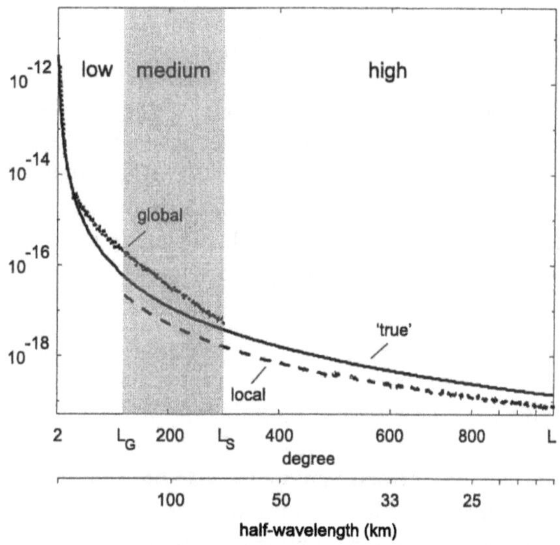

Fig. 1 Degree variances (dimensionless) and frequency zones

The main problem when combining gravity data is that the data are heterogeneous (e.g., Schwarz 1984). Gravity data are functions in space and time. Moreover, they have an individual noise spectrum and resolve a certain frequency band. When using data on the same spatial grid, the main difference between the satellite and the local gravity data is their frequency content. Figure 1 shows global, local and model degree variances in the spectral domain. As the 'true' degree variances are unknown, the Tscherning and Rapp (1974) degree variances model is shown in Figure 1. Satellite data are the best means to recover the long wavelengths of the gravity field. The extent of the medium frequency zone, shown in grey in Figure 1, depends on the area size of the local data and the resolution of the satellite data. Possible correlations between the local data and the global models are strongest in this overlapping zone. The high-frequency part is solely resolved from the local gravity data, representing small-scale features of the gravity field. Figure 1 indicates that a single data set does not resolve the entire spectrum of the gravity field. However, a unified estimate of the entire spectrum could be obtained by combining global and local gravity data in a complementary way.

A combination method that uses the spectral properties of the data seems advantageous. In the following, the spectral combination is proposed for the combination of satellite and local gravity data. After presenting the method in a coherent framework, the main objective is to compare different data weighting methods. The methods are evaluated using their underlying assumptions, requirements and errors. A spherical approximation is used throughout the paper.

2 Spectral combination

Spectral combination is defined as the family of methods that combine heterogeneous gravity data using spectral weights (Moritz 1975). The output of the combination is an estimate of the high-resolution geoidal undulations, $\hat{N}$. Mathematically, this estimate is obtained with (Rapp and Rummel 1975)

$$\hat{N} = \hat{N}^{(1)} + \hat{N}^{(2)} + \delta N , \tag{1}$$

where $\hat{N}^{(1)}$ is the satellite data contribution and $\hat{N}^{(2)}$ is the local gravity data contribution. δN stands for the combination error. It includes discretization and interpolation errors, unmodeled satellite and gravity data errors, truncation or omission errors and residual atmospheric and topographic reduction errors. Obviously, every combination method aims at minimizing the error term δN.

The two contributions $\hat{N}^{(k)}$ are computed in three steps, which are indicated in Eqn. (2).

$$\left.\begin{array}{ll} \text{signal:} & g^{(k)} \xrightarrow{1.} \sigma_l^{(k)} \\ \text{noise:} & e^{(k)} \xrightarrow{2.} \varepsilon_l^{(k)} \end{array}\right\} \xrightarrow{3.} \hat{N}^{(k)}, \; k = 1,2 \tag{2}$$

The first two steps are projections of the signal $g^{(k)}$ and the noise $e^{(k)}$ into the spectral domain. k indicates that either satellite data ($k = 1$) or local gravity data ($k = 2$) are used. The degree variances σ_l and error degree variances ε_l are the spectral equivalents of the signal and the noise. They represent the total power of the signal (or noise) at a certain degree l. The third step involves the mapping of the signal to the output functional. In addition, the signal and the noise information are used to weight the output.

In the following, the last step is further discussed. Satellite and local gravity data are mapped to the output functional by global and local base functions. Spherical harmonic base functions are used to represent the satellite data and radial (harmonic) base functions to map the local gravity data to the geoidal undulations, see for instance Kusche (2002). The spectral weights can be incorporated into the expressions as suggested by Sjöberg (1981) as they only depend on degree l. Consequently, the satellite contribution is computed from

$$\hat{N}^{(1)} \doteq R \sum_{l=2}^{L_S} p_l \sum_{m=-l}^{l} \bar{A}_{lm} \bar{Y}_{lm} , \tag{3}$$

where R stands for the mean Earth radius and L_S is the maximum degree of the satellite model. $\bar{A}_{lm}$ are the satellite spherical harmonic coefficients corresponding to the (Laplace) surface spherical harmonics $\bar{Y}_{lm}$, see Heiskanen and Moritz (1967). $p_l \in \mathbb{R}^L$ are the spectral weights for the satellite data. They control the contribution of the satellite data to the final output $\hat{N}$.

The local data are convolved with a kernel function as follows

$$\hat{N}^{(2)} = \frac{1}{4\pi\gamma} \int_{\alpha=0}^{2\pi} \int_{\psi=0}^{\psi_c} K g^{(2)} \sin\psi \, d\psi \, d\alpha, \tag{4}$$

where the kernel function K is given as

$$K = \sum_{l=2}^{L} (2l+1) \, q_l \, \lambda_l \, P_l(\cos\psi) \, . \tag{5}$$

In Eq. (4), γ stands for the normal gravity evaluated at the ellipsoid, ψ is the spatial angle between the geocentric placement vector of the computation point and the variable integration point, α is the azimuth and ψ_c is the spherical cap. Local gravity data are only used in a limited integration domain σ_c around the computation point.

Corresponding to the local gravity functional $g^{(2)}$ and observation height $H(= r - R)$, different values λ_l are employed in the integral kernel, see Table 1. For instance, gravity anomalies given at the reference sphere, $g^{(2)} = \Delta g|_R$, are convolved with Stokes's kernel function. In Eq. (5), P_l are the l-th degree Legendre polynomials and the spectral weights for the local data $q_l \in \mathbb{R}^L$ are incorporated into the kernel function. The kernel function is frequency limited up to degree L corresponding to the input data $g^{(2)}$.

Table 1 Values λ_l corresponding to the functionals $g^{(2)}$

$g^{(2)}$	λ_ℓ	Name
$\Delta g\|_R$	$\frac{R}{l-1}$	Stokes
$\delta g\|_R$	$\frac{R}{l+1}$	Hotine
$\delta g\|_r$	$\frac{r}{l+1}\left(\frac{r}{R}\right)^{l+1}$	One-step
$T_{rr}\|_r$	$\frac{r^2}{(l+1)(l+2)}\left(\frac{r}{R}\right)^{l+1}$	Gradiometry

Equation (1) can be transformed into the spectral domain, showing the interaction of the spectral weights and the data. The summation is split into the low and medium frequency part (up to L_S) and the high frequency part (from $L_S + 1$ to L).

$$\hat{N} = \sum_{l=2}^{L_S} \left\{ p_l \hat{N}_l^S + q_l \hat{N}_l^G \right\} + \sum_{l=L_S+1}^{L} q_l \hat{N}_l^G + \delta N \,. \tag{6}$$

where $\hat{N}_l^S$ and $\hat{N}_l^G$ are the spectral equivalents of the satellite and local gravity data contributions. Clearly, the spectral weights p_l and q_l are essential to obtain a unified estimate $\hat{N}$. They act as filter functions to smooth the discontinuity between the global and the local spectra.

A special case is derived when the sum of the satellite and the local gravity weights equals one. In addition, setting $q_l = 1$ for $l > L_S$ indicates that the local gravity data are the only means to resolve the higher frequencies. Equation (6) can then be rearranged into (Wenzel 1981)

$$\hat{N} = \sum_{l=2}^{L_S} \hat{N}_l^S + \sum_{l=2}^{L} w_l \left(\hat{N}_l^G - \hat{N}_l^S \right) + \delta N, \tag{7}$$

where

$$w_l = \begin{cases} q_l & , 2 \le l \le L_S \\ 1 & , l > L_S \end{cases} \tag{8}$$

Most of the data weighting methods that will be discussed in the next section can be written in the specific form of Eqs. (7) and (8). However, in order to give a general description of the problem and allow the inclusion of additional local gravity data sets, the formulation is kept in the general form of Eq. (6).

The combination error δN consists of the commission error and the omission error. The commission error is induced by the inaccuracies of the measurements and residual errors from the topographic and atmospheric reductions. It also includes the interpolation and discretization errors. The omission error is due to the spectral truncation in Eq. (3) and the spatial truncation of the integral in Eq. (4). The size of the commission and omission error depends on the quality of the available data and the data weights p_l and q_l. Thus, the combination error will be one of the criteria in the comparison of data weighting methods.

When the error degree variances of the satellite and the local data are provided, the commission variance $\varepsilon_c(\hat{N})$ is given as (Wenzel 1981)

$$\varepsilon_c(\hat{N}) = R^2 \sum_{l=2}^{L_S} \left\{ p_l^2\, \varepsilon_l^{(1)} + q_l^2\, \varepsilon_l^{(2)} \right\} + R^2 \sum_{l=L_S+1}^{L} q_l^2 \varepsilon_l^{(2)} + \delta\epsilon^{(c)} \tag{9}$$

Additional errors, for instance errors due to unmodeled data correlations, are lumped together in $\delta\epsilon^{(c)}$. An approximate value for the omission variance can be obtained by (Heck 1979)

$$\varepsilon_o(\hat{N}) = \frac{1}{\gamma^2} \sum_{l=L_S+1}^{\infty} p_l^2 \lambda_l^2 \sigma_l^M + \frac{1}{4\gamma^2} \sum_{l=2}^{\infty} \kappa_l^2 \sigma_l^M + \delta\epsilon^{(o)} \tag{10}$$

where $\delta\epsilon^{(o)}$ stands for residual errors and

$$\kappa_l = \sum_{n=2}^{L} (2n+1) q_n \lambda_n R_{ln}(\psi_c) \,. \tag{11}$$

σ_l^M is a signal variance model; the Tscherning and Rapp (1974) model is used in the evaluation later on. $R_{ln}(\psi_c)$ are Paul's (1973) coefficients involving the integration of Legendre polynomials as follows ($n \le l$)

$$R_{ln}(\psi_c) = \int_{\psi=\psi_c}^{\pi} P_l(\cos\psi) P_n(\cos\psi) \sin\psi \, d\psi. \tag{12}$$

3 Spectral weight determination

According to Heck and Grüninger (1987), one can roughly classify data weighting approaches as methods that are based on some knowledge of the errors, and methods that neglect a-priori information about the noise and errors. The stochastic methods more or less rely on the quality of the a-priori information, whereas the deterministic approaches might in some cases be too pessimistic. In this contribution, data weighting methods are sorted according to their complexity and difficulty in implementation. Even though the following list is not complete, it gives an overview of available methods. They are as follows:

- deterministic methods
- filtering methods
- least-squares adjustment
- least-squares collocation
- quality measure approach
- quasi-deterministic method
- information measure approach

3.1 Deterministic method (DET)

The simplest way to combine satellite and local gravity data is to neglect possible correlations and a-priori noise knowledge. The method assumes that the satellite and the local gravity data are either error-free or have the same error magnitude. Then, the local gravity data may be convolved with the spherical kernel function. Correspondingly, the weights for the satellite and local gravity data become

$$2 \leq l \leq L_S : \begin{cases} p_l = 1, \\ q_l = 1 \end{cases} \tag{13}$$

As demonstrated in Vaníček and Featherstone (1998), the spherical (Stokes) kernel function may have adverse properties. For instance, the omission error is relatively large and low-frequency local gravity errors leak unattenuated into the final solution. In an attempt to minimize these errors, the Wong and Gore (1969) kernel function - sometimes also denoted as the spheroidal kernel - is often used instead. The weights then yield

$$2 \leq l \leq L_S : \begin{cases} p_l^{\mathrm{DET}} = 1, \\ q_l^{\mathrm{DET}} = 0 \end{cases} \tag{14}$$

The Wong and Gore kernel starts from degree $L_S + 1$, indicating that the lower degrees are removed from the spherical kernel. The size of the omission error is reduced compared to the spherical kernel function, see for instance Vaníček and Featherstone (1998). Other deterministic kernel modifications are discussed in e.g. Evans and Featherstone (2000) or Kern (2003).

3.2 Filtering methods

To ensure a smooth transition between the global and local gravity data, filtering methods can be employed. The data weights are then considered as filter factors. Clearly, a wide range of possibilities and approaches exist, see for instance Oppenheim and Schafer (1999). Haagmans (2000) proposes a spherical Butterworth filter (SBF) for the weighting functions. The satellite data are low-pass filtered from a certain degree l_b and additional weight is assigned to the local gravity data. The spectral weights (or filter) are generated using the formula (Haagmans 2000, Haagmans et al. 2002)

$$2 \leq l \leq L_S : \; p_l^{\mathrm{SBF}} = \sqrt{\frac{1}{1 + \left(\frac{l}{l_b}\right)^{2k}}} \tag{15}$$

and complementary for the local data

$$q_l^{\mathrm{SBF}} = 1 - p_l^{\mathrm{SBF}} \, . \tag{16}$$

The band degree l_b and the filter order k are heuristically determined integer values, depending on the satellite data resolution and local area size. If the spectral weights are properly chosen, the induced errors could be smaller than the ones in the deterministic method.

Another type of filter for the combination of satellite and local gravity data is proposed by Wang (1993). He filters only the satellite data by a Wiener filter of the form

$$2 \leq l \leq L_S : \; p_l^{\mathrm{WNF}} = \frac{\sigma_l^{(1)}}{\sigma_l^{(1)} + \varepsilon_l^{(1)}} \, . \tag{17}$$

The local gravity data are convolved with the spherical kernel, implying $q_l^{\mathrm{WNF}} = 1$ for $2 \leq l \leq L$.

3.3 Least-squares adjustment (LSA)

In the preceding methods, the noise or errors in the satellite and local data are neglected or only partially taken into account. The least-squares spectral combination on the other hand, weights the two contributions according to their errors. The method has been independently introduced by Sjöberg (1981) and Wenzel (1981).

Using the spectral weights derived in Sjöberg (1981), $(2 \leq l \leq L_S)$,

$$p_l^{\mathrm{LSA}} = 1 - q_l^{\mathrm{LSA}}, \quad q_l^{\mathrm{LSA}} = \frac{\varepsilon_l^{(1)}}{\varepsilon_l^{(1)} + \varepsilon_l^{(2)}} \tag{18}$$

the variance in Eq. (9) is minimal and the sum of the weights is one. A larger weight is given to the measurements with smaller errors and vice versa. Unfortunately, the local gravity data errors, $\varepsilon_l^{(2)}$, are largely unknown, which makes the method questionable from a practical point of view. Nevertheless, the method has been successfully applied using stationary error models by Wenzel (1982) and Wichiencharoen (1984).

3.4 Least-squares collocation (LSC)

A least-squares collocation approach for the weight determination has been proposed by Moritz (1975). Further investigations are reported in Sjöberg (1979) and Wenzel (1982). Under the condition of least error variances, least-squares collocation filters the data with the weights (Moritz 1975)

$$p_l^{\mathrm{LSC}} = \frac{(\varepsilon_l^{(1)})^{-1}}{D_l}, \quad q_l^{\mathrm{LSC}} = \frac{(\varepsilon_l^{(2)})^{-1}}{D_l} \tag{19}$$

where D_l is given as

$$D_l = (\sigma_l)^{-1} + (\varepsilon_l^{(1)})^{-1} + (\varepsilon_l^{(2)})^{-1}. \tag{20}$$

The sum of p_l^{LSC} and q_l^{LSC} is always smaller than one. The method requires the knowledge of the error or noise degree variances and the signal degree variances (σ_l). As both requirements are insufficiently known in practice, the quality of the method strongly relies on the chosen models. For higher degrees $l > L_{\mathrm{S}}$, the weights are determined with

$$p_l^{\mathrm{LSC}} = 0, \quad q_l^{\mathrm{LSC}} = \frac{\sigma_l^{(2)}}{\sigma_l^{(2)} + \varepsilon_l^{(2)}}. \tag{21}$$

3.5 Quality measure approach

Quality measures are common tools for representing error characteristics of a process. Although error (degree) variances are the primary error information in most applications, quality measures may represent attractive alternatives.

Signal-to-noise ratio per degree. The (one-dimensional) signal-to-noise ratio (SNR) is a relative measure, relating the signal to the error spectrum. Reliable estimates of the signal-to-noise ratio may be provided for airborne measurements. Generally, the signal-to-noise formulas for unbiased satellite and local gravity data are given as

$$\mathrm{SNR}_l^{(k)} = \sqrt{\frac{\sigma_l^{(k)}}{\varepsilon_l^{(k)}}}, \quad k = 1, 2 \tag{22}$$

The satellite weighting functions are then determined from

$$p_l^{\mathrm{SNR}} = \frac{\mathrm{SNR}_l^{(1)}}{\mathrm{SNR}_l^{(1)} + \mathrm{SNR}_l^{(2)}}. \tag{23}$$

A large signal-to-noise ratio will increase the weight of the data set in the solution. The local data weights are given by $q_l^{\mathrm{SNR}} = 1 - p_l^{\mathrm{SNR}}$.

Gain per degree. Another relative measure is the gain (GAIN). It compares the quality of an error spectrum to another error spectrum. Analogous to the signal-to-noise ratio, it can be given dependent on degree. For $\varepsilon_l^{(1)} \leq \varepsilon_l^{(2)}$, the gain and the weighting functions p_l^{GAIN} are computed from (Kern 2003)

$$\mathrm{GAIN}_l = p_l^{\mathrm{GAIN}} = \sqrt{\frac{\varepsilon_l^{(1)}}{\varepsilon_l^{(2)}}}. \tag{24}$$

Again, $q_l^{\mathrm{GAIN}} = 1 - p_l^{\mathrm{GAIN}}$ determines the local data weights.

3.6 Quasi-deterministic method (QDM)

A quasi-deterministic method has been proposed in Kern et al. (2002). It is based on the assumption that the low-frequency information will be reliable known from satellite missions such as GRACE and GOCE. The basic idea is that the true signal degree variance, σ_l, can be approximated by the degree variance from satellite spherical harmonics.

$$\overline{\sigma}_l^{(2)} = \sigma_l + \varepsilon_l^{(2)} \doteq \sigma_l^{(1)} + \varepsilon_l^{(2)}. \tag{25}$$

The signal degree variances $\overline{\sigma}_l^{(2)}$ are computed from the local gravity measurements $g^{(2)}$. Rearranging Eq. (25), the weights for the local gravity data become (Kern et al. 2002)

$$q_l^{\mathrm{QDM}} = \frac{\varepsilon_l^{(1)}}{\varepsilon_l^{(1)} + |\overline{\sigma}_l^{(2)} - \sigma_l^{(1)}|}. \tag{26}$$

The weights for the global data are given as $p_l^{\mathrm{QDM}} = 1 - q_l^{\mathrm{QDM}}$.

Figure 2 shows the QDM-weights for local gravity data ($5' \times 5'$ spaced terrestrial gravity anomalies in a $3° \times 5°$ area close to Ottawa, Canada; mean(Δg)= -9.7 mGal, std(Δg)= 21.7 mGal) and EGM96 (Lemoine et al. 1998). Possible correlations between the EGM96 and the local data are neglected. Figure 2 clearly shows that both data sets contribute to the medium frequency zone. The local data gain more and more weight as the global data errors increase. Beyond degree $L_{\mathrm{S}} = 360$,

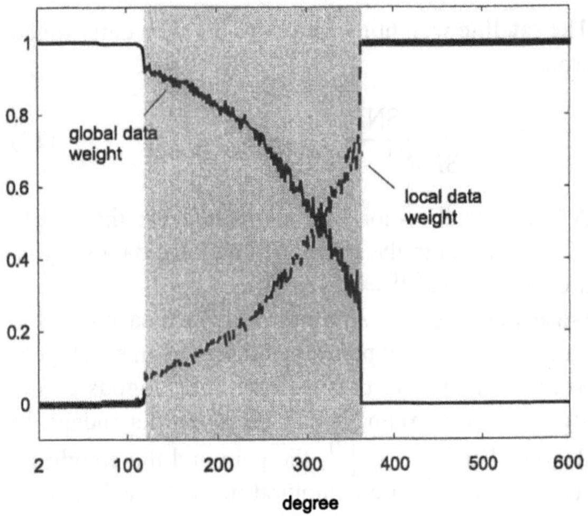

Fig. 2 Weights in the quasi-deterministic approach

the weights are chosen to be $q_l = 1$ and $p_l = 0$. Low-frequency inconsistencies in the local gravity data do not fully propagate into the final solution as their low-frequencies are downweighted. Hence, the method is relatively robust against biases compared to the other methods.

3.7 Information measure approach (IMA)

Yet another method for the combination of satellite and local gravity data sets originates from information theory. Information theory has been developed by Shannon (1948) using information entropy as a measure. In geodesy, the concept of information measures has first been introduced by Blais (1987). Lehmann (1996) uses information measures for the evaluation of geopotential models.

As the information measure is a relative measure, two data sets can be weighted using their inherent information content. In addition, the a-priori information needed for the stabilization of the satellite model can be taken into account.

The information number $I(r;s)$ of two probability density functions $r(x), s(x) \in \mathbb{R}^n$ is given as (Kullback and Leibler 1951)

$$I(r;s) = \int r(x) \log \frac{r(x)}{s(x)} dx \,. \tag{27}$$

A special case is derived when the probability density functions are Gaussian distributed; $r(x)$ is normally distributed with $N(\mu_1, \Sigma_1)$ and $s(x)$ with $N(\mu_2, \Sigma_2)$. The information number in Eq. (27) yields

$$I(r;s) = \frac{\log e}{2} \text{trace W} + \frac{1}{2} \log \frac{\det \Sigma_1}{\det \Sigma_2} \,. \tag{28}$$

The matrix W is given as (Blahut 1987)

$$W = \Sigma_1^{-1} \left(\Sigma_1 - \Sigma_2 + (\mu_1 - \mu_2)(\mu_1 - \mu_2)^T \right) \tag{29}$$

Information theory can be applied to the gravity field, when the gravity quantities are considered as unknowns or a finite number of data affected by measurement noise (Lehmann 1996).

The information measure (in units of bits) of a spherical harmonic model relative to an a-priori degree variance k_l can be obtained as (Lehmann 1996)

$$I = \left(-\frac{L_S^2 + 2L_S - 3}{2} + \sum_{l=2}^{L_S} \frac{l + \frac{1}{2}}{k_l} u_l \right) \log e - \frac{1}{2} \log \det \Sigma + \sum_{l=2}^{L_S} (l + \frac{1}{2}) \log \frac{k_l}{2l+1} \,, \tag{30}$$

where Σ is a diagonal matrix containing the satellite coefficient errors δA_{lm} and

$$u_l = \sum_{m=-l}^{l} (A_{lm}^2 + \delta A_{lm}^2) \,. \tag{31}$$

The information measure for each degree (I_l) can be obtained by summing I degreewise. Replacing k_l by the local gravity degree variance, information measures relating satellite and local data can be computed. Comparing this information measure to a measure computed with a reliable model for k_l defines the weight of the satellite and local gravity data. Obviously, the proper choice of k_l, i.e. of the model, is of vital importance for the method.

4 Discussion

All data weighting methods rely on the quality of the data that are provided. Yet the information that is required for the spectral weight determination varies from method to method, see Table 2. The deterministic methods implicitly imply a certain error behaviour, while the LSA and the LSC explicitly involve the errors of both data sets. If reliable quality measures would be available, the quality measure approaches could be used. Alternatively, the QDM requires precise satellite information. Local gravity data errors are not needed for the spectral weight determination when using the QDM. Some methods model the combination problem with additional parameters, for instance the SBF and the IMA. As the determination of these parameters always involves some a-priori information about the gravity field, these methods may lead to doubtful results.

All methods involve a certain number of assumptions. They are listed in Table 2 in the third column.

Table 2 Comparison of the satellite and local data methods (○ – small error; ◐ – medium error; ● – large error)

Methods	Requirements satellite	Requirements local	Assumptions satellite	Assumptions local	Resulting Errors commission	Resulting Errors omission	Remarks
DET	$\bar{A}_{lm}$	g	$\delta\bar{A}_{lm}$ =?	e =?	●	◒	possibly biased solution
SBF	$\bar{A}_{lm}$ l_b, k	g l_b, k	$\delta\bar{A}_{lm}$ =?	e =?	◒	◒	possibly biased solution
WNF	$\bar{A}_{lm}$ $\delta\bar{A}_{lm}$	g	$\delta\bar{A}_{lm}$	e =?	◒	◒	
LSA	$\bar{A}_{lm}$ $\delta\bar{A}_{lm}$	g e	$\delta\bar{A}_{lm}$	e	○	○	strong assumptions required
LSC	$\bar{A}_{lm}$ $\delta\bar{A}_{lm}$	g e	$\delta\bar{A}_{lm}$	e	○	○	strong assumptions required
SNR	$\bar{A}_{lm}$ $\mathrm{SNR}_l(S)$	g $\mathrm{SNR}_l(g)$	$\mathrm{SNR}_l(S)$	$\mathrm{SNR}_l(g)$	◒	◒	SNR's rarely given
GAIN	$\bar{A}_{lm}$ $\mathrm{GAIN}_l(S)$	g $\mathrm{GAIN}_l(g)$	$\mathrm{GAIN}_l(S)$	$\mathrm{GAIN}_l(g)$	◒	◒	GAIN's rarely known
QDM	$\bar{A}_{lm}$ $\delta\bar{A}_{lm}$	g	$\delta\bar{A}_{lm}$		○	○	relies on satellite data
IMA	$\bar{A}_{lm}$ $\delta\bar{A}_{lm}$	g k_l	$\delta\bar{A}_{lm}$	k_l	◒	◒	strongly relies on the choice of k_l

The question mark indicates that the error is assumed to be unknown and not taken into account. Note that no specific assumption has to be made for the local gravity data in the QDM.

To assess the quality of the data weighting methods, the commission and the omission error are evaluated in each case. The global error degree variances originate from the EGM96 ($L_S = 360$) and the local error degree variances are estimates for the local area ($L = 2160$) described before. White noise of 1.5 mGal standard deviation has been assumed. The same (error) degree variances have been used in all methods. The results are summarized in the fourth column of Table 2. Instead of numbers, which would vary from area to area, the provided data and their errors etc., the (relative) errors are shown in symbols. ○ is a small error, ◒ stands for a medium error and ● indicates a large error. Clearly, the deterministic and the filtering methods involve medium and large commission and omission errors as they do not involve error estimates. However, they could be further improved by modifications of the kernel functions, see for instance Vaníček and Featherstone (1998). On the contrary, the LSA and LSC seek for a minimum of these errors by implying full error knowledge. They would involve the smallest commission and omission error if the correct error estimates are provided. The QDM represents a good alternative to the purely deterministic methods. Small errors are introduced by such a combination with a minimum of assumptions. The QDM is therefore recommended for the combination of satellite and local gravity data.

The spectral combination is a stationary method. Such an approximation may be justified as the satellite errors are global mean values and reliable non-stationary information of the local gravity errors is rarely given. An additional problem are the errors themselves; global and local errors are actually not comparable. A practical solution would be to filter the local errors. Pellinen's smoothing factors may be used for such a purpose (Jekeli 1981). Numerical techniques that can further reduce aliasing and leakage effects have to be found. So far, time-variable effects are not explicitly taken into account. And lastly, the correlations between the satellite and the local gravity data, possibly existing because of a calibration process of the satellite data, may deteriorate the solutions.

5 Conclusions

The analysis of data weighting methods for the combination of satellite and local gravity data was the objective of this paper. Introduced in the framework of spectral combination, the methods are based on different theoretical concepts and requirements. The least-squares adjustment and the least-squares collocation outperform the other methods as they seek for an optimal combination of the satellite and local gravity data. However, they require the a-priori error knowledge for both data sets, which is seldom given. Unless the adopted stochastic models are reliable, methods such as the quasi-deterministic method are recommended for the combination of satellite and local gravity data. It minimizes both commission and omission errors. The deterministic methods could be further modified to produce more reliable results. However, this is outside the scope of the paper.

Acknowledgements

The author wishes to express his sincere gratitude to Dr. KP Schwarz for his continuous inspiration, advice and support. Moreover, Dr. F Sansò is thanked for his encouragement and advice at the symposium. The helpful comments by Dr. WE Featherstone and an anonymous reviewer are gratefully acknowledged.

References

Blahut RE (1987) Principles and practice of information theory. Addison-Wesley, Reading, Mass.

Blais JAR (1987) Information theory and optimal estimation. Manuscr Geod 12: 238-244.

Evans JD, Featherstone WE (2000) Improved convergence rates for the truncation error in gravimetric geoid determination. J Geod 74(2): 239-0248

Haagmans R (2000) A synthetic Earth for use in Geodesy. J Geod 74: 503-511.

Haagmans R, Prijatna K, Omang O (2002) An alternative concept for validation of GOCE gradiometry results based on regional gravity. To be published in the Proc of the 3rd Meet Int Gravity Geoid Com. Thessaloniki, Greece.

Heck B (1979) Zur lokalen Geoidbestimmung aus terrestrischen Messungen vertikaler Schweregradienten. DGK, series C, no. 259.

Heck B, Grüninger W (1987) Modification of Stokes' integral formula by combination of two classical approaches. Proc IAG Symp, XIX IUGG General Assembly, pp 319-337.

Heiskanen WA, Moritz H (1967) Physical geodesy. Freeman and Co., San Francisco.

Jekeli C (1981) Alternative methods to smooth the Earth's gravity field. Department of Geodetic Science. Rep 327. Ohio State University. Columbus.

Kern M, Schwarz KP, Sneeuw N (2002) A study on the combination of satellite, airborne and terrestrial gravity data. To be published by the J Geod.

Kern M (2003) An analysis of the combination and downward continuation of satellite, airborne and terrestrial gravity data. Department of Geomatics Engineering. Rep 20172. The University of Calgary.

Kullback S, Leibler RA (1951) On information and sufficiency. Ann Math Stats. 22: 79-86.

Kusche J (2002) Inverse Probleme bei der Gravitationsfeldbestimmung mittels SST- und SGG- Satellitenmissionen. DGK, series C, no. 548.

Lehmann R (1996) Information measures for global geopotential models. J Geod 70: 342-348.

Lemoine FG, Kenyon SC, Factor JK, Trimmer RG, Pavlis NK, Chinn DS, Cox CM, Klosko SM, Luthcke SB, Torrence MH, Wang YM, Williamson RG, Pavlis EC, Rapp RH, Olson TR (1998) The Development of the joint NASA GSFC and the National Imagery and Mapping Agency (NIMA) Geopotential Model EGM96. Rep. NASA TP-1998-206861.

Moritz H (1975) Integral formulas and collocation. Department of Geodetic Science. Rep 234. The Ohio State University. Columbus.

Oppenheim AV, Schafer RW (1999) Discrete-time signal processing. Prentice Hall signal processing series. Second edition.

Rapp RH, Rummel R (1975) Methods for the computation of detailed geoids and their accuracy. Department of Geodetic Science and Surveying. Rep 233. Ohio State University. Columbus.

Schwarz KP (1984) Data types and their spectral properties. Proc Int Summer School on local gravity field approximation. Ed. KP Schwarz. Bejing. China. pp 1-66.

Shannon CE (1948) A mathematical theory of communication. Bell Syst Techn J 27(379-426): 623-656.

Sjöberg LE (1979) Integral formulas for heterogeneous data in physical geodesy. Bull Géod 53: 297-315.

Sjöberg LE (1981) Least squares combination of satellite and terrestrial data in physical geodesy. Ann Geophys 37: 25-30.

Tscherning CC, Rapp RH (1974) Closed covariance expressions for gravity anomalies, geoid undulations, and deflections of the vertical implied by anomaly degree variance models. Department of Geodetic Science. Rep 208. Ohio State University, Columbus.

Vaníček P, Featherstone WE (1998) Performance of three types of Stokes's kernel in the combined solution for the geoid. J Geod 72: 684-697.

Wang YM (1993) On the optimal combination of potential coefficient model with terrestrial gravity data for FFT computations. Manuscr Geod 18: 406-416.

Wenzel HG (1981) Zur Geoidbestimmung durch Kombination von Schwereanomalien und einem Kugelfunktionsmodell mit Hilfe von Integralformeln. Z Verm. 3: 102-111.

Wenzel HG (1982) Geoid computation by least-squares spectral combination using integral kernels. Proc. IAG General Meet. Tokyo. pp 438-453.

Wichiencharoen C (1984) A comparison of gravimetric undulations computed by the modified Molodenskij truncation method and the method of least squares spectral combination by optimal integral kernels. Bull Géod 58: 494-509.

The rank deficiency in estimation theory and the definition of reference systems

Athanasios Dermanis
Department of Geodesy and Surveying, The Aristotle University of Thessaloniki, Greece

Abstract. The concept of reference frame is examined from the viewpoint of both geophysical and geodetic applications. The concept of parameter estimability in linear models is related to the deterministic concept of determinability in linear or nonlinear improper models without full rank. The geometry of such models is investigated in its linear and nonlinear aspects with emphasis on the common invariance characteristics of observable and estimable parameters and is applied to the choice of datum problem in geodetic networks. The time evolution of the reference frame is investigated and optimal choices are presented from different equivalent points of view. The transformation of a global geodetic network into an estimate of a geocentric Tisserand frame for the whole earth is investigated and a solution is given for the rotational part. The translation to a geocentric frame poses the problem of the estimability of the geocenter coordinates and the more general problem of estimability of coefficients of an unknown function of position, having as domain the frame-dependent coordinates.

Keywords. Reference frames, estimable parameters, earth rotation.

1 The reference frame concept

A reference frame is a mathematical artifice, which is introduced for the sake of convenience only within the framework of Newtonian physics. In a more general setting the study of physical phenomena requires their mathematical modeling in relation to the real number system, so that computations based on the analysis of numerical data may lead to a quantitative realization of the qualitative characteristics described by the relevant modeling equations. This becomes possible with the introduction of coordinates for the representation of points in space and time by real numbers. In addition a smooth field of basis vectors is used in order to represent at every point local vectors (objects of direction and magnitude) and other associated mathematical objects (tensors) by real numbers, namely their components with respect to the local basis.

In such models it is essential that the phenomena described by differential equations are independent of the particular coordinate system and the basis field chosen. Let us recall that in tensor analysis the particular choice of local base vectors having the direction and rate of the coordinate lines, makes the independence more transparent by leading to equations which are coordinate invariant.

Newtonian mechanics is based on the Galilean separation of space and time as well as on the Euclidean model for space. The essential ingredient of the Euclidean model is the possibility of parallel transport of straight lines and vectors, which may be viewed in this case as directed line segments. Thus it is sufficient to introduce a basis only at a particular point, called the origin, and then generate the required field of local bases at each point by parallel transport. In addition the directed line segment from the origin to any point is called the position vector of the point and its components are used as a set of "cartesian" coordinates for the particular point.

A reference frame consists of a particular point O, the origin and an orthonormal triad of vectors $\vec{\mathbf{e}} = [\vec{e}_1 \vec{e}_2 \vec{e}_3]$. Orthogonality and unit length for the base vectors are additionally introduced for the sake of simplicity. Attached to the choice of the reference frame is a parallel triad at every point for the representation of local vectors and tensors as well as the cartesian coordinates of the point. Even if curvilinear coordinates are used, e.g. spherical ones, these are typically defined via their relation to the cartesian coordinates and are therefore attached to the chosen reference frame.

The separation of space and time necessitates the introduction of an additional independent reference frame for time. It consists of an origin (initial epoch) t_0 and a unit vector involving a direction (towards the future) and a particular magnitude, i.e. the definition of a time interval as the unit of time. Let us note that the Galilean separation of space and time is possible when the unit of length for space is independently defined. However in the practice of spatial geodetic techniques only time intervals are measured, while lengths are produced by multiplication with the constant velocity of light c. This means that the scale factor is common for space and time and it cannot be separated by observations. For example it is impossible to determine whether the earth as a whole is expanding (shrinking) or the set of clocks used for the definition of the time unit is getting slower (faster).

Another aspect of the temporal dimension of physical phenomena is the necessity of choosing a reference frame for each epoch, thus facing the problem of not only defining the reference frame at a particular reference epoch but also its evolution with time.

The choice of a reference frame evolving in time is not completely arbitrary. Only a set of particular frames is acceptable, the inertial frames with respect to which the laws of Newtonian mechanics are holding. The choice of two different inertial frames corresponds to a "parallel transport in spacetime" along the time coordinate axis of a frame arbitrarily chosen at the initial reference epoch. Two inertial frames $(O,\vec{\mathbf{e}})$ and $(O',\vec{\mathbf{e}}')$ are dynamically equivalent, when they are "parallel", i.e., the matrix $\mathbf{Q}$ of the transformation $\vec{\mathbf{e}}'=\vec{\mathbf{e}}\mathbf{Q}$ ($\vec{e}_i'=\vec{e}_j Q_i^j$) is constant, and furthermore the origin displacement vector varies linearly with time $\overrightarrow{OO'}=\vec{\mathbf{e}}(\mathbf{a}+t\mathbf{b})$.

2 Reference frames in geophysics

We now come to a different type of frame definition. While inertial frames represent the "absolute" space in which physical phenomena take place, reference frames may be introduced for the convenient representation of a moving and deforming body, in our case the earth. This representation attains an absolute character in the case of rigid bodies, where the time evolution of the representing frame can be chosen by simply imposing the time invariance of the coordinates of the material points of the body. Thus we have only to choose one among "parallel" body frames, each of which represents equally well the moving rigid body. The study of the body motion is equivalent to the study of the motion of the body frame. For the deformable earth it is necessary to choose not only a frame at an original epoch but also to choose its temporal evolution. This must be done in such a way that the total motion of the material points is separated into a common part (motion of the earth frame) and a remaining part (deformation) which is as small as possible. It is the choice of a particular principle, which defines a set of equivalent "parallel" frames (such as the Tisserand axes frame), while a particular frame from this set is chosen through its definition at the reference epoch.

Thus in geophysics we need two frames: an inertial one O^I, $\vec{\mathbf{e}}^I=[\vec{e}_1^I\vec{e}_2^I\vec{e}_3^I]$ and an earth-representing frame O, $\vec{\mathbf{e}}=[\vec{e}_1\vec{e}_2\vec{e}_3]$. The study of the motion of the earth in inertial space is separated into three parts:

(a) The translational motion of the earth, i.e. the variation with time of the vector $O^I O$;

(b) the rotation of the "earth" (in fact of the frame $\vec{\mathbf{e}}$), i.e. the variation with time of the orthogonal matrix $\mathbf{R}(t)$ relating the two frames through $\vec{\mathbf{e}}(t)=\vec{\mathbf{e}}^I\mathbf{R}(t)^T$;

(c) the deformation of the earth with respect to $\vec{\mathbf{e}}$.

In practice the realization of an inertial frame is not easy due to the lack of "motionless" objects around the earth. For its orientation we have to rely on the directions of the extragalactic radio sources involved in VLBI observations, in replacement of the earlier stellar directions of the now abandoned astrogeodetic observations. The choice of origin is problematic since we can hardly find a point in the vicinity of the earth, which has a uniform linear motion with respect to the rest of the universe; the barycenter (center of mass of the solar system) is only an approximation to this requirement. In any case the enormous distances of the radio sources make the exact definition of the origin irrelevant for the study of earth rotation since their directions remain practically unaltered. For this reason the center of mass of the earth may be chosen as the origin of a quasi inertial frame because it has the advantage of best separating translational from rotational motion. The translational motion of the earth is equivalent to the motion of its geocenter as a mass point with the combined effect of all external accelerations applied to it.

Let $\vec{x}$ be the position vector, $\vec{v}$ the velocity, $\vec{f}$ the exerted forces per unit mass, $\vec{h}=\int\vec{x}\times\vec{v}\,dm$ the angular momentum, $\vec{l}$ the exerted torques, while $\mathbf{z}_I$ and $\mathbf{z}$ are the inertial and earth-frame components, respectively, of any vector $\vec{z}=\vec{\mathbf{e}}^I\mathbf{z}_I=\vec{\mathbf{e}}\mathbf{z}$. The laws of translational and rotational motion

$$\vec{a}\equiv\frac{d^2\vec{x}}{dt^2}=\vec{f}\,,\qquad \dot{\vec{h}}=\frac{d\vec{h}}{dt}=\vec{l} \tag{1}$$

take simple forms in the inertial frame

$$\ddot{\mathbf{x}}_I=\mathbf{f}_I\,,\qquad \dot{\mathbf{h}}_I=\mathbf{l}_I\,. \tag{2}$$

The same equations become more complicated in the terrestrial frame because they involve the instantaneous angular velocity vector $\vec{\omega} = \vec{e}^I \omega_I = \vec{e}\omega$, which relates to the rotation matrix $\mathbf{R}$ through the geometric Euler equations

$$[\omega\times] = \mathbf{R}\dot{\mathbf{R}}^T, \qquad [\omega_I\times] = \dot{\mathbf{R}}^T\mathbf{R}, \tag{3}$$

where for any antisymmetric matrix $\mathbf{\Omega} = [\omega\times]$, ω denotes the corresponding axial vector. For any local vector $\vec{z}$ it holds that $\mathbf{z} = \mathbf{R}\mathbf{z}_I$ and the same is true for the position vector. However as a consequence of the relation $\dot{\vec{e}} = \vec{e}[\omega\times]$ the velocity vector has terrestrial components

$$\mathbf{v} = \dot{\mathbf{x}} + [\omega\times]\mathbf{x}, \tag{4}$$

and the translational equations of motion become

$$\mathbf{a} = \ddot{\mathbf{x}} + [\omega\times]^2\mathbf{x} + 2[\omega\times]\dot{\mathbf{x}} + [\dot{\omega}\times]\mathbf{x} = \mathbf{f}. \tag{5}$$

Since the terrestrial frame is not inertial, fictitious accelerations arise in the last equation, namely the centrifugal $[\omega\times]^2\mathbf{x}$, the Coriolis $2[\omega\times]\dot{\mathbf{x}}$ and the gyroscopic acceleration $[\dot{\omega}\times]\mathbf{x}$. The angular momentum has terrestrial components

$$\mathbf{h} = \int[\mathbf{x}\times]\mathbf{v}\,dm = -\left(\int[\mathbf{x}\times][\mathbf{x}\times]\,dm\right)\omega + \int[\mathbf{x}\times]\dot{\mathbf{x}}\,dm \equiv$$

$$\equiv \mathbf{C}\omega + \mathbf{h}_R \tag{6}$$

where $\mathbf{C}$ is the tensor of inertia and $\mathbf{h}_R$ the relative angular momentum. The equations of rotational motion take the form of the Liouville equations

$$\mathbf{C}\dot{\omega} + \dot{\mathbf{C}}\omega + \dot{\mathbf{h}}_R + [\omega\times](\mathbf{C}\omega + \mathbf{h}_R) = \mathbf{l}. \tag{7}$$

In the theory of rotation (Munk & MacDonald, 1960, Moritz & Mueller, 1987) two terrestrial geocentric reference frames have been suggested. The *axes of figure* frame, making $\mathbf{C}$ a diagonal matrix and the *Tisserand axes*, the family of parallel frames for which the relative angular momentum vanishes, $\mathbf{h}_R = \mathbf{0}$, and at the same time the kinetic energy is minimized

$$T \equiv \frac{1}{2}\int \mathbf{v}^T\mathbf{v}\,dm =$$

$$= \frac{1}{2}\omega^T\mathbf{C}\omega + \omega^T\mathbf{h}_R + \frac{1}{2}\int\dot{\mathbf{x}}^T\dot{\mathbf{x}}\,dm = \min. \tag{8}$$

For a Tisserand frame the Liouville equations are readily simplified to

$$\mathbf{C}\dot{\omega} + \dot{\mathbf{C}}\omega + [\omega\times]\mathbf{C}\omega = \mathbf{l}. \tag{9}$$

The practical realization of these frames is difficult for two reasons. First they require complete knowledge not only of the motion ($\dot{\mathbf{x}}$) but also on the mass distribution (dm) within the earth. On the other hand an exact theoretical solution for ω is not possible, in which case the matrix $\mathbf{R}$ resulting from the solution of the Euler geometric equations (3) would transform the inertial frame into the adopted terrestrial one. Therefore the establishment and maintenance of terrestrial frames remains a geodetic responsibility.

3 The estimability concept

3.1 Estimation with improper models

The problem of the establishment of a reference frame arose in geodesy and surveying long before the space era made considerations about the shape of the earth as a whole to be more than a mere theoretical speculation. In surveying the analysis of observations of horizontal or vertical networks, which carry no information on position but only on network shape (or shape and size) becomes problematic when the relevant models involve coordinates referring to an unspecified reference frame.

Different approaches have been developed in order to overcome this problem. The simplest one is the introduction of additional minimal constraints, which introduce a particular reference frame with respect to the network. Such are the trivial constraints of choosing a particular point as the origin and a particular side as the *x*-axis for a planar network or choosing a particular point as the origin (zero height) for a vertical network. These constraints can be introduced a priori, so that the respective coordinates disappear as model unknowns, or a posteriori as additional constraints to be satisfied by any coordinate solution. Although the result will be the same these two choices differ from the methodological point of view.

Another approach is based on the formulation of the normal equations satisfying the least squares principle for the observational errors, which admit more than one solution. A solution can be obtained with the use of a generalized inverse of the coefficient matrix. In fact it was Bjerhammar (1951) who rediscovered the Moore generalized inverse in this context, quite some time before its rediscovery by Penrose (1955).

An approach, which relates to the pseudo-inverse solution, is the introduction of the *inner constraints* by Meissl (1965, 1969), where the network inherits a frame by being best fitted to the network defined by a set of approximate coordinates and, therefore, has a frame attached to it.

The interrelation between least squares adjustment and linear estimation theory through the celebrated Gauss-Markov theorem results in a relation of the "missing" reference frame problem with the notion of parameter estimability. However, the notion of estimability is not a statistical one, but is rather relating to the modeling problem.
Next we will review the fundamentals of the geodetic datum problem, attempting to clarify the relations between its modeling, geometric, algebraic and statistical aspects.

3.2 The datum problem as an improper modeling problem

The problem of data analysis always involves a specific set of parameters, the observables $\mathbf{y}$, while the observation vector $\mathbf{b} = \mathbf{y} + \mathbf{v}$ differs from $\mathbf{y}$ due to the presence of observational errors $\mathbf{v}$. The objective is to exploit the n observations in order to obtain information about a physical system, which can be represented, in more than one ways, by a specific number of parameters r, which we call the parametric rank of the system. We must distinguish between the background physical model consisting of all adopted laws governing the system under study and the specific mathematical model to be used in data analysis (Dermanis, 1991). This model may involve, in addition to the observables $\mathbf{y}$, an additional number of m unknown parameters $\mathbf{x}$, and must consist of a set of equations $\mathbf{F}(\mathbf{y},\mathbf{x}) = \mathbf{0}$. The number of the model equations must be $s = n + m - r$ so that it reduces the total number of parameters $n + m$ to the number $n + m - s = r$ of the actually necessary parameters for the system description. The objective of the analysis is to provide estimates not only for $\mathbf{x}$ and $\mathbf{y}$, but for any system parameter which in general can be expressed as a function $q = q(\mathbf{x},\mathbf{y})$. In fact all system parameters can be actually expressed (at least in principle) as functions $q = q(\mathbf{y})$ of the observables only, because the system is just the part of physical reality covered by the observations. This has been a well-known fact to the surveyors, who without introducing additional parameters ($m = 0$) used $s = n - r$ model equations $\mathbf{g}(\mathbf{y}) = \mathbf{0}$ as conditions to be satisfied by estimates $\hat{\mathbf{y}}$, which also minimize $\mathbf{v}^T\mathbf{P}\mathbf{v} = (\mathbf{b} - \mathbf{y})^T \mathbf{P}(\mathbf{b} - \mathbf{y})$. This is the well-known least-squares adjustment by condition equations, using a suitable weight matrix $\mathbf{P}$.
On the antipodes of this approach is the choice of a set of $m = r$ parameters $\mathbf{x}$ which completely describe the physical system. In this case any system parameter can be described as a function $q = q(\mathbf{x})$. This holds also for the observables, which are expressed as functions $\mathbf{y} = \mathbf{f}(x)$. In this case the $s = n + r - r = n$ model equations have obtained the particular form $\mathbf{F}(\mathbf{y},\mathbf{x}) = \mathbf{y} - \mathbf{f}(\mathbf{x}) = \mathbf{0}$. The determination of estimates $\hat{\mathbf{x}}$, which minimize $\mathbf{v}^T\mathbf{P}\mathbf{v} = [\mathbf{b} - \mathbf{f}(\mathbf{x})]^T \mathbf{P}[\mathbf{b} - \mathbf{f}(\mathbf{x})]$, is the familiar weighted least-squares adjustment by observation equations.
In the more general case (mixed equations) we seek out of all estimates $\hat{\mathbf{y}}$ and $\hat{\mathbf{x}}$, satisfying the model equations $\mathbf{F}(\mathbf{y},\mathbf{x}) = \mathbf{0}$, the ones satisfying the least-squares principle (Schaffrin & Grafarend, 1986, Schaffrin, 2003). Any other system parameter is estimated through its known relation to the parameters by $\hat{q} = q(\hat{\mathbf{x}},\hat{\mathbf{y}})$.
Let us now assume that, either out of ignorance or on purpose, a set of $m > r$ independent parameters is introduced, which describe not the system covered by the observations but a larger system "containing" the original one. Only those parameters $q(\mathbf{x})$ which can also be expressed as functions of only the observables are "determinable" in the sense that we could recover their true values if the true values of the observables were known in an idealized "zero error" situation. If a model of the form $\mathbf{y} = \mathbf{f}(\mathbf{x})$ is used, a function q of the parameters is determinable if it admits a factorization $q = h \circ \mathbf{f}$ so that $q(\mathbf{x}) = (h \circ \mathbf{f})(\mathbf{x}) = h(\mathbf{f}(\mathbf{x})) = h(\mathbf{y})$. This case of *improper modeling* is typical in network adjustment where coordinates $\mathbf{x}$ are used as unknowns, which describe, e.g., the system "network shape, size and position" containing the system "network shape and size" covered by the available observations.
For any set of "adjusted" observables $\hat{\mathbf{y}}$, i.e. such that there exist coordinates $\mathbf{x}$ satisfying $\hat{\mathbf{y}} = \mathbf{f}(\mathbf{x})$, the model equations assume more than one solution. Indeed if $\mathbf{x}$ are any values satisfying the model, the values $\mathbf{x}' = T(\mathbf{x})$ resulting from applying a coordinate transformation leaving the network shape and size invariant also give rise to the same values of the observables $\hat{\mathbf{y}} = \mathbf{f}(\mathbf{x}')$. A way out of this problem is the introduction of a set of additional independent model equations $\mathbf{c}(\mathbf{x}) = \mathbf{0}$, equal in number to the missing $d = m - r$ ones (minimal constraints), so that, after the adjustment, the total set $\hat{\mathbf{y}} = \mathbf{f}(\mathbf{x})$ and $\mathbf{c}(\mathbf{x}) = \mathbf{0}$ admits a unique solution $\hat{\mathbf{x}}$. Since the information contained in the minimal constraints is arbitrary, only the estimates $\hat{q} = q(\hat{\mathbf{x}})$ of determinable parameters $q(\mathbf{x})$ are meaningful.
In practice linear or linearized models are used, in which case the improper model $\mathbf{y} = \mathbf{A}\mathbf{x}$ has an

$n \times m$ design matrix $\mathbf{A}$ with $rank(\mathbf{A}) = r < m < n$. Application of the least squares principle leads to the normal equations $\mathbf{N}\hat{\mathbf{x}} = \mathbf{u}$, with $\mathbf{u} = \mathbf{A}^T\mathbf{P}\mathbf{b}$ and with the $m \times m$ coefficient matrix $\mathbf{N} = \mathbf{A}^T\mathbf{P}\mathbf{A}$ having the same rank deficiency $rank(\mathbf{N}) = r < m$.

3.3 The notion of parameter estimability

When a statistical approach is taken in order to solve the choice of weight matrix, the errors $\mathbf{v}$ appearing in the linear model $\mathbf{b} = \mathbf{y} + \mathbf{v} = \mathbf{A}\mathbf{x} + \mathbf{v}$ are assumed to be random quantities with mean $E\{\mathbf{v}\} = \mathbf{0}$ and covariance matrix $E\{\mathbf{v}\mathbf{v}^T\} = \mathbf{C}_v = = \sigma_0^2\mathbf{P}^{-1}$.
When a <u>proper model</u> is used the design matrix has full (column) rank $rank(\mathbf{A}) = m < n$ and any linear parameter function $q = \mathbf{q}^T\mathbf{x}$ assumes a best linear (uniformly) unbiased estimate (BLUE) $\hat{q}$, i.e., an estimate of the (linear) form $\hat{q} = \mathbf{d}^T\mathbf{b} + k$, which is uniformly unbiased ($E\{\hat{q}\} = q$ for any $\mathbf{x}$) and has minimum mean square error $\text{MSE} = E\{(\hat{q} - q)^2\}$ among all unbiased linear estimates. The unbiasedness condition takes the form $\mathbf{d}^T\mathbf{A}\mathbf{x} + k = \mathbf{q}^T\mathbf{x}$, which holds for any $\mathbf{x}$ when $k = 0$ and $\mathbf{d}^T\mathbf{A} = \mathbf{q}^T$. The solution of this minimization problem leads to an estimate of the form $\hat{q} = \mathbf{q}^T\hat{\mathbf{x}}$ with parameter estimates $\hat{\mathbf{x}} = (\mathbf{A}^T\mathbf{P}\mathbf{A})^{-1}\mathbf{A}^T\mathbf{P}\mathbf{b}$.
The condition $\mathbf{d}^T\mathbf{A} = \mathbf{q}^T$ plays an essential role because it leads to $q = \mathbf{q}^T\mathbf{x} = \mathbf{d}^T\mathbf{A}\mathbf{x} = \mathbf{d}^T\mathbf{y}$ and it means that q is a function of the observables and thus a parameter of the physical system covered by the observations.
In an <u>improper model</u> the estimation is possible only for estimable parameters, i.e. parameters admitting a linear (uniformly) unbiased estimate $\hat{q}$. This requirement which on first sight appears to be of a stochastic character of the form $E\{\hat{q}\} = q$, leads directly to the deterministic condition $E\{\hat{q}\} = = E\{\mathbf{d}^T\mathbf{b} + k\} = \mathbf{d}^T\mathbf{A}\mathbf{x} + k = q = \mathbf{q}^T\mathbf{x}$, $\forall x$, specifically $k = 0$ and $\mathbf{d}^T\mathbf{A} = \mathbf{q}^T$.
Thus estimability of a function of the unknowns is equivalent to the possibility to express this function as a function of the observables. In other words only the parameters belonging to the physical system covered by the observations are estimable.
For any estimable parameters the BLUE estimate is unique and given by $\hat{q} = \mathbf{q}^T\hat{\mathbf{x}}$ where $\hat{\mathbf{x}}$ is any solution of the normal equations $\mathbf{N}\hat{\mathbf{x}} = \mathbf{u}$. Whatever solution is taken the BLUE estimates of estimable parameters remain the same. On the contrary, for the unknowns and any other non-estimable quantities the "estimates" are lucking their usual meaning because they are not unique, depending on which particular solution is taken from the infinitely many solutions of the normal equations.

3.4. The geometry of the linear model

If $\mathbf{x}$ and $\mathbf{x}'$ are mapped by the matrix $\mathbf{A}$ into the same observables $\mathbf{A}\mathbf{x} = \mathbf{A}\mathbf{x}' = \mathbf{y}$ then $\mathbf{A}(\mathbf{x}' - \mathbf{x}) = \mathbf{0}$ and the difference $\mathbf{x}' - \mathbf{x}$ belongs to the null space of $\mathbf{A}$ defined as $N(\mathbf{A}) = \{\mathbf{x} \in R^m \mid \mathbf{A}\mathbf{x} = \mathbf{0}\}$.
The estimability relation $\mathbf{A}^T\mathbf{d} = \mathbf{q}$ asserts that the vector $\mathbf{q}$ belongs to the range $R(\mathbf{A}^T)$ of the transpose matrix $\mathbf{A}^T$, while a theorem of linear algebra asserts that $R(\mathbf{A}^T)$ is the orthogonal complement of $N(\mathbf{A})$, i.e. the space $X = R^m$ of the unknowns admits an orthogonal decomposition

$$X = N(\mathbf{A}) \overset{\perp}{\oplus} R(\mathbf{A}^T), \qquad R(\mathbf{A}^T) = N(\mathbf{A})^{\perp}.$$

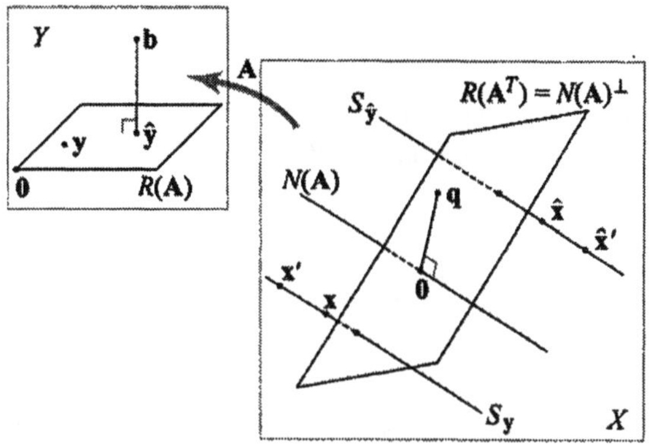

Fig. 1 The geometry of the linear estimation problem

This simply means that, if $\mathbf{q} \in R(\mathbf{A}^T)$, it is orthogonal to any vector $\mathbf{z} \in N(\mathbf{A})$, i.e. $\mathbf{q} \perp \mathbf{z}$ ($\mathbf{q}^T\mathbf{z} = 0$). It also holds that $N(\mathbf{N}) = N(\mathbf{A})$ so that, if $\hat{\mathbf{x}}$ and $\hat{\mathbf{x}}'$ are two solutions of the normal equations, then $\mathbf{N}(\hat{\mathbf{x}}' - \hat{\mathbf{x}}) = \mathbf{0}$, $\hat{\mathbf{x}}' - \hat{\mathbf{x}} \in N(\mathbf{N})$, $\hat{\mathbf{x}}' - \hat{\mathbf{x}} \in \in N(\mathbf{A})$ and since $\mathbf{q} \in N(\mathbf{A})^{\perp}$ it holds that $\mathbf{q} \perp (\hat{\mathbf{x}}' - \hat{\mathbf{x}})$, $\mathbf{q}^T(\hat{\mathbf{x}}' - \hat{\mathbf{x}}) = 0$ and $\hat{q}' = \mathbf{q}^T\hat{\mathbf{x}}' = \mathbf{q}^T\hat{\mathbf{x}} = = \hat{q}$ thus obtaining the same BLUE estimate of q.
For any $\mathbf{x} \in X$ having image $\mathbf{y} = \mathbf{A}\mathbf{x} \in R(\mathbf{A})$ we observe that the linear manifold $\mathbf{x} + N(\mathbf{A})$, resulting by adding to $\mathbf{x}$ any vector from $N(\mathbf{A})$, is mapped by $\mathbf{A}$ to the same vector of observables $\mathbf{y}$.

Thus to any $\mathbf{y} \in R(\mathbf{A})$ corresponds a linear manifold $S_\mathbf{y} = \{\mathbf{x} \in X \mid \mathbf{A}\mathbf{x} = \mathbf{y}\}$ called the solution space of $\mathbf{y}$. The differences of vectors in $S_\mathbf{y}$ are members of $N(\mathbf{A})$ and all manifolds $S_\mathbf{y}$ resulting from the various $\mathbf{y}$ are parallel to the linear subspace $N(\mathbf{A})$. In particular $N(\mathbf{A}) = S_\mathbf{0}$. Thus the space X of the unknowns is separated into disjoint "fibers" $S_\mathbf{y}$, one of which is the optimal one satisfying the normal equations, namely $S_{\hat{\mathbf{y}}}$.

In order to solve the normal equations we might either use a particular generalized inverse of the matrix $\mathbf{A}^T\mathbf{P}\mathbf{A}$ or introduce an appropriate set of minimal constraints $\mathbf{C}^T\mathbf{x} = \mathbf{0}$.

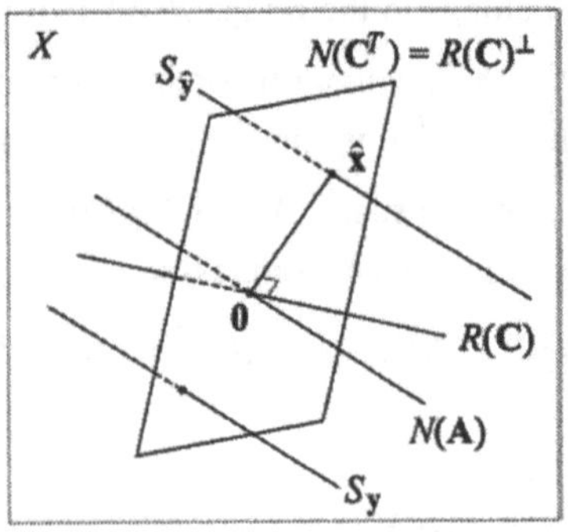

Fig. 2 Obtaining unique solutions with the help of minimal constraints $\mathbf{C}^T\mathbf{x} = \mathbf{0}$.

A generalized inverse $\mathbf{N}^g = (\mathbf{A}^T\mathbf{P}\mathbf{A})^g$ provides a solution $\hat{\mathbf{x}} = (\mathbf{A}^T\mathbf{P}\mathbf{A})^g\mathbf{A}^T\mathbf{P}\mathbf{b} = \mathbf{A}^g\mathbf{b}$, where $\mathbf{A}^g$ is a (reflexive) least-squares generalized inverse of $\mathbf{A}$. As such it maps $\mathbf{b}$ into a $\hat{\mathbf{x}}$ which belongs to the solution space $S_{\hat{\mathbf{y}}}$, where $\hat{\mathbf{y}} = \mathbf{A}\hat{\mathbf{x}} = \mathbf{A}\mathbf{A}^g\mathbf{b}$ is the orthogonal projection of $\mathbf{b}$ onto $R(\mathbf{A})$, in the sense of $<\mathbf{b}-\hat{\mathbf{y}}, \hat{\mathbf{y}}>_Y = \hat{\mathbf{y}}^T\mathbf{P}(\mathbf{b}-\hat{\mathbf{y}}) = 0$. Here Y is R^n equipped with the inner product $(\mathbf{a},\mathbf{b})_Y = \mathbf{a}^T\mathbf{P}\mathbf{b}$ and X is R^m with the usual Euclidean inner product $(\mathbf{u},\mathbf{v})_X = \mathbf{u}^T\mathbf{v}$. More generally a generalized inverse $\mathbf{A}^g$ must map elements $\mathbf{y} \in R(\mathbf{A})$ into corresponding elements of $S_\mathbf{y}$. It is completely characterized by the subspace of R^n which is mapped into $S_\mathbf{0} = N(\mathbf{A})$ and the subspace of R^m into which it maps $R(A)$. This subspace must intersect every $S_\mathbf{y}$ into single element. A least squares generalized inverse maps $R(A)^\perp$ into $N(\mathbf{A})$; a pseudoinverse is a least squares inverse mapping $R(A)$ into $N(A)^\perp$ providing the minimum-norm, least-squares solution.

A set of minimal constraints $\mathbf{C}^T\mathbf{x} = \mathbf{0}$ has the geometric meaning of $\mathbf{x} \perp R(\mathbf{C})$ or $\mathbf{x} \in R(\mathbf{C})^\perp = N(\mathbf{C}^T)$ and must be such that $R(\mathbf{C})^\perp = N(\mathbf{C}^T)$ intersects every $S_\mathbf{y}$ into a single element. This is guaranteed by the condition that $R(\mathbf{C})^\perp \cap N(\mathbf{A}) = \{0\}$. The simultaneous demand that $\mathbf{A}\mathbf{x} = \mathbf{0}$ and $\mathbf{C}^T\mathbf{x} = \mathbf{0}$ should be satisfied only by $\mathbf{x} = \mathbf{0}$, is equivalent to $rank\begin{bmatrix}\mathbf{A}\\ \mathbf{C}^T\end{bmatrix} = m$.

The minimum-norm least-squares solution $\hat{\mathbf{x}}$ satisfying $\|\mathbf{b}-\mathbf{A}\mathbf{x}\|_Y = \min$ ($\Leftrightarrow$ $\mathbf{b}-\mathbf{A}\mathbf{x} \perp R(\mathbf{A})$) and $\|\mathbf{x}\| = \min$ ($\Leftrightarrow$ $\mathbf{x} \perp N(\mathbf{A})$) is realized either by the pseudoinverse solution $\hat{\mathbf{x}} = \mathbf{A}^+\mathbf{b} = \mathbf{N}^+\mathbf{A}^T\mathbf{P}\mathbf{b}$, or by a set of "inner" constraints $\mathbf{E}^T\mathbf{x} = \mathbf{0}$ expressing the requirement $\mathbf{x} \perp N(\mathbf{A})$ through $R(\mathbf{E}) = N(\mathbf{A})$.

If we consider any transformation of the unknowns $\mathbf{x}' = T(\mathbf{x})$ which leaves the observables invariant, i.e., $\mathbf{y}' = \mathbf{A}\mathbf{x}' = \mathbf{A}T(\mathbf{x}) = \mathbf{A}\mathbf{x} = \mathbf{y}$, then an estimable quantity must also be invariant under the same transformation $q' = \mathbf{q}^T\mathbf{x}' = \mathbf{q}^T T(\mathbf{x}) = \mathbf{q}^T\mathbf{x} = q$ (Grafarend & Schaffrin, 1976).

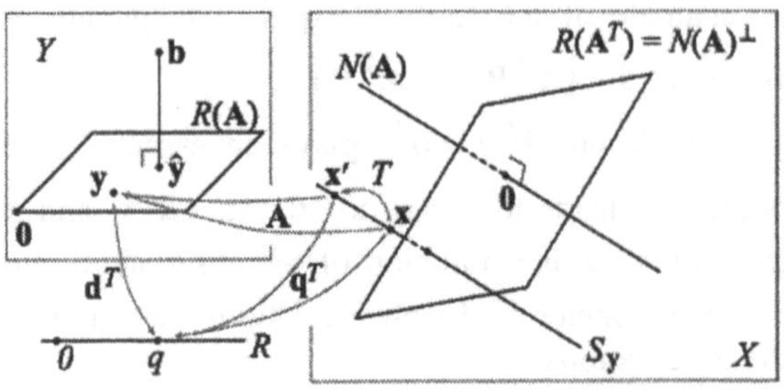

Fig. 3 Common invariance of observables and estimable parameters.

This is guaranteed by either $\mathbf{q}^T = \mathbf{d}^T\mathbf{A}$ leading to $\mathbf{q}^T T(\mathbf{x}) = \mathbf{d}^T\mathbf{A}T(\mathbf{x}) = \mathbf{d}^T\mathbf{A}\mathbf{x} = \mathbf{q}^T\mathbf{x}$, or by $\mathbf{q} \in N(\mathbf{A})^\perp$. Indeed $\mathbf{A}\mathbf{x}' = \mathbf{A}T(\mathbf{x}) = \mathbf{A}\mathbf{x}$ implies $\mathbf{x}' - \mathbf{x} = T(\mathbf{x}) - \mathbf{x} \in N(\mathbf{A})$, $\mathbf{q} \perp \mathbf{x}' - \mathbf{x}$, $\mathbf{q}^T(\mathbf{x}' - \mathbf{x}) = 0$, and thus $q' = \mathbf{q}^T\mathbf{x}' = \mathbf{q}^T\mathbf{x} = q$. If any transformation which leaves the observables invariant also leaves a certain function $q = \mathbf{q}^T\mathbf{x}$ invariant then q is an estimable quantity, provided that for every $\mathbf{x}' \in N(\mathbf{A})$ there exists a transformation T and a $\mathbf{x}$ such that $\mathbf{x}' = T(\mathbf{x}) - \mathbf{x}$. This means that the family of transformations in question must be such that the mapping $\mathbf{x} \to T(\mathbf{x}) - \mathbf{x}$ covers the whole of $N(A)$ and not a subset of it. In the geodetic case where T are coordinate transformations resulting from changes in the reference frame definition this additional requirement is met.

3.5 Invariance and estimability in the non-linear model

We have studied the algebraic properties of the linear model without full rank and arrived at the conclusion that estimable quantities share invariance characteristics with the observables. We shall investigate next the invariance properties at the original non-linear formulation and apply it to the specific geodetic problem of reference frame transformations.

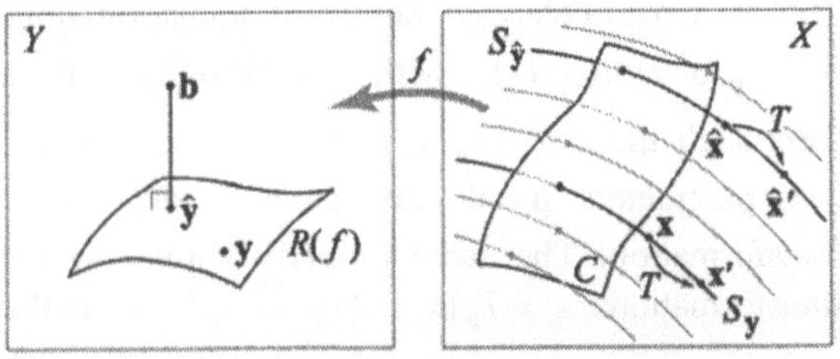

Fig. 4 The geometry of the nonlinear estimation problem

The observables $\mathbf{y} \in Y$ are generally related to the unknowns $\mathbf{x} \in X$ by a nonlinear model $\mathbf{y} = f(\mathbf{x})$. The image of X under f, denoted by $R(f) = f(X)$ is a nonlinear (curved) manifold of Y with dimension $r = m - d < m$ equal to the parametric rank of the model. The solution spaces $S_{\mathbf{y}} = \{\mathbf{x} \in X \mid f(\mathbf{x}) = \mathbf{y}\}$ defined for any $\mathbf{y} \in R(f)$ are d-dimensional nonlinear manifolds of X which constitute a *fibering*, i.e. a separation of X into disjoint subsets (Dermanis, 1998). Estimable from the data $\mathbf{b} = \ = \mathbf{y} + \mathbf{v} \notin R(f)$ are those parameters which are determinable from the observables, i.e. those which can be expressed as functions of the observables $\mathbf{y}$. The null manifold $N(f) = (\mathbf{x} \in X \mid f(\mathbf{x}) = \mathbf{0}\} = S_{\mathbf{0}}$ (kernel of f) is in this case a particular member of the fibering; it is a nonlinear manifold, which does not necessarily contain $\mathbf{0} \in X$.

A function $q = q(\mathbf{x})$ is determinable and thus estimable, if it admits a factorization $q = d \circ f$ so that $q = q(\mathbf{x}) = (d \circ f)(\mathbf{x}) = d(f(\mathbf{x})) = d(\mathbf{y})$. If a transformation $T : X \to X$ leaves the observables invariant $\mathbf{y} = f(\mathbf{x}) = f(T(\mathbf{x}))$ then it also leaves the estimable parameters invariant.

Optimal estimation in this case refers to the satisfaction of the deterministic least squares criterion $\mathbf{v}^T \mathbf{P} \mathbf{v} = (\mathbf{b} - \mathbf{y})^T \mathbf{P} (\mathbf{b} - \mathbf{y}) = \min_{\mathbf{y} \in R(f)}$.

It is satisfied by a particular $\hat{\mathbf{y}} \in R(f)$. When the decomposition $q = d \circ f$ is not readily available we may obtain the optimal estimate $\hat{q} = d(\hat{\mathbf{y}})$ by arbitrarily selecting any element $\hat{\mathbf{x}} \in S_{\hat{\mathbf{y}}}$, i.e. any solution of the compatible equations $\hat{\mathbf{y}} = f(\hat{\mathbf{x}})$, and then compute $\hat{q} = q(\hat{\mathbf{x}})$. The estimate $\hat{q}$ is independent of the particular solution $\hat{\mathbf{x}}$ since $q(\hat{\mathbf{x}}) = d(f(\hat{\mathbf{x}})) = = d(\hat{\mathbf{y}})$.

In order to realize an estimation algorithm, which involves the non-estimable unknown parameters, we need to introduce a "principle of choice" of a single element $\hat{\mathbf{x}}$ out of every solution manifold $S_{\hat{\mathbf{y}}}$. This choice must be smooth in the sense that neighboring $\hat{\mathbf{x}}$ are chosen out of neighboring $S_{\hat{\mathbf{y}}}$. Geometrically this is realized by introducing a section C of the fibering $F = \{S_{\mathbf{y}}\}$, i.e. a manifold C of dimension r, which intersects every solution manifold into a single element $C \cap S_{\mathbf{y}} = \{\mathbf{x}\}$. Functionally C is described as the set of elements of X satisfying the d conditions $c(\mathbf{x}) = \mathbf{0}$, i.e. the kernel of c. Thus the estimation problem is attacked as the problem of minimizing $(\mathbf{b} - f(\mathbf{x}))^T \mathbf{P} (\mathbf{b} - f(\mathbf{x}))$ subject to the conditions (minimal constraints) $c(\mathbf{x}) = \mathbf{0}$.

The obtained values $\hat{\mathbf{x}}$ do not correspond to estimates of the anyway non-estimable unknowns $\mathbf{x}$. They simply serve as representative elements of the corresponding solution spaces $S_{f(\hat{\mathbf{x}})}$, being an intermediate step in the computation of estimates of estimable quantities $\hat{q} = q(\hat{\mathbf{x}})$ where $\hat{q} = q(\hat{\mathbf{x}}')$ for every other $\hat{\mathbf{x}}' \in S_{f(\hat{\mathbf{x}})}$. Since observables are trivially estimable we also have $\hat{\mathbf{y}} = f(\hat{\mathbf{x}}) = f(\hat{\mathbf{x}}')$ for any $\hat{\mathbf{x}}' \in S_{f(\hat{\mathbf{x}})} = S_{\hat{\mathbf{y}}}$, where $\hat{\mathbf{y}}$ is the closest element to the observations $\mathbf{b}$ from $R(f)$.

If T is a transformation leaving the observables invariant, $f(T(\mathbf{x})) = f(\mathbf{x}) = \mathbf{y}$ for any $\mathbf{x}$, then T maps an element $\mathbf{x} \in S_{f(\mathbf{x})}$ into an element $T(\mathbf{x}) \in S_{f(\mathbf{x})}$ belonging to the same solution manifold as $\mathbf{x}$. Such a transformation leaves also estimable parameters invariant $q(T(\mathbf{x})) = q(\mathbf{x}) = q$. Assume that the class G of transformations leaving the observables invariant is such that, for every pair $\mathbf{x}$, $\mathbf{x}'$ belonging to the same solution manifold $S_{\mathbf{y}}$, there exists a transformation $T \in G$ such that $\mathbf{x}' = T(\mathbf{x})$. Then the sets of estimable functions $q(\mathbf{x})$ and functions invariant under G are identical.

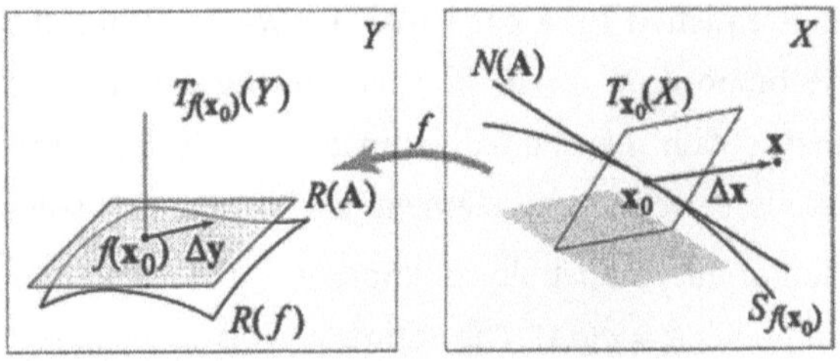

Fig. 5 The linearization of the estimation problem

Linearization in this case provides a means of finding an approximate solution to the nonlinear problem with the help of an "approximate" element $\mathbf{x}_0$, i.e. an element such that $f(\mathbf{x}_0)$ is close to $\mathbf{b}$. The spaces X and Y are replaced by the tangent spaces $T_{\mathbf{x}_0}(X)$ and $T_{f(\mathbf{x}_0)}(Y)$, respectively. The mapping $f: X \to Y$ is replaced by its differential mapping $df: T_{\mathbf{x}_0}(X) \to T_{f(\mathbf{x}_0)}(Y)$, represented by the matrix $\mathbf{A} = (\partial f / \partial \mathbf{x})|_{\mathbf{x}_0}$. The vectors $\mathbf{x}$, $\mathbf{y}$ $\mathbf{b}$ are replaced by $\Delta\mathbf{x} = \mathbf{x} - \mathbf{x}_0$, $\Delta\mathbf{y} = \mathbf{y} - f(\mathbf{x}_0)$, $\Delta\mathbf{b} = \mathbf{b} - f(\mathbf{x}_0)$, respectively. The approximate linear model takes the form $\Delta\mathbf{b} = \Delta\mathbf{y} + \mathbf{v}$, $\Delta\mathbf{y} = \mathbf{A}\,\Delta\mathbf{x}$. The manifold $R(f)$ is replaced by its tangent linear manifold $R(\mathbf{A})$, which is a linear subspace of $T_{f(\mathbf{x}_0)}(Y)$. The solution manifold $S_{f(\mathbf{x}_0)}$ is replaced by its tangent linear manifold $N(\mathbf{A})$, which is a linear subspace of $T_{\mathbf{x}_0}(X)$. Finally the "curved" solution manifolds $S_{f(\mathbf{x})}$ are replaced by the linear manifolds $S_{\mathbf{A}(\mathbf{x}-\mathbf{x}_0)}$, which are parallel to $N(\mathbf{A})$. The estimation of any nonlinear function $q = q(\mathbf{x})$ is replaced by the estimation of the linear function $\delta q = \mathbf{q}^T \delta\mathbf{x}$, where $\delta q = q - q(\mathbf{x}_0)$ and $\mathbf{q}^T = \left.\frac{\partial q}{\partial \mathbf{x}}\right|_{\mathbf{x}_0}$.

4 The geodetic reference frame problem for a single epoch

In the analysis of geodetic observations the class of transformations leaving the observables invariant can be directly determined by simple reasoning. For example in a horizontal surveying network where angles and distances have been measured the permissible transformations are the rigid transformations in the plane consisting of 2 translations along the axes, and 1 rotation around the origin. If no distances have been measured we must add a scaling and consider similarity transformations in the plane. In three-dimensional networks we consider rigid transformation with 6 parameters (3 translations and 3 rotations) or the most general case of a 7-parameter similarity transformation resulting by adding a scaling to the rigid case. The general form of such a transformation for any network point P_i is

$$\mathbf{x}_i' = (1+s)\,\mathbf{R}(\boldsymbol{\theta})\,\mathbf{x}_i + \mathbf{d} \tag{10}$$

where $\mathbf{d} = [d_1 d_2 d_3]^T$ is the translation vector, $1+s$ is the scale factor and $\boldsymbol{\theta} = [\theta_1 \theta_2 \theta_3]^T$ is the vector of the rotational parameters. A usual choice is $\mathbf{R} = \mathbf{R}_3(\theta_3)\mathbf{R}_2(\theta_2)\mathbf{R}_1(\theta_1)$ where $\mathbf{R}_k(\alpha)$ denotes a rotation by an angle α around the kth axis. This is a parameterized family of transformations $\mathbf{x}_i' = t_{\mathbf{p}}(\mathbf{x}_i) = t(\mathbf{p}, \mathbf{x}_i)$, with $\mathbf{p} = [\boldsymbol{\theta}^T \mathbf{d}^T s]^T$, in a form such that, $\mathbf{x}_i' = t_0(\mathbf{x}_i) = t(\mathbf{0}, \mathbf{x}_i) = \mathbf{x}_i$, i.e. the zero parameters $\mathbf{p} = \mathbf{0}$ correspond to the identity transformation. The same property applies to the transformations $\mathbf{x}' = T_{\mathbf{p}}(\mathbf{x}) = T(\mathbf{p}, \mathbf{x})$ induced on the vector of network coordinates $\mathbf{x} = [\mathbf{x}_1^T \cdots \mathbf{x}_N^T]^T$ by

$$\mathbf{x}' = [\cdots \mathbf{x}_i'^T \cdots]^T = T(\mathbf{p}, \mathbf{x}) \equiv [\cdots t(\mathbf{p}, \mathbf{x}_i)^T \cdots]^T . \tag{11}$$

Each solution manifold $S_{f(\mathbf{x}_0)}$ passing through a particular fixed point $\mathbf{x}_0 \in X$, admits a coordinate system by letting the parameters $\mathbf{p}$, which map $\mathbf{x}_0$ into another point $\mathbf{x} = T(\mathbf{p}, \mathbf{x}_0) \in S_{f(\mathbf{x}_0)}$, serve as the curvilinear coordinates of $\mathbf{x}$. A basis for the tangent space $T_{\mathbf{x}_0}\left(S_{f(\mathbf{x}_0)}\right)$ to the manifold $S_{f(\mathbf{x}_0)}$ at $\mathbf{x}_0$, is directly provided by the vectors

$$\mathbf{e}_{p_k} = \left.\frac{\partial \mathbf{x}}{\partial p_k}\right|_{\mathbf{x}_0} = \begin{bmatrix} \vdots \\ \dfrac{\partial \mathbf{x}_i}{\partial p_k} \\ \vdots \end{bmatrix}_{\mathbf{x}_0}, \quad k = 1, \ldots, d, \tag{12}$$

evaluated at $\mathbf{x}_0 = T(\mathbf{0}, \mathbf{x}_0)$, i.e. for $\mathbf{p} = \mathbf{0}$. Direct differentiation of $\mathbf{x}_i = (1+s)\,\mathbf{R}(\boldsymbol{\theta})\,\mathbf{x}_i^0 + \mathbf{d}$ gives

$$\begin{aligned} &\left[\frac{\partial \mathbf{x}_i}{\partial s}\right]_{\mathbf{p}=\mathbf{0}} = [\mathbf{R}(\boldsymbol{\theta})\,\mathbf{x}_i^0]_{\boldsymbol{\theta}=\mathbf{0}} = \mathbf{x}_i^0 , \\ &\frac{\partial \mathbf{x}_i}{\partial \mathbf{d}} = \mathbf{I} \equiv [\mathbf{i}_1 \mathbf{i}_2 \mathbf{i}_3], \quad \left[\frac{\partial \mathbf{x}_i}{\partial d_k}\right]_{\mathbf{p}=\mathbf{0}} = \frac{\partial \mathbf{x}_i}{\partial d_k} = \mathbf{i}_k , \end{aligned} \tag{13}$$

while $\dfrac{\partial \mathbf{R}_k}{\partial \theta_k} = -[\mathbf{i}_k \times]\mathbf{R}_k = -\mathbf{R}_k[\mathbf{i}_k \times]$ leads to

$$\left[\frac{\partial \mathbf{x}_i}{\partial \theta_k}\right]_{\mathbf{p}=\mathbf{0}} = -[\mathbf{i}_k \times]\mathbf{x}_i^0 = [\mathbf{x}_i^0 \times]\mathbf{i}_k , \qquad \left[\frac{\partial \mathbf{x}_i}{\partial \boldsymbol{\theta}}\right]_{\mathbf{p}=\mathbf{0}} = [\mathbf{x}_i^0 \times] . \tag{14}$$

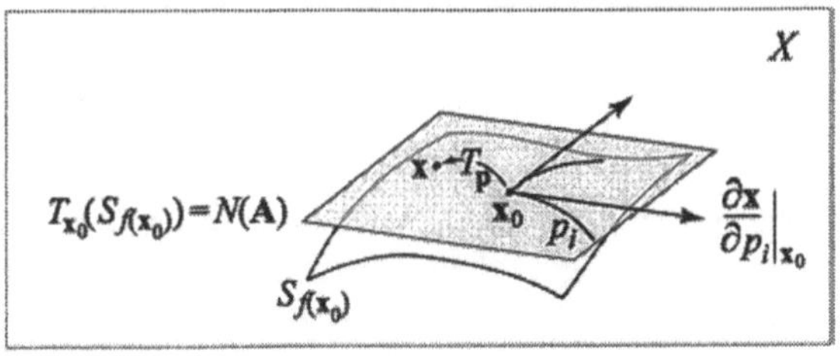

Fig. 6 The determination of a basis for $N(\mathbf{A})$

In the linearized problem the vectors $\mathbf{e}_{p_k}$ form a basis of $N(\mathbf{A})$ and the minimum-norm condition $\mathbf{x} \perp N(\mathbf{A})$, satisfied by $\mathbf{x} \perp \mathbf{e}_{p_k}$, $k = 1,\ldots,d$, takes the form of the inner constraints $\mathbf{e}_{p_k}^T \mathbf{x} = 0$, $k = 1,\ldots,d$, which are combined into

$$\begin{bmatrix} \vdots \\ \mathbf{e}_{p_k}^T \\ \vdots \end{bmatrix} \mathbf{x} = [\cdots \mathbf{e}_{p_k} \cdots]^T \mathbf{x} \equiv \mathbf{E}^T \mathbf{x} = \mathbf{0} , \tag{15}$$

where the matrix $\mathbf{E}$ takes the well-known form

$$\mathbf{E} = \left[\mathbf{e}_{d_1}\mathbf{e}_{d_2}\mathbf{e}_{d_3} \mid \mathbf{e}_{\theta_1}\mathbf{e}_{\theta_2}\mathbf{e}_{\theta_3} \mid \mathbf{e}_s\right] = \begin{bmatrix} \vdots & \vdots & \vdots \\ \mathbf{I} & [\mathbf{x}_i^0 \times] & \mathbf{x}_i^0 \\ \vdots & \vdots & \vdots \end{bmatrix} . \tag{16}$$

Explicitly we have 3 translational, 3 rotational and 1 scale constraints

$$\sum_i \Delta \mathbf{x}_i = \mathbf{0} , \qquad \sum_i [\mathbf{x}_i^0 \times] \Delta \mathbf{x}_i = \mathbf{0} , \qquad \sum_i (\mathbf{x}_i^0)^T \Delta \mathbf{x}_i = \mathbf{0} , \tag{17}$$

which attach to the network a frame having the same "center of mass", the same "angular momentum" and the same "scale" as the network defined by the approximate coordinates.

The inner constraint – minimum norm solution is hardly "optimal" in the real sense of the word. It is just one of the equivalent solutions yielding the same optimal (BLUE) estimates for all estimable parameters expressed as coordinate functions, which are realized by minimal constraints $\mathbf{C}^T \mathbf{x} = \mathbf{0}$. Among these we distinguish a particular class of trivial constraints having the form $\mathbf{C}^T = [\mathbf{0} \ \mathbf{I}_d]$, i.e., $\mathbf{x}_2 = \mathbf{0}$, where $\mathbf{x}_2$ are the last d network coordinates, or any d coordinates after a convenient reordering. Setting $\mathbf{A} = [\mathbf{A}_1 \mathbf{A}_2]$, the constraints can be applied a priori, so that the linear model becomes $\mathbf{b} = \mathbf{A}_1 \mathbf{x}_1 + \mathbf{v}$, with an unbiased coordinate estimate $E\{\hat{\mathbf{x}}_1\} = E\{(\mathbf{A}_1^T \mathbf{P} \mathbf{A}_1)^{-1} \mathbf{A}_1^T \mathbf{P} \mathbf{b}\} = \mathbf{x}_1$!

This estimability of coordinates obtained using minimal constraints is "real" as far as the information contained in the constraints is real and not the result of an arbitrary choice. In this respect the trivial minimal constraints have a peculiar character that needs to be pointed out.

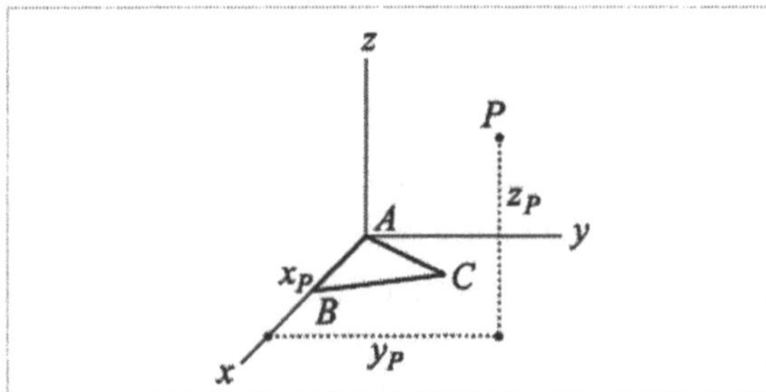

Fig. 7 Reference frame via trivial minimal constraints

Let us assume that they have for example the form $x_A = y_A = z_A = 0$ (origin at A), $y_B = z_B = 0$ (axis x in the AB direction) and $z_C = 0$ (plane ABC as xy plane) for a network where scale is determined from the observations. For any other point P, z_P is its distance from the ABC plane, y_P is the distance of its projection on the ABC plane from the AB line and x_P is the distance of its projection on the AB line from point A. But these are parameters, which depend only on the shape and size of the network and not on its position and hence the coordinates $\mathbf{x}_1$ are estimable quantities so that $E\{\hat{\mathbf{x}}_1\} = \mathbf{x}_1$ is meaningful! The estimability results from the fact that the trivial minimal constraints are incorporated in the model, so that they have a physical meaning, instead of merely being an "algebraic trick" for obtaining one out of many possible solutions.

Let us recall that in surveying we are not only concerned with mapping (conversion of positions to numbers) but also with setting out (conversion of numbers to positions). A reference frame established by trivial constraints can be trivially set out. On the contrary when coordinate estimates $\hat{\mathbf{x}}$ are obtained by other minimal constraints there is no unique way of setting out the (estimated) origin and axes of the frame and we will arrive at different results when starting from different points.

5 The temporal evolution of the reference frame

When the varying shape of a deforming network is known at any epoch t from the analysis of continuous observations, it is necessary to introduce a reference frame at every epoch. The only requirement is that the resulting coordinates $\mathbf{x}(t)$ are smooth functions of time. Another set of reference frames can be introduced by applying a smooth set of transformations $\mathbf{x}'(t) = T_{\mathbf{p}(t)}\mathbf{x}(t) = T(\mathbf{x}(t), \mathbf{p}(t))$ with smooth parameter functions $\mathbf{p}(t)$. Such transformations are induced from the point-wise transformations

$$\mathbf{x}'_i(t) = [1 + s(t)]\,\mathbf{R}(\boldsymbol{\theta}(t))\,\mathbf{x}_i(t) + \mathbf{d}(t)\,. \qquad (18)$$

The question arises whether, by departing from a particular frame choice $\mathbf{x}_0(t)$, it is possible to determine parameter functions $s(t)$, $\boldsymbol{\theta}(t)$, $\mathbf{d}(t)$ leading to a frame $\mathbf{x}(t) = T(\mathbf{x}_0(t), \mathbf{p}(t))$ which is "optimal" in some sense. Such optimality is not real since the estimable network shape remains the same at any epoch. It is, however, natural to select a representation with optimal appearance $\mathbf{x}(t)$ where the apparent motions of the coordinates are minimized. Two reference frame sets are characterized as *parallel,* when the transformation from one to the other is constant in time. Parallel frames are dynamically equivalent and cannot be distinguished by a "minimal motion" optimality criterion. Consequently the initial condition $\mathbf{x}(t_0)$, i.e. the reference frame at an initial epoch t_0, must be arbitrarily chosen.

From a mathematical point of view (Dermanis, 1995), the set of coordinates giving a particular shape at some epoch t is a curved manifold M_t (shape manifold) in the Euclidean space $X = R^{3N}$ of coordinates of the network with N points. The evolution of the reference frame in time is in fact a parameterized curve $\mathbf{x}(t)$ on the manifold of varying shapes $M = \cup_t M_t$.

The same problem arose in surveying where deforming networks are observed at discrete epochs. The solution given was to introduce an arbitrary frame at the first epoch and to use the coordinate estimates of each epoch as approximate coordinates for the next epoch and obtain a solution using the inner constraints of Meissl. Thus the network shape at each epoch is best fitted to the position of the network at the previous epoch thus forming a "*Meissl ladder*". This principle of the form $\Delta\mathbf{x}_i \perp T_{\mathbf{x}_{i-1}}(S_{f(\mathbf{x}_{i-1})})$ can be generalized to continuous time by $\mathbf{v}(t) \equiv \frac{d\mathbf{x}}{dt}(t) \perp T_{\mathbf{x}(t)}(M_t)$ leading to

$$\mathbf{v}(t) \perp \frac{\partial \mathbf{x}}{\partial p_i}(t)\,, \qquad i = 1, \ldots, d\,. \qquad (19)$$

Three optimality criteria have been proposed and shown to be equivalent (Dermanis, 1995, 2000, 2002):

(a) Choose $\mathbf{x}(t)$ as the minimum length geodesic on M passing through a chosen $\mathbf{x}(t_0)$.

(b) Apply the continuous Meissl ladder principle.

(c) Imitate the earth frame choices by applying a discrete Tisserand principle for rotation
$$\mathbf{h}_R(t) = \sum_i [\mathbf{x}_i(t)\times]\,\mathbf{v}_i(t) = \mathbf{0}\,,$$
the preservation of network center for translation
$$\frac{d}{dt}\left(\sum_i \mathbf{x}_i(t)\right) = \mathbf{0}$$
and the preservation of size for scale
$$\frac{d}{dt}\left(\|\mathbf{x}(t)\|^2\right) = \frac{d}{dt}\left(\sum_i \mathbf{x}_i^T(t)\mathbf{x}_i(t)\right) = 0\,.$$

The application of these principles leads to a set of nonlinear differential equations for the transformation parameters $\mathbf{p}(t)$, where the arbitrary initial values $\mathbf{p}(t_0)$ introduce the initial frame $\mathbf{x}(t_0)$.

A somewhat different approach is the modeling of the time evolution of the frame, as well as the shape of the network by $\mathbf{x}(t) = \mathbf{g}(\mathbf{a}, t)$ where $\mathbf{g}$ is a known function involving parameters $\mathbf{a}$ to be estimated from the observations. This approach is followed in the computation of the International Terrestrial Reference Frame (ITRF) where it is assumed that $\mathbf{x}(t) = \mathbf{x}_0 + t\,\mathbf{v}$ where the parameters are the initial coordinates $\mathbf{x}_0 = \mathbf{x}(0)$ and the constant velocities $\mathbf{v}$ (Sillard & Boucher, 2001, Altamini et al., 2002). The optimal solution is obtained by restricting the admissible transformation parameters $\mathbf{p}(t)$ to the particular class $\mathbf{p}(t) = \mathbf{p}_0 + t\,\dot{\mathbf{p}}$ and applying the minimum norm principle with respect to the total set of 14 transformation parameters $\mathbf{p}_0 = [s_0\ \boldsymbol{\theta}_0^T\ \mathbf{d}_0^T]^T$, $\dot{\mathbf{p}} = [\dot{s}\ \dot{\boldsymbol{\theta}}^T \dot{\mathbf{d}}^T]^T$ with extended inner constraints $\left(\partial \begin{bmatrix} \mathbf{x}_0 \\ \mathbf{v} \end{bmatrix} \Big/ \partial \begin{bmatrix} \mathbf{p}_0 \\ \dot{\mathbf{p}} \end{bmatrix}\right)_0^T \begin{bmatrix} \Delta\mathbf{x}_0 \\ \Delta\mathbf{v} \end{bmatrix} = \mathbf{0}$.

The same constraints can be obtained in a much simpler way by applying the conditions of the optimality condition (c) above. In this case $\mathbf{h}_R(t) = = \sum_i [\mathbf{x}_i(0)\times]\mathbf{v}_i + t\sum_i [\mathbf{v}_i\times]\mathbf{v}_i = \sum_i [\mathbf{x}_i(0)\times]\mathbf{v}_i = \mathbf{0}$, and with $\mathbf{x}_i(0) = \mathbf{x}_{i,0} + \Delta\mathbf{x}_i$, $\mathbf{v}_i = \mathbf{v}_{i,0} + \Delta\mathbf{v}_i$ the Tisserand constraints, neglecting second order terms, become

$\sum_i [\mathbf{x}_{i,0}\times]\mathbf{v}_{i,0} + \sum_i [\mathbf{x}_{i,0}\times]\Delta\mathbf{v}_i - \sum_i [\mathbf{v}_{i,0}\times]\Delta\mathbf{x}_i = \mathbf{0}$.

The preservation of the network center gives

$\sum_i (\mathbf{v}_{i,0} + \Delta\mathbf{v}_i) = \mathbf{0}$

The preservation of size gives

$\sum_i (\mathbf{x}_{i,0}^T \mathbf{v}_{i,0} + \mathbf{x}_{i,0}^T \Delta\mathbf{v}_i) + t\sum_i (\mathbf{v}_{i,0}^T \mathbf{v}_{i,0} + 2\mathbf{v}_{i,0}^T \Delta\mathbf{v}_i) = \mathbf{0}$

With the usual choice of $\mathbf{v}_{i,0} = \mathbf{0}$ the above constraints simplify to

$$\sum_i [\mathbf{x}_{i,0}\times]\Delta\mathbf{v}_i = \mathbf{0}, \quad \sum_i \Delta\mathbf{v}_i = \mathbf{0}, \quad \sum_i \mathbf{x}_{i,0}^T \Delta\mathbf{v}_i = \mathbf{0} \quad (20)$$

and must be augmented with the usual inner constraints for the initial epoch

$$\sum_i [\mathbf{x}_{i,0}\times]\Delta\mathbf{x}_i = \mathbf{0}, \quad \sum_i \Delta\mathbf{x}_i = \mathbf{0}, \quad \sum_i \mathbf{x}_{i,0}^T \Delta\mathbf{x}_i = \mathbf{0}. \quad (21)$$

6 Estimating a geocentric Tisserand earth frame

Once a particular time evolution of the reference frame has been established, e.g. the ITRF, it is accompanied by earth rotation parameters relating this frame to the celestial inertial frame. The rotation parameters need to be compared with those derived in theoretical works where the reference frame is a Tisserand frame for the whole earth. To establish such an earth frame we need, in principle, to know the distribution of masses and their motion for the whole earth. An estimate of the earth Tisserand frame can be obtained by considering that the essential contribution to the relative angular momentum is the motion of the tectonic plates. If the motion of each plate in the ITRF is inferred from the motions of stations belonging to the plate, it is possible to compute the total angular momentum of the plates and to determine the frame transformation, which leads to zero plate angular momentum. We present here only the simple case where rigid plates are assumed to rotate around the geocenter. The rotation parameters from the ITRF to the earth Tisserand frame estimate are obtained from the geometric Euler equations (3), where the rotation vector $\boldsymbol{\omega}$ is the weighted mean (Dermanis, 2001)

$$\boldsymbol{\omega} = \left(\sum_{P_k} \mathbf{C}_{P_k}\right)^{-1} \sum_{P_k} \mathbf{C}_{P_k} \boldsymbol{\omega}_{P_k} \quad (22)$$

where $\boldsymbol{\omega}_{P_k}$ and $\mathbf{C}_{P_k}$ is the rotation vector and inertia matrix of each plate P_k. The rotation $\mathbf{R}(\boldsymbol{\theta})$, derived from $\boldsymbol{\omega}$, transforms the ITRF to an almost Tisserand earth frame, which is almost geocentric since the ITRF is centered at an estimate of the geocenter position, obtained from the contribution of satellite laser ranging (SLR) observations.

7 The question of estimability for geocenter coordinates

Despite the routine computation of geocenter positions in the analysis of SLR data, the estimability of the geocenter coordinates has been seriously questioned from a theoretical point of view (Baarda, 1995, § 3.1). The problem is the simultaneous determination of the gravity potential V and station coordinates $\mathbf{x}$ from SLR data, which are invariant to rigid coordinate transformations, assuming that earth rotation is provided by collocated gravity independent VLBI observations. Strictly speaking the problem at hand is underdetermined, the unknown function being an infinite dimensional entity. To overcome this problem either a stochastic character is assigned to V, in which case the inherent datum problem is bypassed, or we assume that V can be sufficiently described by a finite number of parameters, smaller than the available observations.

We examine here the usual truncated representation $V(P) = \sum_{i=1}^{N} a_i \phi_i(\mathbf{x})$, where $\phi_i(\mathbf{x})$ are known basis functions (e.g. spherical harmonics), a_i are unknown coefficients and x are the coordinates of point P in some particular reference frame. If $\tilde{x} = t(\mathbf{p}, x)$ are the coordinates of P in another frame, it will hold that

$$V(P) = \sum_{i=1}^{N} a_i \phi_i(x) = \sum_{i=1}^{N} \tilde{a}_i \phi_i(\tilde{x}) \quad (23)$$

and the transformation $\tilde{\mathbf{x}} = T(\mathbf{p}, \mathbf{x})$ of network coordinates must induce a transformation $\tilde{\mathbf{a}} = S(\mathbf{p}, \mathbf{a})$ on the vector $\mathbf{a}$ of the function coefficients a_i. The total set of unknowns is $[\mathbf{x}^T \mathbf{a}^T]^T$ and the basis vectors of the null space of the estimation problem will be

$$\mathbf{e}_m = \frac{\partial}{\partial p_m}\begin{bmatrix}\mathbf{x}\\ \mathbf{a}\end{bmatrix}_{\mathbf{p}=0} = \begin{bmatrix}\dfrac{\partial \mathbf{x}}{\partial p_m}\\ \dfrac{\partial \mathbf{a}}{\partial p_m}\end{bmatrix}_{\mathbf{p}=0} \equiv \begin{bmatrix}\mathbf{e}_m^x\\ \mathbf{e}_m^a\end{bmatrix} \tag{24}$$

Considering the inverse transformation $x = h(\mathbf{p}, \tilde{x})$ the expansion $\phi_i(x) = \phi_i(h(\tilde{x}, \mathbf{p})) = \sum_k h_i^k(\mathbf{p})\phi_k(\tilde{x})$ with respect to the basis $\phi_i(\tilde{x})$ gives

$$\sum_{i=1}^{N} a_i\phi_i(x) = \sum_{i=1}^{N} a_i \sum_k h_i^k \phi_k(\tilde{x}) = \sum_{k=1}^{N} \tilde{a}_k \phi_k(\tilde{x})$$

and the transformation $\tilde{\mathbf{a}} = S(\mathbf{p}, \mathbf{a})$ becomes $\tilde{a}_k = \sum_k h_i^k(\mathbf{p})\, a_i$. From $\phi_i(x) = \sum_k h_i^k \phi_k(\tilde{x})$ follows that $\partial_{p_m}\phi_i = \sum_k \partial_{p_m} h_i^k \phi_k(\tilde{x})$. The required coefficients $(b_m)_i^k \equiv \partial_{p_m} h_i^k$ can be found by a direct expansion, with respect to the basis $\phi_i(\mathbf{x})$ of the derivatives

$$\frac{\partial \phi_i}{\partial p_m} = \frac{\partial \phi_i}{\partial x}\frac{\partial x}{\partial p_m} = \sum_k (b_m)_i^k \phi_k(x). \tag{25}$$

Then the elements of $\mathbf{e}_m^a$ are given by

$$(\mathbf{e}_m^a)_i = \begin{bmatrix}\dfrac{\partial \tilde{a}_i}{\partial p_m}\end{bmatrix}_{\substack{\mathbf{p}=0\\ \mathbf{a}=\mathbf{a}_0}} = \sum_k \begin{bmatrix}\dfrac{\partial h_i^k}{\partial p_m}\end{bmatrix}_{\mathbf{p}=0} a_k^0 =$$

$$= \sum_k (b_m)_i^k a_k^0 . \tag{26}$$

The geocenter coordinates x_C, y_C, z_C are known scalar multiples of the first-degree spherical harmonic coefficients, which are linear functions of the form $q_C = \mathbf{q}^T\begin{bmatrix}\mathbf{x}\\ \mathbf{a}\end{bmatrix}$, where the vectors $\mathbf{q}$ have only one non-zero element. A necessary condition for estimability is the orthogonality condition $\mathbf{E}^T\mathbf{q} = \mathbf{0}$ where $\mathbf{E}$ is the inner constraint matrix $\mathbf{E} = \begin{bmatrix}\mathbf{E}_x\\ \mathbf{E}_a\end{bmatrix} = \begin{bmatrix}\mathbf{e}_1^x \cdots \mathbf{e}_d^x\\ \mathbf{e}_1^a \cdots \mathbf{e}_d^a\end{bmatrix}$. In general $R(\mathbf{E}) \subseteq N(\mathbf{A})$ and when $R(\mathbf{E}) = N(\mathbf{A})$, i.e., when there are no other deficiencies beyond the one due to the lack of frame definition, the conditions $\mathbf{E}^T\mathbf{q} = \mathbf{0}$ are also sufficient to guarantee estimability.

References

Altamini, Z., P. Sillard, C. Boucher (2002): ITRF 2000: A new release of the International Terrestrial Reference Frame for earth science applications. *Journal of Geophysical Research*, vol. 107, No. B10, p. 2214.

Baarda, W. (1995): *Linking up spatial models in geodesy. Extended S-Transformations*. Netherlands Geodetic Commission, Publ. in Geodesy, New Series, no. 41, Delft.

Bjerhammar, A. (1951): Rectangular reciprocal matrices with special emphasis to geodetic calculations. *Bulletin Géodésique*, 1951, 188-220.

Dermanis, A. (1991): *A Unified Approach to Linear Estimation and Prediction*. Presented at the 20th IUGG General Assembly, Vienna, August 1991.

Dermanis, A. (1995): *The Nonlinear and Space-time Geodetic Datum Problem*. Intern. Meeting *"Mathematische Methoden der Geodäsie"*, Oberwolfach, 1-7 Oct. 1995.

Dermanis, A. (1998): Generalized inverses of nonlinear mappings and the nonlinear geodetic datum problem. *Journal of Geodesy*, 72, 71-100.

Dermanis, A. (2000): Establishing global reference frames. Nonlinear, temporal, geophysical and stochastic aspects. In: M. Sideris (ed.) *"Gravity, Geoid and Geodynamics 2000"*, p. 35-42, IAG Symposia, Vol. 123, Springer, Heidelberg 2001.

Dermanis, A. (2001): Global reference frames: Connecting observation to theory and geodesy to geophysics. IAG Scientific Assembly, 2-8 Sept. 2001, Budapest, Hungary.

Dermanis, A. (2002): On the maintenance of a proper reference frame for VLBI and GPS global networks. In: E.Grafarend, F.W. Krumm, V.S. Schwarze (eds) "*Geodesy – the Chalenge of the 3rd Millenium*", p. 61-68, Springer, Heidelberg.

Grafarend, E. and B. Schaffrin (1976): Equivalence of estimable quantities and invariants in geodetic networks. *Zeitschrift für Vemessungswesen*, 101, 11, 485-491.

Meissl, P. (1965): Über die innere Genauigkeit dreidimensionaler Punkthaufen. *Zeitschrift für Vemessungswesen*, 90, 4, 109-118.

Meissl, P. (1969): Zusammenfassung und Ausbau der inneren Fehlertheorie eines Punkthaufens. *Deutsche Geodät. Komm.*, Reihe A, Nr. 61, 8-21.

Moritz, H. and I.I. Mueller (1987): *Earth Rotation. Theory and Observation*. Ungar, New York.

Munk, W.H. and G.J.F. MacDonald (1960): *The Rotation of the Earth. A Geophysical Discussion*. Cambridge University Press.

Penrose, R (1955): A generalized inverse for matrices. Proc. Cambridge Philos. Soc., 51, 406-413.

Schaffrin, B. and E. Grafarend (1986): Generating classes of equivalent linear models by nuisance parameter elimination - applications to GPS observations. Manuscripta geodaetica, 11 (1986), 262-271.

Schaffrin, B. (2003): Some generalized equivalence theorems for least-squares adjustment. This volume.

Sillard, P. and C. Boucher (2001): A review of algebraic constraints in terrestrial reference frames datum definition. *Journal of Geodesy*, 75, 63-73.

ITRF2000: From Theory to Implementation

Z. Altamimi, P. Sillard
Institut Géographique National, ENSG/LAREG,
6-8 Avenue Blaise Pascal, 77455 Marne-la-Vallée, France

C.Boucher
Institut Géographique National, DAEI, 2, Avenue Pasteur, 94160 Saint Mandé, France

Abstract. One of the main geodetic activities is to determine positions of points or objects over the Earth surface or nearby space. Meanwhile when dealing with geodetic observations, a certain Terrestrial Reference System (TRS) has to be adopted in which point positions will be expressed. Since there are many ways to realize the adopted TRS, the most challenging issue is to select the "best" one that preserves the actual quality of geodesy observations. Conversely, noting that the observations are, and should remain, insensitive to the adopted theoretical TRS, they should not be altered by the added pseudo observations (termed as constraints) permitting to specify a Terrestrial Reference Frame (TRF) as the "best" realization of the TRS. In this paper we will focus on the main important parameters needed to define a TRF (called datum definition), as well as the current achievement of Space Geodesy techniques, through the activities related to the International Terrestrial Reference Frame (ITRF). The use of minimum constraints approach is emphasized both in case of individual TRS realization by space geodesy techniques and in combination, allowing to precisely define the ITRF datum definition. Some important results of the latest ITRF solution, namely ITRF2000, will be used to illustrate the discussion over two axes: accuracy of the datum definition and the current quality of space geodesy solutions used in the ITRF2000 combination.

Keywords. Reference Systems, Reference Frames, rank deficiency, minimum constraints, ITRF

1 Introduction

Geôdaisia, this Greek word, meaning dividing the Earth, has seen its applications extended from the user original (mapping) needs to many other Earth science applications. More than 20 years of space geodesy observations, allow to study a variety of phenomena, including Earth gravity field, geodynamic effects influencing Earth Rotation and geophysical processes such as tectonic plate motion and Earth crust deformation. This is mainly because the uncertainties associated to space geodesy observations initially of decimetric level are now of centimetric, even millimetric level.

Space geodesy products constitute mainly three global references: Celestial Reference Frame (CRF), Terrestrial Reference Frame (TRF) and Earth Orientation Parameters (EOP). Although satellite techniques provide inertial celestial reference frames, implicitly defined by satellite orbits, Very Long Baseline Interferometry (VLBI) is the unique technique producing accessible CRF, through estimated coordinates of radio sources.

These three global references of space geodesy should all be consistent so that our interpretation and the many applications using these products will then refer to the same reference. Ideally, a global combination of all space geodesy observations, using a coherent and unique modelling is a rigorous way to allow consistency between the different products. Before embarking on this ambitious project due to many technical limitations, an alternative way is to combine the individual products (and not the observations) of the different techniques.

In this paper we will focus specifically on the TRS definition and its realization (TRF), with emphasis on the activity related to the International Terrestrial Reference Frame (ITRF). The ITRF is one of the 3 global references of the International Earth Rotation and Reference Systems Service (IERS), the two others being the International Celestial Reference Frame and Earth Orientation Parameters.

Naturally, coordinates (and thus their associated reference system/frame) are not observable, but they are of indispensable utility. The concept of reference systems and frames is in fact a pure mathematical convention (or model) introduced to describe the physical Earth and the temporal variation of its shape. This leads to the simple conclusion that, although geodetic observations provide some of the needed TRS parameters (as described below), the choice of a TRS comprises a large part of convention. In addition, since the notion of reference systems is purely mathematical, thus all the TRS parameters could be arbitrarily (conventionally) selected. What is important after all is that the realization of the TRS (i.e. the TRF) should preserve the original quality of the used observations.

Meanwhile the ITRF, as an international frame used by a variety of users and in particular geophysicists, should preserve the physical properties embedded in space geodesy observations. Consequently, some of the TRS parameters accessed by the observations (such as the scale and the origin, as seen below) should be preserved.

Moreover, a TRF, as a geodetic product, should be associated with its coordinate variances and covariances which should reflect the observations noise, having a stochastic character. The most challenging task is then to be able to separate the observation part from the mathematical (TRS) definition, in order to allow statistical analysis of the coordinate estimates. Minimum constraints approach is found to be an efficient method allowing the de-correlation of the frame and the observation parameters.

2 Definition of a Terrestrial Reference System

In pure mathematics, an ideal Terrestrial Reference System (TRS) is defined as a three-dimensional right-handed cartesian frame whose axes are orthogonal, having an origin (any point in space), a unit of length and a given orientation. In the context of Space Geodesy, we select a TRS as an Euclidean frame (defined in an affine space of dimension 3) with an origin close to the Earth center of mass, a scale satisfying some unit of length (e.g. SI unit) and an equatorial orientation. More details could be found in (Boucher, 2000).

Despite our ideal theoretical definition of a TRS, we should keep in mind that the adopted TRS is an Earth-related system and then, its definition and implementation should take into account the reality of the deformable Earth. This is the case for at least the origin and the orientation time evolution for which some caution should be drawn.

Since we select the Center of Mass (CoM) of the whole Earth (including oceans and the atmosphere) as a natural TRS origin, then a difficulty occurs since this selected point is subject to various geodynamic processes influencing its positions, as observed by space geodesy stations attached to the Earth's surface (effect known as "geocenter motion"). The TRF origin provided by satellite geodesy coincides in practice with the position of the center of mass averaged over the period of the used observations.

On the other hand, as we deal with observations taken from stations located on the Earth's Crust (undergoing various motions, and in particular tectonic plate motion), the TRS theoretical definition should include a clear statement regarding its orientation time evolution.

Unlike the origin and scale, there is no physical property known to allow the orientation determination or its time evolution. Some conventions are then needed to select the initial TRS orientation at a given epoch, as well as a law of its time evolution which should be adopted in such a way that it doesn't affect the estimated parameters.

3 Transformation between TRSs

The general transformation of cartesian coordinates of any point close to the Earth from a TRS (1) to a TRS (2) is a tri-dimensional similarity given by:

$$X_2 = T + \lambda.\mathcal{R}.X_1 \tag{1}$$

where T is a translation vector, λ is a scale factor and $\mathcal{R}$ is a rotation matrix.

A linearized form of equation (1) could also be used, in particular for frames provided by space geodesy techniques (origin difference is about a few hundred meters, and differences in scale and orientation are of 10^{-5} level). In this case, the standard relation of transformation between two frames is a 7 parameter similarity: 3 translations, 1 scale factor, and 3 rotations, designated respectively by, $T1$, $T2$, $T3$, D, $R1$, $R2$, $R3$, allowing to transform a position vector X_1, expressed in frame (1) into a position vector X_2, expressed in frame (2):

$$X_2 = X_1 + T + DX_1 + RX_1 \tag{2}$$

with: $\mathcal{R} = (I + R)$, $\lambda = (1 + D)$, $T = \begin{pmatrix} T1 \\ T2 \\ T3 \end{pmatrix}$

and $R = \begin{pmatrix} 0 & -R3 & R2 \\ R3 & 0 & -R1 \\ -R2 & R1 & 0 \end{pmatrix}$.

If we assume constant linear velocities, X_1, X_2, T, D, R become function of time. Differentiating equation (2) with respect to time gives:

$$\dot{X}_2 = \dot{X}_1 + \dot{T} + \dot{D}X_1 + D\dot{X}_1 + \dot{R}X_1 + R\dot{X}_1 \tag{3}$$

D and R are of 10^{-5} level and $\dot{X}$ is about 10 cm per year, the terms $D\dot{X}_1$ and $R\dot{X}_1$ are negligible which represent about 0.1 mm over 100 years. Therefore, equation (3) could be written as:

$$\dot{X}_2 = \dot{X}_1 + \dot{T} + \dot{D}X_1 + \dot{R}X_1 \tag{4}$$

4 Realization of a Terrestrial Reference System

Using the terminology adopted since the eighties by astronomical and geodetic communities (Kovalevsky et al., 1989), we define a Terrestrial Reference Frame (TRF) by a set of physical points with precisely determined coordinates as a materialization of an ideal

TRS. Therefore, a TRS and its realization, i.e. a TRF, are closely related to each other and so theoretical geodetic formulas which apply to a TRS apply also for the TRF.

In this paper we assume the general case where a TRF contains both station positions as well as linear velocities.

There are mainly two ways to realize a TRS, namely using observations of one or more space geodesy techniques or by combination of various TRS realizations.

4.1 TRS Realization by Space Geodesy Techniques

A TRS realization by a TRF requires, implicitly or explicitly, the adoption of 7 datum parameters (3 translation components, one scale factor and 3 rotation angles) as well as their time derivatives. The adoption of the 14 parameters defines the TRF origin, scale, orientation and their time evolution.

The starting point is to admit that, from the physical point of view, none of space geodesy techniques could provide all the necessary information allowing to completely realize a TRS.

The TRF origin provided by dynamical techniques (such as LLR, SLR, GPS, DORIS) is sensitive to the CoM position, being the point around which the satellite orbits. Given the geocenter motion effect, the TRF origin, in this case corresponds to the CoM mean position, averaged over the period of the used observations. It is in general achieved by fixing to zero the first potential coefficients of the harmonic expansion of the gravity field. The VLBI technique is of course insensitive to the CoM. Although the origin of a TRF could be selected in an arbitrary manner (as a convention), the choice of the CoM, and in particular for the ITRF origin, is to preserve the geophysical information that space geodesy provides naturally.

The TRF scale for all space geodesy techniques is influenced by various parameters, such as the gravitational constant (GM) for dynamical techniques, the speed of the light (c), the troposphere modelling and other technique-specific effects.

The orientation is conventional for all techniques and so additional constraints are needed to relate the TRF orientation to some prescribed definition.

In any case, the 14 parameters should be understood as relative values allowing to transform station positions and velocities from one TRF to another TRF. For notation simplicity, we will designate hereafter by X both position and velocity vectors. Therefore, the two equations (2) and (4) reassembled together could be written as:

$$X_2 = X_1 + A\theta \qquad (5)$$

where $\theta = (T1, T2, T3, D, R1, R2, R3, \dot{T}1, \dot{T}2, \dot{T}3, \dot{D}, \dot{R}1, \dot{R}2, \dot{R}3)^T$ is the vector of the 14 transformation parameters and A is the design matrix of partial derivatives constructed upon elements of X_1 or X_2 used as approximate station positions $(...,x_i^0,...)$, where $1 < i < n$ and n is the number of stations:

$$A = \begin{pmatrix} \cdot & \cdot & \cdot & \cdot & \cdot & \cdot & \cdot & \cdot & \cdot & \cdot & \cdot & \cdot & \cdot & \cdot \\ 1 & 0 & 0 & x_i^0 & 0 & z_i^0 & -y_i^0 & & & & & & & \\ 0 & 1 & 0 & y_i^0 & -z_i^0 & 0 & x_i^0 & & & & 0 & & & \\ 0 & 0 & 1 & z_i^0 & y_i^0 & -x_i^0 & 0 & & & & & & & \\ & & & & & & & 1 & 0 & 0 & x_i^0 & 0 & z_i^0 & -y_i^0 \\ & & & \approx 0 & & & & 0 & 1 & 0 & y_i^0 & -z_i^0 & 0 & x_i^0 \\ & & & & & & & 0 & 0 & 1 & z_i^0 & y_i^0 & -x_i^0 & 0 \\ \cdot & \cdot & \cdot & \cdot & \cdot & \cdot & \cdot & \cdot & \cdot & \cdot & \cdot & \cdot & \cdot & \cdot \end{pmatrix} \qquad (6)$$

Assume that for a given TRF solution s to be provided by a space geodesy technique, the normal equation system is properly reduced to station positions (and velocities), written as:

$$N_s.\Delta X_s = K_s \qquad (7)$$

where N_s is the normal matrix, ΔX_s is the vector of the linearized unknowns (so that $X_s = X_0 + \Delta X_s$, where X_s is the estimated solution), and K_s is the vector of constant terms. When constructing such a normal equation using observations from one or more space geodesy techniques, analysis centers need to add some constraints in order to define the resulting TRF. Mathematically, this reflects the fact that the normal matrix is singular, since it has a rank deficiency corresponding to the datum definition parameters which are not inherent in the observations. Useful discussion on reference systems and the problem of rank deficiency could be found in (Dermanis, this issue).

In practice, all space geodesy techniques need to add 6 pseudo observation (constraint) equations corresponding to the 3 rotation angles and their time derivatives. VLBI technique needs 9 additional equations to define the 3 TRF origin components and their time derivatives and the 3 CRF orientation components.

There are mainly two types of constraints allowing the inversion of the equation system (7):

- classical constraints where the added constraint equation is written as $X_s = X_0$, applied generally over a subset of stations. This kind of constraint allow the inversion of the normal equation system, without any caution to the resulting TRF solution X_s and its variance.

- minimum constraints allowing to complete the rank deficiency of the normal matrix N_s, implying in fact the lacking (sought) datum definition parameters and not more.

In the following, we will focus on the approach of minimum constraints, which represents the "natural" way towards the TRF implementation. This approach is the one currently used in the ITRF combinations.

Many authors contributed to the issue of network adjustment and in particular to the topic of minimum constraints, starting by the pioneering work of Meissl and Blaha (Blaha, 1971). The other authors are too numerous to be cited here, but a large list could be found in a recent review publication on this topic by (Sillard and Boucher, 2001).

The general form of the equation of minimum constraints is directly connected to equation (5). A possible way to understand equation (5) is to use the concept of projectors. Let us consider that (5) corresponds to an estimation problem where X_1 and θ are unknowns and X_2 are observations. It is obvious that the set of parameters (X_1, θ) is unidentifiable since the problem is under-determined. A classical way to treat this problem is to identify $A\theta$ with the component of X_2 on the range of A (designated by $R(A)$) and X_1 with the orthogonal of the range of A (designated by $R(A)^{\perp}$), since

$$\mathbb{R}^{6n} = R(A)^{\perp} \overset{\perp}{\oplus} R(A).$$

The decomposition of X_2 according to the sum of the previous vectorial spaces is unique. With the use of orthogonal projectors, we have

$$\begin{cases} A\theta &= A(A^TPA)^{-1}A^TPX_2 \\ X_1 &= (I - A(A^TPA)^{-1}A^TP)X_2 \end{cases}$$

where P is the scalar product matrix of $\mathbb{R}^{6n}$. It can be chosen through various options, for example :

- unit weights : $P = I$, where I is the identity matrix

- inverse of X_2 covariance matrix

- inverse of X_2 covariance matrix restricted to diagonal terms

And by identification, we can set

$$\theta = (A^TPA)^{-1}A^TPX_2 \qquad (8)$$

We therefore see the key role played by the application $B_P = (A^TPA)^{-1}A^TP$ which makes it possible to work in the space of datum definition parameters.

In terms of TRF datum definition, the estimated TRF solution X_s of a given space geodesy technique, should be related to some specific TRF solution, say X_R, taken as reference. Consequently, the relation between X_s and X_R is of equation (5)-type, where the vector of transformation parameters θ becomes zero.

Therefore using the concept of orthogonal projections developed above, the equation of minimum constraints to be introduced is of the form:

$$B(X_s - X_R) = 0 \qquad (9)$$

where $B = (A^TA)^{-1}A^T$. Equation (9) simply imposes that the reference frames of X_S and X_R are identical. This is done by working in the space of datum definition parameters, as presented above, and nullifying these parameters. Note that the scalar product matrix P is selected here to be the identity matrix in order to avoid any possible contamination that the external solution X_R may introduce in the estimated solution X_s. Moreover, with this kind of minimum constraints approach the estimated TRF takes its full mathematical (or geometrical) meaning.

On the other hand, one can notice that $R(B) = R(A^T)$, so that A^T could be used instead of B. Meanwhile from the numerical computation point of view, we found that using B provides better numerical stability than using A^T.

Finally the matrix A in equation (6) should be reduced to columns corresponding to the datum parameters needed to complete the rank deficiency of the normal equation system (7).

In terms of normal equation, and taking into account the a priori values X_0, this amounts to adding to the normal equation system (7) an equation of the form:

$$(B^T\Sigma_\theta^{-1}B).\Delta X_s = (B^T\Sigma_\theta^{-1}B)(X_R - X_0) \qquad (10)$$

where Σ_θ is a diagonal matrix, used as weights matrix, containing small variances for each one of the transformation parameters (among the 14), corresponding to the rank deficiency of the normal matrix.

Cumulating equations (7) and (10) yields:

$$(N_s+B^T\Sigma_\theta^{-1}B)(\Delta X_s) = K_s+B^T\Sigma_\theta^{-1}B(X_R-X_0) \qquad (11)$$

The introduction of Σ_θ as a weight matrix (which by the way could be selected as the identity matrix) does not change the estimate ΔX_s due to the invariance property, see for instance Schaffrin and Bâki Iz (2001). Meanwhile the estimated variance/covariance matrix resulting from the inversion of equation (11) depends on the selection of Σ_θ which "plays" the role of defining the frame at the level selected by the user. On other words, adding $B^T\Sigma_\theta^{-1}B$ to the singular normal matrix N_s will allow its inversion (even if $\Sigma_\theta = I$) and simultaneously define the underlying TRF at the Σ_θ level.

If the specific TRF solution X_R is initially selected to be the a priori solution X_0, the equation of minimum constraints becomes that of Meissl "*inner constraints*", as quoted by Dermanis (2000), so that from equation (10) we have:

$$(B^T\Sigma_\theta^{-1}B).\Delta X_s = 0 \qquad (12)$$

and then the normal equation system with inner constraints becomes (by accumulating equations (7) and (12)):

$$(N_s + B^T\Sigma_\theta^{-1}B).\Delta X_s = K_s \qquad (13)$$

It should be emphasized here that the approach of minimum constraints described above is an efficient method to:

- avoid rank deficiency of the normal equation system
- provide (complete) the necessary datum definition of the TRF
- preserve (leave unaltered) the original quality of space geodesy observations.

4.2 TRS Realization by Combination

The basic idea behind combining TRF solutions estimated by several space geodesy techniques is to gather the strength of all these techniques in one global frame.

In terms of TRF datum definition, the combination model lies on the relationship connecting two frames. When combining TRF solutions containing station positions as well as velocities, equations (2) and (4) could be used together to form a general combination model, such as the one used for ITRF combination, given by:

$$\left\{ \begin{array}{rcl} X_s^i & = & X_{itrf}^i + (t_s^i - t_0)\dot{X}_{itrf}^i \\ & + & T_k + D_k X_{itrf}^i + R_k X_{itrf}^i \\ & + & (t_s^i - t_k)\left[\dot{T}_k + \dot{D}_k X_{itrf}^i + \dot{R}_k X_{itrf}^i\right] \\ & & \\ \dot{X}_s^i & = & \dot{X}_{itrf}^i + \dot{T}_k + \dot{D}_k X_{itrf}^i + \dot{R}_k X_{itrf}^i \end{array} \right. \qquad (14)$$

where for each individual solution s, and each point i, we have position X_s^i at epoch t_s^i and velocity $\dot{X}_s^i$, expressed in a given TRF k.

The combination consists in estimating:

- Positions X_{itrf}^i at a given epoch t_0 and velocities $\dot{X}_{itrf}^i$ in ITRF
- Transformation parameters (T_k, D_k and R_k) at an epoch t_k and their rates ($\dot{T}_k$, $\dot{D}_k$, $\dot{R}_k$), from the ITRF to each individual frame k.

Similarly to individual TRS realizations by space geodesy techniques, the normal equation system constructed using the above combination model is singular and 14 degrees of freedom need to be selected. There are several ways allowing to define the datum of the combined TRF. A very simple way would be to fix to given values (e.g. zeros) the 14 parameters of a chosen solution s, so that the resulting combined frame is that of the selected solution s. A more interesting approach is the use of minimum constraints, by adding to the combination model a TRF datum equation of type (9), as the one used to define the ITRF2000 orientation and its time evolution (see below).

5 Case of ITRF2000

Starting with the ITRF2000, one of the fundamental concepts adopted is the use of unconstrained individual solutions, but for which we add minimum constraints used solely to accurately define the underlying TRF in origin, scale, orientation and time evolution.

All the ITRF2000 technical description and results are published in (Altamimi et al., 2002). Meanwhile we recall here the most important results to illustrate the discussion of this paper.

The ITRF2000 is intended to be an improved frame in terms of network, quality and datum definition. It includes primary core stations observed by VLBI, LLR, SLR, GPS and DORIS (usually used in previous ITRF versions) as well as regional GPS networks for its densification. The individual solutions included in the ITRF2000 combination are free from any external constraints, reflecting hence the actual quality of space geodesy estimates of station positions and velocities. The implementation of the ITRF2000 datum was achieved as follows:

- Origin: by fixing to zero the translation and translation rate parameters between ITRF2000 and the weighted mean of the SLR solutions of 5 analysis centers: CGS, CRL, CSR, DGFI and JCET.

- Scale: by fixing to zero the scale and scale rate parameters between ITRF2000 and the weighted mean of the above 5 SLR solutions and the VLBI solutions of the analysis centers: GIUB, GSFC and SHA.
- Orientation: by adding to the combination model a TRF datum equation of type (9) (restricted to the orientation and its rate) given by:

$$B(X_0 - X) = 0 \qquad (15)$$

where:

  - X_0: ITRF97 positions at epoch 1997.0 defining the rotation angles and NNR-NUVEL-1A velocities defining the rotation rates. NNR-NUVEL-1A model is adapted from NNR-NUVEL-1 (Argus and Gordon 1991), according to DeMets et al. (1994)
  - X: Estimated station positions and velocities

In order to assess the accuracy of the ITRF2000 scale and origin definition, we computed WRMS values on scale and translation differences (with respect to ITRF2000) upon the 3 VLBI and 5 SLR solutions used in the scale and origin definition. These values listed in Table 1 indicate an accuracy of the scale and origin at epoch 1997.0 at (or better than) the 1 mm level and 0.3 mm/y for their rates.

Moreover, in order to evaluate the long-term stability of the ITRF2000 scale and origin, the WRMS are propagated over 10 years and listed in Table 1. The propagated WRMS values over 10 years suggest a frame stability better than 4 mm in origin and better than 0.5 ppb in scale (equivalent to a shift of approximately 3 mm in station heights). Note however that the WRMS of the Z translation component is larger than those of X and Y. This fact is most likely due to the SLR network configuration having much more stations in the northern hemisphere.

One of the important ITRF2000 results is that there is a statically significant disagreement between ITRF2000 and NUVEL-1A model (DeMets et al., 1990, 1994) in terms of relative tectonic plate motions. Although the ITRF2000 orientation rate is aligned to that of NNR-NUVEL-1A, per-plate residuals between the two corresponding velocity fields appear to be significant as shown in Figure 1. This per-plate residual behavior indicates the inadequacy of NUVEL-1A relative model to describe the current plate motions as seen by ITRF2000 results. Meanwhile it should be emphasized that these differences do not at all disrupt the internal consistency of the ITRF2000, simply because the alignment to NNR-NUVEL-1A defines the ITRF2000 orientation rate and nothing more. On the other hand, it is important to emphasize that the goal of aligning ITRF2000 to the NNR-NUVEL-1A model is to apply, implicitly, the No-Net-Rotation Condition (NNRC) or, equivalently, to realize a Tisserand Frame. The ITRF choice of the No-Net-Rotation Condition is dictated by the fact that the Tisserand axes frame is used in the theory of Earth rotation, having the property to minimize the Earth's crust motion and deformation that would affect the Earth Orientation Parameters. In a recent publication (Altamimi et al., 2003), we demonstrated that the ITRF2000 alignment to NNR-NUVEL-1A model (satisfying the NNRC) is achieved at (or better than) the 1 mm/y level. It is also shown that current knowledge of the NNRC realization is at the level of about 2 mm/y, at the equator of the Earth surface.

Table 1. ITRF2000 Scale and Origin Accuracy at 1997.0 and Long Term Stability

Parameter	WRMS	
	at 1997.0	Over 10 years
D (ppb)	0.2 ($\approx$ 1.2 mm)	0.5 ($\approx$ 3 mm)
$\dot{D}$ (ppb/y)	0.03 ($\approx$ .2 mm/y)	
T_x (mm)	0.4	1.4
$\dot{T}_x$ (mm/y)	0.1	
T_y (mm)	0.5	1.5
$\dot{T}_y$ (mm/y)	0.1	
T_z (mm)	0.9	3.9
$\dot{T}_z$ (mm/y)	0.3	

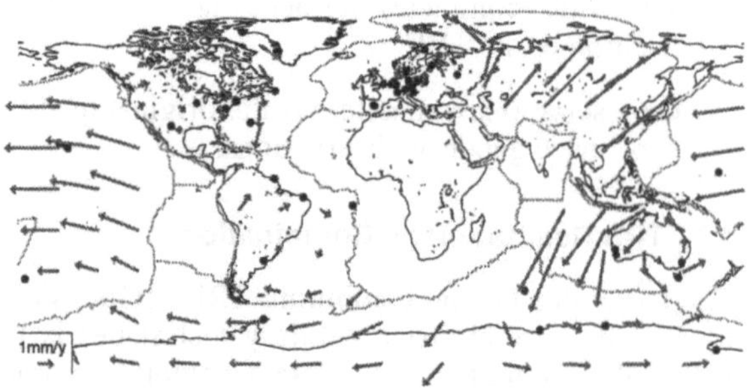

Fig. 1 ITRF2000 and NNR-NUVEL-1A velocity differences

References

Altamimi, Z., P. Sillard and C. Boucher (2002), ITRF2000: A New Release of the International Terrestrial Reference Frame for Earth Science Applications, *J. Geophys. Res.*, 107(B10), 2214, doi:10.1029/2001JB000561.

Altamimi, Z., P. Sillard and C. Boucher (2003). The Impact of a No-Net-Rotation Condition on ITRF2000, *Geophys. Res. Lett.*, 30(2), 1064, doi:10.1029/2002GL016279.

Argus, D.F. and R.G. Gordon (1991). No-net rotation model of current plate velocities incorporating plate motion model NUVEL-1, *Geophys. Res. Lett.18,* 2038-2042.

Blaha G. (1971), Inner adjustment constraints with emphasis on range observations, Department of Geodetic Science, Report 148, The Ohio State University, Columbus.

Boucher, C., (2000). Terrestrial coordinate systems and frames, in *Encyclopedia of Astronomy and Astrophysics,* Version 1.0, pp. 3289-3292, Inst. of Phys. Publ, Bristol, UK.

Dermanis, A., this issue.

Dermanis, A. (2000), Establishing Global Reference Frames. Nonlinear, Temporal, Geophysical and Stochastic Aspects, In: Sideris, (ed.), Gravity, Geoid and Geodynamics 2000, IAG Symposium Vol. 123, pp. 35-42.

DeMets, C., R.G. Gordon, D.F. Argus, and S.Stein (1990). Current plate motions. *Geophys. J. Int., 101,* 425-478.

DeMets, C., R. G. Gordon, D. F. Argus, and S. Stein (1994). Effect of recent revisions of the geomagnetic reversal timescale on estimates of current plate motions, *Geophys. Res. Lett.21(20),* 2191-2194.

Kovalevsky, J., Mueller, I. I., and B. Kolaczek (Eds.),(1989). *Reference Frames in Astronomy and Geophysics,* 474 pp., Kluwer Academic Publisher, Dordrecht.

Schaffrin, B. and H. Bâki Iz (2001). BLIMPBE and its Geodetic Applications. In: Adam and Schwarz (eds.), Vistas for Geodesy in the New Millennium, IAG Symposia, Vol. 125, pp. 377-381.

Sillard, P. and C. Boucher, (2001) Review of Algebraic Constraints in Terrestrial Reference Frame Datum Definition, *Journal of Geodesy , 75,* 63-73.

Problems in the Definition of Vertical Reference Frames

B. Heck,
Geodetic Institute, University of Karlsruhe,
Englerstr. 7,
D – 76128 Karlsruhe, Germany
e-mail: heck@gik.uni-karlsruhe.de
Fax: ++49-721-608-6808

Abstract. Present and future satellite-borne gravity field missions will have a major impact on the unification of regional height systems into a consistent global vertical reference frame. On the background of the high precision and quality of the expected data the scientific concepts of vertical reference frames have to be discussed, before a global frame will be implemented or before international standards and conventions can be fixed.

This paper addresses some major theoretical problems in the definition of vertical reference frames such as difficulties due to the time-dependence of the gravity field and its level surfaces, problems in assigning a geoidal potential value W_0, and the consistent treatment of the permanent tidal effect. Examples from the European Vertical Reference Frame 2000 (EVRF 2000) and the draft of the IERS conventions 2000 prove that these inconsistencies are inherent in several branches of Geodesy and must be taken care of in the practical work involving height systems. Emphasis is given to the formulation of open problems rather than to the presentation of solutions.

Keywords. Vertical reference frames, gravity field, geoid, permanent tidal effects

1 Introduction

Until a few decades ago, the definition of vertical reference systems had been restricted to regional or continental reference frames, due to the exclusive use of levelling data for height determination. In general, the datum of such a national or continental vertical network had been based on tide gauge observations defining a zero level for the heights. Due to the existence of sea surface topography and local phenomena at tide gauge stations, these tide gauge zeros are not situated on one unique level surface globally. This fact was not considered as a drawback until GPS positioning and satellite altimetry have been used in Geodesy. GPS positioning provides (relative) ellipsoidal heights which can be transformed into physical heights by the aid of geoid information. Satellite altimetry on the other hand is able to observe the instantaneous (and, after applying reductions for temporal changes, quasi-stationary) sea surface; by subtracting a global geoid the sea surface topography can be constructed which serves as an input to oceanographic studies.

Both GPS and satellite altimetric heights have to be referred to a unique regional or global equipotential surface, e.g. the Gauß-Listing geoid defining a global vertical reference frame. Due to gaps and systematic effects in terrestrial gravity data and insufficient resolution of satellite-based gravity field models it has not been possible until now to determine the geoid as a global level surface with centimetre accuracy over distances of hundreds to thousands of kilometres. This situation will change in near future when satellite-borne gravity field missions such as CHAMP, GRACE and GOCE will provide high-resolution gravity field information. On the basis of the high precision and quality of the expected data the scientific concepts of vertical reference systems have to be thoroughly discussed before e.g. a global vertical reference frame can be adopted.

The paper addresses some major theoretical problems in the definition of vertical reference frames, such as difficulties due to the time-dependence of the gravity field, problems in assigning a geoidal potential value W_0 as a datum, and the consistent treatment of the permanent tidal effect on the solid Earth. The paper concludes with a formulation of the most important questions to be answered and to be agreed upon by the geodetic community.

2 Heights in the Earth's Time-dependent gravity field

The present status in the precision of geodetic measuring procedures requires the modelling of time-dependent geodynamic phenomena in the terrestrial gravity field, in particular Earth and ocean tides, variations of the angular velocity vector of the Earth's rotation and motions of the lithospheric plates. Therefore, the vector of gravity registered by an earth-fixed observer contains – besides gravitational components – constituents induced by the accelerated motion of the deformable Earth's surface with respect to a (quasi-)inertial or a terrestrial frame.

Based on the general physical model of Newtonian mechanics the motion of any mass point in inertial space is governed by the Newton-Eulerian equation of motion

$$\rho \cdot \ddot{\vec{y}} = \rho \cdot \vec{a}_{grav} + \rho \cdot \vec{a}_{MSP} + \vec{S}(\vec{y}) \tag{1}$$

(Grafarend, 1982, p. 616), where ρ denotes the mass density at the considered mass point P, $\vec{y}$ the position vector of P related to the (quasi-)inertial reference frame $\{0, \vec{e}_i\}$ with origin 0 and orthonormal base vectors $\vec{e}_i, i \in \{1,2,3\}$, $\vec{a}_{grav}$ the gravitational attraction of the Earth at P, $\vec{a}_{MSP}$ the gravitational attraction at P induced by the moon, sun and planets, and $\vec{S}(\vec{y})$ the stress at P related to surface forces; $\ddot{\vec{y}}$ denotes the second derivative of $\vec{y}$ with respect to time τ. Due to the rotation and deformation of the Earth and due to the changing configuration of the bodies in the solar system the constituents $\vec{a}_{grav}$ and $\vec{a}_{MSP}$ are time-dependent.

In a body-fixed frame attached to the Earth additional apparent forces enter the equation of motion (1). Decomposing the position vector $\vec{y}$ into the position vector $\vec{y}_M$ of the Earth's centre of mass M and the topocentric position $\vec{X}$ with respect to M

$$\vec{y} = \vec{y}_M + \vec{X}, \tag{2}$$

it is convenient to represent $\vec{X}$ in the moving frame $\{M, \vec{f}_i\}$; the orthonormal base vectors $\vec{f}_i$ are related to $\vec{e}_i$ via the rotation matrix $\underline{R} = (R_{ij})$

$$\vec{f}_i = R_{ij}\vec{e}_j \tag{3}$$

(application of Einstein's summation convention), such that

$$\vec{X} = X_i \vec{f}_i \tag{4}$$

where X_i can be grouped in the column vector of geocentric coordinates $\underline{X} = (X_i)$. Application of equations (2), (3), (4) to (1) results in the equation of motion related to the moving, body-fixed frame (Heitz, 1980; Schneider, 1981; Grafarend, 1982)

$$\begin{aligned} \rho(D_{\tau\tau}\underline{X}) = \rho \cdot \underline{a}_{grav} + \rho \cdot \underline{a}_{tid} + \underline{S}(\underline{y}) + \\ + \rho(2D_\tau\underline{X} \times \underline{\Omega} + \underline{X} \times D_\tau\underline{\Omega} + (\underline{\Omega} \times \underline{X}) \times \underline{\Omega}). \end{aligned} \tag{5}$$

$D_\tau\underline{X}$ etc. denotes the derivative of the coordinate (column) vector $\underline{X}$ with respect to time τ. Besides the gravitational acceleration $\underline{a}_{grav}$ the tidal acceleration $\underline{a}_{tid}$, the Coriolis acceleration $2 \cdot D_\tau\underline{X} \times \underline{\Omega}$, the Euler acceleration $\underline{X} \times (D_\tau\underline{\Omega})$ and the centrifugal acceleration $(\underline{\Omega} \times \underline{X}) \times \underline{\Omega}$ have to be considered, where $\underline{\Omega} = (\Omega_i)$ is the vector of angular velocity; the coordinates Ω_i are related to the components of the three-dimensional skew-symmetric Cartan matrix

$$\Omega_{ik} = (D_\tau R_{ij}) \cdot R_{kj} \tag{6}$$

by the formula

$$\Omega_i = \frac{1}{2}\varepsilon_{ijk}\Omega_{jk} \tag{7}$$

(ε_{ijk} : ε-tensor). It is well-known that the centrifugal acceleration at any time instant τ forms a conservative vector field which can be expressed by the gradient of the scalar centrifugal potential Z

$$Z := \frac{1}{2}|\underline{\Omega} \times \underline{X}|^2 \quad . \tag{8}$$

Equation (5) forms the basis for the definition of the gravity vector $\vec{\Gamma}_t$ which is related to a conventional terrestrial reference frame (realisation of a body-fixed moving frame)

$$\vec{\Gamma}_t = \Gamma_i^t \vec{f}_i \quad , \quad \underline{\Gamma}_t = (\Gamma_i^t) \tag{9}$$

and can be observed e.g. on the Earth's surface. This results in the basic definition (see e.g. Grafarend, 1982, p. 633)

$$\begin{aligned} \underline{\Gamma}_t := D_{\tau\tau}\underline{X} = \underline{a}_{grav} + \underline{a}_{tid} + |\underline{\Omega} \times \underline{X}| \times \underline{\Omega} \\ + 2(D_\tau\underline{X}) \times \underline{\Omega} + \underline{X} \times (D_\tau\underline{\Omega}) \, . \end{aligned} \tag{10}$$

It can easily be recognized that the Coriolis and Euler terms in (10) are non-conservative, since

$$\text{rot}\left[2(D_\tau \underline{X})\times\underline{\Omega}+\underline{X}\times\dot{\underline{\Omega}}\right]=$$

$$=-2\underline{\Omega}\cdot\text{div}(D_\tau\underline{X})+2\text{grad}(<D_\tau\underline{X},\underline{\Omega}>) \tag{11}$$

$$-2\underline{\Omega}\times\text{rot}(D_\tau\underline{X})-2\dot{\underline{\Omega}}\neq 0$$

(Grafarend, 1982).

The magnitude of the Coriolis and Euler terms as well as the time-variable centrifugal components due to geodynamic phenomena has been estimated in Heck (1984). Coriolis terms induced by lithospheric plate motion and tidal motion of a point on the Earth's surface are always less than a few parts in $10^{-8}\text{m}\cdot\text{s}^{-2}$, similar to the maximum order of magnitude of the Euler terms produced by precession, nutation and length of day variations. The time-variable centrifugal term resulting from polar motion is less than $8\cdot10^{-8}\text{ms}^{-2}$, a magnitude similar to the maximum effects induced by the horizontal displacements of lithospheric plates within 100 years, while the effects of tidal displacements are smaller by two orders of magnitude.

The conservative gravitational acceleration $\underline{a}_{grav}$ is related to the gravitational potential V_{grav}

$$\underline{a}_{grav} = \text{grad}\, V_{grav} \tag{12}$$

where the Newtonian potential

$$V_{grav}(\underline{X}) := G\cdot\iiint\frac{\rho'(\underline{X}')}{\left|\underline{X}-\underline{X}'\right|}d^3\underline{X}' + V_{grav}^{\infty} \tag{13}$$

(G: gravitational constant) depends on the instantaneous mass distribution. V_{grav}^{∞} denotes an arbitrary constant corresponding to the value of the gravitational potential at infinity, i.e. for $|\underline{X}|\to\infty$; mostly this constant (which disappears when the gradient is formed in equation (12)) is fixed to zero. It should be kept in mind that the condition $V_{grav}^{\infty}=0$, producing an "absolute" potential, cannot be checked by any experiment in practice, since this absoute potential refers to a zero point at infinity. Similarly, the spherical harmonic expansion of V_{grav}, used in satellite geodesy, has to be complemented by the constant V_{grav}^{∞}, i.e.

$$V_{grav}(r,\varphi,\lambda)=\frac{GM}{r}\left\{1+\sum_{n=1}^{\infty}\left(\frac{a_\oplus}{r}\right)^n\cdot\sum_{m=0}^{n}(C_{nm}\cdot\cos m\lambda+S_{nm}\cdot\sin m\lambda)\cdot P_{nm}(\varphi)\right\} + V_{grav}^{\infty} \tag{14}$$

(r,φ,λ spherical coordinates of the computation point, $P_{nm}(\varphi)$ associated Legendre functions, C_{nm} and S_{nm} harmonic coefficients). The symbol $a_\oplus$ denotes a fixed radius which is mostly identified with the equatorial radius of a reference ellipsoid approximating the Earth. The series expansion (14) is valid outside a so-called Brillouin sphere around the coordinate origin, enclosing the topographical masses of the Earth.

On the basics of the defining equation (12) a potential is determined up to a constant only. For this reason, only potential differences instead of "absolute" potentials are observable. Of course, V_{grav} is time-dependent, in general; the magnitude of the time-variable constituents in V_{grav}, induced by deformations of the Earth's body and mass shifts due to geodynamic phenomena has been discussed in Heck (1984).

Similar arguments hold for the tidal potential V_{tid}, related to the conservative tidal acceleration

$$\underline{a}_{tid} = \text{grad}\, V_{tid} \quad . \tag{15}$$

The general solution of this equation contains an arbitrary constant V_{tid}^{o}, corresponding to the value of the tidal potential at the origin M, i.e. for $|\underline{X}|\to 0$. Mostly this constant is fixed to zero, but – due to the inherent relativeness of potentials – this condition cannot be checked by any experiment, either.

If $\vec{\Gamma}_t$ is reduced for Coriolis and Euler terms, the resulting vector field

$$\underline{\Gamma}_t + 2\underline{\Omega}\times(D_\tau\underline{X})+(D_\tau\underline{\Omega})\times\underline{X} = \text{grad}(V_{grav}+V_{tid}+Z) \tag{16}$$

is conservative at any time instant τ. But still the potentials V_{grav}, V_{tid} and Z are position– and time-dependent. If the geodynamic processes can be modelled with sufficient precision, a separation between stationary and time-dependent components can be realized. Decomposing the potentials V_{grav} and Z into (quasi-)stationary and time-variable parts

$$\begin{aligned} V_{grav} &= \tilde{V} + \delta V(\tau) \\ Z &= \tilde{Z} + \delta Z(\tau), \end{aligned} \qquad (17)$$

where the time derivative of the (quasi-)stationary components vanishes $\left(D_\tau \tilde{V} = 0,\ D_\tau \tilde{Z} = 0\right)$, the (quasi-stationary) gravity potential

$$W := \tilde{V} + \tilde{Z} \qquad (18)$$

is defined, related to the (quasi-stationary) gravity vector $\vec{g}$, which is described in the moving frame $\{M, \vec{f}_i\}$:

$$\begin{aligned} \underline{g} &= \text{grad} W \\ &= \underline{\Gamma}_t - 2(D_\tau \underline{X}) \times \underline{\Omega} - \underline{X} \times (D_\tau \underline{\Omega}) \\ &\quad - \text{grad}(\delta V + V_{tid} + \delta Z) \ . \end{aligned} \qquad (19)$$

From a practical point of view, W and $\underline{g}$ can be considered as time-independent, if the precision of the reduction models for astronomic tides, Earth, ocean and atmospheric tides, ocean and atmospheric loading, plate tectonic motion, precession and nutation, rotational variations of the Earth and polar motion, post-glacial rebound and other phenomena are sufficient. It has to be pointed out that the problem of defining and fixing V_{grav}^{∞} and V_{tid}^{o} is existent in the realization of the gravity potential W, equation (18), too.

As a matter of fact, equivalent reduction models have to be applied for the definition of a quasi-stationary Earth surface in order to achieve a consistent geodetic model. In the following it is assumed that the reduction procedures are precise enough to provide an Earth surface and gravity field which do not show any significant temporal variations detectable in the observational data for the determination of heights.

The traditional concepts of heights are commonly based on the so-defined (quasi-)stationary gravity field and Earth's surface. Starting from the geopotential number

$$C\,(P) := W_o - W(P) \qquad (20)$$

corresponding to the potential difference between a surface point P and a fixed reference point P_o, several metric height systems, used as national height systems, can be derived (Heiskanen and Moritz, 1967, p. 160; Heck, 1995, p. 292ff), such as the dynamic height

$$H^D(P) := \frac{C(P)}{\gamma_{45}} \qquad (21)$$

(γ_{45}: constant normal gravity at 45° latitude), the normal height

$$H^*(P) := \frac{C(P)}{\bar{\gamma}_P} \qquad (22)$$

($\bar{\gamma}_P$ integral mean of normal gravity along the ellipsoidal normal running through P, between the reference ellipsoid and the telluroid point Q, corresponding to P), which is numerically identical to the ellipsoidal height of the telluroid point Q related to P,
or the orthometric height

$$H(P) := \frac{C(P)}{\bar{g}_P} \qquad (23)$$

($\bar{g}_P$ integral mean of actual gravity along the plumb line running through P, between P and the level surface $W=W_o$), describing the distance between P and the intersection point between the plumb line and the level surface $W=W_o$. Due to the lack of surface gravity measurements, which are necessary for the determination of geopotential numbers C, the normal or orthometric heights have often been replaced by the so-called normal orthometric heights in classical national height systems. It should be worth mentioning that all height concepts are related to potential differences only, whereas no use is made of absolute potentials. Furthermore, H, H* and H^D are completely equivalent to the geopotential number C.

Finally, it should be noted that the determination of temporal height changes in the classical national height systems involving spirit levelling is a strongly time-consuming and laborious task. For this reason the changes in national heights nowadays are often replaced by temporal changes of ellipsoidal heights, determined by GPS observations. The difference between ellipsoidal and national height variations is often neglected, amounting to a maximum of 10% with respect to the total variation (Heck and Mälzer, 1983).

3 Vertical Reference Systems and the Vertical Datum Problem

Vertical reference systems, valid for regional parts of the Earth's surface, can be established in different ways, based on the type and availability of observational data. Any vertical reference system is related to either a fundamental reference point P_o or a reference surface, in general an equipotential surface of the Earth's quasi-stationary gravity field. Having determined the potential differences between this fundamental reference point or reference surface and some points on the Earth's (quasi-stationary) surface by some observational procedure, these potential differences or geopotential numbers can be transformed into the height systems mentioned in section 2. According to the basic definition, the geopotential number at the reference point P_o or at any point situated on the reference surface is zero; equivalently, the orthometric or normal or dynamic height at P_o or on the reference surface vanishes (see Figure 1).

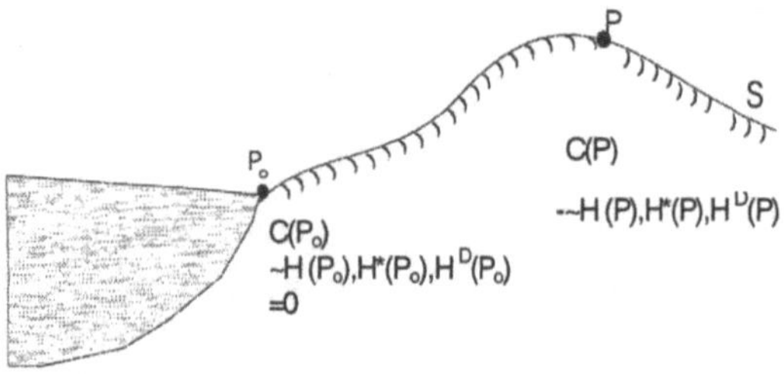

Fig. 1 Vertical datum, referred to a fundamental reference point P_0 (e.g. tide gauge station)

It is evident that only potential differences are relevant for the determination of regional vertical reference systems; the numerical value of the absolute potential W_o at P_o or on the reference equipotential surface is never needed.

Principally, three different approaches are applied in Geodesy and Oceanography for establishing regional vertical reference systems. Over continental regions spirit levelling in combination with gravity measurements along the levelling lines can be used for the determination of potential differences or geopotential numbers C, based on the line integral

$$C(P) = W_o - W(P) = \int_{P_0}^{P} g \cdot dn \; ; \qquad (24)$$

in practical evaluations, the integral is replaced by a discrete sum. In general, a tide gauge station at the sea coast is chosen for the fundamental reference point P_0, such that C(P) refers to a level surface running through P_0. Despite of its very high local precision, spirit levelling is subject to large-scale systematic errors. Furthermore, it is a very time-consuming and expensive procedure, restricted to connected continental areas, and unable to bridge the oceans.

Oceanographic approaches such as steric levelling and dynamic (geostrophic) levelling aim at the determination of potential differences between coastal points or over oceanic regions. Both procedures are based on the approximate evaluation of the general hydrodynamic equations of motion. For a deeper discussion refer to Rummel (2001), Rummel and Ilk (1995) or Heck and Rummel (1990). Steric levelling uses ship measurements of pressure, salinity and temperature along vertical profiles over the oceans, and is restricted to deep ocean areas; problems arise due to the unrealistic assumption of the sea surface being an isobaric surface. Dynamic levelling relies on measurements of the horizontal water velocity; its precision strongly depends on the knowledge of friction forces. The potential differences obtained by steric or dynamic levelling can easily be converted into geometrical heights, just by division by a mean value of gravity, e.g. $\gamma = 9.81 ms^{-2}$, resulting in the sea surface topography (SSTop). In this sense, dynamic levelling is suitable for extrapolating the SSTop information across shallow coastal waters.

A third approach, also applicable over oceanic areas, is based on satellite altimetry and geoid information. Subtracting the geoidal height from the altimetric (stationary) sea surface height, both related to the same ellipsoidal reference surface, the SSTop can be computed. Also this approach suffers from some problems: First, the geoidal heights are not sufficiently well known today (this situation will change significantly after the successful end of the CHAMP, GRACE, and GOCE satellite gravity field missions), and second, the necessary reduction from the observed ocean surface to the quasi-stationary sea surface is still problematic. Furthermore, it is restricted to ocean areas, and it is not usable in coastal areas due to technological difficulties with the altimeter.

It is evident that these three approaches just serve for creating regional vertical datums, which are not completely consistent and cannot be simply merged into a global datum. A unique global vertical datum (Rapp, 1995, Rapp and Balasubramania, 1992) can be established if the respective regional datum zones can be interconnected by determining e.g. the potential differences between regional reference points. Also for the establishment of a global vertical datum it is not necessary to fix an "absolute" value of W_o (Heck and Rummel, 1990; Rummel and Heck, 2000). As a matter of fact, a transition from one to another vertical datum is possible in a mathematically rigorous way by so-called S-transformations (Baarda, 1979).

In principle, it should be possible to interconnect vertical datums e.g. defined on different continents by the oceanographic or altimetric approaches, but still the mentioned error sources obstruct large-scale applications. An alternative procedure for the interconnection of different separated vertical datum zones is based on an extension of the geodetic boundary value problem, proposed by Colombo (1980) and further elaborated by Rummel and Teunissen (1988). Here it is assumed that gravity anomaly data, related to the respective vertical datums, are given over N different datum zones covering the whole globe; furthermore, it is postulated that the ellipsoidal height is given for at least one site in each datum zone, and that the geocentric position of at least one additional site is known. For further discussions of this approach see Heck and Rummel (1990) and Sanso and Usai (1995). Although first accuracy estimates (Xu, 1992) gave promising results, the application of this method suffers from gaps and systematic errors (e.g. induced by horizontal datum inconsistencies) in the gravity anomaly data. It should be noted that in principle "absolute" potential values, related to $V_{grav}^{\infty} = 0$, can be determined by this procedure, imposing the assumption of regularity at infinity

$$\lim_{|\underline{X}\to\infty|}\left(W(\underline{X}) - \frac{1}{2}|\underline{\Omega}\times\underline{X}|^2 \right) = 0 \quad . \tag{25}$$

4 W_o and the geoid

In the quasi-stationary gravity field defined in section 2 a specific level surface can be characterized by selecting a fixed potential value W_o. The traditional, "natural" choice for a global equipotential surface is the Gauß-Listing geoid (Listing, 1873, p. 42) as a best approximation of the mean sea level. For the definition of an "absolute", global vertical datum it is considered necessary to fix a unique level surface. It has to be kept in mind that over continental areas the geoid is mainly situated within the Earth's masses. For this reason, a clear distinction should be made between the "true" geoid running partly within the Earth's interior (and depending on the density in the Earth's crust) and a respective surface in a field harmonically downward continued from external space; this difference is not always taken care of in geodetic literature.

Although at a first glance the Gauß-Listing definition of the geoid seems to be unique, a closer look has to be taken on the basis of present accuracy requirements. At least five main proposals for a precise definition and realisation of the geoid, based on different data sets, can be found in geodetic bibliography (see e.g. Mather, 1978; Heck and Rummel, 1990):

A "physical" definition is based on selecting a fixed numerical value of W_o, i.e.

$$W(\varphi,\bullet,h)=W_o=\text{const.} \tag{26}$$

(Ardalan et al., 2002). This numerical value might be selected as the presently best estimate of W_o from the analysis of various geodetic data sets, e.g.

$$W_o=62\,636\,856.0 \pm 0.5\text{m}^2\cdot\text{s}^{-2}$$

(Bursa et al., 2001). Another choice consists in equating W_o with the constant normal potential U_o at the surface of a level ellipsoid related to a Somigliana-Pizzetti field, i.e. $W_o=U_o$ (Heiskanen and Moritz, 1967; Grafarend and Ardalan, 1999). For the GRS80 this definition results in the postulate

$$W_o=U_{o,GRS80}=62\,636\,860.850\text{m}^2\cdot\text{s}^{-2}$$

(Moritz, 2000). Obviously the level surfaces realized by these two values differ by about 50 cm, if the gravitational part in W tends to zero at infinity, i.e. $V_{grav}^{\infty} = 0$. A drawback of this realization of the Gauß-Listing geoid lies in the fact that the precise position of the surface $W=W_o$ (or, correspondingly, the ellipsoidal height of the geoidal surface $h(\varphi,\lambda)$ by inversion of formula (26)) is not known with sufficient precision, due to the difficulties in validating the condition $V_{grav}^{\infty} = 0$, mentioned in section 2. Equivalently, the potential $W(P_o)$ at a selected point P_o, related to the regularity condition at infinity, cannot be determined with adequate precision. On this background, the respective definition of the European Vertical Reference System (EVRS) has to be strongly questioned. In the definition of the EVRS (EUREF, 2000) it has been proposed to realize the vertical datum by the postulate $W_{NAP}^{REAL} = U_{o,GRS80}$ where W_{NAP}^{REAL} is considered to be identical with the real gravity potential at the fundamental point NAP (Normaal Amsterdams Peil), although such a commitment is not necessary for the definition of a (regional) European vertical reference system. Furthermore, from the above discussion it should have become obvious that the regularity condition at infinity $V_{grav}^{\infty} = 0$ is violated; for a further discussion of this topic see e.g. Rummel and Heck (2000).

A second , "oceanographic", definition considers the geoid as the level surface in relation to which the average permanent sea surface topography (SSTop) is zero when sampled globally over all oceanic regions (Mather, 1978). This definition can be realized by satellite altimetry, providing the quasi-stationary sea surface, and a gravity field model derived from terrestrial gravity data, satellite tracking and dedicated satellite gravity field missions. In principle, it is possible to include tide gauge data, too. The main drawbacks of this approach can be seen in the fact that the precision of the gravity field models is still insufficient (this might change in near future when CHAMP, GRACE and GOCE results will be available), and that tide gauge data are strongly disturbed by local coastal phenomena.

On the basis of the geodetic boundary value problem (GBVP) a third definition (Mather, 1978) is given as follows: The geoid is the surface in relation to which the average quasi-stationary SSTop has a vanishing zero degree harmonic in the solutions of the GBVP. This definition can be realized by satellite altimetry data given in oceanic areas and gravity observations combined with levellings on the continents. The practical application of this approach is hampered by systematic errors in gravity anomaly data over land areas due to horizontal and vertical datum inconsistencies as well as systematic errors in geometric levelling.

The geoid may also be characterized by a geodetic definition (Mather, 1978): The geoid is defined as the level surface in relation to which the average quasi-stationary SSTop has zero mean as sampled at all (discrete) regional vertical datums on the globe. Again, the definition can be realized by satellite altimetry, land and sea gravimetry. A specific difficulty is related to the fact that sea gravimetry should be referred to the geoid which is yet unknown at this stage of evaluation.

Finally, a "natural" definition proposed by Baarda (1979) has to be mentioned: The geoid is the level surface running through a fixed, specific bench mark at a material point P_o, e.g. the NAP. The advantage of this definition is the fact that no further data is needed for the practical realization. On the other hand, this definition requires vertical datum interconnections between different continents, which is still problematic, and a thorough monitoring of vertical motions of the material point P_o on the Earth's surface.

Conceptually, all five definitions of the geoid are equivalent on the level of precision of 1-2 metres. It should be noted that the second, fourth, and fifth definition make no use of a numerical value of W_o, in contrast to the first and third approach relying directly on the condition of regularity at infinity, i.e. $V^{\infty}_{grav} = 0$.

5 The permanent tidal constituent

Another problem to be addressed to in the definition of regional or global vertical reference frames concerns the consistent treatment of the permanent tidal effects. The permanent tidal constituent is the time-independent M_oS_o term in the frequency-domain representation of the (astronomical) tidal potential V_{tid}

$$\begin{aligned} V_{tid} = D \cdot \left(\frac{r}{R}\right)^2 \Big\{ & 0.73807 \cdot \frac{1}{2}\left(1 - 3\sin^2\varphi\right) && M_oS_o \\ & + 0.5300 \cdot \sin 2\varphi \cdot \cos\left(\tau + s + \frac{\pi}{2}\right) && K_1 \\ & + 0.9091 \cdot \cos^2\varphi \cdot \cos(2\tau) && M_1 \\ & + \ldots \Big\} \end{aligned} \tag{27}$$

where φ and r denote the geocentric latitude and radius, τ the mean local lunar time, s the mean lunar longitude, R the mean Earth radius and D=26 277.24 cm^2s^{-2} the Doodson constant (Melchior, 1983; Wenzel, 1997). The permanent tidal constituent M_oS_o just depends on the latitude (and to a lesser extent on the height) of the computation point, but is independent of time, in contrast to the other periodical (diurnal, semidiur-nal, ...) constituents in (27). As indicated in section 2, eventually the constant V^0_{tid} has to be added.

Since the Earth is not a rigid body it deforms under the influence of the tidal force grad V_{tid}. The vertical displacement Δr, the horizontal displacements Δx, Δy at some point on the solid Earth's surface and the incremental potential δw induced by the deformed masses can be related to the tidal potential V_{tid} as follows (Melchior, 1983):

$$\begin{aligned} \Delta x = \ell \cdot \frac{\partial V_{tid}}{\bar{g} \cdot \partial\varphi} \;, \quad & \Delta y = \ell \cdot \frac{\partial V_{tid}}{\bar{g} \cdot \cos\varphi \cdot \partial\lambda} \\ \Delta r = h \cdot \frac{V_{tid}}{\bar{g}} \;, \quad & \delta w = k \cdot V_{tid} \end{aligned} \tag{28}$$

These relations hold for periodical forces acting on a (visco-)elastic model of the Earth. The Love-Shida numbers h, k, ℓ are generally frequency-dependent and can be experimentally determined from the analysis of continuous tidal observations, except the deformation effects potentially enforced by the M_oS_o

permanent constituent. Often the standard values related to the semi-diurnal terms

$$h_2 = 0.6026,\ \ell_2 = 0.0831,\ k_2 = 0.2952$$

are used in practice for the reduction of deformations in the geometry and/or gravity space. These numbers differ considerably from the „fluid" Love numbers

$$h_f = 1.93,\ k_f = h_f - 1 = 0.93$$

which have been proposed for the treatment of the deformation due to the M_oS_o constituent. Since it is principally impossible to determine experimentally the Love-Shida numbers related to the permanent deformation, the "correct" numerical values of h, k, ℓ to be inserted in (28) are unknown.

The difficulties with respect to the calculation of the deformation due to the M_oS_o term have resulted in different concepts for the treatment of the permanent tidal effects. For a thorough discussion see e.g. Heikkinen (1979), Groten (1980), Ekman (1989), Rapp (1989), Heck (1993) and Poutanen et al. (1996).

The mean crust/mean geoid concept relies on the tidal reduction for all periodic direct and indirect tidal effects V_{tid}, δw, Δr, Δx, Δy; the M_oS_o term is not reduced for in the gravity field and in the geometry of the Earth's surface, neither the direct nor the indirect effects. This concept is the "natural" one for oceanography and satellite altimetry, but cannot be used in the framework of geoid determination, since the application of Stokes's integral requires that the disturbing potential is harmonic outside the geoid and regular at infinity; especially the last condition is violated in the "mean" concept.

In the tide-free crust/tide-free geoid concept a complete reduction of both direct and deformation effects, including the M_oS_o tidal constituent, is aimed at. Obviously the tide-free (or non-tidal) case requires that the respective Love-Shida numbers h_o, k_o, ℓ_o are known. From the above discussion it follows that this procedure is unsuitable from theoretical and practical points of view. On the other hand the practical work by institutions such as IERS and IGS is based on adapting conventional semi-diurnal Love-Shida numbers for h_o, k_o, ℓ_o, although the IERS Conventions (McCarthy, 1996) prescribe to formally use $h_o = k_o = \ell_o = 0$.

The zero crust/zero geoid concept tries to circumvent these problems by reducing the direct tidal effects for all constituents of the tidal potential V_{tid} (including the M_oS_o term), but restricting the reduction for the deformation effects to the periodic terms. In this approach, fixed in the resolutions no. 9 and 16 of the IAG at the XVIII General Assembly in 1983 (Rapp, 1983; Tscherning, 1984), the use of dubious numerical values for h_o, k_o, ℓ_o is avoided, since the M_oS_o term is only reduced for in V_{tid}, but not in the induced deformation. It should be noted that the "zero crust" is identical to the "mean crust", but the respective geoids are different; globally the difference between the mean and zero geoidal heights amounts to about 20 cm, while it is on the order of a couple of centimetres at continental scales (Ekman, 1989). Since the M_oS_o term depends only on latitude, the differences produce just a north-south slope, but leave the east-west components unchanged.

The IAG resolutions, recommending the zero crust/zero geoid concept for reductions of all geodetic measurements, have not been followed by the whole geodetic community. They have been consistently applied to terrestrial and sea gravimetric observations and to (quasi-)geoid computation; e.g. the EGG97 quasi-geoid refers to the zero-tidal system (Denker and Torge, 1997). On the other hand, some institutions producing global geopotential models from satellite tracking data have applied the tide-free concept based on a conventional value for k_o. E.g. the widely used EGM96 is based on the non-tidal system, corresponding to a non-tidal value of the coefficient C_{20} (Lemoine et al., 1998).

In the IERS Conventions 1996 (McCarthy, 1996) the zero crust/zero geoid concept is recommended, but in practice the ITRF coordinates refer to the tide-free crust instead. Since the ETRF89 is derived as a partial set from ITRF89, the coordinates of the respective sites have to be considered to refer to the non-tidal crust, too. A conventional tide-free concept, based on standard Love-Shida numbers $h_o = h_2 = 0.6078$, $k_o = k_2 = 0.30190$, $\ell_o = \ell_2 = 0.0874$, has been recommended in the draft IERS Conventions 2000 (not yet published); detailed formulae for the transformation to the "zero" system are provided.

In the European Vertical Reference Network (EUVN) the positions refer to the non-tidal system ETRS89, while the quasi-geoidal heights are taken from the zero-tidal EGG97. This procedure causes discrepancies in the form of a north-south tilt on the order of several centimetres between Catania (φ=37.5°) and Vardoe (φ=70.4°), see Mäkinen (2000) for further discussion.

6 Questions – instead of conclusions

Several aspects concerning the definition of vertical reference frames have been discussed in this paper, in particular the definition of heights in the Earth's time-dependent gravity field, the vertical datum problem, the definition of the geoid as a global height reference surface, and several concepts concerning the permanent tidal potential. Main emphasis has been put on the discussion of present problems instead of providing solutions. Thus, it is considered to be natural to conclude the paper by summarizing open questions:

1. Is there any practically relevant situation for which an absolute potential value (e.g. W_o) is needed?
2. Which combination of data is best suited to interconnect national/continental vertical datums?
3. Is there any definition of the geoid, accepted by both geodesists and oceanographers, which could be used as an international convention?
4. Is there a consistent way for the treatment of the permanent M_oS_o tidal effect which can be accepted by the different branches of Geodesy and by Oceanography?
5. Should the geoid as an equipotential surface refer to the true gravity field (partly inside the Earth's surface) or to the external, downward continued field?

These questions (among others) have to be answered and agreed upon in the geodetic community before a unique, global vertical reference frame can be implemented.

7 References

Ardalan, A, E. Grafarend, and J. Kakkuri (2002). National Height Datum, the Gauss-Listing Geoid Level Value w_o and its Time Variation $\dot{w}_0$. (Baltic Sea Level Project: Epochs 1990.8, 1993.8, 1997, 4). *J. Geod.*, 76, pp. 1-28.

Baarda, W. (1979). A Connection Between Geometric and Gravimetric Geodesy. A First Sketch. *Netherlands Geodetic Commission, Publications on Geodesy*, New Series, Vol. 6, No. 4.

Bursa, M., E. Groten, S. Kenyon, J. Kouba, K. Radej, V. Vatrt, and M. Vojtiskova (2001). Earth's Dimension Specified by Geoidal Geopotential. IAG Scientific Assembly, Sept. 2001, Budapest.

Colombo, O.L. (1980). A World Vertical Network. *Dept. of Geodetic Science, Ohio State University, Columbus, Ohio*, Rep. No. 196.

Denker, H., and W. Torge (1997). The European Gravimetric Quasigeoid EGG97 - An IAG Supported Continental Enterprise. In: Forsberg R, Feissel M, Dietrich R (Eds), *Geodesy on the Move. IAG Symposia*, 119, Springer, pp. 249-254.

Ekman, M. (1989). Impacts of Geodynamic Phenomena on Systems for Height and Gravity. *Bull. Géod.* 63, pp. 281-296.

EUREF (2000). European Vertical Reference System (EVRS). Torres JA, Hornik H (Eds), *Report on the EUREF Symposium Troms∅, June 2000. Veröff. Bay. Komm. f. Internat. Erdmessung*, Astron.-Geod. Arbeiten, Nr. 61, pp. 101-110.

Grafarend, E. (1982). Six Lectures on Geodesy and Global Geodynamics. In: *Geodesy and Global Geodynamics, Mitt. Geod. Inst. TU Graz, Folge 41.* Moritz H Sünkel H (Eds), pp. 531-685.

Grafarend, E.W., and A.A.Ardalan (1999). World Geodetic Datum 2000. *J. Geod.*, 73, pp. 611-623.

Groten, E. (1980). A Remark on M. Heikkinen's Paper „On the Honkasalo Term in Tidal Corrections to Gravimetric Observations". *Bull. Géod.*, 54, pp. 221-223.

Heck, B. (1984). Zur Bestimmung vertikaler rezenter Erdkrustenbewegungen und zeitlicher Änderungen des Schwerefeldes aus wiederholten Schweremessungen und Nivellements. *Deutsche Geodätische Kommission*, Reihe C, Heft Nr. 302, Munich.

Heck, B. (1993). Tidal corrections in geodetic height determination. In: Linkwitz K, Eisele V, Mönicke H-J (Eds), *Applications of Geodesy to Engineering. IAG Symposia*, Vol 108, Springer, pp. 11-24.

Heck, B. (1995). Rechenverfahren und Auswertemodelle der Landesvermessung. 2. Aufl., Wichmann Heidelberg.

Heck, B., and H. Mälzer (1983). Determination of Vertical Recent Crustal Movements by Levelling and Gravity Data. *Tectonophysics, 97*, pp. 251-264.

Heck, B., and R. Rummel (1990). Strategies for Solving the Vertical Datum Problem Using Terrestrial and Satellite Data. In: Sünkel H, Baker T (Eds), *Sea Surface Topography and the Geoid. Proc. of IAG Symposia.* Vol 104, Springer Berlin Heidelberg New York, pp. 116-128.

Heikkinen, M. (1979). On the Honkasalo Term in Tidal Corrections to Gravimetric Observations. *Bull. Géod.*, 53, pp. 239-245.

Heiskanen, W.A., and H. Moritz (1967). Physical Geodesy. W.H. Freeman, San Francisco.

Heitz, S. (1980). Mechanik fester Körper. Mit Anwendungen in Geodäsie, Geophysik und Astronomie. Band 1. Dümmler, Bonn.

Lemoine, F.G. et al. (1998). The Deveploment of the Joint NASA-GSFC and the NIMA Geopotential Model EGM96. NASA/TP-1998-206861.

Listing, J.B. (1873). Ueber unserer jetzige Kenntnis der Gestalt und Grösse der Erde. Nachr. Königl. Gesellschaft der Wissenschaften und der G.A. Universität zu Göttingen, No. 3, pp. 33-98.

Mäkinen, J. (2000). A Note on the Treatment of the Permanent Tidal Effect in the European Vertical Reference System (EVRS). Torres JA, Hornik H (Eds), *Report on the EUREF Symposium Troms∅, June 2000. Veröff. Bay. Komm. f. Internat. Erdmessung*, Astron.-Geod. Arbeiten, Nr. 61, pp. 111-113.

Mather, R.S. (1978). The Role of the Geoid in Four-Dimensional Geodesy. *Marine Geodesy*, 1, pp. 217-252.

McCarthy, D.D. (Ed.) (1996). IERS Conventions. IERS Technical Note 21, CB/IERS, Paris.

Melchior, P. (1983). The Tides of the Planet Earth. Pergamon Press, Oxford.

Moritz, H. (2000). Geodetic Reference System 1980. J. Geod., 74, pp. 128-133.

Poutanen, M., M. Vermeer, and J. Mäkinen (1996). The Permanent Tide in GPS Positioning. *J. Geod.*, 70, pp. 499-504.

Rapp, R.H. (1983). Tidal Gravity Computations Based on Recommendations of the Standard Earth Tide Committee. *Bull. Inf. Marés Terrestres*, 89, pp. 5814-5819.

Rapp, R.H. (1989). The Treatment of Permanent Tidal Effects in the Analysis of Satellite Altimeter Data for Sea Surface Topography. *Manuscr. Geod.*, 14, pp. 368-372.

Rapp, R.H. (1995). A World Vertical Datum Proposal. *Allgemeine Vermessungs-Nachrichten*, 102, pp. 297-304.

Rapp, R.H., and N. Balasubramania (1992). A Conceptual Formulation of a World Height System. *Dept. of Geodetic Science, Ohio State University, Columbus, Ohio*, Rep. No. 421.

Rummel, R. (2001). Global Unification of Height Systems and GOCE. In: Sideris M (Ed), *Gravity, Geoid and Geodynamics 2000. Proc. of IAG Symposia.* Vol 123, Springer Berlin Heidelberg New York, pp. 13-20.

Rummel, R., and B. Heck (2000). Some Critical Remarks on the Definition and Realization of the EVRS. Torres JA, Hornik H (Eds.), *Report on the EUREF Symposium Troms∅, June 2000. Veröff. Bay. Komm. f. Internat. Erdmessung,* Astron.-Geod. Arbeiten, Nr. 61, pp. 114-115.

Rummel, R., and K.H. Ilk (1995). Height Datum Connection – the Ocean Part. *Allgemeine Vermessungs-Nachrichten*, 102, pp. 321-330.

Rummel, R., and P. Teunissen (1988). Height Datum Definition, Height Datum Connection and the Role of the Geodetic Boundary Value Problem. *Bull. Géod.*, 62, p. 477-498.

Sanso, F., and S. Usai (1995). Height Datum and Local Geodetic Datums in the Theory of Geodetic Boundary Value Problems. *Allgemeine Vermessungs-Nachrich-ten*, 102, pp. 343-355.

Sanso, F. and G. Venuti (2002). The Height Datum/Geodetic Datum Problem. *Geoph. J. Int.*, 149, pp. 768-775.

Schneider, M. (1981). Himmelsmechanik. Bibl. Inst. Mannheim/Wien/Zürich, 2. Aufl.

Tscherning, C.C. (Ed.) (1984). The Geodesist's Handbook. *Bull. Géod.* 58, No. 3.

Wenzel, H.-G. (1997). Tide-generating Potential for the Earth. In: Wilhelm H, Zürn W, Wenzel H-G (Eds), *Tidal Phenomena.* Lecture Notes in Earth Sciences 66, Springer-Verlag Berlin Heidelberg, pp. 9-26.

Xu, P. (1992). A Quality Investigation of Global Vertical Datum Connection. *Geophys. J. Int.*, 110, pp. 361-370.

Geodetic Boundary-Value Problems and the Height Datum Problem

Fausto SACERDOTE
Dipartimento di Ingegneria Civile - Universitá di Firenze

Fernando SANSO'
DIIAR - Facoltá di Ingegneria di Como - Politecnico di Milano

Abstract. The analysis of boundary-value problems has always been used in geodesy as a frame to understand the nature of the problem of determining the gravity disturbing potential, with respect to well-posedness (existence, uniqueness, continuous dependence of the solution on boundary data). The first historical problem of this kind is Stokes's, followed by Molodensky's and other more and more sophisticated versions of boundary-value problems of mixed type, in an attempt of approaching more realistic (closer to measurements) conditions. Formulations for realistic problems have invariably displayed "nice" mathematical properties. As shown in this paper, also a realistic boundary-value problem for gravity and geopotential numbers as boundary data on continental areas and marine geoid, as derived from altimetry and stationary sea surface heights, on the oceans, can be demonstrated to have such properties for solutions lying in suitable function spaces. In order to establish realistic boundary conditions for this problem it is necessary to take into account that 1) for different continental areas that are not connected by levelling lines one must assume different reference geopotential values, and even on very large individual continents, because of error propagation in levelling, it is not possible to impose a unique reference value with the required accuracy; 2) ocean circulation models enable us to compute the sea surface topography but for an additive constant; furthermore, such models cannot be applied close to coasts, so the marine geoid cannot be used to connect different continental geoids; 3) moreover, there are oceanic regions (for example, equatorial areas) where stationary SST models do not hold, so that the marine geoid too has to be computed separately for a number of different patches that cannot be directly connected. Consequently, a number of different unknown constants must be introduced in the boundary conditions, and correspondingly suitable constraints have to be imposed. The type of unknowns to account for different height datums and the correspondent conditions to be imposed to the solutions are thoroughly discussed.

1 Introduction

The role of height datum definition and its connection with the solutions of geodetic boundary-value problems has been long recognized and discussed among geodesists; see for example (Balasubramania 1994), (van Onselen 1997), (Rummel and Teunissen 1988), (Sansò and Usai 1995), (Xu and Rummel 1991). Indeed, using gravity measurements and levelling, only potential differences can be obtained, whereas the absolute value of the geopotential cannot be obtained at any point with acceptable accuracy; consequently, the boundary values of the geopotential must be assumed to be known but for an additive constant. Furthermore, if two different regions on the boundary surface cannot be connected by levelling networks, height and geopotential difference between them cannot be measured, and it is not possible to introduce a unique unknown constant valid for both.

In the linearized formulations of geodetic boundary-value problems a reference potential and a reference surface are introduced; yet, the reference surface cannot be defined assuming the reference potential on it to be exactly equal to the geopotential on the earth's surface, but a small unknown constant gap δW must be introduced. The determination of this constant is particularly simple for Stokes's problem (spherical approximation, spherical boundary; see Heiskanen and Moritz 1967, sec.2-19):

$$\begin{cases} \nabla^2 T = 0 & \text{in } \Omega \\ -\frac{\partial T}{\partial r} - \frac{2}{R}T + \frac{2}{R}\delta W = \widetilde{\Delta g} & \text{on } S \\ T = 0(r^{-3}) \end{cases} \quad (1)$$

where S is the boundary and Ω is the exterior domain.

For simplicity, it is assumed that $\widetilde{\Delta g}$ does not contain first degree harmonic components, so that the problem is solvable without further modifications. Owing to the asymptotic condition, it is required that the solution T does not contain the zero-degree harmonic term; consequently, the same must be true for $\widetilde{\Delta g} - \frac{2}{R}\delta W$ in the boundary condition. Hence

$$\frac{1}{4\pi}\int_{S_1} \widetilde{\Delta g} d\sigma = \frac{2}{R}\delta W \qquad (2)$$

where S_1 is the unit sphere. This determines uniquely the value of δW. It is well-known that, in order to obtain a disturbing potential T with vanishing zero-degree harmonic coefficient, it is necessary to assume the mass of the level ellipsoid to be equal to the mass of the earth. More generally, once the earth's mass is known, in order to define a level ellipsoid and its gravity potential, it is necessary to fix the mass of the reference normal potential too; the mass difference between earth and ellipsoid determines the zero-degree harmonic coefficient of the disturbing potential, and consequently the value of δW.

Stokes's problem, as it is well-known, requires in practice the knowledge of reduced gravity anomalies, and consequently gravity and levelling measurements all over the boundary surface. If the boundary is divided into a number of different regions altimetrically disconnected, different unknown constants must be introduced into the boundary conditions, and the constraint imposed by the knowledge of the total mass is no more sufficient to determine them and solve completely the problem.

With the proliferation of precise determinations of ellipsoidal heights by GPS measurements, it has become clear that their role is essential to the unification of regional height datums. Indeed, in the framework of a simple Stokes scheme, using a sort of generalized Stokes's formula containing all unknown regional additive constants, it has been shown (Rummel and Teunissen 1988), (Heck and Rummel 1990) that these constants can be determined algebraically if, in addition to the usual gravity anomaly data and to orthometric heights referred to regional height datums, ellipsoidal heights are determined at a point for each different region at least. The limitation of this approach is in that it is purely algebraic, based on classical Stokes's formula, and does not take into account the developments of the theory of geodetic boundary value problems within the frame of modern functional analysis. Indeed, it will be proved in the sequel that such developments may have consequences on the practical requirements about the correct formulation of the constraints to be imposed in order to obtain a unique solution.

2 A review on altimetry-gravimetry problem 1 and on the functional properties of its solution

The main reason for which Stokes's formulation, and even Molodensky's formulation, cannot be accepted nowadays as a realistic frame for geoid computation is that, actually, modern methods for geoid computation based on gravimetric data make use of a global geopotential model, that is determined from satellite altimetry data on the oceans and gravity anomalies on the continents; therefore, within the frame of boundary-value problem theory, the geoid determination problem is today more suitably classified as a mixed altimetry-gravimetry problem.

Geodetic mixed boundary-value problems have been extensively studied in the eighties. In particular, for the so called Altimetry-Gravimetry Problem 1 (AGP1), whose expression in spherical approximation, is:

$$\begin{cases} \nabla^2 T = 0 & \text{in } \Omega \\ T = f_1 & \text{on } S_S \text{ (sea)} \\ -\frac{\partial T}{\partial r} - \frac{2}{r}T = f_2 & \text{on } S_L \text{ (land)} \\ T = 0(r^{-1}) & \end{cases} \qquad (3)$$

Svensson (1985) proved analytically that, for particular distributions of the continents, the solution, defined in a suitable function space, is not unique. This fact is confirmed by a numerical analysis carried out with a simple model of a continental area covering a spherical cap with varying radius; a radius was found for which uniqueness failed. Sacerdote and Sansò (1987) showed that it is possible to obtain uniqueness introducing an unknown constant in the boundary condition at sea, and adding a constraint to the boundary values of T:

$$\begin{cases} \nabla^2 T = 0 & \text{in } \Omega \\ T = f_1 + a & \text{on } S_S \\ -\frac{\partial T}{\partial r} - \frac{2}{r}T = f_2 & \text{on } S_L \\ T = 0(r^{-2}) \Rightarrow \int_S T d\sigma = 0 & \end{cases} \qquad (4)$$

Such a modified formulation cannot be interpreted as due to the introduction of an additive unknown constant in the potential that would affect both boundary conditions, and was physically justified assuming the altimetric measurements to be affected by a constant bias. This approach, in view of the fact that in recent years the calibration problem of satellite altimetry has been reduced to the level of very few centimetres, has been subsequently criticized by R.Lehmann (2000), who was rather interested in the correct formulation of the problem when the potential is assumed to be known but for an additive constant.

Before discussing the results of Lehmann's work and of further developments, it is useful to introduce some remarks on the functional properties of the solution, which were illustrated in Sacerdote and Sansò (1987) and can be extended to the new formulations.

In problem (4) f_1 is assumed to be the restriction to S_S of a function in $H^{1/2}(S)$, f_2 the restriction to S_L of a function in $H^{-1/2}(S)$. The solution T is proved to belong to $H^1(\Omega)$, so that its trace on the boundary belongs to $H^{1/2}(S)$. The relevant fact is that, by Sobolev's embedding theorem, $T|_S$ is not a continuous function on S, and the evaluation functional $L_P : T \to T(P)$ is not bounded in $H^{1/2}(S)$. Therefore the introduction of pointwise constraints on $T|_S$ is not admissible. In order to make pointwise conditions meaningful, it is necessary to regularize the solution introducing smoother boundary conditions. This step was actually carried out by Sacerdote and Sansò, who proved that it is not sufficient to assume that f_1 and f_2 individually belong to subspaces of smoother functions, but it is necessary to impose constraints across borders between S_S and S_L, i.e. coastlines. Indeed, under suitable conditions of this kind, it is proved that a solution T with $T|_S \in H^{3/2}(S)$ can be obtained. Functions in $H^{3/2}(S)$ are continuous on S and the evaluation functional is bounded, so that no difficulty arises in imposing pointwise conditions, as for example the results of GPS measurements.

3 The introduction of the unknown height datum into AGP1 and its relation with observations

As mentioned in the previous section, a new formulation of AGP1 with the introduction of an additive unknown constant to the potential was given and discussed by Lehmann (2000):

$$\begin{cases} \nabla^2 T = 0 & \text{in } \Omega \\ T = f_1 + \delta W & \text{on } S_S \\ -\frac{\partial T}{\partial r} - \frac{2}{r}T = f_2 - \frac{2}{r}\delta W & \text{on } S_L \\ T = 0(r^{-2}) \end{cases} \tag{5}$$

He showed that problem (5) may not have a unique solution for certain distributions of the continental surface. More precisely, he investigated the problem with numerical methods for an axially symmetric configuration of the boundary, and found that, similarly to the case discussed by Svensson, a singularity occurs in case the continents cover a spherical cap with a certain radius. An analytical proof of the same result, based on the application of Fredholm's alternative and on the investigation of the corresponding homogeneous problem, along the same line followed by Svensson, was given by Sacerdote and Sansò (2002).

The physical assumption underlying problem (5) is that both terrestrial and oceanic observations, i.e. gravity anomaly measurements and levelling on land, satellite altimetry and oceanographic measurements, in addition to reliable circulation models, on sea, enable us to determine the disturbing potential but for a unique additive constant. Indeed, on land potential differences $W(P) - W_0$ are known, where, for instance, W_0 could be considered as the actual gravity potential at a reference tide gauge point; furthermore, the modulus g of the gravity vector is known everywhere. On sea, on the other hand, from a good circulation model the height anomaly $\zeta(P)$ is known with respect to some equipotential reference surface.

As a matter of fact, the simplest and universally accepted stationary model for the ocean is what is called the geostrophic balance model, claiming that

$$\nabla\zeta = \frac{2\Omega \sin\phi}{g}\mathbf{n} \times \mathbf{v} \tag{6}$$

where $\mathbf{n}$ is the local vertical, $\mathbf{v}$ is is the horizontal velocity of the geostrophic current, Ω is the earth's rotation angular rate and ϕ is the latitude. The idea is that a circulation model can provide $\mathbf{v}$ so that (6) can be integrated and ζ obtained but for a constant.

Model (6) is actually known to give a satisfactory approximation of mid-ocean circulation; certainly it does not hold on the western boundary - cf (Pedlosky 1998), (Wunsch 1993). More generally, no universally accepted oceanographic model is available to describe in detail the effects of stationary currents along irregular coasts. It follows then that on the part of ocean described by (6) it can be asserted that, for any two of points,

$$W_1 + \gamma_1\zeta_1 = W_2 + \gamma_2\zeta_2 = \bar{W} \tag{7}$$

and consequently

$$W_1 - W_2 = \gamma_2\zeta_2 - \gamma_1\zeta_1 \simeq \bar{\gamma}(\zeta_2 - \zeta_1) \tag{8}$$

or, in other words, the potential is known but for an unknown additive constant. If such a model could be extended to the tide gauge where the potential reference point for land is located, it could be asserted that one and the same W_0 could be the only unknown constant. The linearization of such a problem is exactly what is given by (4). Nevertheless, if such a connection is not established, the problem should rather be formulated as

$$\begin{cases} \nabla^2 T = 0 & \text{in } \Omega \\ T = f_1 + \delta W_1 & \text{on } S_S \\ -\frac{\partial T}{\partial r} - \frac{2}{r}T = f_2 - \frac{2}{r}\delta W_2 & \text{on } S_L \\ \int_S T d\sigma = 0 \end{cases} \tag{9}$$

Having introduced two unknown constants, however, one should be working with two constraints. So, in addition to $\int_S T d\sigma = 0$, expressing that normal and actual potential are generated by the same mass M, another constraint has to be introduced, which can be chosen to be

$$\int_{S_S} T d\sigma = 0 \tag{10}$$

meaning that the mean equipotential surface interpolating the actual surface of the ocean (given by altimetry) is chosen as reference in the linearization process (Sacerdote and Sansò 2001). In other words, it is acceptable to have different height datums on the oceans and on the continents, provided that in the frame of boundary-value problem theory it is possible to establish a well defined relation between them.

More generally, it must be remarked that

- on land, height datums of different continents are disconnected; furthermore, because of levelling uncertainties over very long profiles, a unique height datum for very large continents cannot be defined with the required accuracy;

- on sea, circulation models cannot be extended to narrow channels between different oceans, and, moreover, instabilities occurring in the equatorial belt do not allow us to connect circulation models of the northern and the southern hemispheres.

Therefore, even the introduction of a unique unknown constant for all continents and of a different unique constant for all oceans provides an unrealistic model for the mixed BVP, and the insertion of a different constant for each disconnected patch of the boundary seems to be more acceptable. Obviously, a corresponding constraint must be added for each unknown constant. Consequently, the correct formulation of the problem resulting from these assumptions is

$$\begin{cases} \nabla^2 T = 0 & \text{in } \Omega \\ T = f_i + \delta W_i & \text{on } S_{S_i} \\ -\frac{\partial T}{\partial r} - \frac{2}{r}T = g_j - \frac{2}{r}\delta W_j & \text{on } S_{L_j} \\ \int_{S_{S_i}} T d\sigma = a_i \, , \ \int_{S_{L_j}} T d\sigma = b_j \end{cases} \tag{11}$$

with the additional constraint

$$\sum a_i + \sum b_j = 0 \Rightarrow T_{00} \equiv \int_S T d\sigma = 0 \tag{12}$$

The proof of the well-posedness of the problem given by eqs. (11) and (12) for a spherical boundary of radius R is very simple; it relies again on Fredholm alternative and on the uniqueness of the solution of the corresponding homogeneous problem:

$$\begin{cases} \nabla^2 T = 0 & \text{in } \Omega \\ T = \delta W_i & \text{on } S_{S_i} \\ -\frac{\partial T}{\partial r} - \frac{2}{r}T = -\frac{2}{r}\delta W_j & \text{on } S_{L_j} \\ \int_{S_{S_i}} T d\sigma = 0 \, , \ \int_{S_{L_j}} T d\sigma = 0 \end{cases} \tag{13}$$

Integrating the boundary values over the spherical boundary surface, and using spherical harmonic expansions yields

$$\frac{1}{4\pi}\int_S -\Big(\frac{\partial T}{\partial r} + \frac{2}{R}T\Big)T d\sigma = \sum \frac{n-1}{R} T_{nm}^2 \tag{14}$$

On the other hand, introducing boundary conditions and constraints from (13) one obtains

$$\begin{aligned} &\tfrac{1}{4\pi}\int_S -\Big(\tfrac{\partial T}{\partial r} + \tfrac{2}{R}T\Big)T d\sigma = \\ &= -\sum_i \tfrac{\delta W_i}{4\pi}\int_{S_{S_i}}\Big(\tfrac{\partial T}{\partial r} + \tfrac{2}{R}T\Big)d\sigma + \\ &-\tfrac{2}{R}\sum_j \tfrac{\delta W_j}{4\pi}\int_{S_{L_j}} T d\sigma = \\ &= -\sum_i \tfrac{\delta W_i}{4\pi}\int_{S_{S_i}} \tfrac{\partial T}{\partial r} d\sigma \end{aligned} \tag{15}$$

From (13) it can be immediately inferred that

$$0 = \int_{S_{S_i}} T d\sigma = \delta W_i \mu(S_{S_i}) \tag{16}$$

where μ indicates the Euclidean measure of the domain, which is obviously a positive quantity. From (16) one can conclude that $\delta W_i = 0$ for all i. Consequently the last term in (15) vanishes, and from (14) and (15) it follows that $\sum \frac{n-1}{R} T_{nm}^2 = 0$. Furthermore from (13) $T_{00} = \int_S T d\sigma = 0$. Hence $\sum \frac{n-1}{R} T_{nm}^2$ must be strictly positive, unless all T_{nm} with $n \neq 1$ vanish. On the other hand no first-degree harmonic function can vanish identically on a portion of a spherical surface, and consequently the coefficients T_{1m} must vanish too. Finally, δW_j must vanish as well for all j as a consequence of the third equation in (13).

4 Concluding remarks

The formulation of the boundary-value problem for the disturbing potential given in the previous sections seems to be quite reasonable and close to the requirements imposed by the actual physical nature of the available data. This formulation suggests the following remarks:

- there is no problem in breaking data into patches, as far as the physics of the problem requires it and the regularity level that can be obtained for the solution (namely $H^{1/2}(\sigma)$) is correctly taken into account;

- this implies that pointwise data (e.g. GPS + levelling) cannot be consistently used to compute the unknown constants, and rather average conditions (e.g. averages over a large number of GPS+levelling points) should be used;

- the reference surface so identified is as a matter of fact a broken surface, but of course this does not mean that T has sharp discontinuities (incompatible with $T \in H^{1/2}(\sigma)$), but just that a broken surface has been used to reduce the data and formulate a unique BVP;

- of course the solution of the problem will provide the unknown constants that, when applied, will unify the datum wordlwide.

References

Balasubramania, N. (1994) - *Definition and Realisation of a Global Vertical Datum*, OSU Report **427**.

Heck, B. and R.Rummel (1990) - *Strategies for solving the vertical datum problem using terrestrial and satellite geodetic data*, in Sea Surface Topography and the Geoid, IAG Symposia **104**, 116-128, Springer Verlag.

Heiskanen, W. and H.Moritz (1967) - *Physical Geodesy*, Freeman and Co.

Lehmann, R. (2000) - *Altimetry-Gravimetry Problems with Free Vertical Datum*, Journal of Geodesy, **74**, 327-334.

van Onselen, K. (1997) - Quality investigation of vertical datum connection, DEOS Rep. **97**.3, TU Delft.

Pedlosky, J. (1998) - *Ocean Circulation Theory*, Springer.

Rummel, R. and P.Teunissen (1988) - *Height datum definition, height datum connection and the role of geodetic boundary value problem*, Bull. Géod., **62**, 477-498.

Sacerdote, F. and F.Sansò (1987) - *Further Remarks on the Altimetry-Gravimetry Problems*, Bull.Géod., **61**, 65-82.

Sacerdote, F. and F.Sansò (2001) - *W_0 : A story of the height datum problem*, in Festschrift W.Torge, Wissenschaftliche Arbeiten der Fachrichtung Vermessungswesen der Universität Hannover, **241**, 49-56.

Sacerdote, F. and F.Sansò (2002) - *Remarks on the role of height datum in altimetry-gravimetry boundary-value problems*, to appear in Proceedings of the Workshop "Earth Gravity from Space - from Sensors to Earth Sciences", Bern.

Sansò, F. and S.Usai (1995) - *Height datum and local geodetic datums in the theory of geodetic boundary value problems*, AVN, **102**, 343-355.

Svensson, L. (1985) - *Some Remarks on the Altimetry-Gravimetry Problem*, in Proc. of the I Hotine-Marussi Symposium on Math. Geodesy, Rome, 559-582.

Wunsch, C. (1993) - *Physics of the Ocean Circulation, in Satellite Altimetry in Geodesy and Oceanography*, R.Rummel and F.Sansò eds., Lecture Notes in Earth Sciences, **50**, Springer-Verlag, 9-98.

Xu, P. and R.Rummel (1991) - *A Quality Investigation of Global Vertical Datum Connection*, Netherland Geodetic Commission, Publications on Geodesy, New Series, **34**.

GPS Permanent Network Solution: the Impact of Temporal Correlations

R. Barzaghi, A. Borghi
DIIAR – Sezione Rilevamento, Politecnico di Milano, p.zza Leonardo da Vinci, 32 - 20133 Milano, Italy

M. Crespi
DITS – Area Topografia, Università di Roma "La Sapienza", via Eudossiana, 18 - 00184 Roma, Italy

G. Pietrantonio, F. Riguzzi
Istituto Nazionale di Geofisica e Vulcanologia, via di Vigna Murata, 605 - 00143 Roma, Italy

Abstract. In this work we analysed the time series of daily solutions of 4 Italian GPS permanent stations with the aim of investigating the presence of temporal correlations and their impact on the estimation of weekly solution and velocity field precisions.

We found that precisions are remarkably lower when temporal correlations are considered; in particular, the mean horizontal precisions of weekly solutions are up to 5 times lower and the horizontal velocity precisions are about 1.5-2 times lower.

This topic has 2 relevant applications: the assessment of the quality of a reference system maintenance by GPS permanent stations and the coordinate differences significance test for geodynamical applications.

Keywords. GPS permanent network, temporal correlations, autocovariance functions

1 Introduction

The correct estimation of GPS-derived coordinate and velocity precisions is of fundamental importance when GPS permanent stations are used to realize and maintain a reference system and to infer geodynamical interpretations.

Despite of the fact that several authors evidenced temporal correlations in the GPS coordinate time series of the daily solutions, in the standard processing procedures for GPS permanent networks weekly solutions and site velocities are estimated by adjusting daily solutions without taking into account the temporal correlations themselves. This fact leads to an overestimation of parameter precisions when daily solutions are least squares adjusted in order to estimate "public" weekly solutions and site velocities.

In the present work, analysing the time series of 4 Italian GPS permanent stations belonging to the IGS global network, we found, as expected, that temporal correlations are not negligible at all.

Without analysing, for the moment, their causes, we just tried to introduce as correctly as possible, but with some limitations, these correlations in the solution of our small "pilot" GPS network (see Figure 1) with the aim of discovering how weekly solution and velocity precisions change when a more correct stochastic model is used.

2 Data set and preliminary processing

We considered GPS data from the IGS permanent stations of Cagliari (CAGL), Matera (MATE), Medicina (MEDI) and Noto (NOTO) collected during a period of 3 years, from the beginning of 1998 to the end of 2000.

During such a period, processing strategy changes and software updating may have occurred and this fact may introduce spurious correlations in daily solution time series. For this reason we completely reprocessed the GPS observations, using the BERNESE software - version 4.2 (Beutler et al., 2000), in order to obtain long enough and homogeneously derived time series, useful for our aims.

We estimated in the processing daily coordinates of MATE, MEDI and NOTO, minimally constrained to CAGL ITRF97 positions.

We followed the recommended strategy for long baselines (about 600 km), estimating troposphere delay parameters every 2 hours and one troposphere horizontal gradient parameter per day, using the Dry-Neill mapping function and considering a cut-off angle of 10 degrees.

The time series of the daily coordinates of MATE, MEDI, NOTO evidenced some outliers that were rejected applying a 3 σ test on the overall time series, previously linear detrended.

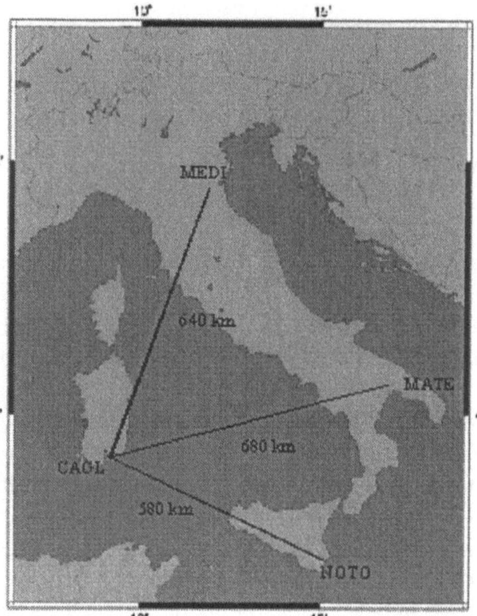

Fig. 1 The "pilot" network

most correct fitting function among eight different autocovariance function models:

1. $y=a \exp(-b|x|)$ exponential
2. $y=a \exp(-bx^2)$ normal
3. $y=a\,(1-cx^2) \exp(-b|x|)$ exp-parable
4. $y=a\,(1-cx^2) \exp(-bx^2)$ normal-parable
5. $y=a \exp(-b|x|) \cos(cx)$ exp-cos
6. $y=a \exp(-bx^2) \cos(cx)$ normal-cos
7. $y=a \exp(-b|x|) \sin(c|x|)/c|x|$ exp-sin x/x
8. $y=a \exp(-bx^2) \sin(c|x|)/c|x|$ normal-sin x/x

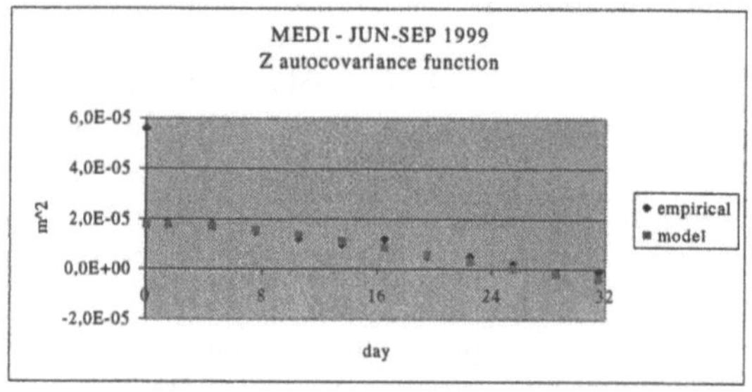

Fig. 2 An example of empirical autocovariance function fitting performed by INTERCOVA software

3 Covariance functions computation

We may assume that, after linear detrending and outlier removal, the time series of (X,Y,Z) station coordinates behave as a second order stationary process and, as a consequence, their autocovariance functions depend on the time span only.

We computed the empirical covariance functions, for the whole analysed period of 3 years, on time intervals of 4 months, by monthly steps (January – April, February – May and so on). Four months are a "right" time interval in order to have more than 100 data, needed for a consistent covariance estimation.

In most cases, we found significant autocovariances generally with a length of 1-2 weeks. Since we were interested in evaluating weekly solutions, we introduced these autocovariances in the stochastic model of our adjustment, whilst, for the moment, we decided to neglect cross-covariances.

Then, when the signal variance was significant with respect to the noise variance, we fitted the empirical autocovariance functions with proper model functions (see Figure 2) by implementing a suitable software (INTERCOVA) able to find the

4 Weekly solutions and site velocities estimation

The focal point of weekly solutions estimation is the correct accounting of the modelled autocovariance functions in the stochastic model of the adjustment (as already said, in the present work cross-covariances are not taken into account).

To evaluate the impact of such autocovariances, we performed also the "standard" adjustment, where daily solutions are assumed as temporally uncorrelated.

In this second case, the covariance matrix representing the stochastic model of the adjustment does have a simple block diagonal structure, where each block is the covariance matrix of one daily solution, obtained as output of daily BERNESE processing (see Figure 3).

When autocorrelations among the daily solutions are taken into account, the stochastic model is quite different. In this case, in fact, the diagonal blocks are represented by the daily BERNESE correlation matrices scaled with the variance of the process; in particular, in the hypothesis of uncorrelated noise, the diagonal elements are scaled with the total variance (signal plus noise), while the extra-diagonal elements are scaled just with the signal variance.

Moreover, in this case, also extra-diagonal blocks appear, containing the model autocovariance functions evaluated at different time intervals (see Figure 4).

For each weekly adjustment, we used the autocovariance functions attaining the 4-months' interval with respect to which the corresponding week is most central.

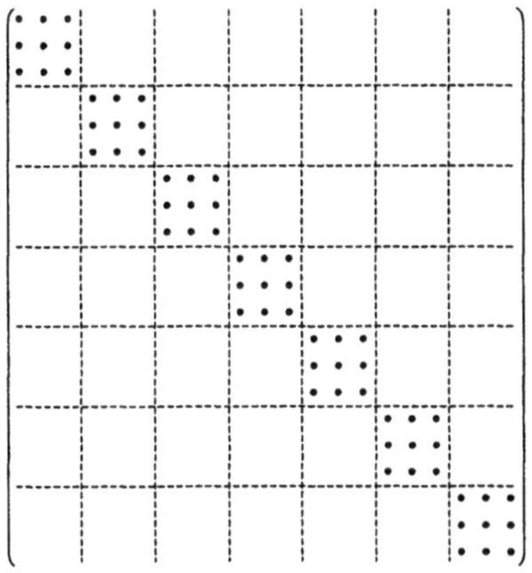

Fig 3 Stochastic model of weekly solution adjustment when autocovariances are neglected

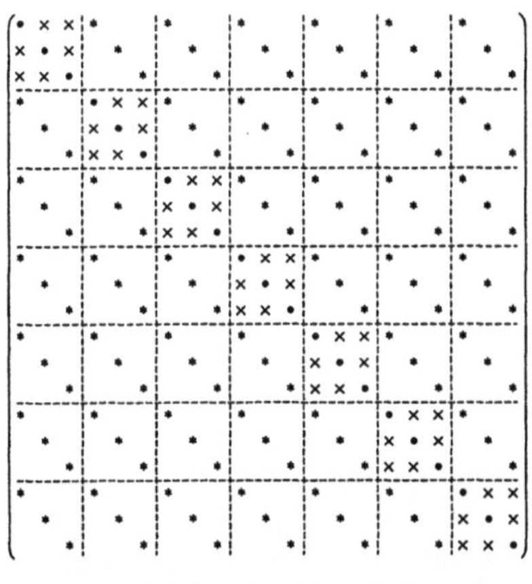

Fig 4 Stochastic model of weekly solution adjustment when autocovariances are considered (cross-covariances are neglected)

Now we focus our attention to the results obtained for the year 1999.

Tables 1 and 2 show weekly solution mean precisions, respectively for the station of NOTO and MEDI for the year 1999: when temporal autocorrelations are considered, the mean horizontal precisions are remarkably lower, up to 5 times for MEDI, while there are no remarkable differences in the vertical ones.

Table 1. Weekly solution mean precisions (year 1999) for NOTO station

Mean 1 σ (mm)	With autocovariances	Without autocovariances
N	4	1
E	3	1
h	4	4

Table 2. Weekly solution mean precisions (year 1999) for MEDI station

Mean 1 σ (mm)	With autocovariances	Without autocovariances
N	5	1
E	4	1
h	5	5

To evaluate the impact of temporal autocorrelations also on velocities estimation, we least squares adjusted all the weekly solutions obtained in both cases, with and without autocovariance introduction.

The results obtained (Tables 3 and 4) show two relevant aspects: first of all, if temporal correlations are completely neglected, the chi-square test on the linear model fails, confirming that in this case the horizontal coordinate precisions are too optimistic and then the stochastic model is inadequate; moreover, if temporal correlations are neglected, horizontal velocity precisions are about 1.5-2 times overestimated.

Table 3. NOTO velocities (year 1999)

WITH AUTOCOVARIANCES:		
Chi-square test ok: $\sigma_0=1$, $\hat{\sigma}_0=0.9$		
V_N (mm/y)	V_E (mm/y)	V_h (mm/y)
12.2±1.2	18.8±1.0	-2.0±1.4
WITHOUT AUTOCOVARIANCES:		
Chi-square test fails: $\sigma_0=1$, $\hat{\sigma}_0=2.1$		
V_N (mm/y)	V_E (mm/y)	V_h (mm/y)
12.7±0.7	19.1±0.6	-1.3±2.9

Table 4. MEDI velocities (year 1999)

WITH AUTOCOVARIANCES:		
Chi-square test ok: $\sigma_0=1$, $\hat{\sigma}_0=0.9$		
V_N (mm/y)	V_E (mm/y)	V_h (mm/y)
20.8±1.4	16.1±1.3	-5.8±1.5
WITHOUT AUTOCOVARIANCES:		
Chi-square test fails: $\sigma_0=1$, $\hat{\sigma}_0=2.7$		
V_N (mm/y)	V_E (mm/y)	V_h (mm/y)
20.8±1.1	17.0±0.7	-4.0±3.7

5 Receiver/antenna change in Matera

In June 1999 (doy 169) a receiver/antenna apparatus change occurred in MATE station.

Time series of daily solutions reveal this change by the presence of a jump, clearly evident in the horizontal components, but not evident in the vertical one, probably because of the greater noise (see Figure 5).

Moreover, the autocovariance functions attaining to time intervals containing receiver/antenna change, show some inconsistencies: in these cases, in fact, the signal variance is almost equal to the total variance of the stochastic process (see Figure 6).

This fact seems to prove that the analysis of solution time series is useful also to assess the data quality and able to highlight eventual anomalous behaviours of GPS permanent stations.

6 Conclusions and future perspectives

The "basic" goal of the present work was the analysis of the impact of temporal autocorrelations among GPS daily solutions on weekly solutions and velocity field estimation.

We considered a small "pilot" network of 4 Italian GPS permanent stations, reprocessed the data of three years with BERNESE software, computed the autocovariance functions of the daily solutions obtained, implemented a new software for autocovariance functions modelling and introduced them in the adjustment.

As expected, we found that when temporal correlations are completely neglected, as in the standard GPS permanent networks processing strategy, horizontal coordinate and velocity precisions are overestimated and the stochastic model results inadequate.

Moreover, we established the efficacy of GPS time series analysis to check data quality and highlight eventual anomalous behaviours of permanent stations (like in the case of receiver/antenna apparatus change in MATE station).

It is evident that this is just a preliminary work, in which cross-correlations among daily solutions were completely neglected. Nevertheless, we showed that the simple introduction of autocorrelations dramatically changes GPS-derived precisions and this fact has clearly relevant geodynamical consequences.

The results obtained encourage us to investigate also the impact of correlations in multibase solutions, what requires a deep investigation on the cross-covariance functions modelling from the empirical data.

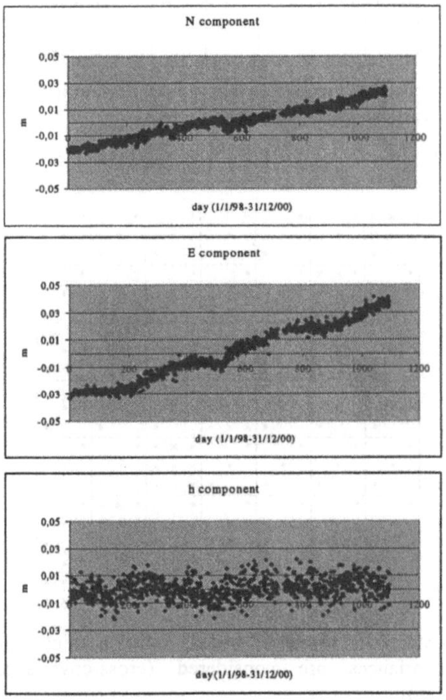

Fig. 5 MATE coordinate time series

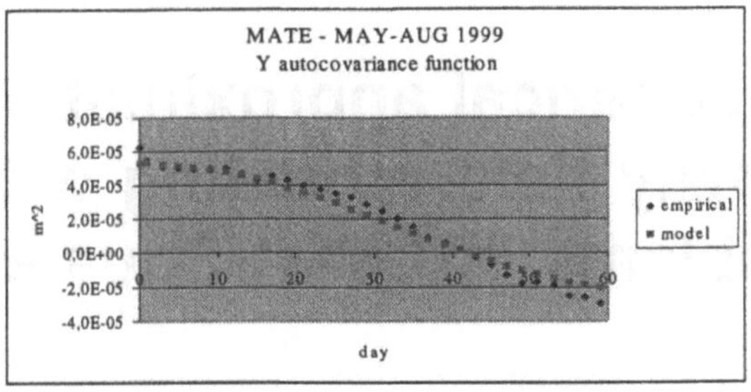

Fig. 6 Example of MATE autocovariance function attaining to a time interval containing receiver/antenna change

Acknowledgements

The authors wish to thank very much Giuseppe Colucci from the Matera Space Geodesy Center of the Italian Space Agency, who supplied GPS data used in this work.

References

Barzaghi, R. and Sansò, F. (1983). *Sulla Stima Empirica della Funzione di Covarianza.* Bollettino di Geodesia e Scienze Affini, n.4.

Barzaghi, R., Borghi, A., Sona, G. (1998). *New Covariance Models for Local Applications of Collocation.* IAG Symposium no.122, IV Hotine-Marussi Symposium on Mathematical Geodesy, Trento, Italy, B. Benciolini edition

Beutler, G., Brockmann, E., Dach, R., Fridez, P., Gurtner, W., Hugentobler, U., Johnson, J., Mervant, L., Rothacher, M., Schaer, S., Springer, T., Weber, R. (2000). *Bernese Software 4.2.* Astronomical Institute, University of Berne.

Crippa, B. (1990). *Un Sistema di Programmi.* Ricerche di Geodesia, Topografia e Fotogrammetria, vol.6, CLUP, Milano.

Mussio, L. (1984). *Il Metodo della Collocazione Minimi Quadrati e le sue Applicazioni per l'Analisi Statistica dei Risultati delle Compensazioni.* Ricerche di Geodesia, Topografia e Fotogrammetria, vol.4, CLUP, Milano.

A discussion of the use of spherical approximation or no approximation in gravity field modeling with emphasis on unsolved problems in Least-Squares Collocation.

C.C.Tscherning
Department of Geophysics, University of Copenhagen, Juliane Maries Vej 30, DK-2100 Copenhagen Ø, Denmark

Abstract: Spherical approximation is widely used in gravity field modeling, but it causes errors of the order of the flattening. Different methods for avoiding using this approximation when linearizing observation functionals are possible due to our improved knowledge of the gravity field and of the surface of the Earth. However a number of problems are not solved today, such as those related to the use of a Bjerhammar sphere in Least-Squares Collocation (LSC). If this concept is used, the associated homogeneous /isotropic kernels in general use will be strongly non-homogeneous on the ellipsoid. A possible solution is to use a "Bjerhammar" ellipsoid inside the Earth.

1 Introduction

Spherical approximation is widely used in gravity field modeling. The best example is the use of the formula

$$\Delta g = g(P) - \gamma(Q) = -\frac{dT}{dn} - \frac{\partial \gamma}{\partial n}\frac{T}{\gamma} \approx -\frac{dT}{dr} - \frac{2}{r}T \qquad (1)$$

where Δg is the gravity anomaly, g is gravity measured in the point P and γ is normal gravity in a point Q on the same ellipsoidal normal as P and where the ellipsoidal height of Q is equal to the orthometric height of P. T is the anomalous gravity potential and r is distance to the origin. The derivative in the direction of the gravity vector, $\frac{\partial}{\partial n}$ is substituted by the radial derivative, the normal gravity is substituted by GM/r^2 and the derivative of gravity with $-2*GM/r^3$ see e.g. Heiskanen and Moritz (1967, section 2-14).

This equation is also the basis for the use of the spherical Stokes equation, solving the boundary problem for gravity data Δg^* on a mean earth sphere with radius R,

$$T(P) = \frac{R}{4\pi}\int_\sigma S(\psi, r)\Delta g^* d\sigma \qquad (2)$$

where S is Stokes kernel, r is the radial distance of P from the origin and ψ is the spherical distance between P and the integration point on the mean earth surface σ.

Note that there are two approximations in use: the first relating the gravity anomaly to the anomalous potential and the second when using the spherical Stokes equation. A further, and most severe problem, is naturally that we have to estimate gravity anomalies Δg^* on the bounding sphere.

The use of these approximations causes errors at least of the order of the flattening. For local applications this seems to be admissible especially if so-called remove-restore methods are used, see e.g. Torge (2001, p. 286).

Earlier, spherical approximation was used when computing spherical harmonic coefficients from mean surface gravity anomalies. Here the use of an ellipsoidal surface of integration and of ellipsoidal harmonic functions has removed most of the error.

However when treating data from the new gravity missions, where data of very high precision is collected and used, we must try to avoid any approximation. In the following we will discuss how we in many cases may be able to avoid using spherical or even ellipsoidal approximation.

The use of better approximations of the observation functionals due to an improved knowledge of positions of points and of the surface of the Earth are discussed in section 2 and 3.

Spherical approximation is also used in the sense that the surface to which a boundary-value problem is referenced is a sphere. Here, the use of an ellipsoidal surface or of the true Earth surface (Meissl, 1981) will possibly give improvements, which are not discussed here.

The use of an ellipsoidal boundary leads to rather complicated numerical procedures, which makes the use of certain methods like least-squares collocation an extremely heavy task. This is discussed in section 4 and 5.

In this paper, I present problems to which I do not have a solution. Other current problems, which I consider important, are listed at http://www.gfy.ku.dk/~cct/topics_for_research.htm

2 Using better approximations for the observation functionals

Spherical approximation is characterized by two approximations: The position of a point in geodetic coordinates, geocentric latitude $\bar{\varphi}$, longitude λ and radial distance r, is put equal to geodetic latitude φ, longitude (unchanged) λ, and r equal to R+h, where R is the mean radius of the Earth (6371 km) and h is the ellipsoidal height. Frequently, one further approximation is made, where h is put equal to the orthometric or normal height.

A second approximation is that the reference ellipsoid is replaced by a sphere with radius R as described in the introduction. Furthermore, observation functionals are used in spherical approximation. Consequently, several approximations are used, so the usual estimate of the error as being of the order of the flattening may not be correct, see e.g. Sansò and Tscherning (2002).

The use of spherical approximation was appropriate earlier, because many other errors (due to data distribution and quality) were also playing a role. But today with modern computers, there is no excuse for using spherical approximation if it is not an absolute necessity. Also, the surface of the Earth is now well known (due to InSAR), and so is the direction of the vertical.

A first improvement is to use "ellipsoidal" approximation; see e.g. Ardalan and Grafarend (2001). We may however get even closer to using no approximation at all.

Using a high degree and order gravity model, contingently enhanced using a regional model, it is possible to come close to making no approximation at all. The gravity vector calculated from the gravity field model must be used when linearizing observation functionals like those associated with the gravity anomaly and the deflection of the vertical. The ellipsoidal height of a point of evaluation must be calculated as the sum of the orthometric height plus the height anomaly calculated using the model. The main point is that the anomalous potential T is now the potential W minus a spherical harmonic model contingently enhanced with a regional model based on local data and topographic information. The linearized observation equation for a gravity anomaly, eq. (1) then becomes

$$\Delta g = -\frac{\partial T}{\partial n} - \frac{1}{\gamma}\frac{\partial \gamma}{\partial n}T \qquad (3)$$

where n is the direction of the gravity vector computed from the model and γ is the magnitude of this gravity vector. The point of evaluation of the partial derivatives and of the gravity vector is a point with altitude computed as the sum of the orthometric height and the height anomaly computed from the model.

3 Consequences for mean anomalies

Mean gravity and mean height anomalies are generally computed from a set of points interpolated to form a regular grid or calculated using least-squares collocation. The related functional has generally been associated with a surface parallel to the mean earth sphere at the mean orthometric height of the block over which the mean value is defined.

Here we have to be much more precise. However, there is today no generally accepted definition of a mean value. One alternative would be to compute the mean value over the surface with a fixed ellipsoidal height. (This height should be equal to or larger than the maximum ellipsoidal height in the area).

4 Spherical approximation in implementations of Least-Squares Collocation

The downward continuation needed when applying Stokes formula eq. (2) may be avoided using different approximation methods such as e.g. LSC in a space of functions harmonic down to a surface inside the Earth. Generally a so-called Bjerhammar sphere is used. Hence LSC has the potential of avoiding completely the use of spherical approximation.

Spherical approximation has to a certain extent been used in the GRAVSOFT (Tscherning et al., 1992) implementations of Least-Squares Collocation. In new versions of the software, the use of spherical approximation is optional.

The solutions are expressed as a linear combination of base functions harmonic outside a Bjerhammar sphere totally enclosed in the Earth.

$$\tilde{T}(P) = \sum_{i=1}^{N} a_i K(L_i, P) \tag{4}$$

where a_i are constants and K is a reproducing kernel (or covariance function) evaluated in P and with respect to the observation functionals L_i. K is rotationally invariant and homogeneous on spheres concentric with the Bjerhammar sphere, such as on the Earth's surface in spherical approximation, cf. eq. (5).

The use of a mean earth sphere as an approximation of the surface of the Earth enables the representation of a large class of kernels using simple closed expressions, see e.g. Tscherning and Rapp (1974).

If no approximation is used, the rotational invariance and homogeneity is lost. This gives problems in global applications, since the gravity anomaly variances calculated from such a kernel may have a large variation between the Equator and the Poles. As a numerical example we have used a typical kernel

$$K(P,Q) = K(r,r',\psi) =$$

$$\sum_{i=2}^{\infty} \sigma_i \left(\frac{R^2}{(rr')}\right)^{2i+2} P_i(\cos\psi) \tag{5}$$

where σ_i is equal to the error-degree variances of EGM96 (Lemoine et al., 1998) for i < 361, equal to the degree-variances of GPM98 (Wenzel, 1998) for i> 360 and i < 721 and

$\sigma_i = \frac{A}{(i-1)(i-2)(i+4)}$ for i > 720. R = 6349.8 km is the radius of the Bjerhammer sphere and the associated variance of gravity anomalies at the surface of the Earth is 123.4 $mgal^2$ for R=6371000. r and r' are the radial distances of P and Q and ψ is the spherical distance. This kernel was used in numerical experiments reported in Sansò and Tscherning (2002).

In Table 1, values of variances for different functionals are given at different distances r from the origin. The functionals are gravity anomalies, Δg, height anomalies, ς, and vertical gravity gradients, T_{zz}, and their corresponding mean-values over a 0.5° x 0.5° equal area block.

Table 1. (Signal) variances of different functionals for different distances from the origin, r, corresponding to points on the ellipsoid with latitude varying from 0° to 90°. Mean values are indicated with an over-bar. Note that the mean-values show (as expected) a much smaller dependence on r than the point values. Mean values were evaluated as point values at 5 km altitude.

r (km)	Δg	$\overline{\Delta g}$	ς	$\overline{\varsigma}$	T_{zz}	$\overline{T_{zz}}$
	$mgal^2$	$mgal^2$	m^2	m^2	EU^2	EU^2
6381	48	31	0.091	0.081	10.8	4.1
6376	75	43	0.112	0.096	21.0	6.8
6371	123	62	0.140	0.116	43.1	11.5
6366	217	93	0.181	0.142	93.6	20.3
6361	409	143	0.246	0.178	214.2	37.2

The variance of the gravity gradients were also calculated at the radial distances (m) r = 340000, 345000, 350000, 355000 and 360000. The variance was in all cases close to 0.00005 EU^2.

If the linear functionals used in eq. (4) are satellite data, and the point of evaluation is on the surface of the Earth, we are implicitly carrying out a downward continuation. The numerical values in Table 1 then show that the results obtained when using LSC for downward continuation will depend on the latitude. One possible method to solve this problem is to use non-isotropic covariance functions or covariance functions based on ellipsoidal harmonics.

The reason for the large variations of the variances is that we use an infinite dimensional space, i.e. the kernels are an infinite series. If we break off the series at a certain, not too low degree, (e.g. i=720), the variances are not changing very much. This may be justified if data at satellite altitude are used to compute spherical harmonic coefficients. However the error-estimates will then not reflect the effect of this. The so-called omission error will be too small.

It would be worthwhile to investigate the magnitude of this type of error.

The use of kernels in a finite dimensional space (degree 720) has been used in numerical experiments with LSC, see Sansò and Tscherning (2002). Data were generated from EGM96, so as to be able to compare estimated spherical harmonic coefficients to the original values. The result of using spherical approximation showed an error much larger than the flattening, up to 10 %. In this experiment, two approximations were used, where in normal practice only one occurs.

In spherical approximation we used r = R+h. The generated gravity anomalies and vertical gravity gradients were evaluated at the correct radial distance, but the derivative was taken in the direction of the radius vector. (In fact the derivative was calculated using the coefficients of degree i multiplied by (i-1) /r). For real observations the derivative is in the direction approximately orthogonal to the bounding surface. This may be the reason why errors of 10 % magnitude normally not are found when using spherical approximation. However, a better understanding of the errors introduced when using spherical approximation is needed.

5 Kernels or covariance functions homogeneous on an ellipsoid

Kernels/covariance functions may be constructed as product sums of ellipsoidal harmonics.

Let E be the excentric anomaly of a "Bjerhammar" ellipsoid with semi minor axis b, smaller than or equal to the semi-minor axis of the Earth. A point P has the ellipsoidal coordinates β (reduced latitude) and u (the semi-major axis of the confocal ellipse on which the point P is located). The Legendre functions of the first kind are denoted P_{ij} and those of the second kind are denoted Q_{ij}, where i is the degree and j the order. Then a general expression for a positive definite kernel in ellipsoidal harmonics is

$$K(P,Q) = K(u,\beta,\lambda,u',\beta',\lambda') =$$

$$\sum_{i=2}^{\infty}\sum_{j=0}^{i} \frac{Q_{ij}(i\frac{u}{E})Q_{ij}(i\frac{u'}{E})}{Q_{ij}(i\frac{b}{E})^2}$$

$$\bullet\, P_{ij}(\sin\beta)P_{ij}(\sin\beta')$$

$$\bullet\,(a_{ij}\cos(j\lambda)\cos(j\lambda') + b_{ij}\sin(j\lambda)\sin(j\lambda')) \tag{6}$$

where a_{ij} and b_{ij} are positive constants. (The "i" occurring in the argument of the associated Legendre function of second order is the imaginary unit).

If the constants all are equal to σ_i for each degree i we have with ψ the "ellipsoidal" distance and u = b

$$K(P,Q) = K(b,b,\psi) =$$

$$\sum_{i=2}^{\infty}\sum_{j=0}^{i} \frac{\sigma_i}{2i+1} P_{ij}(\cos\psi) \tag{7}$$

On the "Bjerhammar" ellipsoid we may find closed expressions similar to those found in Tscherning and Rapp (1974) .

Unfortunately general closed expressions have not yet been found for reproducing kernels in infinite dimensional spaces having such functions as kernels. It may however be possible to express the ratios of Legendre function of the second kind as polynomials in u and u', leading to expressions which are sums of closed expressions.

The use of the ellipsoidal kernels is attractive, but further developments such as the derivations of expressions for the functionals associated with the gradients and the second order derivatives are needed.

6 Conclusion

Using the known position of points of evaluation and a high resolution gravity field model, it is possible to arrive to virtually approximation-free linearised functionals representing the ground data we have. The geodetic points on the bounding surface, the surface of the Earth, are also known with a very small error.

There are a number of problems associated with abandoning spherical approximation. Here the use of an ellipsoidal "Bjerhammar" surface may be a partial solution for the problems occurring in LSC.

There are other areas where even a planar approximation is used. Such approximations may certainly be valid for smaller areas. But what is the limit? And what are the limits of using spherical or ellipsoidal approximation? Numerical studies may tell us something, but theoretical investigations may be a shortcut to obtaining a better handle on the problems.

Approximations used in geodesy have saved a lot of computer time. But now computers are so fast than one may ask: Why do it wrong, when it can be done right ?

References

Ardalan, A.A. and E.W.Grafarend: Ellipsoidal geoid undulations (ellipsoidal Bruns formulae): case studies. Journal of Geodesy, Vol. 75, pp. 544-552, 2001.

Heiskanen, W.A. and H. Moritz: Physical Geodesy. W.H. Freeman & Co, San Francisco, 1967.

Lemoine, F.G., S.C. Kenyon, J.K. Factor, R.G. Trimmer, N.K. Pavlis, D.S. Chinn, C.M. Cox, S.M. Klosko, S.B. Luthcke, M.H. Torrence, Y.M. Wang, R.G. Williamson, E.C. Pavlis, R.H. Rapp, and T.R. Olson, The Development of the Joint NASA GSFC and the National Imagery and Mapping Agency (NIMA) Geopotential Model EGM96, NASA/TP-1998-206861, Goddard Space Flight Center, Greenbelt, MD, July, 1998.

Meissl, P.: The use of finite elements in physical geodesy. Reports of the Dep. Of Geodetic Scoence, no. 313, The Ohio State University, 1981.

Sanso', F. and C.C.Tscherning: Fast spherical collocation: A general implementation. IAG Symposia Vol. 125, pp. 131-137, Springer Verlag, 2002.

Torge, W.: Geodesy. 3. Ed., de Gruyter, 2001.

Tscherning, C.C., R.Forsberg and P.Knudsen: The GRAVSOFT package for geoid determination. Proc. 1. Continental Workshop on the Geoid in Europe, Prague, May 1992, pp. 327-334, Research Institute of Geodesy, Topography and Cartography, Prague, 1992.

Tscherning, C.C. and R.H.Rapp: Closed Covariance Expressions for Gravity Anomalies, Geoid Undulations, and Deflections of the Vertical Implied by Anomaly Degree-Variance Models. Reports of the Department of Geodetic Science No. 208, The Ohio State University, Columbus, Ohio, 1974.

Wenzel, H.G.: Ultra hochaufloesende Kugelfunktionsmodelle GMP98A und GMP98B des Erdschwerefeldes. Proceedings Geodaetische Woche, Kauserslautern, 1998.

Some Topics Related to the Solution of Boundary-Value Problems in Geodesy

P. Holota
Research Institute of Geodesy, Topography and Cartography, 250 66 Zdiby 98, Praha-vychod, Czech Republic
e-mail: holota@pecny.asu.cas.cz, Tel.: +420 323 649235, Fax: +420 2 84890056

Abstract. In geodesy the theory of boundary-value problems has a great number of important applications. It is hardly possible to approach all of them in one single paper. For this reason the exterior Neumann problem is discussed to demonstrate some of the topics which may be encountered in the solution of geodetic boundary-value problems in general. In addition, the problem is closely related to the gravimetric boundary value problem, which has a considerable practical importance. The tie to variational methods and the minimization of the respective quadratic functional are in the focus of the explanation. Within this concept Euler's necessary condition has a form of an integral identity and represents a natural starting point for the interpretation of the solution in terms of function bases. The convergence of Galerkin's approximations is also treated. For the solution domain given by the exterior of a sphere and also of an ellipsoid of revolution the particular attention is paid to the construction of kernels which have a reproducing property with respect to the inner product in the Sobolev space used. They generate suitable function bases for the approximation solution of the boundary-value problem considered. In particular, in the spherical case they allow for a closed expression of the elements in Galerkin's matrix. A relation of the kernels to Dirichlet's and Neumann's function is also shown. Spherical and ellipsoidal harmonics are used as an efficient tool in constructing the respective integral kernels. Finally, an approach to problems with data given on a more complicated boundary is discussed.

Keywords. Earth's gravity field, geodetic boundary-value problems, Green's functions, variational methods, reproducing kernels, ellipsoidal harmonics

1 Introduction

The theory of boundary-value problems has a number of applications in geodesy. They concern both, ordinary and partial differential equations.

A number of these problems may be interpreted as a solution of Euler's differential equation connected with some minimization problem. This in particular is the case if by character the respective application belongs to analytical mechanics, optics, mathematical cartography or in some cases also to other disciplines.

On the other hand, mentioning partial differential equations, we know that Laplace's and Poisson's equations govern applications of the classical theory of the gravity field in geodesy.

Very often the geodetic approach to boundary-value problems for Laplace's (or Poisson's) equation represents what is usually known as a classical solution. We look for a smooth function satisfying the differential equation and the boundary condition "pointwise". Within this concept integral equations and Green's function methods are most frequently used.

Alternatively, we can look for a measurable function satisfying a certain integral identity connected with the boundary-value problem in question. This is the so-called weak solution. Natural function spaces corresponding to this method are Sobolev's spaces. It is surprising that an approach like this is still somewhat marginal in geodesy.

In many cases there exists a possibility to replace the integration of a differential equation under given boundary conditions by an equivalent problem of getting a function that minimizes some integral. This corresponds to variational methods.

In the history the first use of variational methods was in the form of Dirichlet's principle [Peter Gustav Lejeune Dirichlet (1805-1859)]. According to this principle among functions which attain given values on the boundary $\partial\Omega$ of a domain Ω, that and only that function which is harmonic in Ω, minimizes the so-called Dirichlet's integral:

$$\int_{\Omega}\left[\left(\frac{\partial u}{\partial x_1}\right)^2+\left(\frac{\partial u}{\partial x_2}\right)^2+\left(\frac{\partial u}{\partial x_3}\right)^2\right]dx \qquad (1)$$

where x_i, $i=1,2,3$, are rectangular Cartesian coordinates in Euclidean three-dimensional space $\boldsymbol{R}^3$

and $dx = dx_1 dx_2 dx_3$ means the volume element, see Michlin (1970), Rektorys (1974), but also Kellogg (1953) and others. (Note. We write Dirichlet's integral in three dimensional space which, of course, is of a formal meaning only and merely a consequence of our choice.)

Dirichlet's principle was extensively used by Riemann (1826-1866), but critically commented by Weierstrass (1815-1897) and later also by Hadamard (1826-1963). Only in the beginning of the 20th century the principle got a new interest in connection with Hilbert's works. Hilbert (1862-1943) showed that the justification of Dirichlet's principle is much deeper and essentially associated with the notion of the completeness of the metric space. The justification is closely related to Hilbert's contributions to the development of calculus of variations, see Giaquinta (2000).

Now we will be little more specific. We recall that our main intention is to discuss problems which are associated with the determination of the external gravity field of the Earth. In consequence we have to suppose that Ω mentioned above is an unbounded solution domain.

As regards functions spaces, in the sequel we will work with functions from Sobolev's weighted space $W_2^{(1)}(\Omega)$ endowed with inner product

$$(u,v)_1 \equiv \int_\Omega \frac{uv}{|x|^2}dx + \sum_{i=1}^{3}\int_\Omega \frac{\partial u}{\partial x_i}\frac{\partial v}{\partial x_i}dx \qquad (2)$$

see Holota (1997). Note that this product induces the norm

$$\| u \|_1 \equiv (u,u)_1^{1/2} \qquad (3)$$

Finally, also the boundary $\partial\Omega$ of the domain Ω is supposed to have a certain degree of regularity. Putting $\Omega' = \boldsymbol{R}^3 - \Omega \cup \partial\Omega$, we will suppose that Ω' is the domain with Lipschitz' boundary, see Necas (1967), Rektorys (1974) or Kufner et al. (1977). Domains with Lipschitz' boundaries are, e.g., the sphere, ellipsoid, cube, polyhedron, as well as substantially more general domains with smooth or piecewise smooth boundaries. It can be stated that Lipschitz' boundary is already general enough to represent, with a small degree of idealization, the Earth's topography.

2 Exterior Neumann's problem and a quadratic functional

Let $L_2(\partial\Omega)$ be the space of square integrable functions on $\partial\Omega$. Put

$$A(u,v) = \sum_{i=1}^{3}\int_\Omega \frac{\partial u}{\partial x_i}\frac{\partial v}{\partial x_i}dx \qquad (4)$$

which is a bilinear form defined on the Cartesian product $W_2^{(1)}(\Omega)\times W_2^{(1)}(\Omega)$. Consider now the quadratic functional

$$\Phi(u) = A(u,u) - 2\int_{\partial\Omega} uf\, dS \qquad (5)$$

defined on $W_2^{(1)}(\Omega)$, where $f \in L_2(\partial\Omega)$.

One can show that $A(u,u)$ for $u \in W_2^{(1)}(\Omega)$ has all the characteristic properties of a functional norm and that in general

$$A(u,u) \geq \frac{1}{5}\| u \|_1^2 \qquad (6)$$

see Holota (1997). Inequality (6) means that $A(u,v)$ is an elliptic bilinear form.

Similarly, using Schwarz's inequality, we obtain

$$\int_{\partial\Omega} uf\, dS \leq \| u \|_{L_2(\partial\Omega)} \| f \|_{L_2(\partial\Omega)} \qquad (7)$$

which immediately leads to

$$\int_{\partial\Omega} uf\, dS \leq \beta \| f \|_{L_2(\partial\Omega)} \| u \|_1 \qquad (8)$$

in view of $\| u \|_{L_2(\partial\Omega)} \leq \beta \| u \|_1$ guaranteed by the so-called trace theorem, where β is a constant that depends mainly on the geometry of the boundary $\partial\Omega$, see again Holota (1997). Hence it is clear that for $u \in W_2^{(1)}$

$$\lim \Phi(u) = \infty \quad \text{as} \quad \| u \|_1 \to \infty \qquad (9)$$

which (using the standard terminology) means that Φ is a *coercive functional* on $W_2^{(1)}(\Omega)$.

Take now an arbitrary $v \in W_2^{(1)}(\Omega)$ and form the function $\Phi(u+tv)$ of $t \in (-\infty,\infty)$. We can easily compute that

$$\left.\frac{d}{dt}\Phi(u+tv)\right|_{t=0} = 2\left[A(u,v) - \int_{\partial\Omega} vf\, dS \right] \qquad (10)$$

Subsequently, it follows then from the properties of the bilinear form $A(u,v)$ that

$$D\Phi(u,v) \equiv \left.\frac{d}{dt}\Phi(u+tv)\right|_{t=0} \qquad (11)$$

is a *bounded linear functional* of the variable v (Gâteaux' differential of Φ at the point u) and that

$$D\Phi(u+v,v) - D\Phi(u,v) \geq 0 \qquad (12)$$

for all $u, v \in W_2^{(1)}(\Omega)$, see Holota (2000a).

The last inequality and the coercivness of Φ make it possible to conclude that Φ attains its minimum in $W_2^{(1)}(\Omega)$. The proof rests on the theory of non-linear functionals and the solution of an abstract variational problem, as e.g. in Necas and Hlavacek (1981).

On the contrary suppose that at a point $u \in W_2^{(1)}(\Omega)$ the functional Φ has its local minimum. Thus necessarily:

$$\frac{d}{dt}\Phi(u+tv)\bigg|_{t=0} = 0 \tag{13}$$

for all $v \in W_2^{(1)}(\Omega)$ which immediately yields

$$A(u,v) = \int_{\partial\Omega} vf\, dS \tag{14}$$

for all $v \in W_2^{(1)}(\Omega)$. Recall that in calculus of variations this identity represents *Euler's* necessary condition for the functional Φ to have a minimum at the point u.

The last condition has also a classical interpretation, provided that u is sufficiently regular. Indeed, using Green's identity, we have

$$A(u,v) = -\int_{\Omega} v\Delta u\, dx - \int_{\partial\Omega} v\frac{\partial u}{\partial n}\, dS \tag{15}$$

where Δ means Laplace's operator and $\partial/\partial n$ denotes the derivative in the direction of the unit normal $\boldsymbol{n}$. Thus it follows from Eq. (14) that

$$\int_{\Omega} v\Delta u\, dx + \int_{\partial\Omega} v\left(\frac{\partial u}{\partial n} + f\right) dS = 0 \tag{16}$$

is valid for all $v \in W_2^{(1)}(\Omega)$. Supposing in addition that also f and $\partial\Omega$ are sufficiently smooth and applying the usual reasoning to integrals from continuous functions, we obtain

$$\Delta u = 0 \quad \text{in} \quad \Omega \tag{17}$$

and

$$\frac{\partial u}{\partial n} = -f \quad \text{on} \quad \partial\Omega \tag{18}$$

In other words u has to be a solution of (exterior) Neumann's boundary-value problem. As known, Eq. (14) is the basis for its weak formulation.

Remark 1. A more general discussion and further details concerning especially the existence and uniqueness of solutions of Neumann exterior problems in weighted Sobolev spaces may be found in Ambrouche et al. (1997).

3 Interpretation in terms of bases

The integral identity (Euler's condition) represented by Eq. (14) valid for all $v \in W_2^{(1)}(\Omega)$ is a natural starting point for a numerical solution. Indeed, one can approximate u by a linear combination

$$u_n = \sum_{j=1}^{n} c_j^{(n)} v_j \tag{19}$$

where v_j are members of a function basis of $W_2^{(1)}(\Omega)$ and look for the coefficients $c_j^{(n)}$ by solving the respective Galerkin system

$$\sum_{j=1}^{n} c_j^{(n)} A(v_j, v_k) = \int_{\partial\Omega} v_k f\, dS\ , \quad k = 1,\dots,n \tag{20}$$

which is connected with our integral identity in a very transparent way.

The justification of this approach follows from the strong convergence of u_n to u as $n \to \infty$, i.e. from the convergence in terms of the norm $\|\,.\,\|_1$.

To show it we first recall that in $W_2^{(1)}(\Omega)$ the basis functions $v_1, v_2, \dots, v_n, \dots$ generate a sequence of finite dimensional subspaces

$$W_n = \operatorname{span}\{v_i,\ i = 1,\dots,n\} \tag{21}$$

$n = 1, 2, \dots,$ such that

$$\lim_{n\to\infty} \operatorname{dist}(v, W_n) = 0 \tag{22}$$

for all $v \in W_2^{(1)}(\Omega)$, i.e., $\lim_{n\to\infty} W_n = W_2^{(1)}(\Omega)$.

Taking now into consideration the fact that the system (20) has a non-singular determinant, we immediately see that there exists one and only one u_n which satisfies

$$A(u_n, v) = \int_{\partial\Omega} vf\, dS \tag{23}$$

for all $v \in W_n$. Moreover, recalling inequality (6), we have

$$\|u_n\|_1^2 \le 5A(u_n, u_n) = 5\int_{\partial\Omega} u_n f\, dS \tag{24}$$

so that

$$\|u_n\|_1 \le 5\beta\, \|f\|_{L_2(\partial\Omega)} \tag{25}$$

follows from the trace theorem, we have already mentioned in the last section.

It is now relatively easy to show that for $n \to \infty$, u_n converges weakly to u in $W_2^{(1)}(\Omega)$, i.e., in our case that $\lim_{n\to\infty}(u_n,v)_1=(u,v)_1$ for all $v\in W_2^{(1)}(\Omega)$. For this purpose suppose the opposite. The sequence $[u_n]_{n=1}^{\infty}$ is bounded in view of inequality (25). Therefore, there exists a subsequence $[u_{m_n}]_{n=1}^{\infty}$ which converges weakly to $u^*\neq u$. However, for every $v\in W_2^{(1)}(\Omega)$ there exists a sequence $[v_{m_n}]_{n=1}^{\infty}$ ($v_{m_n}\in W_{m_n}$) which converges strongly to v. Now, due to Eq. (23) we have

$$A(u_{m_n},v_{m_n})=\int_{\partial\Omega}v_{m_n}f\,dS \tag{26}$$

and see that

$$\begin{aligned}\int_{\partial\Omega}vf\,dS&=\lim_{n\to\infty}\int_{\partial\Omega}v_{m_n}f\,dS\\&=\lim_{n\to\infty}A(u_{m_n},v_{m_n})=A(u^*,v)\end{aligned} \tag{27}$$

is valid for all $v\in W_2^{(1)}(\Omega)$. In consequence $u^*=u$, which is in contradiction to our supposition.

Finally, we can take $v_n\in W_n$ such that v_n converges strongly to u and write

$$\begin{aligned}A(v_n-u_n,v_n-u_n)=\int_{\partial\Omega}u_nf\,dS-A(v_n,u_n)\\-A(u_n,v_n)+A(v_n,v_n)\end{aligned} \tag{28}$$

From the strong convergence of v_n to u and the weak convergence of u_n to u it then results

$$\lim_{n\to\infty}A(v_n-u_n,v_n-u_n)=0 \tag{29}$$

Thus, due to inequality (6) [the ellipticity of $A(u,v)$], we have

$$\lim_{n\to\infty}\|v_n-u_n\|_1=0 \tag{30}$$

i.e. the strong convergence of u_n to u which completes the proof.

Remark 2. The scheme of the proof is similar to that used for a homogeneous problem in Necas (1967).

Unfortunately, for one essential reason the approximations as above cannot be extremely efficient. Indeed, the expected solution is a harmonic and thus an analytic function, whereas in general, functions from $W_2^{(1)}(\Omega)$ and especially the basis functions have not such a high degree of regularity. For instance, this concerns also approximations based on the frequently used method of finite elements.

4 Inner product and Green's function

For approximation purposes some properties of Hilbert spaces are extremely useful. This concerns e.g. the existence (or non-existence) of a reproducing kernel in a particular Hilber space, see Meschowski (1962) or Zhidkov (1977). Recall also that especially in geodesy function bases are often generated by means of such a kernel, see e.g. Krarup (1969), Moritz (1980), Tscherning (1975), Neyman (1979) and Sanso (1986). As regards $W_2^{(1)}(\Omega)$, however, we can see that there is not too much chance to find a reproducing kernel in $W_2^{(1)}(\Omega)$. This fact may be deduced from the well-known Sobolev lemma on embeddings. In order to make it clear, consider in analogy to $W_2^{(1)}(\Omega)$, Sobolev's space $W_2^{(k)}(\Omega)$ endowed with inner product

$$\begin{aligned}(u,v)_k\equiv\int_\Omega\frac{uv}{|\boldsymbol{x}|^2}\,dx\\+\sum_{1\le|\alpha|\le k}\int_\Omega D^\alpha u(\boldsymbol{x})D^\alpha v(\boldsymbol{x})\,dx\end{aligned} \tag{31}$$

where

$$D^\alpha u=\frac{\partial^{|\alpha|}}{\partial x_1^n\partial x_2^m\partial x_3^l} \tag{32}$$

n,m,l are non-negative integers and $\alpha=(n,m,l)$ is a multiindex of length $|\alpha|=n+m+l$. Note for illustration that $W_2^{(k)}(\Omega)$ is a space produced by functions which are square integrable on Ω under the weight $|\boldsymbol{x}|^{-2}$ and in a generalized sense have partial derivatives up to the order k, which are also square integrable on Ω.

Now it is enough to refer to the theory of Hilbert spaces with reproducing kernel and recall that the existence of a reproducing kernel in $W_2^{(k)}(\Omega)$ may be verified if we show that algebraically and topologically $W_2^{(k)}(\Omega)\subset C(\overline{\Omega})$, where $C(\overline{\Omega})$ is a space of functions defined on Ω, which are continuous up to the boundary $\partial\Omega$ and regular at infinity, cf. also Sanso (1986) for a similar reasoning. (Note here that $\overline{\Omega}$ is the closure of Ω in $\boldsymbol{R}^3$.)

However, in three-dimensional space Sobolev's lemma only guarantees that $W_2^{(k)}(\Omega)\subset C(\overline{\Omega})$ is true for $k>3/2$, see Necas (1967), Kufner et al. (1977) or Rektorys (1974). Unfortunately, in our case we only have $k=1$.

Nevertheless, we can arrive to a conclusion which is still stronger. In $W_2^{(1)}(\Omega)$ consider the inner product

$$(u,v) = A(u,v) = \sum_{i=1}^{3} \int_{\Omega} \frac{\partial u}{\partial x_i} \frac{\partial v}{\partial x_i} \, dx \tag{33}$$

and note that it induces a norm

$$\| u \| \equiv (u,u)^{1/2} \tag{34}$$

which is equivalent to $\| u \|_1$, see Holota (1997).

Let us try now to find a kernel $K = K(x,y)$ which has a reproducing property with respect to the inner product above, i.e. a kernel such that

$$\sum_{i=1}^{3} \int_{\Omega} \frac{\partial K(x,y)}{\partial x_i} \frac{\partial v(x)}{\partial x_i} \, dx \; = \; v(y) \tag{35}$$

holds for all $v \in W_2^{(1)}(\Omega)$. Applying Green's identity (formally in a sense), we get

$$\begin{aligned} &- \int_{\partial\Omega} v(x) \frac{\partial K(x,y)}{\partial n_x} \, d_x S \\ &- \int_{\Omega} v(x) \Delta_x K(x,y) \, dx \; = v(y) \end{aligned} \tag{36}$$

where, using the theory of distributions, one can write

$$v(y) \; = \; \int_{\Omega} \delta(x,y) v(x) \, dx \tag{37}$$

provided that δ is Dirac's delta-function.

Thus K has to meet the following two conditions:

$$\Delta K(x,y) = - \, \delta(x,y) \qquad \text{in} \qquad \Omega \tag{38}$$

$$\frac{\partial K(x,y)}{\partial n} = 0 \qquad \text{on} \qquad \partial\Omega \tag{39}$$

since in Eq. (36) v is an arbitrary function from $W_2^{(1)}(\Omega)$. Hence K has to be a fundamental solution of Laplace's equation which meets Neumann's condition on the boundary $\partial\Omega$, cf. Zhidkov (1977). More precisely, $K(x,y) = (4\pi)^{-1} N(x,y)$, where $N(x,y)$ is Green's function of the second kind often called Neumann's function.

In consequence K has a singularity in Ω and it is obvious that it cannot be an element of $W_2^{(1)}(\Omega)$, i.e., $K \notin W_2^{(1)}(\Omega)$, which confirms the hypothesis from the beginning of the section.

Example. Suppose that Ω is the exterior of a sphere $S_R \equiv \{x \in \boldsymbol{R}^3; |x| > R\}$ of radius R, i.e., $\Omega \equiv S_R$ and that $\bar{x}$ is an image of x given by a transformation of space known as inversion in a sphere, i.e.,

$$\bar{x} = \frac{R}{|x|^2} x \quad \text{and} \quad |x||\bar{x}| = R^2 \tag{40}$$

Then it is relatively easy to show that

$$\begin{aligned} &N(x,y) = \\ &= \frac{1}{|x-y|} + \frac{1}{R} \sum_{n=0}^{\infty} \frac{n}{n+1} \left(\frac{R^2}{|x||y|} \right)^{n+1} P_n(\cos\psi) \end{aligned} \tag{41}$$

or in a closed form

$$\begin{aligned} N(x,y) &= \frac{1}{|x-y|} + \frac{R}{|x|} \frac{1}{|\bar{x}-y|} \\ &- \frac{1}{R} \ln \frac{|\bar{x}-y| + |\bar{x}| - |y| \cos\psi}{|y|(1-\cos\psi)} \end{aligned} \tag{42}$$

where P_n is the usual Legendre polynomial of degree n, and ψ is the angle between the position vectors x and y, see Holota (2002b). Obviously, the function $N(x,y)$ is singular for $x = y$.

5 Restriction to harmonic functions

The situation substantially changes if we consider the space $H_2^{(1)}(\Omega)$ of those functions from $W_2^{(1)}(\Omega)$ which are harmonic in Ω and reformulate the weak setting of Neumann's problem given by Eq. (14), i.e., we look for $u \in H_2^{(1)}(\Omega)$ such that

$$A(u,v) = \int_{\partial\Omega} vf \, dS \tag{43}$$

holds for all $v \in H_2^{(1)}(\Omega)$.

Remark 3. Recall that the boundary behavior of functions from $W_2^{(1)}(\Omega)$ is actually characterized by functions from $H_2^{(1)}(\Omega)$ since

$$W_2^{(1)}(\Omega) = H_2^{(1)}(\Omega) \oplus \dot{W}_2^{(1)} \tag{44}$$

where

$$\dot{W}_2^{(1)} = \{ v \in W_2^{(1)}(\Omega); v = 0 \text{ on } \partial\Omega \text{ in t.s.o.t.} \} \tag{44}$$

see also Neyman (1979). [Note that here "in t.s.o.t." means "in the sense of traces", see Necas (1967), Kufner et al. (1977) or Holota (1997) for the trace theorem.] Conversely, it is clear that the precise characterization of traces given by

$$u \in W_2^{(1)}(\Omega) \iff u|_{\partial\Omega} \in W_2^{(1/2)}(\partial\Omega) \tag{45}$$

where $W_2^{(1/2)}(\partial\Omega)$ is Sobolev-Slobodeckij's space with a fractional derivative, see Necas (1967) or

Kufner et al (1977), is also valid for functions $u \in H_2^{(1)}(\Omega)$.

Applying a constructive approach in the case of some simple domains, we are able to find kernels which in $H_2^{(1)}(\Omega)$ have the reproducing property with respect to the inner product given by Eq. (33).

Note that Eq. (36) reduces to

$$\int_{\partial\Omega} v(x)\frac{\partial K(x,y)}{\partial n_x}\, d_x S = -\, v(y) \tag{46}$$

valid for all $v \in H_2^{(1)}(\Omega)$ since the kernel $K(x,y)$ must be an element of $H_2^{(1)}(\Omega)$ and thus a harmonic function.

6 Spherical case and the reproducing kernel.

Considering the case when $\Omega \equiv S_R$ as in the example above and recalling the well-known integral representation of the solution of (exterior) Dirichlet's problem, we can immediately conclude that in Eq. (46)

$$\frac{\partial K(x,y)}{\partial n_x} = \frac{\partial K(x,y)}{\partial |x|} = -\frac{1}{4\pi}\frac{\partial G(x,y)}{\partial |x|} \tag{47}$$

for $|x| = R$, where

$$G(x,y) = \frac{1}{|x-y|} - \frac{R}{|x|}\frac{1}{|\bar{x}-y|} \tag{48}$$

is the respective Green function (of the first kind). Hence, in our spherical case,

$$\frac{\partial K(x,y)}{\partial n_x} = -\frac{1}{4\pi R}\frac{|y|^2 - R^2}{|y-x|^3} \tag{49}$$

Moreover, the right hand side of Eq. (49) may be easily expressed by means of a series. Indeed,

$$\frac{|y|^2 - R^2}{|y-x|^3} = \frac{1}{R}\sum_{n=0}^{\infty}(2n+1)\left(\frac{R}{|y|}\right)^{n+1} P_n(\cos\psi) \tag{50}$$

see Heiskanen and Moritz (1967) and it is thus clear that

$$K(x,y) = \frac{1}{4\pi R}\sum_{n=0}^{\infty}\frac{2n+1}{n+1} z^{n+1} P_n(\cos\psi) \tag{51}$$

where

$$z = \frac{R^2}{|x||y|} \tag{52}$$

It is also not extremely difficult to find that

$$K(x,y) = \frac{1}{4\pi R}\left(\frac{2z}{L} - \ln\frac{L+z-\cos\psi}{1-\cos\psi}\right) \tag{53}$$

where

$$L = \sqrt{1 - 2z\cos\psi + z^2} \tag{54}$$

see e.g. Tscherning (1975) or Neyman (1979).

In case of $|x| = |y| = R$ the kernel has a singularity for $\psi = 0$ and the series in Eq. (51) definitely does not uniformly converges down to the sphere S_R Nevertheless, for $x, y \in \Omega$ the kernel belongs to $H_2^{(1)}(\Omega)$.

Remark 4. It is worth mentioning explicitly that $K(x,y)$ is a reproducing kernel only for those points x which belong to Ω open [this is indeed enough to identify any $v \in H_2^{(1)}(\Omega)$]; this is because the evaluation functional at $x \in \partial\Omega$ is not bounded when $v \in H_2^{(1)}(\Omega)$, because the trace $v|_{\partial\Omega} \in H_2^{1/2}(\partial\Omega)$, where pointwise values are not well defined. [$H_2^{1/2}(\partial\Omega)$ is an analogue to $W_2^{1/2}(\partial\Omega)$ mentioned above.] By symmetry the same reasoning holds true for $y \in \Omega$.

Remark 5. To have a kernel that converges down to the boundary ∂S_R of the sphere S_R we can modify $K(x,y)$ e.g. as follows

$$\widetilde{K}(x,y) = \frac{1}{4\pi R}\sum_{n=0}^{\infty}\frac{2n+1}{q_n} z^{n+1} P_n(\cos\psi) \tag{55}$$

provided that at least

$$q_n = O(n^3) \qquad \text{as} \qquad n \to \infty \tag{56}$$

This kernel possesses the reproduction property, but with respect to a completely different inner product and generates also a different Hilbert's space (of much smoother functions).

Remark 6. There is an interesting relation between the kernel $K(x,y)$ and the functions $G(x,y)$ and $N(x,y)$. Indeed, recalling Eqs. (42) and (48), we immediately see that

$$\begin{aligned} N(x,y) - G(x,y) &= \frac{R}{|x|}\frac{2}{|\bar{x}-y|} \\ &\quad - \frac{1}{R}\ln\frac{|\bar{x}-y| + |\bar{x}| - |y|\cos\psi}{|y|(1-\cos\psi)} \\ &= \frac{1}{R}\left(\frac{2z}{L} - \ln\frac{L+z-\cos\psi}{1-\cos\psi}\right) \end{aligned} \tag{57}$$

so that

$$K(x,y) = \frac{1}{4\pi}[\, N(x,y) - G(x,y)\,] \tag{58}$$

in view of Eq. (53). Relations which are similar to Eq. (58) may be found (but for a bounded domain) in Garabedian (1964).

Suppose now that $y_i \in \Omega, i = 1, \ldots, \infty$ is a sequence of points which is dense in Ω then the linear manifold

$$H = \operatorname{span}\{K(x, y_i),\ i = 1, \ldots, \infty\} \tag{59}$$

is densely embedded in $H_2^{(1)}(\Omega)$, cf. Sanso (1986). Hence the kernel $K(x, y)$ gives us also a possibility to generate finite dimensional subspaces

$$H_n = \operatorname{span}\{K(x, y_i),\ i = 1, \ldots, n\} \tag{60}$$

in $H_2^{(1)}(\Omega)$ such that $H_n \subseteq H_{n+1}$ and

$$\lim_{n\to\infty} \operatorname{dist}(v, H_n) = 0 \tag{61}$$

for all $v \in H_2^{(1)}$, i.e., $\lim_{n\to\infty} H_n = H_2^{(1)}$.

This is important since in our case it enables us to approximate the solution of the weakly defined Neumann problem given by Eq. (43) by means of the linear combinations

$$u_n = \sum_{j=1}^{n} c_j^{(n)} v_j \tag{62}$$

where

$$v_j(x) = K(x, y_j) \tag{63}$$

In addition, in Galerkin's system given by Eqs. (20) the elements $A(v_j, v_k)$ may be immediately expressed by

$$A(v_j, v_k) = K(y_j, y_k) \tag{64}$$

in view of the reproducing property of the kernel.
Remark 7. Compared with an alternative choice

$$v_j(x) = \frac{1}{|x - y_j|} \tag{65}$$

as in Holota (2000a,b, 2001 and 2002a) the elements given by Eq. (64) have a considerable advantage since (in the spherical cases) they may be expressed in a closed form. In particular for diagonal elements, i.e. for $\psi = 0$, we have

$$A(v_j, v_j) = \frac{1}{4\pi R}\left[\frac{2z}{1-z} - \ln\left(1 + \frac{z}{(1-z)^3}\right)\right] \tag{66}$$

As regards the convergence of Galerkin's approximations we may proceed in a similar way as in Sect. 3. However, problems, which need some care, are the choice of the points y_j (usually under some optimum criteria) and the regularity of the matrix $[A(v_j, v_k)]$.

In fact, the problems are of a similar nature as e.g. that associated with the use of elementary potential [Eq. (65)] in quality of basis functions, cf. Weightman (1967), Balmino (1972), Holota (1979), Heikkinen (1981), Antonov et al. (1983), Kautzleben and Barthelmes (1983), Barthelmes (1986), Lehmann (1992, 1993), Marchenko (1983, 1998), Stromeyer and Ballani (1984), Ihde et al. (1998).
Remark 8. Another choice is $v_j(x) = \widetilde{K}(x, y_j)$. We obtain

$$A(v_j, v_k) = \\ = \frac{1}{4\pi R}\sum_{n=0}^{\infty} \frac{(2n+1)(n+1)}{q_n^2} z^{n+1} P_n(\cos\psi) \tag{67}$$

However, in this case we are in a different Hilbert space, as already mentioned above in Remark 5.
Remark 9. Radially symmetric integral kernels and possibilities of using systems of these functions in quality of basis function are discussed in Freeden et al. (1998) and e.g. also in Freeden and Michel (1999). In both these references multiresolution analysis is essentially used in constructing the sequence of approximations.

7 Ellipsoidal case

The possibility to express the elements $A(v_j, v_k)$ of Galerkin's matrix by means of the kernel $K(x, y)$ leads us to an attempt to construct a kernel $K_{ell}(x, y)$ which has the reproducing property with respect to the inner product (33) also in the case that the solution domain Ω is the exterior Ω_{ell} of an ellipsoid of revolution of semi-axes a and b, $a \geq b$, i.e., $\Omega \equiv \Omega_{ell}$.

In this case it is natural to use ellipsoidal coordinates u, β, λ. They are related to x_1, x_2, x_3 by the equations

$$x_1 = \sqrt{u^2 + E^2}\cos\beta\cos\lambda \\ x_2 = \sqrt{u^2 + E^2}\cos\beta\sin\lambda, \quad x_3 = u\sin\beta \tag{68}$$

where $E = \sqrt{a^2 - b^2}$, see e.g. Heiskanen and Moritz (1967). Clearly, in the coordinates u, β, λ the boundary $\partial\Omega_{ell}$ of Ω_{ell} is defined by $u = b$.

In analogy to Eq. (46) we now write

$$\int_{\partial\Omega_{ell}} v(x)\frac{\partial K_{ell}(x,y)}{\partial n_x}\, d_x S = -v(y) \tag{69}$$

and subsequently also

$$\frac{\partial K_{ell}(x,y)}{\partial n_x} = -\frac{1}{4\pi}\frac{\partial G_{ell}(x,y)}{\partial n_x} \tag{70}$$

for $u_x = b$, where $G_{ell}(x,y)$ is Green's function related to Dirichlet's problem formulated for Ω_{ell}. Thus our first task is to find $G_{ell}(x,y)$.

Following general principles in constructing Green's functions, we start with the fundamental solution

$$J(x,y) = \frac{1}{|x-y|} \tag{71}$$

of Laplace's equation. According to Hobson (1952) or also some related explanations in Hotine (1969), we can express J in terms of the coordinates u, β, λ as follows:

$$\begin{aligned} J(x,y) = \frac{i}{E}\sum_{n=0}^{\infty}(2n+1)\Big[& P_n(z_x)Q_n(z_y)P_n(\sin\beta_x) \\ & \times P_n(\sin\beta_y) + 2\sum_{m=1}^{n}(-1)^m\left(\frac{(n-m)!}{(n+m)!}\right)^2 \\ & \times P_{nm}(z_x)Q_{nm}(z_y)P_{nm}(\sin\beta_x) \\ & \times P_{nm}(\sin\beta_y)\cos m(\lambda_x-\lambda_y)\Big] \end{aligned} \tag{72}$$

where P_{nm} and Q_{nm} are Legendre's functions of the first and the second kind, respectively; while

$$z_x = i\frac{u_x}{E}, \quad z_y = i\frac{u_y}{E} \quad \text{and} \quad i = \sqrt{-1} \tag{73}$$

is the imaginary unit. We put $P_{n0} = P_n$ and $Q_{n0} = Q_n$, and suppose that $u_x < u_y$ which guarantees the convergence of the series.

In the next step we have to look for a function $H(x) = H(u_x, \beta_x, \lambda_x)$ such that

$$\Delta H = 0 \quad \text{for} \quad u_x > b \tag{74}$$

and

$$H = J \quad \text{for} \quad u_x = b \tag{75}$$

From Eq. (74) we see that H is assumed to be a harmonic function, so that it can be expressed as

$$\begin{aligned} H = \sum_{n=0}^{\infty}\Big[& \frac{Q_n(z_x)}{Q_n(z_0)}a_{n0}P_n(\sin\beta_x) \\ & + \sum_{m=1}^{n}\frac{Q_{nm}(z_x)}{Q_{nm}(z_0)}(a_{nm}\cos m\lambda_x \\ & + b_{nm}\sin m\lambda_x)P_{nm}(\sin\beta_x)\Big] \end{aligned} \tag{76}$$

where $z_0 = i(b/E)$ and a_{nm}, b_{nm} are scalar coefficients, cf. e.g. Heiskanen and Moritz (1967) or Hobson (1952). The insertion into Eq. (75) then yields:

$$\begin{aligned} \left.\begin{matrix} a_{nm} \\ b_{nm} \end{matrix}\right\} = & \frac{i}{E}(2-\delta_{0m})(-1)^m \\ & \times(2n+1)\left(\frac{(n-m)!}{(n+m)!}\right)^2 P_{nm}(z_0) \\ & \times Q_{nm}(z_y)\left\{\begin{matrix}\cos m\lambda_y \\ \sin m\lambda_y\end{matrix}\right\} P_{nm}(\sin\beta_y) \end{aligned} \tag{77}$$

where $\delta_{0m} = 1$ for $m = 0$ and $\delta_{0m} = 0$ for $m \neq 0$.

Hence

$$\begin{aligned} H(x,y) = \frac{i}{E}\sum_{n=0}^{\infty}(2n+1)\Big[& A_{n0xy}P_n(\sin\beta_x) \\ & \times P_n(\sin\beta_y) + 2\sum_{m=1}^{n}(-1)^m\left(\frac{(n-m)!}{(n+m)!}\right)^2 A_{nmxy} \\ & \times P_{nm}(\sin\beta_x)P_{nm}(\sin\beta_y)\cos m(\lambda_x-\lambda_y)\Big] \end{aligned} \tag{78}$$

where

$$A_{nmxy} = Q_{nm}(z_x)Q_{nm}(z_y)\frac{P_{nm}(z_0)}{Q_{nm}(z_0)} \tag{79}$$

and we are able to express Green's function. As is well-known,

$$G_{ell}(x,y) = J(x,y) - H(x,y) \tag{80}$$

Now we are in the position to compute the normal derivative of $G_{ell}(x,y)$ on the surface of the ellipsoid of revolution. We have:

$$\begin{aligned} \left.\frac{\partial G_{ell}}{\partial n_x}\right|_{u_x=b} = \frac{i}{E}\sum_{n=0}^{\infty}(2n+1)\Big[& B_{n0y}P_n(\sin\beta_x) \\ & \times P_n(\sin\beta_y) + 2\sum_{m=1}^{n}(-1)^m\left(\frac{(n-m)!}{(n+m)!}\right)^2 B_{nmy} \\ & \times P_{nm}(\sin\beta_x)P_{nm}(\sin\beta_y)\cos m(\lambda_x-\lambda_y)\Big] \end{aligned} \tag{81}$$

where

$$B_{nmy} = Q_{nm}(z_y)\frac{\partial P_{nm}(z_0)}{\partial n_x} - \frac{\partial A_{nmxy}}{\partial n_x}\bigg|_{u_x=b} = \left[\frac{\partial P_{nm}(z_0)}{\partial n_x} - \frac{P_{nm}(z_0)}{Q_{nm}(z_0)}\frac{\partial Q_{nm}(z_0)}{\partial n_x}\right] Q_{nm}(z_y) \quad (82)$$

Going back to the kernel $K_{ell}(\boldsymbol{x},\boldsymbol{y})$, we know that it is a harmonic function. In general, therefore,

$$K_{ell}(\boldsymbol{x},\boldsymbol{y}) = \sum_{n=0}^{\infty}\left[\frac{Q_n(z_x)}{Q_n(z_0)} c_{n0} P_n(\sin\beta_x) + \sum_{m=1}^{n}\frac{Q_{nm}(z_x)}{Q_{nm}(z_0)}(c_{nm}\cos m\lambda_x + d_{nm}\sin m\lambda_x) P_{nm}(\sin\beta_x)\right] \quad (83)$$

where c_{nm}, d_{nm} are scalar coefficients. Thus it follows then from Eq. (70) that

$$\left.\begin{matrix} c_{nm} \\ d_{nm}\end{matrix}\right\} = -\frac{i}{4\pi E}(2-\delta_{0m})(-1)^m \times (2n+1)\left(\frac{(n-m)!}{(n+m)!}\right)^2 \left(\frac{\partial Q_{nm}(z_0)}{\partial n_x}\right)^{-1} \times Q_{nm}(z_0) B_{nmy}\left\{\begin{matrix}\cos m\lambda_y \\ \sin m\lambda_y\end{matrix}\right\} P_{nm}(\sin\beta_y) \quad (84)$$

and in consequence

$$K_{ell}(\boldsymbol{x},\boldsymbol{y}) = \frac{i}{4\pi E}\sum_{n=0}^{\infty}(2n+1)\Big[C_{n0xy} P_n(\sin\beta_x) \times P_n(\sin\beta_y) + 2\sum_{m=1}^{n}(-1)^m\left(\frac{(n-m)!}{(n+m)!}\right)^2 C_{nmxy} \times P_{nm}(\sin\beta_x) P_{nm}(\sin\beta_y)\cos m(\lambda_x-\lambda_y)\Big] \quad (85)$$

where

$$C_{nmxy} = -\left(\frac{\partial Q_{nm}(z_0)}{\partial n_x}\right)^{-1} Q_{nm}(z_x) B_{nmy} = C_{nm0}\, Q_{nm}(z_x) Q_{nm}(z_y) \quad (86)$$

and

$$C_{nm0} = \frac{P_{nm}(z_0)}{Q_{nm}(z_0)} - \frac{\dfrac{\partial P_{nm}(z_0)}{\partial n_x}}{\dfrac{\partial Q_{nm}(z_0)}{\partial n_x}} \quad (87)$$

We try now to express the factors C_{nm0}. For this purpose we approach the derivatives of P_{nm} and Q_{nm} in the direction of the outer unit normal $\boldsymbol{n}$ of our ellipsoid of revolution.

8 Normal Derivatives of Legendre's functions

It is obvious that

$$\frac{\partial P_{nm}(z_0)}{\partial n_x} = \frac{dP_{nm}\left(i\frac{u}{E}\right)}{du}\Bigg|_{u=b}\;\frac{du}{dn_x}\bigg|_{u=b} \quad (88)$$

where

$$\frac{dP_{nm}\left(i\frac{u}{E}\right)}{du} = \frac{i}{E}\frac{dP_{nm}(z)}{dz} \quad (89)$$

and

$$\frac{du}{dn_x}\bigg|_{u=b} = \frac{a}{\sqrt{b^2+E^2\sin^2\beta_x}} \quad (90)$$

as one can easily deduce from differential geometric considerations and from the metric associated with the coordinates u,β,λ, see e.g. Heiskanen and Moritz (1967). In consequence we have

$$\frac{\partial P_{nm}(z_0)}{\partial n_x} = \frac{ia}{E\sqrt{b^2+E^2\sin^2\beta_x}}\frac{dP_{nm}(z_0)}{dz} \quad (91)$$

Moreover, using the definition of Legendre's functions together with recursion formulas valid for these functions, we can deduce that

$$\frac{dP_{nm}}{dz} = -n\frac{z}{1-z^2}P_{nm} - \frac{n+m}{1-z^2}P_{n-1,m} \quad (92)$$

see e.g. Hobson (1952) or Lense (1950), so that

$$\frac{\partial P_{nm}(z_0)}{\partial n_x} = -\frac{n\,b}{a\sqrt{b^2+E^2\sin^2\beta_x}}P^*_{nm}(z_0) \quad (93)$$

where

$$P^*_{nm}(z_0) = P_{nm}(z_0) - \frac{1}{z_0}\frac{n+m}{n}P_{n-1,m}(z_0) \quad (94)$$

Similarly for Q_{nm} we have

$$\frac{\partial Q_{nm}(z_0)}{\partial n_x} = -\frac{(n+1)\,b}{a\sqrt{b^2+E^2\sin^2\beta_x}}Q^*_{nm}(z_0) \quad (95)$$

where

$$Q^*_{nm}(z_0) = Q_{nm}(z_0) + \frac{n-m+1}{z_0\,(n+1)}Q_{n+1,m}(z_0) \quad (96)$$

see Holota (2000b). Note that the expression for the derivative of Q_{nm} in the direction of the normal $\boldsymbol{n}$

can be found, but in another system of coordinates associated with the ellipsoid of revolution, also in Hotine (1969) or in Yurkina (1994).

Finally, combining the results above, we obtain

$$C_{nm0} = \frac{P_{nm}(z_0)}{Q_{nm}(z_0)} - \frac{n}{n+1}\frac{P^*_{nm}(z_0)}{Q^*_{nm}(z_0)} \tag{97}$$

and it is clear at this stage that the next step in the detailed computation of C_{nm0}, and thus also C_{nmxy}, requires a suitable representation of the functions P_{nm} and Q_{nm}.

It seems that among a number of possibilities the following formulas are the most convenient:

$$\begin{aligned} P_{nm}(z) &= \frac{(2n)!}{2^n n!(n-m)!}(z^2-1)^{\frac{n}{2}} \\ &\times F\left(-\frac{n-m}{2}, -\frac{n+m}{2}, -\frac{2n-1}{2}; \frac{1}{1-z^2}\right) \end{aligned} \tag{98}$$

$$\begin{aligned} Q_{nm}(z) &= (-1)^m \frac{2^n n!(n+m)!}{(2n+1)!}(z^2-1)^{-\frac{n+1}{2}} \\ &\times F\left(\frac{n+m+1}{2}, \frac{n-m+1}{2}, \frac{2n+3}{2}; \frac{1}{1-z^2}\right) \end{aligned} \tag{99}$$

Here $F(\alpha,\beta,\gamma;x)$ is the hypergeometric function. The formulas coincide with Bateman and Erdély (1965), where for the construction of P_{nm} and Q_{nm} Legendre's gamma function $\Gamma(x)$ was used and also with Hobson (1952), where on the contrary Gauss' factorial function $\Pi(x)$ was applied.

Passing to the related hypergeometric series expression for $F(\alpha,\beta,\gamma;x)$, we have

$$\begin{aligned} P_{nm}(z) &= \frac{(2n)!}{2^n n!(n-m)!}(z^2-1)^{\frac{n}{2}} \\ &\times \left[1 - \frac{(n-m)(n+m)}{2(2n-1)}\frac{1}{1-z^2} + \ldots\right] \end{aligned} \tag{100}$$

and

$$\begin{aligned} Q_{nm}(z) &= (-1)^m \frac{2^n n!(n+m)!}{(2n+1)!}(z^2-1)^{-\frac{n+1}{2}} \\ &\times \left[1 + \frac{(n-m+1)(n+m+1)}{2(2n+3)}\frac{1}{1-z^2} + \ldots\right] \end{aligned} \tag{101}$$

Thus it is obvious that the subsequent computation is mostly of technical nature, though it is not elementary. However, for its extent (and also the given page limit) we leave it for a next paper.

Remark 9. An alternative way how to express the functions P_{nm} and Q_{nm} may be also found in Sona (1995), subsequently used in Sanso and Sona (2001). It offers a relatively simple and stable approximation of the exact values of ellipsoidal harmonics, but within a limit layer theory.

9 Concluding remarks - A problem on a more complicated boundary

In the determination of the gravity field of the Earth the application of the theory of boundary-value problems has to comply with a rather complex physical reality. Consider e.g. the gravimetric boundary-value problem and try to show its tie to the explanations above.

First recall that W and U denote the gravity and the standard potential of the Earth, respectively. Thus $g = |\,\mathbf{grad}\, W\,|$ is the measured gravity and by analogy $\gamma = |\,\mathbf{grad}\, U\,|$ means the normal gravity. In addition $T(x) = W(x) - U(x)$ is the disturbing potential and $\delta g(x) = g(x) - \gamma(x)$ is the gravity disturbance. We assume that g is corrected for the gravitational interaction with the Moon, the Sun and the planets, for the precession and nutation of the Earth and so on.

The solution domain Ω is the exterior of the Earth and we can write that the problem is to find T such that

$$\Delta T = 0 \quad \text{in} \quad \Omega \tag{102}$$

$$\langle s, \mathbf{grad}\, T\rangle = -\delta g \quad \text{on} \quad \partial\Omega \tag{103}$$

where $s = -(1/\gamma)\,\mathbf{grad}\, U$, $\langle .,. \rangle$ means the inner product of two vectors in R^3. Moreover, T is assumed regular at infinity, i.e., for O denoting Landau's symbol, $T = O(|\,x\,|^{-1})$ as $x \to \infty$.

Following the concept of the so-called weak solution, as in Sect. 2, we assume that on the Cartesian product $W_2^{(1)}(\Omega) \times W_2^{(1)}(\Omega)$ we have a bilinear form $A(u,v)$ such that our boundary-value problem may be written in terms of an integral identity

$$A(T,v) = \int_{\partial\Omega} vf\, dS \tag{104}$$

valid for all $v \in W_2^{(1)}(\Omega)$. Its solution is defined as a function $T \in W_2^{(1)}(\Omega)$ and f is assumed square integrable on $\partial\Omega$, cf. Holota (1997). Note that the tie of problem (104) to that given by Eqs. (102) and (103) requires that $f = \gamma(\partial U / \partial n)^{-1}\delta g$.

Similarly as we have already mentioned in connection with Eq. (14), also Eq. (104) offers a natural starting point for an approximation solution. However, the gravimetric boundary-value problem is an oblique derivative problem and for this reason the bilinear form $A(u,v)$ has a more complex structure: $A(u,v) = A_1(u,v) - A_2(u,v)$, where

$$A_1(u,v) = \int_\Omega \langle \boldsymbol{grad}\, u, \boldsymbol{grad}\, v \rangle \, dx \tag{105}$$

$$A_2(u,v) = \int_\Omega \langle \boldsymbol{grad}\, v, \boldsymbol{a} \times \boldsymbol{grad}\, u \rangle \, dx + \int_\Omega v \langle \boldsymbol{curl}\, \boldsymbol{a}, \boldsymbol{grad}\, u \rangle \, dx \tag{106}$$

and $\boldsymbol{a} = (a_1, a_2, a_3)$ is a vector field such that the components a_i and also $|\boldsymbol{x}|(\boldsymbol{curl}\,\boldsymbol{a})_i$, $i = 1,2,3$, are Lebesgue measurable functions defined and bounded almost everywhere on Ω, see Holota (1997). Moreover, we have to assume that on the boundary $\partial\Omega$ the vector $\boldsymbol{\sigma} = \boldsymbol{s} / \langle \boldsymbol{s}, \boldsymbol{n} \rangle$ and the field $\boldsymbol{a}$ are coupled so that $\boldsymbol{\sigma} = \boldsymbol{n} + \boldsymbol{a} \times \boldsymbol{n}$.

We may then consider a sequence of functions $[u_m]_{m=0}^{\infty}$ defined by the following equations

$$A_1(u_{m+1}, v) = \int_{\partial\Omega} vf \, dS + A_2(u_m, v) \tag{107}$$

which are assumed to hold for all $v \in W_2^{(1)}(\Omega)$ and $m = 0,1,\ldots,\infty$. In Holota (2000a) it has been shown that $[u_m]_{m=0}^{\infty}$ is a Cauchy sequence in $W_2^{(1)}(\Omega)$ and that it converges to a function $u \in W_2^{(1)}(\Omega)$ in the norm $\|.\|_1 = (.,.)_1^{1/2}$.

In addition, one can deduce that under certain regularity assumptions the iterations in Eq. (107) may be interpreted as follows:

$$A_1(u_{m+1}, v) = \int_{\partial\Omega} v f_m \, dS \tag{108}$$

for all $v \in W_2^{(1)}(\Omega)$ while

$$f_m = f - \frac{\partial u_m}{\partial t} \tan(\boldsymbol{s}, \boldsymbol{n}) \tag{109}$$

with $\partial / \partial t$ denoting the derivative in the direction of $\boldsymbol{t} = (\boldsymbol{\sigma} - \boldsymbol{n}) / |\boldsymbol{\sigma} - \boldsymbol{n}|$ which obviously is tangential to $\partial\Omega$ (and exists almost everywhere on Lipschitz' boundary $\partial\Omega$). Details on the computation of the correction term $(\partial u_m / \partial t)\tan(\boldsymbol{s},\boldsymbol{n})$ and the possible use of the components of the deflection of the vertical have been discussed in Holota (2000a).

Formally, Eq. (108) has the same structure as Euler's necessary condition for the quadratic functional connected with Neumann's problem to have a minimum at the point u_{m+1}. Thus every iteration step may be interpreted as a minimization problem, while the iteration process itself makes it possible to solve the problem for the oblique derivative in the direction given by the vector $\boldsymbol{s}$.

Acknowledgements. The work on this paper was supported by the Grant Agency of the Czech Republic through Grant No. 205/01/1463. The presentation of the paper at the 5th Hotine-Marussi Symposium on Mathematical Geodesy, Matera, 17-21 June 2002 was partly sponsored by the Ministry of Education, Youth and Sports of the Czech Republic through Projects No. LA 089 and No. LN 00A005. All this support is gratefully acknowledged. Sincere thanks go also to two anonymous reviewers for very valuable suggestions. They all were instituted in the text. In particular this concerns Remarks 1 and 4.

References

Ambrouche C, Girault V and Giroire J (1997) Dirichlet and Neumann exterior problems for the *n*-dimensional Laplace operator. An approach in weighted Sobolev spaces. J. Math. Pures Appl., 76: 55-81

Antonov VA, Timoshkova EI and Kholshevnikov KV (1983) On the possibility of geopotential representation by means of the point masses system on the basis of satellite observations. In: Holota P (ed.) Proc. Int. Symp. Figure of the Earth, the Moon and other Planets, Prague, Sept. 20-25, 1982. Monograph Series of the Res. Inst. of Geod., Topogr. and Cartogr., Prague: 291-304 (in Russian)

Balmino G (1972) Represeantation of the Earth potential by buried masses. In: Henriksen SW, Mancini A and Chovitz BH. (eds.) The Use of Artificial Satellites for Geodesy. Geophysical Monograph 15, American Geophysical Union, Washington, D.C.

Barthelems F (1986) Untersuchungen zur Approximation des äusseren Gravitationsfeldes der Erde durch Punktmassen mit optimierten Positionen. Veröff. d. Zentralinst. für Physik d. Erde, Nr. 92, Potsdam

Bateman H and A Erdély A (1965) Higer Transcendental Functions, Volume 1. Russion translation of the English original published by Nauka Publishers, Moscow

Freeden W, Gervens T and Schreiner M (1998) Constructive approximation on the sphere – With applications to geomathematics. Claredon Press, Oxford

Freeden W and Michel V (1999) Constructive approximation and numerical methods in geodetic research today – An attempt at a categorization based on an uncertainty principle. Journal of Geodesy, 73: 452-465

Garabedian PR (1964) Partial differential equations. John Wiley and Son, New York

Giaquinta M (2000) Hilbert e il calcolo delle variazioni. Le Matematiche, Vol. LV, Supplemento n. 1: 47-58

Heikkinen M (1981) Solving the shape of the Earth by using digital density models. The Finnish Geod. Inst., Rep. 81:2, Helsinki.

Heiskanen WA and Moritz H (1967) Physical geodesy. W.H. Freeman and Company, San Francisco and London

Hobson EW (1952) The theory of spherical and ellipsoidal harmonics. Inostrannaya Literatura Publishers, Moscow. (Russian translation of the English original published by the University Press, Cambridge, 1931)

Holota P (1979) Representation of the Earth's gravity field by the potential of point masses. Obs. of Artif. Sat. of the Earth, No. 18, Warszawa-Lodz: 159-169
Holota P (1997) Coerciveness of the linear gravimetric boundary-value problem and a geometrical interpretation. Journal of Geodesy, 71: 640-651
Holota P (2000a) Direct method in physical geodesy. In: Schwarz KP (ed.) Geodesy Beyond 2000, The Challenges of the First Decade. IAG Symposia, Vol. 121, Springer-Verlag: 163-170
Holota P (2000b) Variational methods in the recovery of the gravity field – Galerkin's matrix for an ellipsoidal domain. In: Sideris MG (ed.) Gravity, Geoid and Geodynamics 2000, GGG2000 IAG Intl. Symp., Banff, Alberta, Canada, July 31 - August 4, 2000. Intl. Assoc. of Geodesy Symposia, Vol. 123, Springer, Berlin-Heidelberg-New York etc., 2001: 277-283
Holota P (2001) Variational methods in the representation of the gravitational potential. Presented at the AROPA workshop, "Institute d'Europe", Castle of Münsbach, Luxembourg, 23-27 October 2001. Cahiers du Centre Européen de Géodynamique et de Séismologie (to appear)
Holota P (2002a) Model refinements in the recovery of the gravity field of the Earth - Variational approach [CD-ROM]. Proc. IAG 2001 Scientific Assembly – Vistas for Geodesy in the New Millennium, 2-7 September 2001, Budapest, Hungary, IAG and Hungarian Acad. of Sciences
Holota P (2002b) Green's function and external masses in the solution of geodetic boundary-value problem. Presented at the 3rd Meeting of the IAG Intl. Gravity and Geoid Commission, Thessaloniki, Greece, August 26-30, 2002. Ziti Publishers (accepted for publication)
Hotine M. (1969) Mathematical geodesy. ESSA Monographs, Washington, D.C.
Ihde J, Schirmer U, Stefani F and Töppe F (1998) Geoid modelling with point masses. In: Vermeer M and Ádám J (eds.) Second Continental Workshop on the Geoid in Europe, Budapest, Hungary, March 10-14, 1998, Proceedings, Reports of the Finnish Geodetic Inst. No. 98:4, Masala: 199-204
Kautzleben H and Barthelmes F (1983) Point mass representation of the Earth's gravity field. In: Holota P (ed.) Proc. Int. Symp. Figure of the Earth, the Moon and other Planets, Prague, Sept. 20-25, 1982. Monograph Series of the Res. Inst. of Geod., Topogr. and Cartogr., Prague: 307-313
Kellogg OD (1953) Foundations of potential theory. Dover Publications, Inc., New York
Krarup T (1969) A contribution to the mathematical foundation of physical geodesy. Danish Geod. Inst., Publ. No. 44, Copenhagen
Kufner A, John O and Fucik S (1977) Function spaces. Academia, Prague
Lehmann R (1992) Geoid computation in the Gulf of Bothnia by free-positioned point masses. In: Holota P and Vermeer M (eds.) Proc. First continental workshop on the geoid in Europe, Prague, May 11-14, 1992. Research Inst. of Geod., Topog. and Cartog., Prague, in co-operation with IAG-Subcommis. for the Geoid in Europe, Prague: 428-443
Lehmann R (1993) The method of free-positioned point masses – Geoid studies on the Gulf of Bothnia. Bull. Geod., Vol. 67: 31-40
Lense J (1950) Kugelfunktionen. Akad. Vlg., Leipzig
Marchenko AN (1983) On the use of fundamental solutions of Laplace's equation for the determination of the gravitational field and figure of the Earth. In: Holota P (ed.) Proc. Int. Symp. Figure of the Earth, the Moon and other Planets, Prague, Sept. 20-25, 1982. Monograph Series of the Res. Inst. of Geod., Topogr. and Cartogr., Prague: 317-328 (in Russian)
Marchenko AN (1998) Parametrization of the Earth's gravity field – Point and line singularities. Lviv Astronomical and Geodetic Society, Lviv
Meschowski H (1962) Hilbertsche Räume mit Kernfunktion. Springer, Berlin
Michlin AG (1970) Variational methods in mathematical physics. Nauka Publisher, Moscow (in Russian); also in Slovakian: Alpha, Bratislava, 1974 and in English: Pergamon Press, New York, 1964
Moritz H (1980) Advanced physical geodesy. H Wichmann Karlsruhe and Abacus Press Tunbridge Wells Kent
Necas J (1967) Les méthodes directes en théorie des équations elliptiques. Academia, Prague
Necas J and Hlavacek I (1981) Mathematical theory of elastic and elasto-plastic bodies: An introduction. Elsevier Sci. Publ. Company, Amterdam-Oxford-New York
Neyman YuM (1979) A variational method of physical geodesy. Nedra Publishers, Moscow (in Russian)
Rektorys K (1974) Variacni metody v inzenyrskych problemech a v problemech matematicke fyziky. SNTL Publishers of Technical Literature, Prague 1974; also in English: Variational methods. Reidel Co., Dordrecht-Boston, 1977
Sanso F (1986) Statistical methods in physical geodesy. In: Sünkel H (ed.) Mathematical and numerical techniques in physical geodesy. Lecture Notes in Earth Sciences, Vol. 7, Springer-Verlag, Berlin-Heidelberg-New York etc.: 49-155
Sanso F and Sona G (2001) ELGRAM: an ellipsoidal gravity model manipulator. Anno LX / Bollettino di Geodesia e Scienze Affini – N.3: 215-226
Sona G (1995) Numerical problems in the computation of ellipsoidal harmonics. Journal of Geodesy, 70: 117-126
Stromeyer D and Ballani L (1984) Uniqueness of the inverse gravimetric problem for point mass models. Manuscripta geodaetica, Vol. 9, No. 1/2: 125-136
Tscherning CC (1975) Application of collocation. Determination of a local approximation to the anomalous potential of the Earth using "exact" astro-gravimetric collocation. In: Bosowski B and Martensen E (eds.) Methoden und Verfaren der Mathematischen Physik, Vol. 14: 83-110
Weightman JA (1967) Gravity, geodesy and artificial satellites, a unified analytical approach. In: Veis G (ed.) Use of Artificial Satellites for Geodesy, Vol. 2. Natl. Technical University, Athens: 467-486
Yurkina MI (1994) On a more precise Stoke's series. Studia geoph. et geod., Vol. 38, No. 4: 325-332
Zhidkov NP (1977) Linear approximations of functionals. Moscow University Publishers, Moscow (in Russian)

The estimation theory for random fields in the Bayesian context; a contribution from Geodesy

Fernando Sansò, Giovanna Venuti

DIIAR - Sezione Rilevamento, Politecnico di Milano, P.zza Leonardo da Vinci 32, I-20133 Milano, Italy.

Abstract. The paper first reviews the existing mathematical theory for the estimation of a random field T, with known covariance C, from a finite vector of observations, related to T by linear functionals, in the framework of a Bayesian approach.

In particular in Section 2 and Section 3 the equivalence between ordinary collocation formulas, their explaination in terms of generalized random fields and the full probabilistic picture is demonstrated. Then the more general problem of estimating from data both T and its covariance C is tackled; in Section 4 it is shown how to reduce the problem by using prior invariance principles and a prior vague information on the regularity of the field T. In this case it is shown how to construct the posterior distribution of the unknowns. Then in Section 5 the theory of logarithmic derivatives of infinite dimensional distributions is recalled and in Section 6 the corresponding Maximum a-Posteriori equations are constructed. A theorem of existence of at least one solution is proved too. Conclusions follow.

1 Introduction and outline of the paper

Since long time, in Geodesy, collocation theory has been used as a "general" tool for the approximation of the anomalous gravity potential T; schematically one can say that observations are modelled as functionals of T and the prediction of some other functional of T is required. This theory has been interpreted either in a deterministic mode, first by Krarup (1969) who has seen it as an application of the Tychonov approach to improperly posed problems, or in a stochastic mode, as it has been originally conceived by Heiskanen and Moritz (1967) as an application of the Wiener-Kolmogorov theory to spatial random fields.

Due to an obvious formal similarity of the formulas issuing from the two approaches, the two solutions have been considered as equivalent on condition that the interpolation space of the deterministic approach would be equal to the Hilbert Space, H_C, with reproducing kernel identical to the covariance function C.

Subsequently, the two approaches have been considered as incompatible because the norm of $\| T \|^2_{H_C}$, as a random variable, cannot be bounded in the mean (cf. Tscherning (1977)).

In addition the stochastic approach has been criticised as unreliable because, if T is assumed to be Gaussian (as often it has been reasonably done) then one can prove that it is not possible to provide a consistent estimate of the covariance function even if we knew the field over the whole sphere (cf. Lauritzen (1973)). Such problems have been overcome; the latter by proving that a different probabilistic scheme could provide consistent covariance estimates (cf. Moritz (1980); Sansò (1986)), the former by showing that the use of the concept of a generalized random field $\{T\}$, for which $\| T \|^2_{H_C}$ is not necessarily bounded in the mean, can provide a fully logical equivalence between the two approaches (cf. Sansò (1986)).

Even more, the use of a Bayesian concept, which in any way has been extensively applied in geophysics (cf. Tarantola and Valette (1982); Tarantola (1987)), leads to a very nice interpretation of the identification of the covariance with the reproducing kernel, referring to the idea of applying a non-informative prior to a random field. This more or less closes completely the problem at this level with the only proviso that the estimation of the covariance function has then to be performed separately by exploiting a "prior" stochastic invariance principle, which has to be submitted to some testing before it can be accepted as a standard statistical procedure.

Yet one could object that a two-steps approach (estimate the covariance function and then, assuming that the covariance is fixed, predict field functionals) is not satisfactory on a logical ground although one can prove that the result is only weakly dependent on the shape of the covariance (cf. Sansò et al. (2000)). A sound estimation theory, as the Bayesian certainly is, must put on the same ground the estimation of both the covariance function and the predicted functionals of T. This is somewhat analogous to variance-covariance estimation in least squares theory, with the major difference that we have in this case an infinite dimensional problem. The first point here is that in infinite dimensions it is not possible to speak of a probability density because there exists no analogous of the Lebesgue measure; therefore it is not easy to describe the construction of the posterior distribution, according to the Bayesian approach. This however is possible and it has been described in mathematical terms in a way which is also reconstructed in the paper in Section 3. The most important result on the subject is given in Lehtinen et al. (1989), which however is somewhat technical in the use of topological spaces theory; for this reason a simpler version of the solution is reconstructed in this paper.

Moreover, we observe that in infinite dimensional contexts it is not possible to invoke the principle of maximum likelihood, which in Bayesian theory becomes *Maximum a-Posteriori*, because we have no density. Nevertheless a concept similar to maximum likelihood can be constructed by using shifts of the (non-invariant but quasi-invariant) probability measure; these shifts are characterized by a Radon-Nikodym (cf. Bogachev (1996)) density which leads to a direct generalization of the ordinary logarithmic derivative of finite dimensional densities. The corresponding equation is in fact linked to a minimum principle which, at least for Gaussian processes, can be directly studied and proved to have a reasonable solution.

2 Collocation formulas and the theory of generalized random fields when the covariance is known

Let us briefly summarize the context in which Collocation has grown as an estimation method in Geodesy with the two main interpretations, stochastic and deterministic, and look for the logical solution of some inconsistencies that appeared in the comparison of the two.

a) Stochastic approach: we assume that the "unknown field" $u(\underline{t}), \underline{t} \in A$ is in fact random, i.e. $u(\underline{t}, \omega)$ with $\omega \in \Omega$, a set on which some probability distribution is defined. By definition, given any two linear functionals L, G acting on $u(\underline{t}, \omega)$ as function of $\underline{t}$, we define the random variables (r.v.)

$$\begin{cases} U = L_{\underline{t}}\{u(\underline{t}, \omega)\} \\ V = G_{\underline{t}}\{u(\underline{t}, \omega)\} \end{cases} \quad (1)$$

and we assume that $U, V \in \mathcal{L}^2(\omega)$, i.e. that they are endowed with finite mean square values. If in addition we stipulate that

$$E\{u(\underline{t}, \omega)\} \equiv 0 \quad (2)$$

then we put

$$C(\underline{t}, \underline{t}') = E\{u(\underline{t}, \omega)u(\underline{t}', \omega)\} \quad (3)$$

$$C(L, G) = L_{\underline{t}}\{G_{\underline{t}'}C(\underline{t}, \underline{t}')\} = \quad (4)$$
$$= E\{L_{\underline{t}}[u(\underline{t}, \omega)]G_{\underline{t}'}[u(\underline{t}', \omega)]\} .$$

With a small extension of the same notation we put

$$C(\underline{L}, \underline{G}) \equiv [L_i\{G_k C(\underline{t}, \underline{t}')\}] = \quad (5)$$
$$\equiv E\{\underline{L}(u)\underline{G}^+(u)\} .$$

Now assume that we have an observation vector

$$\underline{Y} = \underline{L}(u) \in R^m \quad (6)$$

and from it we want to predict a single variable, expressed as a functional of u

$$V = G(u) ; \quad (7)$$

the criterion is that of minimizing the mean square estimation error (Wiener-Kolmogorov theory) within a class of linear estimators.
If we know $C(\underline{t}, \underline{t}')$ and we propagate the covariance to all the functionals, then the solution is given by

$$\widehat{G(u)} = \widehat{V} = C(G, \underline{L})C^{-1}(\underline{L}, \underline{L})\underline{Y} \quad (8)$$

with the variance of the estimation error, $e = V - \widehat{V}$,

$$\sigma^2(e) = \quad (9)$$
$$C(G, G) - C(G, \underline{L})C^{-1}(\underline{L}, \underline{L})C(\underline{L}, G) .$$

This is nothing but standard linear regression theory and it has been introduced in Geodesy (cf. Moritz (1980)), together with rules for the estimation of the covariance function from data.

b) Deterministic approach: let us switch now to a completely deterministic picture introduced by Krarup (1969). Assume that by definition your unknown u does belong to a Hilbert space with reproducing kernel $K(\underline{t}, \underline{t}')$, i.e. such that

$$< K(t, \cdot), u(\cdot) >_{H_K} \equiv u(t) \ , \ \forall h \in H_K \ . \quad (10)$$

Then if $\underline{L}$ is a vector of observational functionals and G a functional we want to predict, we expect them to be bounded in H_K. Therefore we expect to have Riesz representers of such functionals which, due to (10), can be written as

$$< K(\underline{L}, \cdot), u(\cdot) >_{H_K} = \underline{L}(u) \ , \quad (11)$$
$$< K(G, \cdot), u(\cdot) >= G(u) \ .$$

The observation equations now read

$$\underline{Y} = \underline{L}(u) =< K(\underline{L}, \cdot), u(\cdot) >_{H_K} \quad (12)$$

showing that by (12) we fix the projection of u onto the linear space generated by the functions $K(\underline{L}, \cdot)$, i.e.

$$\begin{cases} u_0 = \arg \underset{\overline{u} \in M_0}{\text{Min}} \parallel u - \overline{u} \parallel_{H_K} \\ M_0 = \text{Span}\{K(\underline{L}, \cdot)\} = \{\underline{\lambda}^+ K(\underline{L}, \cdot), \underline{\lambda} \in R^m\} . \end{cases} \tag{13}$$

Now this u_0 is characterized by (13) or by the dual minimum norm condition

$$\begin{cases} \underline{L}(u_0) = \underline{Y} \\ \parallel u_0 \parallel_{H_K} = \min \end{cases} ; \tag{14}$$

both principles (13) and (14) have in fact as solution

$$u_0 \equiv K(\underline{L}, \cdot)^+ K(\underline{L}, \underline{L})^{-1} \underline{Y} \tag{15}$$

corresponding to the orthogonal projection of u onto M_0.

Now if we apply $G(\ )$ to u_0 to get an "optimal" estimate of $G(u)$, we find

$$\widehat{G(u)} = G(u_0) = K(G, \underline{L}) K(\underline{L}, \underline{L})^{-1} \underline{Y} \tag{16}$$

and this estimator becomes indeed identical to (8), on condition that we identify

$$K(\underline{t}, \underline{t}') = C(\underline{t}, \underline{t}') . \tag{17}$$

At this point it seemed that we were attaining a synthesis of the two points of view, was it not because of the following computation: take any complete orthogonal system $\{e_n(\underline{t})\}$ in $H_K \equiv H_C$ then, as it is known (cf. Sansò (1986))

$$C(\underline{t}, \underline{t}') \equiv K(\underline{t}, \underline{t}') \equiv \sum_n e_n(\underline{t}) e_n(\underline{t}') , \tag{18}$$

so that

$$\begin{aligned} &\parallel u \parallel_{H_K}^2 = \sum_n u_n^2 \\ &u_n = < e_n, u >_{H_C} \\ &E\{u_n\} = 0 \quad E\{u_n^2\} = \sigma^2(u_n) = \\ &= < e_n(\underline{t}), < e_n(\underline{t}'), C(\underline{t}, \underline{t}') >_{H_C} >_{H_C} = \\ &= \parallel e_n \parallel_{H_C}^2 = 1 \end{aligned}$$

and therefore

$$E\{\parallel u \parallel_{H_C}^2\} = +\infty . \tag{19}$$

This was interpreted as a sign of the fact that if we put $C = K$ then $u \notin H_C$ and therefore no stochastic interpretation can be consistent. The dilemma was solved by one of the authors (cf. Sansò (1986)) showing that we really do not need to claim that $u \in H_K$ in order to give a meaning to the random variables of the type

$$< h, u >_{H_K} = V , \ h \in H_K ; \tag{20}$$

on the contrary we can rely on the concept of generalized random field (g.r.f.) where $\{u\}$ is rather seen as a bounded operator

$$h \to < h, u >_{H_K}$$

for instance such that, $E\{< h, u >_{H_K}\} = 0$ and

$$\begin{aligned} E\{< h, u >_{H_K}^2\} &\overset{Def}{=} < h, Kh >_{H_K} \leq \\ &\leq \text{Const} \parallel h \parallel_{H_K}^2 . \end{aligned} \tag{21}$$

It was of help in this generalized interpretation to use the Bayesian concept that any unknown quantity has to be described as a r.v., as well as to consider that we never really need the whole field $u(\underline{t}, \omega)$ but we rather need one functional of u at a time, be it for instance the pointwise value of u at any $\underline{t}$, namely [1]

$$ev_{\underline{t}}(u) \equiv < K(\underline{t}, \cdot), u(\cdot, \omega) >_{H_K} = u(\underline{t}, \omega) . \tag{22}$$

Let us now concentrate a little on the expression (21), providing the definition of the positive selfadjoint operator C, i.e. the covariance operator of the g.r.f. u in H_K. The first remark is that as an operator C does depend on the topology of H_K. By this we mean the following: let us take two reproducing kernels $K(\underline{t}, \underline{t}'), K'(\underline{t}, \underline{t}')$ with the corresponding Hilbert spaces H, H' and assume that the set

$$D \equiv H \cap H'$$

is at least dense in both H and H' (one more usual situation is when one of the two spaces is densely embedded into the other). Then for any $g \in D$ and any linear functional L that is bounded on both H and H', i.e.

$$\max\{K(L, L), K'(L, L)\} < +\infty , \tag{23}$$

we can compute

$$L(g) = < K(L, \cdot), g >_H \equiv < K'(L, \cdot), g >_{H'} \tag{24}$$

that is to say that the same functional L has indeed different Riesz representers in H and H'.

Now if it happens that for a given g.r.f. u the r.v. $L(u)$ is in $\mathcal{L}^2$, L satisfies (23), i.e. $\forall L$ such that $K(L, \cdot) \in D$, we have accordingly the identity

$$< K(L, \cdot), u >_H \equiv < K'(L, \cdot), u >_{H'} \tag{25}$$

(as $\mathcal{L}^2$ r.v.), so that

$$\begin{aligned} E\{L(u)^2\} &\equiv < K(L, \cdot), CK(L, \cdot) >_H \equiv \\ &\equiv < K'(L, \cdot), C'K'(L, \cdot) >_{H'} \end{aligned} \tag{26}$$

[1] It is interesting to note that $u(\underline{t}, \omega)$ for $\underline{t}$ fixed is an $\mathcal{L}^2$ function of ω, which of course needs not to have point values for every ω.

Since $\{K(L,\cdot)\} \in D$ which is dense in H and H', the expression (26) can be extended by continuity to both spaces if both C and C' are bounded. The point is that C and C' cannot be in general the same operator, as they represent the covariance properties of the same g.r.f. on different spaces. In a way choosing a certain Hilbert space H_K is completely free as far as we find that all the functionals L we want to use are bounded $\mathcal{L}^2$ functionals, i.e. are such that

$$E\{L(u)^2\} =< K(L,\cdot), CK(L,\cdot) >_{H_K} < +\infty ,$$

or, said in other words, as far as C is bounded in H_K. A fundamental consequence of this remark is that the solution (8), (9) is exactly the same that we find in the formalism of g.r.f.'s and it is basically independent of K, i.e. of H_K, as long as the covariance function $C(\underline{t},\underline{t}')$ is given. In fact we have for any two bounded $\mathcal{L}^2$ functionals G and L

$$\begin{aligned} &G_t\{L_{\underline{t}'}C(\underline{t},\underline{t}')\} = C(G,L) = \\ &= E\{G_{\underline{t}}(u(\underline{t}))L_{\underline{t}'}(u(\underline{t}'))\} = \\ &= E\{G_{\underline{t}}[< K(\underline{t},\cdot),u(\cdot) >_{H_K}] \cdot \\ &\cdot L_{\underline{t}'}[< K(\underline{t}',\cdot),h(\cdot) >_{H_K}]\} = \\ &= E\{< K(G,\cdot),h(\cdot) >_{H_K} < K(L,\cdot),h(\cdot) >_{H_K}\} \\ &=< K(G,\cdot),CK(L,\cdot) >_{H_K} \qquad (27) \end{aligned}$$

proving that all covariances of the g.r.f. u on H_K can be directly computed from "functionals" and the "covariance function". Then why to introduce this heavier formalism to get on old result? Because in this way we can finally achieve the unification of the two points of view presented at the beginning of the section. In fact, as we said, we are free to choose K as far as C stays bounded on H_K; but then it is natural to consider as the most extended case (i.e. the maximal class of admissible functionals) the particular H_K in which

$$C = I . \qquad (28)$$

To understand the meaning of (28) we notice that taking $G_{\underline{t}} = ev_{\underline{t}}, L_{\underline{t}'} = ev_{\underline{t}'}$ in (27), we find

$$C(\underline{t},\underline{t}') =< K(\underline{t},\cdot), CK(\underline{t}',\cdot) > , \qquad (29)$$

which specialized to the case (28) gives

$$C(\underline{t},\underline{t}') \equiv K(\underline{t},\underline{t}') , \qquad (30)$$

i.e. precisely the unification of (8) and (16).

Another interesting feature of (28) comes from its stochastic interpretation; namely, under (28), for any function $h \in H_K$ of norm equal to 1 ($\| h \|_{H_K} = 1$) we get

$$E\{< h,u >^2_{H_K}\} =< h,h >_{H_K} = 1 ,$$

i.e. $< h,u >_{H_K}$ has the same, unit, variance in all directions of the Hilbert space H. This is in our opinion a good way of translating in the language of g.r.f. the Bayesian concept of non-informative prior, when we know that in any way $< h,u >_{H_K}$ has to have finite variance. So far so good, but now the crucial question comes: if we want to use (30) to define the reproducing kernel $K(t,t')$, i.e. the space H_K, who is giving us $C(\underline{t},\underline{t}')$? For the moment this comes only from an application of some stochastic invariance principles and a separate step of estimation. All that sheds a Bayesian light on the equivalence between stochastic and deterministic approach to Collocation, yet this is still unsatisfactory for at least two reasons:

1) even if we accepted to know $C(t,t')$ (note that this has to be part of the prior knowledge!) the true Bayesian approach would imply that we build a posterior distribution of u and not a simple estimate of $(h,u)_H$ for all $h \in H$. This requires of course the specification of the prior distribution, that we will consider in the next paragraphs by using basically normal infinite-dimensional variates, because these are by far the best known and easier to manipulate, though the theory holds in a much more general context (cf. Rozanov (1971); Mandelbaum (1984); Bogachev (1995));
2) as a matter of fact, to accept that we know $C(t,t')$, as derived empirically from an invariance principle, calls for a more general approach starting only from observations $\{\underline{Y}\}$ and, as prior knowledge, the invariance principle, from which already some consequences on the functional form of $C(t,t')$ can be drawn. This will be developed in the rest of the paper.

3 Generalized random fields and their generalized distributions

Let us go back to our Hilbert space H and see what is the "distributional explanation" of the approach in terms of generalized random fields recalled in §2. Let us assume to define, with the help of some CONS $\{e_n\}$ [2] in H, a ladder of subspaces

$$H_N = \text{Span}\{e_n,\ n = 1\ldots, N\} \qquad (31)$$

and to give a family of normal distributions $\{F_N\}$ on H_N satisfying the famous Kolmogorov condition:

$$\begin{aligned} &P_{N+1}\{\Pi_N h \in A,\ \text{arbitrary } (e_{N+1},h)_H\} \equiv \\ &\equiv P_N\{\Pi_N h \in A\} \qquad (32) \\ &\forall A \in \mathcal{B}(H_N),\ \text{Borel } \sigma\text{-algebra,} \end{aligned}$$

[2] We use the acronym CONS for complete orthonormal system.

where we have denoted with Π_N the orthogonal projection on H_N. Such a scheme is said to provide a weak (or generalized) measure P_* on H. We note that P_* is sufficient to define the covariance operator, namely

$$\begin{aligned} & E_*\{< x,h >_H< x,g >_H\} = \\ & = \int < x,h >_H< x,\ell >_H P_*(dx) = \\ & =< h,Cg >_H \end{aligned} \tag{33}$$

at least

$$\forall h,g \in H_0 = \bigcup_N H_N\,, \tag{34}$$

which is a subspace everywhere dense in H.

According to the Minlos-Sazonov theorem (cf. Skorohod (1974); Mandelbaum (1984); Bogachev (1996)) P_* can be extended to a true probability measure P on H, endowed with the Borel σ-algebra generated by cylinder sets, if and only if C can be extended to a nuclear operator on H, i.e. such that

$$TrC = \sum_{i=1}^{+\infty} < e_i, Ce_i >_H < +\infty\,. \tag{35}$$

When C is not nuclear, for instance when

$$C = \sigma^2 I\,, \tag{36}$$

which is in fact the case we want to consider as we try to define an isotropic (non-informative) prior on H, then the space H is so to say "too small" to accommodate a probability distribution P admitting the family $\{P_N\}$ as marginal distributions on H_N.

This statement is precisely the same as saying that under such conditions one cannot obtain a probability distribution P such that, for every $A \in B(H_N)$, the probability of the cylinder
$\mathcal{C}_A \equiv \{x \in H\ ,\ \Pi_N x \in A\}$ is

$$P(\mathcal{C}_A) = P_N(A) \tag{37}$$

and that at the same time

$$P\{\| x \|_H< +\infty\} \equiv P\{x \in H\} = 1\,; \tag{38}$$

yet, one rather gets

$$P\{\| x \|_H= +\infty\} = 1\,,$$

as indirectly observed long ago by Tscherning (1977); in this respect one can read as well Sacerdote and Sansò (1996).

In order to enlarge the space on which we shall place our distribution, it is useful to exploit the standard isometry that maps all separable Hilbert spaces into the coordinate space ℓ^2

$$x \in H \underset{J^{-1}}{\overset{J}{\rightleftarrows}} \underline{x} = \{< x, e_n >_H= x_n\} = Jx \tag{39}$$

$$\| x \|_H^2= |\underline{x}|_{\ell^2}^2 = \sum_{n=1}^{+\infty} x_n^2 = |Jx|_{\ell^2}^2\,. \tag{40}$$

Now we consider ℓ^2 as a subspace of R^∞, namely the space of all real sequences $\underline{x} \equiv \{x_n\}$, which is known to be a complete linear topological locally convex space, endowed with the metric

$$\rho(\underline{x}, \underline{y}) \equiv \sum_{k=1}^{+\infty} 2^{-k} \frac{|x_k - y_k|}{1 + |x_k - y_k|}\,. \tag{41}$$

The dual of R^∞ with respect to the usual Euclidean coupling is just its subspace of sequences $\{\underline{\lambda}\}$ which have only a finite number of elements different from zero, i.e. for some N

$$\underline{\lambda} \equiv \{\lambda_1, \lambda_2 \ldots \lambda_N, 0, 0 \ldots\} \in R_N^\infty\,. \tag{42}$$

We call this space R_0^∞ and we note that

$$\begin{cases} R_N^\infty = JH_N \\ R_0^\infty = \bigcup_N R_N^\infty = J \bigcup_N H_N = JH_0\,, \end{cases} \tag{43}$$

so that by suitably extending the isometry J^{-1} one has the correspondence

$$(R_0^\infty \subset \ell^2 \subset R^\infty) \overset{J^{-1}}{\rightarrow} (H_0 \subset H \subset H^\infty)\,. \tag{44}$$

The two triples of spaces in (44) are referred to as standard triples in the literature and sometimes they are also called *rigged Hilbert spaces*. Now notice that using the isometry J and considering the cylinders C_A in R^∞

$$C_A = \{\underline{x}; \Pi_N \underline{x} \in A \subset R_N\}$$

one can transfer the family of distributions P_N, originally defined in H_N, onto $R_N^\infty \subset R^\infty$.

On the other hand for R^∞ we have the celebrated Kolmogorov's theorem (cf. Skorohod (1974)) maintaining that to any consistent family $\{P_N\}$ on this space, there does correspond a probability distribution P on the same space, endowed with the Borel σ-algebra defined as the minimal σ-algebra containing all cylinder sets. Indeed in this way we have found a space large enough to support a probability distribution which we were trying to define on ℓ^2.

It is then natural to ask oneself whether in this construction we have not so to say exaggerated and used too large a space. The answer is that typically this is the case, yet the construction is general and it does hold under any circumstances. Before we shed some more light on this point we need the definition of covariance operator for the distribution P on R^∞; this is defined, when it exists, through

$$E_{\underline{x}}\{(\underline{\lambda}^+\underline{x})(\underline{\mu}^+\underline{x})\} = \int (\underline{\lambda}^+\underline{x})(\underline{\mu}^+\underline{x})P(d\underline{x}) \equiv \underline{\lambda}^+ C\underline{\mu}\,, \quad \forall \underline{\lambda}, \underline{\mu} \in R_0^\infty\,. \tag{45}$$

Naturally C is an infinite dimensional matrix and it is related to the covariance operator C_H (cf. (33)) acting originally on H, through the relation

$$C = JC_H J^{-1} \equiv JC_H J^+\,, \tag{46}$$

in other words C is the covariance matrix representing the original covariance operator in the coordinate space. In particular when $C_H = \sigma^2 I_H$ in H, then $C = \sigma^2 I$ (I the identity matrix) in R^∞. More generally when C_H is (or it can be extended to) a bounded operator in H one has (cf. (33))

$$< h, C_H h >_H = E\{< x, h >_H^2\} \le c \parallel h \parallel_H^2 \tag{47}$$

which is then mapped into the standard triple $R_0^\infty \subset \ell^2 \subset R^\infty$ by saying that

$$E\{(\underline{\lambda}^+\underline{x})^2\} \equiv \underline{\lambda}^+ C \underline{\lambda} \le c|\underline{\lambda}|_{\ell^2}^2\,; \tag{48}$$

this relation has to hold ar least on R_0^∞ but it is obvious that it can be extended by continuity to the whole ℓ^2. More precisely if we consider $\underline{\lambda}^+\underline{X}$ as a function of the random variable $\underline{X}$ which has in R^∞ the given distribution P and indeed

$$\varphi_{\underline{\lambda}}(\underline{X}) \equiv \underline{\lambda}^+\underline{X} \tag{49}$$

is a function in $\mathcal{L}^2(\underline{X})$, we see that now $\underline{X}$ can be considered as a bounded operator mapping, through the extension of (49), the space ℓ^2 into $\mathcal{L}_L^2$, i.e. the subspace of $\mathcal{L}^2$ obtained by closing, in the $\mathcal{L}^2$ norm, the subspace of linear functions. In other words $\mathcal{L}_L^2$ can be considered as the subspace of $\mathcal{L}^2$ of functions of the type (49) when $\underline{\lambda} \in \ell^2$, plus all the modifications on sets $\subset R^\infty$ of null P measure. But then $\underline{X}$ is precisely the image, under J, of what we have defined on the original space H to be a generalized random field $\{u\}$ and we understood that this is associated with a probability measure that cannot be supported by the too small H, but can always be defined as a generalized probability measure on the much larger space H^∞.

Remark 3.1. Let us go back to the question whether P can be seen as a distribution on some space smaller than R^∞. As it has been already observed, if $TrC < +\infty$, then $P\{\underline{X} \in \ell^2\} = 1$, i.e. P has support directly in ℓ^2 (cf. (35)); so we have to understand what happens in cases in which C is bounded (cf. (48)) but not nuclear. As a typical example let us take the case that P is Gaussian and

$$C = I\,. \tag{50}$$

Let us define ℓ_w^2 as the Hilbert space with weighted norm

$$|\underline{\lambda}|_{\ell_w^2}^2 = \sum \lambda_n^2 w_n = \underline{\lambda}^+ W \underline{\lambda} < +\infty\,. \tag{51}$$

We think of w_n as a sequence tending to zero, since in this case we have strictly

$$\ell^2 \subset \ell_w^2\,. \tag{52}$$

We see then that, according to the already quoted Minlos-Sazonov theorem (cf. Skorohod (1974)),

$$P\{\underline{X} \in \ell_w^2\} = 1 \tag{53}$$

if and only if the matrix operator W is nuclear,

$$TrW = \sum w_n < +\infty\,. \tag{54}$$

In fact, due to (50), $\forall \underline{\lambda} \in R_0^\infty$,

$$E\{(\underline{\lambda}, \underline{X})_{\ell_w^2}^2\} = E\left\{\left(\sum \lambda_n X_n w_n\right)^2\right\} = \sum \lambda_n^2 w_n^2 = (\underline{\lambda}, W\underline{\lambda})_{\ell_w^2}\,, \tag{55}$$

i.e. W is the covariance operator of $\underline{X}$ in ℓ_w^2 and it is nuclear if and only if (54) holds.

Now that we have set up the right stochastic tools we can show that the approach presented in Section (2) does correspond to the standard Bayesian approach, when the non-informative prior of $\underline{X}$, the image of the g.r.f. u, is taken to be

$$\underline{X} \sim \mathcal{N}[0, I]\,. \tag{56}$$

This is the same as saying that $u = J^{-1}\underline{X}$ is a g.r.f. on the rigged Hilbert space $H_0 \subset H \subset H^\infty$, such that

$$u \sim \mathcal{N}[0, I_H]\,, \tag{57}$$

and it means that the variance of any r.v.

$$v = < h, u >_H \equiv \underline{\lambda}^+\underline{X} \quad (\underline{h} = Jh\,,\ \underline{X} = Ju) \tag{58}$$

along any vector $\underline{\lambda}$ or h of given norm, is the same, i.e.

$$E\{v^2\} = \| h \|_H^2 = |\underline{\lambda}|_{\ell^2}^2$$

irrespectively of the direction of h, or $\underline{\lambda}$. We observe here that I_H is in fact represented by the reproducing kernel $K(\cdot,\cdot)$ through the scalar product in H, and that when the probability P is normal (i.e. P_N are N-dimensional normal variates), H is referred to as the Cameron-Martin space of the distribution P, of the g.r.f. $\{u\}$ (cf. Bogachev (1996)). In the coordinate space R^∞ we can say that the Cameron-Martin space of $\underline{X}$ is ℓ^2.

Assume now that some observations on u, or on $\underline{X}$, are taken so that the variables

$$k = 1, 2 \ldots m, \quad Y_k = < h_k, u >_H \equiv \underline{\lambda}_k^+ \underline{X} \quad (59)$$

attain given values.

If the observation functionals of u are written in the form

$$Y_k = L_k(u) ,$$

then we know that

$$h_k(t) = L_k[K(\cdot, t)] = K(L_k, t) \quad (60)$$

We summarize (59) as

$$\underline{Y} = < \underline{h}, u >_H \equiv \Lambda \underline{X} \quad (61)$$

where Λ is now an (m, ∞) matrix with rows $\underline{\lambda}_k^+$ and we introduce in ℓ^2 the finite subspace

$$M \equiv \text{Span}\{\underline{\lambda}_k\} .$$

We observe also that the orthogonal projection of $\underline{X}$ on M, namely

$$\underline{X}_M = \Lambda^+(\Lambda\Lambda^+)^{-1}\Lambda\underline{X} = \Lambda^+(\Lambda\Lambda^+)^{-1}\underline{Y} , \quad (62)$$

is well-defined and fixed because of (61).

We can now exploit a general statement on normal variates, which is true also in R^∞ (cf. Rozanov (1971)), claiming that the component of $\underline{X}$ along the space $M^\perp$ (the orthogonal complement of M in ℓ^2), $\underline{\xi}$, is normal too and has a covariance operator

$$I_{M^\perp} = [I - \Lambda^+(\Lambda\Lambda^+)^{-1}\Lambda] , \quad (63)$$

i.e. the identity in $M^\perp$, so we maintain

$$\underline{\xi} \sim \mathcal{N}[0, I_{M^\perp}] . \quad (64)$$

Since the posterior variate $\underline{X}$ given $\underline{Y}$ can be written in the form

$$\underline{X}|_{\underline{Y}} = \underline{X}_M + \underline{\xi} , \quad (65)$$

where $\underline{X}_M$ is fixed because of (61), we can claim that $\underline{X}|_{\underline{Y}}$ is a normal variate with mean $\underline{X}_M$ and covariance $I_{M^\perp}$. We have now only to pull back the solution (62) and its covariance matrix (63) to recognize that we have found the standard collocation solution. In fact

$$\begin{aligned} \widehat{u} &= J^{-1}\Lambda^+(\Lambda\Lambda^+)^{-1}\underline{Y} = \quad (66) \\ &= [\ldots J^{-1}\underline{\lambda}_k \ldots] \{\underline{\lambda}_k^+\underline{\lambda}_j\}^{(-1)}\underline{Y} ; \end{aligned}$$

but (cf. (59), (60))

$$J^{-1}\underline{\lambda}_k = h_k = K(L_k, \cdot) \quad (67)$$

and

$$\begin{aligned} \underline{\lambda}_k^+\underline{\lambda}_j &= < h_k, h_j >_H = \\ &= < K(L_k, \cdot), K(L_j, \cdot) >_H = K(L_k, L_j) . \quad (68) \end{aligned}$$

Moreover, even the error covariance of $\underline{X}|_{\underline{Y}}$, namely

$$C_{\underline{X}|\underline{Y}} = C_{\underline{\xi}} = I_{R^\infty} - \Lambda^+(\Lambda\Lambda^+)^{-1}\Lambda , \quad (69)$$

agrees with that of $\{\widehat{u}\}$ since through (66), (67), (68) we see that, putting $\varepsilon(\underline{t}) = u(\underline{t}) - \widehat{u}(\underline{t})$, one has

$$\begin{aligned} C_{\varepsilon\varepsilon}(\underline{t}, \underline{t}') &= K(\underline{t}, \underline{t}') + \\ &- \sum_{k,j} K(L_k, \underline{t}) K^{-1}(L_k, L_j) K(L_j, \underline{t}') . \quad (70) \end{aligned}$$

By using (66) through (70) we find the full equivalence with the standard collocation solution as promised.

In closing the section we remark once more that the estimated field $\widehat{u}$ does belong to H although the original distribution P is indeed spread over the whole H^∞, or, in terms of coordinates, $\underline{X}|_{\underline{Y}} \in \ell^2$ while P is spread on the whole R^∞.

4 The *field and covariance*, Bayesian estimation setup

Now we want to go further and set up a Bayesian machine to estimate a field u and its covariance function $C(\underline{t}, \underline{t}')$, based only on data $\underline{Y}$ and some suitable prior hypotheses.

Let us recall that a Bayesian approach is defined by means of several ingredients:

1) the definition of the "observable vector" $\underline{Y}$, which is a linear (in this context) function of u and, possibly, of the noise $\underline{\nu}$, i.e. the model constituted by observation equations, for instance

$$\underline{Y} = < \underline{h}, u >_H + \underline{\nu} , \quad (\dim \underline{Y} = m) , \quad (71)$$

where $\underline{h}$ is a vector of bounded functionals in H,

2) a prior distribution of u, $P_0(du|\vartheta)$, and possibly for all the nuisance parameters ϑ, so that we can put

$$P_0(du, d\vartheta) = P_0(du|\vartheta)P_0(d\vartheta) \; , \qquad (72)$$

such that u is, at least a priori, a g.r.f. on H and therefore the variables $< h_i, u >_H$ are well-defined and have finite variance

3) the stochastic model of the noise, for instance claiming that

$$\underline{\nu} \sim \mathcal{N}[0, \sigma_\nu^2 I_m] \qquad (73)$$

and that $\underline{\nu}$ and u are independent, so that we can build the likelihood

$$L(\underline{Y}|u) \equiv \frac{1}{(2\pi\sigma_\nu^2)^{m/2}} e^{-\frac{1}{2\sigma_\nu^2}|\underline{Y} - <\underline{h}, u>_H|^2} \qquad (74)$$

To cut it short, we can say that we can then apply the Bayesian concept to derive the posterior distribution of u, namely

$$P(du|\underline{Y}) \propto \int_{(\vartheta)} L(\underline{Y}|u)P_0(du|\vartheta)P_0(d\vartheta) =$$
$$= L(\underline{Y}|u)P_0(du) \; ; \qquad (75)$$

rigorously speaking the only thing we have to verify to be able of carry out a reasonable calculus (e.g. to compute moments) with (75), is that it defines a proper distribution, i.e. that it can be normalized, or

$$\int_{(u)} P(du|Y) = 1 \Leftrightarrow$$
$$\Leftrightarrow 0 < \int_{(u)} \int_{(\vartheta)} L(\underline{Y}|u)P_0(du|\vartheta)P_0(d\vartheta) =$$
$$= \int_{(u)} L(\underline{Y}|u)P_0(du) < +\infty \qquad (76)$$

We observe that $L(\underline{Y}|u)$ is indeed a bounded [3] measurable function of u if $P_0(du)$ is the distribution of a g.r.f. on H, because in this case

$$E_0\{| < \underline{h}, u >_H |^2\} < +\infty \; ,$$

so that

$$P_0\{| < \underline{h}, u >_H | < +\infty\} = 1 \qquad (77)$$

[3] In fact we have

$$|L(\underline{Y}|u)| \le (2\pi\sigma_\nu^2)^{-m/2} \, .$$

and then

$$0 < \int_{(u)} P(du|\underline{Y}) \le (2\pi\sigma_\nu^2)^{-m/2} \int_{(u)} P_0(du)$$
$$= (2\pi\sigma_\nu^2)^{-m/2} \, . \qquad (78)$$

We have therefore to concentrate now on how to build $P_0(du|\vartheta)$ and $P_0(d\vartheta)$. First of all, what kind of nuisance parametrs ϑ do we want to consider? Indeed this depends on how we want to choose $P_0(du|\vartheta)$, but if we stick to the idea of a normal distribution with zero mean, then ϑ can be nothing but the covariance operator. For instance we could fix an $\widetilde{H}$ large enough as to be sure that when we compute the corresponding covariance $C_{\widetilde{H}}$ we know that it is a bounded operator, i.e.

$$\int < h, u >_{\widetilde{H}}^2 P_0(du|C_{\widetilde{H}}) = E_0\{< h, u >_{\widetilde{H}}^2\}$$
$$=< h, C_{\widetilde{H}} h >_{\widetilde{H}} \le c \parallel h \parallel_{\widetilde{H}}^2 \, . \qquad (79)$$

However leaving $C_{\widetilde{H}}$ completely free to vary over the set of symmetric, bounded positive operators in $\widetilde{H}$, is really too vague an information. In order to impose restrictions on $C_{\widetilde{H}}$ we can set up two types of constraints:

4) we can invoke a stochastic invariance principle, as we do when we compute empirical covariance functions; as we shall see through examples this defines the eigenfunctions of $C_{\widetilde{H}}$,
5) we impose a certain asymptotic behaviour of the eigenvalues q_n of $C_{\widetilde{H}}$, when $n \to \infty$, corresponding to the request that a given family of functionals $L(\;)$ generate a family of random variables with finite variance.

As for the point 4) assume that $\widetilde{H} \equiv L^2(B)$, where B is a domain of finite Lebesgue measure ($m(B) < +\infty$) and that we have a certain group of isometric transformations

$$t \in B \to t' = f_\tau(t) \in B \qquad (80)$$
$$m(dt) \equiv m(dt') \qquad ; \qquad (81)$$

then we say that $\{u\}$ is weakly stochastically invariant if

$$E\{u[f_\tau(t)]u[f_\tau(t')]\} = C[f_\tau(t), f_\tau(t')] \qquad (82)$$
$$\equiv C(t, t') \, , \quad \forall t, t' \text{ in } B \, .$$

This fixes the family of the eigenfunctions of C.

Example 4.1. Let $B \equiv S^{(1)}$ (the unit circle in R^2) and

$$f_\tau(t) = (\tau + t) \bmod 1 \; ; \qquad (83)$$

then

$$C(\tau + t, \tau + t') \equiv C(t, t')$$

implies

$$C(t, t') = \widetilde{C}(t - t') \quad (84)$$

and then

$$C(t, t') = \sum_{n=0}^{+\infty} q_n \cos 2\pi n(t - t') \quad (85)$$
$$= \sum_{n=0}^{+\infty} q_n \{e_{n_1}(t)e_{n_1}(t') + e_{n_2}(t)e_{n_2}(t')\} .$$

$$(e_{n_1}(t) = \cos 2\pi nt \ , \ e_{n_2}(t) = \sin 2\pi nt)$$

Example 4.2. Let $B = S^{(2)}$ (the unit sphere in R^3) and

$$f_\tau(\underline{t}) \equiv R_\vartheta \underline{t}$$

where $\underline{t}$ is a point on $S^{(2)}$, and R_ϑ a rotation parametrized by 3 Euler angles. As it is known (cf. Heiskanen and Moritz (1967))

$$C[f_\tau(\underline{t}), f_\tau(\underline{t}')] \equiv C(\underline{t}, \underline{t}')$$

implies

$$C(\underline{t}, \underline{t}') = \widetilde{C}(\psi_{\underline{t}\,\underline{t}'}) , \quad (86)$$

with $\psi_{\underline{t}\,\underline{t}'}$ the spherical angle between the directions $\underline{t}$ and $\underline{t}'$.

But then (cf. Heiskanen and Moritz (1967))

$$C(\underline{t}, \underline{t}') \equiv \sum_{n=0}^{+\infty} q_n (2n + 1) P_n(\cos \psi_{\underline{t}\,\underline{t}'})$$
$$\equiv \sum_{n=0}^{+\infty} q_n \sum_{m=-n}^{n} Y_{nm}(\underline{t}) Y_{nm}(\underline{t}') . \quad (87)$$

Here the eigenfunctions e_{nm} are just the real spherical harmonics

$$e_{nm}(\underline{t}) = \frac{1}{\sqrt{4\pi}} Y_{nm}(\underline{t}) \equiv$$
$$\equiv \frac{1}{\sqrt{4\pi}} \overline{P}_{n,|m|}(\cos\varphi) \begin{cases} \cos m\lambda & n \geq m \geq 0 \\ \sin |m|\lambda & 0 > m \geq -n \end{cases}$$

Note that here any eigenvalue q_n corresponds to a $(2n + 1)$ degenerate eigenspace called the *space of spherical harmonics* of degree n. This concept is particularly useful in Geodesy because in this case $u = T$ (the anomalous gravity potential) is a random field harmonic outside the sphere $\{r \geq R\}$ and once $\{T\}$ is characterized through its covariance operator (87), then its outer harmonic continuation is achieved by the elementary formula

$$C(r_P, r_{P'}, \psi_{PP'}) \equiv$$
$$\sum_{n=0}^{+\infty} q_n (2n + 1) \left(\frac{R^2}{r_P r_{P'}} \right)^{n+1} P_n(\cos \psi_{PP'}) . (88)$$

So if we accept that now the covariance function of u has a shape

$$C(\underline{t}, \underline{t}') = \sum_n q_n e_n(t) e_n(t') \quad (89)$$

with $e_n(t)$ a given sequence, on the basis of the chosen invariance principle, we are left with the problem of defining only a prior distribution for the (still infinite dimensional) nuisance parameter

$$\vartheta \equiv \{q_n\} . \quad (90)$$

In lack of any contrary evidence we shall assume that under the prior $P_0(d\vartheta)$ the components q_n are stochastically independent so that $P_0(d\vartheta)$ will be specified on $R_+^\infty \equiv \{\{q_n\} \in R^\infty , \ q_n \geq 0\}$ by means of one-dimensional distributions

$$P_{q_n}(dq_n) = f_{0n}(q_n) dq_n \ , \ q_n \geq 0 . \quad (91)$$

This is precisely point 5) of our program.

To the aim of better defining the content of (91), let us first notice that typically the functions $e_n(t)$ are bounded and more and more oscillating when $n \to \infty$; at least this is the case for the two Examples 4.1 and 4.2. Therefore the way in which q_n goes to 0 does control the power that the random field u distributes over higher frequencies. If we consider smooth a function with high power at low frequencies and low power at high frequencies, then the way in which q_n goes to 0 is a measure of the smoothness of u.

Remark 4.1. This is certainly the case when $e_n(t)$ are eigenfuntions of differential operators, as in Examples 4.1 and 4.2. In these cases in fact $e_n(t) = e_{n_1}(t) + ie_{n_2}(t)$ are eigenfunctions of $D_\vartheta \equiv i\frac{d}{d\vartheta}$ and $e_{nm}(\underline{t}) = e_{nm}(\underline{t}) + ie_{n,-m}(t)$ are eigenfunctions of $\{M^2 = -\Delta_\sigma, i\frac{\partial}{\partial\lambda}\}$, which are in turn (not accidentally) the generators of the rotation group on the circle (i.e. in R^2) and on the sphere (i.e. in R^3).

At this point of the reasoning we can also observe that there is no hope to derive from data the asymptotic behavior of $\{q_n\}$ because by definition the number of data is finite; therefore this behavior can only be fixed a priori on the basis of some coarse reasoning on the regularity. For instance, as we have already claimed in Section 2, we could first of all fix a sequence $\gamma_n \to 0$ in such a way that, for any linear functional L that we want to apply to u, we give rise to a finite variance r.v., i.e.

$$\sum \gamma_n [L(e_n(t))]^2 < +\infty . \qquad (92)$$

Then we could assign to each q_n a distribution that has mean γ_n and that concentrates around γ_n faster and faster when $n \to \infty$. This should leave the possibility that e.g. the posterior averages of q_n be significantly influenced by the data up to some "degree" n, but then, when $n \to \infty$, they assume a tendency dictated by γ_n.

Remark 4.2. One might think that a good way to build a prior distribution of $\vartheta = \{q_n\}$ would be to follow the classical Bayesian prescription for variances, namely that

$$f_{0n} \propto q_n^{-1} ,$$

i.e. an improper distribution.

On the other hand in this way we would not even constrain q_n to go to zero in the average, so that many functionals like $L = \mathrm{ev_t}$ [4] or, even worse, $\mathrm{ev_t}\partial_\mathrm{t}$, could never be used as observation functionals of u, because for them condition (77) would be violated.

To summarize point 5) we can say that an acceptable prior for $\vartheta = \{q_n\}$ will be of the type product of independent marginal distributions (cf. (91)) such that the two conditions

$$\lim_{n\to\infty} \frac{E_n\{q_n\}}{\gamma_n} = 1 \qquad (93)$$

$$\lim_{n\to\infty} \frac{\sigma\{q_n\}}{\gamma_n} = 0 \qquad (94)$$

are satisfied and that γ_n are chosen in a way that (92) holds true for any functional L we want to be bounded in a mean-square sense.

Example 4.3. We propose as an example a prior distribution for the case of major interest in Geodesy, as well as for other fields of the earth, like the geomagnetic field.

The assumption is that we have a field T on the sphere and we assume it to be at least in L^2 and stochastically invariant under the rotation group. This makes the use of the CONS, in L^2, $\{e_{nm}(\underline{t}) \equiv Y_{nm}(\underline{t})\}$ mandatory and with the help of this system we map $T(\underline{t})$ into a vector

$$\underline{T} \equiv \{\underline{T}_0, \underline{T}_1, \dots \underline{T}_n, \dots\} \in R^\infty \qquad (95)$$

where

$$\underline{T}_0 \equiv T_{00} \equiv (T, Y_{00})_{L^2}$$
$$\underline{T}_1 \equiv \{T_{1,-1}, T_{1,0}, T_{1,1}\},\ T_{1,k} \equiv (T, Y_{1k})_{L^2}$$
$$\dots\dots\dots\dots\dots\dots$$

so that

$$|\underline{T}|^2_{\ell^2} \equiv \sum_{n=0}^{+\infty} |\underline{T}_n|^2 \equiv \sum_{n=0}^{+\infty} \sum_{m=-n}^{n} T^2_{nm}$$

In this way we decompose R^∞ into the direct sum

$$R^\infty = \bigoplus_{n=0}^{+\infty} R_{(n)}$$
$$R_{(n)} \equiv R^{2n+1} .$$

Since every rotation of the sphere produces a rotation in $R_{(n)}$, with respect to the Euclidean metric (see for instance Moritz (1980); Sansò (1986)), it is fairly natural to assume that the prior distribution for $\underline{T}$ is constructed as the product of independent distributions in $R_{(n)}$, in particular we shall assume that

$$P_N(d\underline{T}|\underline{q}) \equiv \prod_{n=1}^{N} P_n(d\underline{T}_n|\underline{q}) \qquad (96)$$

$$P_n(d\underline{T}_n|\underline{q}) = g_{0n}(\underline{T}_n) d\underline{T}_n \qquad (97)$$

$$g_{0n}(d\underline{T}_n|\underline{q}) = \frac{1}{(2\pi q_n)^{n+(1/2)}} e^{-\frac{|\underline{T}_n|^2}{2q_n}} , \qquad (98)$$

so that the resulting distribution in R^∞ is normal and we can write

$$\begin{cases} P_0(d\underline{T}|\underline{q}) \sim \mathcal{N}[0, Q] \\ Q = \mathrm{diag}\{q_n\} . \end{cases} \qquad (99)$$

Remark 4.3. The Cameron-Martin space of the distribution (99) is ℓ^2_q, i.e.

$$H \equiv \{\underline{\lambda} ;\ \underline{\lambda}^+ Q \underline{\lambda} = = \sum_n q_n \sum_m \lambda^2_{nm} < +\infty\} . \qquad (100)$$

It follows that, for instance,

$$\begin{cases} ev_t \equiv \{Y_{nm}(\underline{t})\} \in H \Leftrightarrow \sum_n q_n(2n+1) < +\infty \\ ev_t\partial_r \equiv \{(n+1)Y_{nm}(\underline{t})\} \in H \Leftrightarrow \\ \Leftrightarrow \sum_n q_n(2n+1)(n+1)^2 < +\infty \\ ev_t\Delta_\sigma \equiv \{-n(n+1)Y_{nm}(\underline{t})\} \in H \Leftrightarrow \\ \Leftrightarrow \sum_n q_n(2n+1)n^2(n+1)^2 < +\infty . \end{cases} \qquad (101)$$

[4] We mean

$$\mathrm{ev_t}(\varphi) \equiv \varphi(\mathrm{t})\ \ \mathrm{ev_t}\partial_\mathrm{t}(\varphi) \equiv \varphi'(\mathrm{t}) .$$

Remark 4.4. Note that indeed

$$E_0\{\underline{T}^+Q^{-1}\underline{T}\} = = \sum_{n,m} \frac{E_0\{|T_{nm}|^2\}}{q_n} = \sum_n (2n+1) = +\infty ,$$

as observed by Tscherning; however, if at least the first of (101) is satisfied, it follows that

$$E_0\{\underline{T}^+\underline{T}\} = TrQ = \sum q_n(2n+1) < +\infty \quad (102)$$

and in fact the realizations $\{T\}$ belong to L^2 almost surely, as originally required.

We pass now to construct $f_{0n}(q_n)$; we propose the following prior distribution

$$f_{0n}(q) = \frac{1}{\Gamma(\alpha_n - 1)} \frac{\rho_n^{\alpha_n - 1}}{q^{\alpha_n}} e^{-\frac{\rho_n}{q}} \ , \ q \geq 0 \quad (103)$$

which is an inverse Γ with parameters α_n and ρ_n (cf. Box and Tiao (1973)). We note that the choice of the shape (103) is related to the hypothesis, often used in Bayesian theory, to employ an inverse Wishart distribution for

$$Q_n = \left| \begin{matrix} q_n & & 0 \\ & \ddots & \\ 0 & & q_n \end{matrix} \right| \left.\vphantom{\begin{matrix} q_n \\ \ddots \\ 0 \end{matrix}}\right\} 2n+1 ,$$

in which case we would have

$$\alpha_n - 1 = \frac{1}{2}(2n+1) . \quad (104)$$

The α_n parameters, that can possibly take values different from (104), have been introduced to tune the way in which q_n approach their mean in probability, when $n \to \infty$. This is better understood if we observe that from (103) we can compute

$$E_0\{q_n\} = \frac{\rho_n}{\alpha_n - 2} \ , \ \sigma^2(q_n) = = \left(\frac{\rho_n}{\alpha_n - 2}\right)^2 \frac{1}{\alpha_n - 3} ; \quad (105)$$

so if we put

$$\frac{\rho_n}{\alpha_n - 2} = \gamma_n \quad (106)$$

and we agree that in any way α_n are chosen so that $\alpha_n \to \infty$ for $n \to \infty$, we find that both (93) and (94) are satisfied. Depending on the functional we want to be bounded in mean square sense, we can fix γ_n so that it satisfies (100); this puts a constraint on ρ_n and α_n through (106). On the other hand we note that the joint prior of $\underline{T}$ and $\underline{q}$ can at this point be considered as the infinite product of finite dimensional distributions on R^{2n+2}, of the type

$$P_{0n}(d\underline{T}_n, dq_n) = = \frac{1}{\Gamma(\alpha_n - 1)(2\pi)^{n+(1/2)}} \cdot \frac{\rho_n^{\alpha_n - 1}}{q_n^{\alpha_n + n + (1/2)}} e^{-\frac{|\underline{T}_n|^2}{2q_n} - \frac{\rho_n}{q_n}} d\underline{T}_n dq_n . \quad (107)$$

From (107) we see that

$$E_0\{q_n|\underline{T}_n\} = = \frac{1}{2(\alpha_n - 2) + (2n+1)} \cdot \cdot \left[2(\alpha_n - 2)\gamma_n + (2n+1)\frac{|\underline{T}_n|^2}{2n+1}\right] \quad (108)$$

i.e. if we fix $\underline{T}$ in this distribution the conditional average of q_n becomes a weighted mean between the prior average γ_n and, so to say, the empirical degree variance

$$\sigma_n^2 = \frac{|\underline{T}_n|^2}{2n+1} = \frac{1}{2n+1} \sum_m T_{nm}^2 \quad (109)$$

The parameter α_n is exactly tuning this average. If we put $2(\alpha_n - 2) = w \cdot (2n+1)$ in (108) we see that for $w = 1$ we have exactly

$$E_0\{q_n|\underline{T}_n\} = \frac{1}{2}(\gamma_n + \sigma_n^2)$$

and for $w \to 0$ only the empirical degree variances count, for $w \to \infty$ only the prior γ_n have importance.

Summarizing we can conclude that (106) and (108) can drive us in the choice of the two vectors $\underline{\rho}, \underline{\alpha}$ necessary to fully specify the prior distribution.

Another fundamental consequence of (107) is that, after integrating over q_n, we find

$$P_{0n}(d\underline{T}_n) = \frac{1}{\Gamma(\alpha_n - 1)} \cdot \cdot \frac{\Gamma(\alpha_n + n - (1/2))}{(2\pi\rho_n)^{n+(1/2)} \left(1 + \frac{|\underline{T}_n|^2}{2\rho_n}\right)^{\alpha_n + n - (1/2)}} d\underline{T}_n, \quad (110)$$

so that, after some little computation, we get

$$E_0\{\underline{T}_n\} = 0 \ ; \ C_{0\underline{T}_n\underline{T}_n} = \gamma_n I_{2n+1} \ , \quad (111)$$

proving that for any functional $L(T) = \underline{\lambda}^+\underline{T}$ such that

$$\sum_n \gamma_n \left(\sum_m \lambda_{nm}^2\right) < +\infty \ , \quad (112)$$

i.e. $\forall \underline{\lambda} \in \ell^2_{\underline{\gamma}}$, the corresponding r.v. is bounded in mean square sense. We can conclude therefore that if all the functionals of the observation equations

$$< \underline{h}, u >_H \equiv \Lambda \underline{T} \tag{113}$$

are in $\ell^2_{\underline{\gamma}}$, i.e. if

$$\begin{cases} Tr \Lambda \Gamma \Lambda^+ < +\infty \\ \Gamma \equiv \text{diag}\{\gamma_n\} \end{cases} \tag{114}$$

then we can consistently construct the posterior distributions (cf. (74), (75))

$$\begin{aligned} P(du, dq|Y) &= L(\underline{Y}|u) P_0(du, dq) \\ P_0(du, dq) &= P_0(du|q) P_0(dq) \\ P(du|Y) &= L(\underline{Y}|u) \int_{(q)} P_0(du, dq) \end{aligned} \tag{115}$$

because condition (77) is satisfied. Since the construction of the distribution of u given $\underline{Y}$ is the main target of the Bayesian approach, we can close here and devolve to the next paragraph the study of how to use such a distribution to derive a generalized Maximum a-Posteriori for u.

5 Generalizing the maximum likelihood concept to an R^∞ r.v.

According to (115) we have now, under suitable conditions, the a posteriori distribution of the r.v. $\{\underline{T}\}$ in R^∞; yet, as we do with finite dimensional distributions, we would like to summarize this information in a more concentrated form. This is traditionally done, for instance, by giving the first two moments of the posterior; however, as common in the Bayesian literature is to give the maximum of the posterior distribution (Maximum a-Posteriori or MAP), as well as the "curvature" of this density distribution, to build what is called the normal approximation of the posterior (cf. Box and Tiao (1973)). This approach resembles strictly the usual Kramer-Rao theory of maximum likelihood, although it is conceptually different. Since this is what we shall generalize to our R^∞ case, we consider it useful to repeat it here, keeping an eye to one basic concern. In R^∞ one cannot define probability densities for the very simple reason that there is no analogue of the Lebesgue measure, which is translation invariant. Nevertheless we have in our arsenal probability distributions as well as the corresponding average operators, so if we pay attention to put as far as possible the finite dimensional formulas in the form of averages of measurable functions, these will be most likely generalizable to the infinite dimensional case.

Let us start with the famous definition of score variable $\underline{U}$. We have first (denoting as $m(dx)$ the Lebesgue measure) a probability distribution

$$P(dx) = f(\underline{x}) m(dx) , \tag{116}$$

which does correspond in our case to the posterior of $\underline{x} \equiv (\underline{T}, \underline{q})$ so that all the calculus we will perform has to be thought of as conditional to the observation vector $\underline{Y}$. Classically the score variable is the logarithmic derivative of $f(\underline{x})$, i.e.

$$\underline{U} = \partial_{\underline{x}} \log f(\underline{x}) = \frac{f_{\underline{x}}(\underline{x})}{f(\underline{x})} . \tag{117}$$

Note that with $\underline{U}$ we can compute the derivative of the measure P in a direction $\underline{h}$;

$$\begin{aligned} &\frac{d}{dt} P(dx + t\underline{h})|_{t=0} = \\ &= f_{\underline{x}}(\underline{x} + t\underline{h})^+ \underline{h}\, m(dx + t\underline{h}) + \\ &+ f(\underline{x} + t\underline{h}) m'(dx + t\underline{h})|_{t=0} = \\ &= f_{\underline{x}}(\underline{x})^+ \underline{h}\, m(dx) = \underline{U}(\underline{x})^+ \underline{h}\, f(\underline{x}) m(dx) \\ &= \underline{U}^+(\underline{x}) \underline{h}\, P(dx) . \end{aligned} \tag{118}$$

In (118) we have used the invariance property of Lebesgue measure, so that $m'(dx + t\underline{h}) \equiv 0$. The interesting point in (118) is that the first term and the last of the chain of identities don't involve the Lebesgue measure and the densities, so much so that $\underline{U}(\underline{x})$ could be directly defined in the following interesting way (cf. Bogachev (1996); Skorohod (1974)).

Definition 5.1. Let $P(dx)$ be any (possibly signed) measure defined on a linear topological space L and assume that the set $\{dx + t\underline{h}\}$, translated in direction $\underline{h}$ is still measurable according to P so that $P(dx + t\underline{h})$ as a function of (dx) is absolutely continuous with respect to $P(dx)$, or $P(dx + t\underline{h}) \ll P(dx)$; then according to the Radon-Nicodym theorem (cf. Skorohod (1974)) there exists a density of the first measure with respect to the second, such that

$$P(dx + t\underline{h}) = u(\underline{x} + t\underline{h}) P(dx) \, , \; u \in \mathcal{L}^1_P \, . \tag{119}$$

Assume further that $u(\underline{x} + t\underline{h})$ can be differentiated at $t = 0$ and the (Gateaux) derivative of u in direction h

$$\frac{d}{dt} u(\underline{x} + t\underline{h})|_{t=0} = U(\underline{x})[\underline{h}] \tag{120}$$

is defined at least $\forall \underline{h} \in D$, where D is a linear subspace densely embedded in L

$$D \subset L \, , \; [D]_L \equiv L \, ; \tag{121}$$

then by definition the linear functional $U(x)$ acting on D, which is uniquely defined by (120), is the score or logarithmic derivative of P.

Example 5.1. When $L \equiv R^N$ and the distribution P is generated by a regular (e.g. C^1) density [5], like in (116), then indeed $\underline{h}$ is any direction in R^N, i.e. $D \equiv L \equiv R^N$ and the score variable does correspond to $\underline{U}$ given by (117).

Example 5.2. When $L \equiv R^\infty$ and $P\{dx\}$ is constructed from a family of independent regular distributions

$$P^{(N)}\left(dx^{(N)}\right) \equiv \prod_{n_1}^{N} p_n(x_n) m(dx_n) \quad (122)$$

then we can take $D \equiv R_0^\infty$, i.e. $\underline{h}$ of the form

$$\underline{h}^+ = (h_1, \dots h_M, 0 \dots 0 \dots) , \quad (123)$$

for some M, such that

$$P(dx + t\underline{h}) \equiv P^{(M)}(dx^{(M)} + t\underline{h}) P^{+\infty}_{(M+1)}(dx^{+\infty}_{M+1})$$

where $P^{+\infty}_{M+1}(dx^{+\infty}_{M+1})$ is the distribution generated by the family

$$P^N_{M+1}(dx^N_{M+1}) \equiv \prod_{n=M+1}^{N} p_n(x_n) m(dx_n) . \quad (124)$$

Then we obviously have

$$P'(dx + t\underline{h})|_{t=0} = \sum_{n=1}^{M} \frac{p'_n(x_n)}{p_n(x_n)} h_n \cdot P(dx) , \quad (125)$$

thus showing that

$$\underline{U}(x) = \left\{ \frac{p'_n(x_n)}{p_n(x_n)} \right\} \in R^\infty . \quad (126)$$

For instance if

$$p_n(x_n) \sim \mathcal{N}[0,1]$$

then

$$\frac{p'_n(x_n)}{p_n(x_n)} = -x_n ,$$

i.e.

$$\underline{U}(x) \equiv -\underline{x} \quad (127)$$

[5] Indeed take the case that $P(dx)$ is not regular, for instance $P(dx) = \delta_{x_0}(dx)$, i.e. a probability concentrated at x_0 on the real line R; then no translated $P(dx + th) = = \delta_{x_0 - th}(dx)$ can be absolutely continuous with respect to $P(dx)$, excluding the trivial case $t = 0$.

Remark 5.1. Assume that a probability measure $P(dx)$ is given in the form

$$P(dx) \equiv A \cdot g(x) P_0(dx) , \quad (128)$$

as it happens when we build a posterior distribution with Bayes theorem; assume further that the prior $P_0(dx)$ has a score variable $U_0(x)$. Then the score of $P(dx)$ is given by

$$P'(dx + th)|_{t=0} = = \left[\frac{g_x(x)}{g(x)} + U_0(x) \right] [h] P(dx) \quad (129)$$

or

$$U(x) = \frac{g_x(x)}{g(x)} + U_0(x) . \quad (130)$$

Remark 5.2. In maximum likelihood theory the score variable is used to define the necessary conditions for a point $\widehat{x}$ to be maximum of f; namely

$$f(\widehat{x}) \geq f(x) \Rightarrow U(\widehat{x}) = 0 . \quad (131)$$

Naturally once the definition of $U(x)$ is generalized to an infinite dimensional distribution, the condition (131) can continue to hold to define the maximum a posteriori in a Bayesian environment. For instance for the normal of Example 5.2 we would get (cf. (127))

$$U(\widehat{x}) = -\widehat{x} = 0 \quad (132)$$

which makes indeed a lot of sense. Yet the condition (131) alone is not sufficient, not even to guarantee that $\widehat{x}$ is a relative maximum, for this reason this is usually complemented by the condition

$$\widehat{U}_x = \partial_x U(x)|_{x=\widehat{x}} < 0 . \quad (133)$$

For the Example 5.2 we would have in fact

$$\widehat{U}_x = -I < 0 . \quad (134)$$

The matrix $U_x(\widehat{x}) \equiv \widehat{U}_x$ has also an important meaning in Kramer-Rao theory since, as we know, at least in R^N,

$$-E\{U_x(x)\} \equiv C_{UU} . \quad (135)$$

Now if the probability distribution is assumed to be well-concentrated around its maximum point $\widehat{x}$, we can as well accept the approximation

$$E\{U_x(x)\} \equiv -C_{UU} \cong \widehat{U}_x . \quad (136)$$

With this relation and using the Taylor formula

$$U(x) \cong U(\widehat{x}) + U_x(\widehat{x})(x - \widehat{x}) = \\ = U_x(\widehat{x})(x - \widehat{x}) \qquad (137)$$

which gives us

$$x - \widehat{x} \cong \widehat{U}_x^{-1} U(x)\,, \qquad (138)$$

we see that the "normal approximation" of the posterior

$$x \sim \mathcal{N}\left[\widehat{x}, -\widehat{U}_x^{-1}\right] \qquad (139)$$

can be justified.

Concluding the section we can claim that even in R^∞ the solution of the score equation (131) and the calculation of $\widehat{U}_x^{-1}$, provide us the Gaussian approximation of the posterior itself, defined through (139).

6 The MAP equations for field and covariance estimation

We propose now to build the MAP equations for the posterior distribution of $\{\underline{T}, \underline{q}\}$ given by (115), namely

$$P(dT, dq|\underline{Y}) = \\ = A \cdot L(\underline{Y}|\underline{T}) P_0(dT|\underline{q}) P_0(d\underline{q}) \qquad (140)$$

with

$$L(\underline{Y}|\underline{T}) \propto e^{-\frac{1}{\sigma_\nu^2}|\underline{Y} - A\underline{T}|^2}$$

and $P_0(dT|\underline{q})P_0(dq) = P_0(dT, dq)$ determined by equation (107).

As noticed in Remark 5.1, the score of (140) can be written as

$$\underline{U}(\underline{T}, \underline{q}) \equiv \left| \begin{matrix} \partial_{\underline{T}} \log L \\ \partial_{\underline{q}} \log L \end{matrix} \right| + \underline{U}_0(\underline{T}, \underline{q}) = \qquad (141)$$

$$= \left| \begin{matrix} \frac{1}{\sigma_\nu^2} A^+(\underline{Y} - A\underline{T}) \\ 0 \end{matrix} \right| + \underline{U}_0(\underline{T}, \underline{q})\,;$$

let us observe that for $L(\underline{Y}|\underline{T}) = g(\underline{T})$ smooth, the variation $g(\underline{T} + t\underline{h})$ certainly exists $\forall \underline{h} \in R_0^\infty$, which is sufficient to define $\underline{U}$ univocally.

To compute $\underline{U}_0$ we use the Example 5.2 and, reorganizing the order of $\underline{T}_n$ and q_n variables, we find

$$\underline{U}_0 \equiv \{\underline{U}_{0n}\}$$

where $\underline{U}_{0n}$ are $(2n+2)$ vectors computed from the finite dimensional distributions, i.e. from (107), putting

$$\underline{U}_{0n} \equiv \left| \begin{matrix} \nabla_{\underline{T}_n} \log P_{0n}(\underline{T}_n, q_n) \\ \frac{\partial}{\partial q_n} \log P_{0n}(\underline{T}_n, q_n) \end{matrix} \right|. \qquad (142)$$

In this way we get

$$\underline{U}_{0n} = \qquad (143)$$

$$= \left| \begin{matrix} -\frac{1}{q_n} \underline{T}_n \\ -\frac{(\alpha_n + n + (1/2))}{q_n} + \left[\frac{1}{2}|\underline{T}_n|^2 + (\alpha_n - 2)\gamma_n\right] \frac{1}{q_n^2} \end{matrix} \right|.$$

Going back to (141) we find the two coupled nonlinear MAP equations

$$\begin{cases} \left(\frac{1}{\sigma_\nu^2} A^+ A + Q^{-1}\right) \widehat{\underline{T}} = \frac{1}{\sigma_\nu^2} A^+ \underline{Y} \\ Q \equiv \operatorname{diag}\{\widehat{q}_n\} \end{cases} \qquad (144)$$

and

$$\begin{cases} \frac{1}{\widehat{q}_n} = \frac{\beta_n}{(2n+1)\sigma_n^2 + 2(\alpha_n - 2)\gamma_n} \\ \beta_n = (2n+1) + 2\alpha_n \\ \sigma_n^2 = \frac{|\underline{T}|^2}{2n+1} = \frac{1}{2n+1} \sum_{m=-n}^{n} T_{nm}^2 \end{cases} \qquad (145)$$

where, $\overline{\underline{T}}_n = (0, \ldots, 0, \overline{\underline{T}}_n, 0, \ldots, 0, \ldots)$.

We can observe that (145) with a minor modification becomes equal to (108) and therefore has a quite understandable statistical meaning. Substituting (145) in (144) one gets the reduced non-linear field equation

$$F(\underline{T}) = \frac{1}{\sigma_\nu^2}(A^+ A)\underline{T} + \sum_n \beta_n \frac{\overline{\underline{T}}_n}{|\underline{T}_n|^2 + 2\rho_n} = \\ = \frac{1}{\sigma_\nu^2} A^+ \underline{Y}\,, \qquad (146)$$

and we returned to the use of the parameters

$$\rho_n = (\alpha_n - 2)\gamma_n\,.$$

We study the existence of a solution of (146) by showing that (146) is the variation equation expressing the stationarity condition of a suitable positive functional $\Phi(\underline{T})$ and then proving that $\Phi(\underline{T})$ has at least a minimum.

So let us define

$$\Phi(\underline{T}) = \frac{1}{2\sigma_\nu^2}|\underline{Y} - A\underline{T}|^2 + \qquad (147) \\ + \frac{1}{2}\sum_n \beta_n \log\left(1 + \frac{|\underline{T}_n|^2}{2\rho_n}\right)\,;$$

it is easy to see that the condition

$$\frac{d}{dt}\Phi(\widehat{\underline{T}} + \underline{h})|_{t=0} = 0 \quad \forall \underline{h} \in R_0^\infty \qquad (148)$$

leads exactly to the equation (146), so if (147) has at least one minimum $\widehat{\underline{T}} \in R^\infty$, then (146) has at least one solution.

We note immediately that

$$\Phi(\underline{T}) \geq 0 \tag{149}$$

and that there is no hope that (147) can have a finite value unless $|\widehat{T}_n|^2 \to 0$ in a suitable way, so that $\underline{\lambda}_k^+ \widehat{\underline{T}}$ are finite and the series of logarithms in (147) is converging. This says to us that $\Phi(\underline{T}) = +\infty$ for "most" $\underline{T} \in R^\infty$, so it is interesting to study the domain D of Φ

$$D \equiv \{\underline{T} \in R^\infty \; ; \; \Phi(\underline{T}) < +\infty\} \tag{150}$$

because if a minimum $\widehat{\underline{T}}$ exists it has certainly to be in D. Note that $D \neq \emptyset$ since

$$\Phi(0) = \frac{1}{\sigma_\nu^2} |\underline{Y}|^2 < +\infty \,. \tag{151}$$

Lemma 6.1. We have the identification

$$D \equiv \ell_{\gamma^{-1}}^2 \equiv \left\{ \underline{T} \, ; \, \sum_n \frac{|\underline{T}_n|^2}{\gamma_n} < +\infty \right\} .$$

Proof. In fact if $\underline{T} \in D$ then $\Phi(\underline{T}) < +\infty$ and so we must have

$$\frac{|\underline{T}_n|^2}{2\rho_n} \to 0$$

otherwise the series in (147) diverges. Then

$$\frac{|\underline{T}_n|^2}{2\rho_n} \leq A < 2 \quad \forall n > N$$

for suitable N.

Since

$$x \geq \log(1+x) \geq x - \frac{1}{2}x^2 \tag{152}$$

we find

$$\begin{aligned} \Phi(\underline{T}) &> \frac{1}{2} \sum_{n=N}^{+\infty} \beta_n \log\left(1 + \frac{|\underline{T}_n|^2}{2\rho_n}\right) \geq \\ &\geq \frac{1}{2} \sum_{n=N}^{+\infty} \beta_n \frac{|\underline{T}_n|^2}{2\rho_n} - \frac{1}{4} \sum_{n=N}^{+\infty} \beta_n \frac{|\underline{T}_n|^4}{4\rho_n^2} \geq \\ &\geq \frac{1}{2}\left(1 - \frac{1}{2}A\right) \sum_{n=N}^{+\infty} \beta_n \frac{|\underline{T}_n|^2}{2\rho_n} = \\ &= \frac{1}{2}\left(1 - \frac{1}{2}A\right) \sum_{n=N}^{+\infty} \frac{(2n+1) + 2\alpha_n}{2(\alpha_n - 2)} \frac{|\underline{T}_n|^2}{\gamma_n} \,. \end{aligned} \tag{153}$$

If we assume, as we do, that $\alpha_n = 0(n)$, or even if it diverges faster, then

$$\widetilde{\alpha} \geq \frac{(2n+1) + 2\alpha_n}{2(\alpha_n - 2)} \geq \underset{\sim}{\alpha} > 0 \tag{154}$$

and we have proved

$$\underline{T} \in D \Leftrightarrow \Phi(\underline{T}) < +\infty \Rightarrow \underline{T} \in \ell_{\gamma^{-1}}^2 \,.$$

Vice versa, if $\underline{T} \in \ell_{\gamma^{-1}}^2$ we have, because of (152) and (154)

$$\begin{aligned} +\infty > \widetilde{\alpha} \sum_n \frac{|\underline{T}_n|^2}{\gamma_n} &\geq \sum_n \beta_n \frac{|\underline{T}_n|^2}{2\rho_n} \geq \\ &\geq \sum_n \beta_n \log\left(1 + \frac{|\underline{T}_n|^2}{2\rho_n}\right) \; ; \end{aligned}$$

moreover, since $\underline{\lambda}_k \in \ell_\gamma^2$ by hypothesis (cf. (112)) and $\underline{T} \in \ell_{\gamma^{-1}}^2$, $\Lambda \underline{T}$ is a (finite) vector with finite components $\underline{\lambda}_k^+ \underline{T}$. Therefore

$$\underline{T} \in \ell_{\gamma^{-1}}^2 \Rightarrow \Phi(\underline{T}) < +\infty \Leftrightarrow \underline{T} \in D \,.$$

We now can prove our main theorem.

Theorem 6.1. The functional $\Phi(\underline{T})$ defined by (147) has a minimum at $\widehat{\underline{T}} \in D$.

Proof. Let be $\underline{T}^\ell$ be an extremizing sequence, i.e. $\underline{T}^\ell \in D$ and

$$\lim \Phi(\underline{T}^\ell) = \operatorname*{Inf}_{\underline{T} \in D} \Phi(\underline{T}) = \underset{\sim}{\Phi} \; ; \tag{155}$$

because of (151) we can assume

$$\Phi(\underline{T}^\ell) \leq \Phi(0) \,. \tag{156}$$

Since

$$\Phi(\underline{T}^\ell) \geq \frac{1}{2} \beta_n \log\left(1 + \frac{|\underline{T}_n^\ell|^2}{2\rho_n}\right), \quad \forall n, \ell$$

we see that necessarily

$$\forall n, \ell \;\; \frac{|\underline{T}_n^\ell|^2}{2\rho_n} \leq k(\text{constant})$$

since otherwise (156) could not hold. On the other hand we have for $x \leq k$

$$\log(1+x) = \int_0^x \frac{dt}{1+t} \geq \frac{x}{1+x} \geq \frac{x}{1+k}$$

so that

$$\Phi(0) \geq \Phi(\underline{T}^\ell) \geq \frac{1}{2}\sum_n \beta_n \log\left(1+\frac{|\underline{T}_n^\ell|^2}{2\rho_n}\right) \geq$$
$$\geq \frac{1}{1+k}\frac{1}{2}\sum_n \beta_n \frac{|\underline{T}_n^\ell|^2}{2\rho_n} \geq \tag{157}$$
$$\geq \frac{\underset{\sim}{\alpha}}{1+k}\frac{1}{2}\sum_n \frac{|\underline{T}_n^\ell|^2}{\gamma_n},$$

i.e. $\{\underline{T}^\ell\}$ is a bounded sequence in $\ell^2_{\gamma-1}$. Since $\ell^2_{\gamma-1}$ is a Hilbert space we know that $\{\underline{T}^\ell\}$, possibly reduced to one of its subsequences, is weakly convergent to a limit $\widehat{\underline{T}} \in \ell^2_{\gamma-1}$

$$\underline{T}^\ell \underset{\ell^2_{\gamma-1}}{\overset{*}{\longrightarrow}} \widehat{\underline{T}}. \tag{158}$$

On the other hand $\underline{\lambda}_k \in \ell^2_\gamma$ (i.e. $\underline{\lambda}_k^+ \Gamma \underline{\lambda}_k < +\infty$) so that $\Gamma\underline{\lambda}_k \in \ell^2_{\gamma-1}$ since $(\Gamma\underline{\lambda}_k)^+\Gamma^{-1}(\Gamma\underline{\lambda}_k) =$ $= \underline{\lambda}_k^+\Gamma\underline{\lambda}_k < +\infty$. Therefore from (158) we have

$$\underline{\lambda}_k^+\underline{T}^\ell = (\Gamma\underline{\lambda}_k)^+\Gamma^{-1}\underline{T}^\ell = \tag{159}$$
$$= (\Gamma\underline{\lambda}_k, \underline{T}^\ell)_{\ell^2_{\gamma-1}} \to \underline{\lambda}_k^+\widehat{\underline{T}}.$$

Moreover $\forall n$, $\underline{T}_n$ involves only finitely many components of $\underline{T}$, so that

$$\underline{T}_n^\ell \underset{R^{2n+1}}{\longrightarrow} \widehat{\underline{T}}_n \text{ for } \ell \to \infty. \tag{160}$$

Accordingly, $\forall N$,

$$\underset{\sim}{\Phi} = \lim_{\ell\to\infty} \Phi(\underline{T}^\ell) \geq \tag{161}$$
$$\lim_{\ell\to\infty}\left[\frac{1}{2\sigma_\nu^2}|\underline{Y} - A\underline{T}^\ell|^2 + \right.$$
$$\left. + \frac{1}{2}\sum_{n=1}^N \beta_n \log\left(1+\frac{|\underline{T}_n^\ell|^2}{2\rho_n}\right)\right] =$$
$$= \frac{1}{2\sigma_\nu^2}|\underline{Y} - A\widehat{\underline{T}}|^2 + \frac{1}{2}\sum_{n=1}^N \beta_n \log\left(1+\frac{|\widehat{\underline{T}}_n|^2}{2\rho_n}\right)$$

Taking the limit for $N \to \infty$ of (161) and recalling that $\widehat{T} \in \ell^2_{\gamma-1} \equiv D$ we see that

$$\underset{\sim}{\Phi} \geq \Phi(\widehat{\underline{T}}) \Rightarrow \underset{\sim}{\Phi} = \Phi(\widehat{\underline{T}}) \tag{162}$$

and the theorem is proved.

Remark 6.1. In spite of the deceptive simplicity of the equation (146) we haven't been able to prove the existence of at least one solution without recurring to this variational technique. Even its application was not completely trivial because the functional $\Phi(\underline{T})$ is not convex.

For the very same reason the uniqueness of the solution of (146), if it holds, is still missing. As a matter of fact, if one computes the further derivative of $F(\underline{T})$, given by (146), with respect to $\underline{T}$ to try to gain information on the covariance of $\widehat{\underline{T}}$, one gets

$$\frac{\partial F}{\partial \underline{T}} = \frac{1}{\sigma_\nu^2}A^+A + \tag{163}$$
$$+ \sum_n \frac{\beta_n}{|\underline{T}_n|^2 + 2\rho_n}\left\{\overline{I}_{2n+1} - \frac{2\overline{\underline{T}}_n\overline{\underline{T}}_n^+}{|\underline{T}_n|^2 + 2\rho_n}\right\}$$

where

$$\overline{I}_{2n+1} = \begin{pmatrix} 0 & & 0 \\ & I_{2n+1} & \\ 0 & & 0 \end{pmatrix}$$

Since the matrices in parentheses can be written as

$$\left\{I_{2n+1} - \frac{2|\underline{T}_n|^2}{|\underline{T}_n|^2 + 2\rho_n}\frac{\underline{T}_n\underline{T}_n^+}{|\underline{T}_n|^2}\right\}$$

we immediately see that these are positive definite in the region

$$|\underline{T}_n|^2 < 2\rho_n, \tag{164}$$

while for $|\underline{T}_n|^2$ larger we have non-definite matrices.

Whether condition (164) is met at the estimated $\widehat{\underline{T}}$ or not is a matter that depends on the weighting factors α_n; if these are large enough (164) is likely to be true. Yet the whole subject clearly requires further investigations.

7 Conclusions

In the paper three main issues were addressed, namely

1) is there a precise relation between Collocation theory interpreted in a stochastic mode, with the help of the theory of generalized random fields, and the Bayesian approach applied to the same problem?
2) is it possible to include the covariance operator, or some suitable parameters related to it, into the unknowns and how can we construct then the joint posterior distribution of the field and of its covariance?
3) can we generalize the MAP theory, applied routinely to Bayesian posteriors, to the infinite dimensional cases and in particular what kind of equations do we receive from the prior discussed at point 2)?

As for the issue 1) we have shown that indeed, assuming to know the covariance operator, we can construct a standard triple $H_0 \subset H \subset H^\infty$ of spaces such that the distribution of the unknown random field T in H^∞ appears exactly as the distribution of the generalized random field on H. This result was in fact already known and published in Lehtinen et al. (1989). In this case, assuming that the prior of T is normal, one can show that the "non-informative" hypothesis applied to the prior implies that the covariance operator of T on H is exactly equal to the reproducing kernel K of this space, thus solving an old antinomy.

Moreover, the conditional posterior average and covariance of the field do agree perfectly with Collocation formula.

As for point 2) we have first decided to reduce somehow the dimension of the problem by admitting that the covariance function of T should have a special form according to the invariance principle advocated for the case at hand. For instance for random fields on the sphere, the assumption of rotational invariance fixes the eigenfunctions, i.e. $\{Y_{nm}(\underline{t})\}$, of the covariance, leaving the degree variances, called here q_n, as unknown random variables to be predicted.

Suitable prior distributions for q_n are discussed where an a-priori choice has to be made for their averages, γ_n, which are related to the "prior" degree of smoothness of T and for certain parameters α_n that control the weighting of prior information and empirical degree variances to derive the posterior distribution. The posterior is then constructed properly and it is shown that this is in fact concentrated on the same subspace of H^∞ that we would have at the beginning by fixing $q_n = \gamma_n$.

As for the last point the use of shifts and then directional derivatives of measures in linear topological spaces, allows for the reduction of the MAP problem to the definition of logarithmic derivatives of distributions without using probability densities in the ordinary sense. The result is that MAP equations can be derived for our problem, in which $\widehat{q}_n$ can be eliminated to the effect that a unique non-linear estimation equation for the field T can be written. Such an equation can be studied, as for the existence of at least one solution, by using a variational technique. The question of uniqueness and even of the posterior covariance operator of $\widehat{T}$ are still open.

Basically the theory presented in the paper is making a bridge between geodetic (and geophysical) tradition and the well-established field of infinite-dimensional probability distributions. However, to the knowledge of the authors, the idea of using logarithmic derivatives for statistical estimation seems to be new, although indeed there is a large mathematical literature on differential calculus with infinite dimensional distributions. In specific the equations for predicting $\widehat{T}$ under the hypothesis of rotational invariance are new and the method proposed to study the existence of a solution has been purposely set up.

References

Bogachev VI (1995) Differential measures and the Malliavin calculus. Preprints di Mat no 16, Scuola Normale Superiore di Pisa.

Bogachev VI (1996) Gaussian measures on linear spaces. J Math Sc 79.

Box GEP, Tiao GC (1973) Bayesian inference in statistical analysis. Wiley.

Heiskanen WA Moritz H (1967) Physical geodesy. Freeman, San Francisco and London.

Krarup T (1969) A contribution to the mathematical foundation of physical geodesy. Meddelelse no 44, Geodätisk Inst Köbenhavn.

Lauritzen SL (1973) The probabilistic background of some statistical method in physical geodesy. Meddelelse no 48, Geodätisk Inst Köbenhavn.

Lehtinen MS, Päivärinta L, Somersalo E (1989) Linear inverse problems for generalized random variables. IOP - Inverse Problems 5:599-612.

Mandelbaum A (1984) Linear estimators and measurable linear transformations on Hilbert space. Wahrscheinlichkeitstheorie verw. Gebeite 65:385-397.

Moritz H (1980) Advanced physical geodesy. Wichmann, Karlsruhe.

Rozanov YuA (1971) Infinite-dimensional gaussian distributions. (English trans. AMS). Proceedings of the Steklov Inst. of Math. 108.

Sacerdote F, Sansò F (1996) Optimal linear estimation theory for continuous fields of observations. In: Lecture Notes in Earth Sciences 63, Springer Verlag, Berlin, pp 262-275.

Sansò (1986) Statistical methods in physical geodesy. In: H. Sünkel (ed) Mathematical and numerical techniques in physical geodesy, Springer Verlag, Berlin, pp 49-155.

Sansò F, Venuti G, Tscherning CC (2000) A theorem of insensitivity of the collocation solution to variations of the metric of the interpolation space. In: KP Schwarz (ed) Geodesy beyond 2000, Springer Verlag, Berlin, pp 233-240.

Skorohod AV (1974) Integration in Hilbert space. Ergebnisse der Mathematik und ihrer Grenzgebiete 79, Springer Verlag, Berlin.

Tarantola A (1987) Inverse problems theory; methods for data fitting and model parameter estimation. Elsevier, Amsterdam.

Tarantola A, Valette B (1982) Inverse problems; quest for information. J of Geoph 50:159-70.

Tscherning CC (1977) A note on choice of norm when using collocation for the computation of the approximations to the anomalous potential. Bull Geod 51.

Testing invariance for random field modeling

M. Reguzzoni, G. Venuti
DIIAR, Politecnico di Milano, Polo regionale di Como, Via Valleggio 11, 22100 Como, Italy

Abstract. The application of the collocation theory to the prediction of some random field functional depends on the knowledge of the covariance function. Whether we include the estimation of the covariance into a unique theoretical set up with the prediction of the signal, or we do it separately in a more traditional way, this step can be performed only under the assumption of some stochastic invariance of the field; under such a hypothesis, in fact, we can use a single sample to estimate the empirical covariance function.

However one easily realizes that the empirical estimator provides numerical answers even if applied to signals whose stochastic invariance is out of question. This calls for some criterion to decide, at least a posteriori, whether the hypothesis, on which the covariance estimation is based, is likely to be true or not.

Three different testing procedures were considered and applied to simulated data. The results obtained are discussed in the paper; they lead to a procedure, at least in a simplified context, which verifies whether the sample has or does not have a deterministic linear trend.

A more general result is finally presented, namely how to retrieve an estimate of the empirical covariance function of the signal, correcting the one estimated on the residues of the linear regression.

Keywords. Covariance estimate, invariance hypothesis, testing procedures

1 The invariance hypothesis and a standard testing procedure

The estimation of the random field covariance function from a single sample implies an assumption of some stochastic invariance of the field, Moritz (1989). For a one-dimensional (1D) stochastic process, like a process in time, the classical assumption is that the process is stationary, i.e. with a constant mean and a covariance function depending only on the time lag between two points, Sansò (1986). This hypothesis allows us to obtain several samples of the same process simply by shifting the data in time and, from these samples, to compute an empirical covariance function (ecf). Although formally incorrect, this is possible even if the process is not stationary; therefore a standard testing procedure is to be set up in order to verify a posteriori the likelihood of the invariance hypothesis. In the classical inference theory, an assumption on the process distribution is needed in order to derive the distribution of a suitable statistic; the test rejects the hypothesis if this statistic attains an unlikely value, once a significance level has been fixed. In particular, assuming $s(t)$ to be a normal process with zero mean, and denoting by $\underline{s}$ the sample drawn from it, if the corresponding covariance matrix C_{ss} is known, the statistic

$$\underline{s}^+ C_{ss}^{-1} \underline{s} \tag{1}$$

follows a chi-square distribution. It has to be stressed that, when C_{ss} is the right one, this is true even if the process is not stationary. However the covariance matrix C_{ss} is always unknown; can we expect its empirical estimate $\hat{C}_{ss}$ to be close to C_{ss} and the statistic

$$\underline{s}^+ \hat{C}_{ss}^{-1} \underline{s} \tag{2}$$

to have approximately a chi-square distribution? The guess is that the answer is positive only if the stationarity hypothesis is correctly applied, allowing us to discriminate between stationary and not stationary samples. The subsequent simulations are performed to investigate in this direction.

For the sake of simplicity we work with gridded data, defined on the domain:

$$\underline{t} = [-2 \quad -1.99 \quad \cdots \quad 1.99 \quad 2]^+ . \tag{3}$$

Stationary simulated samples are drawn from a gaussian process $s(t)$ with zero mean and the finite covariance function:

$$C(\tau) = \begin{cases} ke^{-\alpha\tau}\left(1 - \dfrac{\tau}{2}\right) & |\tau| \leq 2 \\ 0 & |\tau| > 2 \end{cases} \tag{4}$$

where τ is the time lag between two points, $k = 0.1$ and $\alpha = 2$ (see Figure 1). To be more precise, they are obtained using the following procedure: first we compute the covariance matrix C_{ss} from $C(\tau)$ for the given grid (3), which has a typical Toeplitz structure; then we produce its Cholesky decomposition, i.e.

$$C_{ss} = T^{+}T \tag{5}$$

where T is an upper triangular matrix; finally, multiplying T^{+} by a normal white noise sample

$$\underline{u} \sim N[\underline{0}, I] , \tag{6}$$

we obtain the sample

$$\underline{s} = T^{+}\underline{u} , \tag{7}$$

which is normally distributed too, with zero mean and, according to the covariance propagation law, with the given covariance matrix

$$C_{ss} = T^{+}C_{uu}T = T^{+}IT = T^{+}T . \tag{8}$$

Starting from the single sample (7) and exploiting the shifting invariance hypothesis, we compute the empirical covariance function by the biased estimator, Box and Jenkins (1976):

$$\hat{C}(|\tau_i|) = \frac{1}{N+1}\sum_{k=0}^{N-i} s_k s_{k+i} \tag{9}$$

where

$$\begin{aligned} i &= 0, 1, \cdots, N = 400 \\ \tau_i &= 0, 0.01, \cdots, 4 \\ s_k &= k^{th} \text{ element of the vector } \underline{s} \end{aligned}$$

and interpolate it with the positive definite parametric model (4) via least square adjustment. The covariance matrix derived from this interpolated model is denoted as $\hat{C}_{ss}$.

It is worth to point out that, even if the stationarity condition is satisfied, the ecf (9) and consequently its interpolated model, could differ a lot from the theoretical one (4), as shown in Figure 2. This makes the chi-square test useless to detect the stationarity of the sample.

Let us compare the empirical distributions of the quadratic forms (1) and (2), computed from a set of 10000 simulated stationary samples (7); by applying the well known approximate relation between standard normal Z and chi-square χ^2 random variables, Sachs (1982):

$$Z \cong \sqrt{2\chi^2} - \sqrt{2\nu - 1} , \tag{10}$$

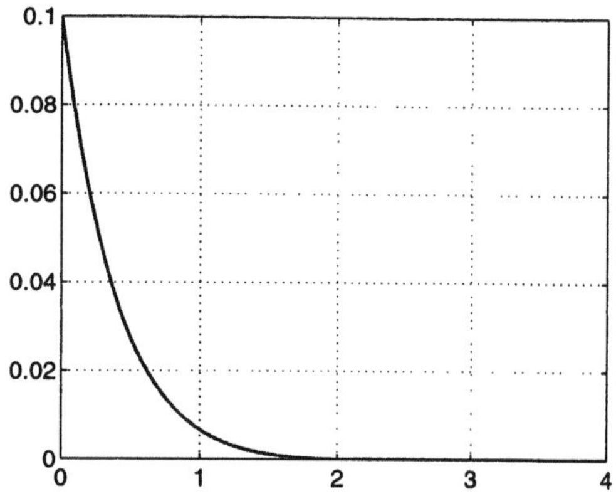

Fig. 1 Covariance function of the stationary process $s(t)$

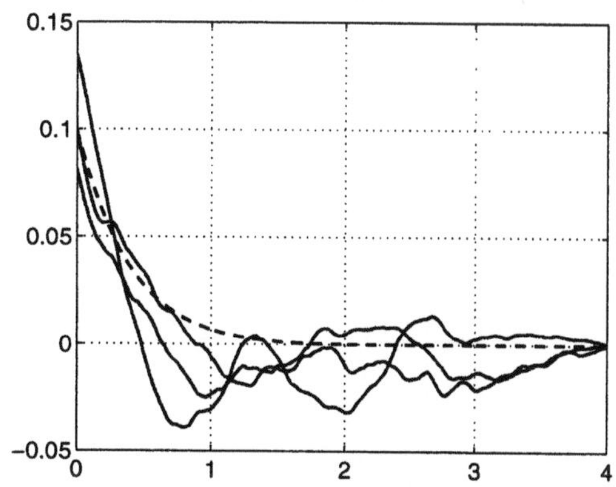

Fig. 2 Variability of the empirical covariance functions (solid line); theoretical covariance function in dashed line

valid for degrees of freedom $\nu > 100$, as in our case ($\nu = N = 400$), we obtained the histograms in Figure 3 and Figure 4. It is clear that only the empirical distribution of the statistic (1), with the theoretical C_{ss}, does fit a standard normal variate, according to (10); in other words the statistical variability of the estimated covariance matrix $\hat{C}_{ss}$ cannot be anyway disregarded, because it heavily affects the distribution of the statistic (2).

We drop the possibility of detecting a generic kind of non stationarity, concentrating on a normal process $y(t)$ with a non constant mean. In particular the non stationary sample is obtained from the stationary one (7) simply by adding a linear trend

$$\underline{y} = \underline{s} + a\underline{t} \tag{11}$$

where $a = \tan\frac{\pi}{6}$ (see Figure 5).

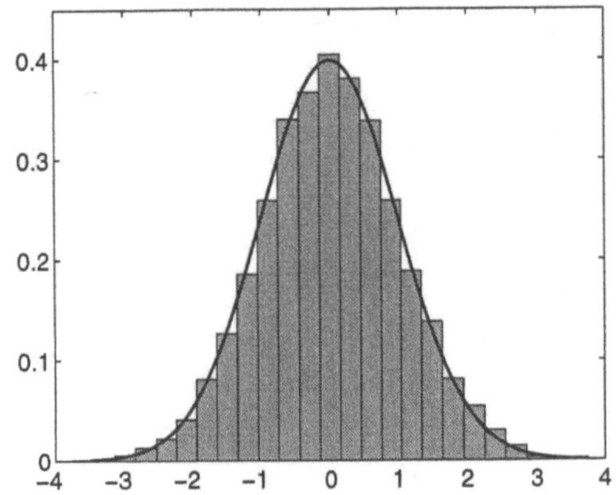

Fig. 3 Histogram of the statistic $\underline{s}^+ C_{ss}^{-1} \underline{s}$, with the correct covariance matrix; the transformation into a standard normal is applied (theoretical Z in solid line)

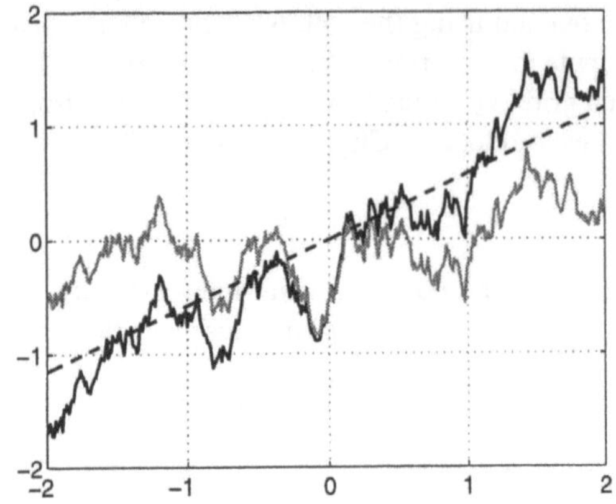

Fig. 5 A stationary sample (in gray) and the corresponding non stationary one (in black); the dashed line is the added linear trend

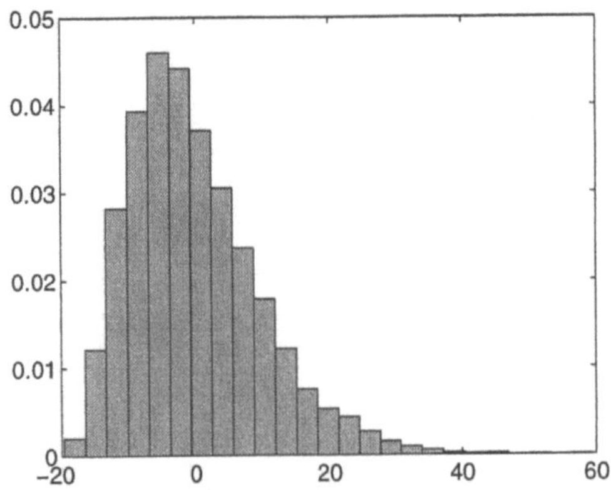

Fig. 4 Histogram of the statistic $\underline{s}^+ \hat{C}_{ss}^{-1} \underline{s}$, with the estimated covariance matrix; the transformation into a standard normal is applied

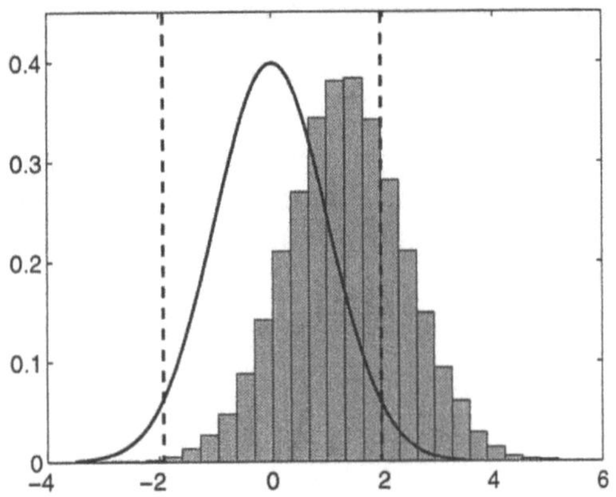

Fig. 6 Histogram of the statistic $\underline{y}^+ C_{yy}^{-1} \underline{y}$; the transformation into a standard normal is applied (theoretical Z in solid line with the acceptance interval at 5% level)

The theoretical covariance matrix C_{yy} is equal to C_{ss}, but the statistic

$$\underline{y}^+ C_{yy}^{-1} \underline{y} \tag{12}$$

is not a chi-square any more, due to the non zero mean of the process $y(t)$. The empirical distribution of this quadratic form, computed from 10000 simulated samples (11), is shown in Figure 6 after applying the transformation (10). Note that the chi-square test statistic (1) or the corresponding Z-test, cannot efficiently reveal the non stationarity of the samples, not even using the theoretical matrix C_{yy}: in fact, according to the histogram in Figure 6 and fixing a significance level of 5%, the non stationary samples are wrongly accepted as stationary with a too high probability (about 73%).

2 The alternative hypotheses test in a bayesian context

We tried to go forward by applying another strategy: the alternative hypotheses test, Cox and Hinkley (1974). The two alternatives to be compared are whether the given signal has or does not have a linear trend in the mean. Let us describe the procedure in the simulated case. Wrongly assuming that the sample (11) is stationary, we can compute its covariance matrix $\hat{C}_{yy}$ as explained in the previous paragraph, replacing $\underline{s}$ with $\underline{y}$. On the contrary, assuming that it has a deterministic linear trend, we first get the residues

$$\underline{r} = \underline{y} - \hat{a}\underline{t} \tag{13}$$

with respect to a standard linear regression, and then compute their covariance matrix $\hat{C}_{rr}$. The two alternatives, for a normal distributed process, can be formalized as follows:

$$\begin{aligned} H_{01} &: \underline{y} \sim N[\underline{0}, \hat{C}_{yy}] \\ H_{02} &: \underline{y} \sim N[\hat{a}\underline{t}, \hat{C}_{rr}] \end{aligned} . \tag{14}$$

We merge these hypotheses, in a bayesian context, by means of a discrete random variable α, whose possible values are 0 and 1, with an equal a-priori probability, obtaining the following likelihood distribution for the data:

$$\underline{y} \sim (1-\alpha)N[\underline{0}, \hat{C}_{yy}] + \alpha N[\hat{a}\underline{t}, \hat{C}_{rr}] . \tag{15}$$

The posterior distribution of α is therefore given by:

$$P(\alpha = 1|\underline{y}) = \frac{1}{1 + e^{\frac{1}{2}\left[Q_r - Q_y + \ln\frac{|\hat{C}_{rr}|}{|\hat{C}_{yy}|}\right]}} \tag{16}$$

where

$$Q_y = \underline{y}^+ \hat{C}_{yy}^{-1} \underline{y} , \qquad Q_r = \underline{r}^+ \hat{C}_{rr}^{-1} \underline{r} . \tag{17}$$

When the sample is not stationary, in the sense indicated, we expect that the posterior probability of $\alpha = 1$ is almost equal to 1, i.e. that H_{02} is likely to be true.

The meaningfulness of this approach has been studied on an elementary example where all the computations can be analytically performed; this example is described in appendix A.

Moreover the method has been numerically tested on a set of 10000 non stationary samples (11), obtaining that the 90% of them has a probability of $\alpha = 1$ greater than 0.99: this means that, when the linear trend is pronounced, the test successfully recognizes the correct hypothesis. Yet, if we apply the test on 10000 stationary samples (7), the wrong decision is taken almost one time out of three. No relevant improvement has been found replacing $\hat{C}_{rr}$ with $\hat{C}_{ss}$, computed using the method described in appendix B.

As in the first paragraph, these non satisfactory results are due to the estimation of the covariance function, in the sense that the likelihood ratio is too sensitive to the variations of the ecf with respect to the theoretical one, especially when the dimension of the samples involved is high. This is confirmed from the results obtained by replacing the estimated covariance matrices $\hat{C}_{yy}$ and $\hat{C}_{rr}$ with the theoretical C_{ss} in the posterior distribution (16): in this way the classification error is null when the test is applied to non stationary samples and it goes down to 8% when it is applied to the stationary cases.

3 A test on the linear regression coefficient

In the previous paragraph the non stationarity due to a linear trend in the mean is verified a posteriori by an alternative hypotheses test. Unfortunately the analytical expression of the posterior distribution involved in the test heavily depends even on small variations of the ecf estimates; in fact it implies the inversion of the ecf matrices and the computation of their determinants. Another approach to verify the same hypothesis, overcoming the described stability problem, is to introduce a test directly on the significance of the estimated tilt parameter. The estimated coefficient is

$$\hat{a} = \frac{\underline{t}^+ \underline{y}}{\underline{t}^+ \underline{t}} , \tag{18}$$

which is distributed as

$$\hat{a} \sim N\left[a, \frac{\underline{t}^+ C_{ss} \underline{t}}{(\underline{t}^+ \underline{t})^2}\right] , \tag{19}$$

so that the null hypothesis $H_0 : a = 0$ can be verified by using the following statistic:

$$Z_1 = \frac{\underline{t}^+ \underline{t}\, \hat{a}}{\sqrt{\underline{t}^+ C_{ss} \underline{t}}} \sim Z . \tag{20}$$

The weaker dependence of this statistic on the ecf estimation makes this approach successful, even using the matrix $\hat{C}_{rr}$ instead of the unknown C_{ss} of the stochastic part of the signal, i.e.

$$Z_2 = \frac{\underline{t}^+ \underline{t}\, \hat{a}}{\sqrt{\underline{t}^+ \hat{C}_{rr} \underline{t}}} \approx Z . \tag{21}$$

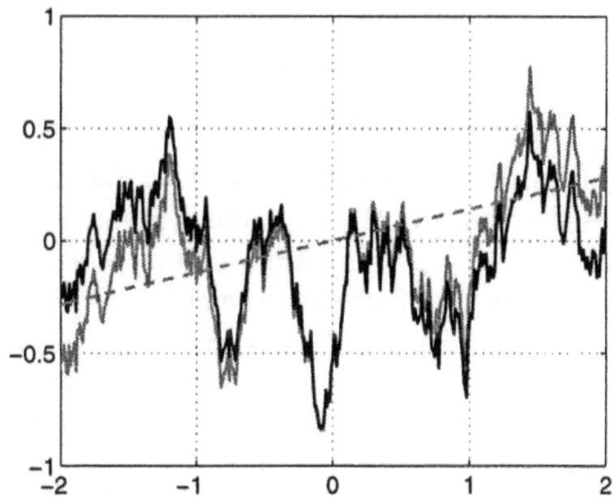

Fig. 7 A stationary sample (gray line), the estimated linear trend (dashed line) and the corresponding residues (black line)

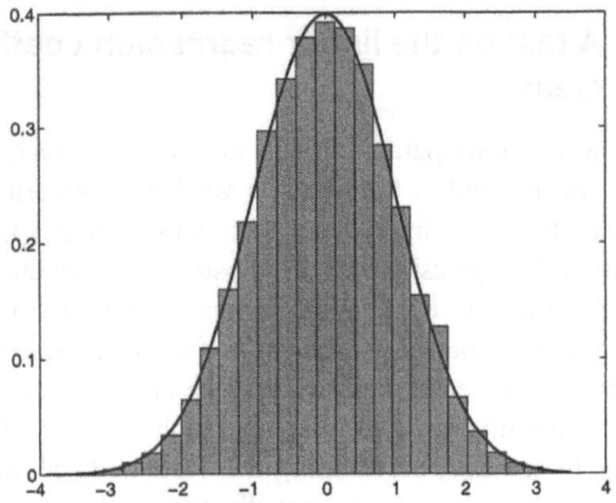

Fig. 8 Histogram of the statistic Z_1 for stationary samples (theoretical Z in solid line)

The method has been verified on several simulated data sets. If applied to the non stationary samples (11), it always correctly detects the presence of the linear trend. When applied to samples (7), the stationarity hypothesis is wrongly rejected in the 20% of the times. This result is due to the fact that the stationary samples can contain a linear trend simply due to the sampling procedure, as shown in Figure 7. If this stochastic trend is not negligible, the test obviously identifies it leading to a wrong conclusion, namely to consider it as deterministic. Things would go better if we knew the theoretical C_{ss}. By using this one in our simulations, the misclassification error is in fact reduced to the fixed significance level of 5%: the empirical distribution of the statistic (20) is indeed a standard normal (see Figure 8). A better approximation of the signal covariance function was therefore searched (see appendix B), obtaining a test error reduction of 3%: Figure 10 shows how the empirical distribution of the statistic

$$Z_3 = \frac{\underline{t}^+ \underline{t}\, \hat{a}}{\sqrt{\underline{t}^+ \hat{C}_{ss} \underline{t}}} \approx Z \qquad (22)$$

approximates the theoretical one better than the empirical distribution (21), displayed in Figure 9.

In Figure 11 the histogram of the parameter $\hat{a}$ of the whole set of stationary samples is displayed; Figure 12 and Figure 13 show the empirical distributions of the values of $\hat{a}$, corresponding to those samples for which the null hypothesis H_0 is rejected by using the statistic (21) and (22) respectively. Note that in both cases these values belong to the tails of the distribution in Figure 11, but they do not coincide with them. The use of $\hat{C}_{ss}$ instead of $\hat{C}_{rr}$ is recommended since the set of $\hat{a}$ misclassified with $\hat{C}_{ss}$ is a strict subset of those misclassified with $\hat{C}_{rr}$. Looking at Figure 12 and Figure 13, one also realizes that there is a threshold a^* such that when $|\hat{a}| < a^*$ the right decision (no trend) is always taken; however it is also clear that when $|\hat{a}| > a^*$ a wrong decision is not necessarily taken. This is due to the non linear dependence of the statistics (21) and (22) on the sample, through the covariance estimation process. In the case considered $a^* \cong 0.075$, which corresponds to a slope of about 4 degrees.

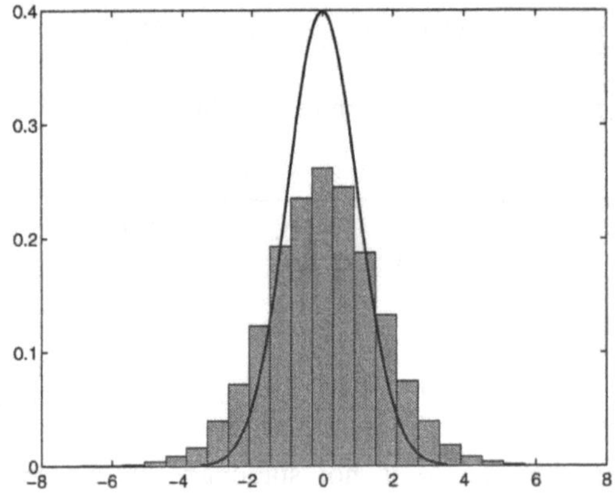

Fig. 9 Histogram of the statistic Z_2 for stationary samples (theoretical Z in solid line)

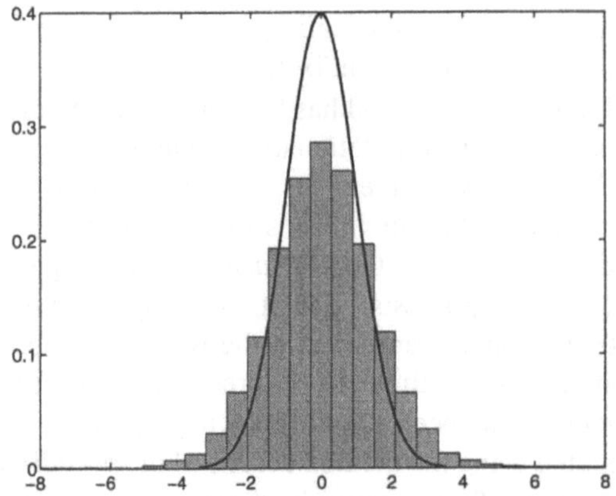

Fig. 10 Histogram of the statistic Z_3 for stationary samples (theoretical Z in solid line)

The same analysis was performed on the results of the alternative hypotheses test too, described in the previous paragraph. As displayed in Figure 14, the values of $\hat{a}$ corresponding to the rejected samples are not only in the tails of the empirical distribution in Figure 11, i.e. the test results are quite independent of the parameter $\hat{a}$. Moreover the use of $\hat{C}_{ss}$ instead of

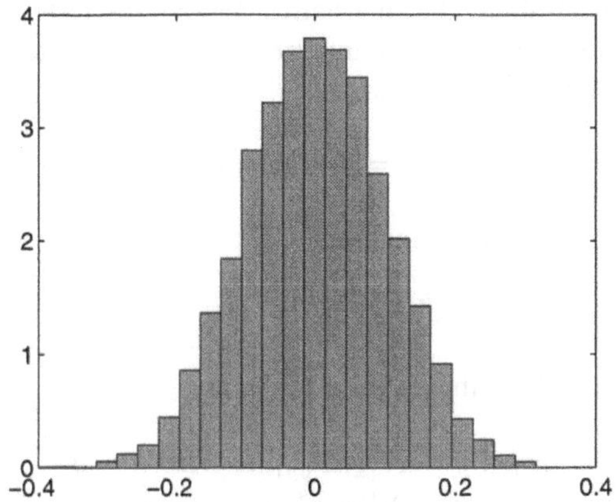

Fig. 11 Histogram of the tilt parameter estimated on 10000 stationary samples

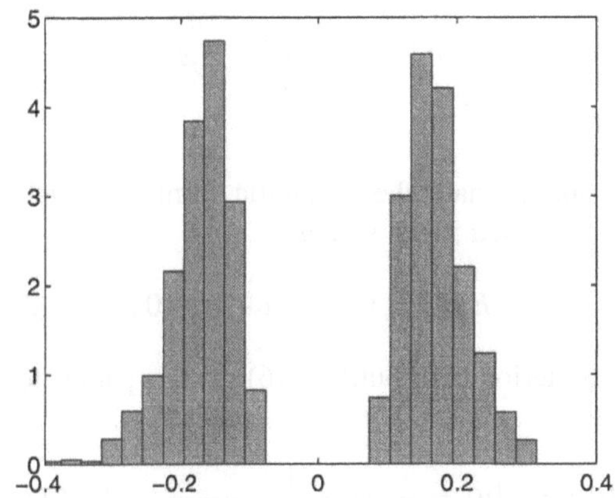

Fig. 12 Histogram of the tilt parameter of the stationary samples rejected using the statistic Z_2 (2019 samples)

$\hat{C}_{rr}$ implies the discrimination of different sets of $\hat{a}$, though the number of the wrongly classified samples is slightly reduced.

4 Conclusions

Three different testing procedures were considered and applied to simulated data. The first one is derived from the standard chi-square test. The covariance matrix, which plays the role of disturbance parameter, however, cannot be replaced with the one estimated from a truncated sample (stationary or not) of the process. It modifies the distribution of the statistic of the test destroying its usefulness.

A test on alternative hypotheses in a bayesian framework was therefore considered. In this case we restrict our analysis on processes invariant but for a linear trend in the mean. The way in which the statistic depends on the variations of the covariance estimates makes this approach not always reliable.

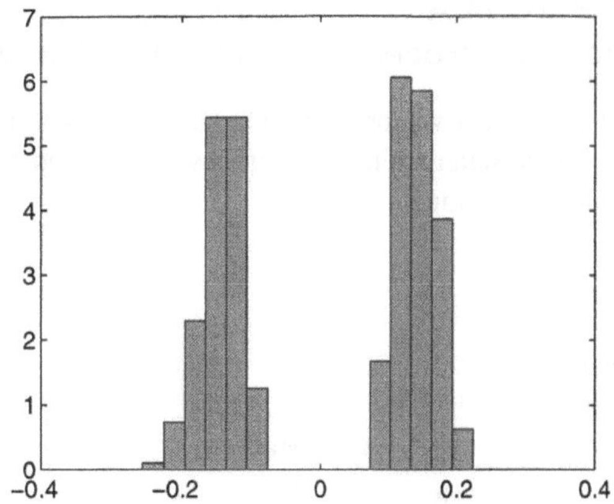

Fig. 13 Histogram of the tilt parameter of the stationary samples rejected using the statistic Z_3 (1700 samples)

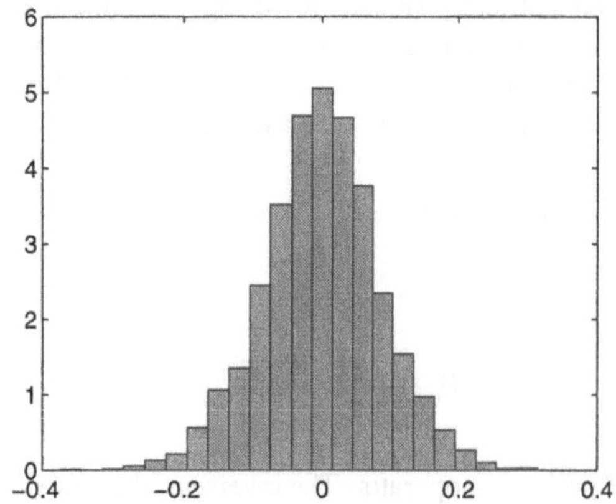

Fig. 14 Histogram of the tilt parameter of the samples rejected using the alternative hypotheses test (3767 samples)

A third successful procedure was defined to test the significance of the tilt parameter of the linear trend. When applied to stationary samples, the test leads to wrong conclusion only if the trend due to the sampling procedure is not negligible. This error, however, can be reduced if a better approximation of the unknown signal covariance function is computed starting from the covariance function of the residues.

This improvement in the covariance matrix estimation seems to be less effective if applied to the alternative hypotheses test.

Let us conclude with a consideration on the next step forward in this work. We will consider different kinds of deterministic trends, testing the significance of the involved parameters. Indeed, the case here considered is typically solved by a kriging approach, Wackernagel (1995). A further generalization of the method to two-dimensional random fields, both on a plane and spherical domain, will be also attempted.

APPENDIX A
Testing a process with a constant mean

We consider a vector $\underline{y}$ (with n components) drawn from a gaussian white noise process with a constant, but non zero, mean:

$$\underline{y} = b\underline{e} + \underline{\nu} \tag{A1}$$

where

$$\begin{aligned} b &= \text{constant} \\ \underline{e} &= [1\ 1\ \cdots\ 1]^+ \\ \underline{\nu} &= N[\underline{0}, \sigma^2 I]\,. \end{aligned}$$

In the same way as we blindly compute the ecf of a non stationary signal, we wrongly assume that the sample (A1) has a zero mean (note that it is anyway stationary). As a consequence we get the following biased estimate of the covariance matrix C_{yy}

$$\hat{C}_{yy} = nb^2 P_e + \sigma^2 I = nb^2 P_e + C_{yy} \tag{A2}$$

where

$$P_e = \frac{\underline{e}\,\underline{e}^+}{\underline{e}^+\underline{e}} = \frac{\underline{e}\,\underline{e}^+}{n}$$

is a projection operator. The inverse of the covariance matrix $\hat{C}_{yy}$ can be written as

$$\hat{C}_{yy}^{-1} = \frac{1}{\sigma^2}\left(I - \frac{\rho}{1+\rho}P_e\right); \quad \rho = \frac{nb^2}{\sigma^2} \tag{A3}$$

and the quadratic form

$$\begin{aligned} \underline{y}^+\hat{C}_{yy}^{-1}\underline{y} &= \frac{1}{\sigma^2}\left\{\frac{nb^2}{1+\rho} + \frac{2b}{1+\rho}\underline{e}^+\underline{\nu} + \right. \\ &\left. + \underline{\nu}^+\left(I - \frac{\rho}{1+\rho}P_e\right)\underline{\nu}\right\} \end{aligned} \tag{A4}$$

behaves, at least in the mean, as a true chi-square with n degrees of freedom

$$\begin{aligned} E[\underline{y}^+\hat{C}_{yy}^{-1}\underline{y}] &= \frac{1}{\sigma^2}\left\{\frac{nb^2}{1+\rho} + \left(n - \frac{\rho}{1+\rho}\right)\sigma^2\right\} \cong \\ &\cong \frac{1}{\sigma^2}\left\{\sigma^2 + (n-1)\sigma^2\right\} = n\,. \end{aligned} \tag{A5}$$

In other words by forcing the hypothesis of zero mean, we estimate a covariance that "eats" the constant b, making the standard chi-square test useless.

What about the alternative hypotheses test? By subtracting the constant mean b to the vector $\underline{y}$, we obtain the residues

$$\underline{r} = \underline{y} - b\underline{e} = \underline{\nu} \tag{A6}$$

with a covariance matrix

$$\hat{C}_{rr} = \hat{C}_{\nu\nu} = \sigma^2 I\,. \tag{A7}$$

The two hypotheses to be tested are

$$\begin{aligned} &H_{01} : \underline{y} \sim N[\underline{0}, \hat{C}_{yy}] && \text{for } \alpha = 0 \\ &H_{02} : \underline{y} \sim N[b\underline{e}, \hat{C}_{rr}] && \text{for } \alpha = 1 \end{aligned} \tag{A8}$$

where α is a discrete classifier. If we compute the determinants of the two covariance matrices:

$$\begin{aligned} \left|\hat{C}_{yy}\right| &= \sigma^{2n}\left(1 + \frac{nb^2}{\sigma^2}\right) \\ \left|\hat{C}_{rr}\right| &= \sigma^{2n} \end{aligned} \tag{A9}$$

and approximate the quadratic forms Q_y and Q_r (17) with their mean values:

$$E[Q_r - Q_y] = n - n = 0\,, \tag{A10}$$

the posterior distribution (16) of the parameter α becomes

$$P(\alpha = 1|\underline{y}) = \frac{1}{1 + e^{-\frac{1}{2}\ln\left(1+\frac{nb^2}{\sigma^2}\right)}} \to 1\,, \tag{A11}$$

leading to choose the right model.

APPENDIX B
The recovery of $\hat{C}_{ss}$ from $\hat{C}_{rr}$

We model the n-dimensional sample $\underline{y}$ as:

$$\underline{y} = \underline{s} + a\underline{t} \tag{B1}$$

where $\underline{s}$ is a stochastic signal with zero mean and covariance matrix C_{ss} and $a\underline{t}$ is a deterministic linear trend in time. The standard least squares estimate of a is:

$$\hat{a} = \frac{\underline{t}^+\underline{y}}{\underline{t}^+\underline{t}} = \frac{\underline{t}^+(\underline{s} + a\underline{t})}{\underline{t}^+\underline{t}} = \frac{\underline{t}^+\underline{s}}{\underline{t}^+\underline{t}} + a\,. \tag{B2}$$

The residues $\underline{r}$ can be related to the signal $\underline{s}$ by:

$$\begin{aligned} \underline{r} &= \underline{y} - \hat{a}\underline{t} = \underline{s} + (a - \hat{a})\underline{t} = \\ &= \underline{s} - \frac{\underline{t}\,\underline{t}^+}{\underline{t}^+\underline{t}}\underline{s} = (I - P_t)\underline{s} \end{aligned} \tag{B3}$$

and their covariance matrix can be derived from the covariance matrix of $\underline{s}$ by the propagation law:

$$C_{rr} = (I - P_t)C_{ss}(I - P_t) , \tag{B4}$$

which indeed cannot be used in general to recover C_{ss} from C_{rr}, since the orthogonal projector $I - P_t$ is not invertible. This problem can be overcome by the following procedure, which exploits the Toeplitz structure of the covariance matrices involved, due to the stationarity hypothesis and to the regular distribution in time of the sample $\underline{y}$. The empirical estimate of C_{rr} is in fact reduced to the recovery of the first row of the matrix itself, namely to the n-dimensional vector $\underline{C}_r$. The i-th element of $\underline{C}_r$, representing the covariance between the residues at distance i ($i = 0, 1, \cdots, n-1$), can be expressed by:

$$C_r(i) = \frac{1}{2N}\mathrm{Tr}[Q_i^s \underline{r}\,\underline{r}^+] \tag{B5}$$

where

$$Q_i^s = Q_i + Q_i^+$$

and Q_i is a n-dimensional square selector matrix, which is null except for the i-th diagonal equal to a sequence of ones:

$$Q_i = \begin{matrix} 0 & 1 & & i & & & n-1 \\ \end{matrix}$$
$$Q_i = \left[\begin{matrix} 0 & 0 & \cdots & 1 & 0 & \cdots & 0 \\ 0 & 0 & \cdots & 0 & 1 & \cdots & 0 \\ 0 & 0 & \cdots & 0 & 0 & \ddots & 0 \\ \cdots & \cdots & \cdots & \cdots & \cdots & \cdots & 1 \\ 0 & 0 & \cdots & 0 & 0 & 0 & 0 \end{matrix}\right] \tag{B6}$$

From the equation (B3), we get

$$\underline{r}\,\underline{r}^+ = (I - P_t)\underline{s}\,\underline{s}^+(I - P_t) \tag{B7}$$

and

$$C_r(i) = \frac{1}{2N}\mathrm{Tr}[\underline{s}\,\underline{s}^+(I - P_t)Q_i^s(I - P_t)] . \tag{B8}$$

At this point, in order to introduce the covariance matrix C_{ss}, we take the mean of the right handside of equation (B8), obtaining

$$C_r(i) \cong \frac{1}{2N}\mathrm{Tr}[C_{ss}(I - P_t)Q_i^s(I - P_t)] . \tag{B9}$$

Thanks to the Toeplitz structure, the unknown C_{ss} can be expressed as

$$\begin{aligned} C_{ss} &= C_s(0) + \sum_{j=1}^{n-1} C_s(j)Q_j^s = \\ &= \sum_{j=0}^{n-1} \overline{C}_s(j)Q_j^s \end{aligned} \tag{B10}$$

where

$$\overline{C}_s(0) = \frac{C_s(0)}{2}$$

and

$$\overline{C}_s(i) = C_s(i) \quad \forall i = 1, 2, \cdots, n-1 .$$

By substituting (B10) in (B9), we end up with the following simply determined linear system:

$$\begin{aligned} C_r(i) &= \frac{1}{2N}\mathrm{Tr}[\sum_{j=0}^{n-1} \overline{C}_s(j)Q_j^s(I-P_t)Q_i^s(I-P_t)] = \\ &= \frac{1}{2N}\sum_{j=0}^{n-1} \overline{C}_s(j)\mathrm{Tr}[Q_j^s(I-P_t)Q_i^s(I-P_t)] = \\ &= \sum_{j=0}^{n-1} a_{ij}\overline{C}_s(j) \end{aligned} \tag{B11}$$

or, in vectorial notation,

$$\underline{C}_r = A\underline{C}_s . \tag{B12}$$

Since the matrix A is invertible, we can obtain an estimate $\hat{C}_{ss}$ of the unknown covariance matrix of $\underline{s}$ from the estimate $\hat{C}_{rr}$ using the system (B12).

References

Box, G.E.P. and G. Jenkins (1976). Time Series Analysis: Forecasting and Control. Holden-Day, San Francisco.

Cox, D.R. and D.V. Hinkley (1974). Theoretical Statistics. Chapman and Hall, London.

Moritz, H. (1989). Advanced Physical Geodesy. Wichmann, Karlsruhe.

Sachs, L. (1982). Applied Statistics. A Handbook of Techniques. Springer-Verlag, New York.

Sansò, F. (1986). Statistical Methods in Physical Geodesy. In: *Mathematical and Numerical Techniques in Physical Geodesy*, Lecture Notes in Earth Sciences, Vol. 7, Springer-Verlag, Berlin, pp. 49-155.

Wackernagel, H. (1995). Multivariate Geostatistics. Springer-Verlag, Berlin.

A formal comparison between Marych-Moritz's series, Sansò's change of boundary method and a variational approach for solving some linear geodetic boundary value problems

J.Otero
Instituto de Astronomía y Geodesia (UCM-CSIC), Facultad de Matemáticas, Universidad Complutense, Plaza de Ciencias 3, Ciudad Universitaria, 28040 Madrid, Spain

A.Auz
GMV, Isaac Newton 11, P.T.M. Tres Cantos, 28760 Madrid, Spain

Abstract. In this paper we are concerned with the simple Molodensky problem and the linearized fixed-boundary gravimetric boundary-value problem in spherical approximation. We find a series solution for these problems from a variational approach using the Molodensky shrinking. These series are compared with the solution by analytical continuation and the change of boundary method.

1 Introduction

Let S_1 be the unit sphere in $\mathbb{R}^3$. If $h \in C^0(S_1)$ and $h \geq 0$, let Σ be the image of S_1 under the map $\phi_h : S_1 \mapsto \mathbb{R}^3$,

$$\phi_h(\sigma) = (R + h(\sigma))\sigma ,$$

where $\sigma \in S_1$ and R is a positive constant.
Let f be a given function defined on the closed surface Σ. We are interested in the boundary-value problems

$$P_k(\Sigma, f) : \begin{cases} \Delta u = 0 & \text{in } \Omega, \\ -\dfrac{\partial u}{\partial r} - \dfrac{k}{r} u = f & \text{on } \Sigma, \\ u(\infty) = 0, & \end{cases}$$

where Ω denotes the domain exterior to Σ and $k \in \{0, 2\}$. If S_R is the sphere of radius R then $\Sigma = S_R$ if, and only if, $h = 0$. $P_2(S_R, f)$ is Stokes' problem and $P_0(S_R, f)$ is Hotine's problem. The function f can be extended throughout the whole space $\mathbb{R}^3$ (with exception of the origin and infinity) so that it remains constant on rays passing through the origin. Below, by f we denote this extension.
At present, $P_2(\Sigma, f)$ is called in geodesy *the simple Molodensky problem* or simply (as some years ago) Molodensky's problem. $P_0(\Sigma, f)$ is the linearized fixed-boundary gravimetric boundary-value problem (or GPS-gravimetric boundary problem as recently called by Moritz (2000)) in spherical approximation. The meaning of Σ and of the functions taking part in $P_k(\Sigma, f)$ is the following: R is a mean radius of the earth, u is the disturbing potential,

- if $k = 2$, Σ is the telluroid and f is the gravity anomaly Δg at ground level,
- if $k = 0$, Σ is the earth's surface and f is the gravity disturbance δg,

and $h(\sigma)$ is the height of $\omega \in \Sigma$ above the sphere S_R where $\sigma = \omega/|\omega|$.
The problem $P_0(\Sigma, f)$ is particularly significant nowadays. Indeed, its solution gives the gravity potential on the earth surface and therefore geopotential numbers (the physical measure of height above sea level), as has been pointed out by Moritz (2000, p.85). By means of approximations such as, for example, the reduction of Poincaré and Prey (see Heiskanen and Moritz, 1967, § 4.4), from these geopotential numbers we can get orthometric heights, and hence the fixed-boundary gravimetric boundary-value problem could be a good alternative to the so called GPS-levelling which needs a very precise determination of the geoid.
From a mathematical point of view the problems $P_0(\Sigma, f)$ and $P_2(\Sigma, f)$ have a different treatment. Whereas $P_0(\Sigma, f)$ is unconditional solvable, the problem $P_2(\Sigma, f)$ may have no solution and if a solution exists it is not unique.
M.S.Molodensky (see Molodensky *et al.*, 1962, Chapter V, §15) solved $P_2(\Sigma, f)$ by means of Fredholm's integral method: he sought to find the solution of $P_2(\Sigma, f)$ in the form of a single layer potential and this leads to the solution of the following integral equation (Molodensky *et al.*, 1962, Eq. V.15.9)

$$2\pi\chi \cos^2 \alpha = (1 + \varepsilon)^2 f + \frac{R}{2}(1 + \varepsilon) \int_{S_1} M(\sigma, \sigma')\chi(\sigma')\, d\sigma' , \quad (1)$$

where χ is an auxiliary unknown function defined on S_1 and $\varepsilon = h/R$. The kernel M is given by

$$M(\sigma,\sigma') = 3L_h^{-1}(\sigma,\sigma') + L_h^{-3}(\sigma,\sigma')(r^2(\sigma') - r^2(\sigma)),$$

where

$$L_h(\sigma,\sigma') = |\phi_h(\sigma) - \phi_h(\sigma')|$$

and $r(\sigma) = |\phi_h(\sigma)| = R + h(\sigma)$. The terrain inclination $\alpha(\sigma)$ at $\phi_h(\sigma)$ is the angle between the outer normal to Σ and the radius vector at $\phi_h(\sigma)$, and it is given by

$$\tan\alpha = \frac{1}{r}|\nabla_s h| = \frac{1}{R}(1+\varepsilon)^{-1}|\nabla_s h|,$$

where $\nabla_s h$ is the gradient of $h(\sigma)$. In terms of χ, the function u at $\phi_h(\sigma)$ is assumed to be

$$u(\phi_h(\sigma)) = R^2 \int_{S_1} L_h^{-1}(\sigma,\sigma')\chi(\sigma')\,d\sigma'.$$

For solving the integral equation (1), Molodensky replaces Σ by $\Sigma_t := \phi_{th}(S_1)$ with $t \in [0,1]$ (*the Molodensky shrinking*). Then, expanding χ, α and the kernel $M(\sigma,\sigma')$ of the integral equation in series of powers of t, he obtains (in planar approximation ($\varepsilon = 0$))

$$u_t \circ \phi_{th} = \sum_{n=0}^{\infty} u_n t^n, \qquad (2)$$

where $u_t(x)$ is the solution of $P_2(\Sigma_t, f)$ and u_n are functions defined on the unit sphere S_1. From (2), setting $t = 1$ we get formally $u \circ \phi_h$.
The function u_0 in (2) is the restriction to S_R of the solution of $P_2(S_R, f)$ given by Stokes' formula

$$u_0 = \frac{R}{4\pi}\int_{S_1} G_0\,[S(\psi) - 1]\,d\sigma,$$

where $S(\psi)$ is the Stokes function and $G_0 = f$. The other functions are also given in terms of Stokes' function and, in words of Molodensky, Eremeev and Yurkina (Molodensky *et al.*, 1962, p. 123) *the topography is taken into account in the second equation* (function) *and the gradients appear clearly from the second equation* (function) *onwards.* For instance, the function u_1 is given by

$$u_1 = \frac{R}{4\pi}\int_{S_1} G_1\,[S(\psi) - 1]\,d\sigma,$$

where

$$G_1(\sigma) = R^2 \int_{S_1} \frac{(h(\sigma') - h(\sigma))}{L_0^3(\sigma,\sigma')}\chi_0(\sigma')\,d\sigma',$$

and

$$\chi_0 = \frac{G_0}{2\pi} + \frac{3}{(4\pi)^2}\int_{S_1} G_0\,[S(\psi) - 1]\,d\sigma.$$

Some refinements to this solution of Molodensky were given later by Brovar (see Brovar, 1964; Moritz, 1980). For the problem $P_0(\Sigma, f)$ a Molodensky-type solution has been derived by Stock (1983).

To our way of thinking, Molodensky's method has a clear connection with the *Poincaré's method of the small parameter* for solving a regularly perturbed Cauchy problem for ordinary differential equations (see, for example, Kuzmina, 2000, § 1 and § 9):

$$\frac{dx}{dt} = F(x,t,\varepsilon); \quad x(0,\varepsilon) = x^0(\varepsilon). \qquad (3)$$

In fact, in Poincaré's method, the solution $x(t,\varepsilon)$ is expanded in powers of ε ($x(t,\varepsilon) = \sum_{k=0}^{\infty} x^k(t)\varepsilon^k$). Then $x(t,\varepsilon) = \sum_{k=0}^{\infty} x^k(t)\varepsilon^k$ is substituted in the equation (3) and both sides of (3) are expanded in powers of ε. Equating the coefficients in terms with equal powers of ε, we finally have the *linear* equations for the coefficients $x^k(t)$.

In 1969, Marych and Moritz independently found *an elementary solution by analytical continuation in the form of a Molodensky series*, using Moritz's own words (see Moritz (2000), where a brief history of the use of analytical continuation can also be found). These Marych-Moritz's series can be derived without needing integral equations and they are nowadays the basis for numerical computations (see, for example, Sünkel (1997)). More details about all these series solutions, their equivalence and the convergence point can be found in (Moritz, 1980). For the linearized fixed-boundary gravimetric boundary-value problem a similar solution has recently be derived by Moritz (2000).

In 1993, Sansò (1993) proposed an iterative method for solving a certain class of linear boundary value problems for Laplace's equation. This method is known as *the change of boundary method* and applied to solve $P_k(\Sigma, f)$ it leads to another series solution. This approach will be outlined in Section 3.

From the basic idea of Molodensky of replacing Σ by Σ_t, in Section 2 we propose an alternative variational approach to solve $P_k(\Sigma, f)$. We show that the variational approach solution and the solution by analytical continuation are *formally* the same. The essential difference is that no use of downward analytical continuation is made in the variational approach. Rather, upward harmonic continuation from S_R is what is needed to explain the terms of the derived series. In this way the Marych-Moritz's solution could be now correctly interpreted. In addition, this upward harmonic continuation allows that one can guess some linking between the change of boundary method and the variational approach, and this relation is also studied in this paper.

We have previously spent some time in describing the method of solution of $P_2(\Sigma, f)$ due to Molodensky since the idea of the variational approach is basically the same: using the Molodensky shrinking, to approximate the solution of $P_k(\Sigma, f)$ by solving boundary problems in the domain Ω_R exterior to S_R (integral equations in Molodensky's method) for which there exist explicit formulae for its solutions (in the case $k = 2$, Stokes' problem with **changing boundary values**). For this approximation to be effective we need that $\Omega \subset \Omega_R$, and for this reason $h \geq 0$ as we are assuming. The variational approach is more direct in the sense that the intermediate step of expressing u as a simple layer potential is avoided. For Dirichlet's problem on nearly circular domains (in the plane), Kautský (1962) follows a method similar to the variational approach described in this paper (see his Theorem 3.1 and compare (Kautský, 1962, formula (9)) with the formula (8) of this paper which gives the boundary values of the approximations of the solution of $P_k(\Sigma, f)$).

The origin of this paper is twofold: firstly, to understand the Marych-Moritz's solution by analytical continuation for Molodensky's problem; and in second place, to look for the connection between the series solutions obtained in the past for this problem and the solution given by the change of boundary method. Since this is the main motivation of this paper we have intentionally not considered solvability conditions for $P_2(\Sigma, f)$. Furthermore, neither regularity of solutions nor convergence problems have been borne in mind. The question of the convergence of the different series solutions presented here is still open.

2 A variational approach for solving $P_k(\Sigma, f)$

Another way of solving $P_k(\Sigma, f)$ is the following variational approach that we formally unfold in this section. We define the function $[0,1] \times \overline{\Omega} \mapsto \mathbb{R}$, $u(t,x) := u_t(x)$ where u_t is the solution of $P_k(\Sigma_t, f)$. For any function $g(t,x)$ we shall use the notation

$$g_n(t,x) = \frac{1}{n!}\frac{\partial^n g}{\partial t^n}(t,x),$$

and $g_n(x) := g_n(0,x)$. If $u_t(x)$ is smooth and depends smoothly on t, we may expand $u(t,x)$ in series of powers of t at $t = 0$ and we get

$$u(t,x) = \sum_{n=0}^{\infty} u_n(x) t^n \quad (x \in \overline{\Omega}).$$

Since $u(1,x)$ is the solution $u(x)$ of $P_k(\Sigma, f)$, then

$$u(x) = \sum_{n=0}^{\infty} u_n(x) \quad (x \in \overline{\Omega}).$$

Note that

$$\Delta(u_n(t,x)) = (\Delta u(t,x))_n,$$

where Δ is the Laplace operator in $\mathbb{R}^3$ given by $\sum_{i=1}^{3} \partial^2/\partial x_i^2$. Hence the functions $u_n(t,x)$ are harmonic in the domain Ω_t exterior to Σ_t and regular at infinity. In particular, the functions $u_n(x)$ are harmonic outside S_R. Note in addition that $u_0(x)$ is the solution of $P_k(S_R, f)$.

In order to obtain a boundary condition for $u_n(x)$, we consider the function $U = r\partial u/\partial r + ku$ which is harmonic in Ω_t, and in Ω_t satisfies

$$U_n(t,x) = r\frac{\partial u_n(t,x)}{\partial r} + k u_n(t,x).$$

Let $F = -r^{-1}U$; then, in points of Ω_t we have

$$-\frac{\partial u_n(t,x)}{\partial r} - \frac{k}{r} u_n(t,x) = F_n(t,x). \tag{4}$$

If $F_n(t,.) \in C^0(\overline{\Omega}_t)$, the functions $u_n(t,x)$ are then solutions of the problems $P_k(\Sigma_t, F_n(t,x))$. Hence, if $t = 0$, we conclude that $u_n(x)$ are solutions of $P_k(S_R, F_n(x))$.

To compute $F_n(x)$ on S_R we note that on Σ_t we have $F = f$, that is to say

$$F\left(t, (R + th(\sigma))\sigma\right) = f(\sigma). \tag{5}$$

Since f does not depend on t, differentiation of (5) with respect to t gives on Σ_t,

$$\frac{\partial F}{\partial t} = -h\frac{\partial F}{\partial r}. \tag{6}$$

Repeated differentiation of (6) with respect to t gives on Σ_t the following recursive formula ($n \geq 1$)

$$\begin{aligned} \frac{\partial^n F}{\partial t^n} &= -h^n \frac{\partial^n F}{\partial r^n} \\ &\quad - \sum_{i=1}^{n-1} \binom{n}{i} h^{n-i} \frac{\partial^{n-i}}{\partial r^{n-i}} \left(\frac{\partial^i F}{\partial t^i}\right). \end{aligned} \tag{7}$$

Since

$$\frac{\partial^j}{\partial r^j}\left(\frac{\partial^i F}{\partial t^i}\right)(0,x) = \frac{\partial^j}{\partial r^j}\left(\frac{\partial^i F}{\partial t^i}(0,x)\right),$$

then from (7) we get on S_R $(t = 0)$

$$F_n = -\sum_{i=1}^{n} \frac{1}{i!} h^i \frac{\partial^i F_{n-i}}{\partial r^i}, \tag{8}$$

where, from (4), if $j = 0, \cdots, n-1$

$$F_j = -\frac{\partial u_j}{\partial r} - \frac{k}{r} u_j .$$

The functions F_j are defined in $\overline{\Omega}_R$ and rF_j are harmonic in Ω_R.

Remark 2.1 Hitherto we have avoided questions of regularity. For instance, to compute u_1 the existence of the radial derivative $\partial F_0/\partial r$ on S_R is required and this depends on the regularity of f. □

Remark 2.2 As it has been pointed out by Sansò (1993, Remark A.1) in the context of the change of boundary method (see Section) *the only convergence we can expect from the above approach is at best on* Σ*, but not certainly in the layer between* S_R *and* Σ*; in fact, such a convergence can happen only if the sought solution* u *has in fact a backward harmonic continuation and this can happen only under very restrictive conditions,* using his own words. □

In order to compare the solution that we have obtained with the solution by analytical continuation we introduce the operators

$$L_n : C^\infty(\overline{\Omega}_R) \mapsto C^\infty(S_R),$$

defined by

$$L_n(g) = \frac{1}{n!}\gamma\left(\frac{\partial^n g}{\partial r^n}\right),$$

where $\gamma(v)$ denotes the trace (restriction) of $v \in C^\infty(\overline{\Omega}_R)$ on S_R. Then we may write (8) in the form $(n \geq 1)$

$$F_n = -\sum_{i=1}^{n} h^i L_i(F_{n-i}). \tag{9}$$

This expression, if compared with the formula given by Moritz (1980, § 45: *Solution by analytical continuation*, Eq. 45-44), shows the *formal* equivalence between the variational approach and the solution by analytical continuation. However, we stress that the equation (9) is meaningful (if f is smooth) and it has been obtained without downward analytical continuation of $f(\omega)$ $(\omega \in \Sigma)$.

Instead of (9) we write $(n \geq 1)$

$$F_n(\sigma) = -\sum_{i=1}^{n} \varepsilon(\sigma)^i H_{i,n-i}(R\sigma)$$

where

$$H_{i,j} = \frac{1}{i!} r^i \frac{\partial^i F_j}{\partial r^i},$$

and $\varepsilon = h/R$.

The functions rF_j are harmonic in Ω_R. Consequently the functions $rH_{i,j}$ are as well harmonic in Ω_R for all $i \geq 1$. In fact, we have

$$H_{i,j} = \frac{1}{i}\frac{\partial}{\partial r}(rH_{i-1,j}) - H_{i-1,j}. \tag{10}$$

Since $rH_{0,j} = rF_j$ is harmonic, multiplying both members in (10) by r, this relation shows recursively that $rH_{i,j}$ are harmonic for all $i \geq 1$.

We now define

$$\widehat{H}_{i,j}(x) = |x| H_{i,j}(Rx).$$

The function $\widehat{H}_{i,j}$ is harmonic outside S_1 and on S_1 we have $\widehat{H}_{i,j}(\sigma) = H_{i,j}(R\sigma)$. In addition

$$\begin{aligned} \frac{\partial \widehat{H}_{i,j}}{\partial r}(\sigma) &= H_{i,j}(R\sigma) + R\frac{\partial H_{i,j}}{\partial r}(R\sigma) \\ &= \frac{\partial}{\partial r}(rH_{i,j})(R\sigma). \end{aligned}$$

From (10) we conclude that

$$H_{i,j}(R\sigma) = \frac{1}{i}\frac{\partial \widehat{H}_{i-1,j}}{\partial r}(\sigma) - \widehat{H}_{i-1,j}(\sigma). \tag{11}$$

Equation (11) has important consequences. It first shows that the function $H_{i,j}(R\sigma)$ $(\sigma \in S_1)$ is the trace on S_1 of the harmonic function

$$\frac{1}{i} r \frac{\partial \widehat{H}_{i-1,j}}{\partial r} - \widehat{H}_{i-1,j}.$$

Since one can define the **trace** on S_1 of an **arbitrary** harmonic function in Ω_1 and regular at infinity (see Dautray and Lions, 1990, Chapter 2, § 6.2) (and therefore also the trace of its radial derivative), equation (11) could be the key to give a solid mathematical foundation to the variational approach that we have followed for solving $P_k(\Sigma, f)$.

From a computational point of view, we recall that if u is harmonic outside S_1 and regular at infinity, its radial derivative on S_1 is given by the well-known formula (see, for example, Heiskanen and Moritz, 1967, Eq. 1-97)

$$\frac{\partial u}{\partial r}(\sigma) = -u(\sigma) + \frac{1}{2\pi}\int_S \frac{1}{|\sigma' - \sigma|^3}\,(u(\sigma') - u(\sigma))\,d\sigma'.$$

Hence we have

$$H_{i,j}(R\sigma) = -\left(\frac{1+i}{i}\right) H_{i-1,j}(R\sigma) + \frac{1}{2i\pi}\int_S \frac{(H_{i-1,j}(R\sigma') - H_{i-1,j}(R\sigma))}{|\sigma' - \sigma|^3}\,d\sigma'. \tag{12}$$

Formula (12) allows a recursive computation of the functions $F_n(\sigma)$.

3 Sansò's change of boundary method

With the notation used in (Sansò, 1993), the series solution of $P_k(\Sigma, f)$ that we get by means of the change of boundary method is the following

$$u = u_0 + \sum_{n=1}^{\infty} \delta u_n \qquad (\text{in } \overline{\Omega}),$$

where u_0 is the solution of $P_k(S_R, f)$. Let B be the operator

$$Bu = -\frac{\partial u}{\partial r} - \frac{k}{r}u\,.$$

The corrections δu_n are computed as follows: δu_1 is the solution of $P_k(S_R, \Delta_1)$ where

$$\Delta_1(\sigma) = f(\sigma) - Bu_0(\phi_h(\sigma))\,.$$

Let $v_1 := u_0 + \delta u_1$, then δu_2 is the solution of $P_k(S_R, \Delta_2)$ where

$$\Delta_2(\sigma) = f(\sigma) - Bv_1(\phi_h(\sigma))\,.$$

We define

$$v_n = \begin{cases} u_0 & \text{if } n = 0, \\ u_0 + \sum_{i=1}^{n} \delta u_n & \text{if } n \geq 1. \end{cases}$$

In general, δu_n ($n \geq 1$) is the solution of $P_k(S_R, \Delta_n)$ where

$$\Delta_n(\sigma) = f(\sigma) - Bv_{n-1}(\phi_h(\sigma))\,.$$

In other words, Δ_n is the pullback to S_1 of the function defined on Σ

$$f(\omega) - Bv_{n-1}(\omega) \quad (\omega \in \Sigma).$$

Note that if $n \geq 1$

$$\Delta_{n+1}(\sigma) = \Delta_n(\sigma) - B(\delta u_n)(\phi_h(\sigma))\,.$$

Hence we may conclude at least formally that on Σ

$$-\frac{\partial u}{\partial r} - \frac{k}{r}u = (f - \Delta_1) + \sum_{n=1}^{\infty}(\Delta_n - \Delta_{n+1}) = f\,.$$

In order to compare the variational approach with the change of boundary method for solving $P_k(\Sigma, f)$ we observe the following. We have seen that the correction u_n to the partial sum $v_{n-1} := \sum_{i=0}^{n-1} u_i$ in the variational approach is the solution of $P_k(S_R, F_n)$. Let

$$(f - Bv_{n-1})(\omega) = (f - \sum_{j=0}^{n-1} F_j)(\omega) \quad (\omega \in \Sigma)\,.$$

The pullback to S_1 of this function is approximated in the following way. Let $\omega \in \Sigma$ and $\sigma = \omega/|\omega|$, then we approximate each F_j ($j = 0, \cdots, n-1$) by the $(n-j)$th partial sum of its Taylor series expansion with respect to r at $r = R$ ($h = 0$), getting in this way a decreasing order of approximation as j increases. Hence

$$F_0(\omega) \approx f(\omega) + \sum_{i=1}^{n} \frac{1}{i!} h^i(\sigma) \frac{\partial^i F_0}{\partial r^i}(R\sigma)$$

$$F_1(\omega) \approx F_1(R\sigma) + \sum_{i=1}^{n-1} \frac{1}{i!} h^i(\sigma) \frac{\partial^i F_1}{\partial r^i}(R\sigma)$$

$$\vdots$$

$$F_{n-1}(\omega) \approx F_{n-1}(R\sigma) + h(\sigma)\frac{\partial F_{n-1}}{\partial r}(R\sigma)\,.$$

Grouping terms we have

$$\sum_{j=0}^{n-1} F_j(\omega) \approx f(\omega) + \underbrace{\sum_{j=1}^{n} \frac{1}{j!} h^j(\sigma) \frac{\partial^j F_{n-j}}{\partial r^j}(R\sigma)}_{=-F_n} + \underbrace{\sum_{j=1}^{n-1}\left(F_j(R\sigma) + \sum_{i=1}^{j} \frac{1}{i!} h^i(\sigma) \frac{\partial^i F_{j-i}}{\partial r^i}(R\sigma)\right)}_{=0}.$$

From equation (8) we finally find

$$\sum_{j=0}^{n-1} F_j(\omega) \approx f(\omega) - F_n(\sigma)\,.$$

Thus we conclude that the pullback of $(f - Bv_{n-1})(\omega)$ to S_1 is approximately given by $F_n(\sigma)$.

4 Gradient solution

Following (Moritz, 1980), we call *gradient solution* of $P_k(\Sigma, f)$ to the function $v_1 := u_0 + u_1$ harmonic in Ω_R. This function can be seen as a first-order approximation to the solution of $P_k(\Sigma, f)$. The function u_1 is the solution of $P_k(S_R, F_1)$, where

$$F_1(\sigma) = -h(\sigma)\frac{\partial F_0}{\partial r}(R\sigma),$$

and

$$F_0 = -\frac{\partial u_0}{\partial r} - \frac{k}{r}u_0\,.$$

Therefore v_1 solves $P_k(S_R, f + F_1)$ and is given by

$$v_1 = \frac{R}{4\pi}\int_{S_1}\left(f - h\frac{\partial F_0}{\partial r}\right)N_k(r,\psi)\,d\sigma,$$

where the kernel $N_k(r, \psi)$ is the spatial Stokes function if $k = 2$ or the spatial Hotine function if $k = 0$ (for these functions see Hotine (1969, Eqs. 29.14, 29.17) or Grafarend *et al.* (1985)). Note that the function $(r/R)F_0$ is harmonic in Ω_R and f is its value on S_R (upward harmonic continuation). Defining momentarily $f(r\sigma) = F_0(r\sigma)$ $(r \geq R)$ the gradient solution can be written in the form

$$v_1 = \frac{R}{4\pi}\int_{S_1}\left(f - h\frac{\partial f}{\partial r}\right)N_k(r,\psi)\,d\sigma.$$

In the geodetic case, if $k = 2$ then $f = \Delta g$ and we have

$$v_1 = \frac{R}{4\pi}\int_{S_1}\left(\Delta g - h\frac{\partial \Delta g}{\partial r}\right)S(r,\psi)\,d\sigma. \quad (13)$$

In this way we have obtained the Moritz gradient solution for Molodensky's problem (Moritz, 1980, Eq. (45-50)) but in an essentially different manner. In fact, in (13) the spatial function $(r/R)\Delta g(r\sigma)$ is the upward harmonic continuation of the surface function $\Delta g(R\sigma)$, whereas Moritz's solution is based in downward analytical continuation of $\Delta g(\omega)$ $(\omega \in \Sigma)$. It should be remarked that in (13), $\Delta g(\sigma) = \Delta g(\phi_h(\sigma))$, that is to say $\Delta g(\sigma)$ is the pullback of Δg to the sphere S_1.

Remark 4.1 Using a different method (based on a change of coordinates transforming the telluroid into an sphere), P.Holota (see Holota (1985) and Holota (1989, p.502)) also found an identical interpretation for the gradient solution of Molodensky's problem without resorting to downward analytical continuation. □

Remark 4.2 Replacing in the equation (13) Δg by δg and $S(r, \psi)$ by $H(r, \psi)$, we have a similar formula for the gradient solution of $P_0(\Sigma, f)$ (see Auz and Otero, 2002). The problem $P_0(\Sigma, f)$ can also be solved by Molodensky's method (Stock, 1983, § 3). See in addition some of the pioneering works on the geodetic boundary value problem using the known surface of the earth, for instance (Koch and Pope, 1972). □

Summing up, the following is the interpretation of the gradient solution as it is obtained by the variational approach.

1. As a zero-order approximation of u we take the solution u_0 of $P_k(S_R, f)$.
2. Now we consider the difference $f(\omega) - F_0(\omega)$ $(\omega \in \Sigma)$ where

   $$F_0 = -\frac{\partial u_0}{\partial r} - \frac{k}{r}u_0\,.$$

   (If u_0 would solve $P_k(\Sigma, f)$ then this difference would be zero.) Since

   $$\begin{aligned} F_0(\omega) &\approx F_0(R\sigma) + h(\sigma)\frac{\partial F_0}{\partial r}(R\sigma) \\ &= f(\omega) + h(\sigma)\frac{\partial F_0}{\partial r}(R\sigma), \end{aligned}$$

   hence

   $$\begin{aligned} f(\omega) - F_0(\omega) &\approx -h(\sigma)\frac{\partial F_0}{\partial r}(R\sigma) \\ &= F_1(R\sigma). \end{aligned}$$

3. Writing $f(\omega) = F_0(\omega) + (f(\omega) - F_0(\omega))$ we then add to u_0 a correction given by the solution of $P_k(S_R, .)$ with boundary data the pullback of $(f - F_0)(\omega)$ $(\omega \in \Sigma)$ to S_1. According to step 2 this pullback is approximately given by F_1. The correction is then the function u_1.

Note that in *the change of boundary method* the approximation made in step 2 is avoided. As we have seen in Section the correction δu_1 is the solution of $P_k(S_R, \Delta_1)$, where Δ_1 is the pullback of

$(f - F_0)(\omega)$ $(\omega \in \Sigma)$ to the sphere S_R. The advantage of this procedure is that $F_0(\omega)$ always exists since $u_0 \in C^\infty(\Omega_R)$. From a computational point of view the disadvantage may lie in that we have to compute u_0 and $\partial u_0/\partial r$ on a non-spherical surface like the telluroid ($k = 2$) or the earth ($k = 0$).
Another interpretation is possible (with $k = 2$, for example), this time in accordance with Moritz's line of thought (see Heiskanen and Moritz, 1967, § 8-10). On the basis of the zero-order approximation u_0 to u we define the gravity anomalies outside S_R by means of

$$\Delta g_0(r\sigma) = -\frac{\partial u_0}{\partial r} - \frac{2}{r} u_0 .$$

Let Δg^* be *a fictitious field of gravity anomalies* on S_R which generate on Σ the measured free-air gravity anomalies Δg (see Heiskanen and Moritz, 1967, p. 317),

$$\Delta g^*(\phi_h(\sigma)) = \Delta g(\sigma)$$

where $\Delta g^* = Bu^*$ and u^* solves $P_2(S_R, \Delta g^*)$. Since

$$\Delta g^*(\phi_h(\sigma)) = \Delta g^*(R\sigma) + h(\sigma)\frac{\partial \Delta g^*}{\partial r}(R\sigma) + \cdots ,$$

as a first approximation we set

$$\Delta g^*(R\sigma) = \Delta g(\sigma) - h(\sigma)\frac{\partial \Delta g_0}{\partial r}(R\sigma) .$$

A first-order approximation to $P_2(\Sigma, \Delta g)$ is then the solution of $P_2\left(S_R, \Delta g - h\frac{\partial \Delta g_0}{\partial r}\right)$. Compare this approach with (Rummel, 1988, § 3)).

5 Conclusion

We conclude this paper noting that the variational method that we have followed to solve $P_k(\Sigma, f)$ can also be applied to solve as well other boundary problems such as the *ellipsoidal Stokes boundary-value problem* (see Martinec and Grafarend, 1997) where a small parameter characterizing the difference between the ellipsoid of revolution and the sphere appears in a natural way (for example, the first eccentricity of the ellipsoid).
More complex problems to which one can apply the methods described in this paper are:

- The *linear gravimetric boundary problem* (see Holota, 1997) which generalizes $P_0(\Sigma, f)$. In this problem the boundary condition is

$$\langle \nabla T, s \rangle = -\delta g ,$$

where $s = -(1/\gamma)\nabla U$, U is the normal potential, $\gamma = |\nabla U|$ is the normal gravity and T the disturbing potential.

- The *linearized scalar boundary problem* which generalizes $P_2(\Sigma, f)$ (see Sacerdote and Sansò, 1986), where the boundary condition on the telluroid is

$$\langle \nabla T, s \rangle - \frac{1}{\gamma \langle s, e_h \rangle}\frac{\partial \gamma}{\partial h} T = -\Delta g ,$$

where e_h is the unit outer vector normal to the reference ellipsoid and $\partial\gamma/\partial h = \langle \nabla\gamma, e_h \rangle$.

Acknowledgements. J.Otero was partially supported by the Ministry of Science and Technology through Project REN2000-0766/CLI. We thank Professor F. Sansò for having made the presentation of this paper.

References

Auz A, Otero J (2002) Gradient solution of the GPS-gravimetric boundary problem (in Spanish). In: Proc III Spanish-Portuguese Assembly on Geodesy and Geophysics, in print

Brovar VV (1964) On the solutions of Molodensky's boundary value problem. Bulletin Géodésique 72: 167–173

Dautrey R, Lions JL (1990) Mathematical analysis and numerical methods for science and technology. Volume 1: Physical origins and classical methods. Springer, Berlin Heidelberg New York

Grafarend EW, Heck B, Knickmeyer EH (1985) The free versus fixed geodetic boundary value problem for different combinations of geodetic observables. Bulletin Géodésique 59: 11–32

Heiskanen WA, Moritz H (1967) Physical Geodesy. W.H. Freeman and Co., San Francisco Londres

Holota P (1985) Boundary value problems of physical geodesy: present state, boundary perturbation and the Green-Stokes representation. In: Proc 1st Hotine-Marussi Symposium on Mathematical Geodesy (Rome), Politecnico di Milano, pp 529–558

Holota P (1989) Higher order theories in the solution of boundary value problems of physical geodesy by means of successive approximations. In: Proc 2nd Hotine-Marussi Symposium on Mathematical Geodesy (Pisa), Politecnico di Milano, pp 471–505

Holota P (1997) Coerciveness of the linear gravimetric boundary-value problem. Journal of Geodesy 71: 640–651

Hotine M (1969) Mathematical geodesy. ESSA Monograph 2, U.S. Department of Commerce, Washington

Kautský J (1962) Approximations of solutions of Dirichlet's problem on nearly circular domains and their application in numerical methods. Aplikace Matematiky 7: 186–200

Koch KR, Pope AJ (1972) Uniqueness and existence for the geodetic boundary value problem using the known surface of the earth. Bulletin Géodésique 106: 467–476

Kuzmina RP (2000) Asymptotic methods for ordinary differential equations. Kluwer Academic Publishers, Dordrecht

Martinec Z, Grafarend EW (1997) Solution of the Stokes boundary-value problem on an ellipsoid of revolution. Studia Geoph. et Geod. 41: 103–129

Molodensky MS, Eremeev VF, Yurkina MI (1962) Methods for study of the external gravitational field and figure of the earth. Israel Program for Scientific Translations, Jerusalem

Moritz H (1980) Advanced physical geodesy. Herbert Wichmann and Abacus Press, Karlsruhe Tunbridge

Moritz H (2000) Molodensky's theory and GPS. In: Moritz H, Yurkina MI (eds) M.S.Molodensky: In memoriam. Mitteilungen der geodätischen Institute der Technischen Universität Graz, Folge 88, Graz, pp 69–85

Rummel R (1988) Zur iterativen Lösung der Geodätischen Randwertaufgabe. In: Deutsche Geodätische Kommission Reihe B, Nr. 287, Munich, pp 175–181

Sacerdote F, Sansò F (1986) The scalar boundary value problem of physical geodesy. Manuscripta Geodaetica 11: 15–28

Sansò F (1993) Theory of geodetic B.V.P.s applied to the analysis of altimetric data. In: Rummel R, Sansò F (eds) Satellite altimetry in geodesy and oceanography. Lecture notes in Earth Sciences 50: 317–371, Springer, Berlin Heidelberg New York

Stock B (1983) A Molodenskii-type solution of the geodetic boundary value problem using the known surface of the earth. Manuscripta Geodaetica 8: 273–288

Sünkel H (1997) GBVP - Classical solutions and implementation. In: Sansò F, Rummel R (eds) Geodetic Boundary Value Problems in View of the One Centimeter Geoid. Lecture notes in Earth Sciences 65: 219–237, Springer, Berlin Heidelberg New York

Potential Coefficients Recovery from the Spectra of the Full Space-Borne Gravity Gradient Tensor

M.S. Petrovskaya, A. N. Vershkov
Central (Pulkovo) Astronomical Observatory of Russian Academy of Sciences,
Pulkovskoe Shosse 65, St.-Petersburg, 196140, Russia; e-mail: petrovsk@gao.spb.ru

Abstract. The solution of the satellite gradiometry problem is studied by the space-wise approach. Simple linear relations are derived between the spectral coefficients of six components of the gravity gradient tensor in the local north-oriented reference frame and the harmonic coefficients of the Earth's potential. These relations are applied for solving numerically two inverse problems. One of them is simulating the spectra of all the gradiometry observables at the altitude $h = 250\ km$ of GOCE satellite on the basis of EGM96 geopotential model. The other problem is recovering a geopotential model from the spectrum of each GOCE observable. As a result, EGM96 model is fully "restored" from all the simulated spectra. From these spectra a number of interesting numerical results are obtained characterizing the behavior of the second order derivatives of the geopotential. The elaborated procedures can be propagated to constructing the expansions of the third and higher order geopotential derivatives from a geopotential model. They can be also utilized for modeling the gravity fields of the other planets and the Moon.

Keywords. Satellite gradiometry, geopotential model

The problem of recovering the geopotential coefficients from a satellite gradiometry mission can be easily solved by the space-wise approach if there are simple linear relations between the spherical harmonic spectra of the second order potential derivatives and the coefficients $\overline{C}_{n,m}$ of the disturbing potential T. However, there is a difficulty in deriving such relations because five of six derivatives contain singularities (quasi-singularities) at the poles, after transition to derivatives with respect to polar coordinates. For this reason alternative base functions were applied instead of conventional spherical harmonic functions in some cases. For instance Albertella et al. (1995) constructed numerically bi-orthogonal truncated expansions for the potential derivatives. Rummel and Van Gelderen (1992) presented these derivatives in the form of infinite series of pure-spin spherical harmonic tensors. Very simple relations were recovered between the spectral coefficients of these series and the coefficients $\overline{C}_{n,m}$ of the conventional spherical harmonic expansion for T. However, since the integral expressions for the coefficients of the tensor harmonics depend on a number of differential operators with respect to the conventional spherical functions, their practical implementation would be difficult.

Alternative principles of constructing a geopotential model from gravity gradiometer data by space-wise approach were proposed by Petrovskaya et al. (2000, 2001). In the present paper these investigations are continued and simple linear relations are derived between the spherical harmonic coefficients of each second order potential derivative and the coefficients $\overline{C}_{n,m}$. Procedures are developed for solving these equations, both with respect to the spectral coefficients of the derivatives and with respect to $\overline{C}_{n,m}$.

For the reference frame the local north-oriented triad $\{x, y, z\}$ is used, where z has the radial direction, x points to the north, and y is directed to the west (the right-handed system). The derivatives of T with respect to these coordinates are simply expressed in terms of the derivatives with respect to polar coordinates $\{r, \theta, \lambda\}$, where r is the geocentric distance, θ is the polar angle and λ is the longitude.

The respective relations are as follows:

$$
\left.
\begin{aligned}
T_{zz} &= T_{rr} \\
T_{xx} &= \frac{1}{r}T_r + \frac{1}{r^2}T_{\theta\theta} \\
T_{yy} &= \frac{1}{r}T_r + \frac{1}{r^2 \tan\theta}T_\theta + \frac{1}{r^2 \sin^2\theta}T_{\lambda\lambda} \\
T_{xy} &= \frac{1}{r^2 \sin\theta}T_{\theta\lambda} - \frac{\cos\theta}{r^2 \sin^2\theta}T_\lambda \\
T_{xz} &= \frac{1}{r^2}T_\theta - \frac{1}{r}T_{r\theta} \\
T_{yz} &= \frac{1}{r^2 \sin\theta}T_\lambda - \frac{1}{r\sin\theta}T_{r\lambda},
\end{aligned}
\right\} \quad (1)
$$

see Koop (1993).

The truncated spherical harmonic series for T is

$$T(r,\theta,\lambda) = \frac{\mu}{a}\sum_{n=2}^{N}\sum_{m=-n}^{n}\left(\frac{a}{r}\right)^{n+1}\overline{C}_{n,m}\,\overline{Y}_{n,m}(\theta,\lambda), \quad (2)$$

$$\overline{Y}_{n,m}(\theta,\lambda) = \overline{P}_{n,|m|}(\cos\theta)\,Q_m(\lambda), \quad (3)$$

$$Q_m(\lambda) = \begin{cases} \cos m\lambda, \; m \ge 0, \\ \sin|m|\lambda, \; m < 0. \end{cases}$$

Here μ is the gravitational constant multiplied by the Earth's mass; a is a scaling factor; $\overline{P}_{n,|m|}(\cos\theta)$ and $\overline{C}_{n,m}$ are the fully normalized associated Legendre functions and the geopotential coefficients, respectively.

After transition in Eq. (2) to the non-normalized quantities, the series is substituted in the right hand sides of Eqs. (1).

From the first of Eqs. (1) the well-known expansion for T_{zz} follows

$$T_{zz} = \sum_{n=2}^{N}\sum_{|m|=0}^{n}\overline{R}_{n,m}\,\overline{Y}_{n,m}(\theta,\lambda) \quad (4)$$

where

$$\overline{R}_{n,m} = \frac{\mu}{a^3}(n+1)(n+2)\left(\frac{a}{r}\right)^{n+3}\overline{C}_{n,m}. \quad (5)$$

Similarly, for the remaining derivatives it then follows:

$$T_{xx} = \frac{\mu}{a^3}\sum_{n=2}^{N}\sum_{m=-n}^{n}\left(\frac{a}{r}\right)^{n+3}C_{n,m} \times \left[-(n+1)P_{n,|m|}(\cos\theta) + \frac{d^2P_{n,|m|}(\cos\theta)}{d\theta^2}\right]Q_m(\lambda), \quad (6)$$

$$T_{yy} = \frac{\mu}{a^3}\sum_{n=2}^{N}\sum_{m=-n}^{n}\left(\frac{a}{r}\right)^{n+3}C_{n,m} \times \left[-(n+1)P_{n,|m|}(\cos\theta) + \frac{\cos\theta}{\sin\theta}\frac{dP_{n,|m|}(\cos\theta)}{d\theta} - \frac{m^2}{\sin^2\theta}P_{n,|m|}(\cos\theta)\right]Q_m(\lambda), \quad (7)$$

$$T_{xy} = \frac{\mu}{a^3}\sum_{n=2}^{N}\sum_{m=-n}^{n}\left(\frac{a}{r}\right)^{n+3}m\,C_{n,-m} \times \left[\frac{1}{\sin\theta}\frac{dP_{n,|m|}(\cos\theta)}{d\theta} - \frac{\cos\theta}{\sin^2\theta}P_{n,|m|}(\cos\theta)\right]Q_m(\lambda), \quad (8)$$

$$T_{xz} = \frac{\mu}{a^3}\sum_{n=2}^{N}\sum_{m=-n}^{n}(n+2)\left(\frac{a}{r}\right)^{n+3}C_{n,m} \times \frac{dP_{n,|m|}(\cos\theta)}{d\theta}Q_m(\lambda), \quad (9)$$

$$T_{yz} = \frac{\mu}{a^3}\sum_{n=2}^{N}\sum_{m=-n}^{n}(n+2)\,m\left(\frac{a}{r}\right)^{n+3}C_{n,-m} \times \frac{P_{n,|m|}(\cos\theta)}{\sin\theta}Q_m(\lambda). \quad (10)$$

Eqs. (4) – (10) are considered on a mean orbital sphere and r means its radius.

The expressions (6) – (10) contain singularities at the poles and the derivatives of the Legendre functions with respect to the polar angle θ. This makes a direct conversion of the right hand sides into truncated spherical harmonic series difficult.

Therefore, we express first the potential derivatives as the truncated series

$$T_{xx} = \sum_{n=0}^{N}\sum_{|m|=0}^{n}\overline{H}_{n,m}\,\overline{Y}_{n,m}(\theta,\lambda), \quad (11)$$

$$T_{yy} = \sum_{n=0}^{N}\sum_{|m|=0}^{n}\overline{G}_{n,m}\,\overline{Y}_{n,m}(\theta,\lambda), \quad (12)$$

$$T_{xy} = \sum_{n=1}^{N}\sum_{|m|=1}^{n}\overline{M}_{n,m}\,\overline{Y}_{n,m}(\theta,\lambda), \quad (13)$$

$$T_{xz} = \sum_{n=2}^{N} \overline{B}_{n,0}\, \overline{P}_{n,1}(\cos\theta) + + \sum_{n=1}^{N} \sum_{|m|=1}^{n} \overline{B}_{n,m}\, \overline{X}_{n,m}(\theta,\lambda), \tag{14}$$

$$T_{yz} = \sum_{n=1}^{N} \sum_{|m|=1}^{n} \overline{A}_{n,m}\, \overline{X}_{n,m}(\theta,\lambda). \tag{15}$$

In the inner sums $-n \le m \le n$.

Note that the right hand sides of Eqs. (14) and (15) represent series which are "slightly shifted" if compared to functions in Eq. (3)

$$\overline{X}_{n,m}(\theta,\lambda) = \overline{P}_{n-1,|m|-1}(\cos\theta)\, Q_m(\lambda). \tag{3*}$$

The coefficients of the harmonics in Eqs. (11) – (15) depend on the indices of summation and on the radius r of a mean orbital sphere. They can be evaluated from gradiometry data by the spherical harmonic analysis, either by the least squares adjustment or by the numerical quadrature technique.

We shall now derive simple linear relations between these coefficients and $\overline{C}_{n,m}$.

At first the above mentioned "peculiarities" are removed from Eqs. (6) – (10). For this purpose, e.g., each of the couple of Eqs. (7) and (12) for T_{yy} is multiplied by $\sin^2\theta$. Eqs. (8) and (13) for T_{xy} are also multiplied by $\sin^2\theta$. The couples of Eqs. (9) and (14) for T_{xz}, as well as Eqs. (10) and (15) for T_{yz}, are multiplied by $\sin\theta$.

As a result, e.g., the function $\sin^2\theta\, T_{yy}$ is presented in two different forms. One of them is derived from Eq. (7) and the other from Eq. (12). By applying the well-known transformation formulas for the Legendre functions each of these two expressions is converted into a conventional spherical harmonic series. The corresponding harmonic coefficients in both series may be denoted by the same quantity $\overline{D}_{n,m}$. By equating the transfer coefficients $\overline{D}_{n,m}$, expressed in terms of $\overline{C}_{n,m}$, on one hand, and in terms of $\overline{G}_{n,m}$, on the other hand, the basic relation is derived between $\overline{G}_{n,m}$ and $\overline{C}_{n,m}$.

The procedure of the subsequent transformations of each couple of equations for T_{xy}, T_{xz} and T_{yz}, constituted from Eqs. (8) – (10) and Eqs. (13) – (15), is similar to that for T_{yy}.

Expression (6) for T_{xx}, containing the second order derivative of Legendre function, is actually not used for deriving the basic relation for this derivative. It will be derived from the relation for T_{yy} and the Laplace equation

$$T_{xx} + T_{yy} + T_{zz} = 0.$$

From Eqs. (4), (11), (12) and the last equation follows

$$\overline{H}_{n,m} = -\overline{R}_{n,m} - \overline{G}_{n,m}. \tag{16}$$

As a result, the following basic relations are derived between the coefficients of the series (11) – (15) and the geopotential coefficients:

$$\begin{aligned} -a_{n,m}\overline{C}_{n-2,m} + b_{n,m}\overline{C}_{n,m} - c_{n,m}\overline{C}_{n+2,m} &= \overline{K}_{n,m}, \\ a_{n,m}\overline{H}_{n-2,m} - b_{n,m}\overline{H}_{n,m} + c_{n,m}\overline{H}_{n+2,m} &= \overline{K}_{n,m}, \end{aligned} \tag{17}$$

$$\begin{aligned} \overline{d}_{n,m}\overline{C}_{n-2,m} - \overline{c}_{n,m}\overline{C}_{n,m} &= \overline{D}_{n,m}, \\ -a_{n,m}\overline{G}_{n-2,m} + b_{n,m}\overline{G}_{n,m} - c_{n,m}\overline{G}_{n+2,m} &= \overline{D}_{n,m}, \end{aligned} \tag{18}$$

$$\begin{aligned} p_{n,m}\overline{C}_{n-1,m} - q_{n,m}\overline{C}_{n+1,m} &= \overline{L}_{n,m}, \\ -a_{n,m}\overline{M}_{n-2,-m} + b_{n,m}\overline{M}_{n,-m} - c_{n,m}\overline{M}_{n+2,-m} &= \overline{L}_{n,m}, \end{aligned} \tag{19}$$

$$\begin{aligned} g_{n,m}\overline{C}_{n-1,m} - h_{n,m}\overline{C}_{n+1,m} &= \overline{S}_{n,m}, \\ \mu_{n,m}\overline{B}_{n,m} - \nu_{n,m}\overline{B}_{n+2,m} &= \overline{S}_{n,m}, \end{aligned} \tag{20}$$

$$\overline{C}_{n,0} = -\omega_{n,0}\overline{B}_{n,0}, \tag{21}$$

$$\overline{C}_{n,m} = u_{n,m}\overline{A}_{n,-m} - v_{n,m}\overline{A}_{n+2,-m}. \tag{22}$$

Eq. (17) for T_{xx} is derived from Eqs. (5), (16) and (18).

The constants $a_{n,m}$, $b_{n,m}$, etc., in Eqs. (17) – (22) depend on the indices n and m and on the radius of the mean orbital sphere. However, their presentation requires considerable space and will be given later in a more specific publication.

Relation (22) for T_{yz} has the most simple form. It is explicitly solved with respect to $\overline{C}_{n,m}$. Eq. (21) for T_{xz} is solved with respect to the zonal harmonic coefficient $\overline{C}_{n,0}$.

Each couple in Eqs. (17) – (22) represents a set of linear equations for a fixed m and variable n. This set can be solved either with respect to $\overline{C}_{n,m}$ or with respect to the coefficients of a geopotential derivative. In particular $\overline{C}_{n,m}$ can be evaluated from the upper equation in (17), after the transfer coefficients $\overline{K}_{n,m}$ are calculated from the lower equation and the known values of $\overline{H}_{n,m}$.

In general, if in an upper or lower equation there are only two unknowns (in particular, two potential coefficients) then it is solved by a recurrence procedure. If there are three unknowns then the corresponding simple matrix equation is solved, providing a unique solution. (In each matrix only three parallel diagonals differ from zero).

On the basis of relations (5) and Eqs. (17) - (22) the coefficients of the potential derivatives are simulated from EGM96 model at the altitude $h = 250\ km$ of GOCE satellite. Since EGM96 model corresponds to the total Earth's potential V, with the exclusion of the zero and first degree harmonics, V_0 and V_1, then it is accepted: $T = V - V_0 - V_1$.

Various spectral characteristics of the geopotential derivatives are presented in Figures 1 – 6. They reveal a very interesting specific regular behavior of the spectrum of each derivative, as well as a similarity of the spectra inside each of two groups, $\{T_{xz}, T_{yz}, T_{zz}\}$ and $\{T_{xx}, T_{yy}, T_{xy}\}$.

On Figures 1.1(a), 2(a) – 5(a) the values of the spectral coefficients of five derivatives are presented (except T_{xy}), for each fixed value of degree n from the range $20 \le n \le 360$ and all the orders $-n \le m \le n$. The high degree coefficients are given in a larger scale on Figures 1.1(b), 2(b) – 5(b). On Figures 1.1(c,d), 2(c,d) – 5(c,d), inversely, the spectra are presented for each fixed value of the order m from the range $-N \le m \le N$ and all the degrees $20 \le n \le N$. It is interesting to observe that the coefficients of T_{xx} and T_{yy} have extreme absolute values in the high frequency part of the spectra ($180 \le n \le 360$) for some small values of $|m|$, in particular for $m = -2$. The pictures for T_{xy}, not presented here, are similar to T_{yy} but with smaller amplitudes of the coefficients.

From Figure 1.2 it may be concluded that for T_{yy} the extreme absolute values of the coefficients of the low order ($m = -2$, etc.) and high degree ($n \ge 180$) characterize the contribution of the gravity gradient anomalies from the upper part of the Earth's crust. A similar conclusion can be drawn for the other horizontal derivatives T_{xx} and T_{xy}.

Each curve on Figure 6, representing the root mean square (RMS) degree variances for the group $\{T_{xx}, T_{yy}, T_{xy}\}$, is divided at $n \approx 180$ into two branches, one of them corresponding to n-even and the other to n-odd. These branches are practically identical for all three components. The RMS variances corresponding to $T_{xx} - T_{yy}$ for the odd n exceed those for all the separate derivatives. It is surprising that the "even" variances for $T_{xx} - T_{yy}$ approximately coincide with the "odd" ones for T_{xx}, T_{yy} and T_{xy}. In the high degree parts of the spectra the curves for the variances of T_{zz}, T_{xz} and T_{yz} are monotonously decreasing and almost parallel, the last two being very close to each other. The above branching does not happen for the derivatives of this group, probably because in these cases at least one differentiation is performed in the radial direction.

The above characteristics of the second order derivatives of the geopotential can be of interest for geophysicists.

The elaborated procedures, after slight modification, can be applied for constructing the expansions of the third and higher order geopotential derivatives. The third derivative will be needed for performing the reduction of the gradiometry data from the satellite orbit to a mean orbital sphere. The elaborated technique can be also utilized for modeling the gravity fields of the other planets and the Moon.

Acknowledgements

M.S. Petrovskaya expresses her deep thanks for a financial support provided by IAG and the Italian Space Agency for participating in V Hotine-Marussi Symposium on Mathematical Geodesy (Matera, Italy, June 17-21, 2002).

The authors appreciate very much very valuable remarks of the reviewers.

References

Albertella, A., F. Migliaccio and F. Sansò (1995) Application of the concept of biorthogonal series to a simulation of a gradiometric mission, IAG Symp. No. 114, *Geodetic Theory Today*, ed. F. Sansò (Third Hotine-Marussi Symposium on Mathematical Geodesy, L'Aquila, Italy, May 29 – June 3, 1994), pp. 350 – 361.

Koop, R. (1993) Global gravity field modeling using satellite gravity gradiometry, *Netherlands Geodetic Commission, Publications on Geodesy, New Series*, Delft, Netherlands, No. 38.

Petrovskaya, M.S., A.N. Vershkov and G.B. Zielinski (2000) Recovering the Earth's potential spectral characteristics from GOCE gradiometry mission. (33rd COSPAR Scientific Assembly, July 16 – 23, 2000, Warsaw, Poland). *Advances in Space Research* (in press).

Petrovskaya, M.S., A.N. Vershkov, G.B. Volfson, J.B. Zielinski (2001) Simulation of the spectra of GOCE observables. Scientific Assembly of the International Association of Geodesy (Budapest, Sept. 2 – 9, 2001), Proceedings in CD, No. 22BD.

Rummel, R. and M. Van Gelderen (1992) Spectral analysis of the full gravity gradient tensor, *Geophysical Journal*, 111, pp. 159 – 169.

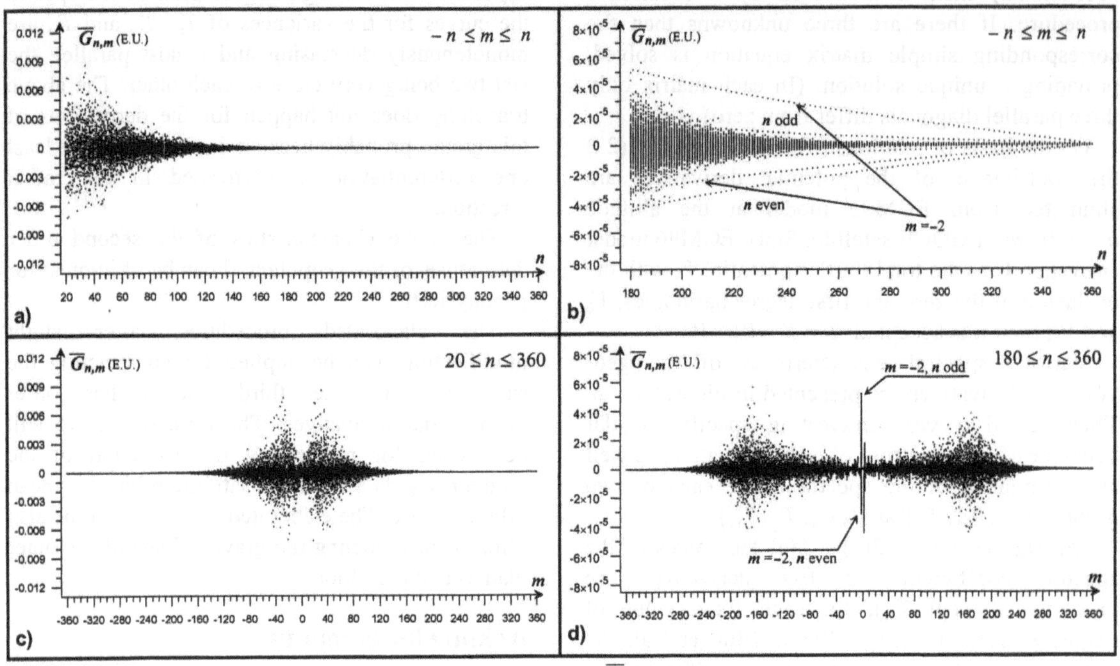

Fig. 1.1 Coefficients $\overline{G}_{n,m}$ of T_{yy}

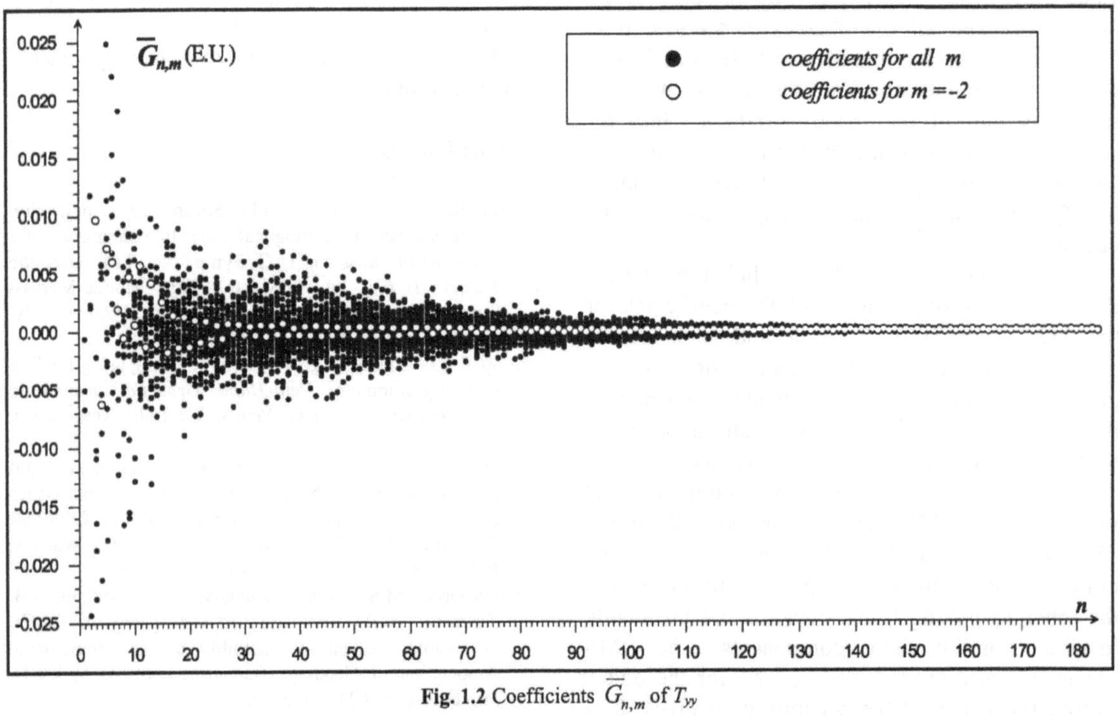

Fig. 1.2 Coefficients $\overline{G}_{n,m}$ of T_{yy}

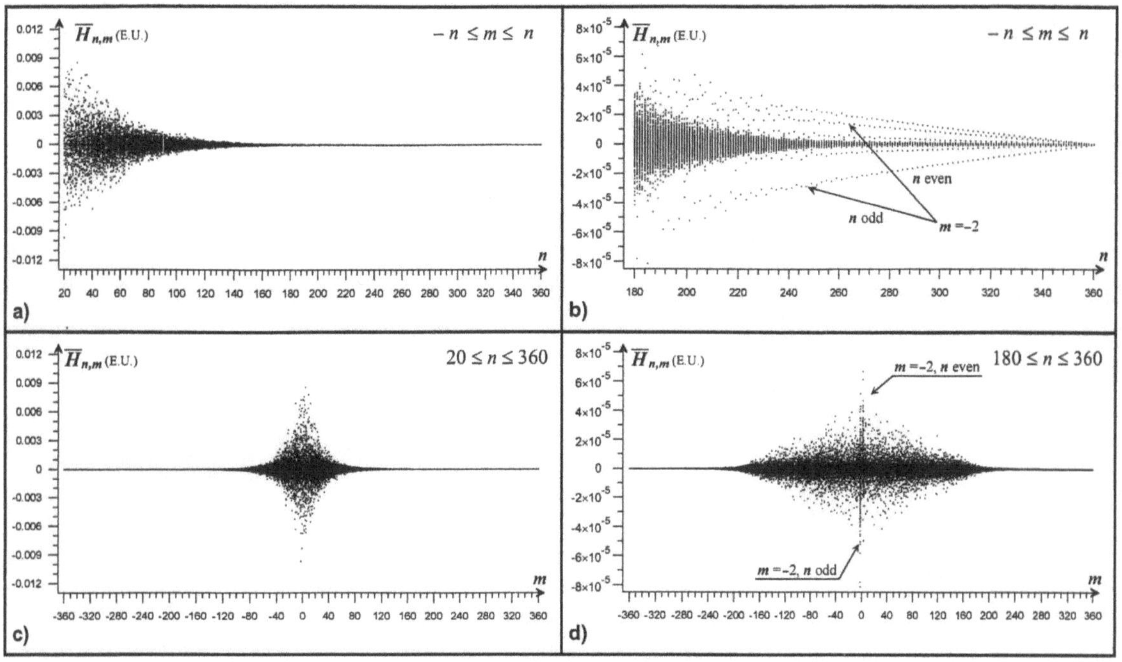

Fig. 2 Coefficients $\overline{H}_{n,m}$ of T_{xx}

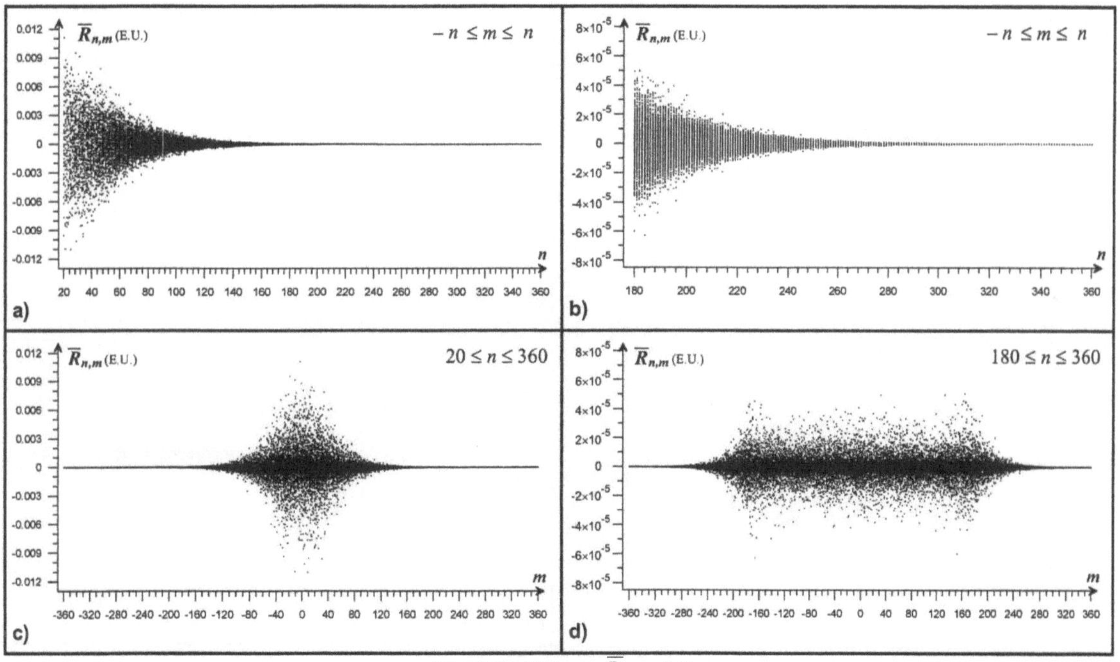

Fig. 3 Coefficients $\overline{R}_{n,m}$ of T_{zz}

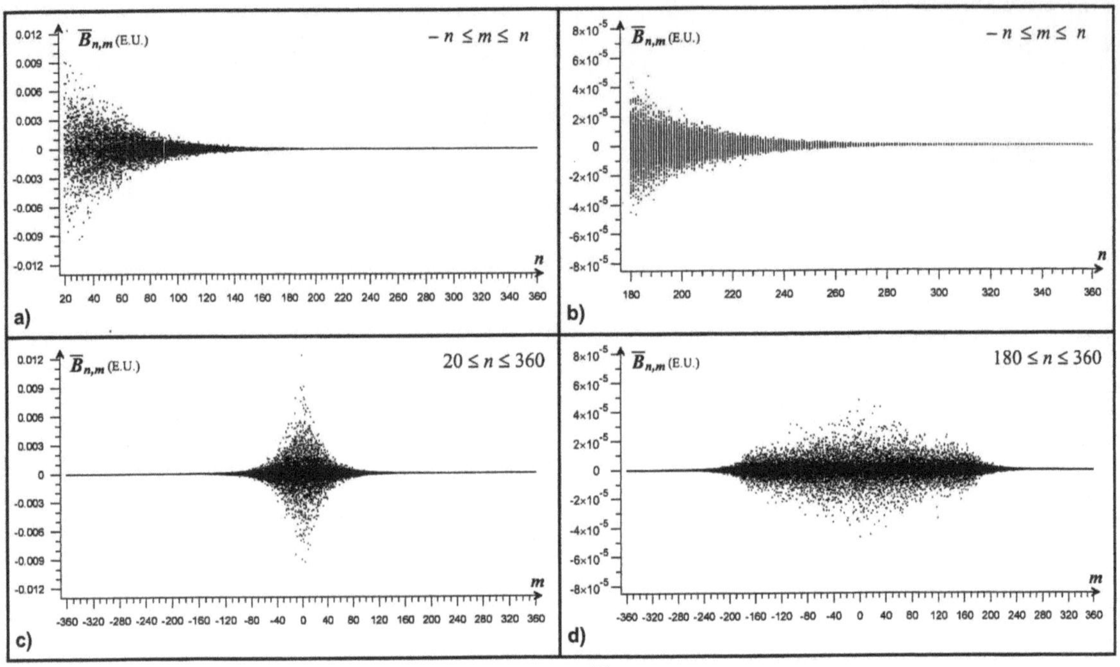

Fig. 4 Coefficients $\overline{B}_{n,m}$ of T_{xz}

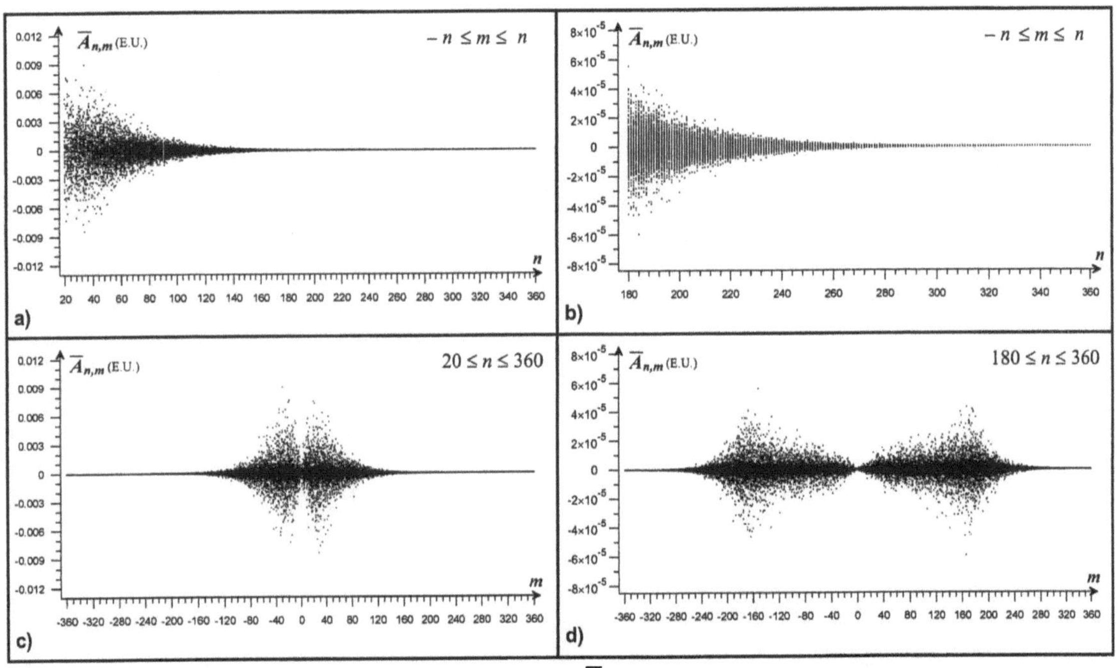

Fig. 5 Coefficients $\overline{A}_{n,m}$ of T_{yz}

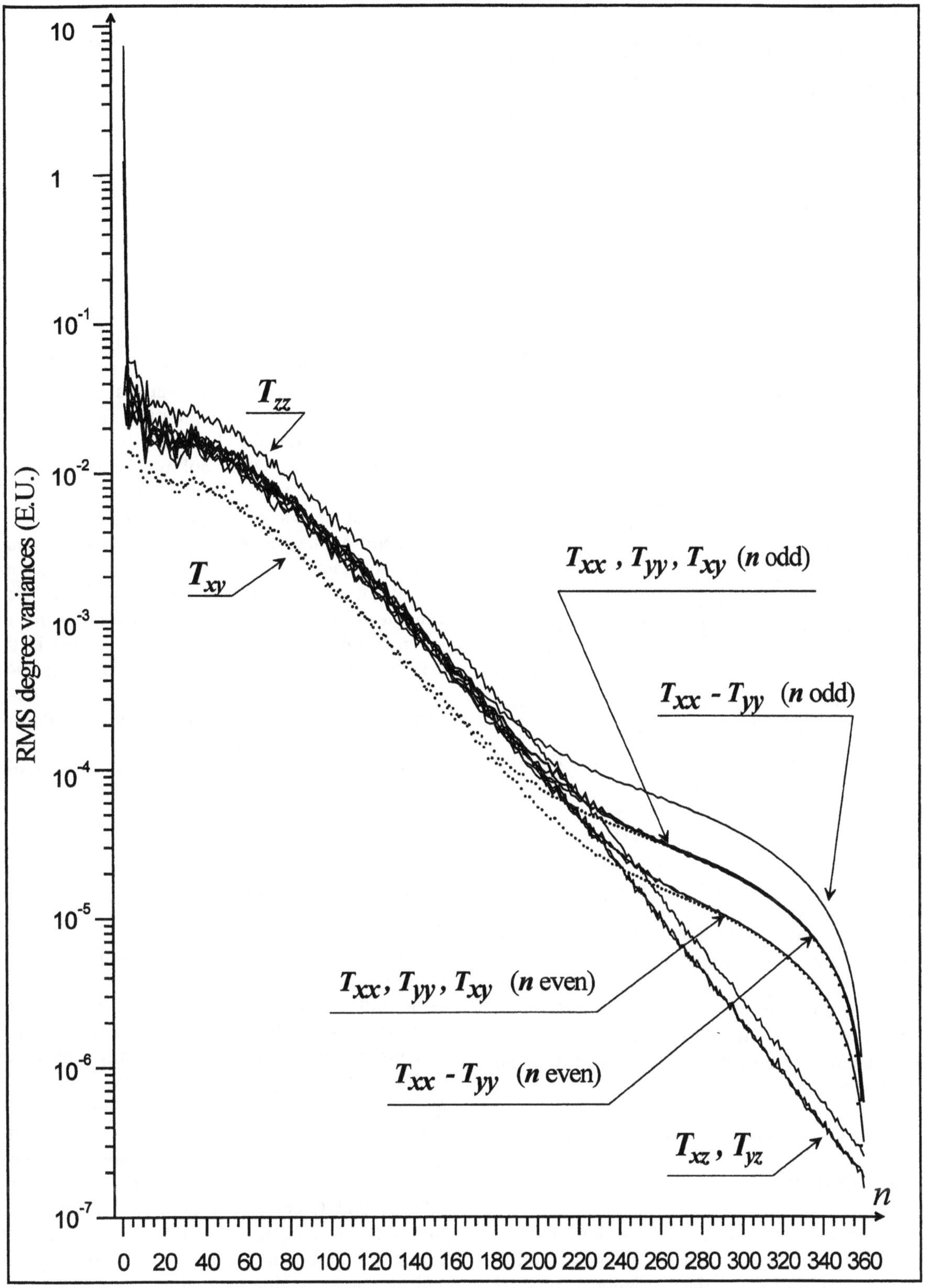

Fig. 6 RMS degree variances for T_{xx}, T_{yy}, T_{zz}, T_{xz}, T_{yz}, T_{xy} and $T_{xx} - T_{yy}$

Application of spherical pseudo-differential operators and spherical wavelets for numerical solutions of the fixed altimetry-gravimetry boundary value problem

R. Grebenitcharsky, M.G. Sideris
Department of Geomatics Engineering, University of Calgary,
2500 University Drive N.W., Calgary, Alberta, Canada, T2N 1N4

Abstract. Two possible solutions of the fixed AGBVP are discussed. Both are based on spherical wavelets generated by the Abel-Poisson kernel as a scale function. This ensures the harmonicity of spherical wavelets, which is necessary in the representation of the harmonic disturbing potential as n-level multiresolution analysis (MRA).

Using GPS/leveling data, the fixed AGBVP II could be transformed into a Neumann boundary value problem. After that, the solution of Neumann's problem in terms of spherical wavelets given in Freeden and Schneider (1998) can be used. The approximation (low-pass filter) and detail coefficients (band-pass filter) in the spherical wavelet representation are spatially distributed and they can be used for establishing smoothness (compatibility) conditions along the coastline.

Another numerical solution of AGBVPs with compatibility conditions is possible with the combined use of spherical harmonics and spherical wavelets (Freeden and Windheuser, 1997). All boundary conditions could be presented as spherical pseudo-differential operators. In this case the application of compatibility conditions in terms of pseudo-differential operators becomes more straightforward. The compatibility conditions in Svensson (1988) are given in an implicit generalized form in terms of pseudodifferential operators as projections between Sobolev spaces. The compatibility conditions are presented in an explicit form.

1 Introduction

One of the most often used methods for the solution of altimetry-gravimetry boundary value problems (AGBVPs) – the variational method (Rektorys, 1977; Holota, 1997) for solving differential equations – requires both the boundary surface and the boundary conditions (observations) to be smooth enough, i.e., to have continuous derivatives up to a certain order, which means they belong to the functional space $C^{(2k)}$, where k shows the order of generalized derivatives for the weak solution or the degree of regularity. The magnitude of k depends on the boundary conditions, the regularity of the weak solution and for a N-dimensional space ($N=2$ for a unit sphere) $2k>N$. Theoretically $k \to \infty$, but in practice k corresponds to the condition the boundary and boundary conditions to be sufficiently smooth. Under such conditions, the weak (generalized) solution will be close enough to the classical (pointwise) solution for the given problem. The complete theory behind this statement can be found in Rektorys, 1977 (p. 546-548). Sobolev embedding theorems have been used to show that if the given data (boundary conditions) are smooth enough, the weak solution is smooth enough, too, and as a result it is the classical solution. In reality, however, irregularities exist on the coastline because different types of data with different observation procedures, accuracy and resolution are used. To overcome these problems, compatibility conditions along the coastline can be used.

Compatibility (smoothness) conditions along the coastline have been discussed by Svensson (1988) as a tool for increasing the smoothness of the solution of AGBVPs. The addition of compatibility conditions along the coastline may cause the problem to become well posed, but it is questionable whether such conditions upon the data are realistic for practical use. To answer this question, two numerical experiments have been conducted; see Grebenitcharsky and Sideris (2001a) and Grebenitcharsky et al. (2001b). The first one investigates the effect of smoothness conditions applied to the same type of data on land and at sea – gravity anomalies from gravity and shipborne gravity measurements. The second one investigates the effect of smoothness conditions along the coastline applied to different types of data

– GPS/leveling and gravity data on land together with altimetry and shipborne gravity data at sea. In both cases, the effect of smoothing results in a geoid change between -5 cm and +5 cm. This magnitude can be considered significant for cm-geoid determination.

The significance of compatibility conditions along the coastline for the solution of AGBVPs was discussed for the first time in Svensson (1988). For AGBVP I and AGBVP II (Sanso, 1993), it has been shown that with the introduction of additional compatibility conditions on the coastline, both problems become normal solvable (satisfying the Fredholm alternative). To achieve this, Svensson (1983b) introduced the new approach of pseudo-differential operators. Using pseudo-differential operators, it is possible not only to reformulate existing AGBVPs but to apply the compatibility conditions along the coastline. Together with new pseudo-differential operators *E, F, L* and *M*, which are the following mappings, $\overline{H}_{1/2}(\delta\Omega_S)\xrightarrow{F}H_{1/2}(\delta\Omega)$, $\overline{H}_{-1/2}(\delta\Omega_L)\xrightarrow{E}H_{1/2}(\delta\Omega)$, $\overline{H}_s(\delta\Omega_L)\xrightarrow{L}H_s(\delta\Omega)$, $\overline{H}_{s-2}(\delta\Omega_S)\xrightarrow{M}H_s(\delta\Omega)$, the new form of AGBVP I and AGBVP II is possible. According to Eskin (1980), page 62, PDOs can be applied only on functions defined on the entire space $\delta\Omega$ on the sphere but we can apply them on the parts $\delta\Omega_L$, $\delta\Omega_S$ only after extending the functions to the entire space $\delta\Omega$. In fact, *E, F, L* and *M* can be considered not only as extensions of function domains but also as extensions from one order of regularity *s-2* to another *s* for AGBVP II and from *-1/2* to *1/2* for AGBVP I representing different type of boundary conditions. More details about those mappings will be given later with respect to the explicit forms of compatibility conditions. H_i and $\overline{H}_i$ are Sobolev spaces of ith order and their closures, and $\delta\Omega$, $\delta\Omega_L$, $\delta\Omega_S$ are the spaces on the entire sphere, land and sea part, respectively and Ω, Ω_L, Ω are the corresponding spaces outside the sphere. The compatibility conditions insure that the boundary conditions (measurements) on land and at sea are consistent along the coastline.

The complete theory of applying pseudodifferential operators (PDOs) for solutions of mixed boundary value problems for elliptic differential equations can be found in Eskin (1980). Because AGBVPs are of the mixed type, and they are strictly related to the elliptic Laplace differential equation, pseudodifferential operators appear to be a very powerful tool for finding numerical solutions for them. The definition of PDOs given in Eskin (1980) allows differential, integral and integrodifferential operators to be presented as pseudo-differential operators. On page 36 of Eskin (1980), it has been shown that PDOs can be considered either as absolutely integrable continuous bounded functions or as bounded infinitely differentiable integro-differential operators. In this way, for both cases they can be easily inverted overcoming some singularity problems. This is the main advantage of PDOs, which can be useful in finding solutions to AGBVPs. After reformulation of a boundary value problem in terms of PDOs, it becomes normal solvable, which means that the inverse PDOs and the solution exist. In the case of "mixed" boundary value problems such as the AGBVPs, the irregularities of data and boundary surface along the coastline could be overcome after reformulating them in terms of PDOs. According to Eskin (1980), for a second order elliptic equation it is possible to select function spaces with weight factor vanishing on the coastline, which will ensure the smoothness of solutions everywhere off the coastline, so the AGBVP will be normal solvable. Combined with spherical harmonics and spherical wavelets, the theory has been applied on the sphere using spherical PDOs (Freeden et al., 1998). Main operators like Stokes, integral operators of single layer potential and double layer potential used in gravity field modeling and some boundary conditions are presented in Freeden et al. (1998) and Freeden and Windheuser (1997) in terms of PDOs.

The properties of wavelets to give not only the frequencies of a signal but also their spatial distribution in different scales can be used to detect discrepancies between different data along the coastline. We can use a wavelet decomposition and reconstruction to put smoothness compatibility conditions on the data and the boundary along the coastline. The main problem of the classical wavelet approach using known wavelets in infinite domain is that they are not harmonic functions. To get a solution of AGBVPs using wavelet transforms, a new wavelet method for approximation of harmonic functions could be used (Freeden and Schreiner, 1995; Freeden and Windheuser, 1996; Freeden and Windheuser, 1997; Freeden and Schneider, 1998).

Using variational methods (Rektorys, 1977), we have a weak solution based on the set of base functions used and on the set of boundary conditions (data). This solution minimizes a functional containing the boundary conditions and it is unique. However, this weak solution can not take into account additional conditions on data like the compatibility conditions along the coastline. To combine variational methods with the theory of spherical pseudo-differential operators and spherical wavelets for a numerical solution of AGBVPs with compatibility conditions, the comparison between bases and frames conducted in Christensen (2001) is used. The frame theory, as a generalization of base functions, offers an opportunity for the application of wavelets. The frame definition corresponding to the stability condition for wavelets allows the wavelet coefficients along the coastline to vary or to be restricted, which is equivalent to smoothing data or establishing compatibility conditions only for the coastal area. The fact that frames do not provide a unique solution seems to be a disadvantage but they can be used when compatibility conditions have to be applied. This flexibility could be useful when we want the weak solution to be close enough to the classical one after smoothing the boundary conditions, boundary surface and even the solution along the coastline. The spherical wavelet packets proposed by Freeden and Windheuser (1996) satisfy this property and could be a good candidate for the application of frame theory to solutions of AGBVPs.

The main task of this paper is to answer two questions: *How the theory of spherical pseudo-differential operators and spherical wavelets can contribute to getting numerical solutions of the fixed altimetry-gravimetry boundary value problem, and how to apply compatibility conditions along the coastline in an explicit form.*

2 Variational methods, base functions and frame theory for solution of AGBVPs

The transition from the unique weak solution given by variational methods to the solution in terms of spherical wavelets used as frame requires more details about the variational methods. Only a brief overview of the application of variational methods for solutions of AGBVPs will be presented here. Having a weak solution by variational methods (using wavelets as base functions), which will correspond to exact decomposition and reconstruction, the wavelet coefficients could be restricted along the coastline using the frame theory. In this way, smoothing of data or applying compatibility conditions in a coastal region become possible. To describe the application of variational methods to AGBVPs, the general formulation given by Lehmann (1999) can be used. In spherical approximation, an altimetry-gravimetry boundary value problem can be represented as

$$\begin{cases} \nabla^2 T = 0 \quad in \quad \Omega \quad (outside \quad the \quad sphere\ \delta\Omega) \\ L_S(T) = b_S \quad at \quad sea \quad \delta\Omega_S \\ L_L(T) = b_L \quad on \quad land \quad \delta\Omega_L \\ T = O(|\xi|^{-2-n}) \quad |\xi| = r \to \infty \end{cases} \tag{1}$$

where $\delta\Omega$ is the sphere as a boundary, ∇^2 is Laplace's operator, L_S and L_L are functionals of the disturbing potential at sea and on land, b_S, b_L are measurements (boundary conditions) at sea and on land, $\xi = (\xi_1, \xi_2, \xi_3) = (\varphi, \lambda, r)$ is the radius vector in spherical coordinates, r is the length of the radius vector ξ, n is the degree of the geopotential model used.

We will discuss the fixed AGBVP II, which means that the boundary surface on land and at sea will be considered as known. The form of this problem is

$$\begin{cases} \nabla^2 T = 0 \quad in \quad \Omega \quad (outside \quad the \quad sphere\ \delta\Omega) \\ L_S(T) = -\dfrac{\partial T}{\partial r} = \delta g_S \qquad (known \quad \delta\Omega_S) \\ L_L(T) = -\dfrac{\partial T}{\partial r} - \dfrac{2}{R}T = \Delta g_L \quad (known \quad \delta\Omega_L) \\ T = O(|\xi|^{-2-n}) \quad |\xi| = r \to \infty \end{cases} \tag{2}$$

where δg_S is the gravity disturbance at sea, Δg_L is the gravity anomaly on land, and R is the Earth radius.

Introducing a bilinear form $B(T,v)$ like in Holota (1983), Keller (1996) and Holota (2000), the problem above could be presented as a variational equation in the form

$$B(T,v) = \int_{\delta\Omega} fv d\sigma = \int_{\delta\Omega_S} f(b_S, b_L) v d\sigma + \int_{\delta\Omega_L} f(b_S, b_L) v d\sigma \tag{3}$$

where the function f is tied to the boundary conditions (data), for example for AGBVPII, $f = f(\delta g_S, \Delta g_L)$; v are base (test) functions which belong to the Sobolev weight space $W_2^{(1)}$; $d\sigma$ is the surface element on the sphere.

The space $W_2^{(1)}$ is a space of functions that are square integrable on $\delta\Omega$ under the weight $|\xi|^{-2}$ and have first derivatives in a generalized sense. It is equipped with the inner product

$$(T,v)_1 = \int_{\delta\Omega}\frac{Tv}{|\xi|^2}d\xi + \sum_{i=1}^{3}\int_{\delta\Omega}\frac{\partial T}{\partial \xi_i}\frac{\partial v}{\partial \xi_i}d\xi \cdot \tag{4}$$

Now, instead of looking for a solution of problem (2) satisfying the differential equation and boundary conditions pointwise (the so-called classical solution) we can have a weak (generalized) solution in space $W_2^{(1)}$, based on certain type of base functions. If we have the so-called Galerkin system of linear equations, the solution can be given as

$$T^{(n)}(\xi) = \sum_{j=0}^{n} c_j^{(n)} v_j(\xi), \tag{5}$$

where $\xi \in \Omega \cup \delta\Omega$ and $c_j^{(n)}$ are numerical coefficients.

The coefficients can be found as a solution of the Galerkin system

$$\sum_{j=0}^{n} c_j^{(n)} B(v_j, v_k) = \int_{\delta\Omega} v_k f d\sigma, \quad k = 0,\dots,n \tag{6}$$

The weak solution will minimize the functional $L(T)$

$$L(T) = (\nabla^2 T, T)_1 - 2(f,T), \tag{7}$$

and will be a solution to the variational equation

$$(\nabla^2 T, v)_1 = (f,v)_1 \cdot \tag{8}$$

The set of coefficients is unique and it has been proven by Svensson (1988) that the weak solution of the free AGBVP II exists under certain conditions (cap size less than 52.2°).

To apply wavelets, the frame theory can be used as a generalization of base functions (Christensen, 2001). Instead of base functions v_n, wavelets ψ_n are used as frames.

Definition 1: A family of elements $\{\psi_n\} \subseteq H$ (H is a separable Hilbert space) is called frame for H if there exist constants $P,Q>0$ such that

$$P\|f\|^2 \le \sum |f, \Psi_n|^2 \le Q\|f\|^2, \quad f \in H, \tag{9}$$

where P,Q are called frame bounds, ψ_n is a family of wavelets, and f is tied to the boundary conditions (data).

The frame bounds are not unique and the wavelet coefficients in this case are not unique either. The optimal frame bounds are the biggest possible P and smallest possible value for Q. If $P=Q$, the frame is called tight and when $P=Q=1$ we have an orthonormal basis. Since the orthonormal basis condition is very strong, it would be difficult to find a basis which satisfies additional conditions like the compatibility conditions at the coastline. From a mathematical point of view, the lack of uniqueness is not desirable but the wavelet frames could provide more flexibility in the application of compatibility conditions at the coastline. The application of spherical pseudo-differential operators and spherical wavelets will be discussed later on.

The relationship between the weak and classical solution of an AGBVP is closely related to the regularity along the coastline of the boundary surface and the boundary conditions. It is known that the classical solution of AGBVPs does not exist in an integral form like Stokes's or Hotine's formula. Using variational methods combined with the theory of spherical pseudo-differential operators and spherical wavelets we can have numerical solutions of AGBVPs.

If all data of the considered problem (including the boundary) are sufficiently smooth (continuous derivatives up to a certain level), then the weak solution is the classical solution of the problem (Rektorys, 1977). According to a theorem in Rektorys (1977), the weak solution of Laplace's equation has derivatives of all orders. The higher the order of derivatives included in the conditions, the closer the weak solution is to the classical solution. It therefore makes sense to investigate the smoothness procedure along the coastline applying compatibility conditions.

3 Compatibility (smoothness conditions) at coastline

For AGBVP I and AGBVP II, it has been shown in Svensson (1988) that with the introduction of additional compatibility conditions at the coastline, both problems become normal solvable. He applied the theory of pseudo-differential operators to impose compatibility conditions at the coastline. For an infinite domain like $\mathbf{R}^3$, the following definition for a pseudo-differential operator is used (Svensson, 1983).

Definition 2: A pseudo-differential operator A of order μ in $\mathbf{R}^3$ is an operator of the form

$$AF(\xi) = (2\pi)^{-3}\int e^{-\xi\omega \cdot i}\hat{A}(\xi,\omega)\hat{F}(\xi)d\omega, \tag{10}$$

where $\mathbf{\omega}=(\omega_1,\omega_2,\omega_3)$ is the vector of frequencies along the three axes, $\hat{F}$ is the Fourier transform of F. $\hat{\Lambda}$, the symbol of Λ, is in the class $S^t(\mathbf{R}^3 X\, \mathbf{R}^3)$ of functions satisfying on each compact set $\mathbf{K}\in\mathbf{R}^3$

$$\left|D_\omega^\alpha D_\xi^\beta \hat{\Lambda}(\xi,\omega)\right| \le C_{\alpha\beta\mathbf{K}}(1+|\omega|)^{t-|\alpha|}, \xi\in\mathbf{K}. \quad (11)$$

$\alpha=(\alpha_1, \alpha_2, \alpha_3)$ and $\beta=(\beta_1,\beta_2, \beta_3)$ are multiindices, and $|\alpha|=\sum|\alpha_j|$. Pseudo-differential operators form an algebra, which is invariant under smooth changes of coordinates. This property is valid for spherical pseudo-differential operators and it will be used later on to get spherical symbols for all pseudo-differential operators involved in formulating compatibility conditions along the coastline.

On a sphere $\delta\Omega$ with radius R, the Laplacean will have the following representation:

$$\nabla^2 = R^{-2}(\partial^2/\partial\varphi^2 + \tan\varphi\,\partial/\partial\varphi + \cos^{-2}\varphi\,\partial^2/\partial\lambda^2) \quad (12)$$

where φ, λ are spherical coordinates. The Laplacean is itself a pseudo-differential operator and it appears in the pseudo-differential operators when imposing compatibility conditions at the coastline. They have the following form:

$$A=(-\nabla^2+1/(4R^2))^{1/2}-3/(2R), \quad (13)$$
$$C=(-\nabla^2+1/(4R^2))^{1/2}+3/(2R),$$

$$AC=CA=-\nabla^2-2/R^2, \quad C=A+3/R. \quad (14)$$

It can be shown that the inverse spherical PDOs like A^{-1}, C^{-1}, $(AC)^{-1}$, and different combinations of them, exist and they have certain physical meaning (see Table 1). For example, A^{-1} is Stokes's operator and $(I-C)^{-1}$ is Hotine's formula. Taking into account that the inverse PDOs exist, AGBVPs can be redefined, compatibility conditions can be established, and the solution can be found numerically. Together with mappings (extensions) L, M from land and sea to the space outside the sphere, a new definition for both AGBVPs containing compatibility conditions along the coastline is possible. The boundary conditions for AGBVP II can be transformed in terms of pseudo-differential operators to

$$\begin{cases} AT=\Delta g_L & on\ \delta\Omega_L \\ (A+2)(A+3)AT=(-\nabla^2-2)\delta g_S & at\ \delta\Omega_S, \\ (A+2)SL\Delta g_L+(A+2)SM(-\nabla^2-2)\delta g_S & \\ +(A+2)T_1=0 & at\ coastline \end{cases} \quad (15)$$

where S is Stokes's operator and T_1 is the first degree term of T. Compatibility conditions in this form have mainly theoretical meaning and it is necessary to discuss their relationship to the regularity of the data along the coastline.

To clarify the meaning of compatibility conditions given in an implicit mathematical form, we can use some simplifications. Although pseudo-differential operators are applied on the sphere, let us assume the planar case, under this assumption, we can have data and their derivatives consistent along the coastline by requiring that

$$\begin{cases} \delta g_S=\Delta g_L & at\ coastline \\ \nabla^2\delta g_S=\nabla^2\Delta g_L & at\ coastline \end{cases} \quad (16)$$

We can prove (Grebenitcharsky and Sideris, 2001a) that this is equivalent to the compatibility conditions at coastline in Eqs. (15). Obviously, although Eqs. (16) can be directly applied only for a planar case they represent the fact that compatibility conditions are closely related to the regularity of the data along the coastline. In planar approximation, it is not necessary for the disturbing potential to be zero at the coastline to satisfy conditions (16) and (2) simultaneously. According to Eqs. (19) for the fixed AGBVPII the Eqs. (16) in spherical approximation on a unit sphere will have the form:

$$\begin{cases} \delta g_S=\Delta g_L+2T & at\ coastline \\ AC\delta g_S=AC\Delta g_L-4T & at\ coastline \\ \nabla^2\delta g_S=\nabla^2\Delta g_L & at\ coastline \end{cases} \quad (16')$$

The best way to apply compatibility conditions along the coastline is to have them in an explicit form. To improve the regularity of data for the spherical approximation we can impose conditions on the higher order derivatives. This can be done with the application of wavelets as multiscale differential operators (Mallat, 1998).

Wavelets can be used to detect discrepancies between different data along the coastline. We can use a wavelet decomposition and reconstruction to place smoothness compatibility conditions on the data and the boundary along the coastline. After the decomposition up to a certain level, we could eliminate irregularities along the coastline in the high frequency part of the decomposition. This is equivalent to putting constrains on the n^{-th} derivatives (i.e., use smoothness conditions). The necessary and sufficient condition for a wavelet transform to be an *n*-order multiscale differential operator is the corresponding wavelet to have *n-1* vanishing moments (Mallat, 1998). It is known that Daubechies wavelets of order *n* have *n-1* vanishing moments. The value for *n* depends on the levels of decomposition and reconstruction. From other side these levels are related to the stripe width along the coastline where the smoothing has to be applied. In wavelet decomposition and reconstruction, we can limit the magnitude of vertical, horizontal and diagonal coefficients up to a certain value for every level of decomposition. This value can be chosen among the statistics of detail coefficients, for example the rms value for every level of wavelet decomposition. These constraints mean that we have implicitly used additional smoothness conditions on the derivatives of the solution. Restricting the detail coefficients on the coastline means that we force the horizontal gradients of the signal both on land and at sea to be equal for every level of decomposition. These restrictions are only for points which are close to the coastline. The signal far from the coastal region remains unchanged. Using this procedure we can expect that the irregularities along the coastline will be smoothed and this smoothing will affect the coastal region only.

4 Spherical pseudo-differential operators and spherical wavelets for the solution of the fixed AGBVP II

Applying spherical harmonic expansion (spherical Fourier transform) for the definition of pseudo-differential operators will give us the so-called spherical PDOs in the finite spherical domain. They have been derived from the theory of singular integrals. For AGBVPs, PDOs can be applied not only for reformulation of AGBVPs in a simple way but for imposing compatibility conditions at the coastline. The theory of spherical PDOs has been described in Freeden and Windheuser (1997) and in Freeden et al. (1998). The definition of spherical PDOs is slightly different from the PDOs in an infinite domain, due to the use of spherical Fourier transforms.

Definition 3. Let $\{\hat{\Lambda}(n)\}_{n=0.1....}$ be a sequence of real numbers $\lim_{n\to\infty}|\hat{\Lambda}(n)|/(n+1/2)^t = const \neq 0$ for some $t\in\mathbf{R}^1$. Then the operator $\Lambda : H_s(\delta\Omega) \to H_{s-t}(\delta\Omega)$, defined by

$$\Lambda(F) = \sum_{n=0}^{\infty}\sum_{j=1}^{2n+1} \hat{\Lambda}(n)\hat{F}(n,j)Y_{n,j}, F \in H_s(\delta\Omega) \tag{17}$$

is called spherical pseudo-differential operator of order *t* and $\{\hat{\Lambda}(n)\}$ is called spherical symbol. Here $H_s(\delta\Omega), H_{s-t}(\delta\Omega)$ are Sobolev spaces, *F* is a function given in $\delta\Omega$, $\hat{\Lambda}, \hat{F}$ are spherical Fourier transforms of PDOs and functions $Y_{n,j}$ are spherical harmonics. Spherical symbols have the following important properties:

$$\begin{aligned} &(\Lambda_1 + \Lambda_2)^\wedge = \hat{\Lambda}_1 + \hat{\Lambda}_2, \\ &(\Lambda_1\Lambda_2)^\wedge = \hat{\Lambda}_1\hat{\Lambda}_2, \\ &(\Lambda^{-1})^\wedge = 1/(\hat{\Lambda}) \end{aligned} \tag{18}$$

Using these properties of spherical symbols and after some derivations, the analogies between PDOs used by Svensson (1988) and known geodetic operators are derived and shown in Table 1.

Table 1. PDOs, spherical symbols and geodetic operators

Name	Symbol	PDO - Spherical symbol	Name	Symbol	Inverse PDO - Spherical symbol
Gravity anomalies	A	(n-1)/R	Stokes's formula	A-1	R/(n-1)
Gravity disturbances	(I-C) =A+2	(n+1)/R	Hotine's formula	$(I-C)^{-1}$ $=(A+2)^{-1}$	R/(n+1)
Gravity gradient	-C	-(n+2)/R		$-C^{-1}$	-R/(n+2)
	AC=CA	(n+2)(n-1)//R2		$(AC)^{-1}$	R2/(n+2)/(n-1)
Second radial derivative	C(I-C)	(n+2)(n+1)//R2		$(C(I-C))^{-1}$	R2/(n+2)/(n+1)

All PDOs are related to the idea of generalized functions, which in fact creates difficulties in the the application of compatibility conditions along the coastline. At the same time, compatibility conditions have to be locally applied. In this case, the combination between spherical PDOs and spherical wavelets can help us overcome this problem.

The application of multiresolution analysis (MRA) and wavelets with well known types of wavelet transforms is not suitable for the solution of AGBVPs because known wavelets for infinite (planar) domain are not harmonic functions. To solve the problem, new wavelet methods for approximating harmonic functions have been suggested in Freeden and Schreiner (1995). Spherical wavelets are useful for the solution of AGBVPs; they are harmonic and have good space and spectral localization properties, which are necessary for the application of compatibility (smoothness) conditions. Spherical wavelets are a result of applying "rotation" and "dilation" operators. In the case of classical wavelets, the operators are "translation" and "dilation" of a mother wavelet. Another characteristic of spherical wavelets is the way of getting the mother wavelet. The kernels of spherical singular integrals are used as spherical scaling functions. As a spherical scaling function for solution of AGBVPs, the most suitable kernel is the Abel-Poisson kernel because it is harmonic (Freeden and Windheuser, 1996). When the harmonic Legendre polynomials are used as base functions, the corresponding spherical scaling function is harmonic, as well.

In general, spherical PDOs, spherical harmonics and spherical wavelets can be combined to solve the fixed AGBVPs. The disturbing potential T and the observations (boundary conditions) can be related trough PDOs, see diagram in Figure 1.

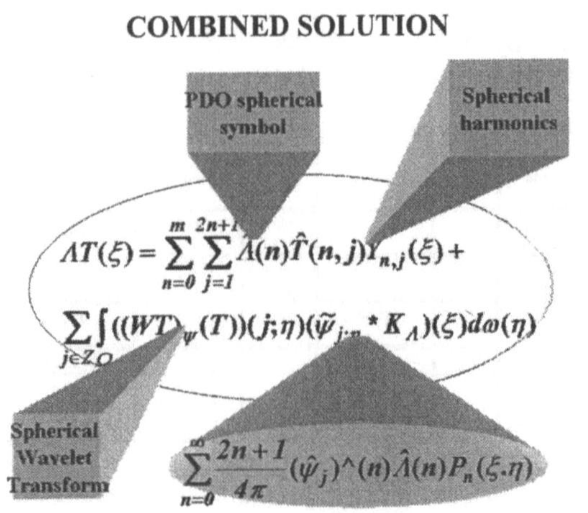

Fig. 1 Combined solution in a general form

ξ and η are points on the sphere, $\widetilde{\psi}_{j;\eta}$ is the dual wavelet, $(\hat{\psi}_j)(n)$ is the spherical wavelet symbol, $\widetilde{\psi}_{j;\eta} * K_\Lambda$ is the convolution of the dual wavelet and the kernel of the corresponding PDO, and j is the level of wavelet decomposition. The combined solution consists of two parts: spherical harmonic expansion for low frequencies and wavelet transformation for the high frequency part of the disturbing potential and the observations.

Now we can proceed to answer the two main questions. How to apply spherical PDOs and spherical wavelets to get two numerical solutions of the fixed AGBVP II, and how to have an explicit form of the compatibility conditions along the coastline.

5 Numerical solution #1 using the solution of Neumann's BVP

We can reformulate the boundary condition on land in Eqs. (2) using geoid heights N from GPS/Leveling. It will take the form

$$\left| -\frac{\partial T}{\partial r} = \delta g_L = \Delta g_L + \frac{2G}{R} N_L \quad on \quad \delta\Omega_L, \right. \tag{19}$$

where G is mean gravity and N_L is the geoid height. After merging gravity disturbances at sea and land, we can apply the wavelet solution of Neumann's problem discussed in Freeden and Schneider (1998). The form of this solution is

$$T_J(\xi) = \sum_{n=1}^{M_0} b_n^0 \sum_{k=0}^{m} \sum_{l=1}^{2k+1} \sum_{s=1}^{M} a_s^{k,l} \hat{\Phi}_0(k) \delta g(\eta_s^M) Y_{-k-1,l}(R;\varsigma_n^{M_0}) \Phi_{0;y_n^{M_0}}(\xi) \tag{20}$$
$$+ \sum_{j=0}^{J} \sum_{n=1}^{MJ} b_n^j \sum_{k=0}^{m} \sum_{l=1}^{2k+1} \sum_{s=1}^{M} a_s^{k,l} \hat{\Psi}_j(k) \delta g(\eta_s^M) Y_{-k-1,l}(R;\varsigma_n^{M_0}) \widetilde{\Psi}_{0;y_n^{M_0}}(\xi)$$

where T is the disturbing potential, $\xi, \eta, \varsigma \in \delta\Omega$, $Y_{-k-1,l}$ are the outer harmonics, $\Phi, \widetilde{\Psi}$ are the scale function and the dual wavelet, $\hat{\Phi}, \hat{\Psi}$ are the spherical symbols for scale function and wavelet function, δg are the boundary conditions (gravity disturbances), and J is the level of wavelet decomposition. For detail information about the notation used, see Eqs. (67) in Freeden and Schneider (1998).

All coefficients b_n^0, b_n^j, a_s^k in front of the scale functions (approximate coefficients) and those in front of the wavelet functions (detail coefficients) are spatially distributed in every J-level of decomposition. These coefficients can be determined separately and independently for a certain configuration and J-level of wavelet approximation. Now the coefficients could be limited up to a certain magnitude using the stability conditions for different types of wavelets. The stability conditions correspond to the definition of the frames (see Eqs. 10) and they will have the form

$$P \leq \sum_{j=-\infty}^{\infty} ((\hat{\Psi}_j)\ (n))^2 \leq Q \quad or \tag{21}$$
$$P \|\delta g\|^2_{L^2(\delta\Omega)} \leq \sum_{j=-\infty}^{\infty} \|(WT)_\Psi(\delta g)(j;.)\|^2_{L^2(\delta\Omega)} \leq Q \|\delta g\|^2_{L^2(\delta\Omega)}$$

Restricting the detail (wavelet) coefficients is equivalent to putting smoothing conditions on the derivatives. Typical for this numerical solution is that the smoothing conditions are imposed on the boundary data, although the conditions are incorporated in the numerical solution.

6 Numerical solution #2 using spherical PDOs and wavelets

This numerical solution is based on the representation of functionals of the disturbing potential as PDOs through their spherical symbols, which can be considered as eigenvalues of those PDOs. The extension and restriction operators used in Eskin (1980) can be applied to define the mappings L and M. In the end, the compatibility conditions can be applied in an explicit way to have a numerical solution of the fixed AGBVP. To combine a spherical harmonic expansion with spherical PDOs and wavelets according to the diagram in Figure 1, the spherical harmonic solution (in the form of geopotential model) has to be subtracted from the data. For the residual part of every functional of the disturbing potential we will have

$$\Lambda T(\xi) = \sum_{j\in z} \int_{\delta\Omega} (WT)_{\psi}^{A}(T)(j;\eta)(\tilde{\Psi}_{j;\eta}^{A} * K_{\Lambda})d\sigma(\eta) \tag{22}$$

$$\Lambda T(\xi) = \sum_{j\in z} \int_{\delta\Omega} (WT)_{\psi}^{A}(T)(j;\eta) \sum_{n=N_{max}}^{\infty} \frac{2n+1}{4\pi}(\hat{\tilde{\psi}}_{j}^{A})(n)\hat{\Lambda}(n)P_n(\xi.\eta)d\sigma(\eta)$$

where A is the type of the wavelet package used. For example, we can use *P, M* or *D* type of wavelet packages described in details in Freeden and Windheuser (1997). Changing the order of integration and summation, yields

$$\Lambda T(\xi) = \sum_{j\in z} \sum_{n=N_{max}}^{\infty} \frac{2n+1}{4\pi}(\hat{\tilde{\psi}}_{j}^{A})(n)\hat{\Lambda}(n) \int_{\delta\Omega} (WT)_{\psi}^{A}(T)(j;\eta)P_n(\xi.\eta)d\sigma(\eta) \tag{23}$$

The integral could be transformed in a discrete form using discrete wavelet representation, given in Freeden and Windheuser (1997, pp. 32-33) in the following way:

$$(WT)_{\Psi}^{A}(T)(j;\eta) = \int_{\delta\Omega} T(\xi)\Psi_{j;\eta}^{A}(\xi)d\sigma(\xi) = \tag{24}$$

$$\sum_{l=1}^{M} c_l^{j;\eta} T(\xi_l) = \sum_{l=1}^{M} \psi_{j;l}^{A}(\eta)T(\xi_l)$$

$$\Downarrow$$

$$\int_{\delta\Omega} (WT)_{\Psi}^{A}(T)(j;\eta)P_n(\xi.\eta)d\sigma(\eta) = \tag{25}$$

$$\int_{\delta\Omega}\sum_{l=1}^{M} (\tilde{\Psi}_{j;l}^{A})T(\xi_l)P_n(\xi.\eta)d\sigma(\eta) =$$

$$\sum_{l=1}^{M} T(\xi_l) \int_{\delta\Omega} \tilde{\Psi}_j(\xi_l.\eta)P_n(\xi.\eta)d\sigma(\eta)$$

After applying the Funk-Hecke formula (Freeden and Windheuser, 1996, pp. 53), we will have for the discrete representation of the integral

$$\int_{\delta\Omega} (WT)_{\Psi}^{A}(T)(j;\eta)P_n(\xi.\eta)d\sigma(\eta) = \tag{26}$$

$$\sum_{l=1}^{M} T(\xi_l)(\hat{\tilde{\Psi}}_j^{A})(n)P_n(\xi_1.\xi) = (\hat{\tilde{\Psi}}_j^{A})(n)\sum_{l=1}^{M} T(\xi_l)P_n(\xi_1.\xi)$$

Now the functional can be represented as

$$\Lambda T(\xi) = \sum_{n=0}^{\infty} \frac{2n+1}{4\pi}\hat{\Lambda}(n)\sum_{l=1}^{M} T(\xi_l)P_n(\xi_1.\xi)\sum_{j=-\infty}^{\infty}((\hat{\tilde{\Psi}}_j^{A})(n))^2 \tag{27}$$

If we assume that

$$\Lambda^{\wedge}(n) = \left(\frac{R}{r}\right)^{n+1} \text{ - upward continuation,} \tag{28}$$

$$\sum_{j=-\infty}^{\infty}((\hat{\tilde{\Psi}}_j^{A})(n))^2 = 1 \text{ - wavelets are orthonormal,} \tag{29}$$

$$\sum_{l=1}^{M} T(\xi_l)P_n(\xi_1.\xi) = \sum_{\lambda'}\sum_{\varphi'} T(\varphi',\lambda',R)P_n(\cos(\psi))\cos\varphi',$$

$$\text{where } \xi = (\varphi,\lambda,r),\ \xi_l = (\varphi',\lambda',R)$$

$$\text{and } P_n(\xi_l.\xi) = P_n(\cos(\psi))$$

then the functional becomes

$$T(\varphi,\lambda,R) = \sum_{n=0}^{\infty} \frac{2n+1}{4\pi}\left(\frac{R}{r}\right)^{n+1} \sum_{\lambda'}\sum_{\varphi'} T(\varphi',\lambda',R)P_n(\cos(\psi))\cos\varphi',$$

But this is exactly the discrete form of Poisson's integral. Eqs. (27) can be successfully used for representing any kind of functional of the disturbing potential as a PDO.

Finally, gravity anomalies on land and gravity disturbances at sea can be represented in the form

$$\Lambda T(\xi) = \sum_{j=0}^{J} \sum_{n=m+M_{j-1}+1}^{m+M_j} \frac{2n+1}{4\pi}\hat{\Lambda}(n)\left[(\hat{\tilde{\Psi}}_j^{A})(n)\right]^2 \sum_{l=1}^{M} T(\xi_l)P_n(\xi_l.\xi)^2 \tag{30}$$

Only the proper spherical symbol has to be chosen from Table 1. For example $\hat{A} = (n-1)/R$ and $(I-C)^{\wedge} = (A+2)^{\wedge} = (n+1)/R$ for gravity anomalies and gravity disturbances, respectively.

7 Explicit form of compatibility conditions

The spherical symbols for PDOs on the unit sphere used in the formulation of compatibility conditions can be written as

$$\hat{A}=(n+1/2)-3/2=(n-1),$$
$$\hat{C}=(A+3)^{\wedge}=n+2 \tag{31}$$
$$(A+2)^{\wedge}=n+1,\quad \hat{S}=1/(n-1)$$
$$(AC)^{\wedge}=(-\nabla^2-2)^{\wedge}=\hat{A}\,\hat{C}=(n-1)(n+2)$$

According to Eskin (1980), the mappings L and M could be simply considered as extensions assuming zeros for the sea and land. Let us introduce two complementary extensions

$$\chi=\begin{cases}1 & on \quad land\\ 0 & at \quad sea\end{cases},\quad \vartheta=\begin{cases}0 & on \quad land\\ 1 & at \quad sea\end{cases} \tag{32}$$

Then the mappings L and M can be represented as the inverse Fourier transforms of Eqs. (33):

$$\hat{L}(\Delta g_L)=\hat{\chi}*\Delta\hat{g}$$
$$\hat{M}(\Delta g_L)=\hat{\vartheta}*\Delta\hat{g} \tag{33}$$

For the specific case of a spherical cap, the Legendre transform will give us the spherical symbol for both extensions as

$$\hat{L}(n)=2\pi\int_{-1}^{1}\chi(t)P_n(t)dt$$
$$\hat{M}(n)=2\pi\int_{-1}^{1}\vartheta(t)P_n(t)dt \tag{34}$$

For a more complicated coastline, a window in the form of a continent-ocean function can be applied and expanded into spherical harmonics (Simons at al., 1997) over the entire unit sphere, and the spherical symbols can be calculated. In this case, the χ and θ functions will not have only zonal non null harmonic coefficients. Possible complications may arise from numerical point of view, when we have to determine these non zero harmonic coefficients.

Assuming the first term T_1 of the disturbing potential to be zero, the compatibility conditions along the coastline can be given in an explicit form in terms of PDOs and wavelets as

$$\Lambda_L(T)+\Lambda_S(T)=0 \tag{35}$$

with spherical symbols

$$\hat{\Lambda}_L(n)=(A+2)^{\wedge}\hat{S}\hat{A}\hat{L}=(n+1)\hat{L}(n)$$
$$\hat{\Lambda}_S(n)=(A+2)^{\wedge}\hat{S}\,\hat{M}(AC)^{\wedge}\hat{A}$$
$$=(n+1)(n+2)(n-1)\hat{M}(n) \tag{36}$$

Now the compatibility conditions can be explicitly applied on the detail coefficients in the coastal area at every J-level of decomposition using the representation of Λ_L, Λ_S in terms of spherical PDOs and spherical wavelets.

In planar approximation, the preliminary results in Grebenitcharsky and Sideris (2001a) and (2002) show that the effect on the geoid of smoothing data along the coastline can have a magnitude from ±5 cm for a flat area to ±30 cm in a mountainous coastline. A numerical solution based on spherical PDOs, wavelets and compatibility conditions given in an explicit form is expected to provide further improvement, and it will be presented in a future paper.

8 Conclusions

Two numerical solutions of the fixed AGBVP II have been suggested using the advantages of spherical PDOs and spherical wavelets.

In numerical solution #1, although the smoothnees conditions are incorporated in the solution, they are applied to the boundary conditions (data).

Difficulties in the processing are expected due to the huge number of coefficients but, for a certain type of wavelets and data configuration, they could be computed once and separately.

In numerical solution #2, the compatibility conditions are applied directly to the solution (disturbing potential) along the coastline.

The main difficulties in this approach are the non-orthogonality of the Abel-Poisson wavelets, which will slow down the processing, and the application of an equidistant grid.

Acknowledgements: The authors wish to thank the two anonymous reviewers for their constructive criticism and suggestions, which significantly improved the quality of this paper. This research has been funded by grants from the Natural Sciences and Engineering Research Council of Canada and by the GEOIDE Network of Centers of Excellence.

References

Christensen, O. (2001). Frames, Riesz bases and discrete Gabor/Wavelet expantions, Bulletin of the American Mathematical Society, Volume 38, Number 3, 273-291.

Eskin, G. (1981). Boundary value problems for elliptic pseudodifferential equations, Tran. of Mathematical Monographs, 52, American Mathematical Society.

Freeden, W. and Schreiner, F. (1995). New wavelet methods for approximating harmonic functions. In: Geodetic theory today (F. Sanso, ed.) International Association of Geodesy Symposia, 114, Springer, 112-121.

Freeden, W. and Windheuser U. (1996). Spherical wavelet transform and its discretization, Adv. Comp. Math., 5, 51-94.

Freeden, W. and Windheuser, U. (1997). Combined spherical harmonic and wavelet expansion – A future concept in Earth's gravitational determination, Appl.,Comp. Harm. Anal., 4, 1-37.

Freeden, W. and Schneider, F. (1998.) Wavelet approximation on closed surface and their application to boundary-value problems of potential theory, Math. Meth in Appl. Sciences, 21, 129-163.

Freeden, W. Gervens, T. and Schreiner, M. (1998). Constructive approximation on the sphere, with applications to Geomathematics, Oxford Science Publications.

Grebenitcharsky, R. and Sideris, M.G. (2001a). Altimetry-gravimetry boundary value problems with smoothness conditions in coastal regions, in Tziavos I. N. and Barzaghi R. (ed.) EGS 2001-G7 Session "Regional and local gravity field approximation", Nice, France, 25-30 March, 2001, Bulletin of IGeS N. 13, special issue, 121-132.

Grebenitcharsky. R. and Sideris, M.G. (2001b). A Comparison of Different Solution Methods for Altimetry-Gravimetry Boundary Value Problems Using Smoothing Conditions Along the Coastline, Presented at the 2001 International Association of Geodesy Scientific Assembly, Budapest, Hungary, September 2-7.

Grebenitcharsky, R., Vergos, G.S., and Sideris, M.G. (2001). Combination of gravity, altimetry and GPS/Leveling data for the numerical solution of the altimetry-gravimetry boundary value problems, In: Adam J., Schwarz K.,P. (eds) Vistas for geodesy in the new millennium, IAG Scientific Assembly, Budapest, Hungary, September 2-7, 150-155.

Grebenitcharsky R., M. G. Sideris (2002) Data smoothing along a mountainous coastline for use in altimetry-gravimetry boundary value problem solutions, in "3rd Meeting of International Gravity and Geoid Commission, Tziavos (ed.), Gravity and Geoid 2002 - GG2002, 2003", 223-228

Holota, P. (1983a). The altimetry gravimetry boundary value problem I: linearization, Friedrich's inequality. Boll Geod Sci Afini, 42, 14-32.

Holota, P. (1983b). The altimetry gravimetry boundary value problem II: weak solution, V-ellipticity. Boll Geod Sci Afini, 42, 70-84.

Holota, P. (1997). Variational methods for geodetic boundary value problems, In: Sanso F., Rummel R. (eds) Geodetic boundary value problems in view of the one centimeter geoid. Lecture Notes in Earth Sciences, vol 65. Springer, Berlin Heidelberg New York, 468-481.

Holota, P. (2000). Variational methods in the recovery of the gravity field – Galerkin's matrix for an ellipsoidal domain, in Sideris (ed.), Gravity, Geoid, and Geodynamics, 2000, IAG Symposia, Vol. 123, 277-283.

Keller, W. (1996). On a scalar fixed altimetry-gravimetry boundary value problem, J. of Geod., vol 70, 459-469.

Lehmann, R. (1999). Boundary-value problems in the complex world of geodetic measurements, *J of Geodesy 73* pp. 491-500.

Mallat, S. (1998). A wavelet tour of signal processing. Academic press, USA.

Rektorys, K. (1977). Variational methods in mathematics, science and engineering. Dordecht-Holand/ Boston-USA.

Sacerdote, F. and Sansò, F. (1983). A contribution to the analysis of altimetry gravimetry problems. Bull Geod, 57, 183-201.

Sacerdote. F. and Sansò, F. (1987). Further remarks on the altimetry-gravimetry problems. Bull Geod, 61, 183-201.

Sansò, F. (1993). Theory of geodetic B.V.P.s applied to the analysis of altimeter data. In: Rummel R, Sanso F. (eds) Satellite altimetry in geodesy and oceanography. Lecture Notes in Earth Sciences, vol 50. Springer, Berlin Heidelberg New York.

Simons, M., S.C. Solomon and B.H. Hager (1997) Localization of gravity and topography: constraints on the tectonics and mantle dynamics of Venus, Geophys. J. Int., 131, 24-44.

Svensson, L. (1983a). Solution of the altimetry-gravimetry problem. Bull Geod, 57, 332-353.

Svensson, L. (1983b). Pseudodifferential operators – New Approach to the boundary problems of physical geodesy, Manuscr Geod, 8, 1-40.

Svensson L (1988) Some remarks on the altimetry-gravimetry problem. Manuscr Geod, 13, 63-74.

Ellipsoidal Vertical Deflections: Regional, Continental, Global Maps of the Horizontal Derivative of the Incremental Gravity Potential

G. Finn, E. W. Grafarend
Department of Geodesy and Geoinformatics,
Stuttgart University, Geschw.-Scholl-Str. 24 D, D-70174 Stuttgart, Germany, e-mail: finn@gis.uni-stuttgart.de

Abstract. A gravity potential field of ellipsoidal type is required for high resolution computation of the deflection of the vertical. With respect to the gravity potential model Spheroidal Earth Gravity Normal (SEGEN) a new theory is given for the prediction of long wave (global) portions of vertical deflections. The vertical deflections are developed into a series of ellipsoidal harmonics up to degree/order 360/360, parameterised in Jacobi spheroidal coordinates, which result in an improvement of accuracy of two orders of magnitude compared with a spherical definition based on a spherical reference gravity potential field. Test computations are presented in regional, continental and global maps with respect to World Geodetic Datum 2000.

Keywords. Ellipsoidal Vertical Deflections, Ellipsoidal Harmonics, Somigliana-Pizzetti Field, SEGEN

1 Introduction

The local *vertical* is defined as the direction or unit vector $\mathbf{\Gamma} \div \|\mathbf{\Gamma}\|_2$, namely the local gravity vector divided by its Euclidean length. Consequently, the local *deflection of the vertical* is the difference between the unit vectors $\mathbf{\Gamma} \div \|\mathbf{\Gamma}\|_2$ and $\boldsymbol{\gamma} \div \|\boldsymbol{\gamma}\|_2$ where the reference local vertical or unit vector $\boldsymbol{\gamma} \div \|\boldsymbol{\gamma}\|_2$ is defined as the reference gravity vector divided by its Euclidean length.

In order to create the reference local vertical we have to choose the optimal gravity potential field. Nearly all contributions in *Mathematical Geodesy* have chosen the *spherical gravity vector* $-\mathbf{e}_r GM/r^2$ as the reference. Such a reference gravity is generated by a mass density distribution, which is *spherically symmetric*. In Geometric Geodesy $(\Lambda_\Gamma, \Phi_\Gamma)$ are experimentally determined from astronomical observations; the reference vertical is described by Gauss ellipsoidal coordinates (L, B) which are materialised by measurements of type global/local positioning system (GPS/LPS). Indeed the ellipsoidal normal already is a very good approximation of the reference local vertical $\boldsymbol{\gamma} \div \|\boldsymbol{\gamma}\|_2$ if an ellipsoidal harmonic reference field is chosen. In *Physical Geodesy*, the local vertical is computed from the *Somigliana-Pizzetti potential field*. There are various good reasons why such a reference potential is chosen and being treated in any geodetic textbook: *First*, such a gravity field is generated as the *ellipsoidal harmonic field of a level ellipsoid* which is identified as the *International Reference Ellipsoid*. Such a perspective has been recently emphasised by *Grafarend and Ardalan* (1999) when the World Geodetic Datum 2000 was presented. Indeed any geodetic reference system applied to LPS as well as to GPS takes reference to the *Somigliana-Pizzetti potential field*. *Second*, any hydrostatic equilibrium figure as well as any isostatic modeling refers to a *level surface of the gravity potential*. A *level surface* bounds any homogeneous or heterogeneous ellipsoidal equilibrium figure of type Maclaurin (1742), Jacobi (1834), Chandrasekhar and Roberts (1963), Chandrasekhar (1969), Grafarend and Hauer (1978) reviewed by Schaffrin et al. (1977), Mikolaiski (1989) and Moritz (1990). Loosely speaking, a disturbing potential or a disturbing gravity, which is defined with respect to the reference potential of Somigliana-Pizzetti type measures the departure from a specific equilibrium figure. *Third, Moritz* (1968 a, b, 1973, 1990) already succeeded to associate a specific mass distribution inside the Earth, which generates the gravity field of a *level ellipsoid*. All these reasons have motivated us to define the incremental gravity potential and its horizontal derivatives in terms of the reference potential field of *Somigliana-Pizzetti type*.

Vertical deflections are caused by density disturbances due to the heterogeneous mass distribution of the Earth. We limit this contribution to the long wave components of the gravity potential of the solid Earth, oceans and atmosphere. The *terrain effect* as well as the effect of isostatic compensation has to be treated separately.

The first topic we deal with here is the representation of vertical deflections in gravity space. In the second step we will introduce the Somigliana-Pizzetti field as the reference gravity potential field w. By means of the high resolution model SEGEN we gain the disturbing gravity potential field δw. Subsequently, the disturbing gravity vector $\boldsymbol{\delta\gamma}$ will be transformed from gravity space to geometry space. Finally, we will succeed to set up equations for ellipsoidal harmonic vertical deflections. Those formulae form the basis of case study computations in the sixth paragraph.

2 Vertical Deflections in Gravity Space

In order to compute vertical deflections synthetically we start from the well-known representation of the deflection of the vertical in gravity space, also called incremental gravity vector $\boldsymbol{\delta\gamma}$, which is defined as the vector-valued difference between the actual gravity vector $\boldsymbol{\Gamma}$ and the reference gravity vector $\boldsymbol{\gamma}$, namely

$$-\boldsymbol{\delta\gamma} = -(\boldsymbol{\Gamma} - \boldsymbol{\gamma}) = [\mathbf{e}_{\lambda_\gamma}\ \mathbf{e}_{\phi_\gamma}\ \mathbf{e}_\gamma] \begin{bmatrix} \gamma\eta \\ \gamma\xi \\ \delta\gamma \end{bmatrix}.$$

The base vectors of the local frame $\{\mathbf{e}_{\lambda_\gamma}, \mathbf{e}_{\phi_\gamma}, \mathbf{e}_\gamma | P\}$ attached at the evaluation point P can be constructed from the partial derivatives of the reference gravity vector $\boldsymbol{\gamma}$ with respect to its spherical coordinates $(\lambda_\gamma, \phi_\gamma, \gamma)$, for instance $\mathbf{e}_{\lambda_\gamma} := \mathrm{D}_{\lambda_\gamma}\boldsymbol{\gamma} \div \|\mathrm{D}_{\lambda_\gamma}\boldsymbol{\gamma}\| = \mathbf{e}_1(-\sin\lambda_\gamma) + \mathbf{e}_2\cos\lambda_\gamma$. The tangent space $\mathbb{T}_P\mathbb{S}^2_\gamma$ is spanned by the base vectors $\mathbf{e}_{\lambda_\gamma}$, $\mathbf{e}_{\phi_\gamma}$, while the normal space $\mathbb{N}_P\mathbb{S}^2_\gamma$ is spanned by $\mathbf{e}_\gamma$. The coordinates of the tangent space divided by the normal gravity $\gamma = \|\boldsymbol{\gamma}\|$ are called vertical deflections (η, ξ), which are directed to (East, North). We obtain $\eta \doteq \cos\lambda_\gamma\,\delta\lambda_\gamma$, $\xi \doteq \delta\phi_\gamma$ in linear approximation employing Eulerian increments $\delta\lambda_\gamma := \Lambda_\Gamma - \lambda_\gamma$, $\delta\phi_\gamma := \Phi_\Gamma - \phi_\gamma$ between astronomical coordinates $(\Lambda_\Gamma, \Phi_\Gamma)$ and reference coordinates $(\lambda_\gamma, \phi_\gamma)$. $\delta\gamma := \Gamma - \gamma$ denotes the gravity disturbance as the third coordinate. This linearisation is necessary in order to represent $\boldsymbol{\Gamma}$ by reference coordinates in the local frame $\{\mathbf{e}_{\lambda_\gamma}, \mathbf{e}_{\phi_\gamma}, \mathbf{e}_\gamma | P\}$. A detailed derivation can be found by Grafarend, Finn and Ardalan (2002). Additionally, we refer to Grafarend (2001) where the spherical case has been treated. Figure 1 illustrates the tangent space $\mathbb{T}_P\mathbb{S}^2_\gamma$.

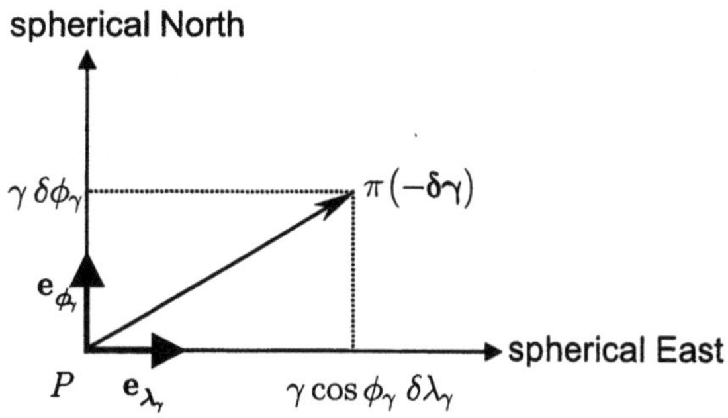

Fig. 1 Negative incremental gravity vector $-\boldsymbol{\delta\gamma}$ in the tangent space $\mathbb{T}_P\mathbb{S}^2_\gamma$ at point P: projection $\pi(-\boldsymbol{\delta\gamma})$ of the negative incremental gravity vector onto the tangent plane $\mathbb{P}_2$ at point P.

3 Actual, Reference and Disturbing Gravity Potential Field

In the case that real observations are not available we can use a model of the actual gravity potential W, which contains the needed physical information. $W = U + V$ consists of the gravitational potential U and the centrifugal potential V. U can be decomposed into two parts: U_{global} and U_{local}. The short wave component U_{local} can be modeled using a digital elevation model. We are focused on the global gravitational part of the gravity potential, namely the long wave component U_{global}. By means of SEGEN (Spheroidal Earth Gravity Normal) the global gravitational part U_{global} of the potential is developed into a series of ellipsoidal harmonics band limited by degree/order 360/360 (Ardalan and Grafarend (2000))

$$W(\lambda,\phi,u) = \tfrac{1}{2}\Omega^2(u^2+\varepsilon^2)\cos^2\phi + \frac{GM}{\varepsilon}\operatorname{arccot}(\tfrac{b}{\varepsilon}) \times \\ \times \sum_{k=0}^{360}\sum_{l=0}^{360} \frac{Q^*_{kl}(\frac{u}{\varepsilon})}{Q^*_{kl}(\frac{b}{\varepsilon})} P^*_{kl}(\sin\phi)(C_{kl}\cos l\lambda + S_{kl}\sin l\lambda).$$

The coefficients C_{kl}, S_{kl} as well as a manual can be downloaded from our web page (http://www.uni-stuttgart.de / gi / research / paper / coefficients / coefficients.zip). The factor $\frac{GM}{\varepsilon}\operatorname{arccot}(\frac{b}{\varepsilon})$ is due to technical reasons. It normalises the coefficients in that way that they are free of units and $C_{0,0} = 1$; all other coefficients $|C_{kl}|$ and $|S_{kl}|$ are smaller. GM denotes the "gravitational mass": G is Newton's gravitational constant, also called the coupling constant of gravitational and inertial forces, M is the symbol for the total mass of the Earth. a/b is the semi-major/semi-minor axis, respectively, of the International Reference Ellipsoid while the linear eccentricity is defined as $\varepsilon := \sqrt{a^2 - b^2}$. The parameter Ω denotes the angular velocity of the Earth. $P^*_{kl}(\sin\phi)$ are the associated Legendre functions of the first kind while $Q^*_{kl}(\frac{u}{\varepsilon}) = i^{k+1} Q_{kl}(i\frac{u}{\varepsilon})$ denote the ones of the second kind. The expansion of the potential field in *ellipsoidal harmonics* can be studied by Asche (1991), Gleason (1988, 1989), Jekeli (1988, 1999), Skrzipek (1998) and Sona (1996).

According to perturbation theory, we decompose the actual gravity potential $W = w + \delta w$ into a reference part w and a disturbing part δw. As the reference we choose an ellipsoidal field of type Somigliana-Pizzetti. The reference local vertical can be generated at the evaluation point P applying the gradient operator to w. It approximates the local vertical significantly better than the spherical vertical as well as the ellipsoidal vertical at the corresponding point on the International Reference Ellipsoid.

$$w(\phi, u) = \frac{1}{2}\Omega^2(u^2 + \varepsilon^2)\cos^2\phi + \frac{GM}{\varepsilon}\operatorname{arccot}(\frac{u}{\varepsilon})$$

$$+\frac{1}{6}\Omega^2 a^2 \frac{(3\frac{u^2}{\varepsilon^2} + 1)\operatorname{arccot}(\frac{u}{\varepsilon}) - 3\frac{u}{\varepsilon}}{(3\frac{b^2}{\varepsilon^2} + 1)\operatorname{arccot}(\frac{b}{\varepsilon}) - 3\frac{b}{\varepsilon}}(3\sin^2\phi - 1).$$

As soon as we take reference to the Somigliana-Pizzetti field the disturbing gravity potential $\delta w := W - w$ does not contain the terms of degree/order (0,0), (2,0) of the series expansion of U_{global} as well as the centrifugal potential; all parts are covered by w. If the origin of the global reference system is situated in the mass center of the Earth, then the coefficients (1,0), (1,1) vanish. The disturbing potential δw contains the remaining series (without the terms (0,0), (1,0), (1,1), (2,0)) which we want to call δw_{global} and $U_{local} =: \delta w_{local}$, namely $\delta w = \delta w_{global} + \delta w_{local}$. In this contribution we investigate only the effect of a global gravitational model. Hence we omit the *terrain effect* as well as the effect of isostatic compensation and set: $\delta w := \delta w_{global}$

$$\delta w(\lambda, \phi, u) = \frac{GM}{\varepsilon}\operatorname{arccot}(\frac{b}{\varepsilon}) \times$$

$$\times \sum_{k=2}^{360}\sum_{\substack{l=0 \\ (k,l)\neq(2,0)}}^{360} \frac{Q^*_{kl}(\frac{u}{\varepsilon})}{Q^*_{kl}(\frac{b}{\varepsilon})} P^*_{kl}(\sin\phi)(C_{kl}\cos l\lambda + S_{kl}\sin l\lambda).$$

The reference gravity γ is required for the computation of vertical deflections. Sufficient accuracy of $0.1\,\mu Gal$ is provided by the following approximation formula, also called the International Gravity Formula, reviewed by Ardalan and Grafarend (2001)

$$\gamma = \frac{\sqrt{u^2 + \varepsilon^2}}{\sqrt{u^2 + \varepsilon^2\sin^2\phi}}|\mathrm{D}_u w| + \mathcal{O}(0.1\,\mu Gal)$$

$$\mathrm{D}_u w = \Omega^2 u\cos^2\phi - \frac{GM}{u^2+\varepsilon^2}$$

$$+\frac{1}{3}\Omega^2 a^2 \frac{3(u^3+u\varepsilon^2)\operatorname{arccot}(\frac{u}{\varepsilon}) - (3\frac{u^2}{\varepsilon}+2\varepsilon)}{(u^2+\varepsilon^2)[(3b^2+\varepsilon^2)\operatorname{arccot}(\frac{b}{\varepsilon}) - 3b\varepsilon]}(3\sin^2\phi - 1).$$

The partial derivative of w with respect to ϕ is neglected. As an alternative, an elegant closed form solution for Somigliana-Pizzetti gravity was recently developed by Grafarend, Finn and Ardalan (2002).

4 Transformation of the Disturbing Gravity Vector from Gravity Space to Geometry Space

W, w and δw are based upon ellipsoidal coordinates (λ, ϕ, u) of type Jacobi, also called mixed elliptic-trigonometric elliptic coordinates which relate to Cartesian coordinates as follows

$$X = \sqrt{u^2 + \varepsilon^2}\cos\phi\cos\lambda$$
$$Y = \sqrt{u^2 + \varepsilon^2}\cos\phi\sin\lambda$$
$$Z = u\sin\phi$$

An extensive review on ellipsoidal coordinates can be found by Thông and Grafarend (1989). The disturbing gravity vector $\delta\gamma$ results from the disturbing gravity potential δw applying the gradient operator

$$\delta\gamma = \operatorname{grad}\delta w$$

$$= \mathbf{e}_\lambda \frac{1}{\sqrt{g_{\lambda\lambda}}}\mathrm{D}_\lambda \delta w + \mathbf{e}_\phi \frac{1}{\sqrt{g_{\phi\phi}}}\mathrm{D}_\phi \delta w + \mathbf{e}_u \frac{1}{\sqrt{g_{uu}}}\mathrm{D}_u \delta w.$$

$g_{\lambda\lambda} = (u^2 + \varepsilon^2)\cos^2\phi$, $g_{\phi\phi} = u^2 + \varepsilon^2\sin^2\phi$, and $g_{uu} = (u^2 + \varepsilon^2\sin^2\phi)/(u^2 + \varepsilon^2)$ denote the metric coefficients of the curvilinear orthogonal coordinates (λ, ϕ, u).

With respect to these coordinates we are able to construct a local frame of reference $\{\mathbf{e}_\lambda, \mathbf{e}_\phi, \mathbf{e}_u | P\}$ attached at point P, which can be obtained from the normalised partial derivatives of the placement vector $\mathbf{x}$, for instance $\mathbf{e}_u := \mathrm{D}_u\mathbf{x} \div \|\mathrm{D}_u\mathbf{x}\| = (u^2+\varepsilon^2\sin^2\phi)^{-\frac{1}{2}}$ $[\mathbf{e}_1 u\cos\lambda\cos\phi + \mathbf{e}_2 u\sin\lambda\cos\phi + \mathbf{e}_3\sqrt{u^2+\varepsilon^2}\sin\phi]$. At this point we have two representations of $\delta\gamma$, one in gravity space containing the unknowns (η, ξ) and one in geometry space containing the physical information. If both representations refer to a unique reference frame, then we can compare the coefficients in order to set up equations for vertical deflections. Hence we have to transform $[\mathbf{e}_\lambda, \mathbf{e}_\phi, \mathbf{e}_u]$ to $[\mathbf{e}_{\lambda_\gamma}, \mathbf{e}_{\phi_\gamma}, \mathbf{e}_\gamma]$.

Both bases are related to the global basis $[\mathbf{e}_1, \mathbf{e}_2, \mathbf{e}_3]$ by rotation matrices $\mathbf{T}_1$ and $\mathbf{T}_2$, outlined by Figure 2. The rotation matrix $\mathbf{T}_2$ is an Eulerian one which rotates first around the 3-axis and subsequently around

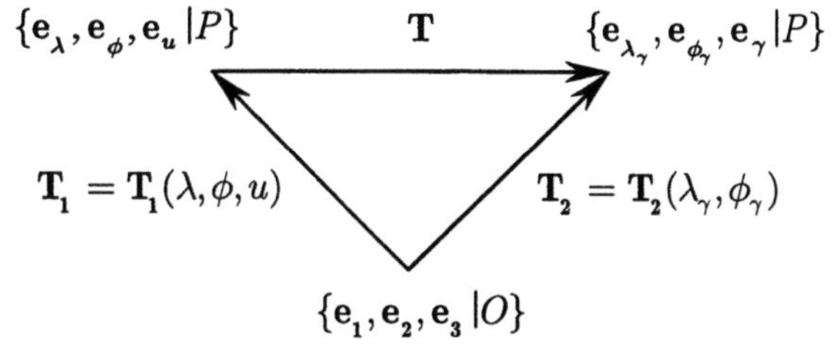

Fig. 2 Basis transformation: geometry space into gravity space.

the 1-axis, namely $\mathbf{T}_2 = \mathbf{R}_3(0)\mathbf{R}_1(\frac{\pi}{2} - \phi_\gamma)\mathbf{R}_3(\lambda_\gamma + \frac{\pi}{2})$. $\mathbf{T}_1$ can be obtained from the Cartesian representation of $\mathbf{e}_\lambda$, $\mathbf{e}_\phi$ and $\mathbf{e}_u$ mentioned above. All matrices belong to the special orthogonal group $SO(3) := \{\mathbf{R} \in \mathbb{R}^{3\times 3} \mid \mathbf{R}^T\mathbf{R} = \mathbf{I}_3 \; and \; \det(\mathbf{R}) = +1\}$. Their inverse is just the transpose. In closing the commutative diagram we are led to the unknown matrix $\mathbf{T} = \mathbf{T}_2\mathbf{T}_1^T$. Employing Eulerian increments $\delta\lambda := \lambda - \lambda_\gamma$, $\delta\phi := \phi - \phi_\gamma$ we arrive at the transformation matrix $\mathbf{T}$ in linear form, namely

$$\mathbf{T} = \begin{bmatrix} 1 & -\sin\phi\delta\lambda & \cos\phi\delta\lambda \\ \sin\phi\delta\lambda & 1 & \delta\phi + \frac{\varepsilon^2}{4u^2}\sin 2\phi \\ -\cos\phi\delta\lambda & -\delta\phi - \frac{\varepsilon^2}{4u^2}\sin 2\phi & 1 \end{bmatrix}.$$

Note that $(\delta\lambda, \delta\phi)$ are properties of the normal field and *not* disturbing quantities like $(\delta\lambda_\gamma, \delta\phi_\gamma)$. They vanish *on* the surface of the reference ellipsoid. Application of this transformation to the representation of $\boldsymbol{\delta\gamma}$ in gravity space leads to

$$\boldsymbol{\delta\gamma} \doteq [\mathbf{e}_\lambda \; \mathbf{e}_\phi \; \mathbf{e}_u]\mathbf{T}^T \begin{bmatrix} -\gamma\eta \\ -\gamma\xi \\ -\delta\gamma \end{bmatrix} = [\mathbf{e}_\lambda \; \mathbf{e}_\phi \; \mathbf{e}_u] \begin{bmatrix} \frac{1}{\sqrt{g_{\lambda\lambda}}} \mathrm{D}_\lambda \delta w \\ \frac{1}{\sqrt{g_{\phi\phi}}} \mathrm{D}_\phi \delta w \\ \frac{1}{\sqrt{g_{uu}}} \mathrm{D}_u \delta w \end{bmatrix}.$$

5 Ellipsoidal Harmonic Vertical Deflections

The comparison of coefficients results in a system of linear equations, which can be solved for the vertical deflections. We gain η, ξ as functionals of the partial derivatives of the disturbing potential δw

$$\eta = -\frac{1}{\gamma}\left(\frac{1}{\sqrt{g_{\lambda\lambda}}}\mathrm{D}_\lambda \delta w + \frac{\sin\phi\delta\lambda}{\sqrt{g_{\phi\phi}}}\mathrm{D}_\phi \delta w - \frac{\cos\phi\delta\lambda}{\sqrt{g_{uu}}}\mathrm{D}_u \delta w\right)$$

$$\xi = -\frac{1}{\gamma}\left(-\frac{\sin\phi\,\delta\lambda}{\sqrt{g_{\lambda\lambda}}}\mathrm{D}_\lambda \delta w + \frac{1}{\sqrt{g_{\phi\phi}}}\mathrm{D}_\phi \delta w - \frac{\delta\phi + \frac{\varepsilon^2}{4u^2}\sin 2\phi}{\sqrt{g_{uu}}}\mathrm{D}_u \delta w\right).$$

These general formulae hold for any choice of the actual as well as the reference gravity potential field. Hence we can use them also for modeling the ellipsoidal gravitational terrain effect, for instance. Finally, by approximating these formulae up to first order by means of the principal terms $[g_{\lambda\lambda}]^{-1/2}\mathrm{D}_\lambda \delta w$, $[g_{\phi\phi}]^{-1/2}\mathrm{D}_\phi \delta w$ we gain explicit equations for vertical deflections as harmonic functions

$$\eta(\lambda, \phi, u) = -\frac{1}{\gamma(\phi, u)} \frac{\frac{GM}{\varepsilon}\operatorname{arccot}(\frac{b}{\varepsilon})}{\sqrt{u^2 + \varepsilon^2}\cos\phi} \times$$
$$\times \sum_{\substack{k=2 \\ (k,l)\neq(2,0)}}^{K} \sum_{l=0}^{k} \frac{Q^*_{kl}(\frac{u}{\varepsilon})}{Q^*_{kl}(\frac{b}{\varepsilon})} P^*_{kl}(\sin\phi)(-lC_{kl}\sin l\lambda + lS_{kl}\cos l\lambda)$$

$$\xi(\lambda, \phi, u) = -\frac{1}{\gamma(\phi, u)} \frac{\frac{GM}{\varepsilon}\operatorname{arccot}(\frac{b}{\varepsilon})}{\sqrt{u^2 + \varepsilon^2\sin^2\phi}} \times$$
$$\times \sum_{\substack{k=2 \\ (k,l)\neq(2,0)}}^{K} \sum_{l=0}^{k} \frac{Q^*_{kl}(\frac{u}{\varepsilon})}{Q^*_{kl}(\frac{b}{\varepsilon})} [\mathrm{D}_\phi P^*_{kl}(\sin\phi)](C_{kl}\cos l\lambda + S_{kl}\sin l\lambda).$$

6 Case Study

Here we present global, continental and regional computations of vertical deflections with respect to Somigliana-Pizzetti modulus of gravity γ (Fig. 3-8). Input data are the coefficients of SEGEN and the parameters of the World Geodetic Datum 2000. Due to numerical reasons we recommend to use the real-valued, stabilised Legendre functions of the second kind $Q^*_{kl}(\frac{u}{\varepsilon})$, which have been introduced by Thông and Grafarend (1989). They have been positively evaluated by Balmino et al. (1991). For a more recent study of the ellipsoidal harmonic modulator we refer to ELGRAM (Sansò and Sona 2001).

We compute the vertical deflections on a $1°\times1°$ grid / $3'\times 3'$ grid for the global / regional maps. It is difficult to interpolate the values. That is why we do the computations *on* the International Reference Ellipsoid. The range of the values for η/ξ is $[-54.2'', 47.3'']$ / $[-75.0'', 49.6'']$. Since 99.37% / 99.33% of all values are in the range of $[-20.0'', 20.0'']$ we plot only this span in order to preserve the detailed structure. Our results are in full agreement with the outcome of Novák, Šimek and Kostelecký (2003). The seismotectonic structure of the Earth is strongly visible in all figures. The statistical properties of the computed vertical deflection components are summarized in Table 1. For a study of the Mollweide projection of the biaxial ellipsoid we refer to Grafarend and Heidenreich (1995).

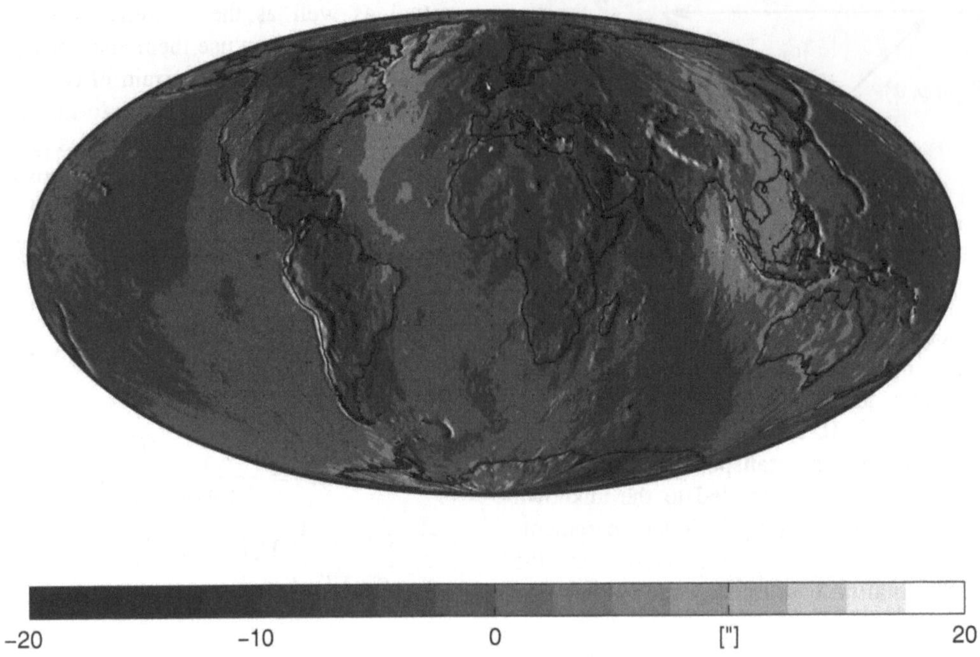

Fig. 3 Global portions of the West-East component η of the deflection of the vertical on the International Reference Ellipsoid with respect to SEGEN and WGD 2000, Mollweide projection of the ellipsoid-of-revolution.

Fig. 4 Global portions of the South-North component ξ of the deflection of the vertical on the International Reference Ellipsoid with respect to SEGEN and WGD 2000, Mollweide projection of the ellipsoid-of-revolution.

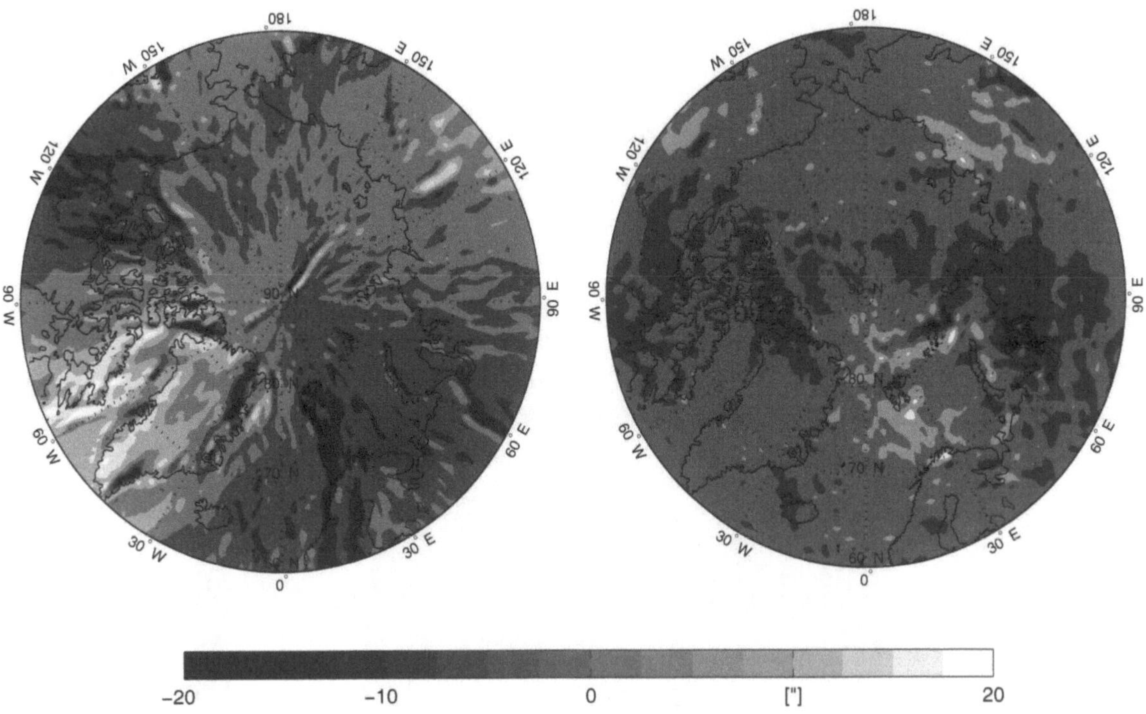

Fig. 5 West-East component η (left) and South-North component ξ (right) of the deflection of the vertical on the International Reference Ellipsoid at the North Pole, equal-area azimuthal projection of the ellipsoid-of-revolution.

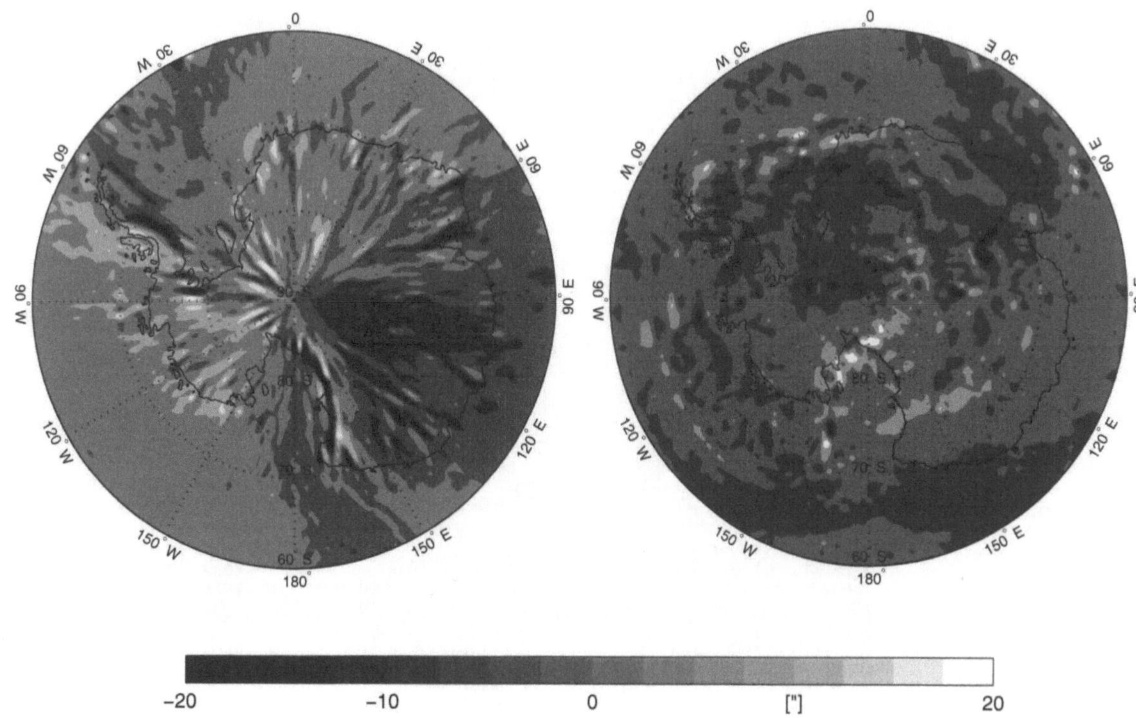

Fig. 6 West-East component η (left) and South-North component ξ (right) of the deflection of the vertical on the International Reference Ellipsoid at the South Pole, equal-area azimuthal projection of the ellipsoid-of-revolution.

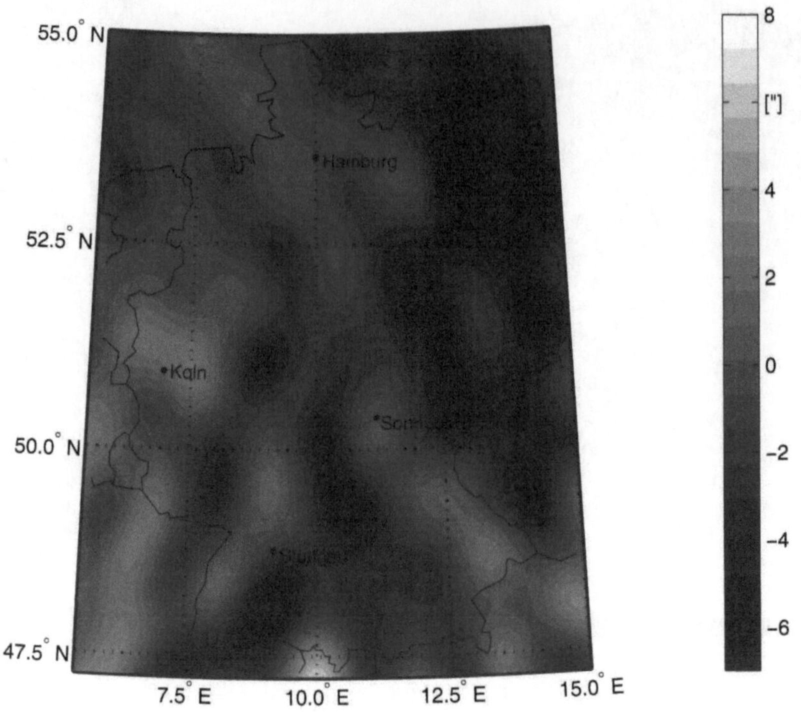

Fig. 7 Global portions of the West-East component η of the deflection of the vertical on the International Reference Ellipsoid of Germany with respect to SEGEN and WGD 2000, Albers projection of the ellipsoid-of-revolution.

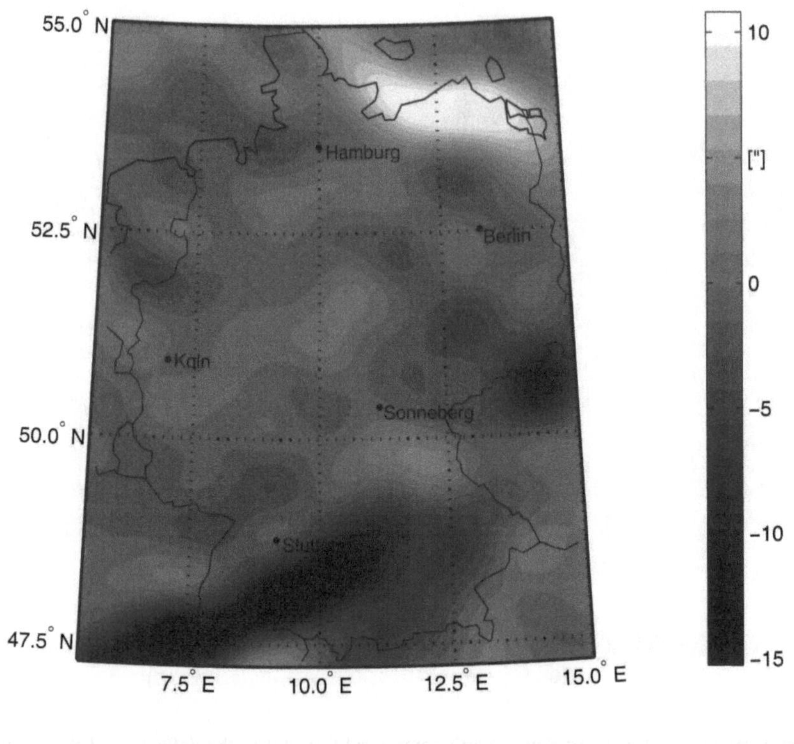

Fig. 8 Global portions of the South-North component ξ of the deflection of the vertical on the International Reference Ellipsoid of Germany with respect to SEGEN and WGD 2000, Albers projection of the ellipsoid-of-revolution.

Table 1. Statistical properties of computed global and regional vertical deflections.

	global		regional	
	η	ξ	η	ξ
max	47.30	49.60	8.90	12.40
min	-54.20	-75.00	-7.10	-15.30
WA	0.00	-0.13	-1.07	-1.27
DWA	1.40	1.26	0.44	0.68

WA, weighted average; *DWA*, deviation of weighted average

At this point we want to highlight, that it is now possible to compute global portions of vertical deflections synthetically which have the same order of magnitude like geometrically observed ones. This results from the excellent approximation of the local vertical by its reference quantity generated by the Somigliana-Pizzetti field. Here we follow the recommendations of the *International Association of Geodesy* (Moritz 1984) to use the Somigliana-Pizzetti potential of a level ellipsoid as the reference potential.

References

Ardalan AA, Grafarend EW (2000) Reference ellipsoidal gravity potential field and gravity intensity field of degree/order 360/360 (manual of using ellipsoidal harmonic coefficients 'Ellipfree.dat' and 'ellipmean.dat'). http://www.uni-stuttgart.de /gi/research/paper/coefficients/ coefficients.zip

Ardalan AA, Grafarend EW (2001) Somigliana-Pizzetti gravity: the international gravity formula accurate to the sub-nanoGal level. J Geod 75: 424-437

Asche W van (1991) Orthogonal polynomials, associated polynomials and functions of the second kind. J Comp Applied Math 37: 237-249

Balmino G, Barriot J, Koop R, Middel B., Thông NC, Vermeer M (1991) Simulation of gravity gradients: a comparison study. Bull Géod 65: 218-229

Chandrasekhar S, Roberts PH (1963) The ellipticity of a slowly rotating configuration. Astrophys J 138: 801-808

Chandrasekhar S (1969) Ellipsoidal figures of equilibrium. Yale University Press, New Haven

Ekman M (1996) The permanent problem of the permanent tide: what to do with it in geodetic reference systems. Mar Terres 125: 9508-9513

Gleason DM (1988) Comparing corrections to the transformation between the geopotential's spherical and ellipsoidal spectrum. Manuscr Geod 13: 114-129

Gleason DM (1989) Some notes on the evaluation of ellipsoidal and spheroidal harmonic expansions. Manuscr Geod 14: 114-116

Grafarend EW (2001) The spherical horizontal and the spherical vertical boundary value problem – vertical deflections and geoidal undulations – the completed Meissl diagram. J Geod 75: 363-390

Grafarend EW, Ardalan AA (1999) World Geodetic Datum 2000. J Geod 73: 611-623

Grafarend EW, Hauer KH (1978) The equilibrium figure of the Earth I. Bull Géod 52: 251-267

Grafarend EW, Heidenreich A (1995) The generalised Mollweide projection of the biaxial ellipsoid. Bull Géod 69: 164-172

Grafarend EW, Finn G, Ardalan AA (2002) Ellipsoidal vertical deflections and ellipsoidal gravity disturbances: case studies. J Geod, received: July 8th 2002

Jacobi CGJ (1834) Über die Figur des Gleichgewichts. Poggendorf Annalen der Physik und Chemie 33:229-238

Jekeli C (1988) The exact transformation between ellipsoidal and spherical harmonic expansions. Manuscr Geod 13: 106-113

Jekeli C (1999) An analysis of vertical deflections derived from high-degree spherical harmonic models. J Geod 73: 10-22

Maclaurin C (1742) A treatise on fluxions. Edinburgh

Mikolaiski HW (1989) Polbewegung eines deformierbaren Erdkörpers. Ph. D. Thesis, Bayerische Akademie der Wissenschaften, Deutsche Geodätische Kommission, Reihe C Heft 354, München

Moritz H (1968a) Density distributions for the equipotential ellipsoid. Report 115, Department of Geodetic Science, Ohio State University, Columbus

Moritz H (1968b) Mass distributions for the equipotential ellipsoid. Bolletino di Geofisica Teorica ed Applicata 10: 59-65

Moritz H (1973) Computation of ellipsoidal mass distributions. Report 206, Department of Geodetic Science, Ohio State University, Columbus

Moritz H (1984) Geodetic Reference System 1980. The geodesist's handbook. Bull Géod 58: 388-398

Moritz H (1990) The figure of the Earth. Wichmann Verlag, Karlsruhe

Moritz H, Mueller I (1987) Earth rotation, theory and observation. Ungar, New York

Novák P, Šimek J, Kostelecký J (2003) Characteristics of the Earth's gravity field for Central Europe in context of some geophysical phenomena. Presented at the 28th EGS General Assembly, Nice, 2003, Geophysical Research Abstracts, Vol. 5, 12011

Sansò F, Sona G (2001) ELGRAM: an ellipsoidal gravity model manipulator. Bolletino di Geodesia e Scienze Affini 60:215-226

Skrzipek MR (1998) Polynomial evaluation and associated polynomials. Num Math 79: 601-613

Sona G (1996) Numerical problems in the computation of ellipsoidal harmonics. J Geod 70: 117-126

Thông NC, Grafarend EW (1989) A spheroidal model of the terrestrial gravitational field. Manuscr Geod 14: 285-304

Harmonic Maps

E.W. Grafarend
Department of Geodesy and Geoinformatics, University of Stuttgart
Geschwister-Scholl-Strasse 24D, 70174 Stuttgart, Germany

Abstract

Harmonic maps are generated as a certain class of optimal map projections. For instance, if the distortion energy over a Meridian Strip of the *International Reference Ellipsoid* is minimized we are led to the *Laplace-Beltrami vector-valued partial differential equation*. Here we construct *harmonic functions* $x(L,B)$, $y(L,B)$ given as functions of ellipsoidal surface parameters (L,B) of type {Gauss ellipsoidal longitude L, Gauss ellipsoidal latitude B} as well as $x(\ell,q)$, $y(\ell,q)$ given as functions of relative isometric longitude $\ell = L - L_0$ and relative isometric latitude $q = Q - Q_0$ gauged to a vector-valued boundary condition of special symmetry. {Easting, Northing} or $\{x(b,\ell),\ y(b,\ell)\}$ and the distortion energy analysis of the new harmonic map are presented as well as case studies for (i) $B \in [-40^\circ, +40^\circ], L \in [-31^\circ, +49^\circ], B_0 = \pm 30^\circ, L_0 = 9^\circ$ *and* (ii) $B \in [46^\circ, 56^\circ]$, $L \in \{[4.5^\circ, 7.5^\circ]$; $[7.5^\circ, 10.5^\circ]$; $[10.5^\circ, 13.5^\circ]$; $[13.5^\circ, 16.5^\circ]\}$, $B_0 = 51^\circ$, $L_0 \in \{6^\circ, 9^\circ, 12^\circ, 15^\circ\}$.

The Analytical Solutions of the Dirichlet and Neumann Boundary Value Problems with Ellipsoidal Boundary

J. Yu, L. Lu
Zhengzhou Institute of Surveying and Mapping, Zhengzhou, Henan, 450052, P.R.China
E-mail: yujinhai@public2.zz.ha.cn

C. Jekeli
Dept. of Civil and Environment Engineering and Geodetic Science, The Ohio State University
Columbus, Ohio, 43210, U.S.A.

Abstract

The analytical solutions of the exterior boundary value problems of the disturbing potential with ellipsoidal boundary are derived with an accuracy of $O(\varepsilon^4)$; for both the Dirichlet and the Neumann boundary value problems, where ε^2 is the square of the second eccentricity of the ellipsoid. In addition, an arithmetic example is implemented to verify that the analytical solutions improve the accuracy from $O(\varepsilon^2)$ to $O(\varepsilon^4)$ compared to the spherical approximation. The solutions are given as integrals of closed-form, analytic Green's functions and are particularly suited to local applications. The singularities of kernel functions for the Dirichlet and the Neumann boundary value problem are equivalent to the Poisson kernel and r^{-1} respectively.

Applications of Normal Mode Relaxation Theory to Solid Earth Geophysics

R. Sabadini
Dipartimento di Scienze della Terra – Sezione Geofisica
Università degli Studi di Milano – Via Cicognara, 7 – 20129 Milano, Italy

Abstract

The normal mode relaxation theory is developed from basic principles for a spherical, self-gravitating and stratified Earth, and is targeted towards new perspectives on the physics of the Earth's interior for graduate students and researchers working in the fields of geophysics and geodesy. This theory allows us to study our planet in an integrated fashion, linking the physics of its interior to the geophysical and geodetic techniques that record, over a broad spectrum of spatial wavelengths, the ongoing modifications in the shape and gravity field of the planet. Basic issues related to the rheological properties of the Earth's mantle and to its slow deformation are introduced, in both mathematical and physical terms, within the framework of an analytical normal mode approach. After the fundamentals of this theory developed in the first, tutorial part, some applications will be presented, in a wide range of applications, ranging from changes in the Earth's rotation to post-seismic deformation and sea-level variations induced by post-glacial rebound. In the study of the physics of the Earth's interior, the normal mode relaxation theory bridges the gap between seismology and geodynamics.

Quantitative rheological Models

In modeling a particular geophysical phenomenon, the choice of the rheology used depends on 1) mathematical difficulty, 2) the quality of the geophysical data which the calculations of the model are required to match and 3) our knowledge of the rheological behavior of the medium at hand. Over the last few decades considerable amount of knowledge has been gained about mantle rheology in terms of the values of the rheological parameters and deformation mechanisms. For instance, what is most important, as far as mantle convection is concerned, is clearly the strong temperature dependence of the viscosity which the laboratory-derived values of the activation energy and volume seem to suggest. This intense interest in understanding convection in a fluid with markedly temperature-dependent viscosity is attested by the recent fundamental studies by geophysicists using analytical, numerical and experimental methods. In what follows, however, rather than discoursing in the topics of mantle rheology and mantle convection we will try to address the main questions that are at issue in attempting to study transient and long time scale geodynamic phenomena in a wide arc of time scales, ranging from years, characteristic of post-seismic deformation, to hundreds of millions of years as in the case of true polar wander driven by subduction, making use of the analytical normal mode theory in viscoelasticity with different models of mantle rheology. One of the key questions is whether one can devise a constitutive law which can satisfactorily model all these phenomena, from the anelastic, transient regime to the steady-state domain. The appropriate constitutive relation which is to be employed in analyzing transient geodynamic phenomena, such as postglacial rebound, is currently a matter of controversy in geophysics. Advocates of the non linear rheology use as their supportive arguments the laboratory data of single-crystal olivine whose power law index is about three. But there is now mounting evidence that at the stress levels obtained in postglacial rebound, the creep mechanism, in fact, may be linear for polycrystalline aggregates, since grain boundary processes, such as Coble creep, may become dominant. There are also recent theoretical studies which indicate that the power law index changes gradually with stress and hence the transition stress which marks the boundary between linear and nonlinear behavior is not as sharply defined as has previously been thought. Indeed, a proper mathematical formulation of the mixed initial and boundary-value problems associated with nonlinear viscoelasticity is a formidable one, fraught with numerical difficulties. It is also important to note that there is no unambiguous evidence in either the postglacial rebound event or in other types of geodynamic data which absolutely requires a nonlinear viscoelastic rheology, in spite of claims to the contrary. For these reasons, geophysicists tend to prefer the simple linear models in viscoelasticity, which allow for a considerably simpler mathematical treatment of the dynamics. Moreover, the linear approach also allows one to study easily the potentially interesting effects of the interaction between transient and steady-state rheologies. The simplest viscoelastic model which can describe the Earth as an elastic body for short time scales and as a viscous fluid for time scales characteristic of continental drift is that of a linear Maxwell solid. The speed for shear wave propagation depends on the square root of the instantaneous rigidity, whereas the strength of mantle convection depends inversely upon the magnitude of the steady-state viscosity. A powerful method of solving transient problems of linear viscoelasticity has been the use of the Correspondence Principle which allows one to employ the elastic solution of a given problem in the Laplace-transformed version of the corresponding viscoelastic problem.

Multi-layer Models

Multi-layer, spherically stratified, self-gravitating relaxation models with a large amount of layers (more than 100) can be dealt with analytically. Relaxation processes are studied for both Heaviside surface loads and tidal forcings. Simulations of the relaxation process of a realistic Earth model with an incompressible Maxwell rheology show that models containing about 30 to 40 layers have reached continuum limits on all timescales and for all harmonic degrees up to at least 150 whenever an elastic lithosphere is present, irrespective of the viscosity profile in the mantle. Especially fine-graded stratification of the shallow layers proves to be important for high harmonic degrees in these models.

Detection of the time dependent gravity field and Global Change

In the dynamical Satellite Laser Ranging data analysis approach the satellite orbit can be used as a gravity probe to monitor the time varying gravity field. Comparison of these zonal rates with the results of the viscoelastic Earth models forced by Pleistocene deglaciation, shows that the SLR retrieved even and odd zonals can be used today to infer the mantle viscosity and lithospheric thickness. In order to explore the capability of our SLR retrieved zonals to infer the upper mantle viscosity and the lithospheric thickness, we model the viscoelastic response of our planet to Pleistocenic deglaciation. These results show that SLR retrieved low degree geoid components can be used in principle to constrain the rheology of the mantle and lithospheric thickness. Discrepancies in the viscosity values necessary to fit the zonal rates when Pleistocenic deglaciation is the only forcing mechanism is a strong indication that mass redistribution is actually occurring over the Earth, eventually associated with mass instabilities in Greenland and Antarctica.

Geodesy and the Problem of Ice Sheets

I. Velicogna,
e-mail: isabella@colorado.edu
Dept. of Colorado & CIRES, Univ. of Colorado

J. Wahr
Dept. of Colorado & CIRES, Univ. of Colorado

Abstract. In recent years, great improvements have been made toward understanding the modern dynamics and recent history of the ice sheets. Several recently-launched satellite missions promise to make geodesy the most powerful tool for investigation of the changing ice sheets, including their past history and their present behavior. Mathematical description of ice sheet behavior from geodetic data requires accurate modeling of all the processes which may affect the measurements. Most geodetic tools measure changes in elevation of the ice sheets, which can include Post Glacial Rebound (PGR), the current Ice Mass Trend (IMT) consisting of both accumulation and glacial outflux, and processes of compaction within the firn column. Consequently it is necessary for mathematical models of geodetic data to separate the effects of IMT, PGR, and compaction. Satellite measurements of the time-variable geoid are insensitive to compaction effects and depend on IMT and PGR differently than do height measurements. Two methodological approaches have been proposed to separate these effects using measurements of height and time-variable geoid: 1- direct inversion for ice mass variability (Wu et al., 2002), which requires *a priori* assumptions about either the Earth's rheology or the ice load history; 2- iterative solution for the fields, which theoretically is more approximate but is computationally much simpler and less dependent on *a priori* assumptions. In this paper we analyze how we can learn about IMT and PGR by combining geodetic measurements, and we assess the conditions required to optimally combine satellite and ground-based data sets.

1 Introduction

Mathematical modeling is an indispensable tool in geodetic problems. However, geophysical problems often are not adequately constrained by modeling a single type of geodetic measurement because of the limitations of measurement uncertainties, spatial and temporal sampling and the large number of different geophysical processes which may contribute to a particular measurement. Often the formal inversion of a complex geophysical problem requires many implicit or explicit assumptions, the adoption of parameters with their own uncertainties, and significant computational time, all of which combine to make it very difficult (if not impossible) to realistically assess the uncertainties of the solution. Consequently, it can be useful to adopt an iterative or convergent approach to solving a geophysical problem, in which the *a priori* assumptions are minimized and a solution can be achieved with significantly less computational expense. In this article we present an example of such a method applied to satellite geodetic measurements of the time-variable geoid, ice sheet elevation, and rock elevation to study changes in the ice sheets. We assess the limitations of the method and explore additional data sets which may improve the results.

The Gravity Recovery And Climate Experiment (GRACE), jointly sponsored by NASA and the Deutsches Zentrum für Luft- und Raumfahrt (DLR), was launched in March, 2002. It will map the Earth's gravity field with unprecedented accuracy and resolution every 30 days during its 5-year lifetime. This will permit monthly variations in gravity to be determined down to scales of a few hundred kilometers and larger. These gravity variations can be used to study a variety of processes involving redistribution of mass within the Earth or on its surface. The expected performance of GRACE and various possible applications are described by Dickey et al. (1997) and Wahr et al. (1998). Among these applications is the use of the GRACE secular gravity signal to constrain models of post glacial rebound (PGR): the viscous adjustment of the solid Earth in response to the removal of the ice loads following the last ice age.

NASA's Ice, Cloud, and Land Elevation Satellite (ICESat) represents an important step toward estimating present-day Antarctic and Greenland ice mass balance. Launched in January of 2003, ICESat carries the Geoscience Laser Altimeter System (GLAS) and will have a mission lifetime of 3-5 years. To study the polar ice sheets, a laser pulse generated by the altimeter will reflect off the snow/ice surface and return to the satellite. Measurements of the round-trip distance, combined with Global Positioning System (GPS) measurements of the geocentric position of the spacecraft, will map changes in the surface elevation of the polar ice sheets at regular time intervals. The exact repeat period of the ground tracks will be 183 days, though there will be a 25-day near-repeat subcycle in which measurement locations shift by 15 km at the equator (and much less over the poles) from those made 25 days previously. Changes in the ice sheet elevation will be determined from crossover differences. Using GLAS data, it will be possible to estimate the rate of change in Antarctic ice mass over the mission lifetime between the ice sheet margins and 88°S latitude.

GLAS measurements of ice sheet elevation will reflect more than just the ice mass change. Uplift and subsidence due to post glacial rebound (PGR) and variable compaction of snow will also influence the height change measured by GLAS. *Velicogna and Wahr* [2002] showed that it is possible to combine the GLAS measurements of change in ice sheet elevation with time variable gravity measurements from GRACE to separate PGR and IMT. Moreover, by adding GPS measurements of vertical rock velocity it is possible to solve for PGR, IMT, and the temporal variation of density due to compaction within the ice column.

In this paper we will examine the state of the art of satellite measurements for investigation of the ice sheets. We will look in particular at methodologies for combining satellite measurements from GLAS, GRACE and GPS and at the limiting factors for IMT and PGR solution uncertainties.

2 Synthetic data

To simulate the recovery of Antarctic post-glacial rebound and ice mass trend from GRACE, GLAS, and GPS, we construct five years of synthetic satellite data as the sum of geophysical signals and measurement errors. Almost all of the contributions to these signals are identical to those described in *Wahr et al.* [2000] and *Velicogna and Wahr* [2002a,b], and a more detailed discussion of the simulated data can be found therein. We briefly summarize the contributions to simulated signals here.

Monthly ice sheet elevations in the simulated GLAS measurements consist of contributions from snow accumulation and horizontal ice flow in Antarctica, from the Earth's elastic response to loading by these mass fields, from Earth's viscoelastic response to past loading (i.e., PGR), and from GLAS measurement errors.

For GRACE, we simulate five years of monthly measurements of the geoid: the equipotential surface corresponding to mean sea level over the oceans. The geoid can be expanded in a spherical harmonic representation as [*Kaula*, 1966]:

$$N(\theta,\phi) = a\sum_{l=0}^{\infty}\sum_{m=-l}^{l}\tilde{P}_{lm}(\cos\theta)\{C_{lm}\cos m\phi + S_{lm}\sin m\phi\}, \quad (1)$$

where a is the Earth's mean radius, θ and ϕ are co-latitude and east longitude, C_{lm} and S_{lm} are dimensionless coefficients, and the $\tilde{P}_{lm}$ are normalized associated Legendre functions [e.g. *Chao and Gross*, 1987]. GRACE will deliver values of C_{lm} and S_{lm}, up to a maximum degree and order (i.e., l and m) of 100, every month. The simulated data include GRACE measurement errors, as well as the gravitational effects of snow accumulation and ice flow on Antarctica, of the elastic response to loading, of PGR, of redistribution of water mass in the ocean and on landmasses other than Antarctica, and of errors in correction for atmospheric pressure.

The simulated GPS data consist of five years of daily height coordinates. The coordinates include a constant vertical velocity contributed by PGR, the Earth's elastic loading response to accumulation and ice flow on Antarctica, and GPS measurement errors estimated to be 1.5 cm root-mean square (RMS) for daily coordinates.

Contributions of Antarctic snow accumulation to our simulated GLAS, GRACE, and GPS data are derived from monthly precipitation in the National Center for Atmospheric Research (NCAR) Climate System Model (CSM-1) [*G. Bonan, personal communication*, 1997; see, e.g., *Boville and Gent*, 1998]. For more detail see *Wahr et al.* [2000] and *Velicogna and Wahr* [2002]. To incorporate snow and ice effects into the GRACE simulated geoid we sum the above (detrended) contribution with *Bentley and Giovinetto*'s [1991] estimate of the spatially varying long-term mean ice mass trend. To simulate changes in GLAS ice heights corresponding to the change in mass fields, we use the density properties of snow and ice modified by the time- and accumulation-dependent compaction model of *Wingham* [2000]. We estimate the contribution from the PGR vertical velocity by convolving viscoelastic load Love numbers, computed as described by *Han and Wahr* [1995], with estimates of the Antarctic deglaciation history. We estimate the elastic vertical displacement U of the solid Earth caused by the mass load, using standard methods (see, e.g., equations (40) and (42) of *Mitrovica et al.* [1994]).

3 Method

To estimate IMT and PGR from the simulated data, we use an iterative method in which we try to avoid dependence on any *a priori* assumptions about Earth viscosity and ice loading history. The iterative approach is analogous to that described by *Velicogna and Wahr* [2002]. Initially we assume that GLAS is sensitive only to the ice thickness change and not to the PGR uplift, and we determine the ice thickness change using only the GLAS elevation data. We refer to this initial estimate as the zeroth iteration.

We compute the secular rate of change in the geoid caused by the zeroth order GLAS mass balance estimate and remove that geoid signal from the GRACE data. The residual secular geoid change is then interpreted as being due entirely to PGR. We use the geoid residual to estimate the corresponding crustal uplift due to PGR, using the method of *Wahr et al.* [2000], in which we let

$$\dot{N}^{PGR}(\theta,\phi) = a\sum_{l=0}^{\infty}\sum_{m=0}^{l}\tilde{P}_{lm}(\cos\theta)[\dot{C}_{lm}^{PGR}\cos(m\phi) + \dot{S}_{lm}^{PGR}\sin(m\phi)] \tag{2}$$

be the secular rate of change in the geoid caused by PGR, where $\dot{C}_{lm}^{PGR}$ and $\dot{S}_{lm}^{PGR}$ are the Legendre expansion coefficients of $\dot{N}^{PGR}$. *Wahr et al.* [2000] found that to a high degree of approximation, the relations between the rates of change of the geoid coefficients $\dot{C}_{lm}^{PGR}$ and $\dot{S}_{lm}^{PGR}$, and those of the uplift rate, $\dot{A}_{lm}^{PGR}$ and $\dot{B}_{lm}^{PGR}$, are

$$\dot{A}_{lm}^{PGR} = \left(\frac{2l+1}{2}\right)\dot{C}_{lm}^{PGR}, \tag{3}$$

$$\dot{B}_{lm}^{PGR} = \left(\frac{2l+1}{2}\right)\dot{S}_{lm}^{PGR}. \tag{4}$$

from which

$$\dot{U}_{GLAS/GRACE}^{PGR}(\theta,\phi) = a\sum_{l=0}^{\infty}\sum_{m=0}^{l}\left(\frac{2l+1}{2}\right)\tilde{P}_{lm}(\cos\theta)[\dot{C}_{lm}^{PGR}\cos(m\phi) + \dot{S}_{lm}^{PGR}\sin(m\phi)] \tag{5}$$

The explanation for (3) and (4) is that the change in the geoid caused by PGR is mostly due to mass anomalies associated with vertical motion of the surface (see *Wahr et al.* [1995]). We estimate PGR uplift from the residual (GRACE minus the IMT estimate of geoid) using equation (5). Subtracting this PGR estimate from GLAS then results in a better estimate of ice balance. We refer to this estimate as the first iteration. We then repeat the process: compute the geoid contributions from this new GLAS ice mass estimate and remove them from GRACE, interpret the secular component of the new GRACE residuals as the effect of PGR, calculate the PGR uplift and remove it from GLAS, and use the GLAS residuals to construct a further improved ice balance estimate. We iterate this procedure eight times, after which the improvement is negligible.

Then, following the last iteration, the PGR error is estimated from the difference between GPS vertical velocities (slope of the daily height coordinates) and the PGR estimate $\dot{U}^{PGR}$ derived by applying equation (5) to the secular GRACE geoid minus the geoid from the estimated IMT. That PGR error is used to estimate the ice compaction trend (see below), and a corresponding correction to the estimate of IMT. In fact we are really trying to solve for three unknowns: IMT, PGR, and the trend of the time-variable density of the ice column. Continuous GPS point measurements of vertical rock velocity add the additional constraint necessary to solve for the third unknown field.

We expect that the GPS measurements available to estimate the PGR correction will be sparse and irregularly distributed, so that assimilating them into the data constraints will require some care. We first interpolate the gridded $\dot{U}_{GLAS/GRACE}^{PGR}$ to each GPS location using two-dimensional optimal interpolation. The PGR correction at a GPS site is simply $\Delta\dot{U}_{compaction}^{PGR} = \dot{U}_{GPS}^{PGR} - \dot{U}_{GLAS/GRACE}^{PGR}$. Then, we interpolate the PGR correction back to the grid points using an optimal interpolation. To downweight the correction in undersampled regions, we multiply the interpolated correction at each grid point by a Gaussian function, $W(\alpha)$, given by (7), where R in (8) is 500 km and α is the angular distance to the nearest GPS site (see section 7 for more details about the choice of R). This effectively localizes the estimate of PGR error near the GPS sites where it is measured. As we will address further in the discussion section, the accuracy of the estimate of PGR error calculated using this method depends fundamentally on the correlation length scale of the time variable compaction effects that cause the error.

Local correction of the GLAS/GRACE PGR estimate with GPS vertical velocities improves the PGR estimate, but subtracting the corrected PGR estimate from GLAS heights does not dramatically improve the estimate of IMT. However, the empirical relation between the IMT compaction error and the compaction error in PGR $\Delta\dot{U}^{PGR}_{compaction} = -0.31\Delta\dot{h}^{IMT}_{compaction}$ (see the discussion at the end of section 4), allows us to use the GPS-derived PGR errors to also estimate the compaction error. In practice, we use the iterative routine that combines GRACE and GLAS (see description in section 3) and after the last iteration we use the GPS to correct the GLAS/GRACE estimate of PGR by removing the difference between GPS and GLAS/GRACE PGR vertical velocities ($\Delta\dot{U}^{PGR}_{compaction}$) estimated as described above. Then we correct the IMT by removing the PGR error and the compaction error estimated as the PGR error divided by 0.31. In this manner we effectively solve for all three unknowns in the Antarctic mass balance (i.e., IMT, PGR, and the trend of time-variable density within ice columns). Once we have obtained the PGR and IMT estimates, we smooth those fields using the 250-km Gaussian averaging function defined in (7).

4 Uncertainties in PGR and IMT

4.1 Time-variable accumulation and IMT

ICESat and GRACE will have lifetimes of about five years. Interannual and interdecadal variations in accumulation rate cause the mass trend on five year time scales to differ from the century scale trend. Part of this difference occurs because the normal variability of climate includes nonsecular components with periods greater than five years. For GRACE, which is directly sensitive to mass change, the difference between the five year and century-scale trends is the only contribution of the time-variable accumulation to the error in estimating the century-scale trend. For GLAS, which is sensitive to ice sheet thickness rather than to mass, there is the additional problem that the relation between thickness and mass is not simple multiplication by ρ_i but is complicated by variable compaction of the snow-ice column. For the application at hand, we will not concern ourselves with the century scale trend but consider only those uncertainties which may be introduced in our estimates of IMT on the five-year timescale over which we will have measurements.

Because the density profile in the upper layers of the snow-ice column depends on prior accumulation rate, the assumption of a constant ice density ρ_i introduces an error in the estimate of IMT from GLAS data. Error in the GLAS estimate of mass balance caused by approximating the time- and accumulation-dependence of snow compaction with $\dot{h} = \dot{m}/\rho_i$, we will refer to as "compaction error". This error affects the GLAS estimate of the five-year trend. *Wahr et al.* [2000] conclude that the compaction error in estimates of the five-year ice mass trend using GLAS alone is likely to be ± 4.5 mm yr^{-1} (water thickness equivalent) when averaged over the entire Antarctic ice sheet, which is equivalent to an error of about ± 0.15 mm yr^{-1} in estimates of global sea level rise.

4.2 Time-variable accumulation and PGR

PGR over Antarctica is poorly known and could contribute on the order of 5 mm yr^{-1} in error to GLAS-only estimates of five-year IMT averaged over Antarctica. The inclusion of GRACE data permits the separation of the PGR and IMT signals, and so reduces the contribution of the unmodeled PGR signal to the IMT error. *Wahr et al.* [2000] find that the GRACE/GLAS iteration described in section 3, without inclusion of the GPS data, would remove all of the PGR contribution to the IMT error if there was no time variable accumulation. However, errors introduced by time variable compaction result in erroneous estimates of mass from GLAS heights, and the mass inconsistency leads to an error in the final estimate of PGR contribution to the surface heights. Thus the GRACE residuals used to infer the PGR signal suffer indirect effects of the compaction error. *Wahr et al.* [2000] find that the final PGR contribution to the IMT error, averaged over the entire ice sheet, is approximately 0.31 times the IMT compaction error. This proportionality factor is an artifact of the method used to combine the GLAS and GRACE data. Specifically, it arises from combining equation (15) of *Wahr et al.* [1998] to estimate the geoid from the GLAS surface mass (i.e. IMT) estimate, with (5) above, relating the GRACE PGR geoid to an inferred uplift rate. The proportionality factor between PGR and IMT after any single iteration is about equal to $d = 0.24$. After N iterations, the proportionality factor = $d(1 - d^N)/(1 - d) \approx 0.31$ when N is large. Hence, errors in the recovery of IMT and PGR from GLAS ice sheet elevations plus the GRACE geoid arise principally because only two observables are used to determine three unknowns (IMT, PGR, and the time-variable density of the ice column).

4.3 Spatial variability

The PGR and IMT estimates derived from GLAS and GRACE data can have large errors at short wavelengths, because the GRACE measurements of $\dot{C}_{lm}^{PGR}$ and $\dot{S}_{lm}^{PGR}$ become increasingly inaccurate as l gets large. To mitigate those short wavelength errors, we smooth the PGR and IMT results by constructing Gaussian averages of those fields. The Gaussian average of the recovered PGR field is [*Wahr et al.*, 1998]:

$$\overline{\dot{U}}_{GLAS/GRACE}^{PGR}(\theta,\phi) = \int \sin\theta'\, d\theta'\, d\phi'\, W(\alpha)\, \dot{U}_{GLAS/GRACE}^{PGR}(\theta',\phi') \quad (6)$$

where $\dot{U}_{GLAS/GRACE}^{PGR}$ is the unsmoothed field, and the averaging function is

$$W(\alpha) = \frac{b}{2\pi} \frac{\exp\left[-b(1-\cos\alpha)\right]}{1-e^{-2b}} \quad (7)$$

where α is the angular distance between (θ,ϕ) and (θ',ϕ'), and

$$b = \frac{ln(2)}{(1-\cos(R/a))} \quad (8)$$

with R the distance along the Earth's surface at which W has dropped to 1/2 its value at $\alpha = 0$ [see *Jekeli*, 1981, eqn. (59)]. We apply this smoothing process to the recovered PGR and IMT fields, using R = 250 km. Thus, the end products of our estimation algorithm can best be described as the PGR and IMT fields smoothed over 250 km scales. To determine the accuracy of these smoothed, recovered fields, we also apply this smoothing process to the input PGR and IMT fields used to construct our simulated data, and we assess the difference between those smoothed input fields and the smoothed recovered fields.

5 PGR and IMT from GLAS, GRACE and GPS Data

Compaction error is by far the largest source of uncertainty in the GLAS/GRACE estimates of PGR and IMT, and GPS measurements of vertical rock velocity will provide an important constraint of this field. In our simulations we begin by assuming that continuous GPS measurements are made at 45 existing and proposed sites, including previous campaign sites along the Antarctic coast [*Dietrich*, 2001] and existing and proposed continuous sites along the Transantarctic Mountains and Marie Byrd Land in West Antarctica [*Raymond et al.*, 1998]. 36 of the 45 sites used are on the Antarctic mainland, while the remainder are on islands at high southern latitudes. Figure 1 shows the spatially-varying recovery of smoothed IMT after assimilation of the GPS vertical velocities, for a five year period. By incorporating the GPS data, we obtain a smoothed PGR estimate with just 2.8 mm yr^{-1} of RMS error (Table 5), a 49% improvement relative to that recovered from GRACE and GLAS data alone. The corresponding estimate of the smoothed IMT has 12.0 mm yr^{-1} of RMS error, a 35% improvement. The similarity of the error reductions for IMT and PGR indicates that the empirical scaling used to relate PGR error to IMT compaction error is justified. Using this distribution of 45 GPS sites, we calculated estimates for all of the independent five year periods of the (300-year long) CSM-1 model. The RMS average of RMS errors in recovery of the simulated fields was 3.4 mm yr^{-1} of PGR error and 15.9 mm yr^{-1} error in IMT. The PGR error ranged from 1.4 to 5.4 mm yr^{-1} and the error in the IMT estimate ranged from 5.8 to 26.9 mm yr^{-1}.

	PGR mm yr^{-1}	IMT mm yr^{-1}
No Time Variable Accumulation – – No GPS	1.3	2.1
No GPS	5.3	19.9
Current+ proposed+ campaign GPS	3.4	15.9
10 GPS sites	2.6	13.2
50 km spacing GPS sites	1.5	7.9

Table 1. RMS error of the smoothed PGR and IMT fields estimated by GLAS/GRACE, with and without GPS.

Because the largest error source in the PGR and IMT estimates is time variable accumulation, we also examine how the estimates of IMT and PGR are affected by using just a few GPS sites located where the compaction error is largest. Figures 2 shows the IMT recovery using only ten GPS sites. The compaction error is largest where the accumulation rates are largest. The results for the PGR after locating the GPS measurements near the largest accumulation centers are shown in Figure 3. With a configuration of only ten GPS sites, the RMS error in the smoothed PGR estimate is 2.3 mm yr^{-1}, for the same example five year period but using 45 GPS, Figure 4 is 2.8 mm yr^{-1}. This is a reduction of 58% from

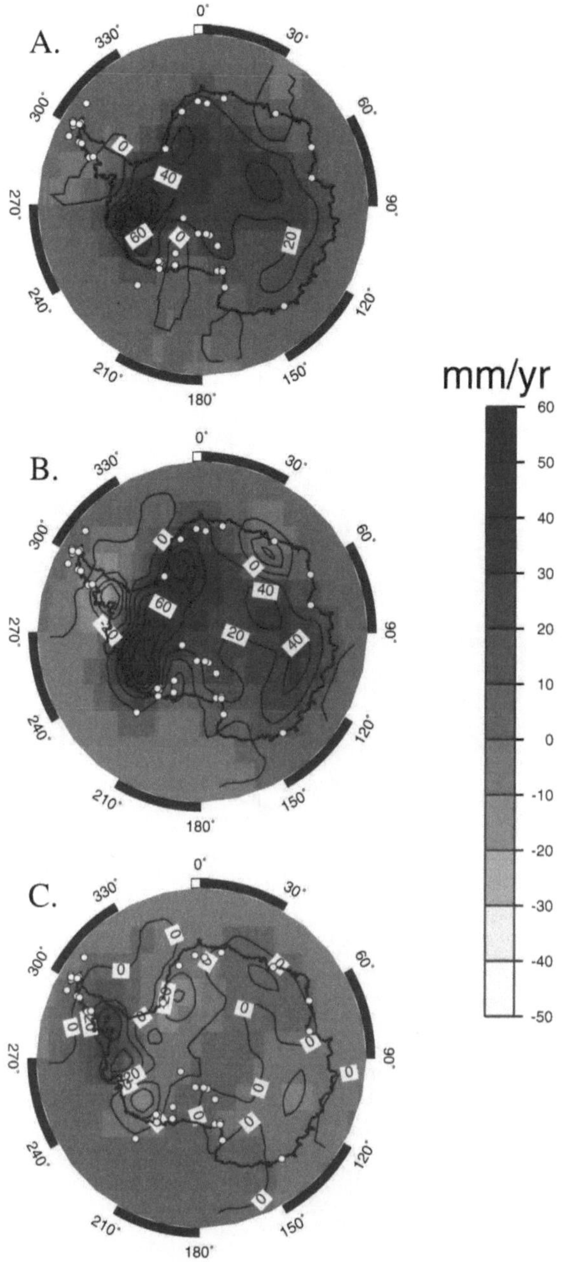

Fig. 1 Smoothed Ice Mass Trend (IMT) signal. (a) Input signal. (b) Recovered signal combining GLAS, GRACE and GPS data from 45 existing and planned continuous and campaign GPS sites. (c) Residual. Contour interval is 20 mm yr^{-1}. Circles denote GPS locations.

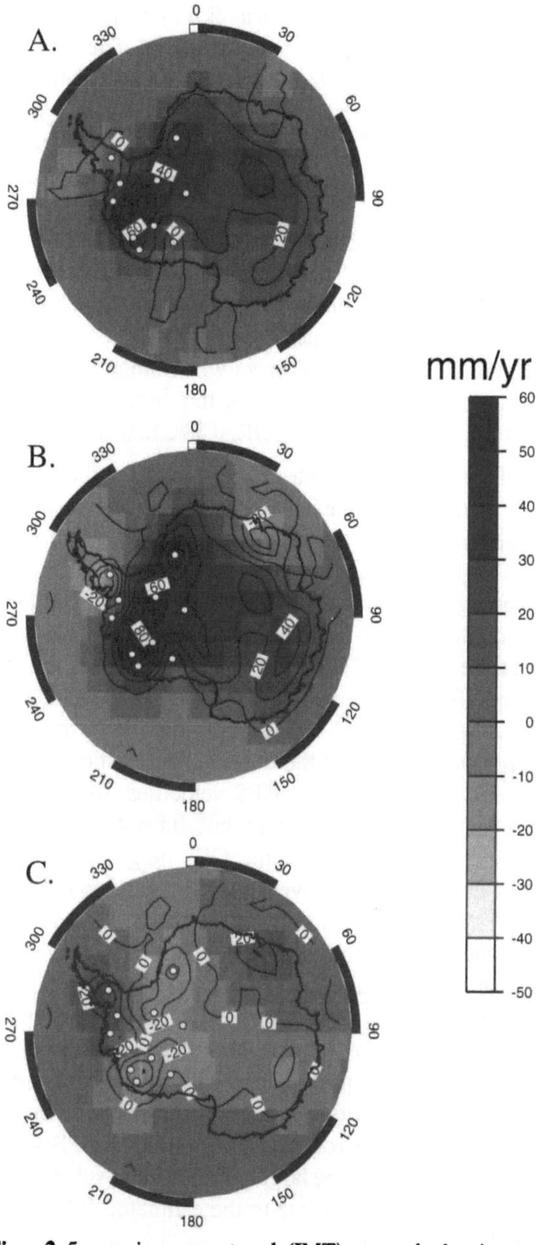

Fig. 2 5-year ice mass trend (IMT), smoothed using a Gaussian averaging function with 250 km radius. (a) Input IMT signal summing two contributions: the century scale IMT from *Bentley and Giovinetto* [1991] plus the 5-year trend from the time-variable accumulation. (b) Recovered signal combining GLAS and GRACE data. (c) Residual. Contour interval is 20 mm yr^{-1}.

the GLAS/GRACE estimate. The RMS error of the smoothed IMT estimate is 10.5 mm yr^{-1}, a similar 43% reduction. When calculated for all of the independent five year periods of CSM-1, the RMS average of the RMS errors was 2.6 mm yr^{-1} of PGR error and 13.2 mm yr^{-1} error in IMT. The PGR error ranged from 1.5 to 3.6 mm yr^{-1} and the error in the IMT estimate ranged from 6.1 to 19.1 mm yr^{-1}. However, even if we ignore the logistical difficulties of siting continuous GPS instruments in the Antarctic interior, the locations where the accumulations are largest are among the least likely places to find exposed bedrock. Moreover, there is no reason to expect that the precipitation rate always will be highest at the locations indicated in Figure 2c. The loci of the maximum compaction errors will change with time, and we cannot predict where the compaction error will be largest during the GRACE/GLAS missions.

To obtain an extreme lower bound for the recovery errors, we also examined the accuracy of signals recovered in the unlikely limit of very dense GPS coverage. By putting GPS sites at a regular 50 km spacing everywhere south of $-60°$, we retrieved the smoothed PGR with RMS error of just 1.5 mm yr^{-1} and the smoothed IMT with 7.9 mm yr^{-1} for the example five year period shown in Figures 2–3 (Table 1). We suspected that a portion of the remaining error was a result of the uncorrected elastic load response contained in the GPS velocities. To test this we ran the simulation again but did not include the elastic load response in the GPS heights. The errors in the signal recovery were reduced to 0.1 mm yr^{-1} (PGR) and 7.7 mm yr^{-1} (IMT). We thus infer that the portion of the PGR RMS error caused by uncorrected secular elastic loading in the GPS signal is $\approx \sqrt{1.5^2 - 0.1^2}$ mm yr^{-1} = 1.5 mm yr^{-1}, and the corresponding portion of the IMT error is $\approx \sqrt{7.9^2 - 7.7^2}$ mm yr^{-1} = 1.8 mm yr^{-1}. We expect that the effects of the uncorrected elastic errors would be similar for any distribution of GPS sites, but that they could be largely removed by subtracting the elastic load response to the estimated IMT from the GPS time series.

6 Discussion

Our simulations show that by combining GLAS and GRACE data it should be possible to recover the spatially-varying Antarctic PGR and IMT signals, smoothed over 250 km scales, to accuracies of about 5 mm yr^{-1} and 20 mm yr^{-1}, respectively. These errors are due almost entirely to the compaction error associated with variability in the accumulation.

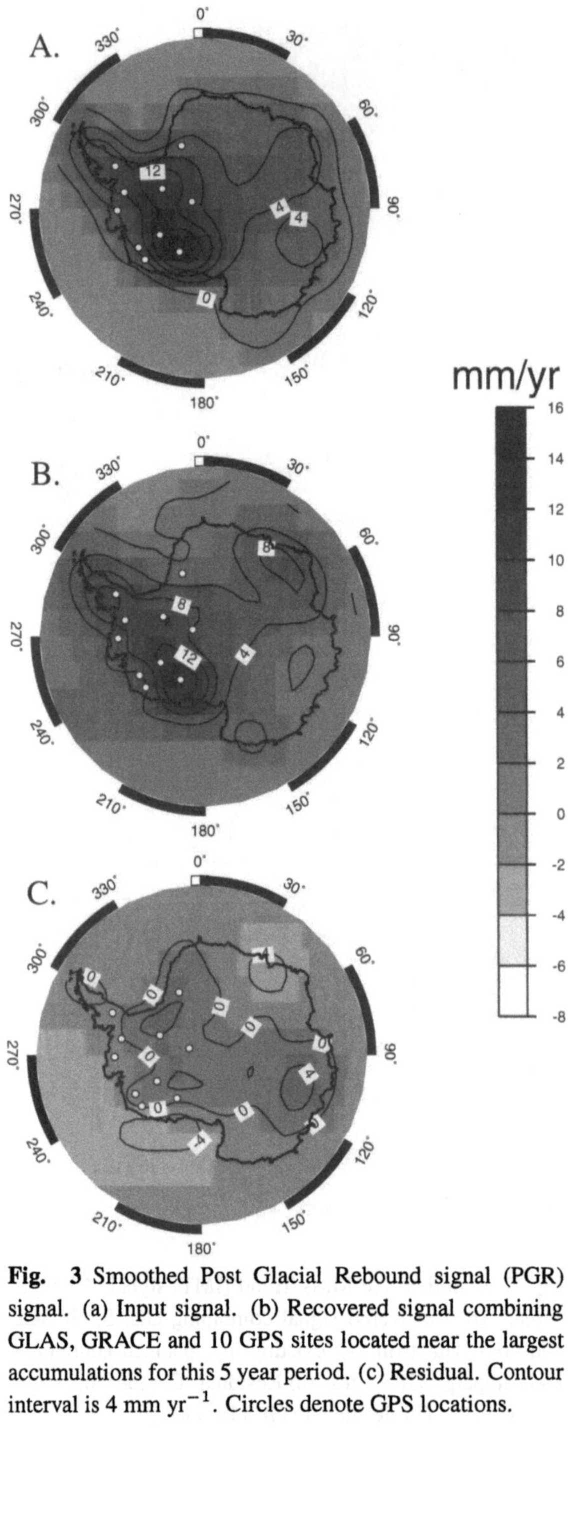

Fig. 3 Smoothed Post Glacial Rebound signal (PGR) signal. (a) Input signal. (b) Recovered signal combining GLAS, GRACE and 10 GPS sites located near the largest accumulations for this 5 year period. (c) Residual. Contour interval is 4 mm yr^{-1}. Circles denote GPS locations.

When GPS measurements are combined with the GLAS and GRACE data, the accuracy of the recovered fields is improved. This improvement depends on the distribution of the GPS receivers. When we assume there are continuous data from 45 existing and proposed GPS sites, the RMS errors in the smoothed PGR and IMT fields drop to about 3 mm yr^{-1} and 16 mm yr^{-1}, respectively.

When GPS data are added to the simulations, compaction error in the IMT is reduced by exploiting the relationship observed in *Wahr et al.* [2000], that compaction errors in PGR are 0.31 times the compaction errors in IMT. However, differences between the GPS vertical velocity and the GRACE/GLAS estimate of PGR that arise from sources other than the compaction trend will cause errors in the correction of IMT, and the correction scales these differences as approximately ϵ_{IMT}=(1+1/.31) ϵ_{PGR}=4.2 ϵ_{PGR}. One example of a difference between the GPS vertical velocity and the GRACE/GLAS estimate of PGR that does not result from the compaction trend arises because of the smoothing which must be applied to GRACE geoid estimates. The PGR signal used in this paper, when smoothed using a Gaussian averaging function with 250 km radius, differs from the unsmoothed PGR signal by ~1.4 mm yr^{-1} RMS. We are able to attenuate the effect of this difference somewhat by applying Gaussian smoothing to the estimate of the PGR correction, $\Delta \dot{U}^{PGR}_{compaction}$, before using it to correct the IMT.

All of the error estimates in this paper depend heavily on the assumed time-variable accumulation field and GPS receiver distribution. It is difficult to assess the accuracy of the CSM-1 accumulation fields [*Wahr et al.*, 2000]. We expect that Antarctic GPS receivers will be sparsely distributed, with the 45 sites representing a best-case scenario. Moreover we expect these sites will be much more sparsely sampled in time than the five years of daily coordinates we simulated. Among existing sites in the Transantarctic Mountains, the coordinate time series from MCM4 (the IGS site at McMurdo station) is quasi-continuous, but remote stations COAT and MTCX have just one to five months of data per year as a result of power and other system failures. Also many of the sites we incorporated are campaign sites, which may have only a few days to weeks of data per year. Vertical velocities estimated from the continuous GPS sites currently have errors of between ~0.4–5.9 mm yr^{-1}, as compared to 0.3 mm yr^{-1} RMS errors in recovery of PGR vertical velocity from the simulated GPS time series. Errors in GPS vertical velocity estimates of more than about 4.4 mm yr^{-1} would introduce errors into the correction of IMT that exceed the RMS errors in GLAS/GRACE recovery without GPS. However, estimates of vertical velocity at continuous sites will improve with more observation, and spatial averaging will reduce the error further in the more densely instrumented regions (e.g., the Transantarctic Mountains and Marie Byrd Land).

Another important point in applying this method relates to the true scales of correlation of the signals we are solving for. The GLAS, GRACE, and GPS data sets sense changes at different length scales, ranging from sparsely distributed point velocity measurements (GPS) to dense altimetric heights with 70-m footprints (GLAS) to surface mass density integrated over scales > 200 km (GRACE). Consequently the accuracy of the estimates of PGR, IMT, and the correction for the time-varying compaction trend depends on the length scales for which these signals are self-correlated. If all three signals self-correlate at scales consistent with the minimum resolution scale of GRACE and the scale of distribution of the GPS measurements, then we expect the results to be quite good. If on the other hand one of the signals is extremely variable at scales of a few hundred kilometers, we expect a poor result.

The PGR signal used in our simulation does have long characteristic length scales. This partially reflects the fact that the Antarctic component of the ICE-3G deglaciation model used to simulate this PGR signal is dominated by these same long scales. But even if the true deglaciation pattern had significant power at short wavelengths, viscoelastic rebound would be dominated by longer wavelengths because the stress induced by short wavelength loading tends to be concentrated near the Earth's surface, and so within the elastic lithosphere. Thus short wavelength variations in loading do not generate a significant viscoelastic relaxation response. *Velicogna and Wahr* [2002] estimate that, at minimum, the PGR signal is significantly self-correlated out to lengths of 250 km and decorrelation occurs at length scales > 700 kilometers, regardless of the scales of ice mass loading present in the deglaciation pattern.

Compaction error is clearly the principle error source in PGR and IMT recovery using this method. Using the same CSM-1 accumulation fields used to generate our simulated GLAS/GRACE/GPS data, we compared the compaction trends with Gaussian averages of those same trends using a 250 km averaging radius. We find that the Gaussian-averaged compaction trends are ~0.5 that of the unsmoothed,

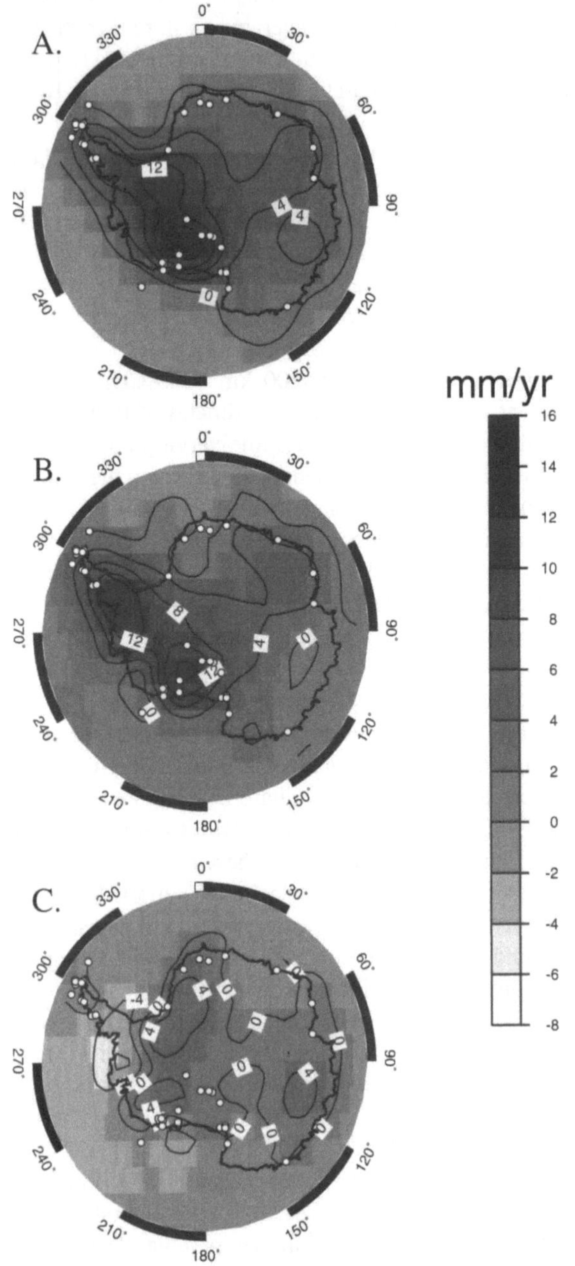

Fig. 4 Post Glacial Rebound (PGR) signal, smoothed using a Gaussian averaging function with 250 km radius. (a) Input signal. (b) aecoveredi signal combining GLAS and GRACE data. (c) Residual. Contour interval is 4 mm yr^{-1}.

suggesting that indeed the compaction predicted from the CSM-1 accumulation is correlated on these length scales. The self-correlation length scale of the compaction error depends on the corresponding scale of the time variable accumulation. It is possible that the spatial correlation properties of the CSM-1 model may not be representative of the true Antarctic accumulation. However, *Wingham et al.* [1998] note that five-year accumulation trends measured by the ERS-1 and ERS-2 satellite radar altimeters are correlated on length scales of $\sim$200 km, with complete decorrelation at $>$ 400 kilometer length scales.

There are several other data sources which might be used to supplement the estimates of the compaction trend in addition to GPS. The *Wingham* [2000] accumulation-dependent compaction model used to simulate ice compaction for this study could be applied to estimates of time variable accumulation during the five years of GRACE and GLAS measurements to improve estimates of compaction trend, but compaction depends significantly on the accumulation history over timescales of decades, so accumulation on five year time-scales will not suffice. The ERS altimetric data and time series of ground-based weather station measurements can be used to extend the record back through time, but assimilating these measurements will be complicated by incomplete sampling in time and/or space. Time variable accumulation measured from firn density in ice cores will provide a very valuable constraint, as these records are temporally complete over the past few centuries [*Oerter et al.*, 2000]. Like GPS measurements, these are point measurements but they can be sampled anywhere (including major accumulation centers) and so may provide even more useful constraint. Figure 5 shows the current sampling of Antarctic ice cores. Microwave remote sensing of ice sheet scattering may eventually help to constrain the compaction trend, but additional research into the microwave signature of firn densification will be needed first.

The simulations presented here assume that glacial outflow is constant with time. This is a questionable approximation, given that glacial flow rates appear to be influenced by basal "lubrication" by liquid water percolating through small fractures in the ice stream. Also, the ice sheet margins support a significant fraction of the gravitational force that drives ice flow. Margins can shift, probably in response to changing thermal and hydrological conditions at the base. These changes in ice-stream width alter the balance of forces and the rate of ice flow. Because we concern ourselves only with the overall trends in

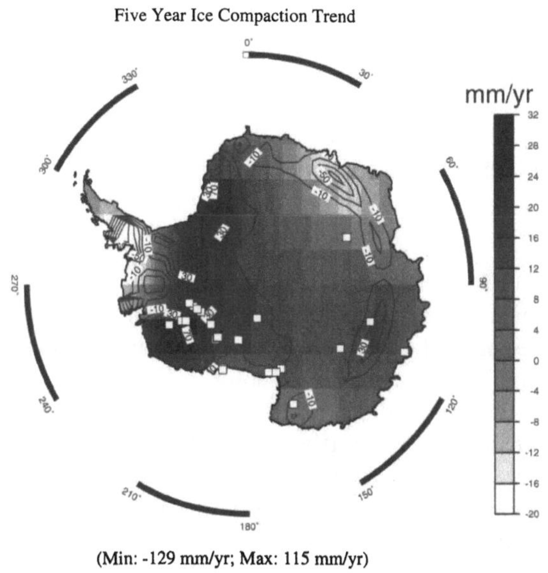

Fig. 5 The compaction trend, which will map directly into error when using GLAS ice heights only to estimate IMT. GRACE data will partially correct this error, but the addition of GPS and ice core data allows estimation of the compaction trend at the observation sites. White squares denote ice core locations.

the methodology presented here, the assumption of constant ice flow contributes negligibly to the overall error. However, as the modeling is augmented to allow estimation of the time-variable accumulation, this assumption will need to be re-evaluated.

7 References

Bentley, C.R., and M.B. Giovinetto, Mass balance of Antarctica and sea level change, in *Proceedings of the international Conference on the Role of the Polar Regions in Global Change*, 1990, edited by G. Weller, C.L. Wilson, and B.A. B. Seeverin, pp.481-488, Univ. of Alaska, Fairbanks, 1991.

Boville, B. A., and P. Gent, The NCAR climate system model, version one, *J. Clim., 11*, 1115-1130, 1998.

Dietrich, R., Present Status of the SCAR GPS Epoch Campaigns, *SCAR Report no. 20*, 15-18, 2001.

Han, D. and J. Wahr, The viscoelastic relaxation of a realistically stratified Earth, and a further analysis of post-glacial rebound, *Geophys. J. Int., 120*, 287-311, 1995.

Jekeli, C., Alternative methods to smooth the Earth's gravity field, *Rep. 327*, Dept. of Geod. Sci. and Surv., Ohio State Univ., Columbus, 1981.

Mitrovica, J.X., J.L. Davis, and I.I. Shapiro, A spectral formalism for computing three-dimensional deformations due to surface loads, 1, Theory, *J. Geophys. Res., 99*, 7057-7073, 1994.

Oerter, H., Wilhelms, F., Jung-Rothenhäusler, F., Göktas, F., Miller, H., Graf, W., Sommer, S., Accumulation rates in *Dronning Maud Land, Antarctica, as revealed by dielectric-profiling measurements of shallow firn cores, Annals of Glaciology*, 30, 27-34, 2000.

Raymond, C. A., E. R. Ivins, T. S. James and M. Heflin, The role of geodesy in assessing the state of the West Antarctic Ice Sheet. *West Antarctic Ice Sheet Chapman Conference*, 1998.

Velicogna, I., and J. Wahr, A method for separating Antarctic post-glacial rebound and ice mass balance using future ICESat Geoscience Laser Altimeter System, Gravity Recovery and Climate Experiment, and GPS satellite data, *J. Geophys. Res. Vol. 107*, 10.1029/2001JB000708, 2002a.

Velicogna, I., and J. Wahr, Post Glacial rebound and Earth's Viscosity Structure From GRACE, In Press *J. Geophys. Res. Vol. 107*, 10.1029/2001JB001735, 2002b.

Wahr, J.M., D. Han, and A. Trupin, Predictions of vertical uplift caused by changing polar ice volumes on a visco-elastic Earth, *Geophys. Res. Lett., 22*, 977-980, 1995.

Wahr, J., M. Molenaar, and F. Bryan, Time-Variability of the Earth's Gravity Field: Hydrological and Oceanic Effects and Their Possible Detection Using GRACE, *J. Geophys. Res., 103*, 30,205-30,230, 1998.

Wahr, J., D. Wingham, C.R. Bentley, A method of combining GLAS and GRACE satellite data to constrain Antarctic mass balance, *J. Geophs. Res., 105*, 16,279-16,294, 2000.

Wingham, D.J., Small fluctuations in the density and thickness of a dry firn column *J. Glaciol.*, 46, 399-411, 2000.

Wu, X., Watkins, M., Ivins, E., Kwok, R., Wang, P.; Wahr, J., Toward global inverse solutions for current and past ice mass variations: Contribution of secular satellite gravity and topography change measurement, *J. Geophys. Res.*, 107, 2001JB000543, 2002.

Angular Momentum in the Earth System

Richard S. Gross
Jet Propulsion Laboratory, California Institute of Technology
Mail Stop 238-332, 4800 Oak Grove Drive, Pasadena, CA 91109-8099, USA

Abstract. Angular momentum is a fundamental conserved property of dynamic systems. Applying the principle of conservation of angular momentum to the Earth allows the causes of the observed changes in the rotation of the solid Earth to be investigated. After reviewing the application of this principle to the study of observed changes in the Earth's rotation, it is used to investigate the influence of the atmosphere and oceans on the Earth's rotation during 1985–1995. Although atmospheric winds are the dominate process causing the Earth's rate of rotation to change on time scales of a few days to a few years, the redistribution of mass within the atmosphere and oceans is shown to be important in causing polar motion on these time scales.

Keywords. Earth rotation, length-of-day, polar motion, atmospheric angular momentum, oceanic angular momentum

1 Introduction

The Earth is a dynamic system—it has a fluid, mobile atmosphere and oceans, a continually changing global distribution of ice, snow, and ground water, a fluid core that is undergoing some type of hydromagnetic motion, a mantle both thermally convecting and rebounding from the glacial loading of the last ice age, and mobile tectonic plates. These dynamic processes change the mass distribution of the Earth (on vastly disparate time scales) and hence change its rotation.

The Earth rotates about its axis once a day, but does not do so uniformly. Instead, the rate of rotation fluctuates by up to a millisecond per day, and the Earth wobbles as it rotates. Much like the wobble of an unbalanced automobile tire, the Earth wobbles because the mass of the Earth is not balanced about its rotation axis. Such changes in the rotation of the solid Earth can be studied by applying the principle of conservation of angular momentum.

The principle of conservation of angular momentum requires that changes in the rotation vector of the solid Earth must be manifestations of (a) torques acting on the solid Earth or (b) changes in the mass distribution within the solid Earth, which alter its inertia tensor. Angular momentum transfers occur between the solid Earth and the fluid regions (the underlying liquid metallic core and the overlying hydrosphere and atmosphere) with which it is in contact; concomitant torques are due to hydrodynamic or magnetohydrodynamic stresses acting at the fluid / solid Earth interfaces.

Here, the principle of conservation of angular momentum as it applies to studying the observed changes in the Earth's rotation is reviewed. It is then used to evaluate the effect on the Earth's rotation of changes in the mass distribution and motion of the atmosphere and oceans.

2 Theory

2.1 Earth Rotation

Within a rotating, body-fixed reference frame, the equation relating changes in the angular momentum $\mathbf{L}(t)$ of a rotating body to the external torques $\boldsymbol{\tau}(t)$ acting on the body is (e.g., Munk and MacDonald 1960; Lambeck 1980):

$$\frac{\partial \mathbf{L}(t)}{\partial t} + \boldsymbol{\omega}(t) \times \mathbf{L}(t) = \boldsymbol{\tau}(t) \tag{1}$$

where $\boldsymbol{\omega}(t)$ is the angular velocity of the body with respect to inertial space. In general, the angular momentum $\mathbf{L}(t)$ can be written as the sum of two terms: (1) that part $\mathbf{h}(t)$ due to motion relative to the rotating reference frame, and (2) that part due to changes in the inertia tensor $\mathbf{I}(t)$ of the body caused by changes in the distribution of mass:

$$\mathbf{L}(t) = \mathbf{h}(t) + \mathbf{I}(t) \cdot \boldsymbol{\omega}(t) \tag{2}$$

The Earth's rotation deviates only slightly from a state of uniform rotation, the deviation being a few parts in 10^8 in speed, corresponding to changes of a few milliseconds (ms) in the length of the day, and about a part in 10^6 in the orientation of the rotation axis relative to the crust of the Earth, corresponding to a variation of several hundred milliarcseconds (mas) in polar motion. Such small deviations in rotation can be studied by linearizing equations (1) and (2). Let the Earth initially be uniformly rotating about its figure axis and orient the body-fixed reference frame so that its z-axis is aligned with the figure axis. Under a small perturbation to this initial state, the relative angular momentum $\mathbf{h}$ will be perturbed to $\mathbf{h} + \Delta\mathbf{h}$, the inertia tensor $\mathbf{I}$ will be perturbed to $\mathbf{I} + \Delta\mathbf{I}$, and the angular velocity vector $\boldsymbol{\omega}$ will be perturbed to $\boldsymbol{\omega} + \Delta\boldsymbol{\omega}$ where:

$$\boldsymbol{\omega} + \Delta\boldsymbol{\omega} = \Omega\,\hat{\mathbf{z}} + \Omega\,(m_x\,\hat{\mathbf{x}} + m_y\,\hat{\mathbf{y}} + m_z\,\hat{\mathbf{z}}) \qquad (3)$$

where Ω is the mean angular velocity of the Earth, the hat denotes a vector of unit length, and the Ωm_i are the elements of the perturbed rotation vector.

Keeping terms to first order in perturbed quantities, the axial component of equation (1) can be written in the absence of external torques as (e.g., Wahr 1982):

$$\Delta\Lambda(t) = \frac{\Lambda_o}{C_m\,\Omega}\left[\Delta h_z(t) + 0.756\,\Omega\,\Delta I_{zz}(t)\right] \qquad (4)$$

where the change $\Delta\Lambda(t)$ in the length of the day is related to $m_z(t)$ by $\Delta\Lambda(t)/\Lambda_o = -m_z(t)$, Λ_o is the nominal length-of-day of 86400 seconds, C_m is the greatest principal moment of inertia of the Earth's crust and mantle, and the factor of 0.756 accounts for the yielding of the crust and mantle to imposed surface loads. The equatorial components of equation (1) can similarly be written in the absence of external torques as (e.g., Wahr 1982):

$$\mathbf{m}(t) + \frac{i}{\sigma_{cw}}\frac{\partial\mathbf{m}(t)}{\partial t} = \boldsymbol{\chi}(t) - \frac{i}{\Omega}\frac{\partial\boldsymbol{\chi}(t)}{\partial t} \qquad (5)$$

where σ_{cw} is the complex-valued frequency of the Chandler wobble, $\mathbf{m}(t) \equiv m_x(t) + i\,m_y(t)$, and the excitation functions $\boldsymbol{\chi}(t) \equiv \chi_x(t) + i\,\chi_y(t)$ are:

$$\chi_x(t) = \frac{1.61}{\Omega\,(I_{zz} - I_{xx})}\left[\Delta h_x(t) + \frac{\Omega\,\Delta I_{xz}(t)}{1.44}\right] \qquad (6)$$

$$\chi_y(t) = \frac{1.61}{\Omega\,(I_{zz} - I_{xx})}\left[\Delta h_y(t) + \frac{\Omega\,\Delta I_{yz}(t)}{1.44}\right] \qquad (7)$$

where the factor 1.44 accounts for the yielding of the crust and mantle to imposed surface loads and the factor of 1.61 additionally accounts for the decoupling of the core and mantle.

Equation (5) relates changes in the location of the rotation pole to changes in angular momentum caused by relative motion and mass redistribution. However, Earth rotation services do not report the location m_x and m_y of the rotation pole, but rather report the location p_x and p_y of the celestial ephemeris pole. Within the rotating, body-fixed terrestrial reference frame, the locations of the rotation and celestial ephemeris poles are related by (Brzezinski 1992; Gross 1992; Eubanks 1993):

$$\mathbf{m}(t) = \mathbf{p}(t) - \frac{i}{\Omega}\frac{\partial\mathbf{p}(t)}{\partial t} \qquad (8)$$

where $\mathbf{p}(t) \equiv p_x(t) - i\,p_y(t)$. The negative sign is required here in the definition of $\mathbf{p}(t)$ since, by convention, the reported $p_y(t)$ is positive towards 90°W longitude. Combining equations (5) and (8) yields the final expression relating the usual reported polar motion parameters $p_x(t)$ and $p_y(t)$ to the excitation functions $\chi_x(t)$ and $\chi_y(t)$:

$$\mathbf{p}(t) + \frac{i}{\sigma_{cw}}\frac{\partial\mathbf{p}(t)}{\partial t} = \boldsymbol{\chi}(t) \qquad (9)$$

Given changes in angular momentum caused by, for example, changes in either the motion or mass distribution of surficial fluid layers of the Earth such as the atmosphere or oceans, equations (4) and (9) can be used to compute the concomitant changes in the length-of-day and polar motion, respectively. In fact, the angular momentum associated with the relative motion and mass redistribution of the atmosphere and oceans is now routinely computed from the horizontal velocities and pressure fields given by numerical models of the general circulation of the atmosphere and oceans.

2.2 Angular Momentum of Surficial Fluids

The angular momentum $\mathbf{L}(t)$ of a surficial fluid layer of the Earth such as the atmosphere or oceans can, as before, be written as the sum of a motion term $\mathbf{L}_{\mathbf{motion}}(t) = \mathbf{h}(t)$ that will change as the strength and direction of the motion of the fluid changes:

$$\mathbf{L}_{\mathbf{motion}}(t) = \int_{V_o(t)} \rho(\mathbf{r}, t)\,\mathbf{r} \times \mathbf{u}(\mathbf{r}, t)\,dV \qquad (10)$$

and a mass term $\mathbf{L}_{\mathbf{mass}}(t)$ that will change as the mass distribution of the fluid changes:

$$\mathbf{L}_{\mathbf{mass}}(t) = \int_{V_o(t)} \rho(\mathbf{r}, t)\,\mathbf{r} \times \boldsymbol{\omega} \times \mathbf{r}\,dV \qquad (11)$$

where the integrals are taken over the entire time-dependent volume $V_o(t)$ of the fluid and $\mathbf{r}$ is the position vector of some mass element of the fluid which has density $\rho(\mathbf{r},t)$ and which is moving with Eulerian velocity $\mathbf{u}(\mathbf{r},t)$ with respect to a terrestrial, body-fixed reference frame which is itself rotating with angular velocity $\boldsymbol{\omega}$ with respect to a celestial, space-fixed reference frame. Since the Earth's rotation deviates only slightly from a state of uniform rotation, to sufficient accuracy here the angular velocity of the body-fixed terrestrial reference frame can be taken to be that of uniform rotation about the z-coordinate axis: $\boldsymbol{\omega} = \Omega\,\hat{\mathbf{z}}$. The Cartesian components of the mass term are then:

$$L_{x,mass}(t) = -\Omega \int_{V_o(t)} \rho(\mathbf{r},t)\, r^2 \sin\phi \cos\phi \cos\lambda \; dV \quad (12)$$

$$L_{y,mass}(t) = -\Omega \int_{V_o(t)} \rho(\mathbf{r},t)\, r^2 \sin\phi \cos\phi \sin\lambda \; dV \quad (13)$$

$$L_{z,mass}(t) = \Omega \int_{V_o(t)} \rho(\mathbf{r},t)\, r^2 \cos^2\phi \; dV \quad (14)$$

where ϕ is North latitude, λ is East longitude, and $dV = r^2 dr \cos\phi \; d\phi \; d\lambda$. From the definition of the inertia tensor:

$$\mathbf{I}(t) = \int_{V_o(t)} \rho(\mathbf{r},t)\, [r^2\, \mathbf{I} - \mathbf{r}\,\mathbf{r}] \; dV \quad (15)$$

where $\mathbf{I}$ is the identity tensor, it is straightforward to show that these components of the mass term are related to the elements of the inertia tensor by:

$$L_{x,mass}(t) = \Omega\, I_{xz}(t) \quad (16)$$

$$L_{y,mass}(t) = \Omega\, I_{yz}(t) \quad (17)$$

$$L_{z,mass}(t) = \Omega\, I_{zz}(t) \quad (18)$$

Thus, it is seen that changes in the components of the mass term of the angular momentum vector are simply related to changes in those components of the inertia tensor that are needed in equations (4 and 6–7) to compute the effect on the Earth's rotation of changes in the distribution of mass within surficial fluid layers of the Earth.

The Cartesian components of the motion term (equation 10) are:

$$L_{x,motion}(t) = \int_{V_o(t)} \rho(\mathbf{r},t)\, [r \sin\lambda \, v(\mathbf{r},t) - r \sin\phi \cos\lambda \, u(\mathbf{r},t)] \; dV \quad (19)$$

$$L_{y,motion}(t) = \int_{V_o(t)} \rho(\mathbf{r},t)\, [-r \cos\lambda \, v(\mathbf{r},t) - r \sin\phi \sin\lambda \, u(\mathbf{r},t)] \; dV \quad (20)$$

$$L_{z,motion}(t) = \int_{V_o(t)} \rho(\mathbf{r},t)\, r \cos\phi \, u(\mathbf{r},t) \; dV \quad (21)$$

where $u(\mathbf{r},t)$ and $v(\mathbf{r},t)$ are the eastward and northward components, respectively, of the velocity of the fluid. Changes in these components of the motion term of the angular momentum vector are just those components of the relative angular momentum $\Delta\mathbf{h}(t)$ that are needed in equations (4 and 6–7) to compute the effect on the Earth's rotation of changes in the motion of surficial fluid layers of the Earth.

Changes in the angular momentum of some surficial fluid layer of the Earth due to the redistribution of mass within that layer can be computed by equations (12–14), and that due to changes in the motion of the fluid can be computed by equations (19–21). The effect on the Earth's rotation of these changes in the angular momentum of the surficial fluid can then be evaluated by comparing the axial component of the angular momentum to the observed length-of-day changes using equation (4) and the equatorial components to the observed polar motion excitation functions using equations (6–7). As an example of this, the influence of the atmosphere and oceans on the Earth's rotation will now be evaluated using this approach.

3 Observed and Modeled Excitation

3.1 Earth Orientation Observations

The observed series of polar motion excitation functions and excess length-of-day values used here is the daily version of the COMB2000 combined Earth orientation series (Gross 2001). COMB2000 is a combination of Earth orientation measurements taken by the techniques of optical astrometry, lunar and satellite laser ranging, very long baseline interferometry, and the global positioning system. The COMB2000 series of polar motion excitation functions and excess length-of-day values spans January 20, 1962 to January 6, 2001 at daily intervals. In order to match the temporal averaging of the oceanic excitation series (see below), 5-day averages of the daily values have been formed.

Prior to comparing the observed excitation series to excitation caused by the atmosphere and oceans,

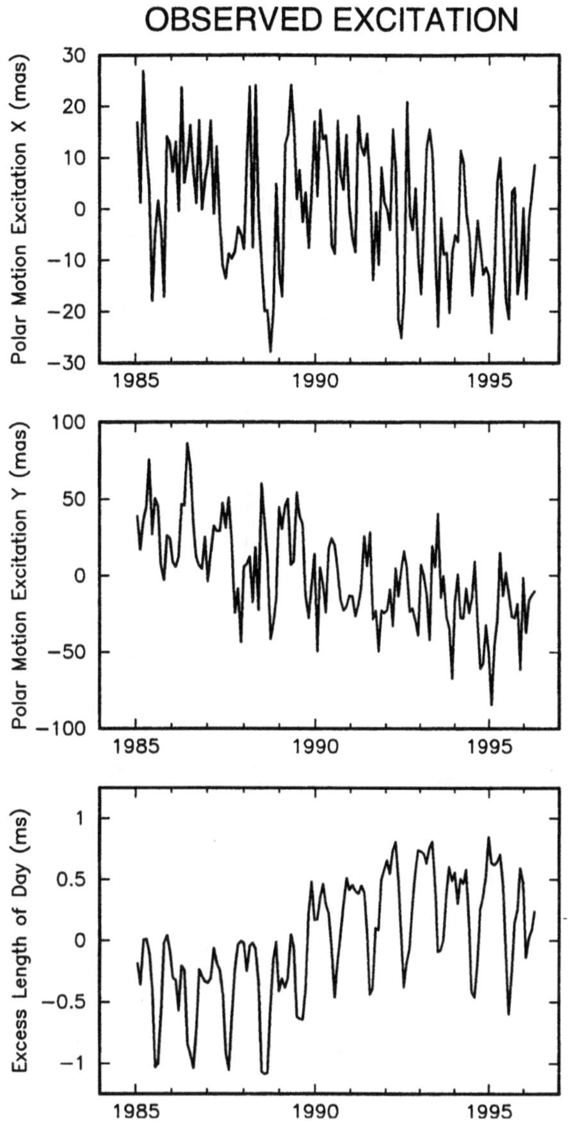

Fig. 1 The COMB2000 series of observed polar motion excitation functions (x-component on top, y-component in the middle) and excess length-of-day values (bottom) from which tidal effects have been removed. For purposes of clarity of display, 30-day averages of the observed values are shown.

non-atmospheric and non-oceanic sources of excitation that can be accurately modeled should be removed from the observations. Solid Earth and ocean tidal effects have thus been removed from the COMB2000 excess length-of-day values using the models of Yoder et al. (1981) and Kantha et al. (1998), respectively. The effect of the termensual (9.12- and 9.13-day), fortnightly (13.63- and 13.66-day), and monthly (27.55-day) ocean tides have been removed from the COMB2000 polar motion excitation series using the model of Gross et al. (1997). The resulting observed polar motion excitation and excess length-of-day series are shown in Figure 1.

3.2 Atmospheric Excitation Functions

The atmospheric excitation functions used in this study are derived from the National Centers for Environmental Prediction / National Center for Atmospheric Research (NCEP/NCAR) reanalysis project (Kalnay et al. 1996) and were obtained from the International Earth Rotation Service (IERS) Special Bureau for the Atmosphere (SBA; Salstein et al. 1993). Both the atmospheric excitation due to winds, computed by integrating the horizontal winds from the surface to the top of the model at 10 hPa (Salstein and Rosen 1997), and that due to surface pressure variations, computed by assuming that the oceans respond as an inverted barometer to the imposed surface pressure variations, are used here. The NCEP/NCAR reanalysis atmospheric excitation series available from the IERS SBA spans January 1, 1948 to the present at 6-hour intervals. Daily averages were formed by summing 5 consecutive values using weights of 1/8, 1/4, 1/4, 1/4, 1/8. In order to match the temporal averaging of the oceanic excitation series (see below), 5-day averages of the daily values were then formed. The resulting atmospheric excitation series are shown in Figure 2.

3.3 Oceanic Excitation Functions

The oceanic current and bottom pressure excitation functions used in this study are those determined by Ponte et al. (1998) and Ponte and Stammer (2000) and were obtained from the IERS Special Bureau for the Oceans. The oceanic general circulation model (OGCM) used by them to determine these excitation functions was forced by 12-hour wind stress fields and daily surface heat and fresh water flux fields from NCEP, but not by surface pressure. Since the OGCM used by them is formulated under the Boussinesq approximation it conserves volume rather than mass. Artificial mass variations can be introduced into Boussinesq ocean models because of the applied surface heat flux. The changing applied heat flux changes the density, which, since volume is conserved, artificially changes the mass. If left uncorrected, this artificial mass change will

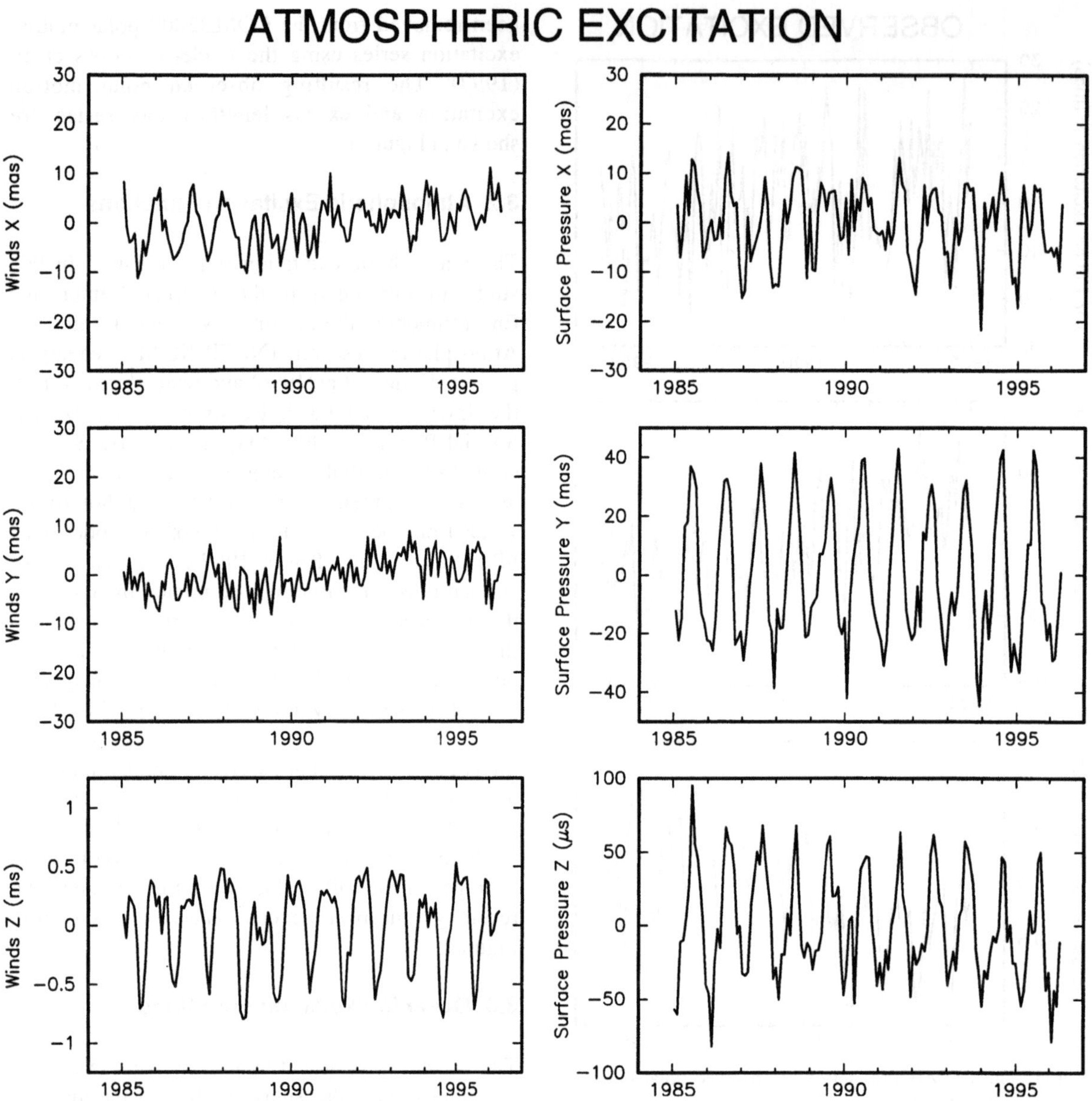

Fig. 2 The *x*-component (top), *y*-component (middle), and z-component (bottom) of the atmospheric excitation functions due to variations in atmospheric winds (lefthand side) and surface pressure (righthand side). The atmospheric excitation function due to surface pressure variations is that computed assuming the oceans respond as an inverted barometer to the imposed atmospheric pressure variations. For purposes of clarity of display, 30-day averages of the atmospheric excitation functions are shown.

cause artificial changes to the ocean-bottom pressure excitation functions. Ponte et al. (1998) and Ponte and Stammer (2000) corrected the ocean-bottom pressure excitation values by computing the effect on them of a spatially uniform sea-surface layer having just the right fluctuating thickness to enforce mass conservation (Greatbatch 1994; Greatbatch et al. 2001). The resulting oceanic excitation series consist of 5-day-averaged values spanning January 1985 to April 1996 and are shown in Figure 3.

3.4 Seasonal Excitation

Seasonal changes in the observed and modeled excitation functions during 1985–1995 are studied here by fitting to them a mean, trend, and periodic terms at the annual and semiannual frequencies.

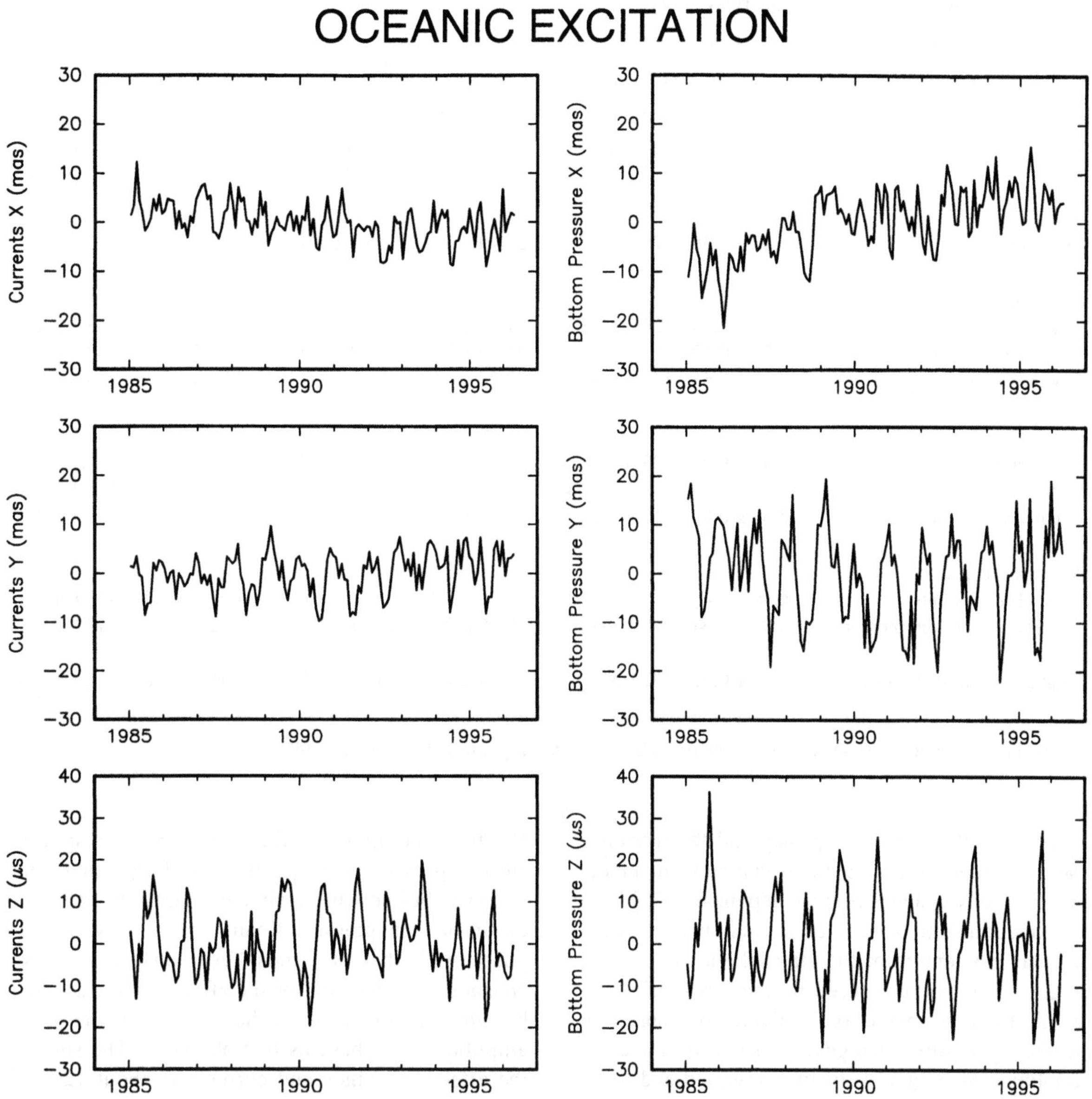

Fig. 3 The *x*-component (top), *y*-component (middle), and z-component (bottom) of the oceanic excitation functions due to variations in oceanic currents (lefthand side) and ocean-bottom pressure (righthand side). For purposes of clarity of display, 30-day averages of the oceanic excitation functions are shown.

Since no uncertainty estimates are available for the atmospheric and oceanic excitation functions, this fit was obtained using unweighted least-squares. For consistency, unweighted least-squares was also used to obtain the fit to the observed excitation function even though uncertainty estimates are available for the observed values. The uncertainties assigned to the fitted parameters and given in Tables 1 and 2 are the 1σ formal uncertainties computed using the standard deviation of the postfit residuals as an estimate of the mean uncertainty of the observed or modeled excitation values. The uncertainties thus determined can be considered to be realistic in the sense that the resulting postfit residuals have a reduced chi-square of one.

Table 1 shows the results of this fit for the amplitude A and phase α of the annual cycles of the observed and modeled excitation series with A and α being defined by the expression:

$$\chi(t) = A \cos[\sigma(t - t_o) + \alpha] \tag{22}$$

Table 1. Annual Excitation During 1985–1995

Excitation process	*x*-component		*y*-component		*z*-component	
	Amplitude (mas)	Phase (degrees)	Amplitude (mas)	Phase (degrees)	Amplitude (μs)	Phase (degrees)
Observed	6.48 ± 0.96	–76.2 ± 8.5	20.49 ± 1.59	–155.5 ± 4.4	398.5 ± 11.7	–32.7 ± 1.7
Atmospheric						
winds	4.15 ± 0.34	–26.0 ± 4.7	1.49 ± 0.34	177.6 ± 13.1	414.2 ± 8.7	–33.3 ± 1.2
surface pressure (i.b.)	7.37 ± 0.50	–175.6 ± 3.9	29.36 ± 0.73	–176.8 ± 1.4	43.5 ± 1.4	147.3 ± 1.8
winds & surface pressure	4.34 ± 0.63	–146.7 ± 8.3	30.85 ± 0.93	–177.1 ± 1.7	370.6 ± 8.8	–33.4 ± 1.4
Oceanic						
currents	2.33 ± 0.28	25.6 ± 6.8	3.95 ± 0.26	–4.9 ± 3.7	6.1 ± 0.5	114.1 ± 4.6
bottom pressure	1.42 ± 0.39	2.4 ± 15.8	9.37 ± 0.54	–17.5 ± 3.3	9.9 ± 0.6	118.1 ± 3.4
currents & bottom pressure	3.64 ± 0.52	–15.1 ± 8.2	13.25 ± 0.72	–13.7 ± 3.1	16.1 ± 1.0	116.5 ± 3.5
Atmospheric plus oceanic						
winds & currents	6.48 ± 0.45	–25.9 ± 4.0	2.46 ± 0.45	–6.4 ± 10.5	409.0 ± 8.7	–32.8 ± 1.2
surface & bottom pressure	5.95 ± 0.64	–175.1 ± 6.1	20.86 ± 0.89	–167.7 ± 2.4	52.4 ± 1.6	142.0 ± 1.8
Total of all atmos. & oceanic	3.33 ± 0.79	–91.8 ± 13.7	18.54 ± 1.17	–165.3 ± 3.6	356.8 ± 9.0	–32.1 ± 1.4

i.b., inverted barometer; atmos., atmospheric; reference date for phase is January 1.0, 1990

where σ is the annual frequency and the reference date t_o is January 1.0, 1990. As can be seen, annual length-of-day changes (the *z*-component in Table 1) are predominantly caused by annual changes in the angular momentum of the atmospheric winds. The angular momentum caused by mass redistribution within the atmosphere, or, equivalently, caused by surface pressure changes, is seen to have an amplitude of only 11% of that observed and to be out-of-phase with the observations. The effects of oceanic currents and of mass redistribution within the oceans, or, equivalently, of ocean-bottom pressure changes, are each less than 3% of that observed and exhibit little agreement in phase.

In contrast, mass redistribution within the atmosphere is seen to be the primary cause of annual changes in the polar motion excitation functions (the *x*- and *y*-components in Table 1). For the *x*-component, even though the effect of atmospheric surface pressure changes is somewhat larger than that observed, it is nearly in quadrature with the observations. The total effect of atmospheric wind and surface pressure and oceanic current and bottom pressure changes has about half the amplitude of that observed with a similar phase. For the *y*-component, the effect of atmospheric surface pressure changes is again larger than that observed but now has a similar phase. The effect of ocean-bottom pressure changes is nearly out-of-phase with that of atmospheric surface pressure changes, with the sum of the effects of surface and bottom pressure changes having nearly the same amplitude and phase as that observed. The sum of the effects of winds and currents is about 12% of that observed and exhibits little agreement in phase.

Table 2 shows the results for semiannual changes in the observed and modeled excitation series. As for the annual frequency, semiannual length-of-day changes are predominantly caused by semiannual changes in the angular momentum of the winds. The effects of semiannual atmospheric surface pressure and oceanic current and bottom pressure changes on the length-of-day are each less than 3% of that observed and exhibit little agreement in phase.

For semiannual changes in the polar motion excitation functions, the effect of atmospheric surface pressure changes are found to be the primary excitation mechanism, as they were found to be at the annual frequency, particularly for the *y*-

Table 2. Semiannual Excitation During 1985–1995

Excitation process	*x*-component		*y*-component		*z*-component	
	Amplitude (mas)	Phase (degrees)	Amplitude (mas)	Phase (degrees)	Amplitude (μs)	Phase (degrees)
Observed	4.56 ± 0.96	177.4 ± 12.0	11.21 ± 1.59	−16.6 ± 8.1	261.5 ± 11.7	117.9 ± 2.6
Atmospheric						
winds	0.92 ± 0.34	96.2 ± 20.9	0.74 ± 0.34	−104.3 ± 26.5	235.3 ± 8.6	110.7 ± 2.1
surface pressure (i.b.)	3.24 ± 0.50	−84.5 ± 8.8	7.66 ± 0.73	−25.3 ± 5.4	7.3 ± 1.4	−100.3 ± 10.6
winds & surface pressure	2.31 ± 0.63	−84.8 ± 15.5	7.84 ± 0.93	−30.6 ± 6.8	229.0 ± 8.8	111.7 ± 2.2
Oceanic						
currents	2.12 ± 0.28	162.1 ± 7.5	1.47 ± 0.26	137.5 ± 10.0	3.5 ± 0.5	−118.2 ± 8.1
bottom pressure	2.52 ± 0.39	143.3 ± 8.9	1.05 ± 0.54	163.7 ± 29.1	5.5 ± 0.6	−156.1 ± 6.1
currents & bottom pressure	4.57 ± 0.52	151.9 ± 6.6	2.46 ± 0.72	148.4 ± 16.7	8.6 ± 1.0	−141.5 ± 6.5
Atmospheric plus oceanic						
winds & currents	2.64 ± 0.45	143.5 ± 9.7	1.30 ± 0.45	167.7 ± 19.9	233.0 ± 8.7	111.4 ± 2.2
surface & bottom pressure	2.42 ± 0.64	−134.8 ± 15.1	6.62 ± 0.89	−26.8 ± 7.7	11.4 ± 1.6	−124.0 ± 8.3
Total of all atmos. & oceanic	3.83 ± 0.80	−177.7 ± 11.9	5.38 ± 1.17	−30.2 ± 12.4	226.7 ± 9.0	113.7 ± 2.3

i.b., inverted barometer; atmos., atmospheric; reference date for phase is January 1.0, 1990

component. For the *y*-component the effects of atmospheric winds, oceanic currents, and ocean-bottom pressure changes are each less than 1/5 that of the effects of atmospheric surface pressure changes and exhibit little agreement in phase with the observations. For the *x*-component, even though the effect of atmospheric surface pressure changes has the largest amplitude, it is nearly in quadrature with the observations, as it was at the annual frequency. The total effect on the *x*-component of atmospheric wind and surface pressure and oceanic current and bottom pressure changes has about 4/5 the amplitude of that observed with a similar phase.

3.5 Nonseasonal Excitation

Nonseasonal changes in the observed and modeled excitation functions during 1985–1995 are studied here by removing from the series the best-fitting mean, trend, annual, and semiannual harmonics that were determined for them and discussed in the previous section. In addition, a highpass filter with a cutoff period of 6 years has been applied to both the observed and, for consistency, the modeled excitation series in order to remove the decadal changes in the length-of-day that are caused by processes acting at the core-mantle boundary (e.g., Jackson et al. 1993) and the decadal changes in polar motion, known as the Markowitz wobble, whose origin is unknown. Thus, the nonseasonal changes studied here consist of signals having periods between 5 days and 6 years excluding signals at the annual and semiannual frequencies.

Table 3 shows the percentage of the observed excitation variance explained by the modeled excitation processes along with the correlation between the observed and modeled excitation series during 1985–1995. The 99% significance level for the correlations is estimated to be 0.10 for the *x*- and *y*-components, and 0.20 for the *z*-component, after accounting for the reduction in the degrees of freedom that was determined from the width of the central peak of the autocorrelation functions.

For nonseasonal length-of-day changes (the *z*-component in Table 3), atmospheric winds are again seen to be the dominant excitation mechanism, as they were at the annual and semiannual frequencies, explaining 75% of the observed variance and having a correlation coefficient of 0.87 with the observations.

Table 3. Nonseasonal Excitation During 1985–1995

Excitation process	% of observed variance explained			correlation with observations		
	x-comp.	*y*-comp.	*z*-comp.	*x*-comp.	*y*-comp.	*z*-comp.
Atmospheric						
winds	20.57	19.25	75.31	0.47	0.56	0.87
surface pressure (i.b.)	19.48	38.89	2.30	0.45	0.65	0.15
winds & surface pressure	36.36	49.42	77.02	0.61	0.72	0.88
Oceanic						
currents	0.15	8.09	2.45	0.15	0.33	0.27
bottom pressure	21.99	14.03	2.93	0.47	0.38	0.27
currents & bottom pressure	17.34	16.01	5.01	0.43	0.40	0.30
Atmospheric plus oceanic						
winds & currents	19.63	26.57	76.77	0.44	0.61	0.88
surface & bottom pressure	41.06	53.58	4.68	0.64	0.76	0.22
Total of all atmospheric & oceanic	57.48	65.64	79.27	0.76	0.81	0.89

%, percentage; comp., component; i.b., inverted barometer; 99% significance level for correlations of the x- and y-components is 0.10, that for the z-component is 0.20

Atmospheric surface pressure, oceanic currents, and bottom pressure are seen to have only a minor effect on nonseasonal length-of-day changes, each explaining no more than 3% of the observed variance. The sum of all atmospheric and oceanic processes can explain 79% of the observed length-of-day variance, and has a correlation coefficient with the observations of 0.89.

In contrast, for nonseasonal changes in polar motion excitation (the x- and y-components in Table 3), atmospheric surface and oceanic bottom pressure changes are seen to be the most important excitation mechanisms with their sum explaining twice as much of the observed variance as does the sum of atmospheric winds and oceanic currents. For the x-component, atmospheric winds, atmospheric surface pressure, and ocean-bottom pressure each explain about 20% of the observed variance, with oceanic currents explaining less than 1% of that observed. The sum of all atmospheric and oceanic processes can explain 57% of the x-component of the observed polar motion excitation variance, and has a correlation coefficient with the observations of 0.76. For the y-component, atmospheric surface pressure changes explain twice as much of the variance as do either atmospheric winds or ocean-bottom pressure changes, with oceanic currents explaining only 8% of that observed. The sum of all atmospheric and oceanic processes can explain 66% of the y-component of the observed polar motion excitation variance, and has a correlation coefficient with the observations of 0.81.

Table 3 shows that oceanic processes, primarily ocean-bottom pressure changes, are an important source of nonseasonal polar motion excitation. For the x-component, adding the effects of oceanic currents and bottom pressure to that of atmospheric winds and surface pressure increases the percentage of the observed variance explained from 36% to 57% and increases the correlation coefficient with the observations from 0.61 to 0.76. For the y-component, adding the oceanic processes to the atmospheric increases the variance explained from 49% to 66% and increases the correlation from 0.72 to 0.81. These variances explained and correlation coefficients for the x- and y-components are still less than those for the z-component, but not by much. The development of oceanic analysis systems that assimilate oceanic data, similar to that of the atmospheric NCEP/NCAR reanalysis system whose results have been used here, can be expected to lead to even closer agreement with the observations (Ponte et al. 2001).

4 Discussion and Summary

The principle of conservation of angular momentum when applied to the Earth system can be used to investigate the origin of observed changes in the rotation of the solid Earth. As an example of this approach, it has been used here to study the effect of the atmosphere and oceans on polar motion and length-of-day changes by computing the changing angular momentum associated with changes in the motion and mass distribution of the atmosphere and oceans. It is shown, in agreement with previous studies, that on timescales of 5 days to 6 years atmospheric winds are the dominant cause of changes in the length-of-day, whereas mass redistribution within the atmosphere and oceans is largely responsible for exciting polar motion.

The oceanic angular momentum series used here was determined from the products of an oceanic general circulation model that was forced by surface wind stress and buoyancy fluxes, but not by surface pressure. At periods greater than a few days, the oceans act predominantly like an inverted barometer and respond statically to changes in surface pressure (Wunsch and Stammer 1997). However, at periods less than a few days, the dynamical response of the oceans to pressure forcing has been shown to be comparable to that of wind stress forcing (e.g., Ponte 1993, 1994). Since the timescales studied here range between 5 days and 6 years, and since the inverted barometer approximation should be valid for periods greater than a few days, neglecting the dynamical response of the oceans to pressure forcing should have minimal impact on the results presented here.

More detailed investigations into the effect of the atmosphere on the Earth's rotation can be found in the reviews and references therein of Hide and Dickey (1991), Eubanks (1993) and Rosen (1993). Recent detailed studies using OGCMs to investigate the effect of the oceans on the Earth's rotation are those of Ponte et al. (1998, 2001), Marcus et al. (1998), Johnson et al. (1999), Ponte and Stammer (1999), Nastula et al. (2000), Thomas et al. (2001), and Gross et al. (2003).

Acknowledgements. The work described in this paper was performed at the Jet Propulsion Laboratory, California Institute of Technology, under contract with the National Aeronautics and Space Administration. Support for this work was provided by the Oceanography Program of NASA's Office of Earth Science.

References

Brzezinski, A. (1992). Polar motion excitation by variations of the effective angular momentum function: considerations concerning deconvolution problem. *manuscripta geodaetica, 17*, pp. 3–20.

Eubanks, T.M. (1993). Variations in the orientation of the Earth. In: *Contributions of Space Geodesy to Geodynamics: Earth Dynamics.* Smith, D.E., and D.L. Turcotte (eds). pp. 1–54, American Geophysical Union Geodynamics Series vol. 24, Washington, D.C.

Greatbatch, R.J. (1994). A note on the representation of steric sea level in models that conserve volume rather than mass. *J. Geophys. Res., 99*, pp. 12767–12771.

Greatbatch, R.J., Y. Lu, and Y. Cai (2001). Relaxing the Boussinesq approximation in ocean circulation models. *J. Atmos. Oceanic Technol., 18*, pp. 1911–1923.

Gross, R.S. (1992). Correspondence between theory and observations of polar motion. *Geophys. J. Int., 109*, pp. 162–170.

Gross, R.S. (2001). Combinations of Earth orientation measurements: SPACE2000, COMB2000, and POLE2000. Jet Propulsion Laboratory Publ. 01-2, Pasadena, Calif.

Gross, R.S., B.F. Chao, and S. Desai (1997). Effect of long-period ocean tides on the Earth's polar motion. *Prog. Oceanogr., 40*, pp. 385–397.

Gross, R.S., I. Fukumori, and D. Menemenlis (2003). Atmospheric and oceanic excitation of the Earth's wobbles during 1980–2000. *J. Geophys. Res.*, in press.

Hide, R. and J.O. Dickey (1991). Earth's variable rotation. *Science, 253*, pp. 629–637.

Jackson, A., J. Bloxham, and D. Gubbins (1993). Time-dependent flow at the core surface and conservation of angular momentum in the coupled core-mantle system. In: *Dynamics of the Earth's Deep Interior and Earth Rotation.* Le Mouël, J.-L., D.E. Smylie, and T. Herring (eds). pp. 97–107, American Geophysical Union Geophysical Monograph Series vol. 72, Washington, D.C.

Johnson, T.J., C.R. Wilson, and B.F. Chao (1999). Oceanic angular momentum variability estimated from the Parallel Ocean Climate Model, 1988–1998, *J. Geophys. Res., 104*, pp. 25183–25195.

Kalnay, E., M. Kanamitsu, R. Kistler, W. Collins, D. Deaven, L. Gandin, M. Iredell, S. Saha, G. White, J. Woollen, Y. Zhu, M. Chelliah, W. Ebisuzaki, W. Higgins, J. Janowiak, K.C. Mo, C. Ropelewski, J. Wang, A. Leetmaa, R. Reynolds, R. Jenne, and D. Joseph (1996). The NCEP/NCAR 40-year reanalysis project. *Bull. Amer. Met. Soc. 77*, pp. 437–471.

Kantha, L.H., J.S. Stewart, and S.D. Desai (1998). Long-period lunar fortnightly and monthly ocean tides. *J. Geophys. Res., 103,* pp. 12639–12647.

Lambeck, K. (1980). The Earth's Variable Rotation. Cambridge University Press, New York.

Marcus, S.L., Y. Chao, J.O. Dickey, and P. Gegout (1998). Detection and modeling of nontidal oceanic effects on Earth's rotation rate. *Science, 281,* pp. 1656–1659.

Munk, W.H., and G.J.F. MacDonald (1960). The Rotation of the Earth. Cambridge University Press, New York.

Nastula, J., R.M. Ponte, and D.A. Salstein (2000). Regional signals in atmospheric and oceanic excitation of polar motion. In: *Polar Motion: Historical and Scientific Problems, IAU Colloq. 178.* edited by Dick, S. D. McCarthy, and B. Luzum (eds). pp. 463–472, Astron. Soc. Pacific Conf. Ser. vol. 208, San Francisco.

Ponte, R.M. (1993). Variability in a homogenous global ocean forced by barometric pressure. *Dyn. Atmos. Oceans, 18,* pp. 209–234.

Ponte, R.M. (1994). Understanding the relation between wind- and pressure-driven sea level variability. *J. Geophys. Res., 99,* pp. 8033–8039.

Ponte, R.M., and D. Stammer (1999). Role of ocean currents and bottom pressure variability on seasonal polar motion. *J. Geophys. Res., 104,* pp. 23393–23409.

Ponte, R.M., and D. Stammer (2000). Global and regional axial ocean angular momentum signals and length-of-day variations (1985-1996). *J. Geophys. Res., 105,* pp. 17161–17171.

Ponte, R.M., D. Stammer, and J. Marshall (1998). Oceanic signals in observed motions of the Earth's pole of rotation. *Nature, 391,* pp. 476–479.

Ponte, R.M., D. Stammer, and C. Wunsch (2001). Improving ocean angular momentum estimates using a model constrained by data. *Geophys. Res. Lett., 28,* pp. 1775–1778.

Rosen, R.D. (1993). The axial momentum balance of Earth and its fluid envelope. *Surveys Geophys., 14,* pp. 1–29.

Salstein, D.A., and R.D. Rosen (1997). Global momentum and energy signals from reanalysis systems. Preprints, 7th Conf. on Climate Variations. pp. 344–348, American Meteorological Society, Boston.

Salstein, D.A., D.M. Kann, A.J. Miller, and R.D. Rosen (1993). The Sub-Bureau for Atmospheric Angular Momentum of the International Earth Rotation Service: A meteorological data center with geodetic applications. *Bull. Amer. Meteorol. Soc., 74,* pp. 67–80.

Thomas, M., J. Sündermann, and E. Maier-Reimer (2001). Consideration of ocean tides in an OGCM and impacts on subseasonal to decadal polar motion excitation. *Geophys. Res. Lett., 28,* pp. 2457–2460.

Wahr, J.M. (1982). The effects of the atmosphere and oceans on the Earth's wobbles — Theory. *Geophys. J. R. astr. Soc., 70,* pp. 349–372.

Wunsch, C., and D. Stammer (1997). Atmospheric loading and the oceanic "inverted barometer" effect. *Rev. Geophys., 35,* pp. 79–107.

Yoder, C.F., J.G. Williams, and M.E. Parke (1981). Tidal variations of Earth rotation. *J. Geophys. Res., 86,* pp. 881–891.

Gravity Field Variability, the Geoid, and Ocean Dynamics

F. Condi
Center for Space Research, University of Texas at Austin

C. Wunsch
Department of Earth, Atmospheric and Planetary Sciences, Massachusetts Institute of Technology

Abstract. With the advent of accurate satellite altimetry, physical oceanography and geodesy have come to have many overlapping problems. The most fundamental of these problems concerns the detailed determination of the geoid. This gravitational equipotential of the earth is central to a description of the solid earth, and appears as the principle reference surface for computing oceanic currents. From the oceanographer's point of view, knowledge of the sea surface elevation relative to the geoid determines the absolute circulation of the ocean. Great progress has occurred in recent years in determining the geoid with much improved accuracy, although much remains to be done for the result to be fully useful for oceanographic purposes. The possibility of measurement of the earth's time variable gravity field and the very high accuracies and precisions which appear possible, raise a myriad of new and interesting challenges for understanding and using the measurements. This present paper summarizes some of the work we have performed (Condi and Wunsch (2003)) in exploring the idea of measuring bottom pressure changes from space, and how these data might be used, but with particular attention paid to basic concepts in ocean dynamics and errors in model approximations.

Keywords. Ocean dynamics, geoid, bottom pressure

1 Introduction

Determination of ocean bottom pressure changes is important for understanding the earth's rotation, diagnosing the physics of oceanic variability and may prove useful in determining the ocean circulation, but the measurements have been difficult to make. The Gravity Recovery And Climate Experiment (GRACE) (Wahr et al. (1998), Tapley and Reigber (2001)), launched in March of 2002, holds the promise of very accurate global measurements of bottom pressure changes from space. The GRACE pair of spacecraft are sensitive to the time-mean gravity field, but will also be able to measure much of the spectral range of the time-dependent components. The Gravity Field and Steady-State Ocean Circulation Explorer Mission (GOCE), planned for launch in 2006, will use a gravity gradiometer to map the static geoid at wavelengths shorter than those of GRACE. The objective of this approach is to better define the static geoid in areas of high sea surface topographic relief over short length scales, such as western boundary currents, where substantial mean transport occurs.

Gravity data will be assimilated into ocean general circulation models (OGCMs) for scientific applications, and OGCMs (or a simplified form neglecting thermodynamics) will be used operationally to correct for the aliasing effect of high frequency motions. In the case of GRACE, the sampling period is long compared to much of the variability of barotropic energy in the ocean, and this higher frequency energy can alias into the signal (Stammer et al. (2000), Tierney et al. (2000)).

To obtain insight into bottom pressure variability and the requirements for assimilating gravity data into an ocean general circulation model, we use the recent results from a constrained, two-degree spatial resolution, global OGCM, contrasting bottom pressure variability to that of surface pressure. The focus is on obtaining a quantitatively useful estimate of the expected variability of the ocean on relevant time scales using an estimated ocean circulation from a constrained general circulation model. Over the spatial scales that we discuss (two-degree resolution), much of this variability will probably fall at the noise level of GOCE measurements

2 The Geoid

For present purposes, we can define the geoid as the particular gravitational equipotential surface of the rotating earth that would coincide with the sea surface if the ocean were brought to rest by removing all external forces, and the density were everywhere set equal to its global mean value. The geoid height, N(θ,φ), where θ and φ are co-latitude and longitude, is measured with respect to the reference ellipsoid (Lambeck (1988)). A map of geoid height, e.g. the EGM96 geoid (Lemoine et al. (1998)) shows that although most values are on the order of tens of meters, high and low extremes of ± 80 meters or greater exist. These extremes are associated with processes such as tectonic subduction zones and deep mantle processes. In contrast, oceanographic processes induce maximum steady changes of the sea surface relative to the geoid of less than two meters and time variable changes of less than a meter. Time variable changes of the geoid itself due to these latter processes are on the order of millimeters.

Figure 1 is a schematic of the relationship of geodetically important quantities. Mass redistribution causes the geoid position to change and the sea floor to deflect. Both atmospheric and oceanic mass redistributions affect geoid height. In general, the variability of geoid position due to both the atmosphere and ocean about a static or time averaged value (schematically the variability of N′ about N) is less than the variability of the sea surface S′ about S, but not necessarily dynamically insignificant. Errors in determining S, S′ are currently larger than the variability of N, N′.

The novel element in the discussion of the geoid is the possibility of determining major elements of its time-dependence through a dedicated gravity mission, in particular GRACE. For oceanographers, this raises the startling possibility of the measurement of ocean bottom pressure changes from space.

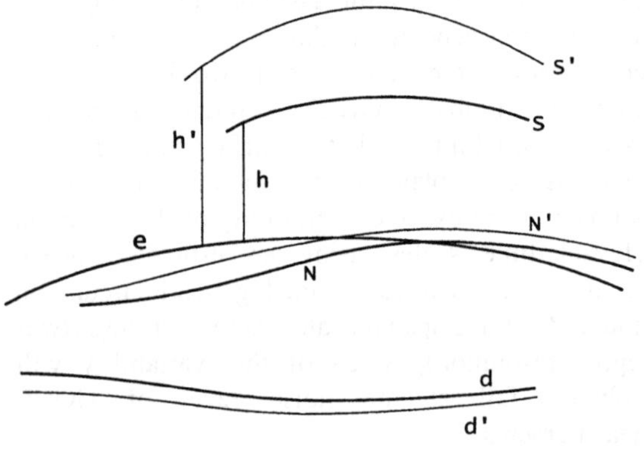

Fig. 1 Schematic of geodetically important quantities. The primed variables represent instantaneous quantities; unprimed, static or time averaged. S, S' - sea surface; N, N' - geoid height; e - ellipsoid; h, h' - geodetic height; η' – sea surface height; d, d' depth of the sea floor.

3 Bottom Pressure and Surface Elevation – Some Basic Theory

The ocean is forced by winds, atmospheric pressure, and buoyancy, and its response to the forcing is reflected in the sea surface elevation and sea floor pressure. As is well-known (Pedlosky, (1987), Wunsch (1996)), on time scales longer than about one day, on spatial scales greater than about 10 km, and at latitudes farther from the equator than a few degrees, the ocean tends, to a very high degree of accuracy, to be in hydrostatic, geostrophic balance. This balance state can be written as,

$$f\,\rho\,u = -\frac{1}{a\,\sin\theta}\frac{\partial p}{\partial\phi} \tag{1}$$

$$f\,\rho\,v = \frac{1}{a}\frac{\partial p}{\partial\theta} \tag{2}$$

$$0 = -g\,\rho - \frac{\partial p}{\partial z} \tag{3}$$

Here, θ is co-latitude, φ is longitude, ρ is density, u and v are southward and eastward velocities, and p is the pressure (the notation is completely standard). $f = 2\,\Omega\cos\theta$, where Ω is the earth rotation rate, is the Coriolis parameter. z is positive upward from the local geoid N. The spherical earth and geoid approximations being used are completely adequate for all known *dynamical* computations (see, e.g., Phillips (1966)), but it is not adequate as a kinematic description. For example, the polar radius is about 21 km shorter than the equatorial radii and the difference dominates altimetric data.

The sea surface, η, does not in general coincide with the geoid (note that in geodetic applications, the sea surface is defined with respect to the earth ellipsoid, while oceanographers refer the sea surface to the geoid - the dynamic topography). The bottom pressure can be written as

$$p_d(\theta,\phi,t) = p_a(\theta,\phi,t) + \rho_s\, g\,\eta(\theta,\phi,t) + \int_{-d}^{0} g\,\rho(z,\theta,\phi,t)\,dz \tag{4}$$

where p_a represents the atmospheric pressure, ρ_s, the density evaluated at the surface, and the upper bound on the integral is taken to be on the local geoid (z = 0). If p_a vanished (no atmospheric pressure forcing), then from the oceanographer's perspective, η would be a proxy for sea surface pressure. In practice, η is in part dependent upon p_a, and one must understand how the sea surface responds to fluctuating atmospheric pressure (reviewed elsewhere; see Wunsch and Stammer (1997), and Model Approximations section below).

The oceanic contribution to the geoid can be obtained in a first approximation using bottom pressure and a thin shell approximation for the mass density σ

$$\sigma(\theta,\phi,t) \approx \int_{z=-d(\theta,\phi)}^{\eta(\theta,\phi,t)} \rho\,(z,\theta,\phi,t)\; dz = \frac{p_d(\theta,\phi,t)}{g} \quad (5)$$

where g is the local gravitational acceleration at the earth's surface. This sheet mass density is then used to calculate the disturbing potential, T, in Bruns's formula

$$N = \frac{T}{\gamma} \quad (6)$$

where γ is normal gravity (Heiskanen and Moritz (1967)), and the disturbing potential is the sum of a static or quasi-permanent and a time variable part. Depending upon the relative accuracies, one can determine the mass distribution from T or vice-versa (Wunsch (1996)). In this context, the geoid measurements could be used to constrain ocean models. Note that actual gravity, g, appears in the dynamical equations since it governs dynamics, but that normal gravity is used in Bruns's formula.

Time variable gravity will allow recovery of bottom pressure variability, but GRACE should also provide a more accurate static geoid than is currently available on length scales larger than those of western boundary currents. On the geoid, by definition, z=0, and the surface velocities are related to the surface elevation by,

$$f\,u_s = -\frac{g}{a\sin\theta}\frac{\partial\eta}{\partial\phi} \quad (7)$$

$$f\,v_s = \frac{g}{a}\frac{\partial\eta}{\partial\theta}, \quad (8)$$

and to the mass field by,

$$\rho u_s(\theta,\phi,t) = \frac{g}{f a \sin\theta}\int_{z_{ref}}^{0} \frac{\partial\rho\,(z,\theta,\phi,t)}{\partial\phi}\, dz + c_1(\theta,\phi,t,z_{ref}) \quad (9)$$

$$\rho v_s(\theta,\phi,t) = -\frac{g}{f a}\int_{z_{ref}}^{0} \frac{\partial\rho(z,\theta,\phi,t)}{\partial\theta}\, dz + c_2(\theta,\phi,t,z_{ref}) \quad (10)$$

where the constants are a function of the reference level, z_{ref}, and are as yet undetermined. Evidently, the ability of altimetry to determine η permits the use of equations (9) and (10) to evaluate the constants directly. The surface velocities on the left hand side of the equations are constrained in current practice by altimetry. Since the variability of the geoid is neglected when considering changes in the sea surface height with respect to the geoid, a more accurate static geoid should improve the estimate of these terms. The first terms on the right hand side containing vertical integrals of horizontal density gradients have traditionally been estimated from hydrography. Time variable gravity will improve the estimate of these terms.

4 Model Approximations

At the level of precision anticipated for GRACE, one must also examine approximations not normally of concern in the large scale general circulation. OGCMs are not formulated with the geostrophic equations as written above but with a more general set of equations called the hydrostatic primitive equations (Marshall et al. (1997). The model being used in ECCO employs the so-called Boussinesq approximation, one that conserves volume and not mass. In addition, density variations are neglected in the momentum equation except in buoyancy terms. When bottom pressure (reflecting the total mass of the system) is integrated globally and expressed as a spatially uniform equivalent height of water about a mean distribution, a seasonal mass exchange superimposed on a drift over time in the model is evident. The drift of about 3.5 cm over six years or about 0.58 cm/ year (Figure 2) appears to be about twice that seen in sea level observations made from altimetry (Nerem et al. (1995)) and is evidence of mass leakage in the model that has no physical basis. The drift is reduced when a constraint is imposed on it (Stammer et al. (2003)).

At this time, atmospheric pressure loading is not included in most OGCMs and was not included in the forcing of the Stammer et al. (2002) state estimates. In practice, an assumption is often made as to how the ocean responds to atmospheric pressure forcing. The so-called inverted barometer response asserts that a 1 millibar increase (decrease) in atmospheric pressure leads to almost exactly a 1 cm. depression (elevation) in the sea surface, allowing load compensation at the sea surface. If the atmospheric load is not compensated at the sea surface, it will be transmitted to the sea floor at high frequencies (see Wunsch and Stammer (1997)), and any lateral gradients in the imposed pressure will generate oceanic flows which one seeks to determine.

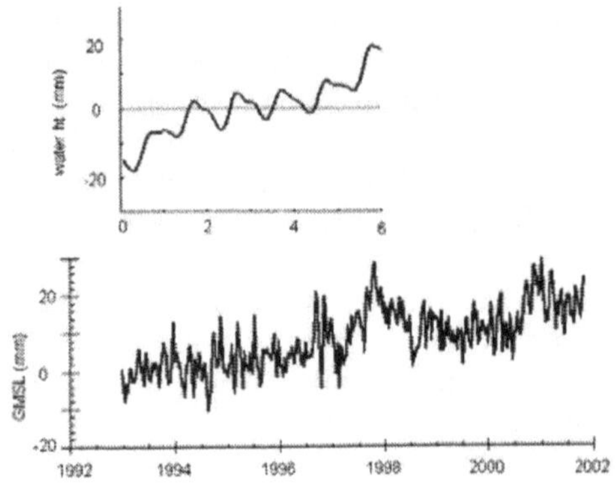

Fig. 2 Upper: total mass variation represented as a uniform change in sea level (scale is in mm); seasonal cycle is superimposed on a drift of about 5 mm/yr. Note the similarity of the trend to that in global mean sea-level determined from altimetry (lower panel). Lower: 10-day average global mean sea-level (GMSL) from the TOPEX altimeter, computed using the method described by Nerem (1995). (D. Chambers, (personal communication, 2002)).

5 Analysis

Ocean bottom pressure has been so difficult to measure for long periods of time (see, e.g., Brown et al. (1975), Luther et al. (1990), Wearn and Baker (1980), Woodworth et al. (1999)) that there is comparatively little experience available in interpreting it and in using it in tests of dynamical ideas. Our strategy has been to use a newly available estimate of the time-varying ocean circulation (Stammer et al. (2002)), obtained from combining six years of altimetric, hydrographic and other data with twice-daily wind fields, and other data and forcing in a least-squares optimization. The model output fields are used to describe the expected global variation of ocean bottom pressure.

We produced variability estimates from the estimated circulation as rms deviations about the six year mean using 30 day averages of the model output. We use a spherical harmonic representation of the bottom pressure and surface elevation in the form

$$S(\theta,\phi,t)=\sum_{n=0}^{N_e}\sum_{m=-n}^{n}\alpha_n^m(t)Y_n^m(\theta,\phi) \qquad (11)$$

where S represents the field to be analyzed, α, the expansion coefficients, Y_n^m , the fully normalized harmonics, and n and m, the degree and order, with a numerical quadrature scheme to determine the coefficients. With this method, the number of known values of the function required is about twice the number of coefficients to be determined, and an exact representation of the function cannot be achieved (unless one imposes solvability conditions that will not be met with real observations). Because the ocean does not cover the globe, difficulties arise in the use of spherical harmonics, including the discontinuities incurred at boundaries. Spurious oscillations are introduced at ocean-continent boundaries because of the abrupt change of properties. We use filtering that is an extension of the Lanczos method (Lanczos (1961)) for reducing the oscillations. The filtering operation attenuates the high wavenumber structures yet maintains the longer wavelength structures that should be detectable by GRACE.

Frequency-degree variance plots of surface and bottom pressure exhibit strong annual and semi-annual signals over a broad wavenumber band (Figure 3); less prominent ridges are evident at other frequencies, such as the 4 month period. The drop in power at high wavenumbers is due to the filtering. While these plots have been constructed from 30 day averages of model output fields, they are still complex and indicate variability on all spatial and temporal scales resolvable by our analysis, with no gaps in the spectrum. Annual, semiannual, 14 month (Chandler wobble frequency), and 4 month periods of bottom and surface pressure were extracted and mapped both as variability amplitude and phase and as time series. Here we consider only the annual period.

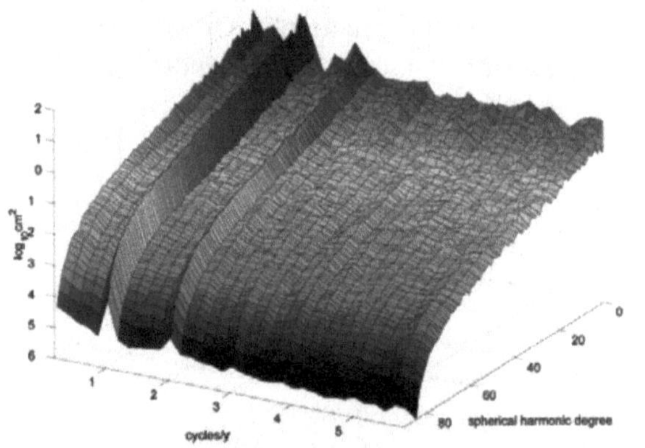

Fig. 3 Frequency-degree variance plot for bottom pressure based on 6 year model.

The prominent regions of annual cycle bottom pressure variability occurring northwest of the Drake Passage, the high latitude Atlantic, and in the Southern Ocean (southwest of Australia) (Figure 4), for the most part, have variability near 2 cm equivalent water thickness. Temporally, the same nearly fixed phase occurs in the North Atlantic, a large region in the North Central Pacific, and the Southern Ocean, just southwest of Australia. The high latitude Atlantic and part of the southern Atlantic peak around the same time as do large parts of the Pacific and Indian oceans. The patterns are quite different from those of surface pressure (Figure 5) for which a notable hemispheric asymmetry in phase is apparent.

In a constant density (homogeneous) hydrostatic ocean, surface and bottom pressure are identical; this state is termed 'barotropic' by oceanographers. In a stratified ocean, the two can be very different. Of the periods considered by Condi and Wunsch (2003), the 4 month period appears to have the greatest barotropic character, but with notable departures from a barotropic structure occurring in the tropical Pacific and Atlantic, and the Indian Oceans. The Southern Ocean is evidently the region of greatest barotropic motion at the annual period; there appear to be significant departures from barotropic structure in the high latitude North Atlantic.

6 Detectability

Sea surface variability (not including that due to pressure forcing) can attain values of up to 10 cm in the filtered fields. The oceanic contribution to the annual component of geoid variability, as measured by a space based observer, can attain values in high variability regions on the order of a millimeter (see Condi and Wunsch (2003)). High variability regions occur in the Pacific, northwest of the Drake Passage, and in the North Atlantic. Because the geoid is an integral over the anomalous mass, its variability is a smoothed representation of the anomalous bottom pressure variability.

A major issue in discussing a satellite mission is the extent to which particular signals can be detected practically. Determination of the contributions from each of the mapped periods can be made by comparing the geoid degree amplitudes for each of the periods with the estimated GRACE errors (Tapley and Reigber (2001), S. Bettadpur (personal communication, 2001)). The present bottom pressure analysis suggests possible recovery of the annual period to about degree 27 and the semi-annual and Chandler wobble period to degree 14 (Figure 6). The steric correction due to mass leakage of the OGCM was calculated as a uniform layer of water over the oceans for both the annual period and full 72 month period, and appears to be detectable to about degree 7 for both. Any dynamical correction would be concentrated at higher wave numbers, and although mission errors become larger at higher wavenumber, the corrections could still be detectable. The length scales at which GRACE will be able to detect changes are larger than typical widths of western boundary currents and GOCE should partially address this shortcoming.

Gravity data and ocean modeling have dual roles. The spatial scale improvement from GRACE for the time variable geoid is expected to be up to about spherical harmonic degree 80 (S. Bettadpur, personal communication, (2001)). Degree 80 corresponds to a half-wavelength of 250 km). Considering that the estimates of bottom pressure variability recovery from GRACE are to degree 27 for the annual, degree 14 for the semi annual and Chandler wobble periods, and somewhat higher degree for the total, the use of the time variable gravity may be limited as a direct constraint, excluding scales with half wavelengths less than about 750 km

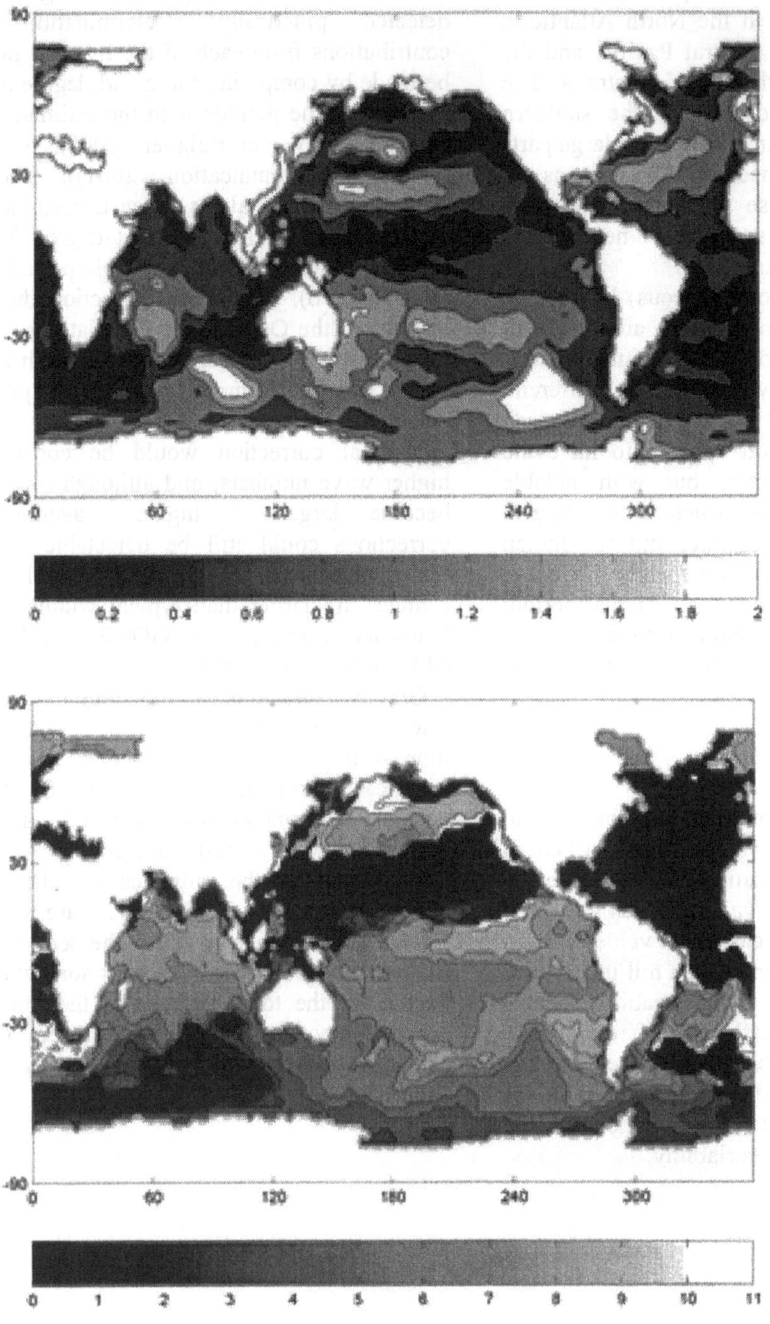

Fig. 4 Upper: annual bottom pressure amplitude (scale in cm). Lower: phase (scale in months).

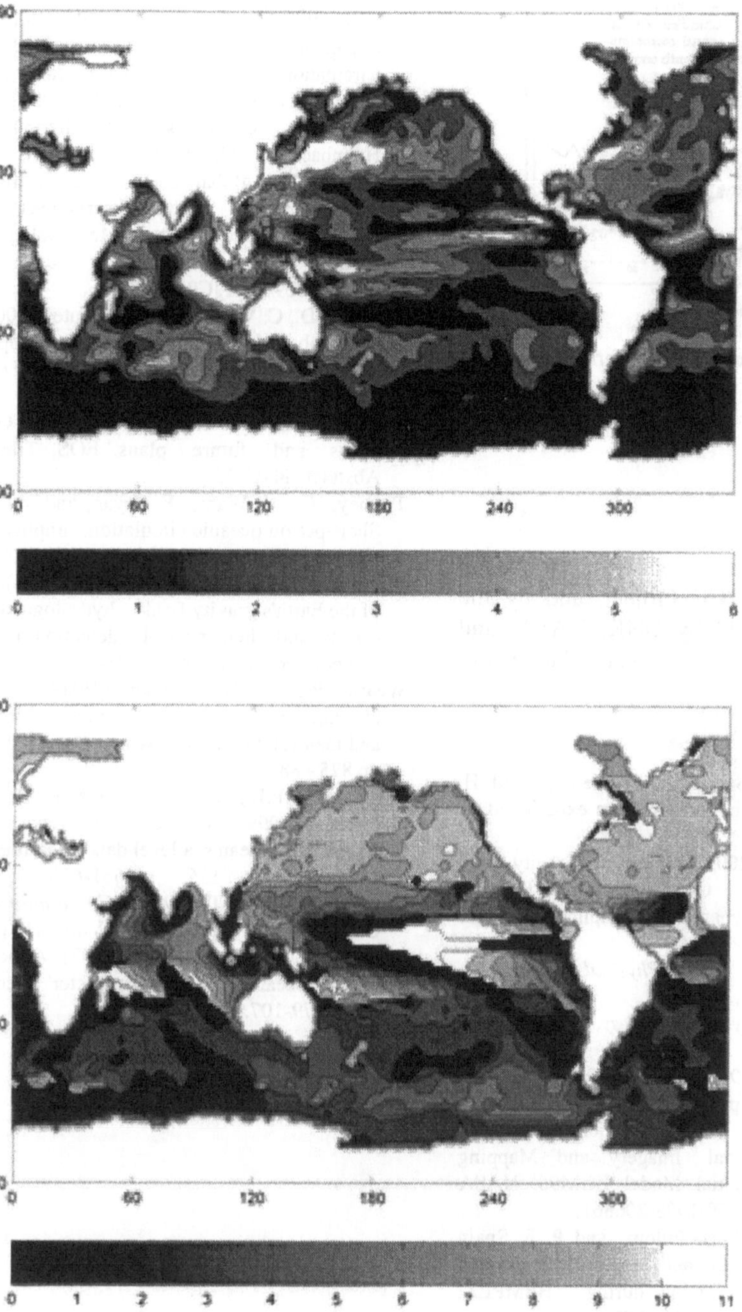

Fig. 5 Upper: annual surface pressure amplitude (scale in cm). Lower: phase (scale in months).

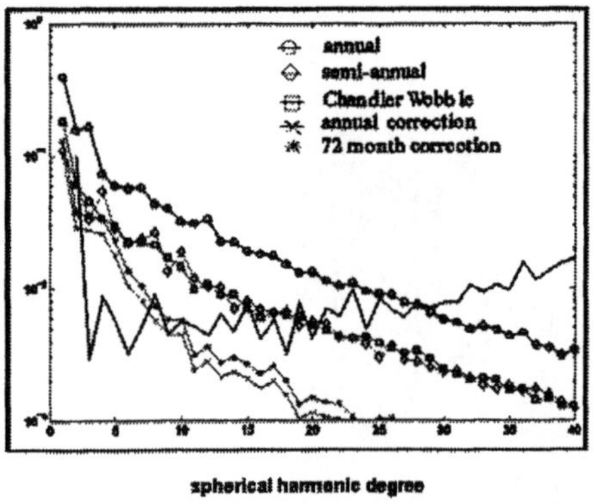

Fig. 6 Detectability of bottom pressure variability (degree amplitude scale in mm). Solid line represents GRACE errors.

Acknowledgments

Supported by the University of Texas at Austin through the NASA GRACE Project, and by the ECCO Consortium funded by ONR, NASA, and NSF through the National Ocean Partnership Program.

References

Brown, W., W. Munk, F. Snodgrass, B. Zetler, and H. Mofjeld (1975). MODE bottom pressure experiment. *J. Phys. Ocean.*, 5, pp.75-85.

Condi, F., and C. Wunsch (2003). Measuring Gravity Field Variability, the Geoid, Ocean Bottom Pressure Fluctuations, and Their Dynamical Implications. *J. Geophys. Res.* (in press).

Heiskanen, W. and H. Moritz (1967). *Physical Geodesy*. W. H. Freeman, San Francisco, 364 pp.

Lambeck, K (1988). *Geophysical Geodesy*. Oxford, New York, 718 pp.

Lanczos, C. (1961). *Linear Differential Operators*. D. Van Nostrand, New York, 564 pp.

Lemoine, F.G., et al. (1998). The Development of the Joint NASA GSFC and the National Imagery and Mapping Agency (NIMA) Geopotential Model EGM96. NASA Technical Report, NASA/TP-1998-206861.

Luther, D.S., A. D. Chave, J. H. Filloux, and P. F. Spain (1990). Evidence for local and nonlocal barotropic responses to atmospheric forcing during BEMPEX, *Geophys. Res. Lett.*, 17, pp. 949-952.

Marshall, J., C. Hill, L. Perelman, and A. Adcroft. (1997). Hydrostatic, quasi-hydrostatic, and nonhydrostatic ocean modeling. *J. Geophys. Res.*, 102, pp.5733-52.

Nerem R. S. (1995). Measuring global mean sea level variations using TOPEX/POSEIDON altimeter data. *J. Geophys. Res.*, 100, C12, pp.25135-25151.

Pedlosky, J. (1987). *Geophysical Fluid Dynamics*, 2nd ed., Springer-Verlag, New York, 710 pp.

Phillips, N. (1966). The equations of motion for a shallow rotating atmosphere and the "traditional approximation". *J. Atmos. Sci.*, 23, pp.626-628.

Stammer, D., C. Wunsch, R. Giering, C. Eckert, P. Heimbach, J. Marotzke, A. Adcroft, C. N. Hill, and J. Marshall (2003). Volume, heat, and freshwater transports of the global ocean circulation 1993–2000, estimated from a general circulation model constrained by World Ocean Circulation Experiment (WOCE) data, *J. Geophys. Res.*, 108(C1), 3007, doi:10.1029/2001JC001115.

Stammer, D., C. Wunsch, R. Giering, C. Eckert, P. Heimbach, J. Marotzke, A. Adcroft, and J. Marshall (2002). The global ocean state during 1992 - 1997, estimated from ocean observations and a general circulation model: Part I: Methodology and Estimated State. *J. Geophys. Res.*, 107(C9)3118, doi:10.1029/2001JC000888.

Stammer, D., C. Wunsch, and R. Ponte (2000). De-aliasing of Global High Frequency Barotropic Motions in Altimeter Observations. *Geophys. Res. Let.*, 27, pp.1175-78.

Tapley, B., and C. Reigber (2001).The GRACE mission: status and future plans. EOS. Trans. AGU,82(47), Abstract G41C-02.

Tierney, C., J. Wahr, F. Bryan, and V. Zlotnicki (2000). Short-period oceanic circulation: implications for satellite altimetry. *Geophys. Res. Lett*, 27, pp.1255-58.

Wahr, J., M. Molenaar, and F. Bryan (1998). Time variability of the Earth's gravity field: hydrological and oceanic effects and their possible detection using GRACE. *J. Geophys. Res.*, 103, pp.30205-229.

Wearn., R., and D. J. Baker (1980). Bottom pressure measurements across the Antarctic Circumpolar Current and their relation to the wind, *Deep Sea Res.*, 27A, pp.875 - 888.

Woodworth, P.L., M. N. Tsimplis, R. A. Flather, and 1. Shennan (1999). A review of the trends observed in British Isles mean sea level data measured by tide gauges. *Geophys. J. Int.*, 136, pp.651-670.

Wunsch, C. (1996). The *Ocean Circulation Inverse Problem*, Cambridge Univ. Press, New York, 437 pp.

Wunsch, C., and D. Stammer (1997). Atmospheric loading and the oceanic 'inverted barometer' effect. *Rev.Geophys.*, 35, pp. 79-107.

Estimating Covariance Parameters in Gravity Downward Continuation

J. Kusche
Department of Physical, Geometrical, and Space Geodesy
TU Delft, Thijsseweg 11, 2629 JA Delft, The Netherlands

Abstract. We discuss same concepts related to the estimation of covariance parameters in the numerical treatment of gravity downward continuation. We aim explicitly at the determination of weighting factors, regularization parameters, and residual data correlations in the inversion of satellite gravity data for spherical harmonic coefficients. Special attention is given to the efficient treatment of large–sized data sets, having in mind data reduction and model validation for e. g. the GOCE mission, and to the derivation of probability density functions.

Keywords. Covariance parameters, variance components, downward continuation, satellite gravity data, pdf's

1 Introduction

In this contribution we discuss same concepts, aspects, and problems related to the estimation of covariance parameters in numerical gravity downward continuation. We aim explicitly at the interpretation of large-sized satellite data sets to be collected for example in the course of ESA's planned GOCE mission (ESA 1999), where the downward continuation process due to satellite altitude and data gaps poses an ill–posed problem, and where pre–specified stochastic models will require validation or even updating from the actual data. Issues such as the assessment of relative weighting factors between different data types (orbit perturbations and gravity gradients, for example), the (possibly degree–dependent) weighting of prior information regarding the Earth's gravity field, and the estimation of residual correlations between observation errors can be viewed as a common problem of determining covariance parameters; that is, estimating a–priori unknown parameters in variance–covariance matrices. We will also discuss an example where the variance–covariance ('regularization') matrix depends non–linear on an unknown parameter; therefore we prefer to speak of 'covariance parameters' rather than of variance–covariance components. Obviously, when we allow for estimating such parameters – more precisely corrections to some pre–specified baseline parameters –, the question of statistical significance becomes relevant, and probability density functions (pdf's) should be applied in testing.

It is known, that the Bayesian theory, among others, provides a suitable framework for this kind of problem. Tikhonov–regularized estimation, for example, can be viewed formally as Bayesian inversion if one agrees to identify the regularization matrix with a prior variance–covariance matrix for the unknown parameters. In this view, the estimation of variance and covariance components, and based on this, the a–posterior determination of regularization and weighting parameters, is inherently linked to the assessment of pdf's for these quantities, so that the interpretation in terms of significance and confidence regions can go beyond what standard procedures provide. However, for large–scale problems like GOCE data inversion (we expect hundred millions of observations to be collected, and the order of 90,000 gravity field unknowns to be estimated), many features of the theory seem to be computationally infeasible. But, with the advent of fast multiprocessor inversion programs and numerical tools that allow to assess for instance functions of elements of large inverse matrices very quickly, these restrictions must be re–evaluated.

This paper is organized as follows: In the second chapter we address briefly the general probabilistic setting for our treatment of linear ill–posed problems with unknown covariance parameters. Thereafter, the determination of variance components and relative weighting factors for disjoint data sets and their pdf's in large–sized problems is discussed. In the fourth chapter, this is further specialized to the estimation of regularization parameters and their assigned pdf. In addition, we give the maximum–likelihood equations which apply to the estimation of a parameter controlling non–linearly the power decrease in the error degree variance model employed in the regularization. Then, in chapter 5 the estimation of covariance components describing residual error correlations is discussed. A summary closes this contribution.

2 Probabilistic Setting

The theory presented in the chapter is not new; it is only meant to serve the algorithmic concepts proposed in the following chapters. Let us begin with the linear model (1), that is with the $u \times 1$ vector $\boldsymbol{\beta}$ of unknown parameters, the $n \times 1$ vector $\boldsymbol{y}$ of observations with $n \times n$ variance–covariance matrix $\boldsymbol{\Sigma}(\boldsymbol{\sigma})$, and the $p \times 1$ vector $\boldsymbol{\sigma} \in \boldsymbol{\Theta}$ of unknown covariance parameters,

$$E(\boldsymbol{y}|\boldsymbol{\beta}) = \boldsymbol{X}\boldsymbol{\beta} \qquad D(\boldsymbol{y}|\boldsymbol{\sigma}) = \boldsymbol{\Sigma}(\boldsymbol{\sigma}) \,. \tag{1}$$

The subspace $\boldsymbol{\Theta}$ must be such that $D(\boldsymbol{y}|\boldsymbol{\sigma})$ is positive definite. $E(\cdot)$ means the mathematical expectation, $D(\cdot)$ the dispersion operator, and | stands for conditional. See Koch (1990) for the notation used throughout this paper. Generally we assume the data being normally distributed, thus

$$p(\boldsymbol{y}|\boldsymbol{\beta},\boldsymbol{\sigma}) = \frac{\exp\left(-\frac{1}{2}(\boldsymbol{y}-\boldsymbol{X}\boldsymbol{\beta})'\boldsymbol{\Sigma}^{-1}(\boldsymbol{y}-\boldsymbol{X}\boldsymbol{\beta})\right)}{(2\pi)^{n/2}|\boldsymbol{\Sigma}|^{1/2}} \,. \tag{2}$$

Bayes' theorem then relates the posterior pdf to the prior pdf being modified by the likelihoodfunction of the observations,

$$p(\boldsymbol{\beta},\boldsymbol{\sigma}|\boldsymbol{y}) = c \cdot p(\boldsymbol{y}|\boldsymbol{\beta},\boldsymbol{\sigma})\, p(\boldsymbol{\beta},\boldsymbol{\sigma}) \tag{3}$$

where c is a normalization constant. In the Bayesian view, all statistical inference is contained in the posterior pdf, eq. (3).

The rationale behind Bayes' theorem, in the light that the unknown random parameter vector $\boldsymbol{\beta}$ in downward continuation of satellite data usually consists of SH coefficients describing a finite–dimensional approximation of the Earth's gravity field,

$$\boldsymbol{\beta} = (\delta C_{20}, \delta C_{21}, \ldots, \delta S_{ll})' \tag{4}$$

plus possibly some additional calibration and state vector parameters, is to describe how our knowledge on $\boldsymbol{\beta}$ and $\boldsymbol{\sigma}$, expressed by prior and posterior pdf, improves when observations are taken and information is absorbed.

For the sake of clarity we first assume that the vector of covariance parameters collected in $\boldsymbol{\sigma}$ is perfectly known. Let us consider a flat (noninformative) prior for $\boldsymbol{\beta}$, that is $p(\boldsymbol{\beta}) = c$; no constraints shall be imposed on $\boldsymbol{\beta}$. The log–posterior becomes then

$$\begin{aligned} &-2 \cdot \log p(\boldsymbol{\beta}|\boldsymbol{y}) \\ &= \log\left((2\pi)^n|\boldsymbol{\Sigma}|\right) + (\boldsymbol{y}-\boldsymbol{X}\boldsymbol{\beta})'\boldsymbol{\Sigma}^{-1}(\boldsymbol{y}-\boldsymbol{X}\boldsymbol{\beta}) \,. \end{aligned} \tag{5}$$

Minimization of this well-known 'cost function' with respect to $\boldsymbol{\beta}$ leads to the common LS normal equations. But unfortunately, and as it is well–known, the downward continuation of satellite gravity data poses an ill–posed problem. This leads to the consequence that even after truncation of the spherical harmonic representation to a finite–dimensional subspace the resulting normal equations tend to be ill–conditioned. To stabilize the solution, one usually applies some kind of regularization, which means nothing else that one introduces smoothness assumptions on the unknowns. On the other hand, it is fair to assume that we have such prior information in stochastic form: From prior determinations of the set of the SH coefficients, and from physics telling us that the energy concentrated in high–degree harmonics must be limited. Whether the deliberate use of this prior information is viewed upon as (stochastic) regularization or considered as a model extension, may be left to the reader. However, seen from a non–Bayesian viewpoint including prior information leads to an unbiased estimate. Assume now as prior pdf the normal distribution

$$p(\boldsymbol{\beta}) = \frac{\exp\left(-\frac{1}{2}(\boldsymbol{\beta}-\boldsymbol{\mu})'\boldsymbol{\Gamma}^{-1}(\boldsymbol{\beta}-\boldsymbol{\mu})\right)}{(2\pi)^{u/2}|\boldsymbol{\Gamma}|^{1/2}} \tag{6}$$

with expectation $\boldsymbol{\mu}$ and covariance matrix $\boldsymbol{\Gamma}$. The particular choice of a normal prior can be motivated from the maximum entropy principle (see e. g. Lehmann 1996). Minimization of the log–posterior

$$\begin{aligned} &-2 \cdot \log p(\boldsymbol{\beta}|\boldsymbol{y}) \\ &= \log\left((2\pi)^n|\boldsymbol{\Sigma}|\right) + (\boldsymbol{y}-\boldsymbol{X}\boldsymbol{\beta})'\boldsymbol{\Sigma}^{-1}(\boldsymbol{y}-\boldsymbol{X}\boldsymbol{\beta}) \\ &\quad + \log\left((2\pi)^u|\boldsymbol{\Gamma}|\right) + (\boldsymbol{\beta}-\boldsymbol{\mu})'\boldsymbol{\Gamma}^{-1}(\boldsymbol{\beta}-\boldsymbol{\mu}) \,. \end{aligned} \tag{7}$$

with respect to $\boldsymbol{\beta}$ leads to the well–known ('regularized') normal equations

$$\left(\boldsymbol{X}'\boldsymbol{\Sigma}^{-1}\boldsymbol{X} + \boldsymbol{\Gamma}^{-1}\right)\hat{\boldsymbol{\beta}} = \boldsymbol{X}'\boldsymbol{\Sigma}^{-1}\boldsymbol{y} + \boldsymbol{\Gamma}^{-1}\boldsymbol{\mu} \tag{8}$$

In what follows, we will always assume $\boldsymbol{\mu} = \boldsymbol{0}$, meaning that the δC_{lm}, δS_{lm} are defined as residual with respect to the best–known reference model.

Problem (8) is linear, though nontrivial for gravity downward continuation from modern satellite missions such as GOCE, since the number of data (n) can be hundreds of millions, the number of unknown gravity field coefficients, nuisance, and state vector parameters (u) tends to be of the order of 90.000. However, current developments (Klees et al., this volume) indicate that the machinery will be at hand to evaluate matrix–vector products $\boldsymbol{\Sigma}^{-1}\boldsymbol{z}$ quickly by filtering techniques, and to solve the system (8) in reasonable time (thus allowing for a number of repetitions), either using direct solvers or the cg–method. Same evidence is that possibly at least a good approximation to the covariance matrix $D(\boldsymbol{\beta}|\boldsymbol{y}) = (\boldsymbol{X}'\boldsymbol{\Sigma}^{-1}\boldsymbol{X} + \boldsymbol{\Gamma}^{-1})^{-1}$ of the posterior pdf will be available.

For the remainder of this article we assume the vector $\boldsymbol{\sigma}$ to be unknown. We focus on maximum–likelihood estimates for the covariance parameters $\boldsymbol{\sigma}$, thus rejecting any prior information on these. Maximum a–posteriori (MAP–) estimates however, which do accommodate for informative prior information about $\boldsymbol{\sigma}$, can be derived as weighted combination of the prior expectation of the $\boldsymbol{\sigma}$ and their ML estimates (Ou and Koch 1994). This means that the main numerical burden will be still in the computation of the ML estimates. We will begin with the normal likelihoodfunction $p(\boldsymbol{y}|\boldsymbol{\beta},\boldsymbol{\sigma})$, eq. (2). As is known, one may form the marginal density w.r.t. $\boldsymbol{\beta}$, and obtains the likelihood function depending only on $\boldsymbol{\sigma}$:

$$p(\boldsymbol{y}|\boldsymbol{\sigma}) \propto \frac{\exp\left(-\frac{1}{2}\boldsymbol{y}'\boldsymbol{W}\boldsymbol{y}\right)}{|\boldsymbol{\Sigma}|^{1/2}|\boldsymbol{X}'\boldsymbol{\Sigma}^{-1}\boldsymbol{X}|^{1/2}} \tag{9}$$

where

$$\boldsymbol{W} = \boldsymbol{\Sigma}^{-1} - \boldsymbol{\Sigma}^{-1}\boldsymbol{X}(\boldsymbol{X}'\boldsymbol{\Sigma}^{-1}\boldsymbol{X})^{-1}\boldsymbol{X}'\boldsymbol{\Sigma}^{-1} \tag{10}$$

with $\boldsymbol{y}'\boldsymbol{W}\boldsymbol{y} = \hat{e}'\boldsymbol{\Sigma}^{-1}\hat{e}$ and $\hat{e} = \boldsymbol{y} - \boldsymbol{X}\hat{\boldsymbol{\beta}}$. Setting the first derivatives of the log–likelihood w.r.t $\boldsymbol{\sigma}$ equal to zero gives

$$\frac{\partial}{\partial\sigma_i}\log\left(|\boldsymbol{\Sigma}|\right) + \frac{\partial}{\partial\sigma_i}\log\left(|\boldsymbol{X}'\boldsymbol{\Sigma}^{-1}\boldsymbol{X}|\right) + \boldsymbol{y}'\frac{\partial}{\partial\sigma_i}\boldsymbol{W}\boldsymbol{y} = 0 \tag{11}$$

Or, as shown e. g. in Koch (1990),

$$\operatorname{trace}\left(\boldsymbol{W}\frac{\partial}{\partial\sigma_i}\boldsymbol{\Sigma}\right) = \boldsymbol{y}'\boldsymbol{W}\left(\frac{\partial}{\partial\sigma_i}\boldsymbol{\Sigma}\right)\boldsymbol{W}\boldsymbol{y} \tag{12}$$

Until now, no special (linear) structure $\boldsymbol{\Sigma}(\boldsymbol{\sigma})$ has been assumed. Together with the common additive variance and covariance components model $\boldsymbol{\Sigma}(\boldsymbol{\sigma}) = \sigma_1^2\boldsymbol{V}_1 + \sigma_{12}\boldsymbol{V}_2 + \cdots$, with $\boldsymbol{\sigma} = (\sigma_1^2, \sigma_{12}, \ldots)'$, a maximum–likelihood estimate leads to the well-known iterative algorithm (see e. g. Förstner 1979) that we are going to exploit in eqs. (18), (20), and eqs. (32), (33).

Estimates according to (12) in the linear variance–covariance component model are numerically identical to those estimates obtained from the iterated MINQUE technique (e. g. Rao and Kleffe 1988), assuming normal statistics. It has been argued, that MINQUE iterations converge faster and that maximum–likelihood estimators in some cases fail and must be handled with care (Rao and Kleffe 1988). On the other hand, the difficulty with the MINQUE technique in the present context is in the computation of the coefficient matrix, whose entries involve trace operations on multiple matrix products that are much more difficult to assess when relying on trace estimation techniques due to the size of the problem.

Finally, we will briefly review the 'regularization' case of eqs. (6) and (7), thus assuming that prior information on the parameter vector $\boldsymbol{\beta}$ has been given in form of a normal distribution with expectation $\boldsymbol{\mu}$ and covariance matrix $\boldsymbol{\Gamma}$, which may very well depend on some unknown parameters collected in the vector $\boldsymbol{\gamma}$. That is, $\boldsymbol{\Gamma} = \boldsymbol{\Gamma}(\boldsymbol{\gamma})$. According to the recursive character of Bayes' theorem, it is clear that one may consider this prior pdf (6) alternatively as a likelihood function $p(\boldsymbol{\mu}|\boldsymbol{\beta},\boldsymbol{\gamma})$, in analogy to (11), (12). Multiplication of the two likelihood functions and integrating out $\boldsymbol{\beta}$, leads then to

$$p(\boldsymbol{y},\boldsymbol{\mu}|\boldsymbol{\sigma},\boldsymbol{\gamma}) \propto \frac{\exp\left(-\frac{1}{2}\boldsymbol{y}'\boldsymbol{W}^{(1)}\boldsymbol{y} - \boldsymbol{y}'\boldsymbol{W}^{(2)}\boldsymbol{\mu} - \frac{1}{2}\boldsymbol{\mu}'\boldsymbol{W}^{(3)}\boldsymbol{\mu}\right)}{|\boldsymbol{\Sigma}|^{1/2}|\boldsymbol{\Gamma}|^{1/2}|\boldsymbol{X}'\boldsymbol{\Sigma}^{-1}\boldsymbol{X} + \boldsymbol{\Gamma}^{-1}|^{1/2}} \tag{13}$$

where

$$\begin{aligned} \boldsymbol{W}^{(1)} &= \boldsymbol{\Sigma}^{-1} - \boldsymbol{\Sigma}^{-1}\boldsymbol{X}(\boldsymbol{X}'\boldsymbol{\Sigma}^{-1}\boldsymbol{X} + \boldsymbol{\Gamma}^{-1})^{-1}\boldsymbol{X}'\boldsymbol{\Sigma}^{-1} \\ \boldsymbol{W}^{(2)} &= \boldsymbol{\Gamma}^{-1}(\boldsymbol{X}'\boldsymbol{\Sigma}^{-1}\boldsymbol{X} + \boldsymbol{\Gamma}^{-1})^{-1}\boldsymbol{X}'\boldsymbol{\Sigma}^{-1} \\ \boldsymbol{W}^{(3)} &= \boldsymbol{\Gamma}^{-1} - \boldsymbol{\Gamma}^{-1}(\boldsymbol{X}'\boldsymbol{\Sigma}^{-1}\boldsymbol{X} + \boldsymbol{\Gamma}^{-1})^{-1}\boldsymbol{\Gamma}^{-1} \end{aligned} \tag{14}$$

which is just a block–matrix extension of (9), (10). Stationarity of the likelihoodfunction (13) requires, in analogy to (11), (12), and evaluated for $\boldsymbol{\mu} = \boldsymbol{0}$,

$$\operatorname{trace}\left(\boldsymbol{W}^{(1)}\frac{\partial\boldsymbol{\Sigma}}{\partial\sigma_i}\right) = \boldsymbol{y}'\boldsymbol{W}^{(1)}\left(\frac{\partial\boldsymbol{\Sigma}}{\partial\sigma_i}\right)\boldsymbol{W}^{(1)}\boldsymbol{y} \tag{15}$$

$$\operatorname{trace}\left(\boldsymbol{W}^{(3)}\frac{\partial\boldsymbol{\Gamma}}{\partial\gamma_i}\right) = \boldsymbol{y}'\boldsymbol{W}^{(2)\prime}\left(\frac{\partial\boldsymbol{\Gamma}}{\partial\gamma_i}\right)\boldsymbol{W}^{(2)}\boldsymbol{y} \tag{16}$$

No special (linear) structure of $\boldsymbol{\Sigma}(\boldsymbol{\sigma})$ and $\boldsymbol{\Gamma}(\boldsymbol{\gamma})$ has been assumed.

3 Weighting in Combination Solutions

In this chapter we will become more specific, with respect to the relative weighting issue in the combination of heterogeneous data sets. We assume that disjoint observation groups are present, uncorrelated with respect to each other. This will be the case for example when combining orbit perturbations derived from GPS–SST with gradiometry data for the GOCE satellite mission.

We consider thus p unknown variance components $\boldsymbol{\sigma} = (\sigma_1^2, \sigma_2^2, \ldots, \sigma_p^2)' \subset \Theta$; such that $D(\boldsymbol{y}|\boldsymbol{\sigma})$ is positive definite:

$$D(\boldsymbol{y}|\boldsymbol{\sigma}) = \boldsymbol{\Sigma}(\sigma_1^2, \sigma_2^2, \ldots, \sigma_p^2) = \begin{pmatrix} \sigma_1^2 \boldsymbol{P}_1^{-1} & 0 & \cdots & 0 \\ 0 & \sigma_2^2 \boldsymbol{P}_2^{-1} & \cdots & 0 \\ \vdots & & & \vdots \\ 0 & 0 & \cdots & \sigma_p^2 \boldsymbol{P}_p^{-1} \end{pmatrix} . \quad (17)$$

The actual number p will be small when compared to the degrees of freedom of the problem; it will be for instance three when one considers orbit perturbations, gradiometry, and a prior gravity model as disjoint observation groups. In what follows, $\boldsymbol{y}_i$, $\boldsymbol{X}_i$, $\hat{\boldsymbol{e}}_i$ are the observation group, the design matrix segment, and the adjustment residuals associated with the i–th data set of length n_i. By $\boldsymbol{N} = \sum_{k=1}^{p} \frac{1}{\sigma_i^2} \boldsymbol{X}_i' \boldsymbol{P}_i \boldsymbol{X}_i$ the weighted combined normal matrix is denoted. The usual iterative algorithm (Förstner 1979), which follows from (12), estimates the variance components in each iteration using

$$\hat{\sigma}_i^2 = \frac{\hat{\boldsymbol{e}}_i' \boldsymbol{P}_i \hat{\boldsymbol{e}}_i}{r_i} \qquad i \in \{1, \ldots, p\} . \quad (18)$$

$$\hat{\boldsymbol{e}}_i = \boldsymbol{X}_i \hat{\boldsymbol{\beta}} - \boldsymbol{y}_i . \quad (19)$$

The numerical bottleneck in a large–size problem is in the computation of the group redundancy numbers

$$r_i = \mathrm{trace}(\boldsymbol{W} \hat{\sigma}_i^2 \boldsymbol{V}_i) = n_i - \frac{1}{\hat{\sigma}_i^2} \mathrm{trace}\left(\boldsymbol{G}_i' \boldsymbol{X}_i \boldsymbol{N} \boldsymbol{X}_i' \boldsymbol{G}_i\right) \quad (20)$$

where $\boldsymbol{P}_i = \boldsymbol{G}_i \boldsymbol{G}_i'$. The symmetric expression on the right–hand side of (20) is amenable to stochastic trace estimation techniques (Kusche, 2003, Klees et al., this volume): Let $\boldsymbol{z}_{ik}$, $\dim(\boldsymbol{z}_{ik}) = n_i$ be a set of K artificially generated realizations of a random vector $\boldsymbol{z}_i$ with $E(\boldsymbol{z}_i) = \boldsymbol{0}$, $D(\boldsymbol{z}_i) = \boldsymbol{I}$. Direct evaluation of (20) can then be replaced by the two–step procedure

1. Solve ($i = \{1 \ldots p\}, k = \{1 \ldots K\}$)

$$\boldsymbol{N} \hat{\boldsymbol{q}}_{ik} = \boldsymbol{X}_i' \boldsymbol{G}_i \boldsymbol{z}_{ik} \quad (21)$$

for the $p \cdot k$ 'random parameter solutions' $\hat{\boldsymbol{q}}_{ik}$

2. Compute the p estimates

$$\hat{r}_i = n_i - \frac{1}{\hat{\sigma}_i^2} \frac{1}{K} \sum_{k=1}^{K} \boldsymbol{z}_{ik}' \boldsymbol{G}_i' \tilde{\boldsymbol{X}}_i \hat{\boldsymbol{q}}_{ik} \quad (22)$$

Step 1 means nothing else than solving the original normal equations with the right–hand sides $\boldsymbol{X}_i' \boldsymbol{G}_i \boldsymbol{z}_{ik}$ i. e. assembled from random observations for observation group i, scaled by $\boldsymbol{G}_i$, and from zeroes for all other observations. By K one may control the accuracy of this estimation; numerical experiments, however, indicate so far that $K = 1$ realization may be sufficient. By taking the expectation of (22) one proofs that the $\hat{r}_i$ are unbiased estimates of r_i. The principle is outlined in the figure 1.

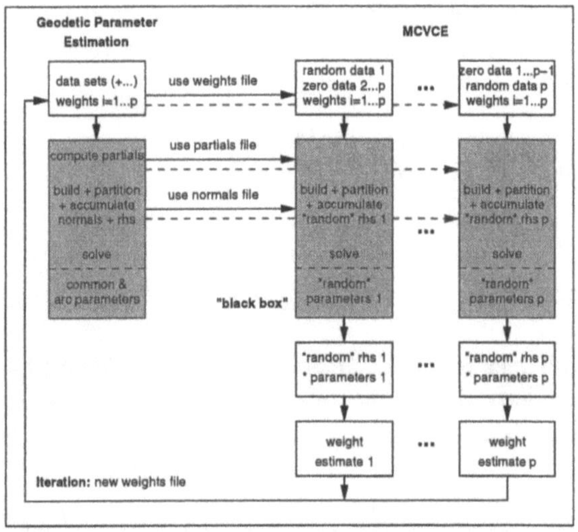

Fig. 1 Use of trace estimation techniques; Monte Carlo variance component estimation (MCVCE); according to (21), (22)

One might, alternatively, work simultaneously with the complete original data set $\boldsymbol{y}$, supplemented by $p \cdot K$ artificially perturbed data vectors $\boldsymbol{y}_{ik}^*$

$$\boldsymbol{y}_{ik}^* = \begin{pmatrix} \boldsymbol{y}_1 \\ \vdots \\ \boldsymbol{y}_i + \boldsymbol{w}_{ik} \\ \vdots \\ \boldsymbol{y}_p \end{pmatrix} = \boldsymbol{y} + \begin{pmatrix} \boldsymbol{0} \\ \vdots \\ \boldsymbol{w}_{ik} \\ \vdots \\ \boldsymbol{0} \end{pmatrix} \quad (23)$$

with K realizations $\boldsymbol{w}_{ik}$ of a random vector $\boldsymbol{w}_i$ with $E(\boldsymbol{w}_i) = \boldsymbol{0}$, $D(\boldsymbol{w}_i) = \sigma_i^2 \boldsymbol{P}_i^{-1}$. In this case we obtain in the first step by accumulation and solution of the normal equations the $p \cdot K$ parameter vectors $\hat{\boldsymbol{\beta}}^*_{ik}$. Thus,

1. Solve ($i = \{1 \ldots p\}, k = \{1 \ldots K\}$)

$$\boldsymbol{N}\hat{\boldsymbol{\beta}}_{ik} = \boldsymbol{X}_i{}'\boldsymbol{P}_i\,\boldsymbol{y}^*_{ik} \tag{24}$$

2. Compute the p estimates

$$\begin{aligned}\hat{r}_i &= n_i - \frac{1}{\hat{\sigma}_i^4}\frac{1}{K}\sum_{k=1}^{K} \boldsymbol{w}'_{ik}\boldsymbol{P}_i\boldsymbol{X}_i(\hat{\boldsymbol{\beta}}^*_{ik} - \hat{\boldsymbol{\beta}}) \\ &= n_i - \frac{1}{\hat{\sigma}_i^4}\frac{1}{K}\sum_{k=1}^{K} \boldsymbol{w}'_{ik}\boldsymbol{P}_i\,(\hat{\boldsymbol{e}}^*_{ik} - \hat{\boldsymbol{e}}_i)\;. \end{aligned} \tag{25}$$

Along these lines the algorithm for deriving $\hat{r}_i$ needs only the residuals $\hat{\boldsymbol{e}}_i$ from the 'unperturbed' data $\boldsymbol{y}$, and those $(\hat{\boldsymbol{e}}^*_{ik})$ from the perturbed data sets. An existing algorithm for accumulation and solution of the normals can be used basically in a 'black–box' fashion. Figure 2 gives an outline of the procedure.

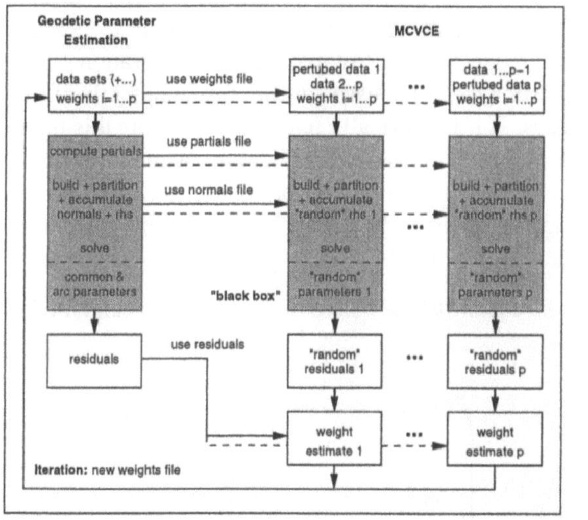

Fig. 2 Use of trace estimation techniques; Monte Carlo variance component estimation (MCVCE); according to (24), (25)

In case of Jeffrey's noninformative priors the variance components are independently distributed like the inverted gamma distribution with the posterior pdf (Ou and Koch 1994)

$$p\left(\sigma_i^2|\boldsymbol{y}_i\right) = \left(\frac{1}{2}\hat{\boldsymbol{e}}_i'\boldsymbol{P}_i\hat{\boldsymbol{e}}_i\right)^{r_i/2} \Gamma(r_i/2)^{-1}\left(\sigma_i^2\right)^{r_i/2+1}$$

$$\times \exp\left(-\frac{1}{2}\sigma_i^{-2}\hat{\boldsymbol{e}}_i'\boldsymbol{P}_i\hat{\boldsymbol{e}}_i\right) \tag{26}$$

with residuals $\hat{\boldsymbol{e}}_i$ computed at convergence. Here Γ is the Gamma function. Relative weighting factors are defined by

$$\omega = \frac{\sigma_i^2}{\sigma_j^2}\,. \tag{27}$$

Their posterior pdf follows (Koch and Kusche 2002)

$$\begin{aligned} p\left(\omega|\boldsymbol{y}_i, \boldsymbol{y}_j\right) &= c\cdot\omega^{-(r_i/2+1)} \\ &\times \left(\frac{1}{2\omega}\hat{\boldsymbol{e}}_i'\boldsymbol{P}_i\hat{\boldsymbol{e}}_i + \frac{1}{2}\hat{\boldsymbol{e}}_j'\boldsymbol{P}_j\hat{\boldsymbol{e}}_j\right)^{-(r_i+r_j)/2} \end{aligned} \tag{28}$$

These pdf's can subsequently be used for testing and for the computation of confidence regions.

4 Regularization

We will now focus on the problem of estimating regularization parameters. In the Bayesian picture regularization parameters are determined by the ratio of variance components (Arsenin and Krianev 1992). Different estimators for variance components, however, consequently lead to different regularization parameter choice rules.

First we consider the standard problem with one regularization parameter α, following as the ratio of the two variance factors of observation covariance matrix and prior information covariance matrix,

$$\boldsymbol{\Sigma}(\sigma^2) = \sigma^2\boldsymbol{P}^{-1} \qquad \boldsymbol{\Gamma}(\gamma^2) = \gamma^2\boldsymbol{Q}^{-1} \tag{29}$$

where $\boldsymbol{P}$ and $\boldsymbol{Q}$ are given, with the resulting normal equations

$$\boldsymbol{N}\hat{\boldsymbol{\beta}} = \left(\boldsymbol{X}'\boldsymbol{P}\boldsymbol{X} + \alpha\,\boldsymbol{Q}\right)\hat{\boldsymbol{\beta}} = \boldsymbol{X}'\boldsymbol{P}\boldsymbol{y} \tag{30}$$

As already mentioned, the estimated regularization parameter is given by ($\boldsymbol{G}\boldsymbol{G}' = \boldsymbol{P}$, $\boldsymbol{F}\boldsymbol{F}' = \boldsymbol{Q}$)

$$\hat{\alpha} = \frac{\hat{\sigma}^2}{\hat{\gamma}^2} \tag{31}$$

where the estimates $\hat{\sigma}^2$, $\hat{\gamma}^2$, according to the maximum–likelihood equations (15) and (16) follow from solving iteratively

$$\hat{\sigma}^2 = \frac{(\boldsymbol{X}\hat{\boldsymbol{\beta}} - \boldsymbol{y})'\boldsymbol{P}(\boldsymbol{X}\hat{\boldsymbol{\beta}} - \boldsymbol{y})}{n - \frac{1}{\hat{\sigma}^2}\mathrm{trace}\left(\boldsymbol{G}'\boldsymbol{X}\boldsymbol{N}^{-1}\boldsymbol{X}'\boldsymbol{G}\right)} \tag{32}$$

$$\hat{\gamma}^2 = \frac{\hat{\beta}' Q \hat{\beta}}{u - \frac{1}{\hat{\gamma}^2}\text{trace}\left(F' N^{-1} F\right)} \,. \tag{33}$$

Note that only one of the two trace terms actually needs to be computed, and that normally the first one will be better suited to trace estimation according to the previous chapter. The posterior pdf for the regularization parameter is given by (Koch and Kusche 2002)

$$p(\alpha|y) = c \cdot \alpha^{-(r/2+1)} \times \left(\frac{1}{2\alpha}(X\hat{\beta} - y)' P (X\hat{\beta} - y) + \frac{1}{2}\hat{\beta}' Q \hat{\beta}\right)^{-n/2} \tag{34}$$

where r is the denominator in (32). It is a special case of (28).

An open problem in the design of regularization strategies for satellite gravity field recovery is the question how an appropriate regularization matrix should be defined and if it can be justified from the given data themselves. In the present context, we will consider this as the question if it is possible to include the power decrease behaviour of an (error) degree variance model by an unknown covariance parameter q in the data analysis. This leads to a model where the covariance matrix depends non–linearly on an unknown parameter.

We therefore consider the following simple extension of the model eq. (29)

$$\Sigma(\sigma^2) = \sigma^2 P^{-1} \qquad \Gamma(\gamma^2, q) = \gamma^2 Q^{-1}(q) \tag{35}$$

where the matrix $Q^{-1}(q)$ is assumed as positive definite and diagonal, with its entries following e. g. a modified Kaula's rule of thumb. This corresponds to letting the entries of Γ be defined by

$$V(\delta C_{lm}) = V(\delta S_{lm}) = \gamma^2 \cdot \frac{1}{f(l,q)} \,. \tag{36}$$

It is clear that this cannot be a proper specification for the low–degree spherical harmonics where we do have accurate degree variances from previous gravity models. However, for an outline we want to keep this as simple as possible.

Stationarity of the likelihood function (13), according to the conditions (15), (16), requires the estimates $\hat{\sigma}$, $\hat{\gamma}$, and $\hat{q}$ to fulfill simultaneously

$$\hat{\sigma}^2 = \frac{(X\hat{\beta} - y)' P (X\hat{\beta} - y)}{n - \frac{1}{\hat{\sigma}^2}\text{trace}\left(G' X N^{-1} X' G\right)} \tag{37}$$

$$\hat{\gamma}^2 = \frac{\sum_l f(l,\hat{q}) \sum_m (\widehat{\delta C^2_{lm}} + \widehat{\delta S^2_{lm}})}{u - \frac{1}{\hat{\gamma}^2}\text{trace}\left(F' N^{-1} F\right)} \tag{38}$$

$$\hat{\gamma}^2 = \frac{\sum_l \frac{df}{dq}(l,\hat{q}) \sum_m (\widehat{\delta C^2_{lm}} + \widehat{\delta S^2_{lm}})}{\sum_l (2l+1)\frac{df/dq}{f(l,\hat{q})} - \text{trace}(H \Gamma^{-1} N^{-1} \Gamma H')} \tag{39}$$

The first two eqs. (37) and (38) are in complete analogy to (32), (33), with the (trivial) decomposition $FF' = Q(\hat{q})$ now referring to the estimate on q. But the third condition (39), with $HH' = d\Gamma/dq$ in its present form, is not suited for an iteration since it does not represent a fixpoint equation for $\hat{q}$.

5 Identification of Correlations

Within this chapter, we assume that only one observation type is present, and that these observations y are taken as an uninterrupted, equi–spaced time series of length n, containing no gaps. Furthermore, we assume that correlations depend only on time difference (stationary noise assumption), that a baseline model describing these correlations is given (correlation function), and that a method exists to convert this model into a suitable filtering operation. The question to be discussed within this chapter therefore reads: Is it numerically feasible to validate this baseline correlation model from the adjustment residuals within the chosen framework; e. g. is there a possibility to improve the model by estimating parameters that describe the influence of residual correlations?

Consequently, one has to deal with the special model

$$\Sigma(\sigma) = \begin{pmatrix} c_0 & c_1 & c_2 & & \cdots \\ c_1 & c_0 & c_1 & c_2 & \cdots \\ & & & & \\ \vdots & & & & \vdots \\ & & \cdots & c_2 & c_1 & c_0 \end{pmatrix} = \sigma_0^2 \Sigma_0 + \sigma_{11} V_1 + \sigma_{22} V_2 + \cdots \tag{40}$$

Here Σ_0 means a given covariance matrix, which is supposed to be Toeplitz. If a power spectral density (PSD) model is to be applied, the rows of Σ_0 contain the Fourier–transformed of the given PSD. It may be a full matrix or simply a unit matrix, if stationary white noise is an appropriate hypothesis. However, we assume that a filtering operation is at hand to compute quickly products $\Sigma_0^{-1} z$.

The role of the matrices V_k is to account for suspected coloured noise that is not already implemented in Σ_0. We call this residual correlation. The frequency behaviour of this suspected residual correlation must be specified within the structure of the V_i (e. g., in multiples of a satellites orbit frequency), the component σ_{kk} is to be estimated. Estimates close to zero, of course, would validate the chosen baseline model Σ_0. A model of type (40) has been first considered by Grafarend and Kleusberg (1980) in estimating the influence of a suspected autoregressive process.

In order to obtain a numerically feasible method at all, we assume: (1) The V_k are banded Toeplitz (say, of width d_k), and (2) if the covariance parameters $\sigma = (\sigma_0^2, \sigma_{11}, \ldots)'$ are estimated in a way that $\hat{\Sigma} = \Sigma(\hat{\sigma})$ is positive definite, we have a machinery that is capable to quickly implement products $\hat{\Sigma}^{-1} z$ (thus replacing $\Sigma_0^{-1} z$).

The V_k need not necessarily be positive definite. But by introducing a positive constant c, it is always possible to decompose $V_k + c\,I$ into Cholesky factors of the same bandwidth d_k, or

$$V_k = C_k C_k' - c\,I \tag{41}$$

Concerning the maximum-likelihood method, according to the general stationarity condition (12) and the model (40),

$$y'W\left(\frac{\partial}{\partial\sigma_i}\Sigma\right)Wy = (\Sigma^{-1}\hat{e})'V_i(\Sigma^{-1}\hat{e}) \tag{42}$$

and

$$\text{trace}\left(W\frac{\partial}{\partial\sigma_i}\Sigma\right) = \text{trace}\left(\Sigma^{-1}V_i\right)$$
$$-\text{trace}\left(\Sigma^{-1}X(X'\Sigma^{-1}X)^{-1}X'\Sigma^{-1}V_i\right) \tag{43}$$

must be equal. The estimates must therefore solve (iteratively) the equations

$$(\Sigma^{-1}\hat{e})'V_i(\Sigma^{-1}\hat{e}) = a_i \tag{44}$$

where by definition

$$a_i = \text{trace}\left(\Sigma^{-1}V_i\right)$$
$$-\text{trace}\left(\Sigma^{-1}X(X'\Sigma^{-1}X)^{-1}X'\Sigma^{-1}V_i\right)\,. \tag{45}$$

The left–hand side expression of (44) requires the computation of a 'generalized' autocovariance function from the filtered residuals $\Sigma^{-1}\hat{e}$. This can be seen from the following special case: Assume e. g. that V_k contains only the kth off–diagonals, with entries of $1/2$; and we identify the unknown component σ_{kk} with a residual correlation at the lag k. It is then just the empirical autocovariance for lag k that we compute by the left–hand side of (44).

For the practical application in a large–scale problem, the trace expressions a_i can be approached using trace estimation techniques and the decomposition (41). This leads to replacing the a_i by

$$\hat{a}_i$$
$$= \frac{1}{K}\sum_{k=1}^{K}\Big((C_i z_k - \sqrt{c}\,z_k)'\Sigma^{-1}(C_i z_k + \sqrt{c}\,z_k)$$
$$+ \quad (C_i z_k + \sqrt{c}\,z_k)'\Sigma^{-1}X$$
$$(X'\Sigma^{-1}X)^{-1}X'\Sigma^{-1}(C_i z_k - \sqrt{c}\,z_k)\Big) \tag{46}$$

where the z_k are K realizations of a standard random vector, $E(z) = 0$, $D(z) = I$, $\dim(z) = n$. Taking the expectation proves that the $\hat{a}_i$ are unbiased estimates of the a_i. Evaluation of (46) requires basically four different operations: (1) generation of a random sequence of the length of the original data vector, (2) application of the band matrices $C_i + \sqrt{c}I$ and $C_i - \sqrt{c}I$, (3) application of Σ^{-1}, (4) estimation of 'random parameter' and synthesis of a new 'random data vector' by $X(X'\Sigma^{-1}X)^{-1}X'\Sigma^{-1}(C_i z_k - \sqrt{c}\,z_k)$.

6 Summary

In this contribution we have discussed some concepts to facilitate the estimation of covariance parameters – mostly variance components – in large parameter estimations as they will occur in the course of gravity downward continuation from, e. g. GOCE satellite data. Such estimates are intended to validate the pre–mission stochastic models; regarding instrument noise variances, PSD functions, or the weighting of prior information and constraints. Also the regularization problem can be treated in this context. As these problems are large–sized, regarding both the number of observations and the number of unknown gravity parameters, existing estimation methods need to be modified. Trace estimation techniques prove to be valuable for this purpose.

Several open questions, however, regard to the convergence and stability behaviour of the ML–based estimators, their reliability, the numerical realization of the derived algorithms, and their application to the satellite data analysis for validation and improvement of the stochastic models.

Our algorithms have been based on the maximum-likelihood estimates; thus rejecting the possibility to specify informative priors on covariance parameters. This is unrealistic since these parameters will not be completely unconstrained for the gravity missions. As an extension, maximum a–posteriori estimation can accommodate such constraints and will be considered in further research.

References

Arsenin VY, Krianev AV (1992). *Generalized maximum likelihood method and its application for solving ill-posed problems.* In Tikhonov A: Ill-Posed Problems in Natural Sciences, pp.3–12, TVP Science Publishers, Moscow

ESA (1999). *Gravity Field and Steady-State Ocean Circulation Mission.* ESA Publications Divisions SP-1233(1), ESTEC, Noordwijk

Förstner W (1979). *Ein Verfahren zur Schätzung von Varianz– und Kovarianzkomponenten.* Allgemeine Vermessungs–Nachrichten 86:446–453

Grafarend, E, Kleusberg A (1980). *Expectation and variance component estimation of multivariate gyrotheodolite observations I.* Allgemeine Vermessungs–Nachrichten 87:129–137

Klees R, Ditmer P, Kusche J (this volume). *Numerical techniques for large–scale least–squares problems with applications to GOCE.*

Koch K–R (1990). *Bayesian Inference with Geodetic Applications.* Springer, Berlin

Koch K–R, Kusche J (2002). *Regularization of geopotential determination from satellite data by variance components.* Journal of Geodesy, 76:259–268

Kusche J (2003). *A Monte–Carlo technique for weight estimation in satellite geodesy.* Journal of Geodesy, 76:641–652

Lehmann R (1996). *Information measures for global geopotential models.* Journal of Geodesy, 70:342–348

Ou Z, Koch K–R (1994). *Analytical expressions for Bayes estimates of variance components.* manuscripta geodetica 19:284–293

Rao CR and Kleffe J (1988). *Estimation of Variance Components and Applications.* North–Holland, Amsterdam

Relating Gravity, Density, Topography and State of Stress Inside a Planet

B. Valette
LGIT, IRD, Université de Savoie, F-73376 Le Bourget du Lac Cedex, France.
e-mail: Bernard.Valette@univ-savoie.fr

F. Chambat
ENS Lyon, LST, 46, Allée d'Italie, F-69364 Lyon Cedex 07, France.
e-mail: Frederic.Chambat@ens-lyon.fr

Abstract. Current interpretations of gravimetric and topographic data rely either on isostasy or on thin plate bending theory. Introducing a fluid rheology constitutes an alternative for global interpretation. In this paper, we present a method that enables to directly relate gravity to deviatoric stresses without any rheological assumption. The relation is obtained by perturbing the equilibrium equation and Poisson's equation around a static spherical configuration, and by introducing a set of suited variables. Namely, we consider the density variation over the equipotential surfaces and the height of interfaces above their corresponding equipotential surfaces. The Backus decomposition of second-order tensors in scalar potentials (Backus 1966) is also found to be very useful. Finally, we show that the method can provide a way to infer strength differences and crustal thickness in a way that generalizes the isostasy approach.

Keywords. Perturbation, topography, Clairaut's equation, gravity, stress, density.

1 Setting the Problem

The relation between the shape of planets and the equilibrium equation has been intensively studied in the hydrostatic context. This has yielded, since Clairaut's work, the classical studies on equilibrium figures. In other respects, the local gravimetric and topographic data are usually interpreted in the framework of isostasy or of plate bending theory. In global approaches, the gravity potential is commonly related to density and to discontinuity topographies through a first-order Eulerian perturbation. Moreover it has become usual to consider a Newtonian fluid rheology in order to relate topography to density variations and to interpret tomographic images in terms of density and gravity.

In this paper we adopt a starting point of view similar to that of Backus (1967) or Dahlen (1981) and refer to the ambient state of stress, without any rheological consideration. Thus we avoid to consider a physical process and to define an initial thermodynamical configuration with the complications it implies. Our aim is only to determine the relations that can be established with the ambient state of stress. More precisely, let us consider a planet occupying the domain V of the space refered to a co-rotating frame located at the centre of mass of the body. Our purpose is to explain how the gravity field of the planet can be written in terms of stress instead of density field, independantly of any rheological law. It consists in solving together Poisson's equation:

$$\Delta\varphi = 4\pi G\rho - 2\Omega^2, \quad \rho(x) = 0 \text{ for } x \in \mathbf{R^3} \setminus V, \tag{1}$$

with the equation of equilibrium:

$$\text{div}\sigma - \rho\text{grad}\varphi = 0, \quad \sigma(x) = 0 \text{ for } x \in \mathbf{R^3} \setminus V, \tag{2}$$

while satisfying the usual boundary conditions:

$$[\varphi] = 0, \quad \varphi(x) + (\Omega^2 x^2 - (\Omega \cdot x)^2)/2 \underset{\infty}{\rightarrow} 0, \tag{3}$$

$$[\text{grad}\varphi \cdot n] = 0, \tag{4}$$

$$[\sigma(n)] = 0, \tag{5}$$

where x, ρ, G, φ, $-\text{grad}\varphi$, Ω, σ denote the position-vector, the density, the gravitational constant, the gravity potential, the gravity vector, the (constant) rotation vector and the Cauchy stress tensor, respectively. [ ] denotes the jump across the closed interfaces Σ, including the outer boundary ∂V, oriented by the unit normal vector field n. In order to solve this system we use a perturbation method because: (a) Equation (2) depends non-linearly on ρ; (b) the shape is involved in the solutions of the equations; (c) planets have a quasi-spherical symmetry.

The paper is structured as follows. In section 2, we set up the shape perturbation formalism. Section 3 is devoted to the perturbation of Poisson's equation and to the generalization of Clairaut's equation by introducing non-hydrostatic variables. Section 4 is devoted to the perturbation of the equilibrium equation and to the expression of the non-hydrostatic variables as functions of the Backus potentials of the deviatoric stress tensor. Finally, in section 5 we outline an inversion scheme of gravity and topography models, considered as data, that relies on the global minimization of the strength difference. Minimizing the deviatoric stress was also considered by Dahlen (1981, 1982) as a possible interpretation of isostasy.

2 Perturbation Formalism

Let us begin by defining the reference hydrostatic spherical configuration as in Chambat & Valette (2001). To this purpose, we first consider a continuous set of surfaces S which interpolate the interfaces Σ from the centre of mass to the boundary ∂V. Secondly, let us define the mean radius r of S as the angular average of the distance of the centre of mass to the points of S. Let us denote by $r = b$ the mean radius of ∂V. We can now define the mean density $\rho(r)$ as the angular average of ρ over S. The potential φ and the pressure p are finally deduced through equations (1-5) with $\sigma = -p\mathrm{I}_d$, $\Omega = 0$ and $V = B(0, b)$. Now, the real configuration must be related to the reference one by introducing a continuous evolution. The physical parameters can be derived from the reference ones through a Taylor expansion which defines the perturbations to the different orders. The deformation of the domain is parameterized by a scalar t ranging from 0, for the reference domain $V_0 = B(0, b)$, to 1 for the real domain V, and which can be thought of as a virtual time. More precisely, let us consider a mapping: $\forall(a,t) \in V_0 \times [0,1]$, $(a,t) \to x(a,t) \in V_t$ with $\forall a \in V_0, x(a,0) = a$, $x(a,1) = x \in V_1 = V$. For any regular tensor field T we consider a mapping: $\forall(a,t) \in V_0 \times [0,1]$, $(a,t) \to T(x(a,t),t)$ with $T(a,0)$ corresponding to the reference field in V_0 and $T(x,1)$ to the real one in V. The nth order Lagrangian displacement is defined as:

$$\xi_n(a) = \left.\frac{\mathrm{d}^n}{\mathrm{d}t^n} x(a,t)\right|_{t=0},$$

and the Eulerian, respectively Lagrangian, n^{th} order perturbation of T as:

$$\delta_{ne}T(a) = \left.\frac{\partial^n}{\partial t^n} T(x(a,t),t)\right|_{t=0},$$

$$\delta_{n\ell}T(a) = \left.\frac{\mathrm{d}^n}{\mathrm{d}t^n} T(x(a,t),t)\right|_{t=0}.$$

Thus, $\delta_{ne}x = 0$ and $\delta_{n\ell}x = \xi_n$. Defining ξ, $\delta_e T$ and $\delta_\ell T$ respectively by:

$$x(a,1) = a + \xi(a), \quad T(a,1) = T(a,0) + \delta_e T(a),$$

$$T(x(a,1),1) = T(a,0) + \delta_\ell T(a),$$

a Taylor expansion of order N yields:

$$\xi(a) = \sum_{n=1}^{N} \frac{\xi_n}{n!}(a),$$

$$\delta_e T(a) = \sum_{n=1}^{N} \frac{1}{n!}\delta_{ne}T(a),$$

$$\delta_\ell T(a) = \sum_{n=1}^{N} \frac{1}{n!}\delta_{n\ell}T(a).$$

>From the definition, it is clear that the Eulerian perturbations commutate with the spatial differentiations. Consider now a scalar field f, a vector field u and a symmetric second order tensor field T. The following usual first-order relations hold:

$$\delta_\ell f = \delta_e f + \mathrm{grad} f \cdot \xi, \tag{6}$$

$$\delta_\ell u = \delta_e u + \nabla u(\xi), \tag{7}$$

$$\delta_\ell(\mathrm{div} T) = \mathrm{div}(\delta_\ell T) - \nabla T : \nabla\xi, \tag{8}$$

where:

$$(\nabla T : \nabla\xi)^j = \nabla_k T^{ij} \nabla_i \xi^k. \tag{9}$$

Finally, we impose that $\delta\Omega^2 = \delta_1\Omega^2 = \Omega^2$, and that the deformation is purely radial, *i. e.*, $\xi = h_\xi\, e_r$ where e_r is the unit radial vector.

3 Generalizing Clairaut's Equation

The purpose of this section is to explain how Poisson's equation can be solved in a way which allows to generalize Clairaut's equation. This is done by introducing new variables which permit to separate topographies from equipotential heights and to identify non-hydrostatic density repartition. The classical way to solve Poisson's equation in a quasi-spherical geometry is to use a perturbation approach between the non-rotating mean model and the aspherical model rotating with angular velocity Ω, and to consider the Eulerian perturbation of potential $\delta_e\varphi$ and of density $\delta_e\rho$. This leads to:

$$\Delta\delta_e\varphi = 4\pi G\delta_e\rho - 2\Omega^2, \tag{10}$$

with the following interface conditions:

$$[\delta_e\varphi] = 0, \quad [\mathrm{grad}(\delta_e\varphi) + 4\pi G\rho\xi] \cdot e_r = 0. \tag{11}$$

Expanding $\delta_e\varphi$ and $\delta_e\rho$ in spherical harmonics yields for each l, m:

$$\left\{\partial_r^2 + \frac{2}{r}\partial_r - \frac{l(l+1)}{r^2}\right\}\delta_e\varphi = 4\pi G\delta_e\rho - 2\Omega^2\delta_l^0, \quad (12)$$

with boundary conditions:

$$[\delta_e\varphi] = 0, \quad [\partial_r\delta_e\varphi + 4\pi G\rho h_\xi] = 0, \quad (13)$$

where we have dropped the indices l and m in the coefficients. The spherical harmonics used here are normalized as

$$\int_0^{2\pi}\int_0^{\pi} Y_l^m(\theta,\lambda)\bar{Y}_{l'}^{m'}(\theta,\lambda)\sin(\theta)\,\mathrm{d}\theta\,\mathrm{d}\lambda = 4\pi\delta_l^{l'}\delta_m^{m'}$$

where θ and λ denote colatitude and longitude, respectively. For instance:

$$Y_0^0(\theta,\lambda) = 1, \quad Y_2^0(\theta,\lambda) = \sqrt{5}(3\cos^2\theta - 1)/2.$$

3.1 Introducing New Variables

For each degree $l \neq 0$, let us now consider the variables:

$$h_\varphi = \frac{\delta_e\varphi}{g}, \; \delta_\varphi\rho = \delta_e\rho + h_\varphi\partial_r\rho, \; h = h_\xi - h_\varphi, \quad (14)$$

where g denotes the (negative) radial gravity in the reference state and satisfies:

$$\partial_r g + 2g/r + 4\pi G\rho = 0. \quad (15)$$

h_φ is the first order equipotential height above the sphere of radius r and h is the height above the equipotential surface. $\delta_\varphi\rho$ represents the lateral variations of density over the associated equipotential surface. Note that applying $\delta_\varphi = \delta_e + h_\varphi\,\partial_r$ corresponds to perturbing to the first order while following the equipotential surfaces and that, for $r = b$, h_φ corresponds to the geoid height and h to the altitude. It is also useful to define:

$$k = \sqrt{(l-1)(l+2)}, \quad (16)$$

$$\gamma = \frac{\rho}{\rho_2} = -\frac{4\pi G\rho r}{3g} = \frac{\partial_r(gr^2)}{3gr}. \quad (17)$$

γ is the ratio of the reference density ρ at radius r to the mean density $\rho_2 = 3\int_0^r \rho(s)s^2\,\mathrm{d}s/r^3$ inside the sphere of radius r.
Using these variables and taking (15) into account, equations (12,13) can be rewritten as:

$$\partial_r^2 h_\varphi - \frac{2}{r}(1-3\gamma)\,\partial_r h_\varphi - \frac{k^2}{r^2}h_\varphi = \frac{4\pi G}{g}\delta_\varphi\rho, \quad (18)$$

or

$$\partial_r\begin{pmatrix} h_\varphi \\ r\partial_r h_\varphi \end{pmatrix} = \frac{1}{r}\begin{pmatrix} 0 & 1 \\ k^2 & 3(1-2\gamma) \end{pmatrix}\begin{pmatrix} h_\varphi \\ r\partial_r h_\varphi \end{pmatrix} - 3\gamma\begin{pmatrix} 0 \\ \delta_\varphi\rho/\rho \end{pmatrix}, \quad (19)$$

with:

$$[h_\varphi] = 0, \quad [\partial_r h_\varphi] = 3[\gamma]\,h/r, \quad (20)$$

$$b\partial_r h_\varphi(b) + (l-1)h_\varphi(b) = 3\gamma h(b) + \frac{\sqrt{5}}{3}\frac{\Omega^2 b^2}{g(b)}\delta_l^2\delta_m^0, \quad (21)$$

$$h_\varphi \underset{r\to 0}{\simeq} \mathrm{cst}\, r^{l-1}. \quad (22)$$

The location of the frame origin at the centre of mass yields the additional condition for $l = 1$:

$$h_\varphi(b) = 0. \quad (23)$$

Condition (21) is derived from (13) by noting that, in virtue of (3) and (12), $\delta_e\varphi$ can be expressed for $r > b$ as:

$$\delta_e\varphi = \frac{\mathrm{cst}}{r^{l+1}} + \frac{\Omega^2 r^2}{3\sqrt{5}}\delta_l^2\delta_m^0,$$

which yields:

$$\partial_r\delta_e\varphi = -(l+1)\frac{\delta_e\varphi}{r} + \frac{\sqrt{5}}{3}\Omega^2 r\,\delta_l^2\delta_m^0.$$

Condition (22) stems from the behaviour of the degree l component of the regular scalar function $\delta_e\varphi$. Let us now consider the homogeneous system corresponding to (19):

$$\partial_r\begin{pmatrix} x \\ r\partial_r x \end{pmatrix} = \frac{1}{r}\begin{pmatrix} 0 & 1 \\ k^2 & 3(1-2\gamma) \end{pmatrix}\begin{pmatrix} x \\ r\partial_r x \end{pmatrix}$$

$$[h_\varphi] = 0, \quad [\partial_r h_\varphi] = 0. \quad (24)$$

Noting that $\gamma(0) = 1$ (see 17), it can be shown that all the solutions of (24) behave like $1/r^{l+2}$ in the vicinity of the centre, except a line of solutions which are proportionnal to r^{l-1}. Let h_1 be such a solution, defined to a multiplicative constant, let h_2 be the solution defined by $(h_2, r\partial_r h_2)(b) = (1, -l+1)$, and let $F(r)$ be the matrix defined as:

$$F = \begin{pmatrix} h_1 & h_2 \\ r\partial_r h_1 & r\partial_r h_2 \end{pmatrix}. \quad (25)$$

From (24), we can deduce that:

$$\partial_r \det(F) = 3\det(F)(1-\gamma)/r, \tag{26}$$

and, taking (17) into account, that:

$$\det(F(r)) = \det(F(b))\frac{bg^2(b)}{rg^2(r)}. \tag{27}$$

Following Poincaré (1902, p. 84), now we will show that h_1 and h_2 remains close to r^{l-1} and to $1/r^{l+2}$ respectively and that F(r) is a fundamental matrix of (24) for $l > 1$.

3.2 Setting Bounds on h_1 and h_2

Let us assume that, for any r, $0 \le \gamma(r) \le 1$, i.e., that $0 \le \rho(r) \le \rho_2(r)$ or that ρ_2 is decreasing with r. Note that this hypothesis is weaker than the one of a decreasing density. Let γ_0 be the minimum of $\gamma(r)$ over [0,b] and define q as:

$$q = \frac{1}{2}\left(3(1-2\gamma_0) + \sqrt{9(1-2\gamma_0)^2 + 4k^2}\right). \tag{28}$$

Under the above hypothesis, h_1 and h_2 satisfy for any r $\in [0, b]$:

$$l-1 \le r\frac{\partial_r h_1}{h_1}(r) \le q \le l+2, \tag{29}$$

$$-(l+2) \le r\frac{\partial_r h_2}{h_2}(r) \le -(l-1), \tag{30}$$

and thus by integration:

$$\left(\frac{r}{b}\right)^{l+2} \le \left(\frac{r}{b}\right)^q \le \frac{h_1(r)}{h_1(b)} \le \left(\frac{r}{b}\right)^{l-1}, \tag{31}$$

$$\left(\frac{b}{r}\right)^{l-1} \le h_2(r) \le \left(\frac{b}{r}\right)^{l+2}. \tag{32}$$

For $l = 1$ ($k = 0$), the proof of (29) and (31) is straightforward, since in this case, h_1 is constant and $h_2 = 1$ on the interval $[0, b]$. For $l \ge 2$, let us define after Poincaré:

$$\alpha_1(r) = r\frac{\partial_r h_1}{h_1}(r), \quad \alpha_2(r) = r\frac{\partial_r h_2}{h_2}(r). \tag{33}$$

The definitions are a posteriori justified by the fact that α_1 and α_2 remain finite, i.e., that h_1 and h_2 do not vanish except for $r = 0$. α_1 and α_2 obey the differential equation:

$$\partial_r \alpha = -\left(k^2 + 3(1-2\gamma)\alpha - \alpha^2\right)/r, \tag{34}$$

which can be reformulated as:

$$\partial_r \alpha = -(\alpha - \alpha_+)(\alpha - \alpha_-)/r, \tag{35}$$

with:

$$\alpha_\pm = \frac{1}{2}\left(3(1-2\gamma) \pm \sqrt{9(1-2\gamma)^2 + 4k^2}\right). \tag{36}$$

Differentiating (36) with respect to γ yields:

$$\partial_\gamma \alpha_+ = \frac{-6\alpha_+}{\sqrt{9(1-2\gamma)^2 + 4k^2}}. \tag{37}$$

Since $\alpha_+ \ge 0$, α_+ is a decreasing function of γ, and thus for any $r \in [0\,,\,b]$:

$$0 < \alpha_+(\gamma = 1) = l - 1 \le \alpha_+(r) \le$$

$$\le \alpha_+(\gamma_0) = q \le \alpha_+(\gamma = 0) = l + 2. \tag{38}$$

The relation $\alpha_-\alpha_+ = -(l-1)(l+2)$ shows that α_- is a negative decreasing function of γ and that:

$$-(l+2) \le \alpha_-(r) \le -(l-1) < 0. \tag{39}$$

At the centre, $\gamma = 1$ so that $\alpha_+ = l-1, \alpha_- = -(l+2)$ and $\alpha_1(0) = l - 1$. Noting that $\partial_r \alpha \ge 0$ for $\alpha \in [\alpha_-, \alpha_+]$ and $\partial_r \alpha < 0$ outside, we conclude that $x(r)$ remains in the interval $[l-1, q]$, that is (29). Noting that (see 25):

$$\det(F(b)) = -h_1(b)(l - 1 + \alpha_1(b)), \tag{40}$$

we can conclude from (27) and (29) that F is a fundamental matrix of (24), i.e., $\det(F) \ne 0$, and consequently that $\alpha_2(0) = -(l+2)$. Remarking once again that $\partial_r \alpha > 0$ for $\alpha > -(l-1)$ and $\partial_r \alpha < 0$ for $\alpha < -(l+2)$ yields (30), which ends up the proof. In order to make completely clear the definition of h_1, inequality (29) shows that for $l > 1$ we can normalize it as:

$$b\partial_r h_1(b) + (l-1)h_1(b) = 1. \tag{41}$$

In this case, (40) yields $\det(F(b)) = -1$ and it follows from (27) that:

$$\det(F(r)) = -\frac{bg^2(b)}{rg^2(r)}, \tag{42}$$

$$F^{-1}(r) = -\frac{rg^2(r)}{bg^2(b)}\begin{pmatrix} r\partial_r h_2 & -h_2 \\ -r\partial_r h_1 & h_1 \end{pmatrix}. \tag{43}$$

3.3 Clairaut's Equation

Under the hydrostatic hypothesis, the level surfaces of density, potential and pressure coincide. In addition these level surfaces include the interfaces. Therefore $\delta_\varphi \rho$ and h, as defined by (14), identically vanish. Equations (19-23) then reduce to the homogeneous system (24) with the boundary conditions (22) and:

$$b\partial_r h_\varphi(b) + (l-1)h_\varphi(b) = \frac{\sqrt{5}}{3}\frac{\Omega^2 b^2}{g(b)}\delta_l^2 \delta_m^0, \quad (44)$$

with the additional condition (23) for $l = 1$. This set of equations is an alternative form of Clairaut's differential equation. Paragraph 3.2 shows that it can only be fulfilled for $l = 2$ and $m = 0$ and thus that, as it is well known (see for instance Jeffreys (1976)), the solution only contains the degree 2 order 0 term:

$$h_\varphi(r) = \frac{\sqrt{5}}{3}\frac{\Omega^2 b^2}{g(b)}\delta_l^2 \delta_m^0 h_1(r). \quad (45)$$

In the general case of a non-hydrostatic repartition of density we must solve the system (19-23) with the use of h_1 and h_2.

3.4 h_φ as a Function of Topographies and Non-hydrostatic Density Repartition

Let us now show that for $l > 1$:

$$h_\varphi(b) = -\frac{4\pi G}{bg^2(b)}\left\{\int_0^b g\, h_1\, \delta_\varphi\rho(r)\, r^2 \mathrm{d}r \right. \quad (46)$$

$$\left. - \sum_{r_\Sigma \le b} r_\Sigma^2 g\,[\rho]\, h\, h_1(r_\Sigma)\right\} + \frac{\sqrt{5}}{3}\frac{\Omega^2 b^2}{g(b)}\delta_l^2 \delta_m^0 h_1(b),$$

and more generally that:

$$h_\varphi(r) = -\frac{4\pi G}{bg^2(b)}\left\{h_1(r)\left(\int_r^b g\, h_2\, \delta_\varphi\rho(s)\, s^2 ds \right.\right.$$

$$\left. - \sum_{r<r_\Sigma \le b} r_\Sigma^2 g\,[\rho]\, h\, h_2(r_\Sigma)\right)$$

$$+h_2(r)\left(\int_0^r g\, h_1\, \delta_\varphi\rho(s)\, s^2 ds \right. \quad (47)$$

$$\left.\left. - \sum_{r_\Sigma \le r} r_\Sigma^2 g\,[\rho]\, h\, h_1(r_\Sigma)\right)\right\} + \frac{\sqrt{5}}{3}\frac{\Omega^2 b^2}{g(b)}\delta_l^2 \delta_m^0 h_1(r),$$

where r_Σ denotes the mean radius of interfaces, including the external boundary.
In order to prove these relations, let us first define:

$$X(r) = \begin{pmatrix} h_\varphi(r) \\ r\partial_r h_\varphi(r) \end{pmatrix}, \; S(r) = \begin{pmatrix} 0 \\ -3\gamma\delta_\varphi\rho/\rho \end{pmatrix}.$$

A solution $X(r)$ of (19, 20, 22) can be written as:

$$X(r) = F(r)\left\{X_0 + \sum_{r_\Sigma<r} F^{-1}(r_\Sigma)[X(r_\Sigma)]\right.$$

$$\left. + \int_0^r F^{-1}(s)S(s)\mathrm{d}s\right\}, \quad (48)$$

where the constant vector X_0 can be taken in the form:

$$X_0 = \begin{pmatrix} (4\pi G/bg^2(b))\, c \\ 0 \end{pmatrix},$$

and where:

$$[X(r_\Sigma)] = \begin{pmatrix} 0 \\ 3[\gamma(r_\Sigma)]h(r_\Sigma) \end{pmatrix}.$$

Applying $F^{-1}(r)$ (see 43) to equation (48) and setting $r = b$ in the resulting second component yields:

$$b^2 g^2 (h_\varphi \partial_r h_1 - h_1 \partial_r h_\varphi)(b) =$$

$$- \sum_{r_\Sigma<b} 3 r_\Sigma g^2 [\gamma] h h_1(r_\Sigma) + 3\int_0^b g^2 \gamma h_1 \frac{\delta_\varphi \rho}{\rho}(s)\, s\, \mathrm{d}s.$$

Taking boundary condition (21) into account in the left hand side of this equation leads to:

$$b^2 g^2 (h_\varphi \partial_r h_1 - h_1 \partial_r h_\varphi)(b) =$$

$$bg^2\left(h_\varphi(b) - 3\gamma h h_1(b) - \frac{\sqrt{5}}{3}\frac{\Omega^2 b^2}{g(b)}\delta_l^2\delta_m^0 h_1(b)\right),$$

and finally to (46) by making use of (17). In order to prove (47), let us start from the first component of equation (48) which can be written as:

$$h_\varphi(r) = \frac{4\pi G}{bg^2(b)}\left\{h_1(r)\left(\int_r^b g\, h_2\, \delta_\varphi\rho(s)\, s^2 ds \right.\right.$$

$$\left. - \sum_{r_\Sigma<r} r_\Sigma^2 g\,[\rho]\, h\, h_2(r_\Sigma) + c\right)$$

$$+h_2(r)\left(-\int_0^r g\, h_1\, \delta_\varphi\rho(s)\, s^2 ds \right. \quad (49)$$

$$\left.\left. + \sum_{r_\Sigma<r} r_\Sigma^2 g\,[\rho]\, h\, h_1(r_\Sigma)\right)\right\}.$$

Putting $r = b$ in this equation and substituting expression (46) of $h_\varphi(b)$ yields the value of the constant:

$$c = \sum_{r_\Sigma \le b} r_\Sigma^2 g\,[\rho]\, h\, h_2(r_\Sigma)$$

$$- \int_0^b g\, h_2\, \delta_\varphi \rho(s)\, s^2 ds + \frac{bg^2(b)}{4\pi G} \frac{\sqrt{5}\,\Omega^2 b^2}{3\; g(b)} \delta_l^2 \delta_m^0.$$

Finally, setting the latter value in (49) leads to (47).

3.5 The Degree $l = 1$

The degree one needs a special treatment since, in this case, the function h_1 and h_2 are constant and F is no longer a fundamental matrix. Besides, a constant degree one h_φ corresponds to a translation of the body and does not affect either the source terms (controlled by $\delta_\varphi \rho$) or the boundary conditions (related to h). Therefore this degree would be undetermined if the centre of mass was not fixed by condition (23).

Taking advantage of $k^2 = 0$, (18) reduces to:

$$\partial_r (r \partial_r h_\varphi) = 3r \partial_r h_\varphi (1 - 2\gamma)/r - 3\gamma \delta_\varphi \rho / \rho \qquad (50)$$

which, with the conditions (20, 22), leads to:

$$r \partial_r h_\varphi(r) = \frac{4\pi G}{r g^2(r)} \left\{ \int_0^r g\, \delta_\varphi \rho(s)\, s^2 \mathrm{d}s - \sum_{r_\Sigma < r} r_\Sigma^2 g\, [\rho]\, h(r_\Sigma) \right\}. (51)$$

Now, condition (21) at $r = b$ implies the following constraint on $\delta_\varphi \rho$ and h:

$$\int_0^b g\, \delta_\varphi \rho(s)\, s^2 \mathrm{d}s - \sum_{r_\Sigma \le b} r_\Sigma^2 g\, [\rho]\, h(r_\Sigma) = 0. \qquad (52)$$

Furthermore, taking (22) into account, (51) yields:

$$h_\varphi(r) = 4\pi G \left\{ \int_0^r \frac{1}{s^2 g^2(s)} \int_0^s g \delta_\varphi \rho(t)\, t^2 \mathrm{d}t\, \mathrm{d}s - \sum_{r_\Sigma < r} r_\Sigma^2 g[\rho]\, h(r_\Sigma) \int_{r_\Sigma}^r \frac{\mathrm{d}s}{s^2 g^2(s)} + c_1 \right\}. (53)$$

The constant c_1 can be determined by the additional condition (23). This yields:

$$c_1 = - \int_0^b \frac{1}{s^2 g^2(s)} \int_0^s g\, \delta_\varphi \rho(t)\, t^2 \mathrm{d}t\, \mathrm{d}s + \sum_{r_\Sigma < b} r_\Sigma^2 g\, [\rho]\, h(r_\Sigma) \int_{r_\Sigma}^b \frac{\mathrm{d}s}{s^2 g^2(s)}.$$

Thus (53) becomes:

$$h_\varphi(r) = 4\pi G \left\{ - \int_r^b \frac{1}{s^2 g^2(s)} \int_0^s g \delta_\varphi \rho(t)\, t^2 \mathrm{d}t\, \mathrm{d}s + \sum_{r_\Sigma < r} r_\Sigma^2 g[\rho]\, h(r_\Sigma) \int_r^b \frac{\mathrm{d}s}{s^2 g^2(s)} + \sum_{r \le r_\Sigma < b} r_\Sigma^2 g[\rho]\, h(r_\Sigma) \int_{r_\Sigma}^b \frac{\mathrm{d}s}{s^2 g^2(s)} \right\}.$$

An integration by part of the first integral yields an expression of $h_\varphi(r)$ similar to (47) provided that we set, for $l = 1$, $h_1 = 1$ and $h_2 = bg^2(b) \int_r^b \mathrm{d}s/(s^2 g^2(s))$. Note that since the condition $h_\varphi(b) = 0$ is conventional, only the compatibility condition (52) is relevant for this degree. Finally, (46) can formally be kept for $l = 1$ with $h_1 = 1$ and $h_\varphi(b) = 0$, since it is equivalent to (52).

4 Taking Stress Into Account

Let us now turn to the equilibrium equation (2). Note that in the spherical reference configuration:

$$\sigma = -p\mathrm{I}_d, \quad \mathrm{grad} p = -\rho \mathrm{grad} \varphi, \quad [p] = 0. \qquad (54)$$

Using (8) the Lagrangian perturbation of (2) yields:

$$\mathrm{div} \delta_l \sigma - \nabla \sigma : \nabla \xi - \rho \delta_l \mathrm{grad} \varphi - \mathrm{grad} \varphi \delta_l \rho = 0. (55)$$

Relations (6, 7, 9) respectively imply that:

$$\delta_l \rho = \delta_\varphi \rho + (h_\xi - h_\varphi)\, \partial_r \rho = \delta_\varphi \rho + h\, \partial_r \rho,$$

$$\delta_l\, \mathrm{grad} \varphi = \mathrm{grad}(g h_\varphi) + \nabla \nabla(\varphi)(\xi),$$

$$\nabla \sigma : \nabla \xi = \rho \nabla \xi^* (\mathrm{grad} \varphi),$$

where the star denotes the adjoint with respect to the usual $\mathbf{R}^3$ scalar product. Substituting these three relations into (55) finally yields:

$$\mathrm{div} \delta_\varphi \sigma + g \delta_\varphi \rho\, e_r = 0 \text{ with } \delta_\varphi \sigma = \delta_l \sigma + \rho g h \mathrm{I}_d. (56)$$

The boundary conditions are directly obtained upon perturbing (5):

$$[\delta_l \sigma](e_r) = -[\sigma(\delta_l n)] = p[(\delta_l n] = 0. \qquad (57)$$

It is useful to apply the Backus scalar potential representation (Backus 1966) to $\delta_l \sigma$. This representation generalizes to second-order tensors the usual representation of vector fields in radial, poloidal and torroidal potentials with respect to the sphere and leads to local relations. In addition, expanding the potentials in spherical harmonics constitutes an alternative to the use of generalized spherical harmonics. Let us first recall the Backus representation in the case of symmetric tensors.

4.1 Backus Representation of Real Second-order Symmetric Tensor Fields

Let T be a regular real valued second-order symmetric tensor field. There exists 6 uniquely determined real potential fields P, Q, R, L, M, N such that:

$$Q_0 = R_0 = M_0 = N_0 = M_1 = N_1 = 0, \qquad (58)$$

and:

$$T = Pe_r \otimes e_r \qquad (59)$$
$$+e_r \otimes (\mathrm{grad}_T(rQ) + \mathrm{grad}_T(rR) \times e_r)$$
$$+ (\mathrm{grad}_T(rQ) + \mathrm{grad}_T(rR) \times e_r) \otimes e_r$$
$$+\mathrm{P}_T.\left\{(L - r^2\Delta_T M)\right.$$
$$\left.+2r^2\left(\mathrm{H}_T(M) + \mathrm{H}'_T(N)\right)\right\}.\mathrm{P}_T,$$

where the indices $_0$ and $_1$ in (58) refer to the spherical harmonic degrees, where the index $_T$ in (59) refers to the sphere of radius r, and where $\times$ denotes the vectorial product. The projector P_T and the tangential gradient grad_T are defined as:

$$\mathrm{P}_T = \mathrm{I}_d - e_r \otimes e_r, \ \mathrm{grad}_T = \mathrm{grad} - e_r\, e_r \cdot \mathrm{grad},$$

and the differential operators over the sphere of radius r, H_T and Δ_T, are related to the covariant derivative ∇_T by:

$$\mathrm{H}_T = \nabla_T \nabla_T, \quad \Delta_T = \nabla_T \cdot \nabla_T.$$

The differential operator H'_T is defined from H_T by the relation:

$$2\mathrm{H}'_T(N)(u_T) =$$
$$(\mathrm{H}_T(N)u_T) \times e_r - \mathrm{H}_T(N)(u_T \times e_r),$$

which holds for any regular scalar field N and any regular vector field u_T tangent to the sphere.
Moreover, it can be verified that (Backus, 1966, 1967):

$$\mathrm{tr}(T) = P + 2L, \qquad (60)$$

$$T(e_r) = Pe_r + \mathrm{grad}_T(rQ) + \mathrm{grad}_T(rR) \times e_r, \quad (61)$$

$$\mathrm{div}T = Ue_r + \mathrm{grad}_T(rV) + \mathrm{grad}_T(rW) \times e_r, \quad (62)$$

with:

$$U = \left\{r\partial_r P + 2P + r^2\Delta_T Q - 2L\right\}/r, \qquad (63)$$

$$V = \left\{r\partial_r Q + 3Q + (r^2\Delta_T + 2)M + L\right\}/r, \quad (64)$$

$$W = \left\{r\partial_r R + 3R + (r^2\Delta_T + 2)N\right\}/r. \qquad (65)$$

Note finally that e_r is an eigenvector of T if and only if $R = Q = 0$ and that T is transversely isotropic with respect to e_r if and only if $R = Q = M = N = 0$. In the latter case: $T = Pe_r \otimes e_r + L\mathrm{P}_T$

4.2 Expression of h and $\delta_\varphi \rho$ as Functions of Stress Potentials

Using relation (62) for $T = \delta_l \sigma$ and identifying the radial, poloidal and toroidal components in the Lagrangian perturbation of the equilibrium equation (56) results in:

$$\delta_\varphi \rho = -(U + \partial_r(\rho g h))/g, \qquad (66)$$

$$\rho g h = -rV, \qquad (67)$$

$$W = 0. \qquad (68)$$

Substituting (67) into (66) and taking (63-65) into account yields:

$$\delta_\varphi \rho = \left\{\partial_r\left(r^2(L - P + (r^2\Delta_T + 2)M\right.\right.$$
$$\left.\left.+\partial_r(rQ))\right) - r(r^2\Delta_T + 2)(Q + 2M)\right\}/gr^2, (69)$$

$$h = -(L + (r^2\Delta_T + 2)M + r\partial_r Q + 3Q)/\rho g, \quad (70)$$

$$r^3 R = -(r^2\Delta_T + 2)\int_0^r s^2 N\, \mathrm{d}s. \qquad (71)$$

Expression (61) shows that boundary conditions (57) can be rewritten as:

$$[P] = [Q] = [R] = 0. \qquad (72)$$

Furthermore, it can be deduced from (70, 72) that:

$$[P - L - (r^2\Delta_T + 2)M - \partial_r(rQ)] = [\rho]\, g\, h, \quad (73)$$

at each interface. Finally, a spherical harmonic expansion yields expressions of $\delta_\varphi \rho$, h, and R similar to (69-71) in which the operator $(r^2\Delta_T + 2)$ is replaced by $-k^2$.

4.3 Expression of $h_\varphi(b)$ as a Function of Stress Potentials

Let us substitute the expression of the harmonic coefficient $\delta_\varphi \rho$ derived from (69) into (46). Integrating by parts and taking boundary conditions (73) into account first yields:

$$h_\varphi(b) = \frac{4\pi G}{bg^2(b)}\int_0^b \left\{-k^2(Q + 2M)h_1\right.$$

$$\left.+(L - P - k^2 M)r\partial_r h_1 + \partial_r(rQ)r\partial_r h_1\right\}(r)\, r\mathrm{d}r$$

$$+\frac{\sqrt{5}\,\Omega^2 b^2}{3\;\;g(b)}\delta_l^2\delta_m^0 h_1(b). \qquad (74)$$

Upon integrating by parts the third term of the previous integral, then taking (72) into account and noting that $(h_1, r\partial_r h_1)$ verifies (24), we finally obtain:

$$h_\varphi(b) = \frac{4\pi G}{bg^2(b)}\int_0^b \{-2k^2(M+Q)h_1$$

$$+\left(L-P-k^2M-2(2-3\gamma)Q\right)r\partial_r h_1\}(r)\,r\mathrm{d}r$$

$$+\frac{\sqrt{5}\,\Omega^2 b^2}{3\;\;g(b)}\delta_l^2\delta_m^0 h_1(b). \qquad (75)$$

The topographies h are not directly involved in the expression. They only intervene through (73). Note also that (75) identically vanishes for $l = 1$. This is due to the fact that the integration of equation (56) over the reference domain also yields the three equations (52) corresponding to the order $m = 0, \pm 1$.

4.4 Taking Strength Lines Into Account

As far as we know, the problem of determining the different kind of strength line patterns - related to the principal stresses - which fulfil the boundary conditions, *i.e.*, normal to any interface where a fluid is involved, is an open problem. Let us assume that the strength lines are quasi-radial or quasi-spherical. We guess that this assumption is very likely even in the context of a convective process. This implies that the rotation R mapping the local spherical frame onto the principal stresses frame can be formally written to the first order as $R = \mathrm{I}_d + A$ where A is an anti-symmetric operator, and that:

$$(\mathrm{I}_d + A).(-p\mathrm{I}_d + \delta_l\sigma).(\mathrm{I}_d - A) = -p\mathrm{I}_d + \delta_l\sigma,$$

correct to first order. Therefore e_r is an eigen-vector field of the field $\delta_l\sigma$ and, according to 4.1 and (71) the stress potentials Q, R and N vanish. As a consequence (75) reduces to:

$$h_\varphi(b) = \frac{4\pi G}{bg^2(b)}\int_0^b \{(L-P-k^2M)r\partial_r h_1(r) -$$

$$-2k^2Mh_1\}(r)\,r\mathrm{d}r + \frac{\sqrt{5}\,\Omega^2 b^2}{3\;\;g(b)}\delta_l^2\delta_m^0 h_1(b), \qquad (76)$$

with:

$$\delta_\varphi\rho = \left\{\partial_r\left(r^2(L-P-k^2M)\right) + 2k^2Mr\right\}/gr^2,$$

$$h = (k^2M - L)/\rho g, \qquad (77)$$

and at the interfaces:

$$[L-P-k^2M](r_\Sigma) = -[\rho]g\,h(r_\Sigma). \qquad (78)$$

5 Inferring Strength Differences

Let us outline how we can get some inference on stress differences in the Earth and extrapolate available information on the crustal thickness. For the sake of simplicity we will now assume that $M = 0$. In this case the only parameter is the stress difference $d = L - P$ between the averaged quasi-horizontal stress and the quasi-vertical one. Equations (76-78) then simplify (omitting the degree two Clairaut term) to:

$$\delta_\varphi\rho = \partial_r(r^2 d(r))/gr^2,\ [d](r_\Sigma) = -[\rho]gh(r_\Sigma), (79)$$

$$h_\varphi(b) = \frac{4\pi G}{bg^2(b)}\int_0^b d(r)r^2\partial_r h_1(r)\;\mathrm{d}r. \qquad (80)$$

Using these last two equations $d(r)$ can be identified from geoid height $h_\varphi(b)$ and topography $h(b)$, both considered as data, through a functional least squares approach. Additional information can be obtained from models of discontinuity height $h(r_\Sigma)$ or density $\delta_\varphi\rho$ through equations (79). The regularization imposes a global minimization of $d(r)$ and corresponds to a simple mechanical criterion. Note that since the kernel $r^2\partial_r h_1(r)$ is close to r^l, the information is localized in the lithosphere, hence only the Moho topography is involved. Seismic models of Moho topography are now available and are reliable for the first harmonic degrees. Moreover, we can scale the control parameters of the inversion with respect to the degree, in order to make the Moho topography and the density a posteriori verify Kaula type laws. It enables to extrapolate the reliable information upon Moho topography to greater degrees. The method also provides a way to estimate the minimum amount of stress difference needed in order to adjust gravimetric and topographic models.

References

Backus, G. E., 1966, Potentials for tangent tensor fields on spheroids, Arch. Ration. Mech. Analysis, 22, 210-252.

Backus, G. E., 1967, Converting vector and tensor equations to scalar equations in spherical coordinates, Geophys. J. R. Astron. Soc., 13, 71-101.

Chambat F., and Valette B., 2001, Mean radius, mass and inertia for reference Earth's models, Phys. Earth. Planet. Inter., Vol 124/3-4, pp 237-253.

Dahlen, F. A., 1981, Isostasy and the ambient state of stress in the oceanic lithosphere, J. Geophys. Res., 86, B9, 7801-7807.

Dahlen, F. A., 1982, Isostatic geoid anomalies on a sphere, J. Geophys. Res., 87, B5, 3943-3947.

Jeffreys, H., 1976, The Earth, 6th Edition, Cambridge University Press, London.

Poincaré, H., 1902, Figure d'équilibre d'une masse fluide, Reprint, Jacques Gabay, 1990, Paris.

The A-optimal regularization parameter in uniform Tykhonov-Phillips regularization - α-weighted BLE -

Jianqing Cai, Erik W. Grafarend
Department of Geodesy and GeoInformatics, University of Stuttgart, Geschwister-Scholl-Str. 24, D-70174 Stuttgart, Germany. E-mail: cai@gis.uni-stuttgart.de; Fax: +49-711-121-3285

Burkhard Schaffrin
Department of Civil and Environmental Engineering and Geodetic Science, The Ohio State University, Columbus, Ohio, U.S.A. E-mail: schaffrin.1@osu.edu; Fax:+1-614-292-2957

Abstract. Numerical tests have documented that the estimate $\hat{\boldsymbol{\xi}}$ of type BLUUE of the parameter vector $\boldsymbol{\xi}$ within a linear *Gauss-Markov model* $\{\mathbf{A}\boldsymbol{\xi} = E\{\mathbf{y}\},\ \boldsymbol{\Sigma}_y = D\{\mathbf{y}\}\}$ is *not* robust against *outliers* in the stochastic observation vector $\mathbf{y}$. It is for this reason that *we give up* the postulate of unbiasedness, but keeping the set-up of a *linear estimation* $\hat{\boldsymbol{\xi}} = \mathbf{Ly}$ of homogeneous type. Grafarend and Schaffrin (1993) as well as Schaffrin (2000) have systematically derived the best linear estimators of type homBLE (*Best homogeneously Linear Estimation*), S-homBLE and α-homBLE of the *fixed effects* $\boldsymbol{\xi}$, which turn out to enhance the best linear uniformly unbiased estimator of type BLUUE, but suffer from the effect being biased. Here the regularization parameter in uniform Tykhonov-Phillips regularization (α-weighted BLE) is determined by minimizing the trace of the *Mean Square Error matrix* $MSE_{\alpha,S}\{\hat{\boldsymbol{\xi}}\}$ (*A-optimal design*) in the general case. In lieu of a case study, both model and estimators are tested and analyzed with numerical results computed from simulated direct observations of a random tensor of type strain rate.

Keywords. A-optimal design, *Tykhonov-Phillips* regularization, α-homBLE, α-HAPS, ridge estimator, biased estimation, Mean Square Error

1 Introduction

Ever since *Tykhonov* (1963) and *Phillips* (1962) introduced the *hybrid minimum norm approximation solution* (HAPS) of a *linear improperly posed problem* there has been left the open problem to evaluate the weighting factor α between the least-squares -norm and the minimum norm of the unknown parameters. In most applications of *Tykhonov-Phillips* type of regularization the weighting factor α is determined by simulation studies, but according to *Table 1* also optimization techniques have been applied. Here we aim at an objective method to determine the *weighting factor* α within α-HAPS.

Alternatively, improperly posed problems which appear in solving integral equations of the first kind or downward continuation problems in potential theory depart from observations which are elements of a probability space. Accordingly, estimation techniques of type BLUUE (best linear uniformly unbiased estimation) have been implemented to estimate $\hat{\boldsymbol{\xi}}$ as an unknown parameter vector $\boldsymbol{\xi}$ ("fixed effects") within a linear *Gauss-Markov model*, such an estimation is *not* robust against *outliers* in the stochastic observation vector $\mathbf{y} \in \mathbb{Y}$. Here we assume that the observation $\mathbf{y}$ is an element of the *observation space* $\mathbb{Y}$, dim $\mathbb{Y}= n$, namely an observation space $\mathbb{Y} = \mathbb{R}^n$ equipped with a Euclidean metric. Due to possibly unstable solutions of type BLUUE with respect to the *"fixed effects"* linear Gauss-Markov model *we give up* the postulate of *unbiasedness*, but keep the set-up of a *linear estimation* $\hat{\boldsymbol{\xi}} = \mathbf{Ly}$ of homogeneous type. According to Grafarend and Schaffrin (1993), updated by Schaffrin (2000), the best linear estimation of type α-homBLE (*α-weighted Best homogeneously Linear Estimation*) which is based on *hybrid norm optimization* of type (i) minimum variance *and* (ii) minimum bias leads us to the *equivalence* of α-homBLE *and* α-HAPS under the following condition. *If* we choose the weight matrix in the least squares norm as the inverse matrix of the variance covariance matrix of the observations *as well as* the weight matrix in the minimum norm acting on the unknown parameter vector as the inverse substitute bias weight matrix, *then* α-homBLE *and* α-HAPS are *equivalent*.

The second method of regularizing an improperly posed problem offers the possibility to determine the regularization parameter α in an optimal way. For instance, by an A-optimal design of type "minimize the trace of the α, S modified *Mean Square Error matrix* $\operatorname{tr} MSE_{\alpha,S}\{\hat{\xi}\}$ of $\hat{\xi}$ (α-hom BLE) to find $\hat{\alpha} = \arg\{\operatorname{tr} MSE_{\alpha,S}\{\hat{\xi}\} = \min\}$ "

we are able to construct the regularization parameter α which *balances the average variance* $\operatorname{tr} D\{\hat{\xi}\}$ *and the average bias* $\operatorname{tr} \mathbf{BSB}'$ for the bias matrix **B:=I-LA.**

Table 1 Commented reference on Tykhonov-Phillips regularization

Author	Topic
Allen, D. M. (1971, 1974)	considered using a generalized ridge estimator for the purpose of selecting variables.
Arslan, O. and Billor, B. (2000)	proposed an alternative class of Liu-type M-estimator (LM-estimators) obtained by shrinking an M-estimator β_M, in stead of the PLS estimator using the matrix $(\mathbf{xx}' + \mathbf{I})^{-1}(\mathbf{xx}' + d\mathbf{I})$.
Bouman, J.(1998)	described different regularization methods and their consequences for the quality of the solution, as well as the determination of optimal regularization parameters.
Chaturvedi, A. and Singh, S.P. (2000)	considered the problem of simultaneous prediction of actual and average values of the dependent variable in a general linear regression model utilizing the philosophy of Stein rule procedure.
Donatos, G.S. and Michailidis, G.C.(1990)	discussed small sample properties of five well known ridge estimators with normal and non-normal disturbances with Monte Carlo method.
Draper, R.N. et al. (1979)	made a broad review and critical comments about ridge regression and James-Stein estimation.
Droge, B. (1993)	determined an optimal selection procedure using the minimax regret principle, which is to behave similarly to the procedure minimizing either an unbiased risk estimator or, equivalently, the C_p – criterion.
Engels, J., et al. (1993)	reviewed the theory of regularization (λ-HAPS-generalized and approximated, λ-HAPS generalized inverse operator) and applied it in the inverse problem to determine the geoid.
Engl, H.(1993)	reviewed various mathematical questions arising in the theory of ill-posed inverse problems and the regularization methods for the stable solution of inverse problems.
EL-Sayed, S.M.(1996)	derived the sampling distribution of the ridge parameter estimator.
Farebrother, R.W. (1975, 1976)	explored the similarity between the minimum mean square linear estimator and Hoerl and Kennard's, and suggested that the subjective method of the ridge trace should be replaced by minimum mean square error estimator.
Farebrother, R. W. (1978)	examined the relationship between the general ridge estimator and the standardized ridge estimator.
Firinguetti, L. (1996)	studied the properties of ridge regression and principle components regression estimator by means of simulation experiments, and found the biased estimator can outperform the ordinary least square estimator.
Firinguetti, L. and Rubio, H. (2000)	investigated the performance of two well known operational ridge regression estimators Hoerl and Kennard (1970), Lawless and Wang (1976).
Gibbons, D. G. (1981)	made a simulation study of 12 promising ridge estimators and systematically evaluated and compared them using Monte Carlo methods.
Goulb, G. H. Heath, M. and Wahba, G. (1979)	used the method of generalized cross-validation (GCV) to determine a value for the ridge parameter. The method does not involve estimation of σ^2.
Grafarend, E. and Schaffrin, B.(1993)	systematically derived the best estimators of type homBLE (best homogeneously linear estimator), S-homBLE and α-homBLE of the fixed effect vector ξ.
Gui, Q. et al. (1998a, b)	studied the robust biased estimation and its applications in geodetic adjustments and the robust universal ridge estimation.
Gui, Q. et al (2000, 2001)	considered the biased estimation in Gauss-Markov models with and without constraints.
Gunst, R.F. and Mason R. L. (1977)	used mean squared error evaluated biased estimations in regression, which include least squares, principal components, ridge regression, latent root, and shrunken estimator. Each of the biased estimations is proved to offer improvements in MSE over least squares.
Gunst, R. F. and Mason, R. L. (1980)	introduced briefly three biased regression techniques: principal component regression, latent root regression and ridge regression (Chapter 10 of their book).

Hanke, M. and Hansen, P. (1993)	comprehensively reviewed the regularization methods for large scale problems.
Hansen, P. (1992)	advocated the use of the graph of the norm or seminorm of the regularized solution verus the norm of the residual vector in the treatment of discrete ill-posed problem and demonstrated that several methods of choosing the regularization parameter are related to locating a characteristic *L*-shaped "corner" of the graph.
Hansen, P. (1993)	presented a unifying characterization of the various regularization method and proposed an a posteriori method of choosing the regularization parameter based on the *L*-curve.
Hemmerle, W. J. (1975)	derived an explicit (non-iterative) solution for generalized ridge regression.
Hemmerle, W. J. and Brantle, T. F. (1978)	derived an alternative explicit closed form solution for selecting the regularization parameter.
Hocking, R. R. (1976)	provided a comprehensive review of the analysis and selection of variables in linear regression, especially to the biased estimation.
Hoerl, A. E. and Kennard, R. W. (1970a, 1970b)	proposed the concept of ridge regression and established the existence of a constant k which yields an estimator which has smaller average distance from β than the least squares estimator.
Hoerl, A.E., et al. (1975)	made some simulation analysis of ridge regression.
Hoerl, R. W.(1985)	reviewed the mathematics of ridge analysis, its literature, practical advantage and relationship to ridge regression.
Hoerl, R. W., Schuenemeyer, J. H. and Hoerl, A. E. (1985)	made a simulation study of biased estimation and subset selection regression techniques.
Ilk, K. H. (1986)	proposed a criterion for the proper choice of an optimal regularizing parameter and applied the procedure based on the singular value decomposition technique to determine an optimal regularization parameter.
Kacirattin, S. Sakalloglu, S. and Akdeniz, F. (1998)	made the comparisons of the modified ridge regression estimator and other restricted ridge regression estimator using the mean squared error matrices.
Lawless, J. F. and Wang, P. (1976)	derived the Lawless and Wang ordinary ridge regression (LOR) from their simulation study of ridge and other regression estimations.
Liu, K. (1993)	proposed a new class of biased estimates in linear regression.
Louis, A. K. and Maass, P.	presented an inverse method for the solution of ill-posed linear problem, which is based on the idea of computing a modified version of the desired solution.
Lowerre, J. M. (1974)	proved that, if the compatible symmetric matrix is positive definite matrix whose eigenvalues are small enough, then the MSE of it is smaller than the MSE of BLUUE.
Mallows, C. L. (1973)	extended the concept of C_p-plots to C_k-plots which may be used to determine the ridge parameter k.
Markiewicz, A. (1996)	proved that the general ridge estimators can be characterized as linearly admissible and linearly sufficient.
Marquardt, W. D.(1970,1974)	noted the relation between ridge estimators and a generalized inverse estimator.
Marquardt, W. D. and Snee, R .D. (1975)	made a review of the theory of ridge regression and its relation to generalized inverse regression along with the results of a simulation experiment and three examples of the use of ridge regression in practice.
Mayer, L. S. and Willke, T. A. (1973)	proved that in any give problem there is at least one member of the class of biased estimators which has total mean square error (TMSE) smaller than the total variance of the corresponding least squares estimator.
McDonald, G. C. and Galarneau, D.I. (1975)	presented the results of simulation studies of some ridge-type estimators with Monte Carlo method.
Nomura, M.(1988)	compared the sampling performance of the minimax generalized ridge regression estimators with that of the ordinary least squares estimator by numerical calculations of the exact MSE of these estimators.
Nomura, M.(1998)	examined the exact small sample properties of ordinary ridge estimators and the biased-corrected ordinary ridge estimator.

Ohtani, K.(1986)	examined the small sample properties of the almost unbiased generalized ridge estimator.
Ohtani, K.(1998)	made an MSE comparison of the restricted Stein-rule and minimum mean squared error estimators in regression.
Phillips, D. L. (1962)	introduced the hybrid minimum norm approximation solution (HAPS) of a *linear improperly posed problem.*
Rao, C.R.(1975)	introduced the general ridge estimator.
Rao, C.R.(1976)	made a survey of contributions to estimation of parameters by linear functions of observations in the Gauss-Markov model and proved that the biased estimators like ridge and shrunken estimators are the special case of Bayes linear estimators.
Schaffrin, B., Heidenreich E. and Grafarend, E.(1977)	introduced the *Tykhonov-Phillips* type of regularization of improperly posed boundary value problem in presenting the standard gravity field.
Schaffrin, B. and B. Middel (1990)	used iterative ridge estimators to approximate generalized inverse estimators such as BLUMBE.
Schaffrin, B.(1995)	made a comparison of inverse techniques considering regularization, weight, estimation and homBLUP.
Schaffrin, B (2000)	introduced repro-BLE (Best Linear Estimate with reproducing property) for the minimum means square error adjustment.
Shalabh (1998)	examined the improved estimation in measurement error models through the stein rule procedure.
Smith, G. and Campbell, F. (1980):	made a critique of some ridge regression methods from the following aspects: (i) Which parameter should be explicitly estimated? (ii) Are the data informative? (iii) what a-priori information should be used? (iv) The relationship among ridge regression , principal components and Marquardt's generalized inverse.
Srivastava, V. K. et al. (1983)	investigated the properties of the sampling distribution of an operational ridge estimator.
Theobald, C. M. (1974)	considered some generalization of mean square error, and established a relationship between mean square error functions and the second order moment matrix, and used it to provide an improved sampling theoretic justification for ridge estimator.
Tykhonov, A. N. (1963)	introduced the regularization of incorrectly posed problem.
Tykhonov, A. N. et al. (1977)	proposed a variational method of regularization in the adjustment of free geodetic nets.
Tykhonov, A.N. and Arsenin, V.Y. (1977)	systematically introduced the solutions of Ill-Posed Problems.
Vinod, H.D. and Ullah, A. (1981)	made a comprehensive review of the advances in regression methods.
Xu, P. (1992a)	reviewed the methods for stabilization of downward continuation problems and introduced biased estimation in determination of surface gravity anomalies using gradiometric observables.
Xu, P. (1992b)	investigated the questions concerning the bias problem of the estimated geopotential fields with respect to the value of minimum norm.
Xu, P. and Rummel, R. (1994a)	investigated the large-sample properties of some smoothness (or ridge) estimators in terms of mean-square error and made a simulation study of the method in recovery of regional gravity fields.
Xu, P. and Rummel, R. (1994b)	generalized the ordinary regularization method by introducing more than one regularization parameter, based on consideration of the minimum mean square error of the estimator and applied it in determination of potential fields.
Xu, P. (1998)	investigated the Truncated Singular Value Decomposition (TSVD) techniques and proposed a new quality-based TSVD estimator with the basic ridge estimate as its initial value, which is shown by simulation to have significant quality advantages.
Wang, Y. and Xiao, T.(2001)	derived a fast realization algorithms for determining regularization parameters in linear inverse problems.
Wenchenko, E.(2000)	presented the estimation of the signal-to-noise ratios in the linear regression model and tested them with a simulation study.

2 The A-optimal design of the regularization parameter in uniform Tykhonov-Phillips regularization (α-weighted BLE)

Let us first introduce the *special Gauss-Markov model* $\mathbf{y} = \mathbf{A}\boldsymbol{\xi} + \mathbf{e}$ specified in *Box 1*, which is given for the *first order moments* in the form of a *consistent system of linear equations relating* the first *non-stochastic* ("fixed"), *real-valued vector* $\boldsymbol{\xi}$ of *unknown* parameters to the expectation $E\{\mathbf{y}\}$ of the *stochastic,* real-valued vector $\mathbf{y}$ of observations, $\mathbf{A}\boldsymbol{\xi} = E\{\mathbf{y}\}$, since $E\{\mathbf{y}\} \in \mathcal{R}(\mathbf{A})$ is an element of the column space $\mathcal{R}(\mathbf{A})$ of the real-valued, *non-stochastic* ("fixed") *"first order design matrix"* $\mathbf{A} \in \mathbb{R}^{n \times m}$. The rank of the fixed matrix $\mathbf{A}$, rk $\mathbf{A}$, equals the number m of unknowns, $\boldsymbol{\xi} \in \mathbb{R}^m$. In addition, the *second order central moments,* the *regular* variance-covariance matrix $\boldsymbol{\Sigma}_y$, also called *dispersion matrix* $D\{\mathbf{y}\}$ constitute the second matrix $\boldsymbol{\Sigma}_y \in \mathbb{R}^{n \times n}$ of *unknowns* to be specified as a linear model furtheron.

Box 1:
Special Gauss–Markov model

$$\mathbf{y} = \mathbf{A}\boldsymbol{\xi} + \mathbf{e}$$

1st moments

$$\mathbf{A}\boldsymbol{\xi} = E\{\mathbf{y}\},\ \mathbf{A} \in \mathbb{R}^{n \times m},\ E\{\mathbf{y}\} \in \mathcal{R}(\mathbf{A}),\ \text{rk}\,\mathbf{A} = m \quad (1)$$

2nd moments

$$\boldsymbol{\Sigma}_y = D\{\mathbf{y}\} \in \mathbb{R}^{n \times n}, \boldsymbol{\Sigma}_y \text{ positive definite, rk}\,\boldsymbol{\Sigma}_y = n$$
$$\boldsymbol{\xi},\ \mathbf{y} - E\{\mathbf{y}\} = \mathbf{e} \text{ unknown} \quad (2)$$
$$\boldsymbol{\Sigma}_y \text{ unknown, but structured.}$$

Here we will focus on best linear estimators of type α-homBLE of *fixed effects* $\boldsymbol{\xi}$. At first let us begin with a discussion of the *bias vector* and the *bias matrix* as well as of the *Mean Square Error matrix* $MSE\{\hat{\boldsymbol{\xi}}\}$ with respect to a *homogeneously linear estimate* $\hat{\boldsymbol{\xi}} = \mathbf{L}\mathbf{y}$ of fixed effects $\boldsymbol{\xi}$ based upon *Box 2.*

Box 2:
Bias vector, bias matrix, Mean Square Error matrix in the special Gauss–Markov model with fixed effects

$$E\{\mathbf{y}\} = \mathbf{A}\boldsymbol{\xi} \quad (3)$$

$$D\{\mathbf{y}\} = \boldsymbol{\Sigma}_y \quad (4)$$

"ansatz"

$$\hat{\boldsymbol{\xi}} = \mathbf{L}\mathbf{y} \quad (5)$$

bias vector

$$\boldsymbol{\beta} := E\{\hat{\boldsymbol{\xi}} - \boldsymbol{\xi}\} = E\{\hat{\boldsymbol{\xi}}\} - \boldsymbol{\xi} \quad (6)$$

$$\boldsymbol{\beta} = \mathbf{L}E\{\mathbf{y}\} - \boldsymbol{\xi} = -[\mathbf{I}_m - \mathbf{L}\mathbf{A}]\boldsymbol{\xi} \quad (7)$$

bias matrix

$$\mathbf{B} := \mathbf{I}_m - \mathbf{L}\mathbf{A} \quad (8)$$

decomposition

$$\hat{\boldsymbol{\xi}} - \boldsymbol{\xi} = (\hat{\boldsymbol{\xi}} - E\{\hat{\boldsymbol{\xi}}\}) + (E\{\hat{\boldsymbol{\xi}}\} - \boldsymbol{\xi}) \quad (9)$$

$$\hat{\boldsymbol{\xi}} - \boldsymbol{\xi} = \mathbf{L}(\mathbf{y} - E\{\mathbf{y}\}) - [\mathbf{I}_m - \mathbf{L}\mathbf{A}]\boldsymbol{\xi} \quad (10)$$

Mean Square Error matrix

$$MSE\{\hat{\boldsymbol{\xi}}\} := E\{(\hat{\boldsymbol{\xi}} - \boldsymbol{\xi})(\hat{\boldsymbol{\xi}} - \boldsymbol{\xi})'\} \quad (11)$$

$$MSE\{\hat{\boldsymbol{\xi}}\} = \mathbf{L}D\{\mathbf{y}\}\mathbf{L}' + [\mathbf{I}_m - \mathbf{L}\mathbf{A}]\boldsymbol{\xi}\boldsymbol{\xi}'[\mathbf{I}_m - \mathbf{L}\mathbf{A}]' \quad (12)$$

$$(E\{\hat{\boldsymbol{\xi}} - E\{\hat{\boldsymbol{\xi}}\}\} = 0)$$

S-modified Mean Square Error

$$MSE_S\{\hat{\boldsymbol{\xi}}\} := \mathbf{L}D\{\mathbf{y}\}\mathbf{L}' + [\mathbf{I}_m - \mathbf{L}\mathbf{A}]\mathbf{S}[\mathbf{I}_m - \mathbf{L}\mathbf{A}]' \quad (13)$$

$\mathbf{S}$ - non-negative definite *substitute* matrix.

Frobenius matrix norms

$$\| MSE\{\hat{\boldsymbol{\xi}}\} \|^2 := \text{tr}\, E\{(\hat{\boldsymbol{\xi}} - \boldsymbol{\xi})(\hat{\boldsymbol{\xi}} - \boldsymbol{\xi})'\} \quad (14)$$

$$\| MSE\{\hat{\boldsymbol{\xi}}\} \|^2 =$$
$$= \text{tr}\,\mathbf{L}D\{\mathbf{y}\}\mathbf{L}' + +\text{tr}[\mathbf{I}_m - \mathbf{L}\mathbf{A}]\boldsymbol{\xi}\boldsymbol{\xi}'[\mathbf{I}_m - \mathbf{L}\mathbf{A}]' \quad (15)$$
$$= \| \mathbf{L}' \|^2_{\Sigma_y} + \| (\mathbf{I}_m - \mathbf{L}\mathbf{A})' \|^2_{\xi\xi'}$$

$$\| MSE_S\{\hat{\boldsymbol{\xi}}\} \|^2 :=$$
$$:= \text{tr}\,\mathbf{L}D\{\mathbf{y}\}\mathbf{L}' + \text{tr}[\mathbf{I}_m - \mathbf{L}\mathbf{A}]\mathbf{S}[\mathbf{I}_m - \mathbf{L}\mathbf{A}]' \quad (16)$$
$$= \| \mathbf{L}' \|^2_{\Sigma_y} + \| (\mathbf{I}_m - \mathbf{L}\mathbf{A})' \|^2_{\mathbf{S}}$$

hybrid minimum variance – minimum bias norm α-weighted

$$\| MSE_{\alpha,S}\{\hat{\boldsymbol{\xi}}\} \|^2 :=$$
$$= \text{tr}\,\mathbf{L}D\{\mathbf{y}\}\mathbf{L}' + \text{tr}[\mathbf{I}_m - \mathbf{L}\mathbf{A}]\frac{1}{\alpha}\mathbf{S}[\mathbf{I}_m - \mathbf{L}\mathbf{A}]'$$
$$= \| \mathbf{L}' \|^2_{\Sigma_y} + \frac{1}{\alpha} \| (\mathbf{I}_m - \mathbf{L}\mathbf{A})' \|^2_{\mathbf{S}} \quad (17)$$

special model

$$\dim \mathcal{R}(\mathbf{S}\mathbf{A}') = \text{rk}\,\mathbf{S}\mathbf{A}' = \text{rk}\,\mathbf{A} = m \quad (18)$$

The *bias vector* $\boldsymbol{\beta}$ is conventionally defined by $E\{\hat{\boldsymbol{\xi}}\} - \boldsymbol{\xi}$ subject to the homogeneous estimation form $\hat{\boldsymbol{\xi}} = \mathbf{L}\mathbf{y}$. Accordingly the bias vector can be represented by (7) $\boldsymbol{\beta} = -[\mathbf{I}_m - \mathbf{L}\mathbf{A}]\boldsymbol{\xi}$. Since the vector $\boldsymbol{\xi}$ of *fixed effects* is unknown, there has been made the proposal to instead use the matrix $\mathbf{I}_m - \mathbf{LA}$ as a *matrix-valued measure of bias*. A measure of the estimation error is the *Mean Square Error matrix* $MSE\{\hat{\boldsymbol{\xi}}\}$ of type (11). $MSE\{\hat{\boldsymbol{\xi}}\}$ can be decomposed into two *basic parts*:

- the dispersion matrix $D\{\hat{\boldsymbol{\xi}}\} = \mathbf{L}D\{\mathbf{y}\}\mathbf{L}'$
- the dyadic bias product $\boldsymbol{\beta\beta}'$

Indeed the vector $\hat{\boldsymbol{\xi}} - \boldsymbol{\xi}$ can be decomposed as well into two parts of type (9), (10), namely (i) $\hat{\boldsymbol{\xi}} - E\{\hat{\boldsymbol{\xi}}\} = \mathbf{Le}$ and (ii) $E\{\hat{\boldsymbol{\xi}}\} - \boldsymbol{\xi} = \boldsymbol{\beta}$ which may be called estimation error (due to observation noise) and bias, respectively. The double decomposition of the vector $\hat{\boldsymbol{\xi}} - \boldsymbol{\xi}$ leads straightforward to the double representation of the matrix $MSE\{\hat{\boldsymbol{\xi}}\}$ of type (12). Such a representation suffers from two effects: *Firstly* the vector $\boldsymbol{\xi}$ of *fixed effects* is unknown, *secondly* the matrix $\boldsymbol{\xi\xi}'$ has only rank 1. In consequence, the matrix $[\mathbf{I}_m - \mathbf{LA}]\boldsymbol{\xi\xi}'[\mathbf{I}_m - \mathbf{LA}]'$ has only rank 1, too. In this situation the proposal has been made to *modify* $MSE\{\hat{\boldsymbol{\xi}}\}$ with respect to $\boldsymbol{\xi\xi}'$ by a *higher rank matrix* **S**. A homogeneously linear α-weighted hybrid minimum variance-minimum bias estimation (α-homBLE) is presented in *Definition 1* which is based upon the weighted sum of two norms of type (17), namely

- *average variance* $\|\mathbf{L}'\|^2_{\Sigma_y} = \operatorname{tr} \mathbf{L}\Sigma_y\mathbf{L}'$
- *average bias* $\|(\mathbf{I}_m - \mathbf{LA})'\|^2_{\mathbf{S}} =$
  $= \operatorname{tr}[\mathbf{I}_m - \mathbf{LA}]\mathbf{S}[\mathbf{I}_m - \mathbf{LA}]'$

  where we expect $\boldsymbol{\xi}$ to belong to the column space $\mathcal{R}(\mathbf{S})$.

The very important estimator α-homBLE *is balancing* variance and bias by the weighting factor α which is illustrated by *Figure 1*.

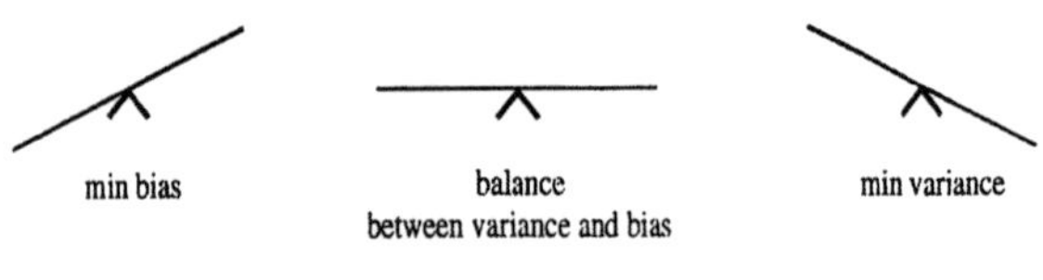

Fig. 1 Balance of variance and bias by the weighting factor α

Definition 1 ($\hat{\boldsymbol{\xi}}$ homBLE of $\boldsymbol{\xi}$):
A $m\times 1$ vector $\hat{\boldsymbol{\xi}}$ is called homogeneous *BLE* of $\boldsymbol{\xi}$ in the *special linear Gauss-Markov model with fixed effects* of *Box 1*, if and only if

(*1st*) $\hat{\boldsymbol{\xi}}$ is a homogeneously linear form

$$\hat{\boldsymbol{\xi}} = \mathbf{Ly}, \tag{19}$$

(*2nd*) in comparison to all other linear estimations $\hat{\boldsymbol{\xi}}$ has the minimum *Mean Square Error* in the sense of

$$\begin{aligned} &\| MSE\{\hat{\boldsymbol{\xi}}\} \|^2 = \\ &= \operatorname{tr} \mathbf{L}D\{\mathbf{y}\}\mathbf{L}' + \operatorname{tr}[\mathbf{I}_m - \mathbf{LA}]\boldsymbol{\xi\xi}'[\mathbf{I}_m - \mathbf{LA}]' = \\ &= \| \mathbf{L}' \|^2_{\Sigma_y} + \| (\mathbf{I}_m - \mathbf{LA})' \|^2_{\boldsymbol{\xi\xi}'} = \min_{\mathbf{L}} \end{aligned} \tag{20}$$

Definition 2 ($\hat{\boldsymbol{\xi}}$ S-homBLE of $\boldsymbol{\xi}$):
A $m\times 1$ vector $\hat{\boldsymbol{\xi}}$ is called homogeneous *S-homBLE* of $\boldsymbol{\xi}$ in the *special linear Gauss-Markov model with fixed effects* of *Box 1*, if and only if

(*1st*) $\hat{\boldsymbol{\xi}}$ is a homogeneously linear form

$$\hat{\boldsymbol{\xi}} = \mathbf{Ly}, \tag{21}$$

(*2nd*) in comparison to all other linear estimations $\hat{\boldsymbol{\xi}}$ has the minimum **S**-modified *Mean Square Error* in the sense of

$$\begin{aligned} &\| MSE_{\mathbf{S}}\{\hat{\boldsymbol{\xi}}\} \|^2 = \\ &= \operatorname{tr} \mathbf{L}D\{\mathbf{y}\}\mathbf{L}' + \operatorname{tr}[\mathbf{I}_m - \mathbf{LA}]\mathbf{S}[\mathbf{I}_m - \mathbf{LA}]' \\ &= \| \mathbf{L}' \|^2_{\Sigma_y} + \| (\mathbf{I}_m - \mathbf{LA})' \|^2_{\mathbf{S}} = \min_{\mathbf{L}} \end{aligned} \tag{22}$$

Definition 3 ($\hat{\boldsymbol{\xi}}$ homlinear hybrid min var-min bias solution, α-weighted, or α- homBLE):

A $m\times 1$ vector $\hat{\boldsymbol{\xi}}$ is called homogeneously linear α-weighted hybrid minimum variance- minimum bias estimate (α-homBLE) of $\boldsymbol{\xi}$ in the *special linear Gauss-Markov model with fixed effects* of *Box 1*, if and only if

(*1st*) $\hat{\boldsymbol{\xi}}$ is a homogeneously linear form

$$\hat{\boldsymbol{\xi}} = \mathbf{Ly}, \tag{23}$$

(*2nd*) in comparison to all other homogeneously linear estimates $\hat{\boldsymbol{\xi}}$ has the minimum variance-minimum bias property in the sense of the α-weighted hybrid norm

$$\begin{aligned} &\| MSE_{\alpha,\mathbf{S}}\{\hat{\boldsymbol{\xi}}\} \|^2 = \\ &= \operatorname{tr} \mathbf{L}D\{\mathbf{y}\}\mathbf{L}' + \frac{1}{\alpha}\operatorname{tr}(\mathbf{I}_m - \mathbf{LA})\mathbf{S}(\mathbf{I}_m - \mathbf{LA})' \\ &= \|\mathbf{L}'\|^2_{\Sigma_y} + \frac{1}{\alpha}\|(\mathbf{I}_m - \mathbf{LA})'\|^2_{\mathbf{S}} = \min_{\mathbf{L}} \end{aligned} \tag{24}$$

in particular with respect to the special model

$$\alpha \in \mathbb{R}^+, \dim \mathcal{R}(\mathbf{SA}') = \operatorname{rk} \mathbf{SA}' = \operatorname{rk} \mathbf{A} = m$$

The estimates $\hat{\boldsymbol{\xi}}$ of type *homBLE*, ***S*** *-homBLE* and α-homBLE can be characterized by the following *Lemma* and the proof of this Lemma is given in the *Appendix A*.

Lemma 4 (homBLE, **S**-homBLE and α- homBLE):

A $m\times 1$ vector $\hat{\boldsymbol{\xi}}$ is *homBLE*, *S -homBLE* and *α -homBLE* of $\boldsymbol{\xi}$ in the special linear Gauss-Markov model with fixed effects of *Box 1*, if and only if the matrix $\hat{\mathbf{L}}$ fulfils the normal equations

(1*st*) *homBLE*:

$$(\boldsymbol{\Sigma}_y + \mathbf{A}\boldsymbol{\xi}\boldsymbol{\xi}'\mathbf{A}')\hat{\mathbf{L}}' = \mathbf{A}\boldsymbol{\xi}\boldsymbol{\xi}' \tag{25}$$

(2*nd*) *S -homBLE*:

$$(\boldsymbol{\Sigma}_y + \mathbf{ASA}')\hat{\mathbf{L}}' = \mathbf{AS} \tag{26}$$

(3*rd*) *α- homBLE*:

$$(\mathbf{SA}'\boldsymbol{\Sigma}_y^{-1}\mathbf{A} + \alpha\mathbf{I}_m)\hat{\mathbf{L}} = \mathbf{SA}'\boldsymbol{\Sigma}_y^{-1}$$
or, if **S** is non-singular, (27)
$$(\mathbf{A}'\boldsymbol{\Sigma}_y^{-1}\mathbf{A} + \alpha\mathbf{S}^{-1})\hat{\mathbf{L}} = \mathbf{A}'\boldsymbol{\Sigma}_y^{-1}$$

For an *explicit representation of* $\hat{\boldsymbol{\xi}}$ as α- homBLE of $\boldsymbol{\xi}$ in the *special Gauss–Markov model with fixed effects* of *Box 1*, we solve the normal equations (25), (26) and (27) for

$$\hat{\mathbf{L}} = \arg\{\mathcal{L}(\mathbf{L}) = \min_{\mathbf{L}}\}.$$

Beside the *explicit representation* of $\hat{\boldsymbol{\xi}}$ of type homBLE, **S** -homBLE and α- homBLE we present you with the related dispersion matrix $D\{\hat{\boldsymbol{\xi}}\}$, the *Mean Square Error matrix* $MSE\{\hat{\boldsymbol{\xi}}\}$, the modified *Mean Square Error matrices* $MSE_{\mathbf{S}}\{\hat{\boldsymbol{\xi}}\}$ and $MSE_{\alpha,\mathbf{S}}\{\hat{\boldsymbol{\xi}}\}$ in

Theorem 5 ($\hat{\boldsymbol{\xi}}$ homBLE):

Let $\hat{\boldsymbol{\xi}} = \mathbf{Ly}$ be *homBLE* of $\boldsymbol{\xi}$ in the *special linear Gauss-Markov model with fixed effects* of *Box 1*. Then equivalent representations of the solutions of the normal equations (25) are

$$\hat{\boldsymbol{\xi}} = \boldsymbol{\xi}\boldsymbol{\xi}'\mathbf{A}'[\boldsymbol{\Sigma}_y + \mathbf{A}\boldsymbol{\xi}\boldsymbol{\xi}'\mathbf{A}']^{-1}\mathbf{y} \tag{28}$$

(if $[\boldsymbol{\Sigma}_y + \mathbf{A}\boldsymbol{\xi}\boldsymbol{\xi}'\mathbf{A}']^{-1}$ exists)

and completed by the dispersion matrix

$$D\{\hat{\boldsymbol{\xi}}\} = \boldsymbol{\xi}\boldsymbol{\xi}'\mathbf{A}'[\boldsymbol{\Sigma}_y + \mathbf{A}\boldsymbol{\xi}\boldsymbol{\xi}'\mathbf{A}']^{-1}\boldsymbol{\Sigma}_y \times \times[\boldsymbol{\Sigma}_y + \mathbf{A}\boldsymbol{\xi}\boldsymbol{\xi}'\mathbf{A}']^{-1}\mathbf{A}\boldsymbol{\xi}\boldsymbol{\xi}', \tag{29}$$

by the *bias vector* (7)

$$\begin{aligned}\boldsymbol{\beta} &:= E\{\hat{\boldsymbol{\xi}}\} - \boldsymbol{\xi}\\ &= -[\mathbf{I}_m - \boldsymbol{\xi}\boldsymbol{\xi}'\mathbf{A}'(\mathbf{A}\boldsymbol{\xi}\boldsymbol{\xi}'\mathbf{A}' + \boldsymbol{\Sigma}_y)^{-1}\mathbf{A}]\boldsymbol{\xi}\end{aligned} \tag{30}$$

and by the *Mean Square Error matrix* $MSE\{\hat{\boldsymbol{\xi}}\}$:

$$\begin{aligned}MSE\{\hat{\boldsymbol{\xi}}\} &:= E\{(\hat{\boldsymbol{\xi}} - \boldsymbol{\xi})(\hat{\boldsymbol{\xi}} - \boldsymbol{\xi})'\}\\ &= D\{\hat{\boldsymbol{\xi}}\} + \boldsymbol{\beta}\boldsymbol{\beta}'\end{aligned} \tag{31}$$

$$\begin{aligned}&MSE\{\hat{\boldsymbol{\xi}}\} :=\\ &D\{\hat{\boldsymbol{\xi}}\} + [\mathbf{I}_m - \boldsymbol{\xi}\boldsymbol{\xi}'\mathbf{A}'(\mathbf{A}\boldsymbol{\xi}\boldsymbol{\xi}'\mathbf{A}' + \boldsymbol{\Sigma}_y)^{-1}\mathbf{A}]\times\\ &\times\boldsymbol{\xi}\boldsymbol{\xi}'[\mathbf{I}_m - \mathbf{A}'(\mathbf{A}\boldsymbol{\xi}\boldsymbol{\xi}'\mathbf{A}' + \boldsymbol{\Sigma}_y)^{-1}\mathbf{A}\boldsymbol{\xi}\boldsymbol{\xi}']\end{aligned} \tag{32}$$

At this point we have to comment what *Theorem 5* tells us. *homBLE* has generated the estimation $\hat{\boldsymbol{\xi}}$ of type (28), the dispersion matrix $D\{\hat{\boldsymbol{\xi}}\}$ of type (29), the bias vector of type (30) and the *Mean Square Error matrix* $MSE\{\hat{\boldsymbol{\xi}}\}$ of type (32) *which all depend on the vector* $\boldsymbol{\xi}$ *and the matrix* $\boldsymbol{\xi}\boldsymbol{\xi}'$, respectively. We already mentioned that $\boldsymbol{\xi}$ and the matrix $\boldsymbol{\xi}\boldsymbol{\xi}'$ are *not* accessible from measurements. The situation is similar to the one in the *theory of hypothesis testing*. As shown later in this section we can produce only an estimator $\hat{\boldsymbol{\xi}}$ and consequently can setup a *hypothesis* $\boldsymbol{\xi}_0$ of the "*fixed effect*" $\boldsymbol{\xi}$. Indeed, a similar argument applies to the *second central moment* $D\{\mathbf{y}\} \sim \boldsymbol{\Sigma}_y$ of the "*random effect*" **y**, the observation vector. Such a dispersion matrix has to be known in order to be able to compute $\hat{\boldsymbol{\xi}}$, $D\{\hat{\boldsymbol{\xi}}\}$, and $MSE\{\hat{\boldsymbol{\xi}}\}$. Again we have to apply the argument that we are only able to construct an estimate $\hat{\boldsymbol{\Sigma}}_y$ and to setup a *hypothesis* about $D\{\mathbf{y}\} \sim \boldsymbol{\Sigma}_y$.

Theorem 6 ($\hat{\boldsymbol{\xi}}$ **S**-homBLE):

Let $\hat{\boldsymbol{\xi}} = \mathbf{Ly}$ be *S-homBLE* of $\boldsymbol{\xi}$ in the *special linear Gauss- Markov model with fixed effects* of *Box 1*. Then equivalent representations of the solutions of the normal equations (26) are

$$\hat{\boldsymbol{\xi}} = \mathbf{SA}'(\boldsymbol{\Sigma}_y + \mathbf{ASA}')^{-1}\mathbf{y} \tag{33}$$

$$\hat{\boldsymbol{\xi}} = (\mathbf{I}_m + \mathbf{SA}'\boldsymbol{\Sigma}_y^{-1}\mathbf{A})^{-1}\mathbf{SA}'\boldsymbol{\Sigma}_y^{-1}\mathbf{y} \tag{34}$$

$$\hat{\boldsymbol{\xi}} = (\mathbf{A}'\boldsymbol{\Sigma}_y^{-1}\mathbf{A} + \mathbf{S}^{-1})^{-1}\mathbf{A}'\boldsymbol{\Sigma}_y^{-1}\mathbf{y} \tag{35}$$

(if $\mathbf{S}^{-1}$ exists)

are completed by the dispersion matrices

$$D\{\hat{\boldsymbol{\xi}}\} = \mathbf{SA}'(\mathbf{ASA}' + \boldsymbol{\Sigma}_y)^{-1}\boldsymbol{\Sigma}_y(\mathbf{ASA}' + \boldsymbol{\Sigma}_y)^{-1}\mathbf{AS} \tag{36}$$

$$D\{\hat{\boldsymbol{\xi}}\} = (\mathbf{A}'\boldsymbol{\Sigma}_y^{-1}\mathbf{A} + \mathbf{S}^{-1})^{-1}\mathbf{A}'\boldsymbol{\Sigma}_y^{-1}\mathbf{A}(\mathbf{A}'\boldsymbol{\Sigma}_y^{-1}\mathbf{A} + \mathbf{S}^{-1})^{-1} \tag{37}$$

(if $\mathbf{S}^{-1}$ exists)

by the *bias vector* (7)

$$\begin{aligned}\boldsymbol{\beta} &:= E\{\hat{\boldsymbol{\xi}}\} - \boldsymbol{\xi}\\ &= -[\mathbf{I}_m - \mathbf{SA}'(\mathbf{ASA}' + \boldsymbol{\Sigma}_y)^{-1}\mathbf{A}]\boldsymbol{\xi}\end{aligned}$$

$$\boldsymbol{\beta} = -[\mathbf{I}_m - (\mathbf{A}'\boldsymbol{\Sigma}_y^{-1}\mathbf{A} + \mathbf{S}^{-1})^{-1}\mathbf{A}'\boldsymbol{\Sigma}_y^{-1}\mathbf{A}]\boldsymbol{\xi} \quad (38)$$

(if $\mathbf{S}^{-1}$ exists)

and by the modified *Mean Square Error* matrix $MSE_s\{\hat{\boldsymbol{\xi}}\}$:

$$\begin{aligned} MSE_s\{\hat{\boldsymbol{\xi}}\} &:= E\{(\hat{\boldsymbol{\xi}} - \boldsymbol{\xi})(\hat{\boldsymbol{\xi}} - \boldsymbol{\xi})'\} \\ &= D\{\hat{\boldsymbol{\xi}}\} + \boldsymbol{\beta}\boldsymbol{\beta}' \end{aligned} \quad (39)$$

$$\begin{gathered} MSE_s\{\hat{\boldsymbol{\xi}}\} = \\ \mathbf{SA}'(\mathbf{ASA}' + \boldsymbol{\Sigma}_y)^{-1}\boldsymbol{\Sigma}_y(\mathbf{ASA}' + \boldsymbol{\Sigma}_y)^{-1}\mathbf{AS} + \\ +[\mathbf{I}_m - \mathbf{SA}'(\mathbf{ASA}' + \boldsymbol{\Sigma}_y)^{-1}\mathbf{A}] \times \\ \times \boldsymbol{\xi}\boldsymbol{\xi}'[\mathbf{I}_m - \mathbf{A}'(\mathbf{ASA}' + \boldsymbol{\Sigma}_y)^{-1}\mathbf{AS}] \end{gathered} \quad (40)$$

$$\begin{gathered} MSE_s\{\hat{\boldsymbol{\xi}}\} = \\ (\mathbf{A}'\boldsymbol{\Sigma}_y^{-1}\mathbf{A} + \mathbf{S}^{-1})^{-1}\mathbf{A}'\boldsymbol{\Sigma}_y^{-1}\mathbf{A}(\mathbf{A}'\boldsymbol{\Sigma}_y^{-1}\mathbf{A} + \mathbf{S}^{-1})^{-1} + \\ +[(\mathbf{A}'\boldsymbol{\Sigma}_y^{-1}\mathbf{A} + \mathbf{S}^{-1})^{-1}\mathbf{S}^{-1}]\boldsymbol{\xi}\boldsymbol{\xi}'[\mathbf{S}^{-1}(\mathbf{A}'\boldsymbol{\Sigma}_y^{-1}\mathbf{A} + \mathbf{S}^{-1})^{-1}] \end{gathered} \quad (41)$$

(if $\mathbf{S}^{-1}$ exists)

The interpretation of *S-homBLE* is even more complex. In extension of the comments to *homBLE* we have to live with another matrix-valued degree of freedom, $\hat{\boldsymbol{\xi}}$ of type (33), (34), (35) and $D\{\hat{\boldsymbol{\xi}}\}$ of type (36), (37) do no longer depend on the inaccessible matrix $\boldsymbol{\xi}\boldsymbol{\xi}'$, $\text{rk}(\boldsymbol{\xi}\boldsymbol{\xi}')$, but on the "*weight of the bias weight matrix*" **S**, rk **S** = *m*. Indeed we can associate any element of the bias matrix with a particular weight which can be "*designed*" by the analyst. Again the *bias vector* $\boldsymbol{\beta}$ of type (38) as well as the *Mean Square Error* of type (39), (40), (41) depend on the vector $\boldsymbol{\xi}$ which is inaccessible. Beside the "*weight of the bias weight matrix* **S**" $\hat{\boldsymbol{\xi}}$, $D\{\hat{\boldsymbol{\xi}}\}$, $\boldsymbol{\beta}$ and $MSE_s\{\hat{\boldsymbol{\xi}}\}$ are vector-valued or matrix-valued functions of the *dispersion matrix* $D\{\mathbf{y}\} \sim \boldsymbol{\Sigma}_y$ of the stochastic observation vector which is inaccessible. By *hypothesis testing* we may decide upon the construction of $D\{\mathbf{y}\} \sim \boldsymbol{\Sigma}_y$ from an estimate $\hat{\boldsymbol{\Sigma}}_y$.

Theorem 7 ($\hat{\boldsymbol{\xi}}$ α-homBLE, also called: *ridge estimator* or *Tykhonov-Phillips regulator*)

Let $\hat{\boldsymbol{\xi}} = \mathbf{Ly}$ be *α- homBLE* of $\boldsymbol{\xi}$ in the *special linear Gauss-Markov model with fixed effects* of *Box 1*. Then equivalent representations of the solutions of the normal equations (27) are

$$\begin{aligned} \hat{\boldsymbol{\xi}} &= (\mathbf{SA}'\boldsymbol{\Sigma}_y^{-1}\mathbf{A} + \alpha\mathbf{I}_m)^{-1}\mathbf{SA}'\boldsymbol{\Sigma}_y^{-1}\mathbf{y} \\ &= (\mathbf{A}'\boldsymbol{\Sigma}_y^{-1}\mathbf{A} + \alpha\mathbf{S}^{-1})^{-1}\mathbf{A}'\boldsymbol{\Sigma}_y^{-1}\mathbf{y} \quad \text{(if } \mathbf{S}^{-1} \text{ exists)}, \end{aligned} \quad (42)$$

completed by the dispersion matrix

$$\begin{aligned} D\{\hat{\boldsymbol{\xi}}\} &= \\ &= (\mathbf{SA}'\boldsymbol{\Sigma}_y^{-1}\mathbf{A} + \alpha\mathbf{I}_m)^{-1}\mathbf{SA}'\boldsymbol{\Sigma}_y^{-1}\mathbf{AS}'(\mathbf{SA}'\boldsymbol{\Sigma}_y^{-1}\mathbf{A} + \alpha\mathbf{I}_m)^{-1} \\ &= (\mathbf{A}'\boldsymbol{\Sigma}_y^{-1}\mathbf{A} + \alpha\mathbf{S}^{-1})^{-1}\mathbf{A}'\boldsymbol{\Sigma}_y^{-1}\mathbf{A}(\mathbf{A}'\boldsymbol{\Sigma}_y^{-1}\mathbf{A} + \alpha\mathbf{S}^{-1})^{-1} \\ &\qquad \text{(if } \mathbf{S}^{-1} \text{ exists)}, \end{aligned} \quad (43)$$

by the *bias vector* (7)

$$\begin{aligned} \boldsymbol{\beta} &:= E\{\hat{\boldsymbol{\xi}}\} - \boldsymbol{\xi} = \\ &= -[\mathbf{I}_m - (\mathbf{A}'\boldsymbol{\Sigma}_y^{-1}\mathbf{A} + \alpha\mathbf{S}^{-1})^{-1}\mathbf{A}'\boldsymbol{\Sigma}_y^{-1}\mathbf{A}]\boldsymbol{\xi} \\ &= -\alpha(\mathbf{A}'\boldsymbol{\Sigma}_y^{-1}\mathbf{A} + \alpha\mathbf{S}^{-1})^{-1}\mathbf{S}^{-1}\boldsymbol{\xi} \\ &= -\alpha(\mathbf{SA}'\boldsymbol{\Sigma}_y^{-1}\mathbf{A} + \alpha\mathbf{I}_m)^{-1}\boldsymbol{\xi} \\ &\qquad \text{(if } \mathbf{S}^{-1} \text{ exists)} \end{aligned} \quad (44)$$

and by the *Mean Square Error matrix* $MSE\{\hat{\boldsymbol{\xi}}\}$

$$\begin{aligned} MSE\{\hat{\boldsymbol{\xi}}\} &:= E\{(\hat{\boldsymbol{\xi}} - \boldsymbol{\xi})(\hat{\boldsymbol{\xi}} - \boldsymbol{\xi})'\} \\ &= D\{\hat{\boldsymbol{\xi}}\} + \boldsymbol{\beta}\boldsymbol{\beta}' \end{aligned} \quad (45)$$

as well as the hybrid α-weighted variance-bias norm $MSE_{\alpha,s}\{\hat{\boldsymbol{\xi}}\}$

$$\begin{gathered} MSE_{\alpha,s}\{\hat{\boldsymbol{\xi}}\} = \\ = (\mathbf{A}'\boldsymbol{\Sigma}_y^{-1}\mathbf{A} + \alpha\mathbf{S}^{-1})^{-1}\mathbf{A}'\boldsymbol{\Sigma}_y^{-1}\mathbf{A}(\mathbf{A}'\boldsymbol{\Sigma}_y^{-1}\mathbf{A} + \alpha\mathbf{S}^{-1})^{-1} + \\ +[(\mathbf{A}'\boldsymbol{\Sigma}_y^{-1}\mathbf{A} + \alpha\mathbf{S}^{-1})^{-1}\alpha\mathbf{S}^{-1}]\boldsymbol{\xi}\boldsymbol{\xi}'[\alpha\mathbf{S}^{-1}(\mathbf{A}'\boldsymbol{\Sigma}_y^{-1}\mathbf{A} + \alpha\mathbf{S}^{-1})^{-1}] \\ \text{(if } \mathbf{S}^{-1} \text{ exists)}. \end{gathered} \quad (46)$$

The interpretation of the very important estimator α-homBLE $\hat{\boldsymbol{\xi}}$ of $\boldsymbol{\xi}$ is as follows: $\hat{\boldsymbol{\xi}}$ of type (42), also called *ridge estimator* or *Tykhonov-Phillips regulator*, contains the *Cayley inverse* of the normal equation matrix *which is additively decomposed* into $\mathbf{A}'\boldsymbol{\Sigma}_y^{-1}\mathbf{A}$ and $\alpha\mathbf{S}^{-1}$. The weight factor α balances the first observational weight and the second bias weight within the inverse. While the experiment informs us of the variance-covariance matrix $\boldsymbol{\Sigma}_y$, say $\hat{\boldsymbol{\Sigma}}_y$, the *weight of the bias weight matrix* and the *weight factor* α are at the disposal of the analyst. For instance, by the choice **S**= Diag$(s_1, \ldots, s_m)$ we may emphasize increase or decrease of certain bias matrix elements. The choice of an equally weighted bias matrix is $\mathbf{S} = \mathbf{I}_m$. In contrast, the weight factor α can be alternatively determined by the *A-optimal design* of type

- $\text{tr}\, D\{\hat{\boldsymbol{\xi}}\} = \min_\alpha$, or
- $\text{tr}\, \boldsymbol{\beta}\boldsymbol{\beta}' = \min_\alpha$, or
- $\text{tr}\, MSE_{\alpha,s}\{\hat{\boldsymbol{\xi}}\} = \min_\alpha$.

In the *first case* we optimize the *trace of the variance-covariance matrix* $D\{\hat{\boldsymbol{\xi}}\}$ of type (43). Alternatively by means of $\text{tr}\, \boldsymbol{\beta}\boldsymbol{\beta}' = \min_\alpha$ we optimize the *quadratic bias* where the bias vector $\boldsymbol{\beta}$ of type (44) is chosen, regardless of the dependence on ξ. Finally for the *third case* – the most meaningful one – we minimize the trace of the *Mean Square Error matrix* $MSE_{\alpha,S}\{\hat{\xi}\}$ of type (46), despite of the dependence on $\xi\xi'$. In this paper we concentrate on the third case and the main result is summarized in

Theorem 8 (A-optimal design of α):

Let the average hybrid α-weighted variance-bias norm $MSE_{\alpha,S}\{\hat{\xi}\}$ of $\hat{\xi}$ (α- homBLE) with respect to the linear Gauss Markov model be given by

$$\mathrm{tr}MSE_{\alpha,S}\{\hat{\xi}\} =$$
$$= \mathrm{tr}(A'\Sigma_y^{-1}A+\alpha S^{-1})^{-1}A'\Sigma_y^{-1}A(A'\Sigma_y^{-1}A+\alpha S^{-1})^{-1}+$$
$$+\mathrm{tr}[(A'\Sigma_y^{-1}A+\alpha S^{-1})^{-1}\alpha S^{-1}]\xi\xi'[\alpha S^{-1}(A'\Sigma_y^{-1}A+\alpha S^{-1})^{-1}],$$

then $\hat{\alpha}$ follows by A-optimal design in the sense of

$$\mathrm{tr}MSE_{\alpha,S}\{\hat{\xi}\} = \min$$

if and only if

$$\hat{\alpha} = \frac{\mathrm{tr}\, A'\Sigma_y^{-1}A(A'\Sigma_y^{-1}A+\alpha S^{-1})^{-2}S^{-1}(A'\Sigma_y^{-1}A+\alpha S^{-1})^{-1}}{\xi' S^{-1}(A'\Sigma_y^{-1}A+\alpha S^{-1})^{-2}A'\Sigma_y^{-1}A(A'\Sigma_y^{-1}A+\alpha S^{-1})^{-1}S^{-1}\xi} \qquad (47)$$

The proof of *Theorem 8* is given in the *Appendix B*. The subject of *optimal design* within Mathematical Statistics has been studied since nineteen sixties. For more detail let us refer to *R.B. Bapat* (1999), *D.R. Cox* and *N. Reid* (2000), *E. P. Liski*, et al. (2001), *A. Pazman* (1986) and *F. Pukelsheim* (1993).

For the independent, identically distributed (i.i.d) observation *Theorem 8* will be simplified as:

Corollary 9 (A-optimal design of α for the special Gauss-Markov model with i.i.d. observations):

For the special Gauss-Markov model

$$A\xi = E\{y\},\ I_n\sigma^2 = \Sigma_y = D\{y\},\ I_m\sigma^2 = S \qquad (48)$$

of independent, identically distributed (i.i.d.) observations with variance σ^2 and an analogous substitute weight matrix **S** scaled by the variance σ^2, an A-optimal choice of weighting factor α is

$$\hat{\alpha} = \frac{\mathrm{tr}\, A'A(A'A+\alpha I_m)^{-3}\sigma^2}{\xi'(A'A+\alpha I_m)^{-2}A'A(A'A+\alpha I_m)^{-1}\xi} \qquad (49)$$

For the i.i.d. observations of a random scalar parameter case (direct observation) the α-homBLE of ξ and BIQUUE of σ^2 are summarized in *Box 3* and *Corollary 10*.

Box 3:
Special Gauss-Markov model: direct observations

$$\tau\xi = E\{y\},\ I_n\sigma^2 = \Sigma_y = D\{y\},\ \tau = [1,...,1]'$$

"α-homBLE of ξ "

$$\hat{\xi} = \frac{1}{n+\alpha}\tau' y = \frac{n}{n+\alpha}\Big(\frac{1}{n}\sum_i y_i\Big) = \frac{n}{n+\alpha}\bar{\xi}$$

"$MSE_{\alpha,S}\{\hat{\xi}\}$ of $\hat{\xi}$"

$$S = 1\sigma^2 = [\sigma^2]$$

$$MSE_{\alpha,S}\{\hat{\xi}\} = D\{\hat{\xi}\} + \beta\beta' =$$
$$= \sigma^2\frac{n}{(n+\alpha)^2} + \frac{\alpha^2}{(n+\alpha)^2}\xi^2$$
$$= \gamma_1(\alpha) + \gamma_2(\alpha).$$

"BIQUUE of σ^2"

$$\hat{\sigma}^2 = \frac{1}{n-1}y'(I_n - \frac{1}{n}J_n)y,\ J_n = \tau\tau'.$$

In case of the *special Gauss-Markov model of direct observations* the first order design matrix **A** is of full rank. Accordingly an estimation $\bar{\xi}$ of type *BLUUE* (Best Linear Uniformly Unbiased Estimation) exists and may be used to replace ξ. Although we have so far treated $S = [\sigma^2]$ as known, we note that in the particular special case we may treat the variance factor σ^2 as a common unknown and resort to a classical estimation $\bar{\sigma}^2$ of type *BIQUUE* (Best Invariant Quadratic Uniformly Unbiased Estimation), which is a useful substitute of σ^2 in computing the weight α.

Corollary 10 (A-optimal design of α for the special Gauss-Markov model with direct i.i.d. observations):

Let us replace (i) ξ by $\bar{\xi}$ (BLUUE) and (ii) σ^2 by $\bar{\sigma}^2$ (BIQUUE) within the A-optimal choice of the weighting factor $\hat{\alpha}$, Eq. (49), with respect to the special Gauss- Markov model Eq. (48). Then an approximation $\tilde{\alpha}$ of the A-optimal choice $\hat{\alpha}$ of type

$$(50)\quad \tilde{\alpha} = \frac{\bar{\sigma}^2}{\bar{\xi}^2} \qquad\qquad \lim_{n\to\infty}\tilde{\alpha} = \hat{\alpha} = \frac{\sigma^2}{\xi^2} \quad (51)$$

is achieved with

$$\mathrm{tr}\, MSE_{\alpha,S}\{\hat{\xi}\}|_{\bar{\sigma}^2,\bar{\xi}} = MSE_{\alpha,S}\{\hat{\xi}\}|_{\bar{\sigma}^2,\bar{\xi}} = = \bar{\sigma}^2\frac{n}{(n+\alpha)^2} + \frac{\alpha^2}{(n+\alpha)^2}\bar{\xi}^2 \qquad (52)$$

3 Case study: 2-dimensional strain rate tensor

With these models developed above we were able to successfully perform the α-homBLE estimation to determine the weighting factor α for a special case . In lieu of a case study, the model is applied to simulated observations of a random tensor of type strain rate based on the real estimate of one station in the Finnish Primary Geodetic Network (Kakkuri and Chen 1992).

Box 4 provides the real estimated random strain rate tensor and related standard deviations and the simulated observations in 11 epochs.

Using these observations we can estimate the BLUUE of distinct parameters t_{11}, t_{12}, t_{22} with *Box 3* and the sample variance $\bar{\sigma}^2_{y_A}$, $\bar{\sigma}^2_{y_B}$, $\bar{\sigma}^2_{y_C}$ of type BIQUUE (*Box* 3), which are also presented in *Box 4*.

Box 4:

Observations of a random tensor of type strain rate (epoch 1: Kakkuri and Chen (1992))

$$\mathbf{t} = \begin{bmatrix} t_{11} & t_{12} \\ t_{12} & t_{22} \end{bmatrix} = \begin{bmatrix} 0.236 & -0.049 \\ -0.049 & 0.148 \end{bmatrix} (\mu\,\text{strain/year})$$

$\bar{\sigma}_{11} = 0.094$, $\bar{\sigma}_{12} = 0.054$ and $\bar{\sigma}_{22} = 0.065$

"the vectorized form "

$\mathbf{y} = [y_A \ y_B \ y_C]' = [t_{11} \ t_{12} \ t_{22}]'$ with

$\bar{\sigma}_A = 0.094$, $\bar{\sigma}_B = 0.054$, $\bar{\sigma}_C = 0.065$

"Observations of the distinct elements y_A, y_B, y_C in 11 epochs"

epoch i	y_A (μstrain/year)	y_B (μstrain/year)	y_C (μstrain/year)
1	0.1513	-0.0305	0.1615
2	0.2913	-0.0081	0.1624
3	0.2881	-0.0864	0.0826
4	0.1970	-0.0123	0.1186
5	0.2418	-0.1069	0.2390
6	0.2790	-0.0004	0.1180
7	0.2547	-0.1636	0.1501
8	0.2602	-0.0336	0.1999
9	0.4316	-0.0886	0.2063
10	0.0219	-0.0908	0.1570
11	0.2679	-0.0408	0.0428

" the BLUUE of t_{11}, t_{12}, t_{22} (*Box 3*)"

$$\begin{bmatrix} \bar{\xi}_A \\ \bar{\xi}_B \\ \bar{\xi}_C \end{bmatrix} = \begin{bmatrix} \bar{t}_{11} \\ \bar{t}_{12} \\ \bar{t}_{22} \end{bmatrix} = \begin{bmatrix} 0.2441\ (\mu\,\text{strain/y}) \\ -0.0602\ (\mu\,\text{strain/y}) \\ 0.1489\ (\mu\,\text{strain/y}) \end{bmatrix}$$

"the sample variance $\bar{\sigma}^2_{y_A}, \bar{\sigma}^2_{y_B}, \bar{\sigma}^2_{y_C}$ of type BIQUUE(*Box 3*)"

$\bar{\sigma}^2_{y_A} = 0.010168 (\mu\,\text{strain/y})^2$, $\bar{\sigma}^2_{y_B} = 0.002585\ (\mu\,\text{strain/y})^2$, $\bar{\sigma}^2_{y_C} = 0.003195\ (\mu\,\text{strain/y})^2$

"the dispersion matrix of $\bar{\xi}_A$, $\bar{\xi}_B$, $\bar{\xi}_C$"

$\bar{\sigma}^2_{\bar{\xi}_A} = 0.000924\ (\mu\,\text{strain/y})^2$, $\bar{\sigma}^2_{\bar{\xi}_B} = 0.000235 (\mu\,\text{strain/y})^2$, $\bar{\sigma}^2_{\bar{\xi}_C} = 0.000290 (\mu\,\text{strain/y})^2$, no correlation.

By means of *Figure* 2 we compare α-homBLE (*dashed line*) and BLUUE estimates (*full line*), in particular document the dependence on the uniform regularization parameter α. Based upon the results summarized in *Box 3* we have computed *at first* the trace of the *Mean Square Error* (*MSE*) *matrix*, in particular the functions $\gamma_1(\alpha)$ as variance term and $\gamma_2(\alpha)$ as bias term squared. Within { $\gamma_1(\alpha)$, $\gamma_2(\alpha)$} we have substituted $\{\sigma, \xi\}$ by $\{\bar{\sigma}(\text{BIQUUE}), \bar{\xi}\,(\text{BLUUE})\}$, i.e. the true value σ and ξ with their estimates $\bar{\sigma}$ and $\bar{\xi}$ and plotted in *Figure* 3. The interrelation between the variances, squared biases and the weighting factor is evident. The variance term $\gamma_1(\alpha)$ (*dashed line*) decreases as α increases while the squared bias term $\gamma_2(\alpha)$ (*dotted line*) increases with α . The *dash-dotted line* which represents $\gamma_1(\alpha) + \gamma_2(\alpha)$ as $\text{trMSE}_{\alpha,S}\{\hat{\xi} : \alpha\text{-homBLE}\}$ is under the level of $\text{tr}\,MSE\{\hat{\xi} : \text{BLUUE}\}$ as expected. In summary, these estimates determine approximate values of the weighting factor α of type A-optimum with respect to $MSE_{\alpha,S}\{\hat{\xi}\}$, *Eq.* (50), in particular

$$\tilde{\alpha}_A = 0.171,\ \tilde{\alpha}_B = 0.714 \text{ and } \tilde{\alpha}_C = 0.144$$

Figure 4 is a "zoom-in" version of *Figure 3*, which illustrates the optimal values $\tilde{\alpha}$ consistent with these curves, respectively $\text{trMSE}_{\alpha,S}\{\hat{\xi}\}$ as well as $\text{trMSE}_{\alpha,S}\{\hat{\xi}\}$ at minimal points.

Secondly, by an iterative procedure we have updated every element of the first approximate $\{\tilde{\alpha}_A, \tilde{\alpha}_B, \tilde{\alpha}_C\}$ by means of

$$\tilde{\alpha}_{k+1} = \frac{\hat{\sigma}^2_k}{\hat{\xi}^2_k},\quad \hat{\xi}_k = \hat{\xi}(\tilde{\alpha}_k),\quad \hat{\sigma}_k = \hat{\sigma}(\hat{\xi}(\tilde{\alpha}_k)) \quad (53)$$

$$\begin{aligned} \hat{\sigma}^2_k &= \frac{1}{n-1}\mathbf{y}'(\mathbf{I}_n - \frac{1}{n+\tilde{\alpha}_k}\mathbf{J}_n)^2\,\mathbf{y} \\ &= \frac{1}{n-1}\sum_{i=1}^{n}(y_i - \frac{n}{n+\tilde{\alpha}_k}\bar{\xi})^2 \end{aligned} \quad (54)$$

The sequential optimization ends at the *reproducing point* $\tilde{\alpha}_{k+1} \doteq \tilde{\alpha}_k$ in computer arithmetic subject to $\text{tr}\,\text{MSE}_{\alpha,S}\{\hat{\xi}_k\}$at minimum. Such an iterative procedure illustrated in *Figure 5* supports the optimization procedure to generate $\hat{\alpha} = \arg\{\text{tr}\,MSE_{\alpha,S}\{\hat{\xi}\} = \min\}$.

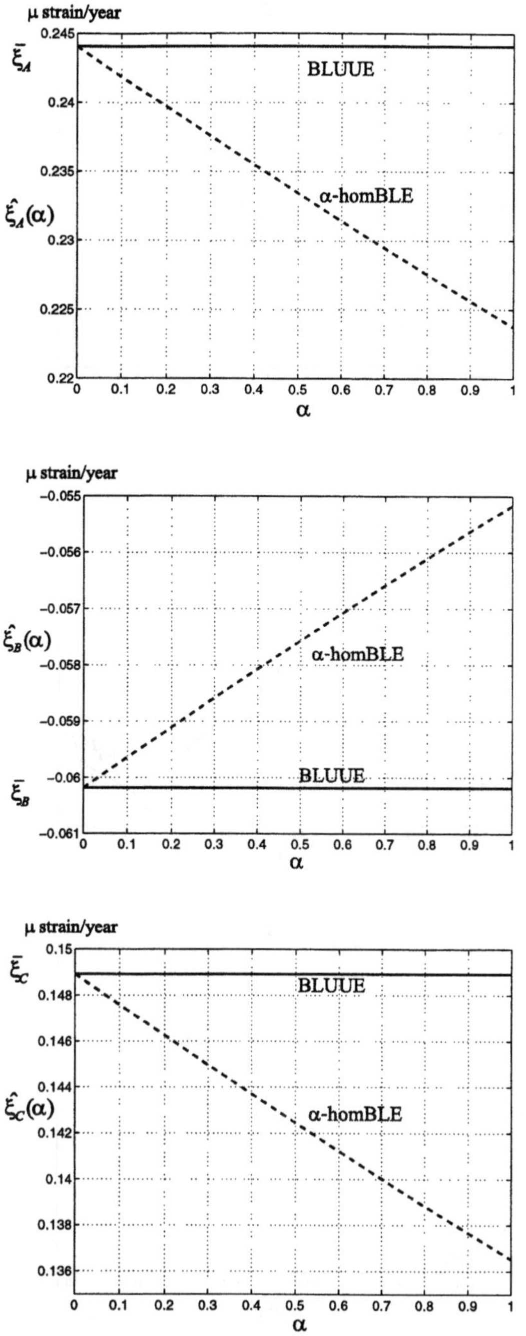

Fig.2 $\bar{\xi}$ (BLUUE) versus $\hat{\xi}$ (α -homBLE) of the direct observations { t_{11}, t_{12}, t_{22} }, symmetric random tensor of type strain rate, as functions of balancing parameter α .

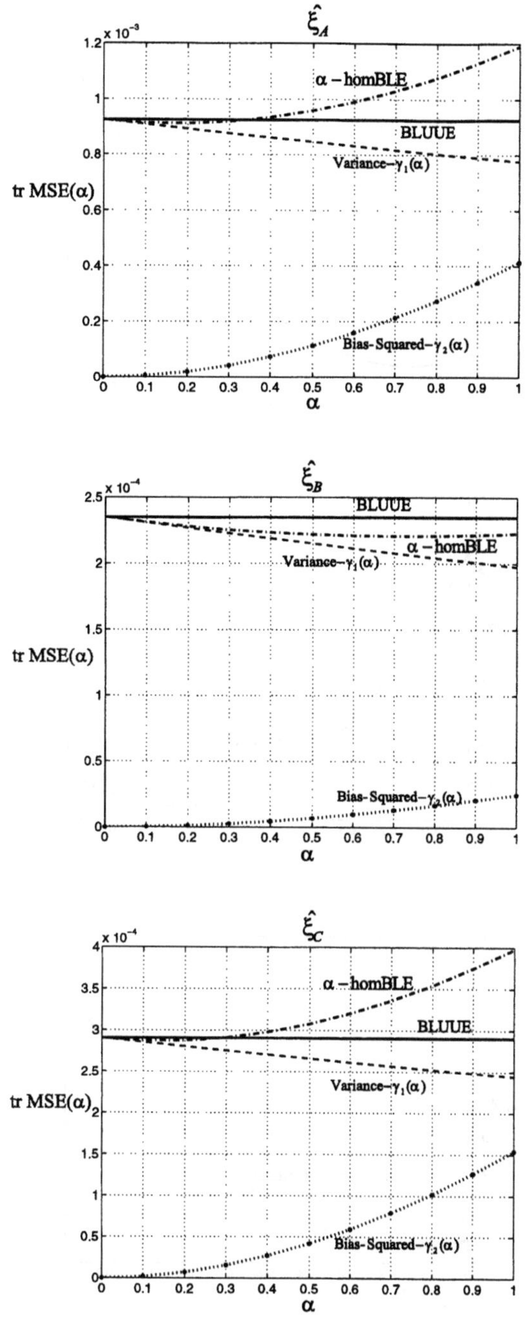

Fig.3 The trace of the *Mean Square Error (MSE)* functions for the BLUUE and α-homBLE estimates of unknowns (ξ_A, ξ_B, ξ_C) as functions of α. Depicted are $\operatorname{tr} MSE\{\bar{\xi}\} = \hat{\sigma}^2_{\bar{\xi}}$ for BLUUE and $\operatorname{tr} MSE_{\alpha,S}\{\hat{\xi}\} = \gamma_1(\alpha) + \gamma_2(\alpha)$ for α-homBLE, as well as the separate parts for variance $\gamma_1(\alpha)$ and bias squared $\gamma_2(\alpha)$.

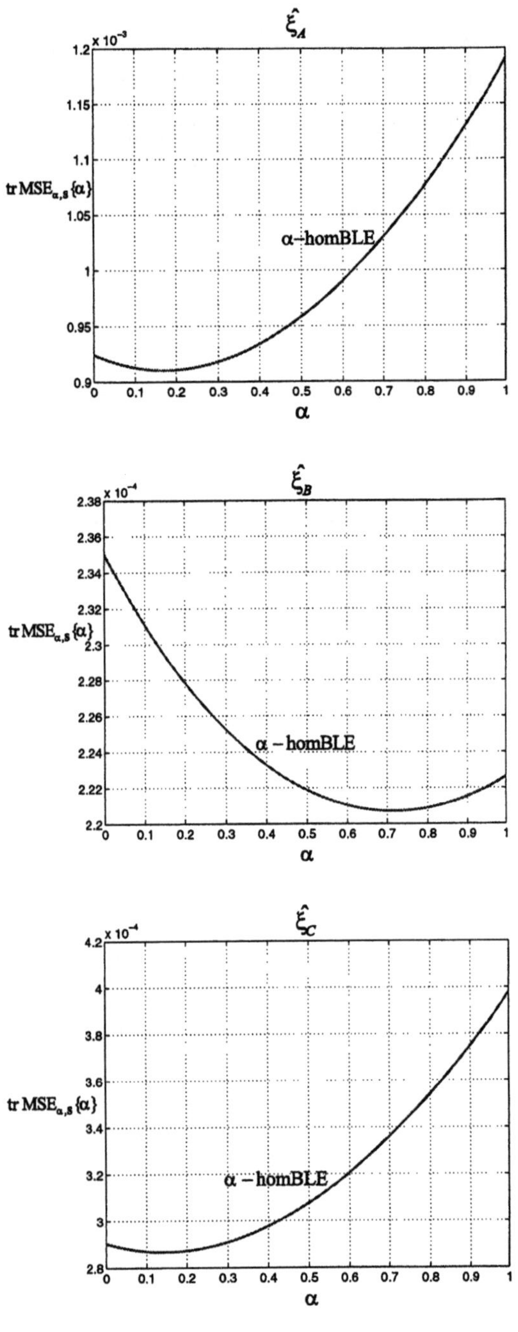

Fig.4 The trace of the *Mean Square Error* (*MSE*) functions $\operatorname{tr} MSE_{\alpha,S}\{\hat{\xi}\}$ for α-homBLE estimates of (ξ_A, ξ_B, ξ_C), also depicted in figure 3. Here the function graph is "zoomed-in", in order to emphasize the behavior of the function around its minimal point.

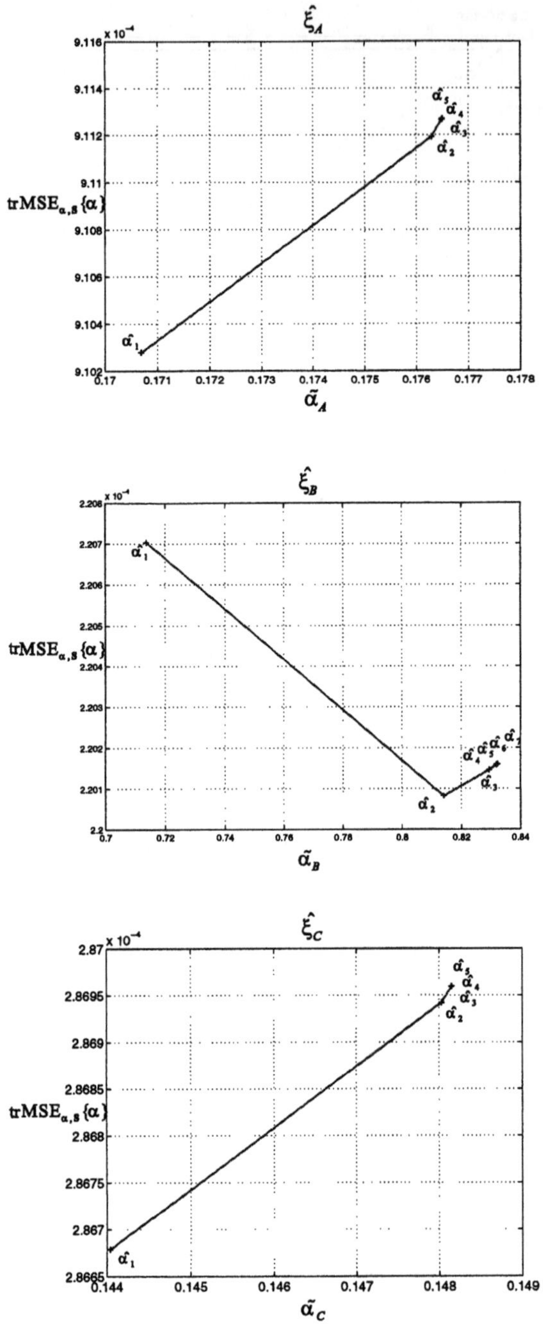

Fig. 5 Iteration steps for A-optimal α, three sets of direct observations

(i) upper graph: the trace of $MSE_{\alpha S}\{\alpha\}$ for the estimate $\hat{\xi}_A(\alpha)$ of type α-homBLE of first set of direct observations;

(ii) middle graph: the trace of $MSE_{\alpha S}\{\alpha\}$ for the estimate $\hat{\xi}_B(\alpha)$ of type α-homBLE second set of direct observations;

(iii) lower graph: the trace of $MSE_{\alpha S}\{\alpha\}$ for the estimate $\hat{\xi}_C(\alpha)$ of type α-homBLE of third set of direct observations.

Acknowledgements. This work relates to research results of the DFG project GR 323/36-1 *"Hypothesis tests and evaluating statistics of the eigenvalues and orientations of a random deformation tensor"*, which is sponsored by the Deutsche Forschungsgemeinschaft (DFG). This support is gratefully acknowledged. The authors thank an anonymous reviewer for his many constructive comments, which helped to clarify a number of points in this revised version of the paper.

References

Allen, D.M.(1971): Mean square error of prediction as a criterion for selecting variables, Technometrics *13* (1971) 469-481

Allen, D.M.(1974): The relationship between variable selection and data argumentation and a method for prediction, Technometrics *16* (1974) 125-127

Arslan, O. and Billor, B.(2000): Robust Liu estimator for regression based on an M-estimator, Journal of Applied Statistics *27* (2000) 39-47

Bapat, R.B. (1999): Linear Algebra and Linear Models, 2nd Edition, Springer, New York, Berlin, 1999

Bouman, J. (1998): Quality of regularization methods, DEOS Report no. 98.2, Delft University Press 1998

Chaturvedi, A. and Singh, S.P.(2000): Stein rule prediction of the composite target function in a general linear regression model, Statistical Papers *41* (2000) 359-367

D. R. Cox and N. Reid (2000): The Theory of the Design of Experiments, Chapman & Hall/CRC, 2000

Donatos, G.S. and Michailidis, G.C.(1990): Small sample properties of ridge estimators with normal and non-normal disturbances, Commun. Statist.-Simula. *19* (1990) 935-950

Draper, R.N. and Van Nostrand, C.R.(1979): Ridge regression and James- Stein estimation : Review and comments, Technometrics *21* (1979) 451-466

Droge, B.(1993): On finite-sample properties of adaptative least squares regression estimates, Statistics *24* (1993) 181-203

Engels, J., et al.(1993): The geoid as an inverse problem to be regularized, Proceedings of the international conference "Inverse Problems: Principles and Applications in Geophysics, Technology, and Medicine" Anger, G et al. (eds). Akademie Verlag, Berlin 1993

Engl, H.(1993): Regularization methods for the stable solution of inverse problems, Surveys on Math. for Industry *3* (1993) 371-143

EL-Sayed, S.M.(1996): The sampling distribution of Ridge parameter estimator, The Egyptian Statistical Journal *40* (1996) 211-219

Farebrother, R.W.(1975): The minimum mean square error linear estimator and ridge regression, Technometrics *17* (1975) 127-128

Farebrother, R.W.(1976): Further results on the mean square error of ridge regression, Journal of the Royal Statistical Society *B.38* (1976) 248-250

Farebrother, R.W.(1978): Partitioned ridge regression, Technometrics *20* (1978) 121-122

Firinguetti, L.(1996): A simulation study of ridge regression in an incorrectly specified model, Estadistica *48* (1996) 115-154

Firinguetti, L. and Rubio, H. (2000): A note on the moments of stochastic shrinkage parameters in ridge regression Commun. Statist.-Simula. *29(3)* (2000) 955-970

Gibbons, D.G(1981): A simulation study of some ridge estimators, Journal of the American Statistical Association *7* (1981) 131-139

Golub, G, Heath, M. and Wahba, G (1979): Generalized cross-validation as a methods for choosing a good ridge parameter, Technometrics *21* (1979) 215-223

Grafarend, E. and Schaffrin B.(1993): Ausgleichungsrechnung in linearen Modellen, Bibliographisches Institut Wissenschaftsverlag, Mannheim 1993

Gui, Q. and Zhang, J.(1998a): Robust biased estimation and its applications in geodetic adjustments, Journal of Geodesy *72* (1998) 430-435

Gui, Q., Huang, W., and Zhang, J.(1998b): Robust universal ridge estimation, Arta geodetica et cartographica sinica *3* (1998)

Gui, Q. and Liu, J.(2000): Biased estimation in Gauss-Markov model, AVN *3* (2000) 104-107

Gui, Q., Yang, Y. and Guo, J.(2001): Biased estimation in Gauss-Markov model with Constraints, AVN *1* (2001) 28-30

Gunst, R.F. and Mason R.L.(1977): Biased Estimation in Regression: An Evaluation Using Mean Squared Error, Journal of the American Statistical Association *72* (1977) 616-628

Gunst, R.F. and Mason, R.L.(1980): Regression Analysis and its Application, Marcel Dekker, New York 1980

Hansen, P.(1992): Analysis of discrete ill-posed problems by means of the *L*-curve, SIAM Review *34* 561-580

Hansen, P.(1993): The use of the *L*-curve in the regularization of discrete ill-posed problems, SIAM J. Sci. Comput. *14* (1993) 1487-1503

Hanke, M. and Hansen, P.(1993): Regularization methods for large scale problems, Surveys on Math. for Industry *3* (1993) 253-315

Hemmerle, W.J.(1975): An explicit solution for generalized ridge regression, Technometrics *17* (1975) 309-314

Hemmerle, W.J. and Brantle, T.F.(1978): Explicit and constrained generalized ridge estimation, Technometrics *20* (1978) 109-120

Hocking, R.R.(1976): The analysis and selection of variables in linear regression, Biometrics *32* (1976) 1-49

Hoerl, A. E. and Kennard, R.W.(1970a): Ridge regression: Biased estimation for nonorthogonal problems, Technometrics *12* (1970) 55-67

Hoerl, A.E. and Kennard, R.W.(1970b): Ridge regression: Applications to nonorthogonal problems, Technometrics *12* (1970) 69-82

Hoerl, A.E., Kennard, R. W. and Baldwin, K.F.(1975): Ridge regression: Some simulations Analysis, Communications in Statistics *4* (1975) 105-123.

Hoerl, R. W.(1985): Ridge Analysis 25 Years Later, The American Statistician *39* (1985) 186-192

Hoerl, R. W., Schuenemeyer, J. H. and Hoerl, A. E. (1985): A simulation of biased estimation and subset selection regression techniques, Technometrics *28* 369-380

Ilk, K. H. (1986): On the regularization of ill-posed problems, Proc. Int. symp. Figure and Dynamics of the Earth, Moon and Planets, Prague 1986, 365-383

Kacirattin, S. Sakalloglu, S. and Akdeniz, F. (1998): Mean squared error comparisons of the modified ridge regression estimator and the restricted ridge regression *27* (1998) 131-138

Kakkuri J. and Chen R. (1992): On Horizontal Crustal Strain in Finland. Bull. Géod. *66* (1992) 12-20

Lawless, J.F. and Wang, P.(1976): A simulation study of ridge and other regression estimations. Communications in Statistics *A5* (1976) 307-322

Liski, E. P., et al. (2001) Topic in Optimal Design, Springer, New York and Berlin, 2001

Liu, K (1993): A new class of biased estimates in linear regression, Communications in Statistics *A22* (1993) 393-402

Louis, A. K. and Maass, P. (1990): A mollifier method for linear operator equation of the first kind, Inverse Problems, *6* (1990) *427-440*

Lowerre, J. M.(1974): On the mean square error of parameter estimates for some biased estimators, Technometrics *16* (1974) 461-464

Mallows, C.L.(1973): Some comments on Cp, Technometrics *15* (1973) 661-675

Markiewicz, A.(1996): Characterization of general ridge estimators, Statistics & Probability Letters *27* 145-148

Marquardt, W.D.(1970): Generalized inverses, ridge regression, biased linear estimation, and nonlinear estimation, Technometrics *12* (1970) 591-612

Marquardt, W.D.(1974): Discussion No.2, Technometrics *16* (1974) 189-192

Marquardt, W.D. and Snee, R.D.(1975): Ridge regression in practice, The American Statistician *29* (1975) 3-20

Mayer, L.S. and Willke, T.A.(1973): On biased estimation in linear models, Technometrics *15* (1973) 497-508

McDonald, G.C. and Galarneau, D.I.(1975): A Monte Carlo evaluation of some ridge-type estimators, Journal of the American Statistical Association *70* (1975) 407-416

Nomura, M.(1988): On exact small sample properties of ordinary ridge estimators, Economics letters *27* (1988) 251-255

Nomura, M.(1998): On exact small sample properties of the minimax generalized ridge regression estimators, Commun. Statist.-Simula. *17* (1988) 1213-1229

Ohtani, K.(1986): On smalll sample properties of the almost unbiased generalized ridge estimator, Commun. Statist.-Simula. *15* (1986) 1571-1578

Ohtani, K.(1998): An MSE comparison of the restricted Stein-rule and minimum mean squared error estimators in regression, Sociedad de Estadistica e Investigacion Operativa Test 7 (1998) 361-376

Phillips, D.L.(1962): A technique for the numerical solution of certain integral equations of the first kind, J. Ass. Comput. Mach. *9* (1962) 84-96

Pazman, A. (1986): Foundations of Optimum Experimental Design, Kluwer Academic, Dordrecht, 1986

Pukelsheim, F. (1993): Optimal Design of Experiments, Wiley, New York 1993

Rao, C.R.(1975): Simultaneous estimation of parameters in different linear models and applications to biometric problems, Biometrics *31* (1975) 545-554

Rao, C.R.(1976): Estimation of parameters in a linear model (The 1975 Wald Memorial Lectures), The Annals of Statistics *4* (1976) 1023- 1037

Schaffrin, B., Heidenreich, E. and Grafarend, E.(1977): A representation of the standard gravity field, Manuscipta geodaetica 2 (1977) 135-174

Schaffrin, B. and Middel B.(1990): Robust predictors and an alternative iteration scheme for ill-posed problem, in Geophysical Data Inversion-Methods and Applications, A. Vogel, Ch. O. Ofoegbu, R. Gorenflo and Bj. Ursin (Eds.), Vieweg, Braunschweig (1990) 33-51

Schaffrin, B.(1995):A comparision of inverse techniques: regularization, weight estimation and homBLUP, IUGG General Association 1995

Schaffrin, B (2000): Minimum mean square error adjustment, Part I: The empirical BLE and the repro-BLE for direct observations, Journal of the Geodetic Society of Japan *46* (2000) 21-30

Shalabh (1998): Improved estimation in measurement error models through Stein rule procedure, Journal of Multivariate Analysis *67* (1998) 35-48

Smith, G. and Campbell, F.(1980): A Critique of Some Ridge Regression Methods, Journal of the American Statistical Association *75* (1980) 74-81

Srivastava, V.K. et al.(1983): Some properties of the distribution of an operational ridge estimator, Metron *30* (1983) 227-237

Theobald, C.M.(1974): Generalizations of mean square error applied to ridge regression, Journal of the Royal Statistical Society *B.36* (1974) 103-106

Tykhonov, A.N.(1963): The regularization of incorrectly posed problem, Soviet Math. Doklady *4* (1963) 1624-1627

Tykhonov, A.N. et al. (1977): A variational method of regularization in the adjustment of free geodetic nets, Geodesy, Mapping and Photogrammetry *19* (1977) 127-139

Tykhonov, A.N. and Arsenin, V.Y.(1977): Solutions of Ill-Posed Problems, Wiley, New York 1977

Vinod, H.D. and Ullah, A.(1981): Recent advances in regression methods, Marcel Dekker, New York 1981

Xu, P.(1992a): Determination of surface gravity anomalies using gradiometric observables, Geophys. J. Int. *110* (1992) 321-332

Xu, P.(1992b): The value of minimum norm estimation of geopotential fields, Geophys.J.Int. *111* (1992) 170-178

Xu, P.(1998): Truncated SVD methods for discrete linear ill-posed problems, Geophys.J.Int. *135* (1998) 505-514

Xu, P. and Rummel, R(1994a): A simulation study of smoothness methods in recovery of regional gravity fields, Geophys.J.Int. *117* (1994) 472-486

Xu, P. and Rummel, R(1994b): Generalized ridge regression with applications in determination of potential fields, manuscripta geodaetica *20* (1994) 8-20

Wang, Y. and Xiao, T.(2001): Fast realization algorithms for determining regularization parameters in linear inverse problems, Inverse Problems *17* (2001) 281-291

Wenchenko, E.(2000): Estimation of the signal-to-noise in the linear regression model, Statistical Papers *41* 327-343

Appendix A: Proof of *Lemma 4*

(i) homBLE:

The hybrid norm $\| MSE\{\hat{\xi}\} \|^2$ establishes the Lagrangean

$$\mathcal{L}_1(\mathbf{L}) := \operatorname{tr} \mathbf{L}\mathbf{\Sigma}_y\mathbf{L}' + \operatorname{tr}(\mathbf{I}_m - \mathbf{L}\mathbf{A})\xi\xi'(\mathbf{I}_m - \mathbf{L}\mathbf{A})' = \min_{\mathbf{L}}$$

for $\hat{\xi}$ *homBLE* of ξ. The *necessary conditions* for the minimum of the *quadratic Lagrangean* $\mathcal{L}_1(\mathbf{L})$ are

$$\frac{\partial \mathcal{L}_1}{\partial \mathbf{L}}(\hat{\mathbf{L}}) := 2[\mathbf{\Sigma}_y\hat{\mathbf{L}}' + \mathbf{A}\xi\xi'\mathbf{A}'\hat{\mathbf{L}}' - \mathbf{A}\xi\xi']' = 0$$

which agree to the normal equations (25). The theory of derivative of a scalar-valued function of a matrix: *trace* is reviewed in *Appendix A* of the book by Grafarend and Schaffrin (1993).

The second derivatives

$$\frac{\partial^2 \mathcal{L}_1}{\partial(\operatorname{vec}\mathbf{L})\partial(\operatorname{vec}\mathbf{L})'}(\hat{\mathbf{L}}) > 0$$

at the "point" $\hat{\mathbf{L}}$ constitute the *sufficiency conditions.*

In order to compute such a $mn \times mn$ matrix of second derivatives we have to vectorize the matrix normal equation

$$\frac{\partial \mathcal{L}_1}{\partial \mathbf{L}}(\hat{\mathbf{L}}) := 2\hat{\mathbf{L}}(\mathbf{\Sigma}_y + \mathbf{A}\xi\xi'\mathbf{A}') - 2\xi\xi'\mathbf{A}',$$

$$\frac{\partial \mathcal{L}_1}{\partial(\operatorname{vec}\mathbf{L})}(\hat{\mathbf{L}}) := \operatorname{vec}[2\hat{\mathbf{L}}(\mathbf{\Sigma}_y + \mathbf{A}\xi\xi'\mathbf{A}') - 2\xi\xi'\mathbf{A}'],$$

where the notation vec is the vec operation (vectorization of an array). With the property of vec opertation $\operatorname{vec}\mathbf{BC} = (\mathbf{C}' \otimes \mathbf{I}_n)\operatorname{vec}\mathbf{B}$ for suitable matrices **B** and **C**, $\forall \mathbf{B} \in \mathbb{R}^{n \times m}$, $\forall \mathbf{C} \in \mathbb{R}^{m \times q}$ and the *Kronecker-Zehfuss product* $\mathbf{B} \otimes \mathbf{C}$ of two arbitrary matrices as well as $(\mathbf{B}+\mathbf{C}) \otimes \mathbf{D} = \mathbf{B} \otimes \mathbf{D} + \mathbf{C} \otimes \mathbf{D}$ of three arbitrary matrices subject to dim **B** = dim **C** we have

$$\frac{\partial \mathcal{L}_1}{\partial(\operatorname{vec}\mathbf{L})}(\hat{\mathbf{L}}) = 2[(\mathbf{\Sigma}_y + \mathbf{A}\xi\xi'\mathbf{A}') \otimes \mathbf{I}_m]\operatorname{vec}\hat{\mathbf{L}} - 2\operatorname{vec}(\xi\xi'\mathbf{A}').$$

With the theory of matrix derivatives: Derivatives of a matrix-valued function of a matrix, namely $\partial(\operatorname{vec}\mathbf{X})/\partial(\operatorname{vec}\mathbf{X})'$, we are now prepared to compute the second derivatives

$$\frac{\partial^2 \mathcal{L}_1}{\partial(\operatorname{vec}\mathbf{L})\partial(\operatorname{vec}\mathbf{L})'}(\hat{\mathbf{L}}) = 2[(\mathbf{\Sigma}_y + \mathbf{A}\xi\xi'\mathbf{A}') \otimes \mathbf{I}_m].$$

Since $\mathbf{\Sigma}_y + \mathbf{A}\xi\xi'\mathbf{A}'$ is a *positive-definite matrix* the second derivatives constitute to the sufficiency conditions

$$\frac{\partial^2 \mathcal{L}_1}{\partial(\operatorname{vec}\mathbf{L})\partial(\operatorname{vec}\mathbf{L})'}(\hat{\mathbf{L}}) = 2[(\mathbf{\Sigma}_y + \mathbf{A}\xi\xi'\mathbf{A}') \otimes \mathbf{I}_m] > 0.$$

The *vec operation*, the *Kronecker-Zehfuss product* and the *derivatives of a matrix-valued function of a matrix* are also reviewed in *Appendix A* of the book by Grafarend and Schaffrin (1993).

(ii) S-homBLE:

The hybrid norm $\| MSE_S\{\hat{\xi}\} \|^2$ establishes the Lagrangean

$$\mathcal{L}_2(\mathbf{L}) := \operatorname{tr} \mathbf{L}\mathbf{\Sigma}_y\mathbf{L}' + \operatorname{tr}(\mathbf{I}_m - \mathbf{L}\mathbf{A})\mathbf{S}(\mathbf{I}_m - \mathbf{L}\mathbf{A})' = \min_{\mathbf{L}}$$

for $\hat{\xi}$ *S- homBLE* of ξ. Following the first part of the proof we are led to the *necessary conditions* for the minimum of the *quadratic Lagrangean* $\mathcal{L}_2(\mathbf{L})$

$$\frac{\partial \mathcal{L}_2}{\partial \mathbf{L}}(\hat{\mathbf{L}}) := 2[\mathbf{\Sigma}_y\hat{\mathbf{L}}' + \mathbf{A}\mathbf{S}\mathbf{A}'\hat{\mathbf{L}}' - \mathbf{A}\mathbf{S}]' = 0$$

as well as to the *sufficiency conditions*

$$\frac{\partial^2 \mathcal{L}_2}{\partial(\operatorname{vec}\mathbf{L})\partial(\operatorname{vec}\mathbf{L})'}(\hat{\mathbf{L}}) = 2[(\mathbf{\Sigma}_y + \mathbf{A}\mathbf{S}\mathbf{A}') \otimes \mathbf{I}_m] > 0.$$

The *normal equations* of *S-homBLE* $\partial\mathcal{L}_2/\partial\mathbf{L}(\hat{\mathbf{L}}) = 0$ agree to (26).

(iii) α-homBLE:

The hybrid norm $\| MSE_{\alpha,S}\{\hat{\xi}\} \|^2$ establishes the Lagrangean

$$\mathcal{L}_3(\mathbf{L}) := \operatorname{tr} \mathbf{L}\mathbf{\Sigma}_y\mathbf{L}' + \frac{1}{\alpha}\operatorname{tr}(\mathbf{I}_m - \mathbf{L}\mathbf{A})\mathbf{S}(\mathbf{I}_m - \mathbf{L}\mathbf{A})' = \min_{\mathbf{L}}$$

for $\hat{\xi}$ *α-homBLE* of ξ. Following the first part of the proof we are led to the *necessary conditions* for the minimum of the *quadratic Lagrangean* $\mathcal{L}_3(\mathbf{L})$

$$\frac{\partial \mathcal{L}_3}{\partial \mathbf{L}}(\hat{\mathbf{L}}) = 2[\frac{1}{\alpha}\mathbf{A}\mathbf{S}\mathbf{A}'\hat{\mathbf{L}}' + \mathbf{\Sigma}_y\hat{\mathbf{L}}' - \frac{1}{\alpha}\mathbf{A}\mathbf{S}]' = 0$$

as well as to the *sufficiency conditions*

$$\frac{\partial^2 \mathcal{L}_3}{\partial(\operatorname{vec}\mathbf{L})\partial(\operatorname{vec}\mathbf{L})'}(\hat{\mathbf{L}}) = 2[(\mathbf{\Sigma}_y + \frac{1}{\alpha}\mathbf{A}\mathbf{S}\mathbf{A}') \otimes \mathbf{I}_m] > 0$$

The *normal equations* of *α-homBLE* $\partial\mathcal{L}_3/\partial\mathbf{L}(\hat{\mathbf{L}}) = 0$ agree to (27).

Appendix B: Proof of *Theorem 8*

Before we prove *Theorem 8* let us introduce auxiliary results which are used subsequently.

Lemma B1 (Cayley matrix inverse differentiation):

$$d(\mathbf{A}'\boldsymbol{\Sigma}_y^{-1}\mathbf{A}+\alpha\mathbf{S}^{-1})^{-1} = -(\mathbf{A}'\boldsymbol{\Sigma}_y^{-1}\mathbf{A}+\alpha\mathbf{S}^{-1})^{-1}\mathbf{S}^{-1}(\mathbf{A}'\boldsymbol{\Sigma}_y^{-1}\mathbf{A}+\alpha\mathbf{S}^{-1})^{-1}d\alpha \tag{B.1}$$

Proof:

$$\mathbf{M}\mathbf{M}^{-1}=\mathbf{I}\Rightarrow (d\mathbf{M})\mathbf{M}^{-1}+\mathbf{M}(d\mathbf{M}^{-1})=0\Rightarrow$$
$$d\mathbf{M}^{-1}=-\mathbf{M}^{-1}d\mathbf{M}\,\mathbf{M}^{-1}$$

Example B1: $\mathbf{M}:=\mathbf{A}'\boldsymbol{\Sigma}_y^{-1}\mathbf{A}+\alpha\mathbf{S}^{-1} \Rightarrow$

$$\Rightarrow \quad d\mathbf{M}=\mathbf{S}^{-1}d\alpha$$

$$d\mathbf{M}^{-1}=-(\mathbf{A}'\boldsymbol{\Sigma}_y^{-1}\mathbf{A}+\alpha\mathbf{S}^{-1})^{-1}\mathbf{S}^{-1}(\mathbf{A}'\boldsymbol{\Sigma}_y^{-1}\mathbf{A}+\alpha\mathbf{S}^{-1})^{-1}d\alpha$$

q.e.d.

Lemma B2 (differentiation of a scalar function of a matrix, such as the trace):

$$\begin{aligned}&\mathrm{tr}\,(\mathbf{A}+\mathbf{B})=\mathrm{tr}\,\mathbf{A}+\mathrm{tr}\,\mathbf{B}\\&d(\mathrm{tr}\,(\mathbf{A}+\mathbf{B}))=\mathrm{tr}\,d\mathbf{A}+\mathrm{tr}\,d\mathbf{B}\\&d(\mathrm{tr}\,(\mathbf{X}\mathbf{A}\mathbf{X}'))=\mathrm{tr}(\mathbf{A}+\mathbf{A}')\mathbf{X}'d\mathbf{X}\end{aligned} \tag{B.2}$$

Lemma B3 (Cayley inverse: sum of two matrices)

$$\begin{aligned}&(\mathbf{A}'\boldsymbol{\Sigma}_y^{-1}\mathbf{A}+\alpha\mathbf{S}^{-1})^{-1}\mathbf{A}'\boldsymbol{\Sigma}_y^{-1}\mathbf{A}=\\&=[\mathbf{I}_m+\alpha(\mathbf{A}'\boldsymbol{\Sigma}_y^{-1}\mathbf{A})^{-1}\mathbf{S}^{-1}]^{-1}=\\&=\mathbf{I}_m-\alpha(\mathbf{A}'\boldsymbol{\Sigma}_y^{-1}\mathbf{A}+\alpha\mathbf{S}^{-1})^{-1}\mathbf{S}^{-1}=\\&=\mathbf{I}_m-\alpha(\mathbf{S}\mathbf{A}'\boldsymbol{\Sigma}_y^{-1}\mathbf{A}+\alpha\mathbf{I}_m)^{-1}\end{aligned} \tag{B.3}$$

Lemma B4 :

$$\begin{aligned}&\mathrm{tr}\,(\boldsymbol{\beta}\boldsymbol{\beta}')=\boldsymbol{\beta}'\boldsymbol{\beta}\\&d(\boldsymbol{\beta}'\boldsymbol{\beta})=(d\boldsymbol{\beta})'\boldsymbol{\beta}+\boldsymbol{\beta}'(d\boldsymbol{\beta})=2\boldsymbol{\beta}'(d\boldsymbol{\beta})\end{aligned} \tag{B.4}$$

Example B2:

$$\begin{aligned}\boldsymbol{\beta}&:=-[\mathbf{I}_m-(\mathbf{A}'\boldsymbol{\Sigma}_y^{-1}\mathbf{A}+\alpha\mathbf{S}^{-1})^{-1}\mathbf{A}'\boldsymbol{\Sigma}_y^{-1}\mathbf{A}]\boldsymbol{\xi}=\\&=-\alpha(\mathbf{A}'\boldsymbol{\Sigma}_y^{-1}\mathbf{A}+\alpha\mathbf{S}^{-1})^{-1}\mathbf{S}^{-1}\boldsymbol{\xi}=\\&=-\alpha(\mathbf{S}\mathbf{A}'\boldsymbol{\Sigma}_y^{-1}\mathbf{A}+\alpha\mathbf{I}_m)^{-1}\boldsymbol{\xi}\end{aligned}$$

$$\begin{aligned}d\boldsymbol{\beta}&=-d\alpha(\mathbf{A}'\boldsymbol{\Sigma}_y^{-1}\mathbf{A}+\alpha\mathbf{S}^{-1})^{-1}\mathbf{S}^{-1}\boldsymbol{\xi}-\\&\quad-\alpha d[(\mathbf{A}'\boldsymbol{\Sigma}_y^{-1}\mathbf{A}+\alpha\mathbf{S}^{-1})^{-1}]\mathbf{S}^{-1}\boldsymbol{\xi}\\&=-d\alpha(\mathbf{A}'\boldsymbol{\Sigma}_y^{-1}\mathbf{A}+\alpha\mathbf{S}^{-1})^{-1}\mathbf{S}^{-1}\boldsymbol{\xi}+\\&\quad+\alpha(\mathbf{A}'\boldsymbol{\Sigma}_y^{-1}\mathbf{A}+\alpha\mathbf{S}^{-1})^{-1}\mathbf{S}^{-1}(\mathbf{A}'\boldsymbol{\Sigma}_y^{-1}\mathbf{A}+\alpha\mathbf{S}^{-1})^{-1}d\alpha\mathbf{S}^{-1}\boldsymbol{\xi}=\\&=-(\mathbf{A}'\boldsymbol{\Sigma}_y^{-1}\mathbf{A}+\alpha\mathbf{S}^{-1})^{-1}[\mathbf{I}_m-\alpha\mathbf{S}^{-1}(\mathbf{A}'\boldsymbol{\Sigma}_y^{-1}\mathbf{A}+\alpha\mathbf{S}^{-1})^{-1}]\mathbf{S}^{-1}d\alpha\boldsymbol{\xi}\\&=-(\mathbf{A}'\boldsymbol{\Sigma}_y^{-1}\mathbf{A}+\alpha\mathbf{S}^{-1})^{-1}\mathbf{A}'\boldsymbol{\Sigma}_y^{-1}\mathbf{A}(\mathbf{A}'\boldsymbol{\Sigma}_y^{-1}\mathbf{A}+\alpha\mathbf{S}^{-1})^{-1}\mathbf{S}^{-1}d\alpha\boldsymbol{\xi}\\&=-(\mathbf{S}\mathbf{A}'\boldsymbol{\Sigma}_y^{-1}\mathbf{A}+\alpha\mathbf{I}_m)^{-1}\mathbf{S}\mathbf{A}'\boldsymbol{\Sigma}_y^{-1}\mathbf{A}(\mathbf{S}\mathbf{A}'\boldsymbol{\Sigma}_y^{-1}\mathbf{A}+\alpha\mathbf{I}_m)^{-1}d\alpha\boldsymbol{\xi}\end{aligned}$$

$$\begin{aligned}d(\boldsymbol{\beta}'\boldsymbol{\beta})&=2\boldsymbol{\beta}'(d\boldsymbol{\beta})=2[-\alpha(\mathbf{A}'\boldsymbol{\Sigma}_y^{-1}\mathbf{A}+\alpha\mathbf{S}^{-1})^{-1}\mathbf{S}^{-1}\boldsymbol{\xi}]'\times\\&\quad\times[-(\mathbf{A}'\boldsymbol{\Sigma}_y^{-1}\mathbf{A}+\alpha\mathbf{S}^{-1})^{-1}\mathbf{A}'\boldsymbol{\Sigma}_y^{-1}\mathbf{A}(\mathbf{A}'\boldsymbol{\Sigma}_y^{-1}\mathbf{A}+\alpha\mathbf{S}^{-1})^{-1}\mathbf{S}^{-1}d\alpha\boldsymbol{\xi}]\\&=2\alpha\boldsymbol{\xi}'\mathbf{S}^{-1}(\mathbf{A}'\boldsymbol{\Sigma}_y^{-1}\mathbf{A}+\alpha\mathbf{S}^{-1})^{-2}\mathbf{A}'\boldsymbol{\Sigma}_y^{-1}\mathbf{A}(\mathbf{A}'\boldsymbol{\Sigma}_y^{-1}\mathbf{A}+\alpha\mathbf{S}^{-1})^{-1}\mathbf{S}^{-1}\boldsymbol{\xi}d\alpha\\&=2\alpha\boldsymbol{\xi}'(\mathbf{S}\mathbf{A}'\boldsymbol{\Sigma}_y^{-1}\mathbf{A}+\alpha\mathbf{I}_m)^{-2}\mathbf{S}\mathbf{A}'\boldsymbol{\Sigma}_y^{-1}\mathbf{A}(\mathbf{S}\mathbf{A}'\boldsymbol{\Sigma}_y^{-1}\mathbf{A}+\alpha\mathbf{I}_m)^{-1}\boldsymbol{\xi}d\alpha\end{aligned}$$

Lemma B5: (A-optimum):

$$\mathrm{tr}\,MSE_{\alpha,\mathbf{S}}\{\hat{\boldsymbol{\xi}}\}=extr$$
$$\Leftrightarrow$$
$$\begin{aligned}&\frac{d}{d\alpha}(\mathrm{tr}\,MSE_{\alpha,\mathbf{S}}\{\hat{\boldsymbol{\xi}}\})=\\&=-2\,\mathrm{tr}[\mathbf{A}'\boldsymbol{\Sigma}_y^{-1}\mathbf{A}(\mathbf{A}'\boldsymbol{\Sigma}_y^{-1}\mathbf{A}+\alpha\mathbf{S}^{-1})^{-2}\mathbf{S}^{-1}(\mathbf{A}'\boldsymbol{\Sigma}_y^{-1}\mathbf{A}+\alpha\mathbf{S}^{-1})^{-1}]+\\&+2\alpha\boldsymbol{\xi}'\mathbf{S}^{-1}(\mathbf{A}'\boldsymbol{\Sigma}_y^{-1}\mathbf{A}+\alpha\mathbf{S}^{-1})^{-2}\mathbf{A}'\boldsymbol{\Sigma}_y^{-1}\mathbf{A}(\mathbf{A}'\boldsymbol{\Sigma}_y^{-1}\mathbf{A}+\alpha\mathbf{S}^{-1})^{-1}\mathbf{S}^{-1}\boldsymbol{\xi}\\&=0\end{aligned} \tag{B.5}$$
$$\Leftrightarrow$$
$$\hat{\alpha}=\frac{\mathrm{tr}[\mathbf{A}'\boldsymbol{\Sigma}_y^{-1}\mathbf{A}(\mathbf{A}'\boldsymbol{\Sigma}_y^{-1}\mathbf{A}+\alpha\mathbf{S}^{-1})^{-2}\mathbf{S}^{-1}(\mathbf{A}'\boldsymbol{\Sigma}_y^{-1}\mathbf{A}+\alpha\mathbf{S}^{-1})^{-1}]}{\boldsymbol{\xi}'\mathbf{S}^{-1}(\mathbf{A}'\boldsymbol{\Sigma}_y^{-1}\mathbf{A}+\alpha\mathbf{S}^{-1})^{-2}\mathbf{A}'\boldsymbol{\Sigma}_y^{-1}\mathbf{A}(\mathbf{A}'\boldsymbol{\Sigma}_y^{-1}\mathbf{A}+\alpha\mathbf{S}^{-1})^{-1}\mathbf{S}^{-1}\boldsymbol{\xi}} \tag{B.6}$$

Proof:

$$\begin{aligned}&\mathrm{tr}\,MSE_{\alpha,\mathbf{S}}\{\hat{\boldsymbol{\xi}}\}=\\&=\mathrm{tr}(\mathbf{A}'\boldsymbol{\Sigma}_y^{-1}\mathbf{A}+\alpha\mathbf{S}^{-1})^{-1}\mathbf{A}'\boldsymbol{\Sigma}_y^{-1}\mathbf{A}(\mathbf{A}'\boldsymbol{\Sigma}_y^{-1}\mathbf{A}+\alpha\mathbf{S}^{-1})^{-1}+\\&+\mathrm{tr}[(\mathbf{A}'\boldsymbol{\Sigma}_y^{-1}\mathbf{A}+\alpha\mathbf{S}^{-1})^{-1}\alpha\mathbf{S}^{-1}]\boldsymbol{\xi}\boldsymbol{\xi}'[\alpha\mathbf{S}^{-1}(\mathbf{A}'\boldsymbol{\Sigma}_y^{-1}\mathbf{A}+\alpha\mathbf{S}^{-1})^{-1}]=\\&=\mathrm{tr}(\mathbf{A}'\boldsymbol{\Sigma}_y^{-1}\mathbf{A}+\alpha\mathbf{S}^{-1})^{-1}\mathbf{A}'\boldsymbol{\Sigma}_y^{-1}\mathbf{A}(\mathbf{A}'\boldsymbol{\Sigma}_y^{-1}\mathbf{A}+\alpha\mathbf{S}^{-1})^{-1}+\mathrm{tr}\,(\boldsymbol{\beta}\boldsymbol{\beta}')\\&=\mathrm{tr}(\mathbf{A}'\boldsymbol{\Sigma}_y^{-1}\mathbf{A}+\alpha\mathbf{S}^{-1})^{-1}\mathbf{A}'\boldsymbol{\Sigma}_y^{-1}\mathbf{A}(\mathbf{A}'\boldsymbol{\Sigma}_y^{-1}\mathbf{A}+\alpha\mathbf{S}^{-1})^{-1}+\boldsymbol{\beta}'\boldsymbol{\beta}\end{aligned}$$

$$\begin{aligned}&d\,\mathrm{tr}\,MSE_{\alpha,\mathbf{S}}\{\hat{\boldsymbol{\xi}}\}=\\&=\mathrm{tr}\{d[(\mathbf{A}'\boldsymbol{\Sigma}_y^{-1}\mathbf{A}+\alpha\mathbf{S}^{-1})^{-1}\mathbf{A}'\boldsymbol{\Sigma}_y^{-1}\mathbf{A}(\mathbf{A}'\boldsymbol{\Sigma}_y^{-1}\mathbf{A}+\alpha\mathbf{S}^{-1})^{-1}]\}+d(\boldsymbol{\beta}'\boldsymbol{\beta})\end{aligned}$$

"the first term"

$$\begin{aligned}&\mathrm{tr}\{d[(\mathbf{A}'\boldsymbol{\Sigma}_y^{-1}\mathbf{A}+\alpha\mathbf{S}^{-1})^{-1}\mathbf{A}'\boldsymbol{\Sigma}_y^{-1}\mathbf{A}(\mathbf{A}'\boldsymbol{\Sigma}_y^{-1}\mathbf{A}+\alpha\mathbf{S}^{-1})^{-1}]\}=\\&=\mathrm{tr}\,2\mathbf{A}'\boldsymbol{\Sigma}_y^{-1}\mathbf{A}(\mathbf{A}'\boldsymbol{\Sigma}_y^{-1}\mathbf{A}+\alpha\mathbf{S}^{-1})^{-1}d(\mathbf{A}'\boldsymbol{\Sigma}_y^{-1}\mathbf{A}+\alpha\mathbf{S}^{-1})^{-1}\\&=-2\,\mathrm{tr}[\mathbf{A}'\boldsymbol{\Sigma}_y^{-1}\mathbf{A}(\mathbf{A}'\boldsymbol{\Sigma}_y^{-1}\mathbf{A}+\alpha\mathbf{S}^{-1})^{-2}\mathbf{S}^{-1}(\mathbf{A}'\boldsymbol{\Sigma}_y^{-1}\mathbf{A}+\alpha\mathbf{S}^{-1})^{-1}]d\alpha\le 0\end{aligned}$$

"the second term"

$$\begin{aligned}&d(\boldsymbol{\beta}'\boldsymbol{\beta})=\\&=2\alpha\boldsymbol{\xi}'\mathbf{S}^{-1}(\mathbf{A}'\boldsymbol{\Sigma}_y^{-1}\mathbf{A}+\alpha\mathbf{S}^{-1})^{-2}\mathbf{A}'\boldsymbol{\Sigma}_y^{-1}\mathbf{A}(\mathbf{A}'\boldsymbol{\Sigma}_y^{-1}\mathbf{A}+\alpha\mathbf{S}^{-1})^{-1}\mathbf{S}^{-1}\boldsymbol{\xi}d\alpha.\end{aligned}$$

"differentiation"

$$\begin{aligned}&\frac{d}{d\alpha}(\mathrm{tr}\,MSE_{\alpha,\mathbf{S}}\{\hat{\boldsymbol{\xi}}\})=\\&-2\,\mathrm{tr}[\mathbf{A}'\boldsymbol{\Sigma}_y^{-1}\mathbf{A}(\mathbf{A}'\boldsymbol{\Sigma}_y^{-1}\mathbf{A}+\alpha\mathbf{S}^{-1})^{-2}\mathbf{S}^{-1}(\mathbf{A}'\boldsymbol{\Sigma}_y^{-1}\mathbf{A}+\alpha\mathbf{S}^{-1})^{-1}]+\\&+2\alpha\boldsymbol{\xi}'\mathbf{S}^{-1}(\mathbf{A}'\boldsymbol{\Sigma}_y^{-1}\mathbf{A}+\alpha\mathbf{S}^{-1})^{-2}\mathbf{A}'\boldsymbol{\Sigma}_y^{-1}\mathbf{A}(\mathbf{A}'\boldsymbol{\Sigma}_y^{-1}\mathbf{A}+\alpha\mathbf{S}^{-1})^{-1}\mathbf{S}^{-1}\boldsymbol{\xi}\\&\frac{d}{d\alpha}(\mathrm{tr}\,MSE_{\alpha,\mathbf{S}}\{\hat{\boldsymbol{\xi}}\})=0\quad\Rightarrow\end{aligned}$$

$$\hat{\alpha}=\frac{\mathrm{tr}[\mathbf{A}'\boldsymbol{\Sigma}_y^{-1}\mathbf{A}(\mathbf{A}'\boldsymbol{\Sigma}_y^{-1}\mathbf{A}+\alpha\mathbf{S}^{-1})^{-2}\mathbf{S}^{-1}(\mathbf{A}'\boldsymbol{\Sigma}_y^{-1}\mathbf{A}+\alpha\mathbf{S}^{-1})^{-1}]}{\boldsymbol{\xi}'\mathbf{S}^{-1}(\mathbf{A}'\boldsymbol{\Sigma}_y^{-1}\mathbf{A}+\alpha\mathbf{S}^{-1})^{-2}\mathbf{A}'\boldsymbol{\Sigma}_y^{-1}\mathbf{A}(\mathbf{A}'\boldsymbol{\Sigma}_y^{-1}\mathbf{A}+\alpha\mathbf{S}^{-1})^{-1}\mathbf{S}^{-1}\boldsymbol{\xi}}$$

where $\hat{\alpha}=\hat{\alpha}(\boldsymbol{\xi},\alpha,\mathbf{S})$ reaches a minimum indeed.

Response of the Earth's Crust due to Topographic Loads Derived by Inverse Isostasy

Hussein A. Abd-Elmotaal
Civil Engineering Department, Faculty of Engineering,
Minia University, Minia 61111, Egypt

Abstract. Let us propose that the density anomaly is proportional to the earth's radius vector in such a way that it is linearly related to the topography by a convolution of the topography and an isotropic kernel function. Hence, one can derive that the attraction of the compensating masses is also a convolution of the topography and an isotropic isostatic response function. Such an isostatic response function of the earth's crust can be determined by deconvolution. The paper gives the necessary derivation of such a deconvolution by means of global spherical harmonics. A practical determination of the isotropic isostatic response of the earth's crust needs the harmonic analysis of both the topography and the attraction of the compensating masses. To avoid the assumption of an isostatic model, the principle of inverse isostasy, by which we aim to have zero isostatic anomalies, has been employed. The harmonic analysis of the Bouguer anomalies is thus a combination of the harmonic analysis of the topographic potential and the already existed global (free-air) reference models. Two global reference models have been used. They are EGM96 and GPM98CR models complete to degree and order 360 and 540, respectively. The needed harmonic analysis of the topography has been carried out using TUG87 and TBASE digital height models after smoothing to $20'$ and $30'$ resolutions. The results show that the isostatic response of the earth's crust derived by inverse isostasy behaves in the same sense as those given by the exact solution of the Vening Meinesz isostatic model.

Keywords. isostasy, inverse problems, response function, density anomaly.

1 Introduction

Geophysical evidences already prove that the topographic masses are somewhat isostatically compensated. Let us assume that the compensating density anomaly is linearly related to the topography by a convolution of the topographic height and an isotropic kernel function. One can prove that the attraction of the compensating masses is also a convolution of the topography and an isotropic isostatic response function. Such an isostatic response function can be determined by deconvolution.

A practical determination of the isotropic isostatic response function demands the harmonic analysis of both the topography and the attraction of the compensating masses. The principle of inverse isostasy (forcing the isostatic anomalies to be zero) has been employed. Thus the Bouguer anomalies replace the attraction of the compensating masses. Combining the harmonic analysis of the topographic potential and the free-air reference models perform the harmonic analysis of the Bouguer anomalies.

The paper gives the necessary derivation of the isostatic response function by means of spherical harmonics by a deconvolution of the vertical derivative of the compensating potential. A practical computation of the isotropic isostatic response function has been carried out using EGM96 model (complete to degree and order 360) and GPM98CR model (complete to degree and order 540) representing the geopotential model. TUG87 ($30' \times 30'$ resolution) and TBASE ($20' \times 20'$ and $30' \times 30'$ resolutions) digital height models have been used for computing the harmonic coefficients of the topography and of the topographic potential. A broad comparison of the isostatic response functions estimated within this investigation has been made. These estimated isostatic response functions are also compared with the Vening Meinesz isostatic model.

It should be noted that the assumption employed here that the density anomaly is linearly related to the topography by a convolution was given originally by Dorman and Lewis (1970). Moritz (1990) has also treated this topic in terms of spherical harmonics. Numerical results may be found, e.g., in (Lewis and Dorman, 1970; Bechtel et al., 1987; Hein et al., 1989).

2 Isostatic Response of the Earth's Crust

Let us consider a vertical column starting at the earth's center ($r = 0$) till the earth's surface ($r = R$). For each column, the density anomaly $\Delta\rho$ is a function of the radius vector r (or rather the depth), i.e.,

$$\Delta\rho \equiv \Delta\rho(r,\theta,\lambda)\,, \tag{1}$$

where θ and λ denote the spherical co-latitude and longitude, respectively. Let us assume that the density anomaly $\Delta\rho$ is linearly related to the topography h by a convolution (Dorman and Lewis, 1970)

$$\Delta\rho(r',\theta',\lambda') = \iint_\sigma h(\theta'',\lambda'')\,K(r',\psi')\,d\sigma\,, \tag{2}$$

where the kernel function $K(r',\psi')$ is isotropic, i.e., it depends *only* on the spherical distance ψ' between the point (r',θ',λ'), for which $\Delta\rho$ is referring, and the point (θ'',λ''), for which the topography h is referring, cf. Fig. 1. The spherical distance ψ' is given by the cosine theorem

$$\cos\psi' = \cos\theta'\cos\theta'' + \sin\theta'\sin\theta''\cos(\lambda''-\lambda')\,. \tag{3}$$

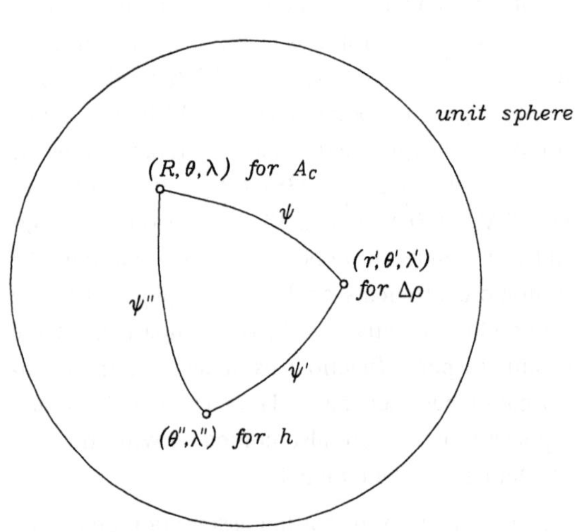

Fig. 1 The used point notations for density anomaly $\Delta\rho$, height h and compensating masses A_C.

Equation (2) can be written as

$$\Delta\rho(r',\theta',\lambda') = h(\theta'',\lambda'') * K(r',\psi')\,, \tag{4}$$

where $*$ stands for the spatial convolution. We emphasize the two assumptions made so far:

- the linearity of $\Delta\rho$ in h, cf. (2).
- the isotropy of the kernel $K(r',\psi')$.

The potential of the compensating masses V_C at point (r,θ,λ) is given by Newton's integral as (Heiskanen and Moritz, 1967, p. 3)

$$V_C(r,\theta,\lambda) = G\iiint_v \frac{\Delta\rho(r',\theta',\lambda')}{l}\,dv\,, \tag{5}$$

where G stands for Newton's gravitational constant, l is the spatial distance between (r,θ,λ) and (r',θ',λ'), given by

$$l^2 = r^2 + r'^2 - 2rr'\cos\psi\,. \tag{6}$$

and dv is the volume element, given by

$$dv = r'^2 dr' d\sigma\,, \tag{7}$$

and $d\sigma$ refers to the point (θ'',λ''). The spherical distance ψ is again given by the cosine theorem

$$\cos\psi = \cos\theta\cos\theta' + \sin\theta\sin\theta'\cos(\lambda'-\lambda)\,. \tag{8}$$

The attraction of the compensating masses A_C is given by the vertical derivative of the potential of the compensating masses V_C as:

$$\begin{aligned} A_C(r,\theta,\lambda) &= -\frac{\partial V_C}{\partial r} \\ &= -G\iiint_v \Delta\rho(r',\theta',\lambda')\frac{\partial}{\partial r}\left(\frac{1}{l}\right)dv\,, \end{aligned} \tag{9}$$

which can be re-written as

$$\begin{aligned} A_C(r,\theta,\lambda) &= -G\iint_\sigma \int_{r'=0}^{R} \Delta\rho(r',\theta',\lambda')\,\cdot \\ &\quad \cdot\frac{\partial}{\partial r}\left(\frac{1}{l}\right) r'^2 dr' d\sigma\,. \end{aligned} \tag{10}$$

We put

$$D(r',\psi) = -Gr'^2\frac{\partial}{\partial r}\left(\frac{1}{l}\right)\,. \tag{11}$$

The dependence of D on r is eliminated by computing A_C at the sea level, for which

$$r = R\,. \tag{12}$$

Thus, the attraction of the compensating masses $\Delta\rho$ is given by

$$A_C(R,\theta,\lambda) = \iint_\sigma \int_{r'=0}^{R} D(r',\psi)\,\Delta\rho(r',\theta',\lambda')\,dr'd\sigma, \tag{13}$$

or, symbolically

$$A_C(R,\theta,\lambda) = \int_{r'=0}^{R} D(r',\psi) * \Delta\rho(r',\theta',\lambda')\,dr'. \tag{14}$$

Substituting (4) into (14) gives

$$A_C(R,\theta,\lambda) = h(\theta'',\lambda'') * \int_{r'=0}^{R} D(r',\psi) * K(r',\psi')\,dr'. \tag{15}$$

Note that $h(\theta'',\lambda'')$ doesn't depend on r' (see Fig. 1).

The *isostatic response function F* is defined by

$$F(\psi'') = \int_{r'=0}^{R} D(r',\psi) * K(r',\psi')\,dr'. \tag{16}$$

This can be written explicitly as

$$F\left[(\theta,\lambda),(\theta'',\lambda'')\right] = \int_{r'=0}^{R} \iint_\sigma D(r',\psi)\cdot \cdot K(r',\psi')\,d\sigma dr', \tag{17}$$

where ψ is given by (8) and ψ'' is given by (see Fig. 1)

$$\cos\psi'' = \cos\theta\cos\theta'' + \sin\theta\sin\theta''\cos(\lambda''-\lambda). \tag{18}$$

The isostatic response function F depends *only* on the spherical distance ψ'' between the points (θ,λ) and (θ'',λ'') for reasons of symmetry.

Substituting (16) into (15) gives

$$A_C(R,\theta,\lambda) = h(\theta'',\lambda'') * F(\psi''). \tag{19}$$

Since the points (θ',λ') and (θ'',λ'') are freely moving points all over the sphere, we can replace the point (θ'',λ'') by (θ',λ') in (19) to simplify the notations. We thus can write

$$A_C(R,\theta,\lambda) = h(\theta',\lambda') * F(\psi), \tag{20}$$

where ψ denotes the spherical distance between the points (θ,λ) and (θ',λ'), given by (8). Given the attraction of the compensating masses A_C and the topographic height h, the isostatic response function F can be determined by deconvolution.

3 Deconvolution

To compute the isostatic response function F by deconvolution of (20), we confine ourselves to the global problem, which can be solved by means of spherical harmonics.

Let us expand the three components of (20)

$$A_C(R,\theta,\lambda) = \sum_{n=0}^{\infty}\sum_{m=-n}^{n} A_{nm}Y_{nm}(\theta,\lambda), \tag{21}$$

$$h(\theta,\lambda) = \sum_{n=0}^{\infty}\sum_{m=-n}^{n} H_{nm}Y_{nm}(\theta,\lambda), \tag{22}$$

$$F(\psi) = \sum_{n=0}^{\infty} F_n P_n(\cos\psi), \tag{23}$$

where $P_n(\cos\psi)$ stands for the standard Legendre polynomial and the spherical harmonics Y_{nm} are given by (Moritz, 1990, p. 195)

$$Y_{nm}(\theta,\lambda) = P_{nm}(\cos\theta)\begin{cases}\cos m\lambda, & m=0,1,\ldots,n,\\ \sin m\lambda, & m=-1,\ldots,-n,\end{cases} \tag{24}$$

where $P_{nm}(\cos\theta)$ are the standard Legendre functions. Since F depends only on ψ, its expansion is purely zonal, cf. (23).

Equation (20) can be written explicitly as

$$A_C(R,\theta,\lambda) = \iint_\sigma h(\theta',\lambda')\,F(\psi)\,d\sigma. \tag{25}$$

Substituting F from (23) gives

$$A_C(R,\theta,\lambda) = \sum_{n=0}^{\infty} F_n \iint_\sigma h(\theta',\lambda')\,P_n(\cos\psi)\,d\sigma. \tag{26}$$

Now by using the well-known integral formula (Heiskanen and Moritz, 1967, eq. 1–71), one can write

$$A_C(R,\theta,\lambda) = 4\pi\sum_{n=0}^{\infty}\frac{F_n}{2n+1}H_n(\theta,\lambda). \tag{27}$$

We express the Laplace harmonic $H_n(\theta,\lambda)$ of the topography h in terms of the spherical harmonics Y_{nm}

$$H_n(\theta,\lambda) = \sum_{m=-n}^{n} H_{nm}Y_{nm}(\theta,\lambda). \tag{28}$$

Thus A_C is given by inserting (28) into (27)

$$A_C(R,\theta,\lambda) = 4\pi\sum_{n=0}^{\infty}\sum_{m=-n}^{n}\frac{F_n}{2n+1}H_{nm}Y_{nm}(\theta,\lambda). \tag{29}$$

Comparing (21) with (29) gives immediately

$$A_{nm} = \frac{4\pi}{2n+1} F_n H_{nm} . \tag{30}$$

Equation (30) is the so-called the *spherical convolution theorem* (Moritz, 1990, p. 251). Equation (30) can be written for F_n as

$$F_n = \frac{2n+1}{4\pi} \frac{A_{nm}}{H_{nm}} , \tag{31}$$

which is *independent* of the order m. This condition should be satisfied by the topography h and the attraction of the compensating masses A_C if the assumption of isotropy is justified. Finally, the isostatic response function is given by (23) with (31).

4 Practical Considerations

Equation (31) computes F_n as a ratio of the conventional harmonic coefficients of the attraction of the compensating masses A_{nm} and of the topography H_{nm}. The conventional harmonic coefficients are related to the fully normalized ones by, e.g., (Heiskanen and Moritz, 1967, p. 32)

$$\begin{aligned} \bar{A}_{n0} &= \frac{A_{n0}}{\sqrt{2n+1}} , \\ \bar{A}_{nm} &= \sqrt{\frac{1}{2(2n+1)} \frac{(n+m)!}{(n-m)!}} A_{nm} \quad (m \neq 0) . \end{aligned} \tag{32}$$

Replacing the conventional harmonic coefficients by the fully normalized ones, the *constant* factors in the nominator and dominator of (31) will cancel, and F_n is then given by

$$F_n = \frac{2n+1}{4\pi} \frac{\bar{A}_{nm}}{\bar{H}_{nm}} , \tag{33}$$

where $\bar{A}_{nm}$ and $\bar{H}_{nm}$ are the fully normalized harmonic coefficients of the attraction of the topographic masses and of the topography, respectively.

Practical computation of F_n by (33) may face a problem of imperfect isotropy. This leads to lose of power of F_n, especially for higher degrees n, if one computes F_n by, e.g.,

$$F_n = \frac{2n+1}{4\pi} \left(\frac{1}{2n+1} \sum_{m=-n}^{n} \frac{\bar{A}_{nm}}{\bar{H}_{nm}} \right) = \frac{1}{4\pi} \sum_{m=-n}^{n} \frac{\bar{A}_{nm}}{\bar{H}_{nm}} . \tag{34}$$

Alternatively, F_n may be computed by

$$F_n = \frac{2n+1}{4\pi} \frac{\bar{A}_n}{\bar{H}_n} , \tag{35}$$

where $\bar{A}_n$ and $\bar{H}_n$ is given by

$$\bar{A}_n = \sqrt{\sum_{m=-n}^{n} \bar{A}_{nm}^2} , \tag{36}$$

$$\bar{H}_n = \sqrt{\sum_{m=-n}^{n} \bar{H}_{nm}^2} . \tag{37}$$

Practical comparison between (34) and (35) will be given in sec. 8.

The summation on n in (23) is infinite. Examining (31) shows that using a certain practical upper limit N_{max}, representing infinity, will change the values of the isostatic response function (the higher the upper limit, the larger are the values of the isostatic response function). Hence, no direct comparison of isostatic response functions could be made. To solve this problem, we define the *normalized* isostatic response function $\bar{F}(\psi)$ as

$$\bar{F}(\psi) = \frac{F(\psi)}{F(0)} , \tag{38}$$

where $F(0)$ stands for the value of the isostatic response function at the origin ($\psi = 0$).

5 Principle of Inverse Isostasy

A practical determination of the isotropic isostatic response function needs the harmonic analysis of both the topography $\bar{H}_{nm}$ and the attraction of the compensating masses $\bar{A}_{nm}$, cf. (33). Creating compensating masses by means of an isostatic hypothesis already implies an assumption of the earth's isostatic response. Instead, one may wish to estimate a more realistic isostatic response of the earth without postulating an isostatic hypothesis. This may be achieved by the principle of inverse isostasy, by which we aim to have zero isostatic anomalies Δg_I, i.e.,

$$\Delta g_I = 0 . \tag{39}$$

The reader who is interested in more details about inverse problems in isostasy is referred, e.g., to (Moritz, 1990, sec. 8.3). Equation (39) can alternatively be written in the form

$$A_C = -\Delta g_B , \tag{40}$$

where A_C stands for the attraction of the compensating masses, as before, and Δg_B stands for the Bouguer anomalies. Thus, the Bouguer anomalies will be used instead of the attraction of the compensating masses.

6 Harmonic Analysis of the Compensating Masses and Topography

The Bouguer anomaly Δg_B is defined by

$$\Delta g_B = \Delta g_F - A_T, \tag{41}$$

where Δg_F stands for the free-air anomaly and A_T refers to the attraction of the topographic masses. Using (40), the attraction of the compensating masses may be written as

$$A_C = A_T - \Delta g_F. \tag{42}$$

Accordingly, the fully normalized harmonic coefficients of the potential of the compensating masses $\bar{C}_{nm}$ can be defined as

$$\bar{C}_{nm} = \bar{t}_{nm} - \bar{T}_{nm}, \tag{43}$$

where $\bar{t}_{nm}$ are the fully normalized harmonic coefficients of the potential of the topographic masses and $\bar{T}_{nm}$ are the fully normalized harmonic coefficients of the free-air disturbing potential. $\bar{T}_{nm}$ are provided by the global geopotential models (e.g., EGM96). The harmonic analysis of the topographic potential $\bar{t}_{nm}$ will be given in the next section. Finally, the fully normalized harmonic coefficients of the attraction of the compensating masses $\bar{A}_{nm}$ (in gravity anomaly unites) is related to the (unitless) fully normalized harmonic coefficients of the potential of the compensating masses $\bar{C}_{nm}$ by (Heiskanen and Moritz, 1967, p. 89)

$$\bar{A}_{nm} = \frac{GM}{R^2}(n-1)\bar{C}_{nm}, \tag{44}$$

where GM is the geocentric gravitational constant and R is the mean earth's radius.

Expand the topography h in terms of fully normalized harmonic coefficients $\bar{H}_{nm}$

$$h(\theta,\lambda) = \sum_{n=0}^{\infty}\sum_{m=-n}^{n} \bar{H}_{nm}\bar{Y}_{nm}(\theta,\lambda), \tag{45}$$

where the fully normalized spherical harmonics $\bar{Y}_{nm}$ are given by (analogous to (24))

$$\bar{Y}_{nm}(\theta,\lambda) = \bar{P}_{nm}(\cos\theta)\begin{cases}\cos m\lambda, & m = 0,1,\ldots,n, \\ \sin m\lambda, & m = -1,\ldots,-n,\end{cases} \tag{46}$$

and $\bar{P}_{nm}(\cos\theta)$ are the fully normalized Legendre functions. The fully normalized harmonic coefficients $\bar{H}_{nm}$ are orthogonal, and given by (ibid., p. 31)

$$\bar{H}_{nm} = \frac{1}{4\pi}\iint_{\sigma} h(\theta,\lambda)\,\bar{Y}_{nm}(\theta,\lambda)\,d\sigma. \tag{47}$$

The practical determination of the fully normalized harmonic coefficients $\bar{H}_{nm}$ is carried out using the HRCOFITR program written by Abd-Elmotaal (2001) based on HARMIN and SSYNTH subroutines, written originally by Colombo (1981).

7 Harmonic Analysis of the Topographic Potential

The topographic potential T_T can be easily defined as the potential of all topographic masses outside the geoid and ocean water inside the geoid. The topographic potential T_T can be written as:

$$T_T(P) = G\iiint_{v} \frac{\rho_Q}{\ell_{PQ}}\,dv_Q, \tag{48}$$

where G is the Newton's gravitational constant, ρ_Q denotes the density at Q, ℓ_{PQ} is the spatial distance between P and Q and dv_Q is the volume element at Q.

It is known that T_T is harmonic outside the earth's surface and its spherical harmonic series is convergent outside a sphere completely enclosing the earth (Heiskanen and Moritz, 1967, p. 60). Outside that sphere, the convergent series representation of the reciprocal distance can be used (ibid., p. 33)

$$\frac{1}{\ell_{PQ}} = \sum_{n=0}^{\infty} \frac{r_Q^n}{r_P^{n+1}} P_n(\cos\psi_{PQ}), \tag{49}$$

where r is the modulus of the radius vector, $P_n(\cos\psi)$ is the conventional Legendre polynomial of degree n and ψ_{PQ} is the spherical distance between P and Q.

The Legendre polynomial may be expressed as

$$P_n(\cos\psi_{PQ}) = \frac{1}{2n+1}\sum_{m=-n}^{n} \bar{R}_{nm}(P)\bar{R}_{nm}(Q) \tag{50}$$

with the fully normalized spherical harmonics

$$\bar{R}_{nm}(P) = \sqrt{2^{1-\delta_{m0}}(2n+1)\frac{(n-m)!}{(n+m)!}}\cdot P_{nm}(\cos\theta_P)\begin{cases}\cos m\lambda_P & \text{for } m \le 0 \\ \sin m\lambda_P & \text{for } m > 0\end{cases} \tag{51}$$

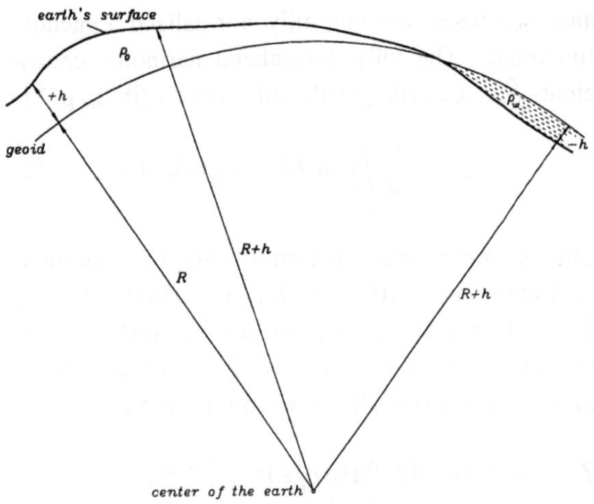

Fig. 2 Potential of the topographic masses.

with

$$\delta_{ij} = \begin{cases} 1 & \text{if } j = i \\ 0 & \text{if } j \neq i \end{cases}, \quad (52)$$

where θ is the polar distance, λ is the geodetic longitude, δ_{ij} is the Kronecker symbol and $P_{nm}(\cos\theta)$ is the conventional Legendre function of degree n and order m. Thus the topographic potential can be represented by

$$T_T(P) = G \sum_{n=0}^{\infty} \frac{1}{(2n+1) r_P^{n+1}} \sum_{m=-n}^{n} \bar{R}_{nm}(P) \cdot \left[\iiint_v \rho_Q r_Q^n \bar{R}_{nm}(Q) \, dv_Q \right] \quad (53)$$

with

$$dv = r^2 dr d\sigma, \quad (54)$$

where $d\sigma$ is the spherical surface element.

To calculate the volume integral inside the brackets of (53), we confine ourselves to the spherical approximation. Then its contribution due to the topographic masses outside the geoid and ocean water inside the geoid, see Fig. 2, is

$$\iiint_v = \iint_\sigma \int_{r=R}^{R+h} \rho_Q r_Q^{n+2} dr_Q \bar{R}_{nm}(Q) \, d\sigma_Q, \quad (55)$$

where R is the radius of the mean earth's sphere and h is the topographic height (+) or ocean bottom depth (−).

The integration of (55) with respect to r is straightforward and is expressed as (Sünkel, 1985, p. 5)

$$\iiint_v = \frac{R^{n+3}}{n+3} \iint_\sigma \rho_Q \left[\left(1 + \frac{h_Q}{R}\right)^{n+3} - 1 \right] \bar{R}_{nm}(Q) \, d\sigma_Q. \quad (56)$$

Hence, the harmonic coefficients of the topographic potential $\bar{t}_{nm}$ and the harmonic series expansion of the topographic potential can be expressed, by inserting (56) into (53), as

$$T_T(P) = \frac{GM}{r_P} \sum_{n=0}^{\infty} \left(\frac{R}{r_P}\right)^n \sum_{m=-n}^{n} \bar{t}_{nm} \bar{R}_{nm}(P), \quad (57)$$

where

$$\bar{t}_{nm} = \frac{R^3}{M(2n+1)(n+3)} \iint_\sigma \rho_Q \cdot \left[\left(1 + \frac{h_Q}{R}\right)^{n+3} - 1 \right] \bar{R}_{nm}(Q) \, d\sigma_Q, \quad (58)$$

where M denotes the mass of the earth, given by

$$M \doteq \frac{4\pi R^3}{3} \rho_M, \quad (59)$$

where ρ_M denotes the mean earth's density

$$\rho_M = 5.517 \text{ g/cm}^3.$$

For the practical determination of the harmonic coefficients of the topographic potential $\bar{t}_{nm}$, (58) may be written as

$$\bar{t}_{nm} = \frac{3 \Delta\phi \Delta\lambda}{4\pi\rho_M (2n+1)(n+3)} \sum_i^{\phi} \sum_j^{\lambda} \rho_{ij} \cdot \left[\left(1 + \frac{h_{ij}}{R}\right)^{n+3} - 1 \right] \begin{Bmatrix} \cos m\lambda_j \\ \sin m\lambda_j \end{Bmatrix} \cdot \bar{P}_{nm}(\cos\theta_i) \cos\phi_i, \quad (60)$$

where $\sum$ denotes the summation along ϕ and λ, $\Delta\phi$ and $\Delta\lambda$ are the grid sizes of the used digital height model in the latitude and the longitude directions, respectively, and ρ is given by

$$\begin{aligned} \rho &= \rho_\circ && \text{for } h \geq 0, \\ \rho &= \rho_\circ - \rho_w && \text{for } h < 0, \end{aligned} \quad (61)$$

where $\rho_\circ$ is the constant density of the topography and ρ_w is the density of the ocean's water.

8 Results

TUG87 ($30' \times 30'$ resolution) and TBASE ($20' \times 20'$ and $30' \times 30'$ resolutions) digital height models have been used in this investigation. EGM96 and GPM98CR models (complete to degree and order

360 and 540, respectively) have been used representing the geopotential model.

Figure 3 shows a comparison of computing F_n by (34) and (35) using TUG87 DHM and EGM96 geopotential model. It shows that for the lower degrees (till around $n = 40$) both curves are nearly coincident. For higher degrees, however, eq. (35) saves the power of F_n with less variations between degree-to-degree values.

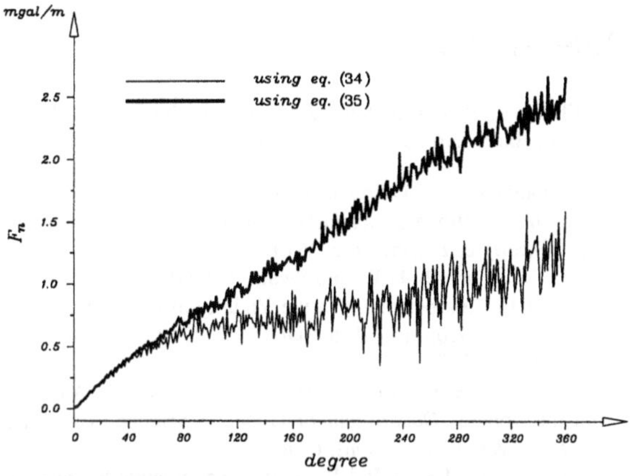

Fig. 3 F_n computed by (34) and (35) using TUG87 DHM and EGM96 geopotential model.

Figure 4 shows F_n computed by (35) using TBASE DHM ($30' \times 30'$ resolution for $N_{max} = 360$ and $20' \times 20'$ resolution for $N_{max} = 540$) and GPM98CR geopotential model. The behaviour of F_n shows, more or less, a linear function, which follows directly the definition of F_n (note the factor (2n+1) appearing in (33) which is an equation of straight line). Figure 4 shows clearly that using a higher value for N_{max}, representing infinity, will change the values of the isostatic response function F. Hence, no direct comparison of isostatic response functions could be made. To solve this, we defined the normalized isostatic response function $\bar{F}(\psi)$ (cf. (38)).

Figure 5 shows the normalized isostatic response functions compared to the Kelvin function kei$x$. Two normalized isostatic response functions are considered here. The first uses EGM96 geopotential model and TUG87 DHM with $N_{max} = 360$. The second uses GPM98CR geopotential model and TBASE DHM ($20' \times 20'$) with $N_{max} = 540$. The Kelvin function kei$x$ (or rather the *scaled* Kelvin function) is the solution of the equation of equilibrium of a thin plate bent by a concentrated load. The Kelvin function kei$x$ then gives the bending curve of the earth's crust subject to the topographic loads. This follows the Ven-

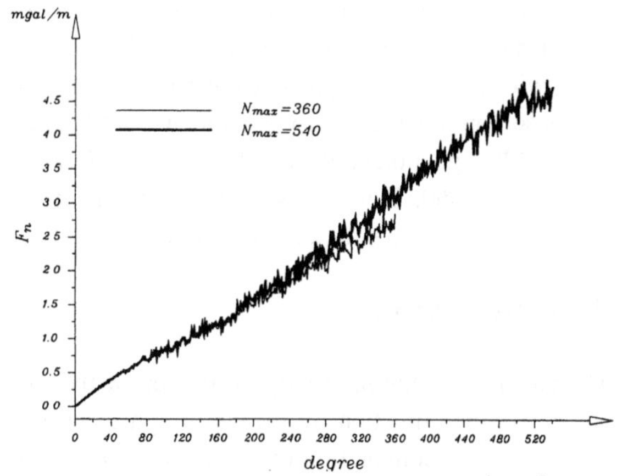

Fig. 4 F_n computed by (35) using TBASE DHM ($30' \times 30'$ resolution for $N_{max} = 360$ and $20' \times 20'$ resolution for $N_{max} = 540$) and GPM98CR geopotential model.

ing Meinesz regional isostatic model (Abd-Elmotaal, 1993). For the Kelvin function, a value of the so-called *degree of regionality l* of 20 km has been chosen. The degree of regionality l is a function of the changeable physical parameters of the earth's crust. Hence, an exact estimation of l is obviously hard, and usually one tries to reasonably assume it.

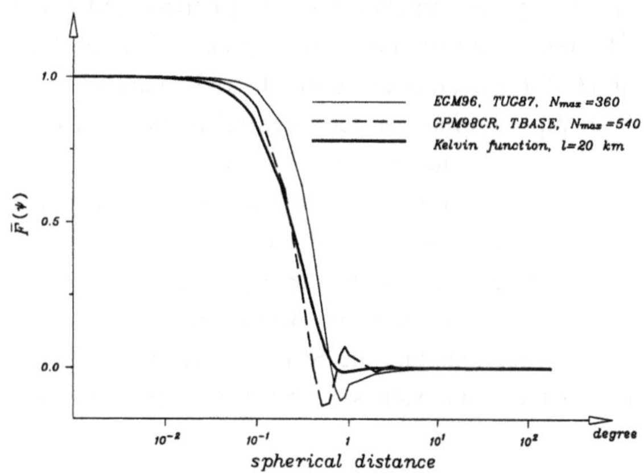

Fig. 5 Normalized isostatic response functions compared to Kelvin function kei$x$.

Figure 5 shows that the normalized isostatic response functions and the Kelvin function kei$x$ give nearly the same values. For the Kelvin function kei$x$, the degree of regionality l may play a role of a fitting parameter. Figure 5 shows that the isostatic response of the earth's crust computed by *direct* isostasy using the most realistic isostatic model (expressed by the *exact* solution of the earth's crust subject to the topographic loads (Abd-Elmotaal, 1993)) matches the isostatic response of the earth's crust derived by *inverse* isostasy.

It should be noted that using GPM98CR geopotential model and TBASE DHM ($30' \times 30'$ resolution) with $N_{max} = 360$ gives a normalized isostatic response function practically coincident with that using EGM96 geopotential model and TUG87 DHM with $N_{max} = 360$. This shows the insignificant effect of the used digital height and geopotential models.

9 Conclusion

The paper gives the necessary derivation of the isostatic response function by means of spherical harmonics by a deconvolution of the vertical derivative of the isostatic potential. A practical determination of the isostatic response function needs the harmonic analysis of both the topography and the attraction of the compensating masses. To avoid the assumption of any isostatic model, the principle of inverse isostasy (forcing the isostatic anomalies to be zero) has been employed. The harmonic analysis of the Bouguer anomalies is thus a combination of the harmonic analysis of the topographic potential and the already existed global (free-air) reference models.

A practical computation of the isotropic isostatic response function has been carried out using EGM96 model (complete to degree and order 360) and GPM98CR model (complete to degree and order 540) representing the geopotential model. TUG87 ($30' \times 30'$ resolution) and TBASE ($20' \times 20'$ and $30' \times 30'$ resolutions) digital height models have been used for computing the harmonic coefficients of the topography and of the topographic potential. The results show that the normalized isostatic response functions and the Kelvin function $\mathrm{kei}x$ give nearly the same values. This shows that the isostatic response of the earth's crust computed from direct isostasy using the most realistic isostatic model (Vening Meinesz isostatic model with Kelvin function $\mathrm{kei}x$ as the exact bending curve (Abd-Elmotaal, 1993)) matches the isostatic response of the earth's crust derived by inverse isostasy.

Acknowledgement. The author would like to thank Prof. Helmut Moritz for his useful comments, valuable discussions and great encouragement.

References

Abd-Elmotaal, H. (1993) Vening Meinesz Moho Depths: Traditional, Exact and Approximated, *manuscripta geodaetica*, **18**, 171–181.

Abd-Elmotaal, H. (2001) Enhanced Approach for Harmonic Analysis on the Ellipsoid, *Scientific Bulletin, Ain Shams University, Faculty of Engineering*, **36 (3)**, 53–65.

Bechtel, T.D., Forsyth, D.W. and Swain, C.J. (1987) Mechanisms of Isostatic Compensation in the Vicinity of the East African Rift, Kenya, *Geophysical Journal of Royal Astronomical Society*, **90**, 445–465.

Colombo, O. (1981) Numerical Methods for Harmonic Analysis on the Sphere, *Department of Geodetic Science, The Ohio State University, Columbus, Ohio*, **310**.

Dorman, L.M. and Lewis, B.T.R. (1970) Experimental Isostasy: 1. Theory of the Determination of the Earth's Isostatic Response to a Concentrated Load, *Journal of Geophysical Research*, **75**, 3357–3365.

Hein, G.W., Eissfeller, B., Ertel, M., Hehl, K., Jacoby, W. and Czerwek, D. (1989) On Gravity Prediction Using Density and Seismic Data, *Preprint, Institute of Astronomical and Physical geodesy, University FAF Munich*.

Heiskanen, W.A. and Moritz, H. (1967) *Physical Geodesy*, Freeman, San Francisco.

Lewis, B.T.R. and Dorman, L.M. (1970) Experimental Isostasy: 2. An Isostatic Model for the USA Derived from Gravity and Topographic Data, *Journal of Geophysical Research*, **75**, 3367–3386.

Moritz, H. (1990) *The Figure of the Earth: Theoretical Geodesy and the Earth's Interior*, Wichmann, Karlsruhe.

Sünkel, H. (1985) An Isostatic Earth Model, *Department of Geodetic Science, The Ohio State University, Columbus, Ohio*, **367**.

SAR interferometry for deformation control

B. Crippa
Department of Earth Sciences, University of Milan, Via Cicognara 7, 20129 Milan, Italy

M. Crosetto, M. Blázquez
Institute of Geomatics, Campus de Castelldefels, 08860 Castelldefels (Barcelona), Spain

R. Barzaghi
DIIAR, Politecnico di Milano, P. Leonardo da Vinci 32, 20133 Milan, Italy

Abstract. A quantitative control of deformations using the differential interferometric SAR (DInSAR) technique may be achieved when multiple observations and suitable modelling and analysis tools are employed. The paper begins with a description of the main characteristics of the DInSAR data. Then, it discusses a new modelling and filtering strategy, which takes advantage of the specific properties of the DInSAR observations. The core of the procedure is the least squares collocation filtering and prediction, which exploits the correlation properties of the DInSAR data. The proposed procedure was tested on simulated DInSAR data that reproduce the characteristics of a small scale subsidence, and that include the main components of the interferometric data: the atmospheric contribution, the phase noise component, and the outliers due to the unwrapping related errors.

Keywords. SAR, monitoring, modelling, estimation, simulation.

1 Introduction

The differential interferometric SAR technique (DInSAR), based on spaceborne SAR data, has been successfully employed in different application fields: glacier dynamics, earthquakes, volcanoes, landslides, and the deformations related to water exploitation, mining activity, and construction works. For a general review, see Hanssen (2001). The above applications require different quality levels to the estimates provided by DInSAR. A quite qualitative use of the DInSAR results seems to be sufficient for the purposes of some applications. This is the case of several studies where the geophysical interpretation may be simply based on the qualitative information derived from DInSAR observations. However, this is not the case for some other important applications, e.g. the subsidence monitoring in urban areas, which need to be characterised by high quality standards like those usually achieved by the geodetic techniques. These applications need a fully quantitative DInSAR monitoring, based on multiple observations (i.e. multiple interferograms).

The DInSAR technique may provide a quantitative monitoring tool only if suitable data modelling and analysis procedures are employed. It is important to note that the use of multiple DInSAR observations requires 3D (2D in space, plus the time) modelling and data analysis tools. In literature these procedures have received little attention. Furthermore, little importance has been usually given to important quality aspects, like the precision, accuracy and reliability of the DInSAR estimates.

This paper describes new modelling strategies based on 2D and 3D adaptative models, which fully take advantage of the specific properties of the DInSAR observations. This work only focuses on vertical displacement phenomena. The paper begins with a description of the basic properties of the DInSAR observations. This is followed by the discussion of a modelling strategy for multiple DInSAR observations and the results obtained on 3D simulated datasets. Finally, some possible evolutions of the described research are described.

2 Properties of the DInSAR data

The D-InSAR technique exploits the phase difference (interferometric phase) $\Delta\Phi_{Int}$ of two SAR images (hereafter referred to as the master, M, and the slave, S, images). Let us consider a point P on the ground, which remains stable in the time interval between the image acquisitions. $\Delta\Phi_{Int}$ is related to the distance difference $SP-MP$, which is the key element for the InSAR DEM (Digital

Elevation Model) generation. When the point moves from P to P^1 between the two image acquisitions, besides the component due to terrain topography, Φ_{Topo}, $\Delta\Phi_{Int}$ includes the terrain movement contribution, Φ_{Mov}. In the general case $\Delta\Phi_{Int}$ consists of the following components:

$$\Delta\Phi_{Int} = \Phi_S - \Phi_M = \frac{SP - MP}{\frac{\lambda}{4\cdot\pi}} + \frac{SP^1 - SP}{\frac{\lambda}{4\cdot\pi}} + \Phi_{Atm} + \Phi_{Noise} =$$

$$= \Phi_{Topo} + \Phi_{Mov} + \Phi_{Atm} + \Phi_{Noise},$$

where Φ_S, Φ_M are the phases of S and M; Φ_{Atm} is the atmospheric contribution; Φ_{Noise} is the phase noise; SP^1 is the slave-to-P^1 distance; and λ is the radar wavelength. If the terrain topography is known (i.e. a DEM of the imaged area is available), Φ_{Topo} can be computed (Φ_{Topo_Sim}) and subtracted from $\Delta\Phi_{Int}$, obtaining the D-InSAR phase $\Delta\Phi_{D-Int}$:

$$\Delta\Phi_{D-Int} = \Phi_{Mov} + \Phi_{Atm} + \Phi_{Res\_Topo} + \Phi_{Noise}$$

where Φ_{Res_Topo} represents the residual component due to DEM errors. In order to derive information on the terrain movement, Φ_{Mov} has to be separated from the other phase components. When multiple DInSAR observations are available, the following important properties may be exploited to this purpose (see next section):

- Φ_{Mov} is highly correlated, spatially and temporally;
- Φ_{Atm} is correlated spatially, and uncorrelated temporally;
- Φ_{Noise} and Φ_{Res_Topo} can be considered spatially decorrelated (hereafter they are referred to as the noise). As far as Φ_{Res_Topo} is concerned, this assumption holds for high quality DEMs, e.g. the high resolution photogrammetric DEMs, while it does not hold for the InSAR DEMs, potentially affected by atmospheric effects.

3 A model for multiple observations

The fusion of multiple DInSAR observations involves a 3D modelling of the subsidence at hand: two dimensions in space, plus the time. As already mentioned in the introduction, the models proposed in literature are sometimes only limited to one single dimension of the phenomenon under analysis, see the 1D models adopted by Berardino et al. (2001) and Ferretti et al. (2000): they provide a pointwise description of the subsidence temporal evolution. Other authors employs a very simple strategy based on averaging the observations of multiple interferograms, see Strozzi et al. (2001) and Williams et al. (1998). This last procedure, which is quite popular in literature, is often named interferogram stacking technique.

In this section we describe a procedure based on an adaptative modelling, which fully takes advantage of the correlation properties mentioned in the previous section. The proposed procedure employs a "2.5D modelling", which couples two different models: a 2D model, which is used to assess the subsidence velocity map of the observed field (assuming a constant deformation rate over the observed time interval); and a 1D model, which is employed to estimate the temporal evolution of the subsidence in some selected points of the deformation field. The main features of the proposed procedure are briefly outlined below.

The assumption of a constant deformation rate over the observed time interval allows a 2D model to be adopted, instead of a quite more complex 3D one. This assumption should be quite appropriate to different applications (at least as a first approximation).

The procedure requires multiple interferograms of the same subsidence area. The interferograms are processed separately, deriving from each interferogram the corresponding velocity field. The velocity fields are then put together as input observations of the 2D estimation procedure (stacked velocity fields). These fields contain the contributions of the four phase components described in the previous section, where in this case Φ_{Mov} represents the stationary signal; Φ_{Atm} is temporally uncorrelated but can be spatially correlated within each velocity field; and Φ_{Noise} and Φ_{Res_Topo} are white noise.

The subsidence velocity map is estimated through the classical LS collocation procedure, which involves the estimation of the autocovariance function of the stacked velocity fields, the separation of the signal from the noise (filtering), and the prediction of the signal over location not covered by the input data, Moritz (1978). The proposed procedure provides an adaptative filtering, which is only driven by the autocovariance function of the input data, without requiring any explicit modelling of the subsidence at hand: it is quite more flexible than the classical interpolation techniques, e.g. based on polynomials, etc. The collocation filtering may only work if the stationary signal (related to Φ_{Mov}) is "strong enough" to be separated from the Φ_{Atm}. This is not the case over the stable areas, which have to be excluded before running the collocation filtering. For this purpose we run a pre-processing of the stacked velocity fields based on a robust filter (a 3D median filtering).

The analysis of the temporal evolution of the subsidence is run on the residuals of the 2D collocation filtering (i.e. the differences between the observations and the corresponding estimated signals). Different analysis tools may be implemented: LS linear regression, 1D collocation filtering, etc. In this analysis the data redundancy plays a critical role. In fact, unlike the 2D collocation filtering, where usually large data sets are filtered, this analysis concerns very limited sets of data (say, 15-20 observations, one for each stacked interferogram). In order to gain redundancy, the temporal profiles of neighbour points may be put together. The analysis is typically performed on few selected locations of the deformation field under analysis (for instance, the location of the maximum deformation rate and some points of particular interest). This step is fundamental to check the goodness of the 2D model assumption, i.e. the constant deformation velocity.

4 Discussion of the results

The performances of the proposed procedure were tested on simulated DInSAR data that roughly reproduce the characteristics of a real phenomenon: the urban subsidence of Sallent (North Catalunya, Spain), which has been extensively analysed at the Cartographic Institute of Catalonia, see Crosetto et al. (2002a) and Crosetto et al. (2002b). This subsidence has slow deformation rate and small spatial extent, and represents a difficult case for the usage of DInSAR. The simulated data include different components, as it is detailed below.

- A 3D deformation field, $d = F(X,Y,t)$, whose main characteristics, like the shape, extension, deformation rate and temporal evolution, are similar to those of the Sallent subsidence. This is an "error-free" field, which was used in the analysis as a reference to assess the performances of the proposed filters. A deformation map, which corresponds to an observation interval of 5 years, is shown in Fig. 1. The maximum deformation rate is 20 mm/yr.

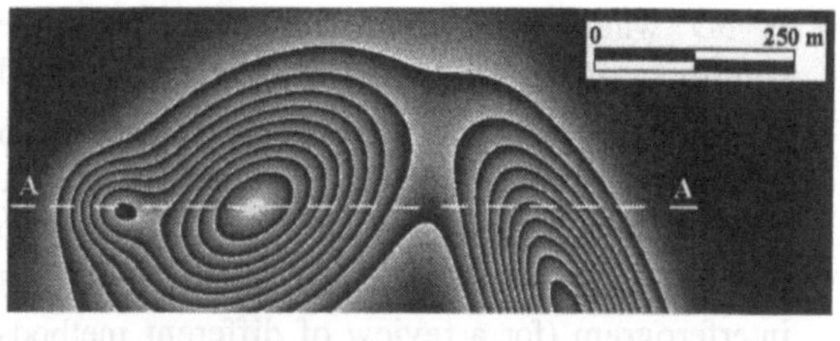

Fig. 1 Simulated deformation map, which corresponds to a 5 year time interval. Each fringe corresponds to a movement of 10 mm.

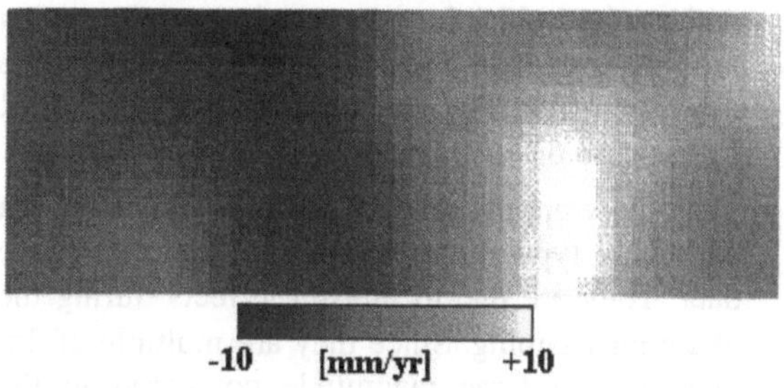

Fig. 2 Example of an atmospheric field with 20-m grid spacing, which covers the same area from Fig. 1.

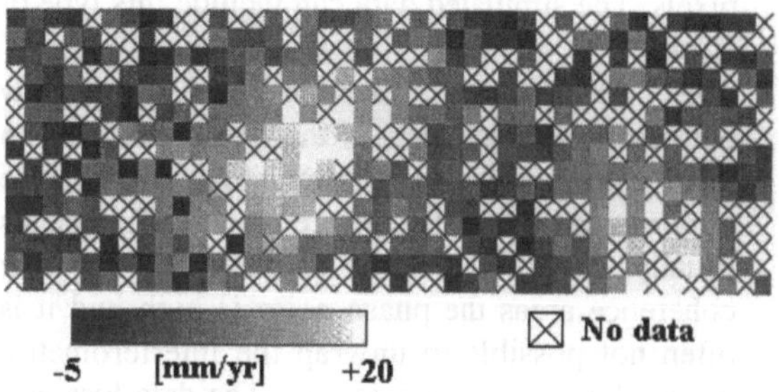

Fig. 3 A noise field superposed to a deformation filed, which covers the same area from Fig. 1. The grid spacing is 20 m; the data loss is 32.6 %.

- A 3D atmospheric field, which reproduces the contribution of the atmospheric effects. This field is spatially correlated, i.e. within a given interferogram (at a fixed t) it is correlated in the $[X,Y]$ domain. The data simulation was designed to work in a Monte Carlo framework. Therefore all the data (with the exception of the deformation field) were generated randomly. For each interferogram the generation of the atmospheric fields is based on the Cholesky decomposition and is driven by the standard deviation σ_{Atm} of the atmospheric component and its correlation length L_C. The values of σ_{Atm} and L_C can be estimated by analysing the $\Delta\Phi_{D-Int}$ over stable areas, see Crosetto et al. (2002a). An example of an atmospheric field is shown in Fig. 2. This field covers approximately the same area shown in Fig. 1 and has a grid spacing of 20 m. In this case σ_{Atm} is 5.7 mm/yr and L_C is about 210 m.

- A 3D white noise field, which includes the contributions of Φ_{Noise} and Φ_{Res_Topo}. This field is randomly generated assuming a zero-mean normally distributed noise. For each pixel, the noise standard deviation σ_{noise} is derived from the coherence map of the associated interferogram (for a review of different methods to estimate the phase noise as a function of the coherence, see Hanssen (2001)). An example of white noise, which is superposed to a deformation filed is shown in Fig. 3. This field covers the same area from Fig. 2, with a 20-m grid spacing. In this case the average σ_{noise} is 3.4 mm/yr.
- The unwrapping-related errors represent a particular type of errors in the interferometric data. They are due to aliasing effects during the phase unwrapping. Since they are multiple of 2π, they have a large magnitude compared to the noise. They are usually associated with low coherence areas and affect clusters of neighbour pixels. The simulated data can include this type of errors.
- The subsidences characterized by slow deformation rates, such as the one considered in this work, are only detectable over large observation intervals, where the SAR images usually have very low coherence. In low coherence areas the phase noise is high and it is often not possible to unwrap the interferometric phase, thus causing a data loss. The data loss was taken into account in the data simulation. An example is shown in Fig. 3, where the data loss is 32.6 %.

In order to test the proposed procedure, different data simulations were performed. Firstly, the 2D LS collocation was run on a 3D dataset which consists of 10 stacked velocity fields and includes the reference deformation field, the white noise and the data loss effect. Its total variance σ_T^2 is 47.91 mm^2/yr^2, while the variance of the signal is 64.5 % of σ_T^2. The average data loss is 32.6 %. This dataset does not include a temporal evolution of the deformation (i.e. it has a constant deformation rate). In Fig. 4 it is shown a profile of the observations (i.e. the input data of the estimation procedure), the estimated deformation field (estimated signal) with the associated confidence bands (signal ± standard deviation). The location of the profile is illustrated in Fig. 1. One may notice that the estimated signal follows rather well the reference profile. It is important to note that the reference profile was not used in the estimation of the signal. An exception occurs in the left part of the profile, where the reference values have a localized variation: in this case the LS collocation method, due to the structure of the autocovariance function, does not "recognize" this variation as a signal and therefore performs a smoothing of the surface.

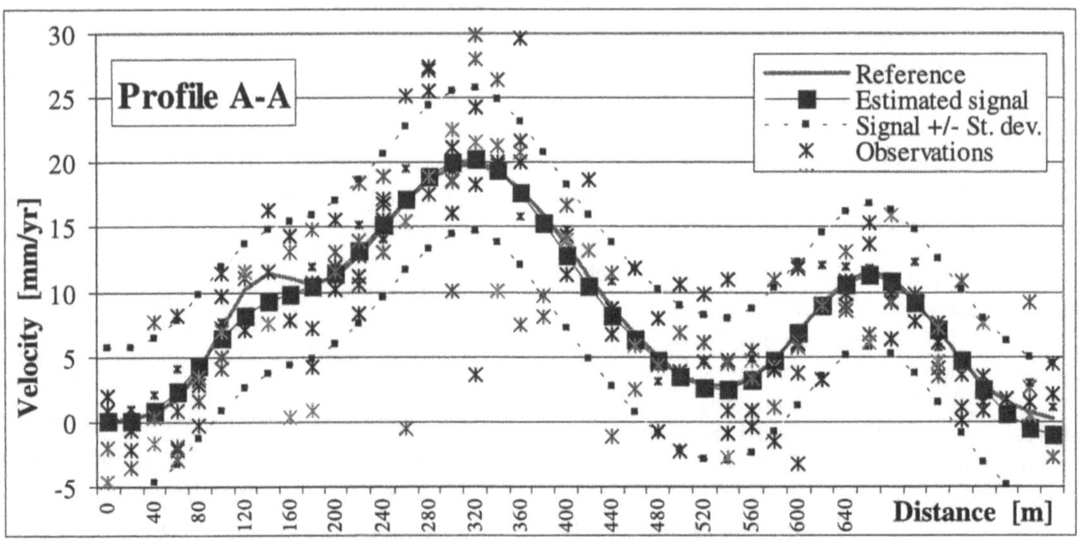

Fig. 4 Filtering on a set of 10 stacked velocity fields, which includes the white noise and the data loss effect. Profile of the observations and the estimated deformation (estimated signal) with the associated confidence bands (signal ± standard deviation).

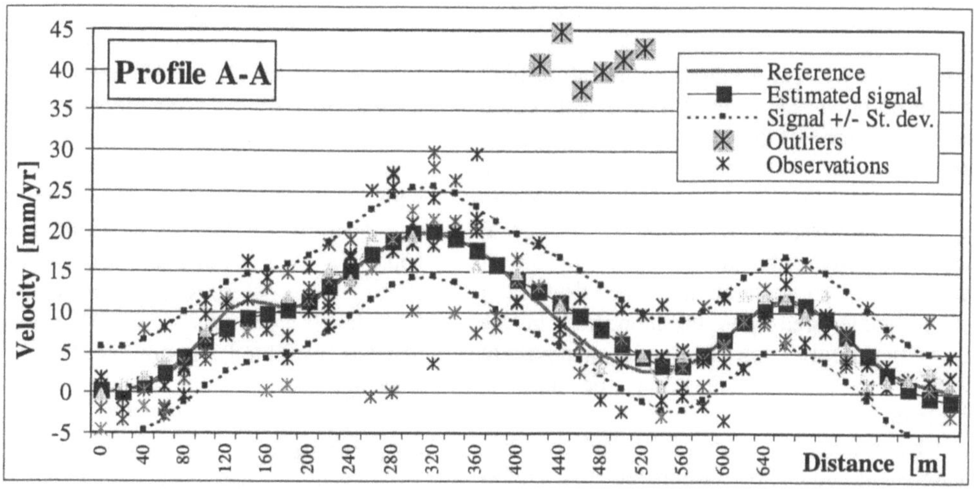

Fig. 5 Filtering on a set of 10 stacked velocity fields, which includes outliers due to unwrapping errors, whose magnitude is 30 mm/yr.

A second dataset was generated to check the robustness of the method against the "outliers" due to unwrapping related errors. In Fig. 5 it is shown a profile of the observations, the estimated signal and its associated confidence band. The noise and data loss were the same of in the previous dataset. The considered cluster of outliers includes 6 by 6 pixels (six of them are visible in the profile) with a magnitude of 30 mm/yr (this corresponds to a phase error of 2π for an interferogram with 1-year observation interval). One may notice a localized effect (in the interval 400-520 m) on the estimated signal. However, the maximum amplitude of this effect is less than 3 mm/yr, i.e. less than one tenth of the magnitude of the input errors. The method is quite robust against the unwrapping-related errors, which can be easily detected by analysing the residuals of the LS collocation estimation. This may appropriately be performed using the classical Baarda data snooping, Baarda (1968).

The filtering of the atmospheric component represents the most delicate step of the proposed procedure. In fact, the LS collocation is only able to filter out the uncorrelated part (noise) of the input observations from their correlated part (signal). As already discussed in the previous section, the collocation filtering is expected to work on a stacked velocity fields if the signal related to Φ_{Mov} is enough strong to overcome the signal due to Φ_{Atm}. This should reasonably happen because the component due to Φ_{Mov} is stationary in time (therefore it has a strong correlation in the stacked velocity fields), while Φ_{Atm} is temporally uncorrelated: increasing the number of interferograms implies Φ_{Atm} to be considered as a noise component in the stacked signal. We have however to recognize that this is a quite heuristic approach. In practice, given a set of interferograms the strength of the signal due to Φ_{Atm} will depend on the magnitude of the atmospheric effects on each interferogram, on their spatial correlation length and on the total number of interferograms.

Two scenarios may be foreseen. In the most favourable one, which is most likely to occur using large numbers of interferograms, the atmospheric effects tend to cancel out in the mean, leaving a weak signal due to Φ_{Atm}. In this case the signal estimated with the LS collocation method will be unbiased, or only slightly affected by the atmospheric effects, like in the case shown in Fig. 4. In this case the interferograms affected by severe atmospheric effects can be identified by analysing the residuals of the collocation: the corresponding residuals will show a very strong correlation, while in absence of (or with weak) atmospheric effects the residuals will tend to behave as white noise. These interferograms could be eliminated to perform a more refined estimation of the signal (second iteration). What happens in the second scenario, when the atmospheric effects do not cancel out completely? In this case, the average of the atmospheric effects might significantly different from zero and might result in a residual trend. In this case it might be not possible to identify the interferogram(s) affected by severe atmospheric effects. In fact, the bias in the estimated signal might corrupt the residuals of the collocation, which might exhibit correlation even if the associated interferograms have no atmospheric effects.

The possibility to get a biased deformation field due to non compensated atmospheric effects could represent a limitation of the procedure. It is however important to underline that the actual possibility to get, in real data, important residual atmospheric effects has to be assessed by analysing sets of real interferograms over stable areas. The influence of the atmospheric component may be drastically reduced if stable areas can be identified in the vicinity of the deformation area under analysis. In this case, in fact, these areas may be used to classify, from the viewpoint of atmospheric effects, reliable and potentially degraded interferograms. The latter ones could be eliminated (or under-weighted, in case a weighted collocation procedure is used) before running the collocation filtering.

5 Conclusions

In this paper an adaptative modelling, which takes advantage of the properties of multiple DInSAR observations has been described. In the proposed procedure two models are coupled: a 2D model, which is used to assess the velocity map of the observed field, and a 1D model, which is employed to estimate the temporal evolution of the subsidence in some selected points. In this paper the analysis of the results has been only focused on the 2D collocation filtering. The proposed filtering provides a flexible modelling tool for the deformation fields with low spatial frequency characteristics. An important advantage is that it provides the filtered and predicted signals with their associated variances, thus describing the stochastic features of the estimated deformation field. The procedure works properly with quite noisy observations. It accepts in input evenly distributed observations: all the results described in this paper were obtained on irregular grids of data, which include an important percentage of "holes" due to data loss. Furthermore, the procedure is robust against outliers, like those due to unwrapping related errors, which may be easily eliminated by analysing the residuals. The procedure can separate the deformation from the atmospheric component, unless there are strong non-compensated atmospheric effects.

Three future developments of the present work are foreseen: the analysis of subsidences based on real SAR data, the implementation of a fully 3D modelling, and the fusion of data coming from different sources. The use of real data will be fundamental to compare the above results with those obtained with real DInSAR observations. The extension from 2D to 3D would be of particular interest for all the subsidences characterized by important temporal variability. It would require an adequate number of interferograms. In fact, dealing with 3D fields involves a much larger number of unknown in the LS collocation. The data fusion could be used to estimate the deformation fields starting from heterogeneous sources, like DInSAR, GPS, levelling networks, etc. This could be done by using a weighted version of the LS collocation filtering.

References

Baarda, W. (1968). A testing procedure for use in geodetic networks. Netherlands Geodetic Commission – Publications on Geodesy, Vol. 2, No. 5, Delft, Holland.

Berardino, P., G. Fornaro, A. Fusco, D. Galluzzo, R. Lanari, E. Sansosti, and S. Usai (2001). A new approach for analysing the temporal evolution of earth surface deformations based on the combination of DIFSAR interferograms. *Proc. of IGARSS 2001*, Sidney, Australia.

Crosetto, M., C.C. Tscherning, B. Crippa, and M. Castillo (2002a). Subsidence monitoring using SAR interferometry: reduction of the atmospheric effects using stochastic filtering. *Geophysical Research Letters*, Vol. 29, No. 9, pp. 26-29.

Crosetto, M., M. Castillo, and R. Arbiol (2002b). Urban subsidence monitoring using radar interferometry: Algorithms and validation. *Photogramm Eng Rem Sens* (in press).

Ferretti, A., C. Prati, and F. Rocca (2000). Non-linear subsidence rate estimation using permanent scatterers in differential SAR interferometry. *IEEE TGRS*, Vol. 38, No. 5, pp. 2202-2212.

Hanssen, R. (2001). Radar interferometry. Kluwer Academic Publishers, Dordrecht, Holland.

Moritz, H. (1978). Least-squares collocation. *Rev. Geophys. Space Phys.*, Vol. 16, pp. 421-430.

Strozzi, T., U. Wegmüller, L. Tosi, G. Bitelli, and V. Spreckels (2001). Land subsidence monitoring with differential SAR interferometry. *Photogramm Eng Rem Sens*, Vol. 67, No. 11, pp. 1261-1270.

Williams, S., Y. Bock, and P. Fang (1998). Integrated satellite interferometry: Tropospheric noise, GPS estimates and implications for interferometric synthetic aperture radar products. *Journal of Geophysical Research*, Vol. 103, B11, pp. 27051-27067.

Changes in the Antarctic Ice Sheet Mass and the Instability of the Earth's Rotation over the Last 110 Years

N. S. Sidorenkov
Hydrometcenter of Russia, 11-13 Bolshoi Predtechensky pereulok, Moscow, 123242, Russia
E-mail: sidorenkov@rhmc.mecom.ru

Abstract. The redistribution of water masses on the Earth entails changes in the components of the Earth's inertia tensor and causes the motion of poles and changes of the Earth's rotation speed.

N.S. Sidorenkov (1982) has deduced equations which connect the fluctuations of the World Ocean water mass ζ_O or the ice mass of Antarctica ζ_A, Greenland ζ_G, and "the rest part of land" ζ_C with the parameters of the Earth's rotation (coordinates of the North Pole v_1, v_2 and velocity of rotation v_3).

Presently, parameters v_1, v_2, and v_3 of the Earth's rotation are measured with a very high accuracy and can be used for the calculation of water exchange (values ζ_O, ζ_A, ζ_G, and ζ_C).

According to this idea, we have calculated the unknown changes of the specific amount of water in the World Ocean ζ_O and the accumulation of ice in Antarctica ζ_A, Greenland ζ_G, and "the rest part of land" ζ_C for 1891-2000 with a one-year discretion.

The computed (theoretical) series ζ_O, ζ_A, ζ_G are compared with the empirical values ζ_O, ζ_A, ζ_G. A close correlation between theoretical and empirical series ζ_A for Antarctica is revealed. The possibility of using the theoretical series for the analysis of temporal variations in the geoid parameters is found. It has been estimated that because of the decadal fluctuations of the Antarctic ice sheet mass, the polar diameter of the geoid varied within ±25 cm over the last century.

Keywords. Antarctic and Greenland ice sheets, sea level, accumulation of ice via precipitation, decade instability of the Earth's rotation, geoid

1 Introduction

Apart from all other reasons, the parameters of the geoid depend on the distribution of water over the planetary surface. Most considerable changes in the geoid parameters can result from the redistribution of water between the World Ocean and the polar ice sheets. In cold glacial epochs, when some portion of water was accumulated in the polar ice sheets, the geoid ellipticity was minimal. In warm interglacial epochs, when almost all the water went in the World Ocean, the geoid ellipticity was increased up to its maximal value.

The purpose of this paper is to call attention to a close correlation of the decade variations in the Earth rotation with the mass changes in the Antarctic ice sheets.

2 Theory

The redistribution of water over the earth's surface causes changes in the components of the Earth's inertia tensor and, as a consequence, variations in the vector's components of instantaneous angular velocity. This relationship is described by the following system of equations (Munk & MacDonald, 1960)

$$-\frac{1}{\sigma}\frac{dv_2}{dt}+v_1=\frac{n_{13}}{C-A}=-\frac{R^2}{2(C-A)}\iint_S \zeta(\theta,\lambda,t)\sin 2\theta\cos\lambda ds, \quad (1)$$

$$\frac{1}{\sigma}\frac{dv_1}{dt}+v_2=\frac{n_{23}}{C-A}=-\frac{R^2}{2(C-A)}\iint_S \zeta(\theta,\lambda,t)\sin 2\theta\sin\lambda ds, \quad (2)$$

$$\delta v_3=-(1+k')\frac{\delta n_{33}}{C}=-(1+k')\frac{R^2}{C}\iint_S \zeta(\theta,\lambda,t)\sin^2\theta ds. \quad (3)$$

Here, we use the Cartesian coordinate system Ox_i fixed with Earth. Its centre O is in the centre of the Earth's mass, and the axes are directed as follows: x_1 - to the prime meridian, x_2 - to a meridian of 90^0 E, and x_3 - along the earth's spin axis averaged for 1900 -1905. Dimensionless values $v_1=\omega_1/\Omega$, $v_2=\omega_2/\Omega$ and $1+v_3=\omega_3/\Omega$ are the direction cosines of the instantaneous Earth's rotation axis; ω_1, ω_2 and ω_3 are the components of the vector of instantaneous angular velocity; Ω is the mean angular velocity of the Earth equal to 7.29 x 10^{-5} radians per sidereal second; $\sigma=2\pi/1.18$ is the Chandler frequency; t is the time; C and A are the polar and equatorial planetary moments of inertia of the Earth; n_{13}, n_{23} and n_{33} are the variable components of the Earth's inertia tensor; $k'=-0.3$ is the load deformation coefficients; $\zeta(\theta,\lambda)$ is the deviation of the specific (i.e. per unit area) amount of water or ice at point $\{\theta,\lambda\}$ and at time instant t ($\zeta(\theta,\lambda)$=0, when $v_1=v_2=v_3$=0); θ is the colatitude; λ is the east longitude; R is the mean radius of the Earth; $ds=R^2\sin\theta d\theta d\lambda$; S is the total area of the Earth.

It is convenient to combine equations (1-2)

$$\frac{d}{dt}(v_1+iv_2)-i\sigma(v_1+iv_2)=\frac{-i\sigma}{C-A}(n_{13}+in_{23})=\underline{\psi}. \quad (4)$$

Let the mass of ice in Antarctica and Greenland vary under the linear law $\underline{\psi}=(z_1+iz_2)t$ and let at instant $t=0$ $v_1+iv_2=0$. Then the solution of equation (4) will be (Sidorenkov, 2002, p. 334)

$$v_1+iv_2=(z_1+iz_2)t+\frac{z_1+iz_2}{i\sigma}\left(1-e^{i\sigma t}\right) \quad (5)$$

This means that the pole makes the circular motion with Chandler frequency and the linear drift in the direction of the meridian $\lambda=arctg\frac{z_2}{z_1}$ with the velocity $\sqrt{z_1^2+z_2^2}$. At averaging of expression (5) over interval of time multiple to Chandler period (1.2 year), the periodic term disappears leaving only the linear terms that describe the drift of the pole. As we are interested in time scale over several years, further on we shall use only the averaged coordinates of the so-called secular polar motion, which do not contain periodic terms any more. Averaging of equations (1-3) substantially removes the restrictions superimposed on the choice of the model of the Earth.

Let us confine our attention to the computation of integrals on the right side of equations (1-3). We divide the integration domain S (the area of the entire earth's surface) into four natural parts: the World Ocean S_O, Antarctica - S_A, Greenland -

S_G and "the rest part of land" - S_C, (i.e. $S = S_O + S_A + S_G + S_C$).

The World Ocean has a tendency to equalize the spatial nonhomogeneity of the specific water mass distribution on the long-period variations (>2 year). Therefore, we may assume that $\zeta(\theta,\lambda) = \zeta_O =$ constant over the ocean area. In addition, the dimensions of Antarctica S_A and Greenland S_G are small in comparison with S. Thus, we may approximately assume that the specific mass of ice $\zeta(\theta,\lambda)$ is independent of θ, λ and use the mean value:

$$\zeta_N(t) = \frac{1}{S_N}\iint_{S_N} \zeta(\theta,\lambda,t)ds,$$

denoting with S_N any of the above specific areas. On the continents the water is primarily in the liquid phase and consequently it rapidly down to the World Ocean (T<1 year). The mass of continental glaciers constitutes only 2 % from a lump of ice on the Earth. Therefore we can suppose, that the same assumption is valid for the S_C area.

Thus, any integral on the right side of equation system (1-3) can be represented as the sum of four integrals

$$\iint_S \zeta(\theta,\lambda,t)\Phi(\theta,\lambda)ds = \sum_{N=1}^{4}\zeta_N(t)\iint_{S_N}\Phi(\theta,\lambda)ds \cdot \quad (6)$$

The integrals of the type $\iint_{S_N}\Phi(\theta,\lambda)ds$ have fixed bounds of integration, which are independ of time. Such integrals may be computed by means of numerical integration (and we have carried out these calculations – (Sidorenkov, 1982, 2002).

Considering only the long-term fluctuations of the Earth's rotation and the secular polar motion, we may ignore the terms $\sigma^{-1}d\nu_i/dt$, because $\sigma^{-1}d\nu_i/dt << \nu_i$. The system of equations (1-3) is therefore reduced ultimately to the system of three algebraic equations, to which a fourth is added representing water mass conservation

$$\nu_1 \cdot 10^{13} = 79965\zeta_O + 4426\zeta_A - 6960\zeta_G - 77440\zeta_C,$$

$$\nu_2 \cdot 10^{13} = 124880\zeta_O + 12650\zeta_A + 6320\zeta_G - 143850\zeta_C,$$

$$\nu_3 \cdot 10^{14} = -88562\zeta_O - 287\zeta_A - 93\zeta_G - 30688\zeta_C, \quad (7)$$

$$0 = 71436\zeta_O + 2820\zeta_A + 414\zeta_G + 25330\zeta_C.$$

Here, ζ_O, ζ_A, ζ_G and ζ_C - are the averaged (in accordance with (6)) fluctuations of the specific amount of water or ice, g/cm^2, in the World Ocean, Antarctica, Greenland and "the rest part of land", respectively.

The simple algebraic system (7) enables us to solve not only direct problems, take known time series of ζ_O, ζ_A, ζ_G and ζ_C, that is to define the secular polar motion and variations of the earth's rotation, but also to solve the reverse problems, i.e. using the pole's coordinates ν_1 and ν_2 and of the long-term Earth's rotation velocity components ν_3, to compute the unknown global water exchange parameters ζ_O, ζ_A, ζ_G and ζ_C. This approach can be used in the case of a combined problem as, where partly ζ_N and partly ν_i may be considered as unknown.

The Earth's rotation irregularities have been recorded since XVIII and the secular polar motion since XIX century (Fedorov, 1972; Vondrak, 1999). Unfortunately, the accuracy of the records of the secular polar motion is extremely unreliable (Yatskiv, 1976). We have made an assumption that values ν_1, ν_2 and ν_3 are precisely known and have calculated the time series of values ζ_O, ζ_A, ζ_G and ζ_C for the period of 1891-2000 (Sidorenkov, 1982, 2002). The graph ζ_A is given at Fig. 1.

3 Glaciolological and meteorological observations

Let's compare the obtained theoretical time series ζ_A of the specific mass of ice in Antarctica with the available empirical data ζ'_A based on glaciolological and meteorological observations.

As it is known, the increment Z of the ice mass of any ice sheet is determined by the sum of values of the resultant water balance B over the previous years:

$$Z = \sum_i B_i \, , \qquad (8)$$

where i - the number of a year. The balance $B = P - R$ is evaluated as the difference between the amount of precipitation P falling on the ice sheet area and the total discharge R of its water resources. The total water discharge R includes the rates of the iceberg-calving, melting, liquid precipitation, losses through vaporization, bottom melting, blowing of snow by wind, and other factors. It is clear that to determine the value R and its long-term changes is virtually impossible.

The long-term changes in precipitation over the Antarctica or Greenland can be investigated on the basis of meteorological measurements of the sums of precipitation, or determined from the data on the annual layers of snow accumulation in ice cores.

Through the efforts of the Antarctic expeditions of a number of countries, some factual material on the income part of the budget of ice sheet (accumulation of snow in Antarctica) has been collected. This material is generalized in V.N. Petrov's monograph (1975). He analysed the annual layers of snow accumulation on the ice sheet using the ice cores from nine stations. Three of them (Amundsen-Scott, Little America, and Wilkes) provide data series going back to 1880. Other stations have been used in investigations since the XX century. V.N. Petrov has determined the values of P, which characterize the alteration of speeds of snow accumulation on the Antarctic ice sheet for the period of 1885 -- 1953 with an annual discretization. He has not estimated the changes in the losses of the ice mass (ablation) of the Antarctic ice sheet.

As a result of calculations by a set of equations (7), we obtain time series of the specific mass $\zeta_A(t)$. The study of the ice core records or the annual sum of atmospheric precipitation gives the accumulation rates P. To determine the values of the equivalent for series of $\zeta_A(t)$, it is necessary to calculate the cumulative sums $\sum_{j=1}^{k} P_j$. Because $P_j > 0$, there is an uncommonly high linear trend hampering the comparison of calculated and observed values $\zeta_A(t)$. Actually, the mass accumulation is compensated substantially by the ice discharge and therefore, the trend is small. It is impossible to determine exactly the value of the ice discharge R. Therefore, we assumed in our study that the magnitudes of the calculated and observed trends should be equal. From this requirement we have found, that in Antarctica, the losses of ice mass R exceeds the mean value of snow accumulation $\overline{P}$ (averaged over the whole period of observations, i.e. 1885--2000) by 3%. Thus $B = P - R = P - 1.03\overline{P}$, where $\overline{P}$=15 g cm^{-2} year^{-1}. Then, having determined the annual values of B, we calculated the cumulative sums of the deviation of the specific mass of ice ζ'_A in Antarctica.

$$\zeta'_A(n) = \frac{Z_A}{S_A} = \frac{1}{S_A}\sum_{j=1}^{n} B_j\,, \qquad (9)$$

where S_A is the area of Antarctica; n is the number of a year.

The comparison of the theoretical and empirical curves ζ_A and ζ'_A for 1891-1953 (Fig. 1) reveals a good qualitative agreement. Only at the end of the 1940s, the decrease in the specific mass of the ice sheet ζ'_A occurred more rapidly than it is predicted by the theoretical curve ζ_A. This discrepancy could arise because of an inadequate registration of ice density with depth. In the upper layer, where snow dominates, the ice density has an order of 0.35 g/cm^3. With depth, in the process of snow subsidence, packing, and transformation into ice, the ice density is gradually increased approaching to the normal value of ice density (0.9 g/cm^3).

Unfortunately, we have not found the data on the accumulation of snow in Antarctica for the last decades (since 1954 till the present time). Dr. N.N. Brjazgin, of the Arctic and Antarctic Research Institute, Sankt Petersburg, has kindly given us the series of the annual sums of precipitation for 11 Antarctic meteorological stations since 1958 till the present time (Brjazgin, 1990). O.V. Luzenko and V.G. Zakharov have used these data to calculate the accumulated anomalies of the annual sums of precipitation over the Antarctic ice sheet. Then we have calculated the series of the mean specific mass of snow over the Antarctic area ζ'_A. This empirical curve ζ'_A for 1958--2000 is shown in Fig. 1. It is also coincides well with the theoretical curve ζ_A/30,2 (in Fig. 1, the value ζ_A is reduced 30,2 times). The coefficient of correlation between series ζ'_A and ζ_A is r=0.91±0.08.

4 Discussion

So, the theoretical values ζ_A coincide rather well with the observed values ζ'_A. However, there are some significant differences between the observed and calculated magnitudes. The observed values of the redistribution of water are 30 times smaller than the theoretical ones. There are some reasons for these discrepancies. The most important reason is likely to be the noncorrespondence of the actual and theoretically assumed properties of the Earth. When developing the theory, it was assumed that the Earth rotates as a unit and therefore, its planetary moments of inertia C and A were used. In the case of the short-periodic effects (not more than two years), the subcrustal material of the Earth completely satisfies this assumption. In the case of water redistribution, the effects of the same sign last for decades. It is not precluded that under the effect of such long impacts, the subcrustal material of the Earth behaves not as a solid but flows like a viscous fluid. Then, the redistribution of water affects not the entire Earth but only its uppermost layer – the lithosphere. The lithosphere can slide over the asthenosphere; in this case, the theory uses only the moments of inertia only of the lithosphere. These moments are 30 times smaller than the moments of the total Earth's inertia. Consequently, all theoretical estimates of values ζ_O, ζ_A, ζ_G and ζ_C should be reduced by a factor of 30. In this case, the theoretical and empirical values would agree both qualitatively and quantitatively (Sidorenkov, 2002).

The sliding of the lithosphere over the asthenosphere is possible in the case when the action duration T is many times longer as compared with the characteristic relaxation time within the asthenosphere τ. It is known that the relaxation time τ is determined by the relationship: $\tau = \eta / \mu$, where η - the viscous coefficient and μ - the rigidity. In the case of the asthenosphere, $\eta \approx 10^{18} - 10^{23}$ P and $\mu \approx 10^{12}$ dyn/cm^2 (Fowler, 1996). As a result, $\tau = \eta / \mu = 10^6 - 10^{11}$ s or 0.03--3000 years. Clearly, the hypothesis mentioned above is reasonable if we take the lower interval of the permissible values of η. This hypothesis is also conformed with the fact of the existence of a significant correlation between the seismic activity and the irregularities of the Earth's rotation.

With the help of the set of equations (7), it is impossible to receive any absolute values of the water exchange parameters (ζ_A, ζ_G, ζ_O). Equations (7) allow us to study only temporal variations of these parameters. The configuration of temporal variations $\zeta(t)$ does not depend on the model of the Earth (i.e., on parameters C, A and k'); it is completely defined by the configuration of temporal variations of the Earth rotation parameters $\nu_1(t)$, $\nu_2(t)$ and $\nu_3(t)$. Parameters of the model C, A and k' are constant (independent of time *t*) and do not influence the configuration of variations in parameters $\zeta_A(t)$, $\zeta_G(t)$, $\zeta_O(t)$. They serve as the scale factors and influence the magnitude of the oscillation amplitude of $\zeta_A(t)$, $\zeta_G(t)$, $\zeta_O(t)$. The higher are the values of C and A and the lower value of k', the greater is the amplitude of variations in $\zeta_A(t)$, $\zeta_G(t)$, $\zeta_O(t)$.

The trend $\zeta_A(t)$ was obtained with the use of Earth orientation parameters $\nu_1(t)$, $\nu_2(t)$ and $\nu_3(t)$. It is equal -1.03 $\overline{P}$ for an interval of 1891-2000. This trend includes combined effect of the accumulation mass of the ice sheet, the postglacial rebound (PGR), the tidal deceleration of the Earth rotation. Magnitude of the trend depends on duration of the explored interval of the time. PGR affects variability $\zeta_A(t)$ on thousand-year time scales. In our study we explored the decadal oscillations $\zeta_A(t)$, which are excited mainly by the decadal change of the climate. That is why PGR does not distort a configuration $\zeta_A(t)$ on fig. 1.

5 Conclusions

The synthesis of the series of the theoretical and empirical values ζ'_A demonstrates the following variations in the specific mass of ice in Antarctic.

The specific mass of ice ζ'_A decreased from 7 to -51 g/cm^2 over the period of 1891—1903; by 1934, it gradually increased up to 1 g/cm^2; during 1936--1972, it decreased to -38 g/cm^2, and increased again up to -7 g/cm^2 during 1973--2000.

Let the ice density in Antarctica be equal to 0.69 g/cm^3 and take into account the subsidence of the lithosphere by introducing the load deformation coefficients $k' = -0.31$. Then, the increments of the altitude h of the Antarctica surface will coincide with the values ζ'_A ($h=((1-0,31)/0,69)\zeta'_A=\zeta'_A$).

Thus, during the last 110 years, the altitude of the Antarctica surface varied as follows: it decreased by 58 cm from 1891 to 1903, increased by 52 cm from 1904 to 1934, decreased by 39 cm from 1935 to 1972, and increased by 31 cm in 1973--2000. Evidently, the polar radius of the Earth in the Southern Hemisphere changed in the same way.

Thus, the redistribution of water between the World Ocean and the Antarctic ice sheet causes variations in the length of the Southern Hemisphere polar radius; they have a decadal time period and a range of about 50 cm. The variations in the ice mass of the Antarctic and Grenland ice sheets and in the level of the World Ocean lead to constant changes in the geoid parameters and acceleration of gravity. The geoid parameters depend on time. The more considered period of time, the more amplitude of the variation of geoid parameters. To correctly determine these variations, constant monitoring is needed. Calculations of the ice mass in Antarctica and Greenland, and the amount of water in the World Ocean performed by the data on the nonstability of the Earth's rotation with the use of the equation system (3) allow us to carry out an indirect control over the time variations of certain parameters of the geoid.

The state of ice sheets in Antarctica and Greenland depends on climatic variations. Therefore, irregularities in the geoid parameters can also correlate with the fluctuations of the climatic characteristics and indices. To find these correlations, it is necessary to perform the monitoring of the geoid parameters and gravitational field, as well as of glaciological and climatic characteristics.

Studies of this problem promise a significant advance in understanding the decade variations of the Earth rotation, oscillations of the sea level, and some global changes.

Acknowledgements

This work was supported by the Russian Foundation for Basic Research (project code 02-02-16178) and the IAG fellowship

References

Bryazgin N.N. (1990). Atmoshpheric precepitation in Antarctic and theirs multi-year change. In collected articles "Meteorological Research in Antarctic". Part 1, Leningrad. Gidrometeoizdat. pp. 30-34.

Fedorov E.P., Korsun A.A., Maior S.P., Panchenko N.I.,Taradee B.K., Yatzkiv Ya.S., (1972). Dvizenie poljusa Zemli s 1890 po 1969. Izdatelstvo "Naukova Dumka", Kiev, 264 pp.

Fowler, C.M.R. (1996). The Solid Earth: An Introduction to Global Geophysics. Cambridge University Press, New York.

Klique R.K., (1980). Uroven okeana v geologicheskom proshlom. Nauka, Moskva, 110pp.

Munk, W.H., G.J.F. MacDonald. 1960. The rotation of the Earth. Cambridge University Press. Cambridge

Sidorenkov N.S. (1980). Irregularites of the Earth's rotation as possible indices of global water exchange. Soviet Meteorology and Hydrology, No.1, 39-45.

Sidorenkov N.S. (1982). Global water exchanges and irregularities of Earth Rotation. Vodnye Resursy (Water Resources) No 3, 39-46.

Sidorenkov N.S. (1987) The nature of seasonal and interannual variations of the Earth's rotation. //Proc. Int. Symp. "Figure and Dynamics of the Earth, Moon, and Planets." Prague. 1987. P. 947-960.

Sidorenkov N.S. (2002). Physics of the Earth's rotation instabilities. Nauka, Publishing Company Fizmatlit, Moscow. 384 p.

Vondrak J. (1999). Earth rotation parameters 1899.7-1992.0 after reanalysis within the Hipparcos frame. //Surveys in Geophysics. 1999. Vol. 20. pp. 169-195.

Yatzkiv Ya.S., Mironov N.T., Korsun A.A., Taradii V.K., (1976). Dvizenie poljusov i neravnomernost vrachenija Zemli. "Astronomija" ("Itogi nauki i techniki"), Vol. 12 (chasti I, II). VINITI. Moscow.

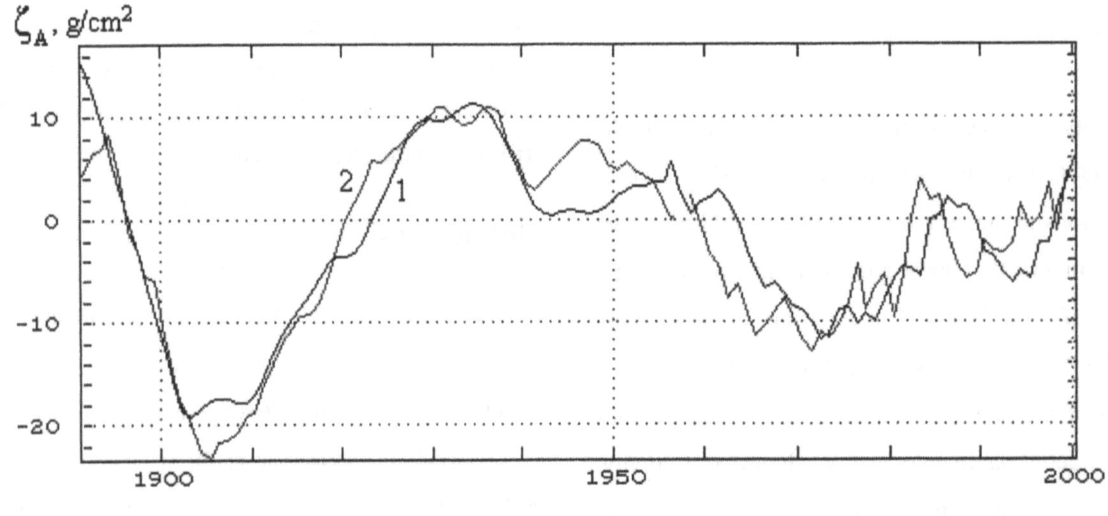

Fig. 1 Temporal variations of the specific mass of ice ζ_A in Antarctica:

1 – the theoretical value ζ_A/30,2;

2 – the empirical value ζ'_A.

Index of Authors

List of Participants

Abate Giovanni; Italy
Abd-Elmotaal Hussein; Egypt
Allasio Andrea; Italy
Altamimi Zuheir; France
Barzaghi Riccardo; Italy
Beinat Alberto; Italy
Benciolini Battista; Italy
Biagi Ludovico; Italy
Bianco Giuseppe; Italy
Borghi Alessandra; Italy
Borre Kai; Denmark
Cabrucci Andrea; Italy
Caprioli Mauro; Italy
Condi Francis; USA
Coratella Duilio; Italy
Crippa Bruno; Italy
De Franco Roberta; Italy
De Lacy Clara; Italy
Dermanis Athanasios; Greece
Devoti Roberto; Italy
Ditmar Pavel; The Netherlands
Dominici Donatella; Italy
Faccioli Andrea; Italy
Ferraro Lina; Italy
Finn Gunter; Germany
Freeden Willi; Germany
Gandolfi Stefano; Italy
Gatti Marco; Italy
Grafarend Erik; Germany
Grebenitcharsky Rossen; Canada
Gross Richard; USA
Heck Bernhard; Germany
Holota Petr; Czech Republic
Keller Wolfgang; Germany
Kern Michael; Canada
Klees Roland; The Netherlands
Kusche Juergen; The Netherlands
Kutterer Hansjoerg; Germany
Lanotte Roberto; Italy
Luceri Vincenza; Italy
Marotta Anna Maria; Italy
Mayer Carsten; Germany
Nardi Antonio; Italy
Nardini Marco; Italy
Negusini Monia; Italy
Pacione Rosa; Italy
Petrovskaya Margarita; Russia
Pietrantonio Grazia; Italy
Reguzzoni Mirko; Italy
Riguzzi Federica; Italy
Rutigliano Paolo; Italy
Sabadini Roberto; Italy
Sacerdote Fausto; Italy
Sansò Fernando; Italy
Sarti Pierguido; Italy
Schaffrin Burkhard; USA
Schuh Wolf-Dieter; Germany
Sideris Michael; Canada
Sidorenkov Nikolai; Russia
Sillard Patrick; France
Sorgente Mauro; Italy
Teunissen Peter J.G.; The Netherlands
Tscherning Carl Christian; Denmark
Valette Bernard; France
Velicogna Isabella; USA
Venuti Giovanna; Italy
Watkins Michael; USA
Yu Jinhai; P.R.China

MIX
Papier aus verantwortungsvollen Quellen
Paper from responsible sources
FSC® C105338

If you have any concerns about our products,
you can contact us on
ProductSafety@springernature.com

In case Publisher is established outside the EU,
the EU authorized representative is:
Springer Nature Customer Service Center GmbH
Europaplatz 3, 69115 Heidelberg, Germany

Printed by Libri Plureos GmbH
in Hamburg, Germany